XX International Scientific Symposium of Students, Postgraduates and Young Scientists on "Problems of Geology and Subsurface Development"

IOP Conference Series: Earth and Environmental Science
Volume 43

Tomsk, Russia
4 - 8 April 2016

ISBN: 978-1-5108-3100-1
ISSN: 1755-1307

Printed from e-media with permission by:

Curran Associates, Inc.
57 Morehouse Lane
Red Hook, NY 12571

Some format issues inherent in the e-media version may also appear in this print version.

Printed by Curran Associates, Inc. (2016)

For permission requests, please contact the Institute of Physics at the address below.

Institute of Physics
Dirac House, Temple Back
Bristol BS1 6BE UK

Phone: 44 1 17 929 7481
Fax: 44 1 17 920 0979

techtracking@iop.org

Additional copies of this publication are available from:

Curran Associates, Inc.
57 Morehouse Lane
Red Hook, NY 12571 USA
Phone: 845-758-0400
Fax: 845-758-2633
Email: curran@proceedings.com
Web: www.proceedings.com

Table of contents

Volume 43

XX International Scientific Symposium of Students, Postgraduates and Young Scientists on "Problems of Geology and Subsurface Development"

4–8 April 2016, Tomsk, Russia

Accepted papers received: 7 September 2016
Published online: 11 October 2016

Preface

Papers

Mineralogy, Geochemistry and Petrography

Ore Deposits: Current Prospecting and Exploration Methods

Petroleum Geology: Current Exploration and Extraction Methods

Geophysical Surveying Methods for Ore Deposit Prospecting and Geoinformation Systems in Geophysical Surveys

Hydrogeology and Engineering Geology

Hydrogeochemistry and Hydrogeoecology

Geoecology and Environmental Protection

Latest Research in Reservoir Engineering

Latest Research in Hydrocarbon Refining

Integrated Use and Processing of Mineral Resources

Latest Technologies in Oil and Gas Driling

Mining, Ore and Non-metallic Minerals Development

Modern Oilfield Equipment and Upgrading

Latest Technology in Oil-and-Gas Pipelines and Petroleum Storage

Energy Efficiency and Automation Systems for Oil and Gas Facilities

Petroleum Engineering Economics and Mining Law

Problems of Arctic Region development

IOP Conf. Series: Earth and Environmental Science **43** (2016) 011001 doi:10.1088/1755-1315/43/1/011001

XX INTERNATIONAL SCIENTIFIC SYMPOSIUM OF STUDENTS, POSTGRADUATES AND YOUNG SCIENTISTS DEVOTED TO THE 120TH ANNIVERSARY OF THE FOUNDING OF TOMSK POLYTECHNIC UNIVERSITY

XX International Scientific Symposium in honor of Academician M.A. Usov "Problems of Geology and Subsurface Development" (for students and young scientists) was organized under the guidance of the Ministry of Education and Science of the Russian Federation and the Russian Foundation for Fundamental Research within the National Research Tomsk Polytechnic University (NR TPU). TPU is one of the oldest technical higher education institutions in Russia, training specialists in the domain of geoscience and enhancing their further research in this area. The Institute of Natural Resources, National Research Tomsk Polytechnic University (INR TPU) was chosen to hold the International Scientific Symposium. In 2016 the Institute of Natural Resources celebrated its 115th anniversary. It was founded by V.A. Obruchev in 1901, first Siberian geologist, member of USSR Academy of Sciences, Hero of Socialist Labor, and first Laureate of the Lenin Prize. He was recognized as a prominent scientist in the area of geology all over the world.

INR is the first institute of geological education and geosciences in Asian Russia. Even today the Siberian Mining and Geological School, established by V.A. Obruchev and M.A. Usov, has retained its significance in the discovery, exploration and development of mineral resources not only in Siberia, the Far East and North-East of Russia, but also in Central Asia. There are numerous outstanding scientists and engineers among alumni of The Institute of Natural Resources. There are over 35,000 graduates of the institute, 450 of whom became explorers of mineral deposits including one Nobel laureate, 50 laureates of Lenin and State Prizes, more than 250 doctors (DSc) and 1600 associate professors (PhD), 15 academicians and corresponding members of the USSR Academy of Sciences.

Within the framework of the XX International Scientific Symposium, there were 18 panels and workshops in four days. The Symposium was unique in the fact that it embraced all scientific fields of geology and subsurface development, including mineral and hydrocarbon prospecting and exploration methods, space geological research and geoinformation systems, as well as ecological problems of the planet and humanity, rational utilization of mineral resources, land management, natural resource law and economics.

The participants were students, post graduates, Master degree students and young scientists from Russia, foreign and CIS countries. The research issues presented at the Symposium highlighted the latest achievements via modern technology and comprehensive interpretation methods; revealed the experimental research results and computer technology in geology and geoecology; provided the analysis of theoretical and experimental research on various geological and environmental protection problems.

The reports embraced the vital issues and latest scientific achievements in stratigraphy, paleontology, tectonics, historical and regional geology, mineralogy, geochemistry, petrography, lithology, metallogeny, hydrogeology and engineering geology, geophysics, petroleum geology, oil and gas field development and processing of hydrocarbon and mineral resources, geoinformation systems (GIS) in geology, geospace research, oilfield equipment upgrading; modern technological achievements in mineral exploration, oil and gas production, transportation and storage, drilling, mining engineering, geoecology, hydrogeoecology, environmental protection engineering, rational utilization of mineral resources, land management, mining and natural resources law, and economical problems of mineral resource sectors in Russia and CIS countries, problems of the Arctic Region development. There was a special panel for those who wanted to present reports in English and / or German.

The presented reports were of scientific importance due to the fact that new concepts and original ideas were suggested by the authors. A number of research topics introduced fundamentally new discoveries. There is also a practical aspect: the findings of the young scientists' research, both fundamental and experimental, could result in DSc and PhD theses. Some authors were granted patents and licenses for their significant discoveries.

Within the framework of the XX International Scientific Symposium, 1043 reports were presented (including poster presentations), including 332 non-residents from other cities and 119 from foreign countries (12 - non-CIS countries and 107 – CIS countries, correspondingly).

All in all, 637 students and 217 young scientists from Russia, CIS countries and foreign countries participated in this Symposium.

Within the 18 different panels and RT 854 reports were presented (including 590 from TPU) of 1043 submitted reports, as well as 189 poster presentations.

The representatives of different organizations were as follows: higher education institutions - 835, Russian Academy of Science - 42, R&D Institutes - 11, industrial enterprises - 6.

There were 48 participants from eleven non-CIS countries: Germany, the Netherlands, the USA, China, Vietnam, Venezuela, Cameroon, Iraq, the Republic of Côte d'Ivoire, Indonesia, and India.

There were 92 participants from seven CIS countries: National Academies of Sciences and Universities of Belarus, Kazakhstan, Uzbekistan, Azerbaijan, Moldova, Ukraine and Tajikistan.

The research results of the Symposium can be found in a two-volume edition which is available at http://portal.tpu.ru/science/konf/usovma/eng and /or www.usovma.tpu.ru.

The Editorial Boards of this Symposium and IOP Conference Series: Earth and Environmental Science consider that the articles would be of great interest for university researchers and young scientists, research and academic institutes, academies of sciences and their branches, engineering and technical staff of ministries and government departments - for those who explore and develop the Earth subsurface.

Many people have contributed in a variety of ways in the preparation of this edition. We would like to thank the administration of the Institute of Physics and Professor P.S. Chubik, Rector of National Research Tomsk Polytechnic University.

Tributes to colleagues

Editorial board of the collection (journal) thanks the organizers of the XX International scientific symposium of students, postgraduates and young scientists in honour of academician M.A. Usov "Problem of geology and subsurface development" devoted to the 120th anniversary of the founding of Tomsk Polytechnic University:

A.Yu. Dmitriev, chairman of the programme and organizing committee, PhD (Engineering Sciences), associate Professor, TPU, Tomsk, Russia
G.M. Ivanova, secretary of the conference, associate Professor, PhD (in Mineralogy), TPU, Tomsk, Russia
S.V. Seryakov, 1th Section co-Chairman, associate Professor, PhD (in Mineralogy), TPU, Tomsk, Russia I.V. Kucherenko, 2th Section co-Chairman, Professor, Dsc (in Mineralogy), TPU, Tomsk, Russia
R.Yu. Gavrilov, 3th Section co-Chairman, associate Professor, PhD (in Mineralogy), TPU, Tomsk, Russia
N.M. Nedolivko, 4th Section co-Chairman, associate Professor, PhD (in Mineralogy), TPU, Tomsk, Russia
Yu.V. Kolmacov, 5th Section co-Chairman, associate Professor, PhD (in Mineralogy), TPU, Tomsk, Russia
V.K. Popov, 6th Section co-Chairman, Professor, Dsc (in Mineralogy), TPU, Tomsk, Russia
S.L. Shvartzev, 7th Section co-Chairman, Professor, Dsc (in Mineralogy) TPU, Tomsk, Russia
A.A. Potseluev, 8th Section co-Chairman Professor, Dsc (in Mineralogy) TPU, Tomsk, Russia
O.A. Pasko, 9th Section co-Chairman, Professor, Dsc (Agricultural Sciences) TPU, Tomsk, Russia
Ye.G. Yazikov, 10th Section co-Chairman, Professor, Dsc (in Mineralogy) TPU, Tomsk, Russia
O.S. Chernova, 11th Section co-Chairman, associate Professor, PhD (in Mineralogy) TPU, Tomsk, Russia
V.I. Yerofeyev, 12th Section co-Chairman, Professor, Dsc (Physico-Mathematical Sciences) TPU, Tomsk, Russia
Ye.D. Ivanchina, 12 (1)th Section co-Chairman, Professor, Dsc (Engineering Sciences), TPU, Tomsk, Russia
V.V. Korobochkin, 12 (2)th Section co-Chairman Professor, Dsc (Engineering Sciences) TPU, Tomsk, Russia
V.I. Vereshchagin, 13th Section co-Chairman, Professor, PhD (Engineering Sciences) TPU, Tomsk, Russia
A.Yu. Dmitriyev, 14th Section co-Chairman, associate Professor, PhD (Engineering Sciences) TPU, Tomsk, Russia
E.N. Pashkov, 15th Section co-Chairman, associate Professor, PhD (Engineering Sciences) TPU, Tomsk, Russia
L.A. Saruyev, 15th Section co-Chairman, Professor, Dsc (Engineering Sciences) TPU, Tomsk, Russia

Peer review statement

All papers published in this volume of *IOP Conference Series: Earth and Environmental Science* have been peer reviewed through processes administered by the proceedings Editors. Reviews were conducted by expert referees to the professional and scientific standards expected of a proceedings journal published by IOP Publishing.

Paleogeographic and litho-facies formation conditions of Mid-Upper Jurassic sediments in S-E Western Siberia (Tomsk Oblast)

M Shaminova, I Rychkova and Ju Sterzhanova

Institute of Natural Resources, Tomsk Polytechnic University, Tomsk, Russia
E-mail: [1]mshaminova@mail.ru, [2]irina.rychkova@mail.ru, [3]usterzhanova@yandex.ru

Abstract This paper describes the criteria to identify Tumen (lower Bathonian) and Naunak (upper Bathonian-Callovian- Oxfordian) suites within S-E Western Siberia (Tomsk Oblast). The specific paleogeographic and litho-facies formation conditions of sediments and numerous vegetable remains and ichnofossils indicated the fact that this territory was the location of sedimentogenesis transition during Tumen and Naunak suite formation. Based on integrated survey oil-gas potential litho-facies groups were defined in Mid-Upper Jurassic sediments within S-E Western Siberia.

1. Introduction

Mid-Upper Jurassic sediments in S-E Western Siberia are the most potential oil-gas deposits. However, it should be noted that due to existing polyfacies cross-sections it is rather difficult to identify complex-structured reservoirs [7]. Mid-Upper Jurassic sediments in S-E Western Siberia include Vasugan horizon embracing marine Vasugan suite and coastal-marine Naunak suite [4] being underlain by Tumen suite. Today, there are two controversial questions which are under discussion: what are the distinct and visible division between these suites and how to trace the reservoirs within these suites. Hard-to-identify stratigraphic units are conditioned by the lenticular lithological bodies and insignificant suite rock component contrast intensity.

Traditionally, production logging data is applied in sectional layering. However, log data interpretation of geological bodies with different facies features scarcely ever seems to be distinct. In this case, direct methods are more significant in the biostratigraphic control [3].

Backstripping and effective forecasting of natural petroleum and gas reservoirs were based on core sampling involving paleobotanic (macro-plant remains), lithological-petrographic, geochemical, fluorescence microscopy and X-ray diffraction methods, as well as logging interpretation.

2. Data and methods

Investigation target is Mid-Upper Jurassic sediments in S-E Western Siberia (Tomsk Oblast)-Sneznaya field (wells 135, 170, 301, 430, 446) and Mayskaya (wells 228, 309, 393, 400, 698).

The investigated fields embraced rocks of different facies genesis: coastal-marine (aleurolites, silt-sandstones), lacustrine-boggy (silt-claystones, argillites), lacustrine-alluvial (silt-sandstones, sandstones) and channel facies (sandstones) [2]. Obtained results are based on calculated geochemical models of Yudovich and Ketris [8] and plotted asymmetry-excess diagrams (according to grain size measurement data).

3. Results and discussion

Based on the integrated analysis the lithological-petrographic and palaeontological criteria were established for the diagnosis of Tumen and Naunak suites.

IOP Conf. Series: Earth and Environmental Science **43** (2016) 012001 doi:10.1088/1755-1315/43/1/012001

Tumen suite is mainly composed of argillites and silt-argillites, including thin layers of sandstones and silt-sandstones (fig. 1).

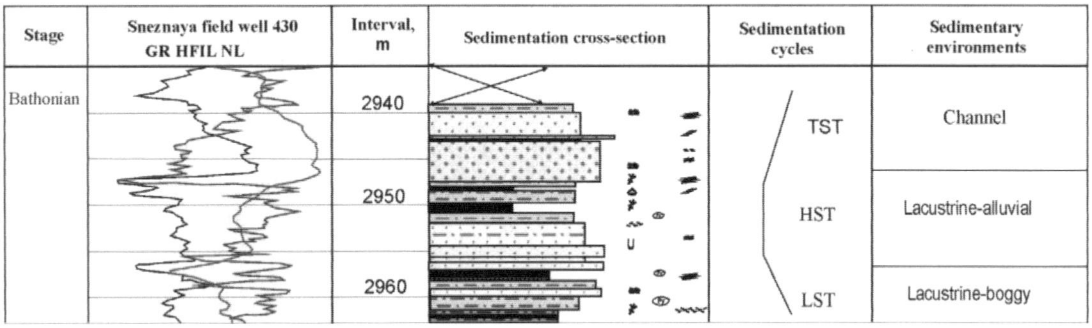

Figure 1. Vertical sedimentation model of Tumen suite

*HFIL- High Frequency Induction Logging GR- gamma ray NL- neutron logging

Characteristic features of Tumen suite

Argillites and silt-agrillites are of silicious hydromica and insignificant amount of fossilized vegetable remains. The rocks are oriented- structured governed by flattened silt-to-quartz sized fragments alongwise the bedding plane. One can also observe interrupted bituminous-filled cracks in this bedding direction.

Studied polished thin sections under fluorescent microscope MIKMED II reflected numerous bitumen migration traces. Bitumen composition varied from greasy-resinous (yellow-brownish luminescence) to resinous- asphaltene (dark-brownish luminescence) and predominately resinous bitumen (dark yellow-brownish luminescence). Bitumen content was 0.02-0.04% (fig. 2).

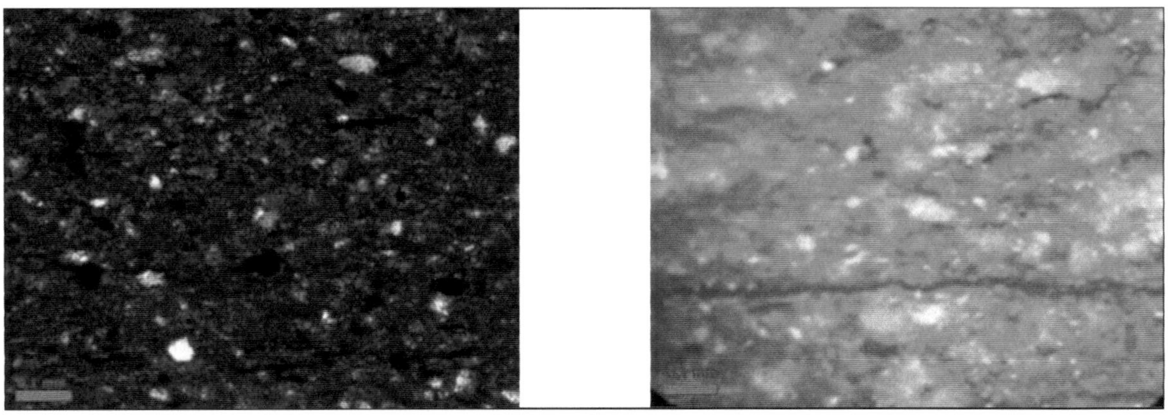

Figure 2. Silt-, bitumen argillite
(left: nicol "+"; right: under fluorescent microscope MIKMED II)

The fragment material composition in sandstones and silt-sandstones was predominately quartz and feldspar of average roundness and sorting. The cementing matter included mainly hydromica, chlorite, muscovite (more often plastically deformed), which, in its turn, indicates the fact of intensive catagenesis. The postdiagenesis alteration in Tumen suite was more intensive than in Naunak suite (intensive cataclasis, fracturing, flatness of fragments alongwise the bedding plane).

Numerous and various plant traces pertaining to upper Tomsk phytohorizon (Lower Bathonian) could be observed on the bedding surface [5]. This flora included equisetum (horsetail), fern and ginkgophyta, of which the predominate type was Coniopteris (fern). The following could also be found throughout Tumen suite profile: *Equisetites sp., Coniopteris vialovae, C. simplex, Raphaelia diamensis* and *Nilssonia urmanica*. Based on detailed studies of the taxons several ferns such as *Coniopteris vialovae* and *Raphaelia diamensis* were revealed as limited stratigraphic bands. This, in its turn, could be applied in stratigraphic breakdown and correlation of the continental sediments in S-E Western Siberia.

Characteristic features of Naunak suite

Naunak suite is mainly composed of thick medium-coarse grained sandstone and silt-sandstone interbeds (fig.3).

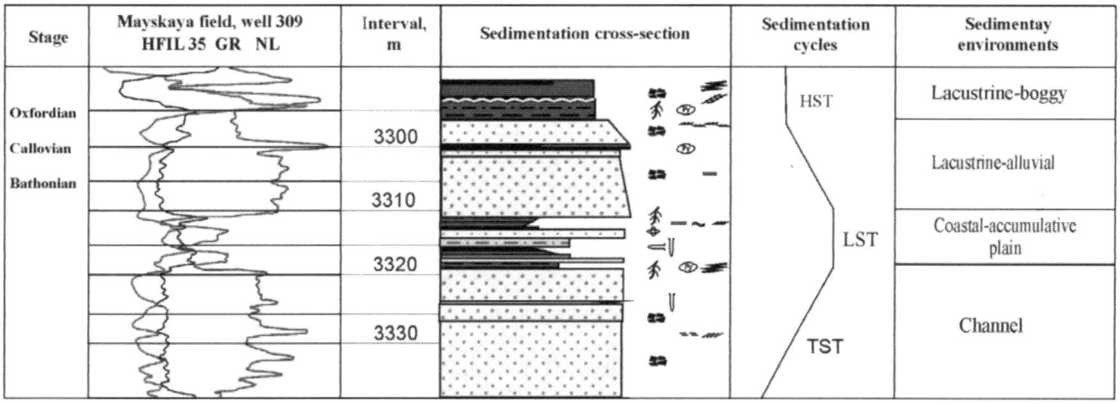

Figure 3. Vertical sedimentation model of Naunak suite

*HFIL- High Frequency Induction Logging GR- gamma ray NL- neutron logging

The terrigenous varieties in Naunak suite comparable to those in Tumen suite involve weakly sorted and less rounded fragment material. The cementing matter includes not only micaceous minerals but also carbonates (siderite, calcite) and authigenous glauconite. Post-diagenesis alteration predominately embraces the corrosion and regeneration of the quartz fragments and feldspar kaolinization; however, to a lesser extent, it also involves cataclasis and fracturing.

Studied polished thin sections under fluorescent microscope MIKMED II reflected numerous bitumen migration traces. Bitumen composition varied from greasy (light yellow luminescence) to resinous- asphaltene (dark-brownish luminescence) and predominately greasy-resinous bitumen (light yellow-brownish luminescence). Bitumen content was 0.04-0.06% (fig. 4) [9].

A few horizon cross-sections revealed a high bitumen content being confined to the pores and cementing matter. Naunak phytohorizon plant traces (upper Bathonian-Callovian-Oxfordian) could be observed on the bedding surface involving an insignificant paleoflora composition comparable to underlying Tomsk phytohorizon. This composition predominately includes gymnosperms (naked-seeded): czekanowskiales, cycadales, ginkgoales and coniferales. Lower ranking sequence involves ferns. Sporadically equisetum types can be found. Czekanowskia types are few in number-only 3-4. However, they form monodominant accumulations composing interbeds within terrigenous sediments. The following plants have been identified: equisetopsida *Equisetostachus sibiricus*, ferns *Coniopteris hymenophylloides, C. burejensis C. latilobus, C. simplex, C. burejensis*, czekanowskiales *Czekanowskia tomskiensis*, ginkgoales *Sphenobaiera* sp., cycadales *Nilssonia urmanica, N. maiskaja* and coniferales *Podozamites lanceolatus*, *Podozamites* sp. [6]. Fine-pinnulate *Coniopteris* is the major fern type, i.e. *Coniopteris latilobus, Nilssonia maiskaja* and *Podozamites eichwaldii* are predominate in Naunak suite.

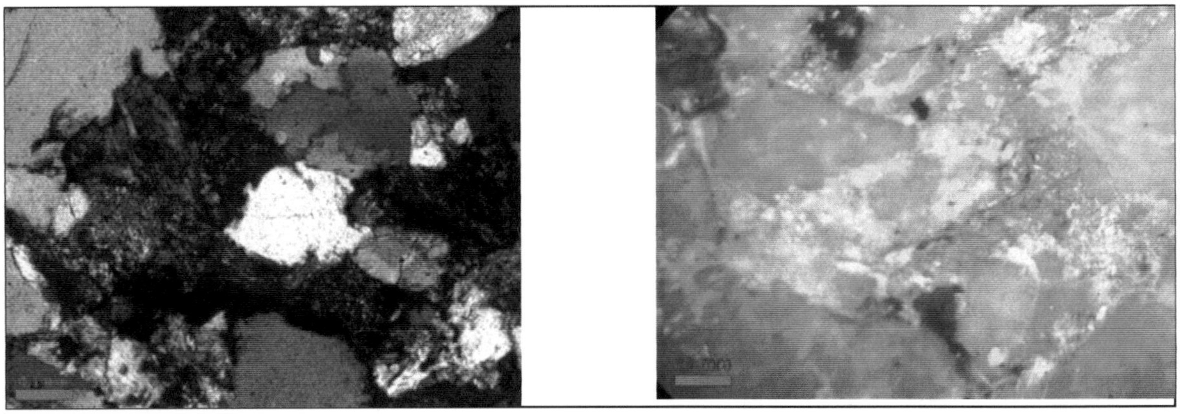

Figure 4. Feldspar-quartz graywacke with hydromica cement indicating decomposition of feldspar grains and quartz and bituminous grain co-ordination

(left: nicol "+"; right: under fluorescent microscope MIKMED II)

4. Conclusion

Chemical-petrographic study of coals from Tumen and Naunak suites via Rock-EVAL method did not only indicate different biochemical coal formation environment conditions but could also be applied in identifying the stratigraphic units.

The analysis of geophysical data based on gamma ray, high-frequency induction logging 35, neutron logging and induction electrical logging in correlation with complex lithological and biofacies studies made it possible to isolate the sedimentation cycles in Tumen and Naunak suites. The most potential petroleum-gas reservoirs (irrespective of age) are sandstone sediments of channel facies formed during TST [1]. Horizons enriched by migrated bitumen of greasy-resinous (light yellow-brownish luminescence) and greasy (light yellow luminescence) composition were identified under the fluorescent microscope. Such horizons having porous and cemented bituminous structure could be potential hydrocarbon reservoirs.

Therefore, detailed integrated study of hard-to-identify Mid-Upper Jurassic sediments in S-E Western Siberia furthered possible stratigraphic breakdown and intraregional correlation, as well as the prediction of potential reservoirs [10].

References

[1] Baraboshkin E 2011 *Terrigennye kollektory Moscow.* Sedimentologija kerna. pp.157.

[2] Botvinkina L 1965 *RAS USSR Publishing House Moscow Nauka.* Metodicheskoe rukovodstvo po izucheniju sloistosti Proceedings of Geological Institute. Vol. **119** pp. 260

[3] Kirichkova A, Chizhova V, Stashkova E, Fortunatova N, Shurigin B 2007 *Petroleum Geology. Theory and Practice.* Stratigrafija v neftjanoj geologii: metodologija issledovanij i aktual'nye problem. Vol. **2** URL: http://www.ngtp.ru/rub/2/007.pdf.

[4] *Reshenija 6-go Mezhvedomstvennogo stratigraficheskogo soveshhanija po rassmotreniju i prinjatiju utochnennyh stratigraficheskih shem mezozojskih otlozhenij Zapadnoj Sibiri* 2003 Novosibirsk 114

IOP Conf. Series: Earth and Environmental Science **43** (2016) 012001 doi:10.1088/1755-1315/43/1/012001

[5] Rychkova I 2013 *Paleobotanicheskij vremennik // Neperiodicheskoe prilozhenie kzhurnalu "Lethaea rossica"*. Paleobotanicheskaja harakteristika tjumenskoj svity srednej jury (Tomskaja oblast). Vol. **1** pp. 132-136.
[6] Rychkova I, Bistritskaya L 2015 *Uspehi sovremennogo estestvoznanija.* Naunak plant variety in Ust-Tim depression within S-E Western Siberia. Vol. **5** pp.175 – 177.
[7] Shurigin B, Nikitenko B, Devyatov V et al 2000 *Novosibirsk SB RAS Jurskaja sistema.* Stratigrafija neftegazonosnyh bassejnov Sibiri. pp. 480
[8] Yudovich J, Ketris M 2000 *Nauka.* Fundamentals of lithochemistry. pp. 479.
[9] Rychkova I, Shaminova M, Sterzhanova U , Baranova A 2015 *IOP Conference Series: Earth and Environmental Science*. Lithologic-facies and paleogeographic features of Mid-Upper Jurassic oil-gas bearing sediments in Nurolsk depression (Western Siberia). Vol. **27** pp. 1 – 6.
[10] Shaminova M, Rychkova I, Sterzhanova U and Dolgaya T 2014 *IOP Conference Series: Earth and Environmental Science*. Lithologo-facial, geochemical and sequence-stratigraphic sedimentation in Naunak suite (south-east Western Siberia). Vol. **21** pp. 012001.

Trace metals in soils of the main geomorphological units in the southwestern part of Western Siberia

E Yu Konstantinova

Tyumen State University, Volodarskogo Street 6, Tyumen, 625003, Russia
E-mail: konstantliza@gmail.com

Abstract. Total concentrations of Ti, Cr, Mn, Ni, Cu, Zn, Rb, Sr, and Pb as well as soil granulometric texture were studied for three plot sites representing different geomorphologic units of the southwestern part of Western Siberia: periphery of the upland Tobolsky Mainland, Ishim plain, Turinskaya plain. Interregional difference in the relationship by and among the content of trace elements and particle size distribution of soil horizons is established. Thus, for the soils of Turinskaya plain such interrelations are not observed. For the soils of Ishim Plain moderate negative correlation between Pb concentrations and medium silt, as well as average positive correlations between Zn and fine sand, coarse sand and Pb are found. For the soils of the high terraces of the Irtish and periphery of Tobolsky Mainland interface zone moderate positive correlations between contents of Ti, Zn, Sr and fine sand, weak positive ones between Rb and medium sand, moderate negative ones between Zn and clay, Ti, Ni, Cu, Zn, Rb, Sr, and fine dust, Ti, Cu, Zn, Rb, Sr and medium silt are observed. Consequently, properties and genesis of local parent rocks are significant factors for distribution and accumulation of trace elements in the soils of the southern taiga; at the same time, the processes of bioaccumulation in thick humus horizons of dark gray soils and chernozems apparently play an important part in accumulation and migration of trace elements in forest-steppe soils of Ishim and Turinskaya plains.

1. Introduction

The southwestern part of Western Siberia can be described as a territory with a diverse geomorphological structure, a variety of individual morphological units with different history of formation, combinations of geomorphological processes, and genesis of near-surface deposits.

Soil resistance to different types of industrial impact is largely determined by their elemental composition and granulometry inherited from subsurface unconsolidated sediments [1-4]. The particle size distribution, chemical and mineralogical composition of the soil largely specify migration and accumulation of trace elements in the landscape, the background concentrations of inorganic pollutants in soils.

The main purpose of this research is to determine the differences in total concentrations of Ti, Cr, Mn, Ni, Cu, Zn, Rb, Sr, and Pb in soils of several geomorphological regions of the south-west of Western Siberia associated with the lithological characteristics of the parent rocks.

2. Materials and methods

The main geomorphological units for the territory of the south of Tyumen region and the northern Trans-Urals (figure 1) are Turinskaya base-level dissected clinoplain, Predturgayskaya base-level occasionaly dissected clinoplain, Ishim denudation clinoplain, base-level constructional low dissected upland Tobolsky Mainland, Sredneirtyshskaya alluvial terraced lowland, Kondinskaya terraced alluvial lowland, Irtysh-Kondinskaya lacustrine-alluvial low dissected plain [5]. Deposits of these geomorphological units differ in chemical composition and properties, which determines the diversity of soil-forming rocks.

Subaeolian quaternary loessy pale or yellow-brown light and medium loams and sandy loams of lacustrine-alluvial genesis with thickness from one to eight meters present near-surface deposits of upland Tobolsky Mainland [5-6]. These deposits are characterized by the high content of coarse silt (up to 35-45%), the absence of carbonates and enrichment in Ti, Mn and Al [7].

Soil-forming rocks of Ishim plain are confined to subaerial sediments – typical for Western Siberia carbonate loess loams with a thickness of 3-8 m or more (in local depressions) [8]. Loess soil-forming rocks of the region are characterized by high content of coarse silt and there are thin layers and lenses of sand in some sections.

Soil-forming rocks of Turinskaya plain are rather similar to the loess-like loams of Ishim plain. Calcareous loams of this geomorphologic structure are characterized by low content of silt fraction and high content of fine sand [7, 9].

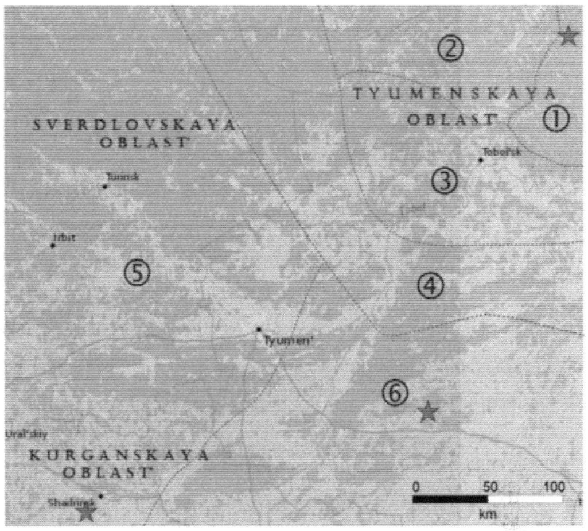

Figure 1. Main geomorphologic units of southwestern part of Western Siberia [5]: 1. upland Tobolsky Mainland; 2. Kondinskaya lowland; 3. Sredneirtyshskaya lowland; 4. Irtysh-Kondinskaya lowland; 5. Turinskaya plain; 6. Ishim plain. Asterisks show the locations of plot sites.

Field studies were carried out on three plot sites (figure 1): high terraces of the Irtysh river, passing into the watershed surfaces of the upland Tobolsky Mainland (outskirts of village Turtas), the central part of Ishim plain (an area near Okunevka village) and river valley part of Turinskaya plain near Shadrinsk. Selected areas are located rather far from the major sources of pollution. Soil profiles (14, 18 and 11, respectively) were placed in basic landscape positions within each site. Samples for further chemical analysis were taken from each genetic horizon of the studied soils; soils were described in accordance with Russian soil classification [10].

Grain size distribution was determined by laser microanalyzer Fritsch «Analysette 22» with the pyrophosphate method [11], total concentrations of trace elements - on the spectrometer Rontec PicoTAX (72, 80 and 51 samples, respectively) (TXRF method) [12]. Classification of particle size distribution in studied soils was carried out in accordance with [13].

Statistical processing of the data and its visualization were performed in STATISTICA software: calculation of the average, minimum and maximum values, standard error of the mean, correlation between the content of trace elements and soil granulometry.

3. Results and discussion

Soils of the upland Tobolsky Mainland are presented by sod-podzolic-gleys with the second humic horizon in elevated landscape positions and gleysols in subordinate positions. In forest-steppe landscapes of Ishim plain predominant soils are dark gray forest soils under aspen-birch groves and arable land on the watersheds and solods in the bottom parts of local depressions. Soils of Turinskaya plain are ploughed clay-illuvial chernozems and solods.

Soils of the studied sites vary in the particle size distribution. Particle size distribution in mineral samples from Tobolsky Mainland ranges from medium to heavy clay loams, higher content of clay was found in organomineral horizons (figure 2). The soils of Ishim and Turinskaya plains are poorly differentiated by particle size distribution, heavy clay dominates (figure 2).

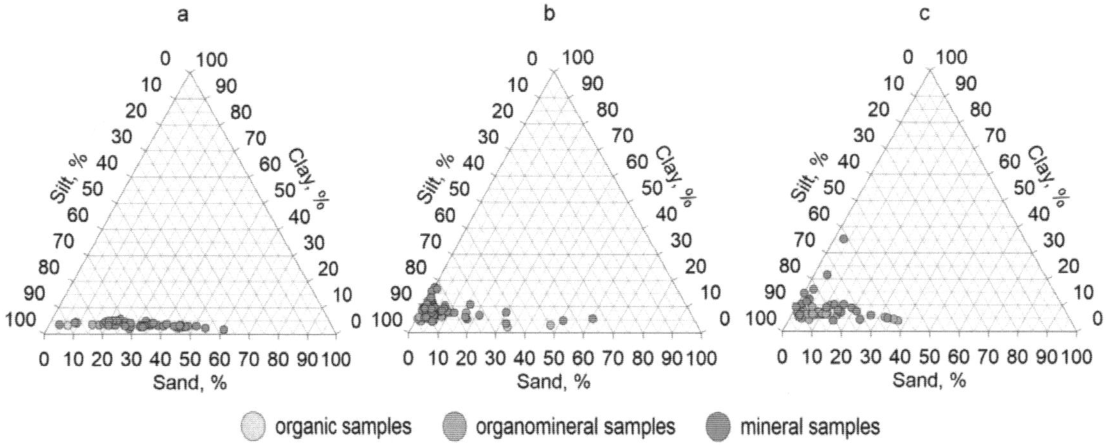

Figure 2. Particle size distribution in soils of three plot sites: *a* – upland Tobolsky Mainland (n=48), *b* – central part of Ishim plain (n=80), *c* – Turinskaya plain (n=51). Particle sizes: clay (<0.001 mm), silt (0.001- 0.05 mm), sand (0.05-1 mm).

The results of the total elements concentrations analysis in the studied soils have led to the following conclusions. Ti is a typical lithophylous element and one of the main components of the silicates, so its maximum average values are characteristic for the soils formed on the most weathered rocks (figure 3b). The behavior of Cr is largely dependent on landscape position and local climatic conditions. In acid environment of the southern taiga Cr intensively precipitates (figure 3a). The maximum Mn concentrations are observed in soils of the southern taiga landscapes due to its accumulation in the form of nodules, intensively formed under hydromorphic conditions (figure 3a). The highest concentration of Ni is typical for loamy soils rich in organic matter (figure 3b). The average Cu values in soils are low in chernozem soils of Turinskaya plain formed on calcareous loams (figure 3c). Highs values of Cu in the soils of Ishim Plain can be probably explained by the intensive processes of its bioaccumulation and relatively low mobility of the metal in steppe landscapes (figure 3b). The average Zn content reduces, while the content of clay particles in the examined soils increases (figure 3c). Low average concentration of Rb is characteristic for sandy soils of terraces (figure 3a); the maximum concentrations are approximately the same, which is associated with the analysis of samples including organic horizons. Sr is concentrated in the most enriched in silt calcareous soils of Turinskaya plain (figure 3c). Intensive removal of Sr is observed in sandy soils of the Irtysh terraces (figure 3a). Pb is less mobile, the content of this element in soils is the same in all plot sites, but it tends to accumulate in soils which are the most enriched in silt (figure 3c).

The relationship between total concentrations of trace elements and soil particle size distribution was studied using the nonparametric Spearman coefficient ($p < 0.5$). Moderate negative correlation between Pb and medium silt (-0.32), average positive relationship between Zn and fine sand (0.3), Pb and coarse sand (0.23) were observed in soils in the high terraces of the interface of Ishim plain. Moderate positive correlation between Ti, Zn, Sr and fine sand (0.34, 0.32 and 0.32, respectively), weak positive ones between Rb and medium sand (0.29), moderate negative ones between Zn and clay (-0.32), Ti, Ni, Cu, Zn, Rb, Sr, and fine silt (-0.34, -0.31, -0.40, -0.32 and -0.36, respectively), Ti, Cu, Zn, Rb, Sr and medium silt (-0.34, -0.31, -0.40, -0.32 and -0.36, respectively) were observed in soils of the upland Tobolsky Mainland. In soils of Turinskaya plain, such correlations were not identified.

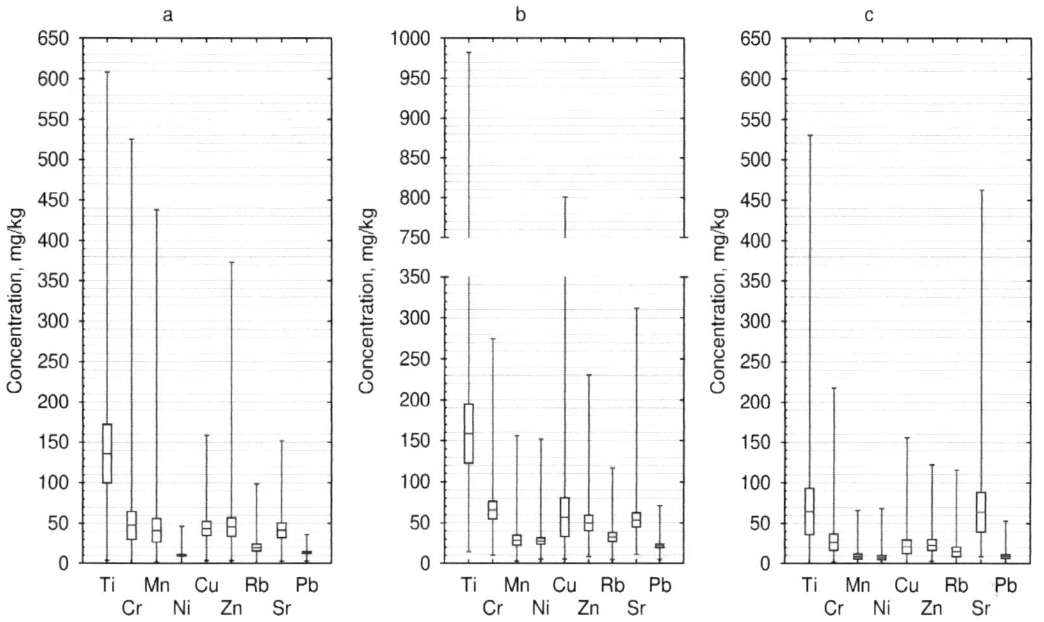

Figure 3. Average concentrations and variance of trace element concentrations in soils of studied plot sites: *a* – upland Tobolsky Mainland (n=72), *b* – Ishim plain (n=80), *c* – Turinskaya plain (n=51). Concentration of Mn and Ti are given in mg/100 g of soil.

Noteworthy is the fact that the content of elements such as Ti, Cr, Ni, Rb, and Sr in the studied soils is lower than their average content in the soils of the world, and the most common soil-forming rocks in Western Siberia (table 1). Primarily, some features of the sampling procedure can explain these results. We analyzed samples collected not only from the upper horizon, but also throughout the whole profile depth, which lead to the sample heterogeneity. In total, the average total concentrations of trace elements in soils of the studied geomorphological regions were significantly different (table 1), which, perhaps, may result from differences in particle size distribution of soils.

Table 1. Mean total concentrations of trace elements in soils of different geomorphologic units of Western Siberia, mg/kg

	Tobolsky Mainland (n=72)	Ishim plain (n=80)	Turinskaya plain (n=51)	Southwestern part of Western Siberia	Loess loams of Western Siberia [2]	Average concentrations in world soils [14]
Ti	1360	1590	650	1270	4100	4600
Cr	48	66	27	50	61	200
Mn	410	290	90	280	835	850
Ni	10	28	7	16	40	40
Cu	44	57	21	43	43	20
Zn	46	50	23	42	92	50
Rb	20	33	15	24	-	100
Sr	42	53	64	52	595	300
Pb	13	22	9	16	18	10

4. Conclusion

The content of trace elements in soils of the main geomorphological units of the south-west of Western Siberia is not always determined by the characteristics of particle size distribution and genesis of local parent rocks. The intensity of the biological cycle and the water regime are the main factors determining their content in soils and landscapes of a particular territory. Properties and genesis of local parent rocks are significant factors for distribution and accumulation of trace elements in the soils of the southern taiga; at the same time, the processes of bioaccumulation in thick humus horizons of dark gray soils and chernozems apparently play an important part in accumulation and migration of trace elements in forest-steppe soils of Ishim and Turinskaya plains.

References

[1] Il'in V B, Syso A I, Konarbaeva G A, Baidina N L and Cherevko A S 2000 *Eurasian Soil Sci.* Heavy Metal Contents of Soil-Forming Rocks in the South of Western Siberia. Vol. **33(9)** pp. 950-953.

[2] Il'in V B, Syso A I, Baidina N L, Konarbaeva G A, Cherevko A S 2003 *Eurasian Soil Sci.* Background Concentrations of Heavy Metals in Soils of Southern Western Siberia. Vol. **36(5)** pp. 494-500.

[3] Acosta J A, Martínez-Martínez S, Faz A and Arocena J 2011 *Geoderma.* Accumulations of major and trace elements in particle size fractions of soils on eight different parent materials. Vol. **161** pp. 30–42.

[4] Qian J, Shan X, Wang Z and Tu Q 1996 *Sci. Total Environ.* Distribution and plant availability of heavy metals in different particle-size fractions of soil. Vol. **187(2)** pp. 131–141.

[5] Varlamov I P 1972 *Geomorfologiya Zapadno-Sibirskoy Ravniny* (Novosibirsk: Zapadno-sibirskoe knizhnoe izdatel'stvo).

[6] Karetin L N 1987 *Eurasian Soil Sci.* Genetic features of sod-podzolic soils of the «Tobol'sk Continent». Vol. **11** pp. 34-43.

[7] Karetin L N 1990 *Novosibirsk Nauka.* Pochvy Tyumenskoy Oblasti.

[8] Karetin L N 1982 *Novosibirsk Nauka SB.* Chernozemnye i Lugovye Pochvy Tobol-Ishimskogo Mezhdurech'ya.

[9] Chizhikova N P, Grakovskiy V G and Gradusov B P 1982 . *Eurasian Soil Sci.* Some features of the mineral soil composition of the Transurals forest-steppe. Vol. **8** pp. 84-96.

[10] Shishov L L, Tonkonogov V D, Lebedeva I I and Gerasimova M I 2004 Klassifikatsiya i Diagnostika Pochv Rossii (Smolensk: Oykumena)

[11] Shein E V, Milanovskii E Yu and Molov A Z 2006 *Eurasian Soil Sci.* The effect of organic matter on the difference between particle-size distribution data obtained by the sedimentometric and laser diffraction methods. Vol. **39(1)** pp. 84–90

[12] Towett E K, Shepherd K D, Cadisch G 2013 *Sci. Total Environ.* Quantification of total element concentrations in soils using total X-ray fluorescence spectroscopy (TXRF). Vol. **463-464** pp. 374-388

[13] Kachinskiy N A 1965 *Fizika Pochvy* (Moscow: Vysshaya shkola)

[14] Vinogradov A P 1959 *New York: Consultants Bureau Enterprises.* The Geochemistry of Rare and Dispersed Chemical Elements in Soils.

Environmental formation conditions of Kimmeridgian-Oxfordian Upper Jurassic sediments in Aleksandrov mega-swell

G Il'ina[1], Ye Schmidt[2] and R Abramova[3]

[1 2 3]Institute of Natural Resources, Tomsk Polytechnic University, Tomsk, Russia
E-mail: ilgf@sibmail.com[1], lena_shmidt@bk.ru[2]

Abstract The research target is Upper Jurassic sediments within Vasugan horizon. Identification of Vasugan and Naunak suites based on the differentiation and well log correlation and detailed lithological core analysis of Vakhsk, Ob and Protochni fields revealed the existing paleo-environments. It was established that continental facies significantly influenced sedimentation within the investigated territory.

1. Introduction

Detailed paleogeographic differentiation describing the formation conditions of thicknesses and isolated reservoirs (1980-1990) revealed and defined the transition zone of Vasugan and Naunak suites as v-interpenetrating coastal-marine and continental sedimentary units (according to the Law of Facies- Walther's). Ranging formation age of separate petrographic horizons, i.e. facies types and their boundaries, is governed by coastline movements. The overlying sediments within the sedimentary thickness profile extend on the lithosphere surface and/or sedimentation basin bottom. In the case of sea level transgression and / or regression the horizontal sediment zones (facies) merge into sedimentary thickness vertically. As a result sediments in one and the same facies are of different ages towards land-sea direction [1].

2. Geological setting

The investigation target is Aleksandrov mega-swell embracing the formation conditions of Upper Jurassic suite. Aleksandrov mega-swell sheath (mantle) formed during Triassic age with the formation of rift structures intersecting geosyncline structural-formational zones [2, 3], and the formation of basement blocks uplifting during the Mesozoic and Cenozoic periods dominating as regional oil and gas accumulation zones. Jurassic sediments compose up to 1/7 of the terrigenous mantle functioning both as caprocks (Upper Jurassic Bazhenov suite rocks) and as reservoir rocks of petroleum deposits (Vasugan, Naunak and Tumen suite sediments).Vasugan and Naunak suite sediments were exposed to the entire Aleksandrov mega-swell and includes interbedded sandstones, siltstones and claystones. The sandstones are characterized as gray or light-gray, polymistic texture, aleuro-psammo and psammo-aleuro structures, random, massive; total amount of clastic particles ranges from 88 to 92 % and cement content -8-12%. Clastic matter includes quartz (32-65 %), feldspar (20-50 %), rock fragments (10-47 %), mica (1-3 %). Although the sediment composition in Vasugan and Naunak suites is identical, the rocks have different assortment of particles (sandstone in Vasugan suite- 2.44 and in Naunak- 1.72) and different content of pellite fraction (in Vasugan suite- 21.15% and in Naunak- 13.26%). Siltstones are predominately fine-grained, from dark gray to black, subconchoidal fracture, finely horizontal bedding of clays with isolated glauconite grains. Total amount of clastic material- 60 %. Claystones are dark gray to black, coarsely tabulated, horizontally bedded. Major clay matter is composed of strongly montmorillonite hydromica aggregates with undirectional clay particle orientation.

According to well logging data

a) high apparent resistivity (AR) values of enclosing rocks (4.0-28.0 Ohm) were determined, especially for carbonized argillaceous-silt interlayers;

b) self-potential (SP) curve revealed values of up to 80mV in sand interlayer intervals;

c) spontaneous potential reduction value ($a_{пс}$) is directly proportional to rock porosity (permeability) value.

The cumulative thickness of Vasugan and Naunak suites within the mega-swell is uniformly distributed. Thickness ranges from 33m in Saimovsk area to 89m in Traigorodsk area. This thickness is more contrastively revealed within the wells: from 0 m (well № 217 in Chebachya area) to 94m (well № 1 in Pankovskaya area). The thickness decreases westward to eastward in Aleksandrov mega-swell. Maximum values were revealed in the central and north-east areas of this structure. Sand content(net-to-gross system) is coherent with corresponding total suite thickness data, i.e. the maximum net-to-gross system in N-E megaswell (Vakhskaya, Traigorodsk and Migitinsaya areas) and central area (Obskaya) is relevant to the zone of maximum developed thickness suite zone (north-eastern area) and reduced thickness zone (central area of the structure).

This could be conditioned by the facial differences in sand material deposition. This can be explained by the fact of traceable changing distribution of organic remains in vertically Oxfordian – Kimmeridgian sediments. Marine fauna remains (belemnite layers, pelecypoda shells, foraminifer *Globulina*) were found in the sediments of the upper suite section, i.e. layer U_1^1 located within the central and south-western areas of Aleksandrov mega-swell (Obskaya area-wells №1,2; Kondakovskaya area- wells № 32, 33; Chebachya area- well № 220 and Protochaya area- well № 1). According to the organic remains in Obskaya area – well № 1 (core sampling interval 2271,1-2276,1 m) the sediments were defined as Oxfordian –lower Kimmeridgian (J_3o-km); according to *Globulina obskaensis Dain* remains in Obskaya area – well № 2 (core sampling interval 2312,8-2314,9m) the sediments were defined as Late Oxfordian (J_3o_3); pelecypoda shells *Entolium sp. indet., Meleagrinella sp. Indet* found in Kondakovskaya area- well № 32 (core sampling interval 2045,9-2052,8m) are typical of Oxfordian sediments (J_3o); pelecypoda shells *Buchia cf. Concentrica* found in Kondakovskaya area- well № 33 (core sampling interval 2138,9-2140,9m) are typical of Early Kimmeridgian sediments (J_3 km_1). Downwardly, fauna has not been found in the layers (U_1^2, U_1^3, U_1^4). However, plant remains (*Coniopteris sp.*), pollen and vegetal spores, characteristic of Upper Jurassic sediments (J_3) in Naunak suite, are pervasive in these layers. Fauna remains identical to those found in layers $U_1^{2\text{-}4}$ (J_3o-km_3) can be found in the north-eastern area of the mega-swell in the upper suite section[7].

3. Research methods

It is rather difficult to identify correlating Vasugan and Naunak suite analogues as they are genetically different: Vasugan suite- coastal-marine while Naunak suite- coastal plain continental sediments. Based on the traces of marine sediments, the composition of which is related to pollen-spore spectrum, I. Nesterov (1966) defined a sovereign stratigraphic unit- Milidzhinskaya suite. Later (in 1967) this suite was renamed Naunak [I.Nesterov, et al., 1975].

Passega C-M diagram (1981) was applied in dividing Naunak and Vasugan suites and identifying the sedimentation genesis (fig. 1) which involved the sand and silt cross-sections of upper Vasugan sediments.

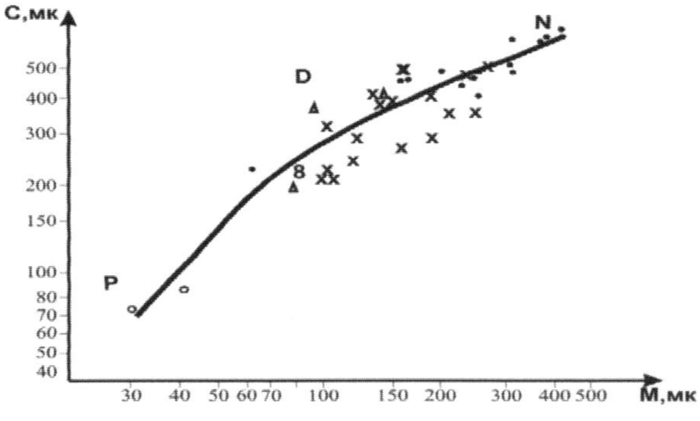

x – 1 • – 2 ○ – 3 Δ – 4

Figure 1. Passega C-M diagram for sand and silt cross-sections of Upper Jurassic sediments in Aleksandrov mega-swell. Suite sandstones: 1- Naunak, 2- Vasugan; suite silts: 3- Naunak, 4- Vasugan

It is noted that the application of genetic diagrams, including Passega C-M diagram, accurately indicate only the features of this or that deposition environment. The diagrams could show either identical and / or approximately close features in different facies, as the movement of water and material transport are similar. Re-deposited rocks could inherit the properties of the breakdown rocks which could be noted on the diagrams [4,5]. Above-mentioned facts indicate that it is necessary to apply a multi-discipline study, involving several methods (fig.2).

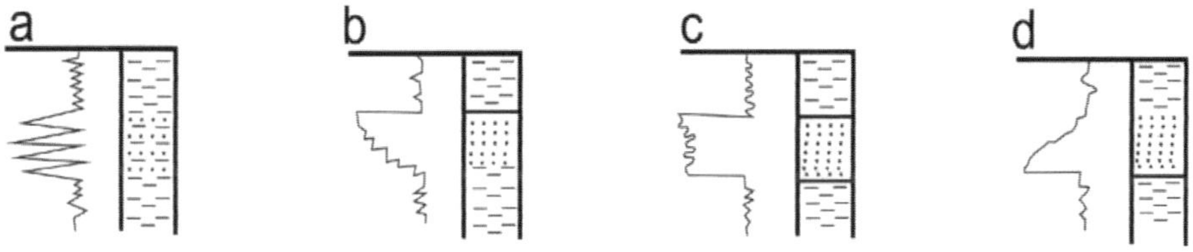

Figure 2. Four types of gamma logging curves: a- thin interbedded sandstones and slates; b- cross-section, revealing upward coarse material, sharp sandstone-slate contrast; c – homogeneous sandstone with distinct upper and lower contacts; d- thickness with gradual upward fine-grained material, i.e. from sandstone to slate, including abrupt bottom formation.

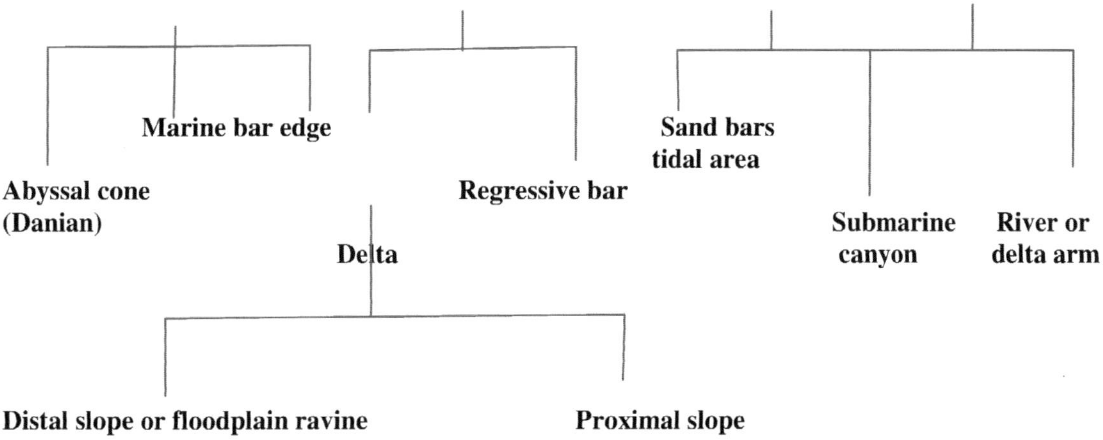

Viewing the deposition conditions, it is important to consider the rock component particle size-deposition system alteration dependence. This fact plays an important role in the case of limited core recovery [6]. The configuration and type of well logs indirectly reflect the changes of sediment particle sizes, which, in its turn, stipulates the possibility of applying different well logging methods to develop the models showing the changes of vertical grain sizes. In this case, it is viable to use both SP curves reflecting reservoir properties and radioactivity logging (RL) identifying clay content. For example, V. Murotsev method (1984) is lab-referenced to rock material excluding overall genetic description of the rocks obtained from low core recovery or without any core recovery.

If applying gamma log and SP curve regularity patterns in accordance with Selley method (1989), then this problem could be solved (fig. 2). It should be noted that none of the above-mentioned features could be a major one for identifying this or that deposition environment. However, these features in combination with content values of glauconite (typical for deep-water facies) or carbon-bearing detritus (typical for coastal and continental facies) could be auxiliary identification criteria for physic-geographical formation conditions of sand bodies.

4. Results and discussion

Based on the analysis results of deposition conditions, reservoir properties and petrographic composition of Upper Jurassic rocks in Vasugan and Naunak suites, the most potential areas with lithologically screened accumulations could be in the transition zone (fig.3).

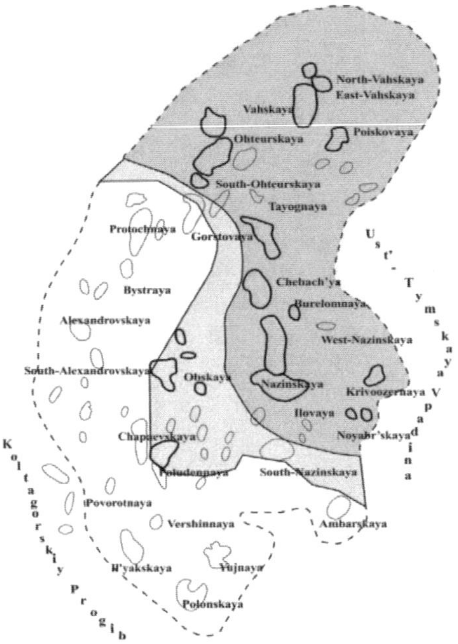

Figure 3. Diagram of paleogeographic formation conditions of Vasugan and Naunak suite sediments in Aleksandrov mega-swell

- Formation zone of Vasugan suite sediments (marine sediments of littoral slope)
- Formation zone of Naunak suite sediments (continental sediments of coastal plain)
- Formation zone of transition sediments (sediments of beach sands)

5. Conclusion

The division of Vasugan and Naunak suites (Aleksandrov mega-swell) based on Selley method, Passega C-M diagram and analysis of organic remains in core samples, provides a rather detailed and complete genesis characteristic description of the sand layers in Upper Jurassic sediments [7, 8]. However, there still remains one relevant question- what is the influence of rock genesis on HC saturation and filtration properties of the rock itself? Core sampling and logging data indicated the fact that there is an insignificant difference between Vasugan and Naunak suites regarding their reservoir properties. Increasing clay mineral content in Vasugan suite sand-silt rocks is conditioned by the slow transport rate of the clastic material, which, in its turn, influenced the suite aggregate thickness [9, 10].

.

References

[1] Danenberg E, Belozerov V and Brilina N 2006 *Tomsk Publishing House.* Geologicheskoe stroenie i neftegazonosnost' verhnejursko-melovyh otlozhenij jugo-vostoka Zapadno-Sibirskoj plity (Tomsk Oblast) pp. 291.
[2] Surkov V, Zhero O 1981 *Moscow Nedra.* Fundament i razvitie platformennogo chehla Zapadno Sibirskoj plity. pp.62-91.
[3] Kontorovich A, Nesterov I, Salmanov F 1975 *Moscow Nedra.* Geologija nefti i gaza Zapadnoj Sibiri. pp. 76-112.
[4] Proshlyakov B, Kuznetsev V 1981 *Moscow Nedra.* Litologija i litologo-facial'nyj analiz. pp. 167-213.
[5] Selley Richard 1988 Ancient sedimentary environments and their sub-surface diagenesis. pp. 31-159.
[6] Murotsev V 1984 *L Nedra.* Jelektrometricheskaja geologija peschanyh tel - litologicheskih lovushek nefti i gaza. pp. 24-110.
[7] Il'ina G 2000 *Geology-mineralogical sciences Tomsk.* Geologija i uslovija formirovanija jurskih otlozhenij Aleksandrovskogo megavalaDissertation abstract. pp. 25
[8] Alekseev V P 2007 *Ekaterinburg: publishing house of UGGU.* Atlas of facies from the Jurassic terrigenous sediments (coal-bearing strata of Northern Eurasia). pp .209.
[9] Logutenkova N, Sorokina I E 1980 *lithofacies and paleogeographic criteria of petroleum potential.* lithofacies features of oil and gas formations of biohermal. pp. 64-68.
[10] Korolyuk I K, Krylov N, Maltsev A K 1987 *Nauka.* Petroleum potential of sedimentary formations.

IOP Conf. Series: Earth and Environmental Science **43** (2016) 012004 doi:10.1088/1755-1315/43/1/012004

Identification of clay minerals in reservoir rocks by FTIR spectroscopy

Vu Cong Khang, Mikhail V. Korovkin, Ludmila G. Ananyeva

Institute of Natural Resources, National Research Tomsk Polytechnic University, Russia
E-mail: congkhangvu@gmail.com, mvk@tpu.ru, lga@tpu.ru

Abstract. Clay minerals including kaolinite, montmorillonite and bentonite in oil and gas reservoir rocks are identified by absorption spectra obtained via Fourier Transform Infrared (FTIR) spectroscopy. Bands around 3695, 3666, 3650 and 3630 cm^{-1} and bands around 3620 and 3400 cm^{-1} are the most diagnostically reliable for kaolinite and montmorillonite, respectively; also absorption bands in the region of 1200…955 cm^{-1} are equally diagnostic for all the clay minerals studied.

1. Introduction

Currently the issue of studying the oil and gas reservoir rocks and clay minerals composing them becomes very important and relevant to in geology and oil and gas fields development. Clay minerals are formed as a result of alteration of such primary minerals as feldspars, micas, chlorides, etc., and as a result of their deposition as gels from colloid and true solutions. Clay minerals of montmorillonite, kaolinite and hydromica groups and mixed-layer formations prevail in a lot of clay rocks [1].

The structure of clay rocks including their size, shape, surface texture, structural elements quantitative and spatial relation and structural relation type is the main factor that determines clay rock properties. The clay rocks microstructure is a sensitive indicator of the rock formation environment and various combinations of microstructure features are in strong relation with the rock properties. The most important property of clay rocks is their swelling ability, i.e. an ability to expand when interacting with water and water solutions [2]. Clay rocks composed by clay minerals with a more elastic crystalline structure (like montmorillonite) and a less elastic crystalline structure (like kaolinite) show off the most swelling and the least swelling ability, respectively. Such clay minerals significantly impact the filtration properties of reservoir rocks and this has to be taken into account when designing oil and gas production methods.

It is a rather complicated task to identify clay minerals that compose reservoir rocks. For this purpose various methods including Mossbauer spectroscopy, thermo luminescence, thermal analysis, SEM investigation, X-ray diffraction, infrared (IR) spectroscopy and other are used [3]. Such methods are mutually complimentary and should be used in combination.

In this paper an IR spectroscopy was used to identify minerals by their characteristic absorption spectra [4].

2. Research methodology

The IR absorption spectra were recorded with a spectrophotometer Specord M40 in a region of 400…4000 cm^{-1} with a resolution of 0.01 cm^{-1} and a Shimadzu spectrophotometer IRPrestige-21 with Fourier Transform (FTIR-8400S) in a region of 300…4000 cm^{-1} with a resolution of 0.001 cm^{-1} (FT-IR), using the software IRsolution [5]. Study samples were taken from training collections and Mineralogical Museum collections of the Tomsk Polytechnic University and Tomsk region oil and gas fields. A sample was finely ground in an agate mortar to the powder with a particle size of 2 μm and less. Then it was mixed with KBr powder of Very Pure and Analytical Reagent (AR) grades, and then the powder mixture was put in a press mould and pressed at a high pressure to form translucent thin disks. When measuring IR spectra by a spectrophotometer FTIR – 8400S, a sample weight varied in the range of 0.1 – 0.5 mg.

3. Experimental results and discussion

The clay mineral samples shown in fig.1 can be attributed to kaolinite by IR absorption spectra. Their IR spectra demonstrate fully-represented and well-resolved Si-O-Si and Al-O-Si bands of the mineral lattice around 1102, 1033, 1014, 938, 755, 693, 541, 462 and 432 cm^{-1} and also clearly-seen absorption bands of OH stretching. The main distinctions of kaolinite-group minerals is observed in OH stretching absorption; particularly, there are two narrow and intensive absorption peaks around 3695 and 3630 cm^{-1} and weak band peaks around 3650 and 3666 cm^{-1} which are characteristic of an ordered variety of kaolinite [6, 7].

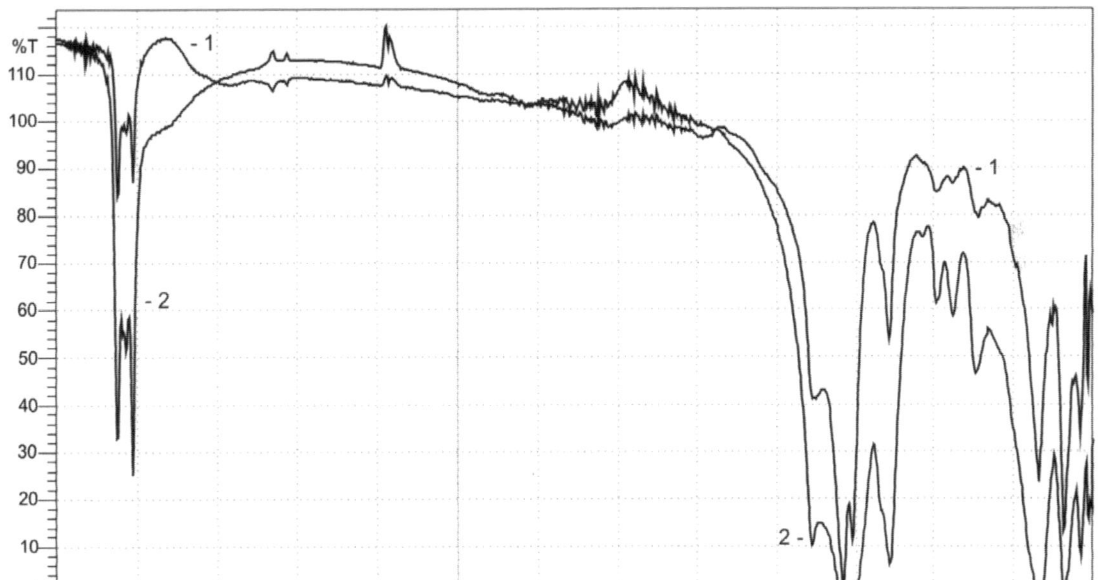

Figure 1. Panoramic spectra of minerals in the region of 4000...400 cm^{-1}: 1 – kaolinite(Tomsk Polytechnic University's training collection), 2 – kaolinite (Russia, Krasnoyarsk region, Field Verkhoturovskoye)

IR absorption spectra of montmorillonite-group minerals illustrated in fig. 2 are characterized by a double absorption peak around 3620 and 3400 cm^{-1}, an intensive peak around 1032 cm^{-1} and weak peaks around 920 and 842 cm^{-1} [6, 8]. Spectra of montmorillonite and bentonite, being a clay mineral whose main constituent is montmorillonite, are rather the same. The bentonite quality depends on the content and ionic form of montmorillonite. The difference is that there appear intermediate absorption bands in the region of 1679…1600 cm^{-1} and that a weak splitting of the peak around 1032 cm^{-1} is and, at the same time, a weak peak is not observed around 1105 cm^{-1}.

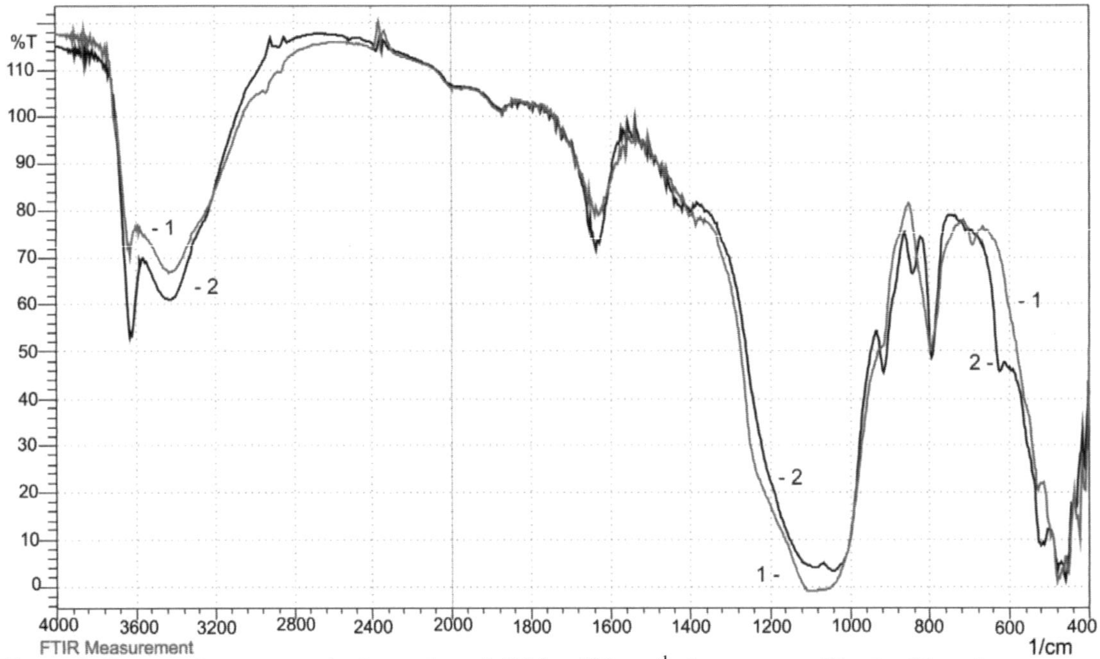

Figure 2. Panoramic spectrum in the region of 4000…400 cm^{-1}: 1 – montmorillonite (Tomsk Polytechnic University's training collection), 2 – bentonite (Republic of Tadjikistan)

4. Conclusion

Spectra of clay minerals are rather complicated; the intensity and position of IR absorption peaks are dependent on conditions of mineral formation and presence of various impurity ions and elements including Mg, Ca, Fe in a lattice of the mineral; the larger the ionic radius is, the more peaks shift to a longer-wave region. Comparison of IR spectra of various modifications of a clay mineral allows us to identify spectra of the mineral structure and to assess the impact of such mineral structure spectra on the structure itself. Bands around 3695, 3666, 3650 and 3630 cm^{-1} and bands around 3620 and 3400 cm^{-1} are the most diagnostically reliable for kaolinite and montmorillonite, respectively; also absorption bands in the region of 1200…955 cm^{-1} are equally diagnostic for the all examined clay minerals. Thus, an FTIR spectroscopy makes it possible to reliably identify unknown clay minerals. Currently, with the help of atlases and computer-assisted files of spectra, it is possible to identify any compounds, minerals and substances, provided that they and their vibration spectra have been already recorded and held in storage.

References

[1] Burlin Yu, Ivanov M, Kalmykov G, Karnyushina E, Korobova N 2008 *Moscow University publishing house* . Petrophysical methods in core studies. Vol. **1** pp. 112.
[2] Osipov V, Sokolov V, Rumyantseva N 1989 *Moscow Nedra Pabl.* Clay rocks microstructure. pp. 211
[3] Madejová J 2003 *J. Vibrational spectroscopy*. Vol. **31 (1)** pp. 1–10.

[4] Korovkin M 2012 *Tomsk Polytechnic University publishing house.* Infrared spectroscopy of carbonated minerals. pp. 80.
[5] Razva O, Anufrienkova A, Korovkin M, Ananyeva L, Abramova R 2014 *IOP Conference Series: Earth and Environmental Science*. Vol. **21 (1)** pp. 012006.
[6] Madejová J, Kečkéš J, Pálková H, Komadel P 2002 *J. Clay Minerals*. Vol. **37 (2)** pp. 377.
[7] Madejová J, Komadel P 2001 *J. Clays and Clay Minerals*. Vol. **49 (5)** pp. 410.
[8] Bishop J, Pieters C, Edwards J. 1994 *J. Clays and Clay Minerals*. Vol. **42 pp.** 702.

Petrographic-geochemical characteristics of granitoids and their epigenetic alteration products in paleovalley fields (Vitim uranium-ore site)

E S Kuznetsova, V A Domarenko, I A Matveenko

Tomsk Polytechnic University, 30 Lenin Avenue, Tomsk, 634050, Russia

E-mail: usoltseva_es@mail.ru, viktor_domarenko@ mail.ru, mia2046@yandex.ru

Abstract. The study describes the results of the mineral and element composition of granitoids in basement and weathering crust of Khiagdinsk ore field in Vitim uranium ore site. It has been stated that granitoids in basement consist of leucocratic biotite granite of subalkaline group. The major rock-forming, accessory (apatite, zircon, sphene (titanite), magnetite, monazite, xenotime), and uranium-bearing minerals have been determined. Weathering crust is composed of unlithified or weakly lithified sediments, among which sandy and sandy medium gravel deposits have been distinguished in terms of mineralogical and granulometric texture. High radioactivity of granitoids was revealed in thorium-uranium basement and natural uranium. The combination of the specified factors presupposes that granitoids of Vitim uranium ore site may be a source of uranium in the fields of the paleovalley type.

1. Introduction

The interest in studying uranium and thorium in granitic rocks started in the 1950's due to two facts: with understanding the fact that these rocks have, on average, higher uranium and thorium contents than any other rock types and weathering of granitic rocks may be a source of uranium in some deposits [1]. For example, as early as in 1969 Rosholt and Bartel showed a clear relationship of uranium deposits and surrounding outcrops of granitic rocks [2].

Specialized survey of uranium within Vitim plateau by Sosnov expedition began in 1947. Khiagdinsk ore field is located in the Republic of Buryatia, Central Transbaikal. Ore field includes 8 uranium deposits of paleovalley type (Khiagdinsk, Vershinnoye, Istochnoye, Namaru-Koretkonde, Kolichikan, Tetrakhskoye, Dybryn, Gilindinskoye) (fig. 1).

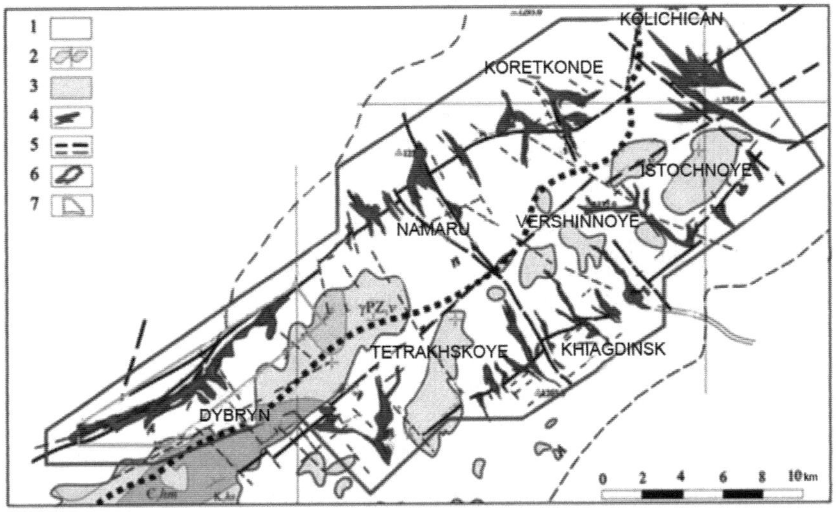

Figure 1. Overview geologic map of Khiagdinsk ore field (according to the data presented by JSC "Rusburmash"): 1 – Neogene deposits; 2 - 3 basement rocks: sedimentary (a) and metamorphic (b) (2), granitoids (3); 4 – uranium-ore deposits; 5 – discontinuous faults of different types; 6 – the contour of Khiagdinsk ore field.

Geologically, within the zone, a large Caenozoic paleoriver system of Bol'shoy Amalat with major N-E flow was exposed under Miocene basalt mantle of 4000 sq. km in area. Paleovalleys are predominantly intersected by granitoid crystalline basement consisting of grey sedimentary rocks. Three units were distinguished in Neogen-Quaternary formations (bottom-up): terrigenous arkosite (80 m), igneous-sedimentary (120 m) and igneous (250 m); two lower units are comprised of a paleovalley system, upper – a superposed sheath (mantle) [3].

The geological structure of Khiagdinsk ore field is conditioned by the formation of two structural levels consisting of heterogeneous basement, predominantly granitic, and mantle of Cainozoic igneous- terrigenous formation.

Paleovalley basement and its margins consist of 90% highly-radioactive granitoids of Vitim-Kansk complex (25…40 mcR/h). In Pre-Neogene period all basement rocks were exposed to intensive chemical weathering resulting in the formation of overlying clay coarse-gravel weathering crust, the thickness of which was from 1 meter to ten meters. Transported clastic material forms Neogene productive sedimentary sequences in Dzhilinda suite.

Localized mineralization conditions within Khiagdinsk ore fields were considered by many researchers [4, 5, 6, 7, 8]. The question - what is the source of ore matter - remains is still open, and this defines the relevance and practical application of the study.

The purpose of the given paper is to study the composition of basement granitoids and their derivative products and determine their geochemical differentiation and potential ore source in paleovalley deposits.

2. Methods and material

The study is based on the materials of specialized mineralogical-geochemical survey of basement rocks and derivative products in Namaru-Koretkonde and Dybryn fields.

Rock composition was studied by traditional methods of optical diagnostics on transmitted -light and reflected -light microscopes in thin and polished sections by X-ray difractometer D2 Phaser (Bruker AXS GmbH, Tomsk, analyst – D.G. Usol'tsev).

Scanning electron microscope (SEM) Hitachi S-3400N with energy-dispersive spectrometer Bruker XFlash was used to determine the composition from Li to U with an element content more than 0.1 wt.

% in accessory microinclusions and ore mineralization (Training and Research Laboratory of Optical and Electron Microscopy, Uranium Geology International Center, Tomsk Polytechnic University).

To determine the chemical composition highly sensitive method of inductively coupled plasma mass spectrometry (ICP MS) was used to examine 60 elements in accredited laboratory of LLC Chemical-Analytical Center "Plasma", Tomsk (supervisor – N.V. Fedyunina).

3. Results and discussion

Based on optical diagnostics and X-ray difractometry, the mineral composition of basement granitoids and weathering crust were investigated (not transported, slightly transported, transported).

Basement granitoids are characterized by mature mineral composition and structural features. Porphyric leucocratic (light-colored) biotite granites of subalkaline group are predominate. The major rock-forming minerals: quartz 20...35%, plagioclase 15...25%, K-feldspar 25...35%, biotite 1...5%, muscovite 1...7%. Accessory minerals (1...3%) includes apatite, zircon, sphene, magnetite, monazite, xenotime.

Weathering crust includes unlithified or weakly lithified sediments. Mineralogically and granulometrically, the following petrographic types of deposits are distinguished: sandy and sandy medium gravel and differ in large rock fragments.

They are light-gray rocks with gray, yellowish or brownish sections due to organic remain impurities and pyroclastics. Core samples revealed sand, granitic subsoil or gravel of weakly lithified irregular coarse rocks.

The microscopic study revealed their similar mineral composition. Based on this fact several types of sands can be distinguished: 1) quartz-feldspar sand without organic impurities and pyroclastics which could be a product of slightly transported redeposited weathering crust along subalkaline leucogranites; 2) quartz-feldspar sand with insignificant organic impurities and pyroclastics (up to 1...2%) which could be transported weathering crust along subalkaline leucogranites; 3) quartz-feldspar sand with organic impurities and pyroclastics (up to 7...10%) which could be a displacement product of transported weathering crust along subalkaline leucogranites and pyroclastic ash flows prior to the formation of the basalt plateau.

The electron-microscopic analysis of granitoid and weathering crust thin sections has revealed rare-earth minerals including rare-earth phosphates, presumably monazite (Ce, La, Nd, Th) [PO_4] (fig. 2) and xenotime (YPO_4) with low impurities of Ga, Dy, Er, Yb. These minerals are rather prevalent and include crystalline fines (2...4 mkm). Uranium (1 %) may be isomorphic impurities.

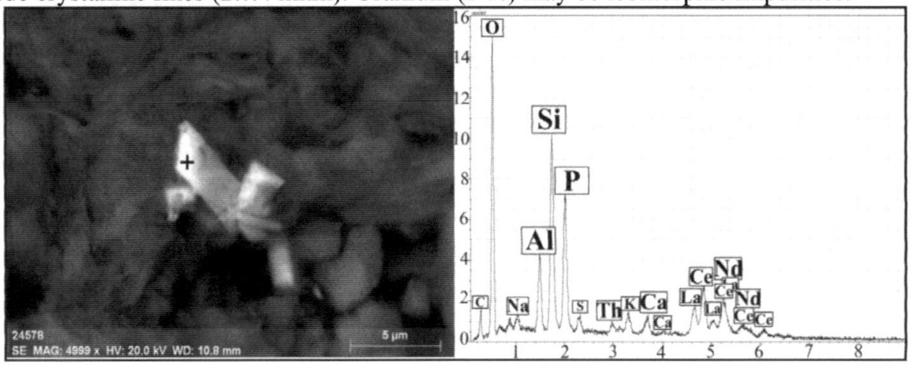

Figure 2. Koretkonde uranium deposit. C-5648, interval: 266.7...267.2 m. Monazite and its energy-dispersive spectrum

Electron-microscopic analysis showed the presence of aggregates and some zircon crystals of up to 10...40 mkm in diameter. There are few uranium-bearing (U about 8.5%) variations.

We have found intrinsic uranium minerals. Granitoid basement involves calcium phosphate (U 24%) (fig. 3) and titanate uranium (U 45%) (fig. 4). Examination of thin sections showed that calcium phosphate occurs as veinlets of up to 15 mkm or small grains in the quartz intergranular space. These veins are most likely to have developed in the postmagmatic stage of granitoid transformation.

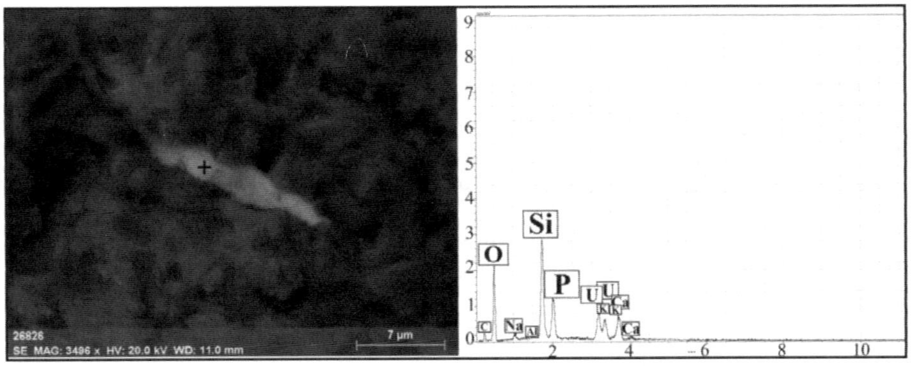

Figure 3. Koretkonde uranium deposit. C-5648, interval: 266.7…267.2 m. Calcium phosphate of uranium and its energy-dispersive spectrum

The energy-dispersive spectrum (fig. 3) revealed silicon which is governed by the siliceous crystal matrix (SiO), similar to uranium titanate.

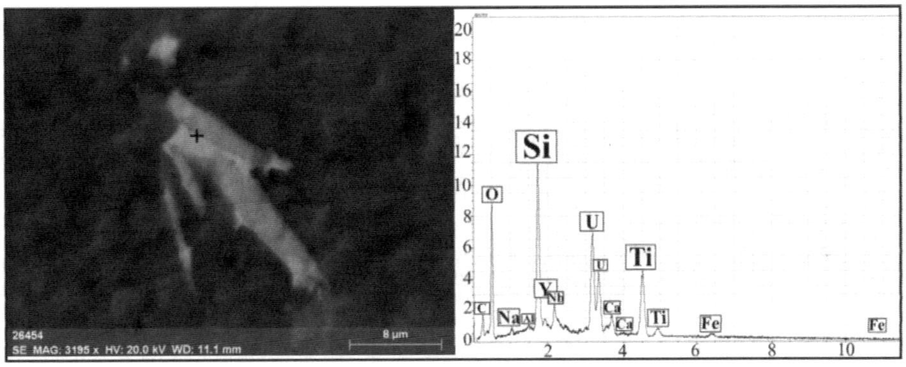

Figure 4. Namarus uranium deposit. C-5459, interval: 327.5…327.8 m. Uranium titanate (brannerite) and its energy-dispersive spectrum

In case of weathering crust natural uranium mineralization was found as calcium-uranium-phosphorous composition (U 32 %). These compounds are mostly finely-dispersed amorphous concentration within intergranular space of clay sands or granite fragments.

Another remarkable feature is the occurrence of iron sulfides (pyrite?) in granitoids and weathering crust. In granitoid basement there are iron sulfides as small crystal impregnations and aggregates. However, investigated weathering crust revealed concentrations of framboid pyrite everywhere which could be the result of the decomposition of organic matter due to bacterial sulfate reduction [9, 10].

In addition, we found significant impregnations of zinc, barium, silver, lead sulfides, as well as iron, titanium, copper, tungsten, silver, tin oxides and gold with impurities of silver, iron, copper, titanium. Intermetallic compounds of Sn, Cu, Zn, Fe, SiO, as well as native metals of W and Ni are revealed.

Analysis of uranium and thorium variations (fig. 5) based on ICP-MS data shows that granitoid basement and weathering crust are characterized by high radioactive thorium-uranium and natural uranium.

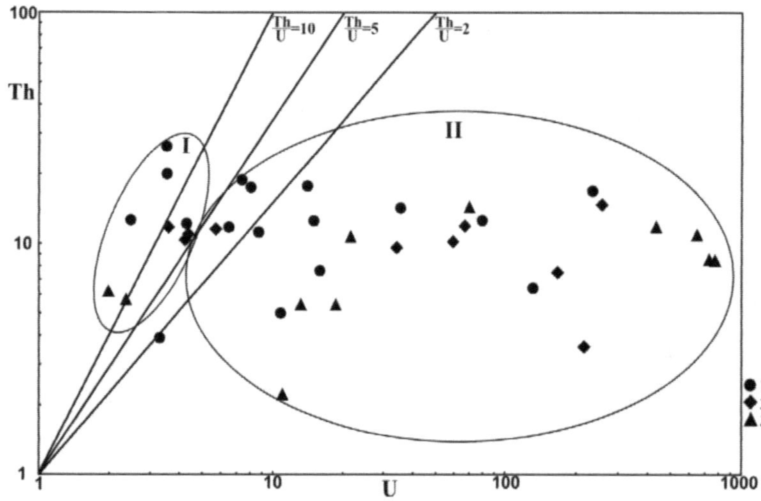

Figure 5. Diagram of content variations in uranium and thorium (according to L.V. Komlev; ed. by A.A. Smyslov): I – highly radioactive granites; II – highly radioactive substantially uranium-bearing granites. 1 – basement granitoids; 2 – slightly transported weathering crust; 3 –transported weathering crust.

4. Conclusion

Based on conducted analysis the following conclusions:

- basement granitoids are leucocratic (light-colored) biotite granite of subalkaline group;
- major accessory minerals of prevailing granitoids are apatite, zircon, sphene, magnetite, monazite, xenotime, single grains of calcium uranium phosphate and brannerite;
- increased concentrations of uranium and thorium indicate uranium – based granitoid basement;
- in the process of hypergene transformations weathering crust develops on granitoids composed of the following prevailing minerals: quartz and feldspar by 1:2, K-feldspar, granite fragments, pyroclastics, organic remains that demonstrate the arid conditions of its development;
- accessory and ore mineral composition in weathering crust include sphene, zircon, monazite, ningyoite;
- as a result of the areal uplifting during denudation a significant part of mantle was washed out and redeposited in paleovalleys;
- due to outwash and roundness of matter the following sediments can be distinguished: slightly transported and transported weathering crust;
- mineral composition of rock-forming components, ore minerals (sphene, zircon, monazite, ningyoite), uranium-thorium relation (from 0.01 to 3) of the weathering crust show that sediment composition of paleovalley and granitoids inherited from the sources of those sediments.

Based on the research presented above it can be concluded that granitoid basement of Vitim site could be a source of uranium-forming material of paleovalley type.

References

[1] Thomas T Tieh, Ernest B Ledger and Marvin W Rowe 1980) *Chemical Geology.* Release of uranium from granitic rocks during in situ weathering and initial erosion central Texas. Vol. **29** pp. 227-248.

[2] Rosholt J N and Bartel A J 1969 *Earth and Planetary Science Latters.* Uranium, thorium, and lead systematics in Granite Mountains, Wyoming. Vol. **7** pp. 141—147.

[3] Mashkovtsev G Konstantinov A Miguta A Shumilin M and Schetochkin V 2010 *Moscow. VIMS Press.* Uranium Russian subsoil. pp. 820

[4] Luchinin I, Peshkov P and Dementiev P 1992 *Prospect and protection of mineral resources* Uranium deposits in paleovalleys of Trans-Ural and Trans-Baikal. Vol. **5** pp. 12-15.

[5] Kochenov A and Khaldey A 1997 *Lithology and mineral resources.* On the question of uranium sources in the exogenous epigenetic deposits. Vol. **4** pp. 371-378.

[6] Nikitina Ye and Prokhorov D 2012 *Moscow University Bulletin.* Regularities of ore formation localization, mineral composition of ore and ore-bearing rocks of Koretkonde uranium deposits (Vitim uranium-ore site). Vol. **6** pp. 44-50.

[7] Domarenko V Kramorenko S Martynenko V Novgorodtsev A Rudenko A Mitrofanov E Derevenets E 2013 *Conf. on Uranium: geology, resources, production* Moscow: VIMS Press. Vvol **66** pp. 50-51.

[8] Kochkin B Novgorodtsev A Tarasov N and Martynenko V 2014 *Geology of Ore Deposits*. The morphological characteristics of the ore bodies and genesis of deposits of uranium ore field Khiagda Vol. **6** pp. 539-555.

[9] Astafiev M, Rozanov A and Khuver R 2005 *Paleontologic Journal.* Flamboids: their structure and origin. Vol. **5** pp. 1-7.

[10] Borkow P and Babcock L 2003 *The Sedimentary Record.* Turning Pyrite Concretions Outside-In: Role of Biofilms in Pyritization of Fossils. Vol. **4** pp. 4-7.

Structural features and formation of lower Cretaceous AV_1 layer in the Soviet oil field (Tomsk Oblast)

A B Zhamsaranova, E N Osipova, T A Gaydukova[1] and N V Aksenova

National Research Tomsk Polytechnic University, Tomsk
E-mail: [1]Zhamsaranova.ab@mail.ru, [1]osipovaen@list.ru, [2]polozova15@rambler.ru

Abstract. The analysis of the collected geological and geophysical information on AV1 layer known as Ryabchik formation is carried out. The facial conditions of this formation which define structural features of «Ryabchik» sandstones formations are considered. Maps characterizing permeability and porosity of reservoir are plotted. Areal tracking technique of sand streaks is given.

1. Introduction

Oil industry development in Tomsk Oblast began in 1962 with discovering of the Soviet field located in the northwestern part of the region, 10 km from Strezhevoy (figure 1). In 1966 it was included into pilot well program, 75 exploration and 1397 producing and injection wells from 1682 to 2795 m in depth were drilled here. The field is multilayer, oil-bearing potential is proved by 18 reservoirs and it belongs to the category of the largest fields by its reserves. The field is at the fourth stage of extraction, water influx is more than 80%.

The Alymsky AV_1 layer represents a periodic, thin-layer, lens-shaped interlayering of clayey sandstones, siltstone and clays, it is characterized by the distinctive low permeability and porosity, high structural discontinuity, which results in low well yields and low well flow index [1]. The AV_1 layer oil reserves contain more than 70.9% of the residual extractable C1 reserves. Oil-bearing strata range from 4.8 to 24.9 m between wells, averaging at 15.6 m. In the AV_1 horizon five oil-bearing sandy rider seams are allocated: $AV_1{}^1$, $AV_1{}^{2a}$, $AV_1{}^{2b}$, $AV_1{}^3$ and $AV_1{}^4$.

Production geologists call the AV_1 layer «Ryabchik» formation due to its distinctive appearance and a lenticular structure (reminds of hazel grouse feathers, ryabchik being the name of the bird in Russian). Research conducted in Reservoir Engineering Laboratory has established low values: permeability from 0 to 107 10^{-3} micron^2, porosity from 0 to 10%; which practically identifies this layer as non-collector and therefore it was considered as no-prospect [2]. In addition, there were no techniques to estimate oil reserves and the field was not developed. However, with advances in hydraulic fracturing, there appeared an opportunity to extract oil from this reserve, thus determining the timeliness of this research.

Oil production in Tomsk Oblast has decreased since 2012 [3]. Development of underexplored Alymsky AV_1 layer, as well as exploration and development of the Neocomian reservoir [4] [5], is urged to stabilize a source of raw materials.

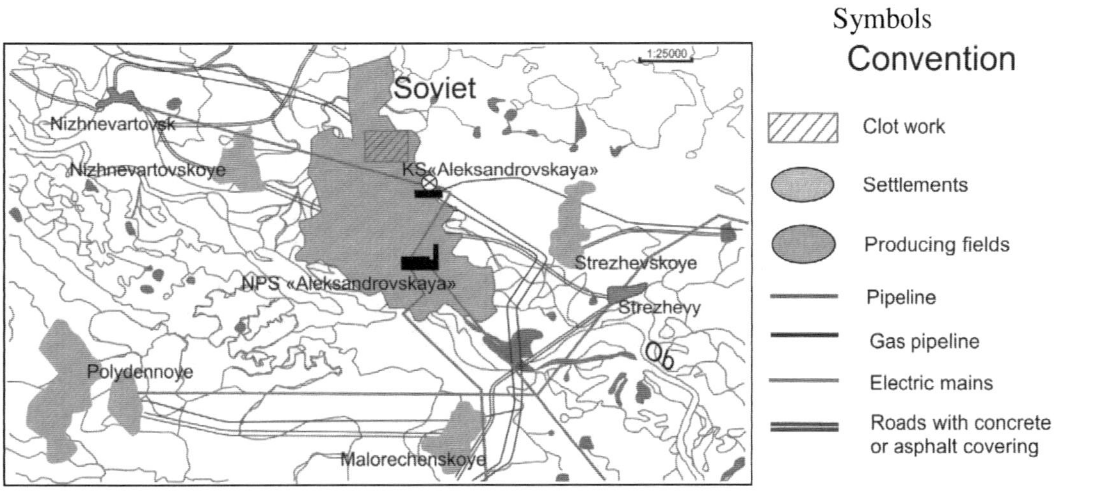

Figure 1. Works site location on the survey map of hydrocarbons fields in Tomsk Oblast

2. AV_1 layer characteristics

Formation genesis of the «Ryabchik» formations of sandstones can be explained in terms of their structural features. The typical structural distinctive features of the sandstones allow classifying them with faltering lamination. Sedimentology links formation of such lamination with consecutive frequent alternating stages of quiet and vigorous hydrodynamic activity resulting in deformation of primary layer structure [6].

Considering that the complex of deposits in question was formed in the sea pool, the model of "storm" sandstones is the most acceptable to it. On the one hand, it explains extensive development of the «Ryabchik» sandstones, on the other hand, it explains their structural features. Storm sandstones are one of widespread elements of sandy component in a shelf zone. Their formation is linked to sites subjected to coastal storms [7]. Destruction of barrier coastlines due to strong hydrodynamic activity promotes transportation of coarse-grained materials for considerable distances to remote shelf sites of the sea basin (figure 2).

Alternation of the storm periods with quiet stages of sedimentation are favorable for interlayering of well-sorted sandy and clay material. However, clay interlay in "storm" sandstones of a shelf zone does not form a homogeneous lithological layer. It is due to the fact that the thin clay cover formed during quiet sedimentation is subjected to the subsequent storm situation with strong hydrodynamic impact. At the same time clay material mixes up with sandy varieties, forming a sandy-argillaceous band with wavy faltering, the typical «Ryabchik» structure. One of the storm sandstones important features is considerable lateral development of separate thin sandy band [8].

As showed in figure 2, packs of storm layers are formed as a result of periodic destruction of barrier structures, when well-sorted sandy material is transported to considerable distance from the coastline towards the sea in short time, covering the sea bottom with a thin but extensive (tens of square kilometers) sandy cover. Subsequent calm period promotes accumulation of the blocking band of clays, and further activation of wave activity resumes process of coarse-grained material transportation. Hydrodynamic activity promotes crushing of the previously formed clay sublayer and introduction of separate parts into the spreading sandy reserves, which leads to faltering lamination formation [6].

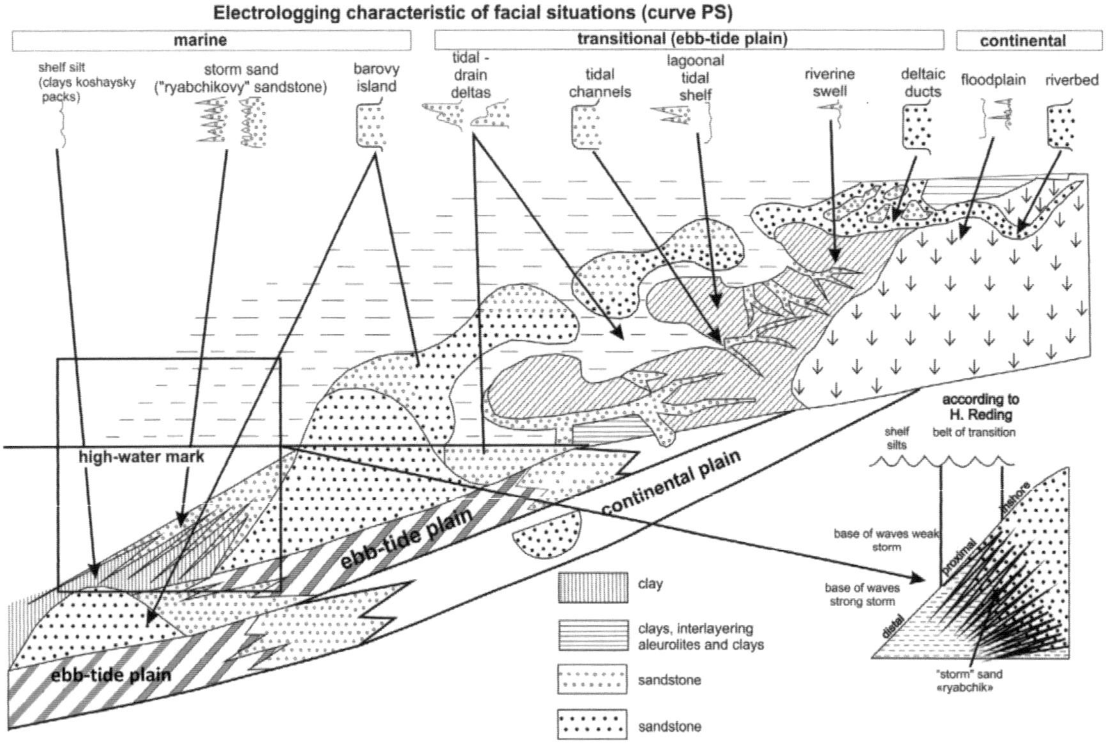

Figure 2. Paleogeographic facial conditions of «Ryabchik» sandstones formation [7].

3. AV$_1$ layer porosity and permeability analysis

The studied AV$_1$ layer permeability and porosity have been determined from data of geophysical surveys, carried throughout the history of the Soviet field development. In our research, the analysis of porosity and permeability of the northeast part has been carried out.

Research has shown (figure 3) that porosity is practically identical throughout the site, while the permeability coefficient significantly increases around wells 379, 381, 646, i.e. filtration characteristics of rocks increase in the northeast direction (figure 4).

Analysis of the permeability map shows:

• The layer is mainly sandy and has graded layering in a coastal zone in close proximity to barrier structures. While it advances towards the sea, the number of clay sublayers increases and in the deepest zones accumulation of clays prevails.

• Coastal storm sands represent lithological uniform section of storm deposits limited by erosion surfaces with thickness from 5 to 130 cm without slate interlayers.

• Proximal storm sands (layers from 5 to 100 mm thick) usually keep the lithologic and structural characteristics demonstrating their single-shot formation from a uniform stream.

• Distal storm sands are evident in the zone of shelf silt facies. They are fine-grained formations usually with thickness of less than 50 mm, with horizontal or slanting lamination [2].

The studied deposits being storm sandstones is confirmed by the following:

- broad development and close similarity of the geophysical characteristic among the composing packs all over the studied area;

- «Ryabchik»-type appearance of separate packs that can be linked to faltering lamination development;

- considerable content of clay in the section testifies to remoteness of the area from the coastline during the deposit formation [9].

According to logging charts there are post-sedimentation carbonated sandstones within low thickness sublayers, which being the uniform stratigraphic level of dissociation of a collector, can be used for distinguishing the horizons to correlate sandy layers of sea, coastal and continental genesis [10]. Together with clay sublayers, they represent the reliable intra-reservoir confining beds dividing a reservoir into hydrodynamically isolated systems. Repeated change of thickness and weakening of hydrodynamic processes promote formation of the interstratifying layers of considerable thickness and extent in the form of a "puff pie" which can also have various lithologic "stuffing" (figure 2).

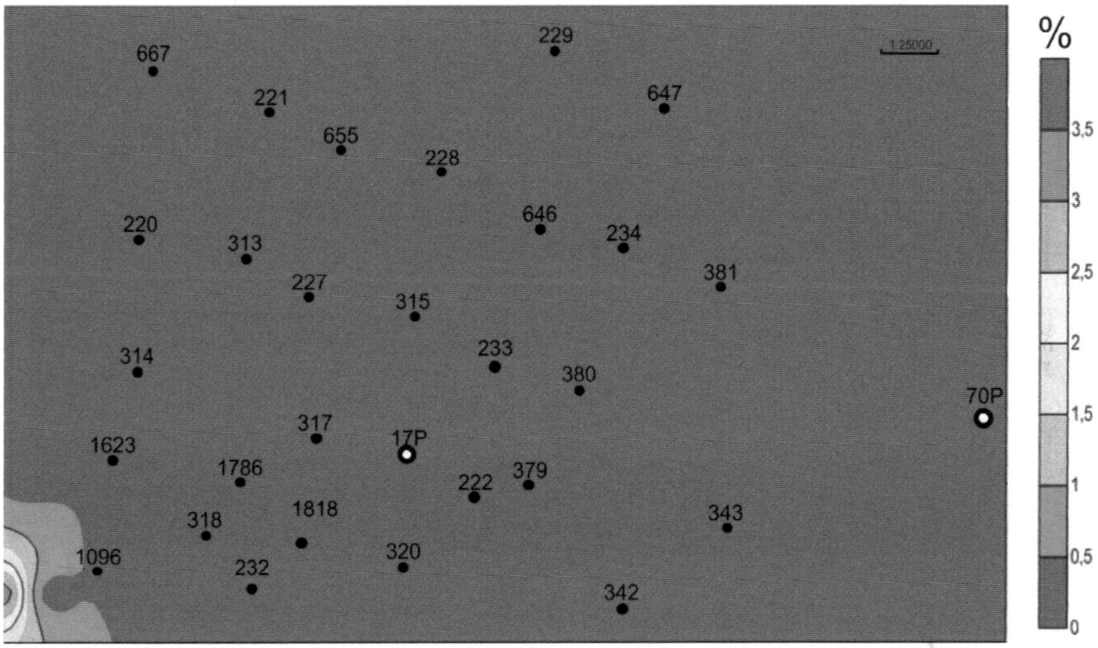

Figure 3. Porosity of the study area

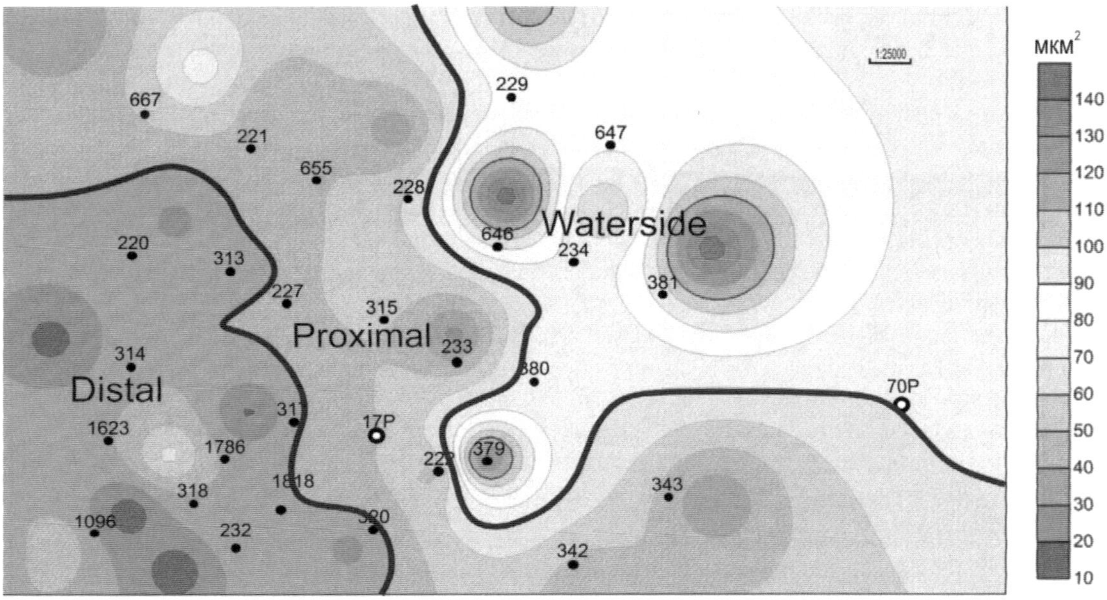

Figure 4. Permeability of the study area

4. Results

Proceeding from the considered model of AV_1 layer structure it is possible to draw a conclusion that the allocated inter-reservoir covers presented by both clay sublayers and sublayers of carbonated sandstone can serve as local hydrodynamic divisions for AV_1^1 sublayers AV_1^{2a} and AV_1^{2b}. In this regard each of the allocated sublayers should be considered as independent hydrodynamic system.

The studied AV_1 layer reserves in the Soviet field are difficult to recover. Therefore, development of the appropriate technologies allowing increasing oil recovery is required to bring these new areas into development. Well interference testing may be one of the most effective techniques. This method will allow conducting research on the unsteady mode of filtration and to define connectivity of the layer between wells.

To obtain reliable data on characteristics of layer and to reveal their heterogeneity, it is important to select the wells for the interference test correctly. For this purpose, it is necessary to consider the actual azimuth of a wellbore direction, data on special arrangement of wells, changes in vertical and radial permeability of layer.

5. Conclusion

The AV_1 layer formative conditions analysis in the Soviet field has revealed major sedimentation factors proving complex lithologic structure of the reservoir, and allowed evaluating sandy oil-saturated sublayer area.

References

[1] Osipova E N, Lobova G A, Isaev V I and Starostenko V I 2015 *News of Tomsk Polytechnic University.* Petroleum potential of the Lower cretaceous reservoirs of Nyurolsky depression. Vol. **326** **(1)** pp.14 – 33.

[2] Gorobets E A, Arzhilovsky A B, Volkov L A, Glebov A S and Chuprov A A 2010 *Neftyanoe hozyajstvo.* Development of hardly recoverable stocks of AV1 1-2 Samotlor field. Vol. **11** pp. 54 – 57.

[3] Kazanenkov V A 2016 *Drilling and oil.* A source of raw materials of hydrocarbons and regional features of deposits distribution in the Tyumen suite and its analogs in Western Siberia. Vol. **3** pp. 3 – 11.

[4] Isaev V I, Lobova G A and Osipova E N 2014 *Russian Geology and Geophysics.* The oil and gas contents of the Lower Jurassic and Achimovka reservoirs of the Nyurol'ka megadepression. Vol. **55** pp. 1418 – 1428.

[5] Lobova G, Osipova E, IsaevV and D Terre 2015 *Scientific and Technical Challenges in the Well Drilling Progress IOP Conf. Ser.: Earth and Environ. Sci.* Petroleum potential of Lower-Jurassic deposits in Nurolsk megadepression. Vol. **21** pp. 012001.

[6] Reding H 1977 Situations of accumulation and facies. Vol. **1** pp. 245.

[7] Danenberg E E, Belozyorov V B and Brylina N A 2006 *Tomsk: TPU publishing house.* Geological structure and oil-and-gas cretaceous deposits contents of the southeast of the West Siberian plate (Tomsk region). pp. 291.

[8] Kontorovich A E, Nesterov I I, Salmanov F K et al. 1975 Geology of oil and gas of Western Siberia. pp. 561.

[9] Nezhdanov A A 1985 *Works of ZAPSIBNIGNI, issue 201, Novosibirsk.* Types of carbonate concretions and their role in studying of oil-and-gas formations of Western Siberia. pp. 95

[10] Kalmykov G A, Korotkov K V, Yampolskaya E N, Akopyan G A, Belokhin V S 2012 *Moscow University Geology Bulletin* . Estimation of expected inflow in the AV1 bed by GWI data on the basis of C/O logging. Vol. **67** **(4)** pp. 259 – 264.

Features of lithological and granulometric composition of bottom sediments in the northern part of Laptev Sea

T Perevertailo, N Nedolivko, M Kazhumukhanova

Department of Geology and Mineral Exploration, Institute of Natural Resources, National Research Tomsk Polytechnic University, Russia
Email: [1]ptg@tpu.ru, [2]nedolivkonm@yandex.ru, [3]kazhumukhanova@gmail.com

Abstract. Bed-sediment samples were collected from the northern part of the Laptev Sea to assess particle-size distribution, mineralogical composition as well as to examine organic matter content. The organic carbon distribution map is constructed to reveal the trends of C_{org} content and distribution in the surface of the bottom sediments.

1. Introduction

Since the 30-s of the 20th century the intensive development of oil and gas exploration industry has led to the necessity of exploration of such unique gas fields as Medvezhye, Urengoy, Yamburg, Zapolyarnoye, Bovanenko on the continental area of the Russian region of the Arctic. According to prospecting surveys, approximately a quarter (22 %) of the world oil and gas reserves is located within this territory.

It is obvious that the Arctic zone of the Earth will be the primary source of the oil and gas reserves, not only for Russia, but also for other countries. Thus, the interest of Arctic countries in the exploration of the Arctic natural resources tends to increase, and, according to experts' opinion, the competition for these resources among the countries will become stronger in future.

However, mineral resource development of the Arctic areas is associated with high risks, such as specific environmental conditions, lack of infrastructure and lack of experience in eliminating the impact of oil spill on the Arctic system, insufficient knowledge of the shelf. While organizing and performing the offshore drilling operations, a serious risk is represented by the gas-hydrates, which are underexplored in our seas in the Arctic. The Laptev Sea shelf is of primary interest in terms of hydrates [4]. The Laptev Sea is a submarine cryolithic zone [3], where the powerful and multiple methane releases [2, 3] (concentration up to 700 nM) in a form of flowing bubbles (known as plumes) were discovered. They rise from the bottom of the sea (from the depth of 60-110 m) through the water layer, and sometimes go up to the atmosphere. It is this depth interval in which the most significant thawing of the submarine cryolithic zone has been anticipated. Such an interval provides gas flow to gas outlet [2].

Gas hydrate is a solid clathrate compound in which a large amount of gas is trapped within a crystal structure of water, forming a solid similar to ice. It should be added that the methane content in gas hydrates is being lowered by a factor of about 200. In case of decreasing pressure or increasing temperature, the destruction of gas hydrates takes place and high gas volume is released. On the one hand, gas hydrates are involved in the process of oil and gas accumulation as they are considered to be impermeable layers. On the other hand, in case of fractured zones activation, the intense migration of the gas from gas hydrates deposits occurs, oil and gas pool collapse and decomposition of oil and gas bearing rocks take place. During the hydrate destruction, the gas flows through talik zones into the air and further to the atmosphere, which results in significant redistribution of hydrocarbon gases and gas hydrates formed at different times and at different depth. This fact poses a serious ecological problem of a global scale.

According to available data, seas of Eastern Arctic are the source of methane released into the atmosphere of the Arctic Region. Contribution of the latter is comparable with that of the entire global ocean, and it is conditioned by degradation of subsea and near-shore permafrost.

The facts of temperature jumps as compared to the past century (about 5°C) [1], the highest coastal erosion rate, annual methane discharge, enormously intense, in contrast with the other objects, have revealed ocean anomalies of the dissolved methane, including big gas plumes (up to 500 m in diameter) [2-4] and induced to perform a more detailed research.

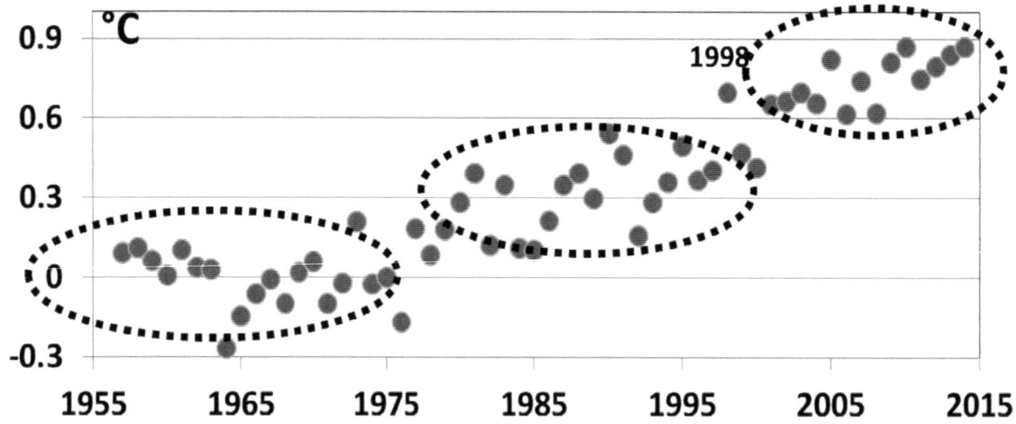

Figure 1. Global land-ocean temperature index (°C) [4]

2. Unit characteristics and methods

24 bed-sediment (0-5 cm) samples were collected from the northern part of the Laptev Sea to assess particle-size distribution and mineralogical composition as well as to examine organic matter content. The study area (figure 2) was explored by the Russian and US scientists in September-October 2011 on the board of research vessel Academician Lavrentiev.

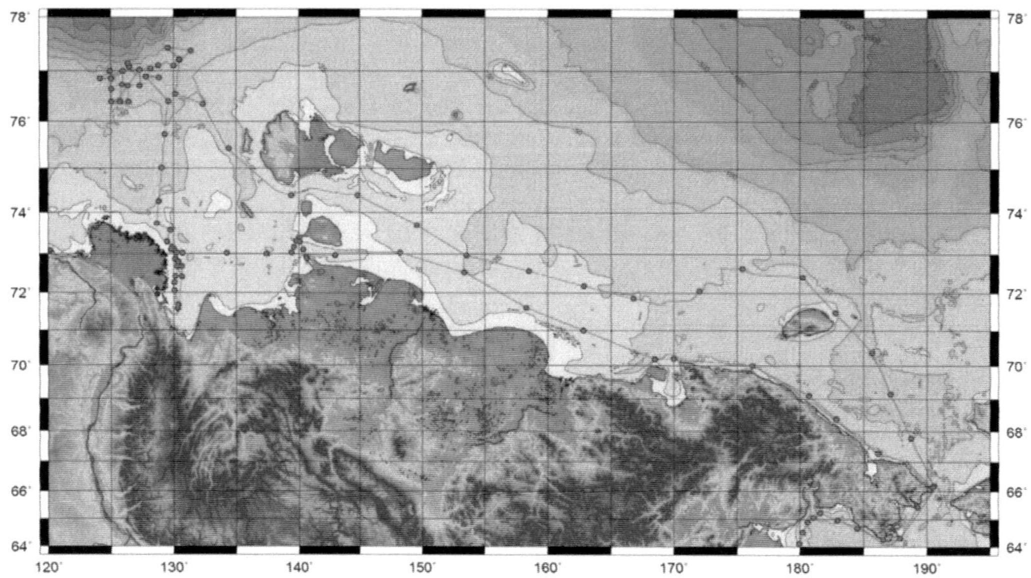

Figure 2. The map of study area

Particle size distribution was examined by laser diffraction method. Laser diffraction measures particle size distributions by measuring the angular variation in intensity of light scattered when laser beam passes through a dispersed particulate sample. Mineralogical analysis was carried out using the binocular microscope to reveal the different mineral species within sandy and silt fractions. Organic matter content was determined in the sediments by the Rock Eval pyrolysis. The analysis was carried out in the Arctic Sea's Carbon Research International Laboratory, Tomsk Polytechnic University.

In terms of the granulometric composition, the fractions with dimensions from 0.01 mkm to 3.08 mm were obtained. The range of fractions consisted of predominant pellite fraction (<0.01 mm) with the content from 24 to 77 %, aleurite (form 22 to 45 %), and psammitic (from 0.3 to 47 %) fractions were minor. According to the classification of clastic sedimentary rocks, the samples were primarily in the form of aleuritic-clays, less frequently in the form of silty clays and sands. Using the granulometry data, the cumulative curves were built (figure 3), the calculated sorting coefficient changed within the range of 1.18-3.65, which indicated the mean sorting ratio for aulerite clastic material, and good sorting ratio for sand clastic material.

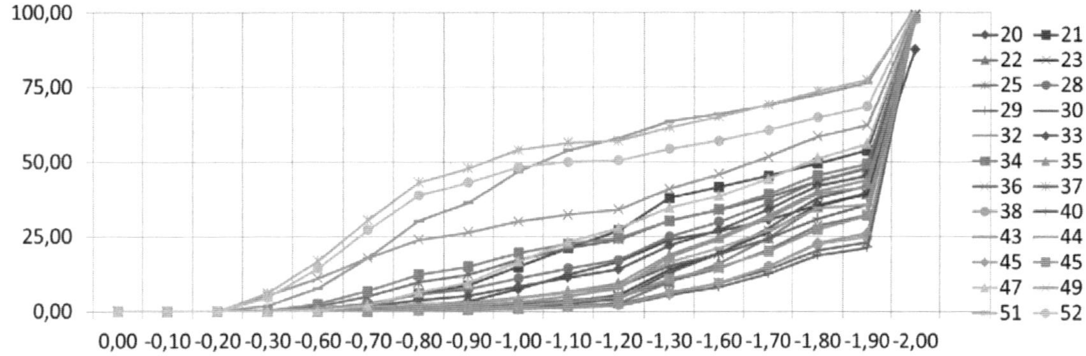

Figure 3. Particle-size accumulation curve

Using the granulometric analysis results, the distribution maps of pellitic, aleurite and sandy fraction content were constructed, and correlation between fractions content and surface of submarine relief was determined (figure 4).

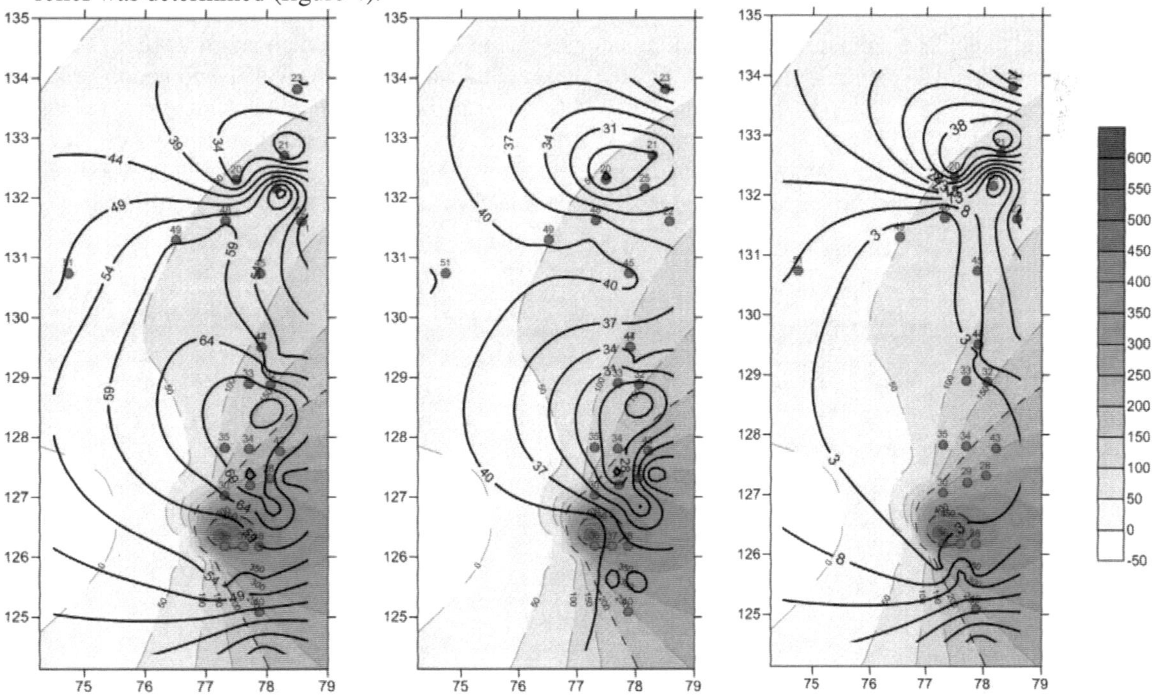

a) b) c)

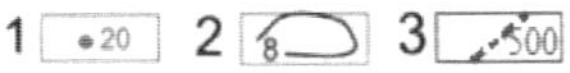

Note: 1 – № of the samples; 2 –a) pellitic fraction content, %; b) aleurite fraction content, %; c) sand fraction content, %; 3 – height contour, m.

Figure 4. Particle-size distribution map

Based on the information obtained in the course of study, the relation between water depth and grain size distribution has been established: pelitic sediments tend to the most low-lying areas, whereas the psammit depositions, on the contrary, are confined to the shallow part (figure 4).

Mineral analysis by physical method revealed the presence of quartz, feldspar, muscovite, biotite, garnet, and some of the other sediment minerals in sediment samples. Quartz and feldspars predominated in all samples. Quartz was the most common mineral which was found to be about 30-60% of total sediment, grains being colorless and transparent or translucent, ranging from rounded to angular. They were colorless with rare traces of hematite (figure 5 (a)).

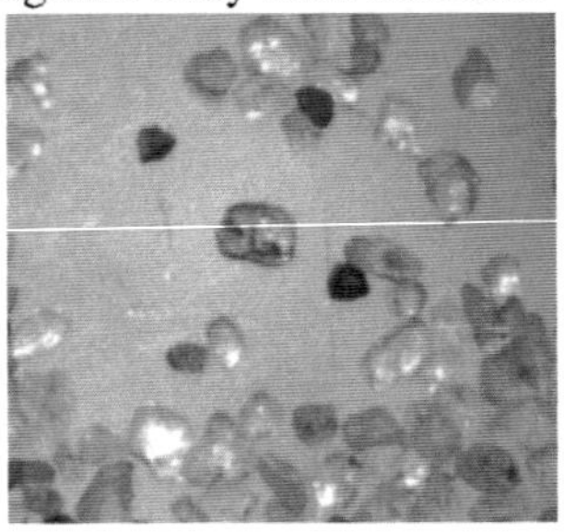

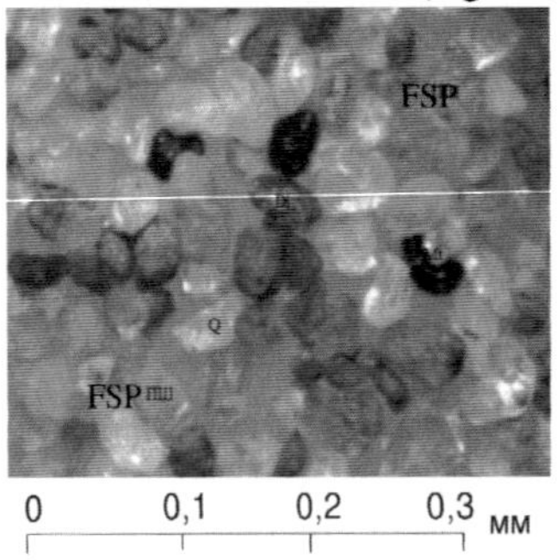

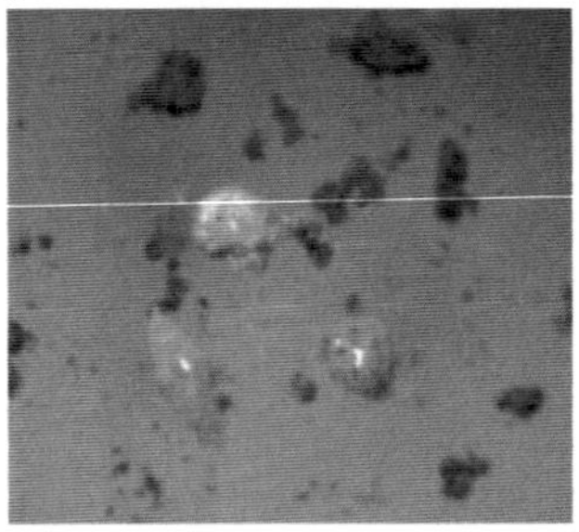

A) – The trace of hematite in quartz

B) – The grains of quartz (Q), feldspar (FSP), diopside (Di) and magnetite (Mt)

C) – The remains of clamshell of various forms

Figure 5. Features of the mineralogical composition of sands

Feldspar (25-35 %) is elongated, translucent, colorless, gray, pink, and pale yellow angular-rounded and angular fragments. Micas (biotite and muscovite) are found in small amounts (5-15 %) in all samples (figure 5 (b)). Carbonate minerals are presented by calcite brown siderite and are observed in samples No. 21, 22, 28, 30 in the amount of 10-15 %. The magnetite (to 10 %) is widespread among the ore minerals which has specific iron-black color and strong magnetic properties. Accessory minerals (to 5 %) include epidoty, diopside, apatite, chlorite, and grenades. Also, there are carbonized vegetable fragments and the remains of clamshell of various forms (figure 5 (c)).

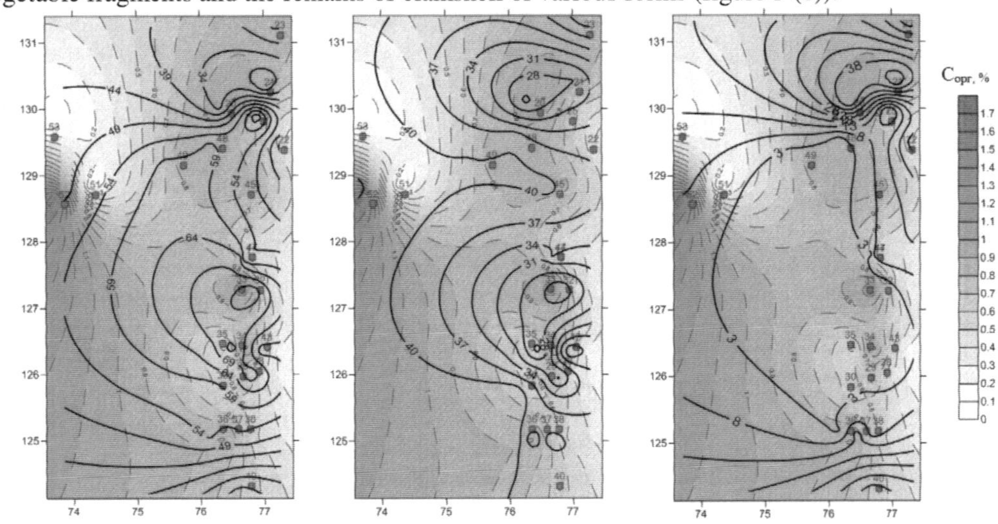

a) b) c)

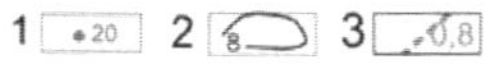

Note: 1 – № of the samples; 2 – a) clay fraction content, %; b) silt fraction content, %; c) sand fraction content, %; 3 – organic carbon content (C_{org}), %.

Figure 6. Organic carbon distribution map

Organic substance study is the most important aspect of hydrochemical, hydrobiological and geological research. Besides, organic carbon is its most primary characteristic. In the studied samples the C_{org} content varies from 0.03 to 1.61 %. The organic carbon distribution map is constructed to reveal the trends of C_{org} content and distribution on the surface of the north Laptev Sea's bottom sediments (figure 6). It is compared with the granulometric composition distribution map. Therefore, the relative saturation of north Laptev Sea's pelitic sediments with organic carbon is revealed. The correlation between organic carbon and sediment granulometry can be partly explained by the sorption of organic carbon by the thin pelitic fraction.

3.Conclusion

Based on the obtained data, it has been found that depositions of pelitic sediments tend to the lower areas, whereas the psammit depositions are confined to the shallow areas.

The concentration of north Laptev Sea's pelitic sediments with organic carbon is revealed. The probable reason for this process is the increased sorption of C_{org} by fine pelitic fraction.

References

[1] Karlsson E, Bruchert V, Tesi T, Charkin A, Dudarev O and Semiletov I 2015 Marine Chemistry. **170** pp. 11-12.
[2] Selver A, Sparkes R, Bischoff J, Talbot H, Gustafsson O and Semiletov I 2015 Organic Geochemistry. Vol. **83-84** pp. 16-26.
[3] Rachold V and Bolshiyanov D 2007 Eos. Vol. **13** pp. 149-156.
[4] Romanovsky V and Hubberten H 2001 Izvestiya RAS. Vol. **3** pp. 15-28.
[5] Salvado J, Tesi T, Andersson A and Ingri J 2015 Geophysical Research Letters. Vol. **42** pp. 8122–8130.
[6] Leifer I, Rekant P, Salyuk A, Nicolsky D, Romanovsky V and Romanovskii N 2010 Journal of Geophysical Research. pp. 115.
[7] Shakhova N, Sergienko V and Semiletov I 2009 Vestnik RAS. Vol. **6** pp. 507–518.
[8] Sparkes R, Dogrul S, Bischoff J and Talbot H 2015 Biogeosciences. Vol. **12** pp. 1726-4189.
[9] Sward H, O'Regan M, Ampel L, Ananyev R and Chernykh D 2016 GFF. Vol. **138** pp. 336-354.
[10] Vetrov A, Ponyaev M, Belyaev M and Romankevich E 2015 Oceanologiya. Vol. **55** pp. 387–394.

Fracture system influence on the reservoirs rock formation of Ordovician-Devonian carbonates in West Siberia tectonic depression

A E Koveshnikov[1,2], A C Nesterova[1], T F Dolgaya[3]

[1] National Research Tomsk Polytechnic University, Tomsk, Russia
[2] Tomsk Division of Trofimuk Institute of Petroleum-Gas Geology and Geophysics of the Siberian Branch of the RAS, Tomsk, Russia
[3.] National Research Tomsk Polytechnic University, Tomsk, Russia

E-mail: [1,2]Kovesha@mail.ru, [1]As_nesterova@mail.ru

Abstract. During the Paleozoic period from the beginning of the Cambrian to the end of the Carboniferous in the boundaries of the West Siberia tectonic depression there occurred the sea, where the carbonate platforms were formed by the limestones accumulation. All the area at the end of the Carboniferous period was turned to land. Resulting from Gertsynskaya folding in the times of Permian – Triassic the formed deposits were folded and denudated to a considerable extent. Besides, the reservoir rocks of the crust of weathering including redeposited one, were formed as a result of hypergenesis, during the continental stand of the area in the near-surface zone. A new geological prospecting unit has been suggested which underlies these crusts of weathering and formed during fracture tectonic processes with hydrothermal-metasomatic limestones reworking and the processes of hydrothermal leaching and dolomitization. So, in the carbonate platforms the system of fissure zones related to tectonic disturbance was formed. This has a dendrite profile where the series of tangential, more thinned fractures deviate from the stem and finish in pores and caverns. The carbonate platforms formation in the West Siberia tectonic depression has been analyzed, their dynamics and gradual increasing from the minimal in Ordovician and Silurian to maximal at the end of the Late Devonian has been shown.

1. Introduction

West Siberia tectonic depression is one of the most promising sources of oil and gas stock addition for the whole territory of Russia. Here, the carbonate formations are thought to be promising but are not explored properly [1]. They are characterized by the following varieties: carbonate, carbonate-flinty and carbonate-clayey rocks, which have complex history of formation and secondary processes of transformation.

The concept of Paleozoic reservoir rock formation in the near-surface zone, where the Paleozoic deposits occurred on land is a generally accepted point of view. In this area during the Permian – Triassic period or the continental stand time in the near-surface zone which is also called the oil-bearing horizon of the contact zone for the Paleozoic and overlapping Mesozoic deposits [9] there formed the crusts of weathering including redeposited ones.

They occurred resulting from hypergenesis processes [3], and, according to this point of view, all newly - formed Paleozoic reservoir rocks and oil\gas deposits relate to the above mentioned crusts of weathering.

Reservoir rocks formation mechanism has been suggested [2]. This is provided in the unweathered near-surface zone limestones, located lower the oil-gas bearing horizon of the contact zone. These became the reservoir rocks of the fissured – porous - cavernous type as a result of fissure-hydrothermal-metasomatic processes.

The oil-bearing horizon of the contact zone was formed in the following way. After Gertsynskaya folding phenomena in the boundaries of the West Siberia tectonic depression [2] the vast synclinorium and antinclinorium zones were formed [2], which are characterized by the syncline and anticline folds of the second and third order (like Mezhovskiy medial massif) – (Fig. 1B). These are mostly surrounded by the clayey-flinty rocks which are modified, magmatic and metamorphosed sedimentary terrigeneous rocks of the Proterozoic period (Fig. 1B). Synclinorium zones of the major order (Fig.1) were formed in the pre-Jurassic exposure areas of the spatially linked syncline areas (like Mezhovskiy medial massif). Antinclinorium zones were formed in the areas characterized by Proterozoic period rocks. These had less structures like Mezhovskiy medial massif had (generally more uplifted). Then, with the further formation of the tectonic depression in the boundaries of the West Siberia depression and sediment accumulation in the sea of the Jurassic-Paleogene period, these Paleozoic modified rocks due to the sea bottom subsidence were transformed by the hydrothermal secondary-katagenesis processes [3, 4] (resulting in reservoir rock formation of the fissured hydrothermal – metasomatic genesis).

The described processes relate primarily to the carbonate rocks. So, knowledge of their formation territory and the age interval in the boundaries of the West Siberia allow to determine the most promising areas of the reservoir rocks likely to be formed and oil\gas deposits in the carbonate rocks formed by the above mentioned scheme.

2. Research methods

To identify the Paleozoic deposits of various ages likely to be expanded through the West Siberia territory, the following approach has been applied [8], where the Paleozoic deposits are characterized as the carbonate-clayey formations of the Paleozoic age, folded into syncline and anticline major and the third-order folds. As described in [2] the whole territory of the West Siberia depression with the Gertsynskaya folding had great synclinorium and antinclinorium major folds to be formed (Fig. 1A) and described in [2]. The areas with the prevalence of synclinorium minor and the third-order structures in the territory make synclinoria; the Paleozoic profile is mostly preserved (Fig.1A). The areas with the exposures of the modified magmatic and the metamorphosed terrigeneous Proterozoic alumosilicate rocks are of antinclinoria (Fig.1A). So, we show the areas, where the Paleozoic profile is mostly preserved in the West Siberia boundaries. It is the area where oil\gas fields are most likely to be opened. These were formed by the above described scheme for reservoir rocks of the fissured hydrothermal – metasomatic genesis.

To study the extent area of the carbonate formations in West Siberia depression that were formed in Middle-Late Devonian period, the paleogeography reconstruction method has been applied. Based on the identified data [10] on lithological composition and the thickness of the Paleozoic deposits opened-up by drilling, the deposits expansion paleomaps for Ordovician and Silurian, Low-Middle-Late Devonian periods (Famennian and Frasnian) have been made (Fig.2). While correlating the data in Fig.1 and Fig.2 we drew a conclusion that there is potential for reservoir rocks identification in some areas of the West Siberia depression. These were formed according to the general point of view, i.e. in oil-gas bearing horizon of the contact zone (synclinorium zones), as well as in the carbonate rocks due to the fissured hydrothermal – metasomatic processes of Paleozoic rocks (pre-Jurassic carbonate rocks exposure zone).

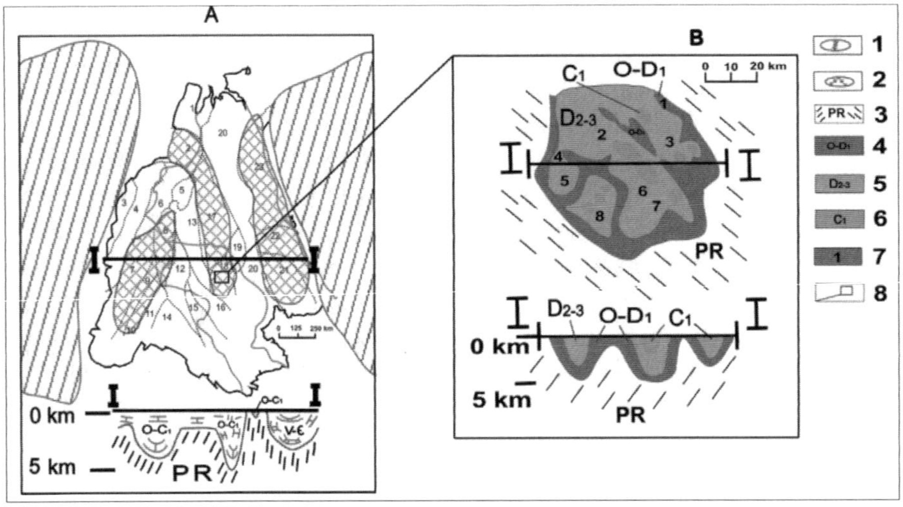

Fig1.A. *Paleozoic deposits of West Siberia depression - synclinorium and antinclinorium folded. Profile schematic according to I – I line. Legend: 1. Areas of Paleozoic deposits with thickness of more than 500 m, 2. Baikalite development areas. Structurally -facial areas: 1 – Bovanenkovskiy; 2 – Novoportovskiy; 3 – Tagilskiy; 4 – Beryozovo-Sartynyinskiy; 5 – Yarudeiskiy; 6 – Sherkalinskiy; 7 – Shaimskiy; 8 – Krasnoleninskiy; 9 – Tyumenskiy; 10 – Kosolapovskiy; 11 – Uvatskiy; 12 – Salymskiy; 13 – Ust-Balykskiy; 14 – Ishimskiy; 15 – Tevrizskiy; 16 – Tuisko-Barabinskiy; 17 – Varyeganskiy; 18 – Nyurolskiy; 19 – Nikolskiy; 20 – Kolpashevskiy; 21 – Vezdekhodniy; 22 – Tyiskiy; 23 –Yermakovskiy .*

B. *Paleozoic deposits* of *Mezhovskiy medial massif. Nyurolskiy structurally-facial areas synclinorium folded. Legend: 1. Baikalite development areas. 2. Synclinorium zones of Gertsynskaya folding. Rocks: 3. Cambrian – Low Devonian; 4. Middle-Upper Devonian; 5. Low-Middle Carboniferous; 6. Proterozoic. 7. Mezhovskiy medial massif location in the boundaries of Nyurolskiy structurally-facial area. 8. Oil-gas fields: 1 – Severo-Ostaninskoye, 2 – Gerasimovskoye, 3 – Ostaninskoye, 4 – Urmanskoye, 5 – Archinskoye, 6 – Severo-Kalinovoye, 7 – Kalinovoye, 8 – Nizhne-Tabaganskoye.*

3. Reservoir rock formation scheme

The whole territory of the West Siberia depression according to [10] is subdivided into 23 structurally-facial areas which have their own scheme of Paleozoic rocks formation. This includes the lithological composition of rocks, fundamental secondary processes of transformation and the identified data of paleontologically characterized age (Fig.1, 2).

Reservoir rocks of Paleozoic deposits are characterized by the limestones transformed by the secondary processes of hydrothermal dolomitization as well as of hydrothermal leaching resulting in voids formation. Limestone accumulations produce large platforms (which are

IOP Conf. Series: Earth and Environmental Science **43** (2016) 012008 doi:10.1088/1755-1315/43/1/012008

called carbon-bearing platforms) in terms of area and thickness. To analyze Paleozoic complex potentials for oil\gas field exploration which are assigned to these carbon-bearing platforms, the series of paleogeographical maps for various rock ages have been made (Fig.2) [6, 7].

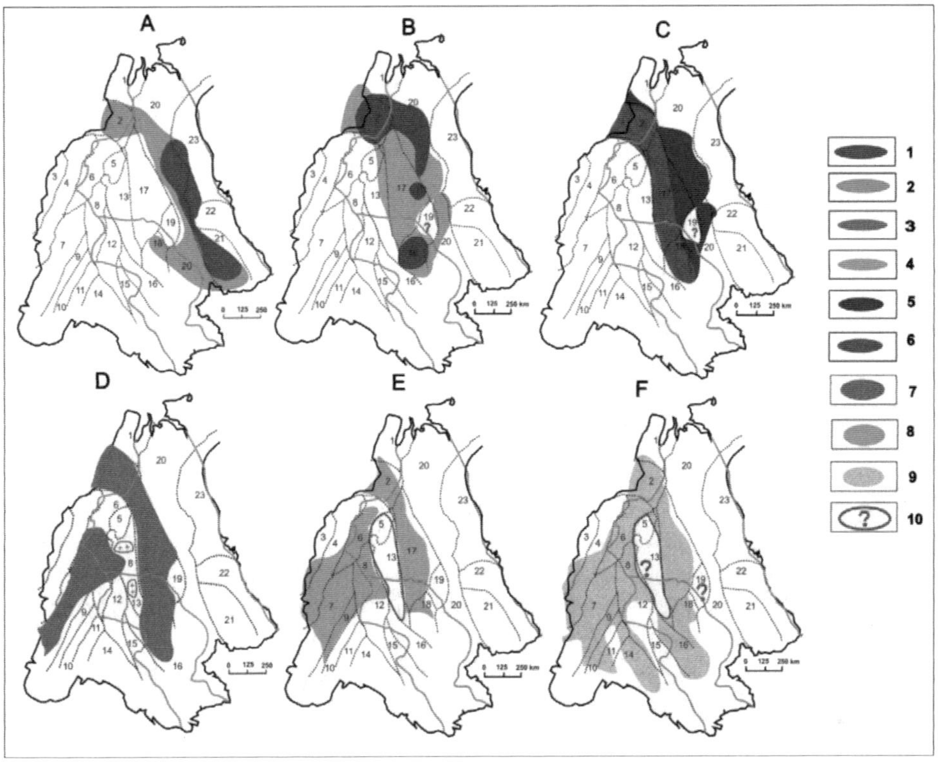

Fig. 2. *Carbon-bearing platforms accumulation in the boundaries of the West Siberia depression in: A – Ordovician; B –Silurian, C – Low Devonian; D – Middle Devonian; E – Middle Devonian, Frasnian, F – Late Devonian, Famennian. Legend: Deposits: 1– Low Ordovician; 2 –Middle and Upper Ordovician; 3 – Low Silurian; 4 – Upper Silurian; 5 – Low Devonian; 6 – Low Devonian, Emsian; 7 – Middle Devonian; 8 – Upper Devonian, Frasnian; 9 – Upper Devonian, Famennian, 10 – deposits are not identified by drilling.*

Synclinorium zones alternate with antinclinorium ones (Fig.1A). Three synclinorium folds separated by antinclinorium zones have been identified. These will be largely destructed by the denudation processes (Kolpashevskiy structurally-facial area). All Paleozoic deposits are dissected by the fault lines striking to the north-east and north-west.

Paleogeographical maps have been made to identify the territory for carbon-bearing platforms accumulation of various ages. These show the expected development of the platforms in Ordovician (Fig.2A), Low Devonian (Fig.2C), Middle Devonian (Fig.2D), Upper Devonian (Frasnian) (Fig.2E), and Upper Devonian (Famennian) (Fig.2F).

The carbonate formations by their petrophysical properties (shown above) [5], from

Ordovician to Late Devonian have allied parameters and differ in details, which characterize the organic remains complex forming limestones of various ages. Integrally, the void geometrization is subjected to the secondary processes activity spatially related to the faults. The tangential fractures, feathering the faults, are concluded by the crazes, effective fractures surrounded in the adjacent rock by the pores and caverns resulting from hydrothermal leaching and dolomitization. Such void geometrization can be called "fault-zone detached-and-fractured" (dendrite) [6, 7].
The profile of the described system will look like reservoir rocks development zones, alternating with the non-reservoir development zones. When the fissured-metasomatic limestones are adjacent underneath the limestones altered in the oil-gas bearing horizon of the contact zone, they will combine the uniform fracture-cavern-pore system. The most important is the fact that the extended spatially-fractured zones will combine the discreted areas for reservoir rocks development in oil-gas bearing horizon of the contact zone. It's necessary to mention that these areas occurred on the increased paleorelief. Generally, the fractured zones combine that sort of discreted areas into the uniform hydrodynamic system, through which both hydrothermal waters and fluids (oil and gas) can migrate.

Carbon-bearing platforms formation from Ordovician to the end of the Devonian
Beginning from the Ordovician period to the end of the Devonian in some areas of the West Siberia depression the carbonate formations accumulated sequentially and continuously, but somewhere else they accumulated occasionally.
In the first case the carbonate formations are not separated among themselves. They seem to be the uniform massif in the event of tectonic disturbance. This increases their overall potential for the reservoir rocks formation and future oil\gas fields occurrence. The characteristic feature of the sedimentation in Ordovician, Silurian and Early Devonian periods (Fig.2A, 2B, 2C), is the following. When correlating the extent area of the carbon-bearing platforms, it has been identified that they gradually decrease or increase.
The sedimentation pattern changed in the Middle Devonian period. When there was a certain decrease in the carbonate accumulation area in the central part of the West Siberia depression, a new, not very vast area of the carbonate rocks accumulation occurred (Fig.2D) in its western extreme point. In the Late Devonian period (Frasnian) the carbonate accumulation area was decreasing in the central part of the West Siberia depression, when the carbonate accumulation in the western area was increasing. At last, in the Late Devonian (Famennian), two areas became the uniform carbon-bearing platform. This expanded its borders along the territories where the carbonate accumulation processes had never occurred before.

4. Conclusion

1. Paleozoic carbonate rocks, producing carbon-bearing platforms on the West Siberia depression, were formed continuously from the beginning of the Cambrian to the end of the Carboniferous. This was the prolonged formation of the carbonaceous rock mass being either uniform or separated layers by the other rocks.
2. The crusts of weathering in the oil-gas bearing horizon of the contact zone were formed. These occurred along the pre-Jurassic surface as a result of Gertsynskaya folding processes followed by the continental stand of the area, and had the advanced

reservoir rock properties. During sedimentation in the sea of Jurassic-Paleogene period, the Paleozoic formations that had been altered by the hypergenesis processes underwent the secondary transformation of katagenesis processes (secondary katagenesis) which relate to the system of renovated faults.

3. Secondary-katagenesis processes of the fissured hydrothermal–metasomatic genesis through the carbonate formations resulted in the occurrence of the secondary-katagenesis reservoir rocks, being a zoneform of the reservoir rocks, alternating with the stationary zones of the non-reservoir rocks. Their superposition in the reservoir rock areas formed in the oil-gas bearing horizon of the contact zone created the uniform hydrodynamic system for oil and gas movement and oil-gas accumulations and fields.
4. The Ordovician, Silurian and Early Devonian witnessed the carbonate accumulation in the central part of the West Siberia depression boundaries. In the Middle Devonian along with the central zone there formed the isolated western carbonate accumulation zone. The central carbon-bearing platform area was reduced in Late Devonian (Frasnian), while the western one – expanded; they became convergent. The carbon-bearing platform was uniform in Late Devonian (Famennian); the carbon accumulation occurred along the vast territory of the West Siberia depression. The Carboniferous suffered the carbon-bearing platform area decrease, while the end of the period saw the carbon accumulation decline.

References

[1] Ezhova AV 2007 *Geologiya Nefti i Gaza.*Pore space genesis and rock-fluid properties of Paleozoic reservoirs of Tomsk area fields. Vol. **3** pp. 20-24.

[2] Koveshnikov A E 2013 *Izvestiya* (TPU Tomsk). Gertsynskaya folding influence on Paleozoic formations preservation in West Siberia depression. Vol. **323** pp. 148-151.

[3] Koveshnikov A V, Nedolivko N M 2012 *Izvestiya* (TPU Tomsk). Crusts of weathering of Pre-Jurassic deposits in West Siberia depression. Vol. **320** pp. 77-81.

[4] Koveshnikov A E, Nedolivko N M 2012 *Izvestiya* (TPU Tomsk). Secondary-katagenesis transformations of Pre-Jurassic rocks in West Siberia depression. Vol. **320** pp. 82-86.

[5] Koveshnikov A E 2012 *Izvestiya* (TPU Tomsk). Reservoir rocks formation in Pre-Jurassic deposits in West Siberia depression. Vol. **321** pp. 132-137.

[6] Koveshnikov A E, Nesterova A C 2016 *"Problemy geologii i osvoyeniya nedr" Proceedings of XX International symposium of Academician M.A.Usov* (TPU Tomsk) Fissured reservoir rocks formation in Ordovician-Low Devonian carbonate formations of West Siberia depression. Vol. **1** pp. 342-344.

[7] Koveshnikov A E, Merkulova A A 2016 *"Problemy geologii i osvoyeniya nedr" Proceedings of XX International symposium of Academician M.A.Usov* (TPU Tomsk) Fracture system during reservoir rock formation in Middle- Upper Devonian carbonate accumulations of West Siberia depression. Vol. **1** pp. 340-342.

IOP Conf. Series: Earth and Environmental Science **43** (2016) 012008 doi:10.1088/1755-1315/43/1/012008

[8] Kontorovich A E *et al* 1991 *"Theoretical and regional geological problems of oil\gas"* (Novosibirsk) Oil\gas-bearing geological conditions for the upper part of the Paleozoic Profile in West Siberia (like Mezhovskiy medial massif). pp. 152–171.

[9] Kontorovich V A 2007 *Geologiya i Geofizika.*Seismogeological criteria of the oil\gas-bearing contact zone for Paleozoic and Mesozoic deposits of West Siberia (like Chuzik-Chizhapka oil\gas accumulation zone) Vol. **48** pp. 538–547.

[10] *Resolutions of inter-departmental commission on consideration and acception of the regional stratigraphic scheme of Paleozoic formations in the territory of West Siberia Plain* 1999 (Novosibirsk, SNIIGGIMS) 80.

Influences of Neo-Pleistocene permafrost on thermal history of petromaternal Lower Jurassic Togur suite (Tomsk region)

A A Iskorkina[1], V I Isaev[2], A N Fomin[3]

[1-2] Department of Geophysics, Institute of Natural Resources, National Research Tomsk Polytechnic University, Tomsk
[3] Laboratory of oil and gas geochemistry, Institute of oil and gas geology and geophysics of A.A. Trofimuk of Siberian Branch of the Russian Academy of Sciences, Novosibirsk.
E-mail: [1]iskorkina.a@mail.ru, [2] isaevvi@tpu.ru, [3]FominAN@ipgg.sbras.ru

Abstract. Based on paleotemperature modeling, evaluation of Neo-Pleistocene permafrost rock thickness impact on geothermal regime of the petromaternal Togur deposits has been performed within the territory of oil fields of the Tomsk region (the southeast of Western Siberia). It has been stated that paleopermafrost with the thickness of about 300 m must be considered for appropriate reconstruction of geothermal history of petromaternal rocks in the south-east areas of West Siberia. This condition is relevant to a consistent consideration of thermal history of maternal deposits in course of assessment of resources by a volumetric-genetic method.

1. Introduction

The previous researches have studied the impact of secular trend in the temperature on the surface of the Earth, as well as, the influence of neopleystotsen permafrost on the thermal history and oil-generation potential of Bazhenov deposits in the south-east of West Siberia [1]. The article presents the assessment of significant influence of secular temperature trend and paleopermafrost on implementation extent of oil-generation potential of Bazhenov suite forming deposits of hydrocarbons in the lower Jurassic and upper Jurassic oil-and-gas bearing complexes in the south-east of West Siberia. The purpose of the present study is to estimate the influence of Neo-Pleistocene permafrost thickness on geothermal regime of the Togur deposits forming deposits of hydrocarbons of the lower Jurassic and pre-Jurassic oil-and-gas bearing complexes in the south-east of West Siberia.

2. Research methods

Based on data from sedimentary cross-section of deep well 1 in the North Festival field, modeling of paleogeothermal conditions in Togur deposits was performed (fig. 1). Generative potential of the Togur deposits within this territory is caused by the high content of dispersed organic matter of gumic-sapropelic type and rather high content of organic C (to 10%) [2]. In the North Festival field hydrocarbon deposits are mainly associated with medium-Lower Jurassic reservoirs (table 1).

The evaluation of Neo-Pleistocene permafrost rock thickness impact on the geothermal regime and degree of oil-generation potential implementation of Togur deposits is performed on the basis of the results of four optional paleotemperature reconstructions variability analysis. Reconstruction 1 considers both secular trend in the temperature on the Earth surface [3], and Neo-Pleistocene permafrost sequence of about 300 m in thickness [4]. Reconstruction 2 involves secular trend in the Earth surface temperature analysis without considering permafrost rock sequence. Reconstruction 3 gives analysis regardless of secular trend of surface temperature and permafrost rock sequence. Reconstruction 4 refers to secular trend of surface temperature and Neo-Pleistocene permafrost which is assumed to be up to 1000 m thick [5].

In *Reconstruction 1* (table 2) permafrost sequence is considered to be of 300 m thick. Formalized calculation of permafrost thickness is provided beginning from 240 thousand years ago when an

IOP Conf. Series: Earth and Environmental Science **43** (2016) 012009 doi:10.1088/1755-1315/43/1/012009

"immediate"(according to the standards of geological time, for the period of 1,5+ 3,0 thousand years) replacement of "normal" sedimentary deposits by permafrost sequence with particular thermophysical parameters – thermal conductivity, temperature conductivity occurred [6]. This sequence of permafrost rocks has overlaid sedimentary mantle for 179 thousand years. Hereafter, "immediately" (1,5+3,0 thousand years) permafrost sequence is substituted for "normal" sedimentary deposits and since that time "normal" sedimentary mantle has been retained up to the present moment for over the recent 52 thousand years.

Reconstruction 4 deals with permafrost thickness of 1000 m, other procedure being the same as in Reconstruction 1.

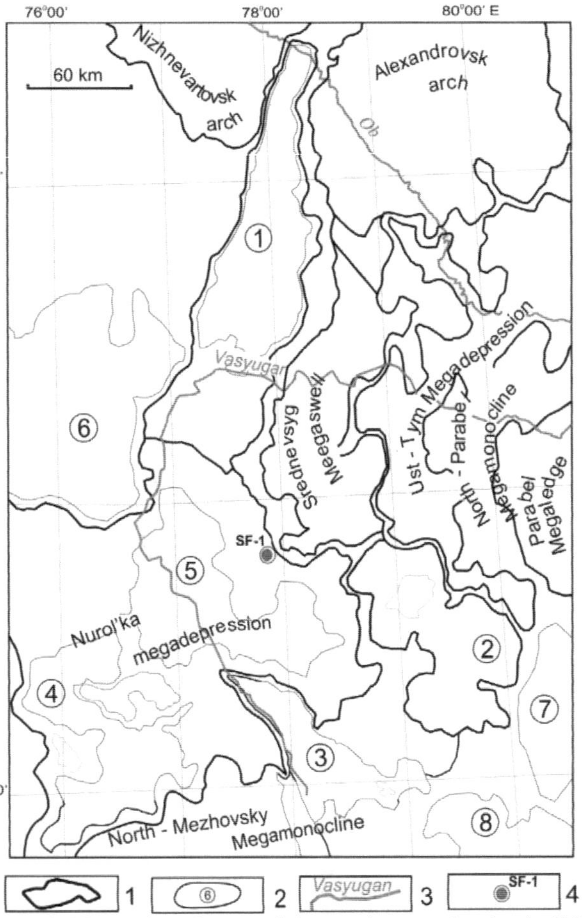

Figure 1. Tectonic scheme of the study area (on basis [7]): 1) – structures of the I order 2) – structures of the II order: 1 – Koltogorsky mesodeflection, 2 – Pudinskoe mesoraising, 3 – Lavrovsky mesoledge, 4 – Southern Nyurol`ka megadepression, 5 – Tsentralnonyurolsk mesohollow, 6 – Novovasyugansk mesoshaft, 7 – Goreloyarsk mesoraising, 8 – Kalgachsky mesoledge; 3) – rivers; 4) – the studied well: SF-1– North Festival 1

Simulation of the thermal history of the Togur suite deposits has been carried out on the basis of paleotectonic and paleotemperature reconstructions. In the present research paleotemperature modeling has been applied [8]

Table 1. Characteristic of the well North Festival 1 section (northeast board of the Nyurolsky megahollow)

Characteristics	Value
Face, m	3270
Deposits on a face (suite)	Paleozoic (PZ)
Base of Lower Jurassic deposits, m	3234
Top of togur suite, m	3165
Power of togur suite, m	30
Top of Bazhenov suite, m	2705
Power of Bazhenov suite, m	23
Power of paleogenov deposits, m	704
Power of neogenov deposits, m	-
Power of quarternary deposits, m	35
Results of tests (suite; layer; fluid type; output)	Bazhenov-vasyugan (Yu$_1$); chilly. (an oil smell in a core) Tyumen; Ю$_{13-15}$; oil; 2,57 m^3/d. Tyumen; Ю$_{13-14}$; oil; 0,28 m^3/d. Urmansk; Ю$_{16}$; oil; 0,13 m^3/d.. Urmansk+Paleozoic; Ю$_{16}$+PZ; oil, gas; 1,54 m^3/d.890 m^3/d. Paleozoic; PZ; gas; 410 m^3/d.
Measured rock temperatures (suite; measurement depth; rock temperature)	Tyumen; 3130 m; 118 °C. Tyumen; 3145 m; 123 °C.
«Measured temperatures " according to vitrinite reflectance (suite; selection depth; temperature)	Urmansk; 3232 m; 124 °C.

Table 2. Description of sedimentation history and thermophysical properties of the sedimentary sequence tapped with the North Festival well (Neo-Pleistocene permafrost thickness is 300 m).

Suite, sequence (stratigraphy)	Thick ness, m	Age, Ma ago	Accumu lation period, Ma	Density, g/cm^3	Thermal Conductivit y, W/m K	Temperature Conductivity , m^2/s	Temperatur e Conductivit y, m^2/s
Quaternary *Q*	-	0,052-0,00	0,052	-	-	-	-
Quaternary *Q*	300	0,055-0,052	0,003	2,10	1,3	7e-007	1,22e-006
Quaternary *Q*	-300	0,0565-0,055	0,0015	2,10	2,09	1,05e-006	1,22e-006
Quaternary *Q*	-	0,2355-0,0565	0,179	-	-	-	-
Quaternary *Q*	300	0,2385-0,2355	0,003	2,10	2,09	1,05e-006	1,22e-006
Quaternary *Q*	-300	0,24-0,2385	0,0015	2,10	1,3	7e-007	1,22e-006
Quaternary *Q*	35	1,64-0,24	1,4	2,02	1,27	6,5e-007	1,1e-006
Pliocene N_2	–	1,64–4,71	3,07	–	–	–	–
Miocene N_1	–	4,71–24,0	19,29	–	–	–	–
Nekrasovskaya *nk* Pg_3	154	24,0–32,3	8,3	2,09	1,35	7e-007	1,2e-006
Cheganskaya *hg*Pg_{3-2}	70	32,3-41,7	9,4	2,09	1,35	7e-007	1,2e-006
Lyulinvorskaya *ll* Pg_2	240	41,7-54,8	13,1	2,09	1,35	7e-007	1,2e-006
Talitskaya *tl* Pg_1	70	54,8-61,7	6,9	2,09	1,35	7e-007	1,2e-006
Gankinskaya $P_{1\text{-}K2gn}$	170	61,7-73,2	11,5	2,11	1,37	7e-007	1,25e-006
Slavgorodskaya *sl* K_2	130	73,2-86,5	13,3	2,11	1,37	7e-007	1,25e-006
Ipatovskaya *ip* K_2	-	86,5-89,8	3,3	-	-	-	-
Kuznetsovskaya *kz* K_2	15	89,8-91,6	1,8	2,18	1,43	8e-007	1,25e-006
Pokurskaya *pk* $K_{1\text{-}2}$	800	91,6–114,1	22,5	2,26	1,49	8e-007	1,25e-006
Alymskay a_2 K_1	24	114,1-116,3	2,2	2,39	1,6	8e-007	1,25e-006
Alymskay a_1 K_1	17	116,3-120,2	3,9	2,39	1,6	8e-007	1,25e-006
Kiyalinskay *kls* K_1	613	120,2–132,4	12,2	2,39	1,6	8e-007	1,25e-006
Tarskay *tr* K_1	54	132,4–136,1	3,7	2,44	1,62	8e-007	1,25e-006
Kulomzinskay *lm*K_1	313	136,1–145,8	9,7	2,44	1,64	8e-007	1,25e-006
Bazhenov *bg* J_3	23	145,8–151,2	5,4	2,42	1,62	8e-007	1,3e-006
Georgiev *gr* J_3	5	151,2–156,6	5,4	2,42	1,62	8e-007	1,3e-006
Vasyugan *vs*$J_{3\text{-}2}$	70	156,6–162,9	6,3	2,42	1,6	8e-007	1,3e-006
Tyumen *tm* $J_{1\text{-}2}$	362	162,9–200,8	37,9	2,46	1,64	8e-007	1,3e-006
Togur *tg* J_1	**30**	**200,8-203,9**	**3,1**	**2,46**	**1,64**	**8e-007**	**1,3e-006**

Urmanskoye *ur* J_1	39	203,9-208,0	4,1	2,46	1,64	8e-007	1,3e-006

Note. Grey shading indicates geological time intervals of "immediate" formation and degradation of NeoPleistocene permafrost sequence. Dark shading indicates time interval of permafrost sequence existence.

3. Results and discussion

A number of observations can be deduced from the analysis of computational values of mantle basement heat flow density (table 3). In *Reconstructions 1, 3 and 4* heat flow increases by 1,1–1,6–4,7 mW/m^2 relatively to computational value of heat flow in *Reconstruction 2* which is – 55.7 mW/m^2. In *Reconstructions 1 and 4* the increase of computational density of heat flow is due to the increase of heat diffusion throughout daylight surface caused by abnormally high thermal conductivity and temperature conductivity of the permafrost sequence present in the model. In this case more heat is dissipated through the daylight surface; therefore, higher value of computational density of mantle basement heat flow is required.

Provided that the secular trend of surface temperature (*Reconstruction 3*) is not taken into consideration, there is also an increase in computational heat flow – 57,3 mW/m^2, which is due to absence of heat solar source in the paleotemperature reconstruction model of this type.

The comparison of calculated and measured geotemperatures in the borehole is presented in table 4. Since the measured temperatures (including those defined against vitrinite reflectance) and calculated geotemperatures can have uncertainty of ±2°C, results of *Reconstructions 3 and 4* are unacceptable. In these reconstructions true error exceeds optimal [9] rate more than four times, while the difference from vitrinite reflectance («maximum paleotermometer») data is 7–13°C. Thus, neglect of paleoclimate (*Reconstruction 3*) does not allow producing a precise physic-mathematical model of geothermal regime of Togur source rock. In the same way the hypothetical assumption about Neo-Pleistocene permafrost being 1000 m thick in the latitudes of 57-61 °is not confirmed by paleotemperature modeling.

Table 3. Calculated geotemperatures of the Togur suite in North Festival well 1 cross-section.

Time, million years ago	Secular trend of surface temperature («local),°C	Togur suite basement depth, m	Suite geotemperatures, °C			
			Reconstruction 1	Reconstruction 2	Reconstruction 3	Reconstruction 4
0	0	3183	119	120	125	115
0.001	+1	3182	119	120	125	115
0.003	+2	3182	119	120	125	115
0.005	+3	3182	119	120	125	115
0.018	+1	3182	119	120	125	115
0.03	–2	3182	119	120	125	115
0.05	–1	3181	119	120	125	115
0.052	–1	3181	119	120	125	115
0.055	–1	3181	119	120	125	115
0.0565	–2	3181	119	120	125	115
0.07	–4	3181	119	119	125	115
0.09	–1	3180	119	119	125	115
0.11	–4	3180	119	119	125	115
0.13	–1	3179	119	119	125	115
0.15	–4	3179	119	120	125	115
0.19	–9	3178	119	120	125	115
0.21	–6	3177	119	121	125	115
0.222	–7	3177	119	120	125	115
0.225	–8	3177	119	121	125	115
0.235	–10	3177	119	121	125	115
0.2355	–9	3177	119	121	125	115
0.2385	–2	3177	119	121	125	115
0.24	0	3177	119	122	125	115
1.4	+1	3158	124	122	124	131
1.64	+1	3158	124	122	124	131
3.1	+2	3147	126	124	124	133

3.2	+2	3147	126	124	124	133
3.8	+12	3147	126	124	124	133
4.7	+3	3147	130	128	124	138
5.2	−3	3146	126	124	124	133
5.7	+7	3146	126	124	124	134
6.3	+10	3146	132	129	124	139
7	+4	3146	127	124	124	134
20	+15	3146	138	135	124	145
24	+16	3141	138	136	124	146
31,5	+17	3011	133	131	117	140
32,3	+16	2993	132	130	117	139
34	+15	2977	130	128	116	137
37,6	+14	2955	128	126	115	135
41,7	+12	2925	125	123	114	131
42	+11	2913	123	121	113	130
46	+8	2838	117	115	110	124
54,8	+19	2687	120	118	103	126
58	+24	2647	124	122	101	130
61,7	+22	2616	121	119	100	127
73	+15	2441	107	105	92	113
73,2	+16	2441	107	105	92	113
86,5	+22	2314	107	105	87	113
89,8	+22	2311	107	105	86	113
90	+23	2307	107	105	86	113
91,6	+22	2299	107	105	86	112
114,1	+21	1491	74	73	54	77
118	+19	1463	71	70	53	74
120,2	+19	1453	71	70	52	73
132,4	+19	850	49	48	30	50
136,1	+19	783	46	46	28	48
145,8	+19	493	36	36	17	37
Computational basement heat flow, mW/m^2			56.8	55.7	57.3	60.3

Note. Shaded areas indicate temperatures of major oil generation zone (OGZ) [10], dark-colour shading indicates absolute OGZ paleotemperature maxima. Threshold OGZ geotemperature is 95 °C.

Table 4. Comparison of measured and calculated geotemperatures (North Festival well 1).

Depth, m	Measured (observed) temperature, °C	Measure ment method	Reconstruction 1, °C		Reconstruction 2, °C		Reconstruction 3, °C		Reconstruction 4, °C	
			Calcu lated tempera tures	Dis crepa ncy	Calcu lated tempe ratures	Discre pancy	Calcu lated tempe ratures	Dis crepan cy	Calcu lated tempe- ratures	Dis crepanc y
3130	118	Sheeted	117	−1	119	+1	124	+6	114	−4
3145	123	Sheeted	118	−5	119	4	124	+1	114	−8
3232	124	On OSV	130	+6	128	+4	117	−7	137	+13
Mean squared error («true error»), °C			±4		±3		±5		±9	

In case of the paleoclimate accounting (*reconstructions 1 and 2*) the true error of inverse modelling is close to optimal and is equal. Thus, comparison of the measured and assumed geotemperatures, allows concluding that according to the criteria of the true error results of *reconstructions 1 and 2* are acceptable and equal.

Existence of intensive Togur petroleum paleocharge in results of *reconstructions 1* (table 3) well explains exposed oil accumulation in the Lower Jurassic deposits with the North Festival 1 well (table 1). Moreover, in *reconstructions 1* during the longest period 34-6 million years ago (table 3) catagenetic conditions of gas generation deep zone (geotemperatures reach 138^{0}C) took place; the given fact conforms with gas bearing of Lower Jurassic and Paleozoic oil-and-gas bearing complexes (table 1).

4. Conclusion

By the example of Mesozoic-Cainozoic section of the south-east of West Siberia it has been stated that the neglect of secular trend of temperature and Neo Pleistocene permafrost thickness do not make it possible to reconstruct thermal history of Togur source rocks appropriately.

To estimate hydrocarbon resources in south-east areas of West Siberia using volumetric-genetic method [10] it is advisable to apply "local" secular trend of surface temperature [3] and deal with permafrost thickness of 300 m, that provides more consistent consideration of the main oil generation phase history and prevent from underestimating.

The present reaearch is conducted with financial support from the Russian Foundation of Basic Research as a part of the scientific project № 16-35-00080 mol_a

References

[1] Iskorkina A, Isaev V, Terre D 2015 *IOP Conf. Series: Earth and Environmental Science.* Assessment of Mesozoic-Kainozoic climate impact on oil-source rock potential (West Siberia). Vol. **27** pp. 012023.
[2] Isaev V I, Fomin A N 2006 *Russian Geology and Geophysics.* Loki of generation of bazhenov- and togur-type oils in the southern Nyurol`ka megadepression. Vol. **47 (6)** pp. 734-745.
[3] Isaev V I, Iskorkina A A 2014 *Journal of Geophysics.* Mezozojsko-kajnozojskij khod temperatur na poverkhnosti Zemli i geotermicheskij rezhim Yurskikh neftematerinskikh otlozhenij (Uyzhnaya paleoklimaticheskay zona Zapadnoj Sibiri. Vol. **36 (5)** pp. 64–80.
[4] Efimenko S V, Badina M V, Efimenko V N 2013 *TGASU bulletin.* K obosnovaniyu territorialnogo rasprostraneniya granici I–II dorojno_klimaticheskih zon v Zapadno-Sibirskom regione. Vol. **4** pp. 295–303.
[5]Pavlov A V, Gravis G F 2000 *Nature.* Vechnaya merzlota i sovremennii klimat. Vol. **4** pp. 10-18.
[6] Ivanov N S, Gavrilev R I 1965 *Nauka.* Teplofizicheskie svoistva merzlih gornih porod. Pp. 74.
[7] Kontorovich V A, Belyaev S Yu, Kontorovich A E, Krasavchikov V O, Kontorovich A A, Suprunenko O I 2001 *Geology and Geophysics.* Tektonicheskoe stroenie i istoriya razvitiya Zapadno-Sibirskoi geosineklizi v mezozoe i kainozoe. Vol. **42** pp. 1832–1845.
[8] Gulenok R Yu, Isaev V I, Kosygin V Yu, Lobova G A, Starostenko V I 2011 *Russian Journal of Pacific Geology.* Estimation of the Oil-and-Gas Potential of Sedimentary Depression in the Far East and West Siberia Based on Gravimetry and Geothermy Data. Vol. **5 (4)** pp. 273–287.
[9] Isaev V I 2013 *Russian Journal of Pacific Geology.* Interpretation of High-Accuracy Gravity Exploration Data by Mathematic Programming. Vol. **7 (2)** pp. 92–106.
[10] Kontorovich A E, Burshtein L M, Malishev N A, Safronov P I, Guskov S A, Ershov S V, Kazanenkov V.A. Kim N.S. Kontorovich V.A. Kostireva E.A. Melenevskii V.N. Livshic V.R. Polyakov A A, Skvorcov M B 2013 *Geology and Geophysics.* Istoriko-geologicheskoe modelirovanie processov naftidogeneza v mezozoisko-kainozoiskom osadochn. Vol. **54 (8)** pp. 1179–1226.

Petrophysical rock properties of the Bazhenov Formation of the South-Eastern part of Kaymysovsky Vault (Tomsk Region)

A M Gorshkov[1,3], L K Kudryashova[2,4] and O S Lee-Van-Khe[2]

[1] Petroleum Learning Center, National Research Tomsk Polytechnic University, Russia
[2] Department of Geology and Minerals Prospecting, National Research Tomsk Polytechnic University, Russia
[3]gorshkovam89@mail.ru, [4]KudryashovaLK@tpu.ru

Abstract. The article presents the results of studying petrophysical rock properties of the Bazhenov Formation of the South-Eastern part of Kaymysovsky Vault with the Gas Research Institute (GRI) method. The authors have constructed dependence charts for bulk and grain density, open porosity and matrix permeability vs. depth. The results of studying petrophysical properties with the GRI method and core description have allowed dividing the entire section into three intervals each of which characterized by different conditions of Bazhenov Formation rock formation. The authors have determined a correlation between the compensated neutron log and the rock density vs. depth chart on the basis of complex well logging and petrophysical section analysis. They have determined a promising interval for producing hydrocarbons from the Bazhenov Formation in the well under study. Besides, they have determined the typical behavior of compensated neutron logs and SP logs on well logs for this interval. These studies will allow re-interpreting available well logs in order to determine the most promising interval to be involved in Bazhenov Formation development in Tomsk Region.

1. Introduction

Oil companies have recently started paying greater attention to unconventional reserves due to the depletion of conventional hydrocarbon resources. The biggest unconventional oil reservoir in Russia is the Bazhenov Formation located in the West Siberian Basin. The oil pools of this formation are considered one of the most important targets to replenish hydrocarbon resources in Russia [1, 2]. The uniqueness of the Bazhenov Formation consists in the fact that it, on the one hand, is the main source rock of the West Siberian Basin and, on the other hand, it is reservoir. Thus, Bazhenov Formation rocks are a source of both existing and potential oil, which indubitably should be taken into account during the selection of oil production technologies for such complicated reservoirs [3].

The main technology used to produce movable oil from the Bazhenov Formation is horizontal drilling with a subsequent multistage hydraulic fracturing [3]. The thermal gas treatment technology is now being tested for producing potential oil from kerogen-containing rocks in Russia [4, 5].

One of the most promising areas for J_0 development in Tomsk Region is the eastern part of Kaymysovsky Vault. This structure was submerged for a long time under sea basin transgression during the Late Jurassic – Early Cretaceous time, which allowed forming thick Bazhenov Formation rocks (about several dozen meters). Several major oil fields have been discovered in the eastern part of

Kaymysovsky Vault and they have been commercially developed for a long time. There are hundreds of wells drilled in this territory that penetrate Bazhenov Formation deposits. Besides, they have obtained an oil flow on the South-Eastern slope of Kaymysovsky Vault from J_0 at the Fedyushkinskoe field. All these facts together with the developed oilfield infrastructure will allow significantly reducing expenses on bringing the Bazhenov Formation into development in this territory.

However, despite the proven oil-bearing capacity of the Bazhenov Formation, there is a range of problems connected with production of hydrocarbons from it. First, these hydrocarbon reserves are hard-to-recover and require an individual approach to development. Second, the deposits of this formation are not well-studied enough now. Thus, the main problem of the geological-geophysical studies of Bazhenov Formation deposits at most fields in Tomsk Region is connected with the absence of a complete range of well logging methods and poor core recovery. Besides, there is the absence of a sufficient number of laboratory core studies for the unambiguous localization of areas with improved reservoir properties of the Bazhenov Formation. Due to the ultra-low reservoir properties of the siliceous-argillaceous-carbonate deposits of the Bazhenov Formation, it is impossible to determine the petrophysical properties of these rocks with laboratory equipment using standard methods [6].

Gas Research Institute (GRI) in Chicago has made significant progress in developing methods used to determine the petrophysical properties of shale formations. The Institute has developed a comprehensive method of determining shale reservoirs: "Development of laboratory and petrophysical techniques for evaluating shale reservoirs." [7].

The aim of this work was to study the petrophysical properties of Bazhenov Formation deposits in the South-Eastern part of Kaymysovsky Vault with the GRI method.

2. Method of studying petrophysical properties of Bazhenov Formation rocks

The GRI method [7] was developed to determine the main petrophysical properties (porosity, permeability, saturation) of shale formations. The method consists in crushing core into fractions of an equal size and determining parameters on crushed core. The GRI method allows significantly reducing the time of permeability experiments (2,000 seconds) and excluding the impact of natural fracturing on the values of the parameter under consideration. However, despite all the benefits of this method, its main drawback is making experiments in atmospheric conditions (without pore and ground pressure).

The paper [8] shows that the determined permeability values of the core plugs are $10\text{-}10^6$ times higher vs. rock matrix permeability due to the developed natural fracturing of Bazhenov Formation rocks. Thus, in order to determine the true values of shale reservoir permeability, experiments should be made on crushed core.

The petrophysical properties (bulk and grain density, open porosity, matrix permeability) of Bazhenov Formation deposits were determined on cores with natural saturation the SMP-200 shale permeameter. Before the start of the experiment, whole core was crushed and sieved through two- and one-millimeter cells to obtain a fraction of a homogeneous size. The mass of one weighed portion amounted to 30±0.01 g.

The bulk and grain density of the samples was determined on crushed core via the gas saturation method. The experiment on determining bulk and grain density consisted in calculating a sample volume at the baseline (before helium penetration into pores) and at the end of the experiment (after sample saturation with helium) correspondingly for the pressure drawdown curve. Pressure drops as gas (helium) penetrates into the micropores of separate rock particles [9]. Since SMP-200 works on the basis of the Boyle's law, we can calculate a corresponding value of a core volume for every pressure value in case of the isothermal process during the whole experiment.

The open porosity coefficient was calculated on the basis of bulk and grain density values [9].

The permeability of the Bazhenov Formation samples was determined via the Pressure Pulse Permeability method. The method consists in the approximation of the theoretical curve of experimental data on a pressure drop that occurs due to helium penetration into the micropores of separate particles in crushed core, as well as the subsequent calculation of matrix permeability values on the basis of the obtained curve parameters [7, 10].

The insignificant pressure change during the experiments (0.1-0.5% of the absolute value) imposes heightened requirements to device calibration, the absence of leaks and thermal stabilization during the process of determining petrophysical properties [9].

3. Petrophysical and geological-geophysical study of Bazhenov Formation deposits in the South-Eastern part of Kaymysovsky Vault

The authors used rock samples from Bazhenov Formation deposits (J_0) at one of the fields in the South-Eastern part of Kaymysovsky Vault. The Bazhenov Formation is exposed at this field and drilled with core sampling in an exploration well. In the well section, the Bazhenov Formation with its top located at the depth of 2,663.0 m is characterized by conformable bedding on deep-sea dark-grey Georgiev Formation clays and is interbedded by a unit of marine, predominantly argillaceous Kulomzin Formation deposits. The thickness of this formation is 13 m.

Figure 1 shows the results of the petrophysical studies of the Bazhenov Formation rock samples from the exploration well under consideration.

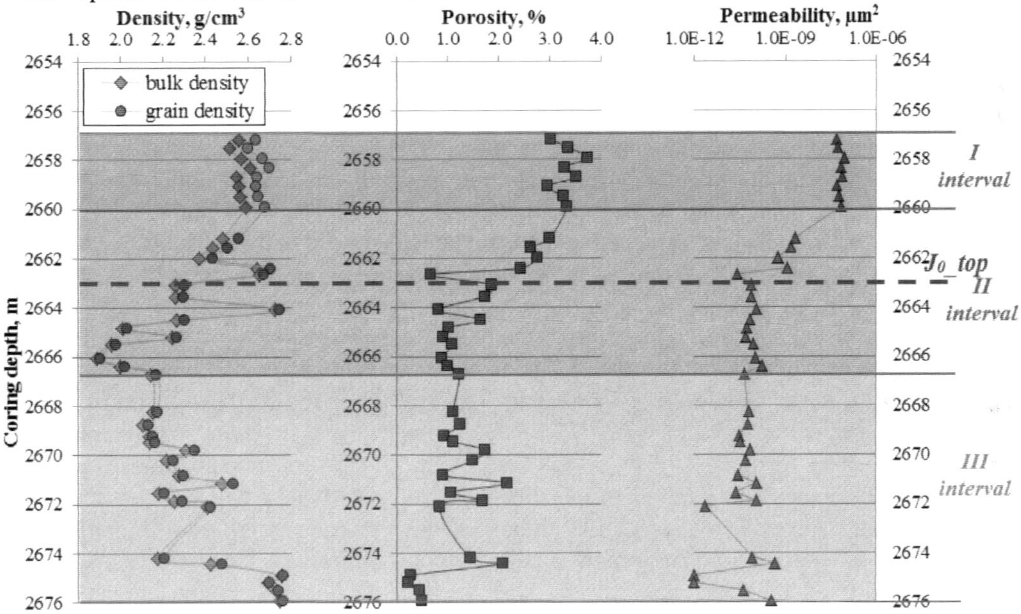

Figure 1. Results of the petrophysical studies of Bazhenov Formation rocks in the South-Eastern part of Kaymysovsky Vault.

A chart featuring the dependence of petrophysical rock properties vs. depth is constructed on the basis of processing 13 Kulomzin Formation samples and 28 Bazhenov Formation samples (interval – 2,657.2-2,675.9 m).

The petrophysical data (Figure 1) and core description allow dividing the entire section into three intervals each of which characterized by different conditions of rock formation.

In compliance with the core description, the first interval (2,657.2-2,659.9 m) is formed by dark-grey, silty, sometimes slabby shalestone having homogeneous composition. Composition homogeneity is confirmed by the insignificant change in the grain density of rocks varying from 2.60 g/cm^3 to 2.70 g/cm^3 and close values of the open porosity coefficient (from 3.00% to 3.73%). The significant changes in the grain density values and open porosity coefficient are connected with the horizontal microlamination of the deposits. Besides, this interval is characterized by high matrix permeability coefficient values – about $6 \cdot 10^{-8}$ μm^2. It is possible to conclude for the first interval that these deposits were formed in a relatively quiet depositional environment (Kulomzin Formation).

The second interval (2,659.9-2,666.7 m) is formed by brownish shalestones. The samples are characterized by a big amount of pyrite in the form of microscopic inclusions, which testifies to deep-water depositional conditions. The interval demonstrates significant changes in the mineral composition of its rocks: varying density from 1.91 g/cm^3 (at the depth of 2,666.1 m) to 2.75 g/cm^3 (at the depth of 2,664.1 m). The low rock density values are connected with the high content of kerogen and bitumen in the shalestones. The content of total organic carbon in this interval reaches 25% in compliance with pyrolytic core studies. The high density values are connected with the presence of big broken thick-walled shells composed of white calcite. Besides, this interval has two dense beds: one in the top of the Bazhenov Formation and another in the bottom of the Kulomzin Formation. These beds can be mapped on the basis of the high rock density (2.71 g/cm^3 and 2.75 g/cm^3, respectively) and low open porosity coefficient values (0.66% and 0.82%). All these factors testify to a heterogeneous and changeable deposition process in the South-Eastern part of Kaymysovsky Vault. The shape of the curve in the open porosity coefficient distribution chart for this interval repeats the shape of the curve of rock grain density vs. depth, but has a more flattened look. Sample porosity gradually drops from 3.35% (at the depth of 2,659.9 m) and reaches a mean value of about 0.88% by the end of the interval under study. Analogous behavior is also typical of matrix permeability. The matrix permeability of the rocks drops with depth to reach the value of about $6 \cdot 10^{-11}$ μm^2 and is practically invariable up to the bottom of the Bazhenov Formation. The research findings show that the second interval corresponds to a transient depositional process from sea basin transgression to regression. The deep-water shalestones of the Bazhenov Formation are gradually interbedded with more shallow-water Kulomzin Formation deposits. The density and porosity chart does not allow clearly determining the boundary of the Bazhenov Formation top – both on well logs and on the basis of petrophysical studies.

The third interval (2,666.7-2,675.9 m) is formed by homogeneous dark-brown to black shalestones. The density of its rocks makes 2.17 g/cm^3 on average, which testifies to the relatively quiet depositional environment of Bazhenov Formation rocks. The beds of carbonatized shalestones characterized by the density values of up to 2.53 g/cm^3 and open porosity coefficient values making up to 2.16% start appearing at the depth of 2,669.8 m. The bottom of the Bazhenov Formation has an argillaceous-carbonate bed at the depth of 2,674.2 m. It is characterized by almost equal high values of grain density amounting to up to 2.76 g/cm^3 and low open porosity coefficient values reaching less than 0.43%. The carbonate rocks found in this interval have zero matrix permeability (emphasized with white color in Figure 1), whereas, in compliance with the core description, this argillaceous-carbonate interval is characterized by a high degree of rock fracturing and the core is crushed into separate pieces.

Figure 2 shows the results of the petrophysical studies of the Bazhenov Formation samples vs. the results of the geological-geophysical studies for the exploration well under consideration.

The analysis of the range of well logging methods in the exploration well of the field under consideration makes it possible to conclude that the thickness of the Bazhenov Formation is determined on the sections on the basis of the abnormally high electric resistance and high gamma ray logging values, therefore the Bazhenov Formation shalestones are one the basic benchmarks in Western Siberia. Besides, the section shows several negative SP log anomalies that testify to the presence of a reservoir in the Bazhenov Formation. However, the negative anomaly of the SP log also testifies to carbonate beds. In this connection, there is a need for a priori information on the formation (geochemical core analysis, field tests, etc.) to map reservoir rocks in the Bazhenov Formation in a more detailed way.

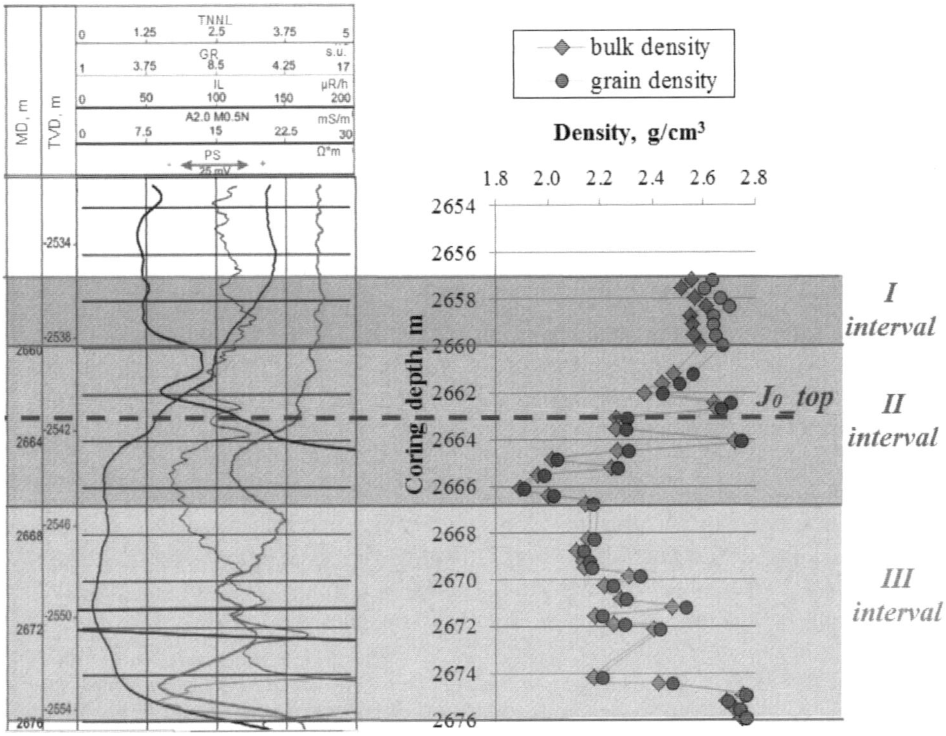

Figure 2. Results of geologic-geophysical and petrophysical studies of Bazhenov Formation rocks in the southeastern part of Kaymysovsky Vault.

It should be noted that the intervals determined with the help of petrophysical studies and core descriptions are invisible on well logs. However, the chart featuring the dependence of the bulk and grain density of the rocks vs. depth reliably correlates with the compensated neuron log on well logs. This relationship can be used in well logging interpretation to determine Bazhenov Formation lithotypes.

4. Conclusion

Thus, the GRI method was first applied in Russia at the laboratory of Tomsk Polytechnic University to study reservoir rocks with ultra-low reservoir properties. The petrophysical analysis of the Bazhenov Formation rocks in the South-Eastern part of Kaymysovsky Vault shows that the GRI method with the use of the SMP-200 shale permeameter allows dismembering deposits depending on rock density distribution in a more detailed way. This dismembering allows determining the inhomogeneity of depositional conditions.

The geological-geophysical and petrophysical study of the section has allowed determining a correlation between the compensated neutron log and the rock density vs. depth chart. This relationship can be used in the future in well logging interpretation at the fields of Tomsk Region to determine Bazhenov Formation lithotypes. Besides, compensated neutron log values can be used for the preliminary evaluation of mechanical rock properties forming the section in order to design a multistage hydraulic fracturing.

The complex study of the section and the geochemical analysis have allowed concluding that one of the intervals under consideration (the second interval) is the most promising bed of the Bazhenov Formation for development. This interval is formed by bituminous shalestones characterized by the high content of movable hydrocarbons and kerogen (TOC amounting to up to 25% in compliance with the pyrolytic research data). The interval has the minimum values of compensated neutron logging on well logs and density in compliance with the petrophysical analysis and a negative anomaly on its SP log.

Finally, the present studies and the relationships obtained will allow re-interpreting the existing well logging data in order to determine the most promising intervals for further Bazhenov Formation development.

5. Thanks

We would like to thank the employees of the Reservoir Engineering Laboratory of OJSC «TomskNIPIneft» for the disintegrated core samples that it has provided. We are thankful to our colleagues V. E. Baranov for the equipment provided and T. A. Gaydukova for her valuable recommendations and assistance in data interpretation. The study is supported by the Ministry of Education of the Russian Federation and performed within the framework of the grant VIU-IPR-50/2016 "Hard-to-recover natural resources".

References

[1] Afanasjev I S, Gavrilova E V, Birun E M, Kalmykov G A and Balushkina N S 2010 *Scientific and technical journal OJSC «Rosneft».* Bazhenov Formation. Overview, problems. Vol. **4** pp. 20–25.

[2] Lobusev A V, Lobusev M A, Vertievec Iu A and Kulik L S 2011 *Territory of Neftegaz* . Bazhenov Formation – supplementary source of hydrocarbons materials in West Siberian. Vol. **3** pp. 28–31.

[3] Lazeev A N, Kashik A S, Bilibin S I, Valova L V, Kalmykov G A, Bachin S I, Diakonova T F, Isakova T G and Iukanova E A 2015 *Geophysics.* Main problems of studying of the Bazhenov Formation deposits. Vol. **3** pp. 2–4.

[4] Kokorev V I, Darischev V I, Akhmadeishin I A, Schekoldin K A and Bokserman A A 2014 *Drilling and crude oil.* Field test results and prospects of thermal gas treatment technologies for the Bazhenov Formation development in OJSC «RITEK». Vol. **11** pp. 26–28.

[5] Kokorev V I 2010 *Petroleum engineering.* Basic aspects of controlling of thermal gas impact on rocks of Bazhenov Formation as to geological conditions of Sredne-Nazymsky and Galyanovsky fields (part 1). Vol. **6** pp. 29–31.

[6] Tonkonogov Iu M and Muler P B 2013 *Oil and gas vertical.* Innovative methods of petrophysical investigation of low-permeability rocks. Vol. **9** pp. 34–36.

[7] Guidry K, Luffel D and Curtis J 1995 *Development of Laboratory and Petrophysical Techniques for Evaluating Shale Reservoirs. Gas Research Institute Final report. GRI-95/0496* (Des Plaines) pp. 1–49.

[8] Skripkin A G, Parnachev S V, Baranov V E and Zakharov S V 2014 *Oil industry.* Experience of using different methods for evaluation of reservoir properties of nanopermeable rocks. Vol. **11** pp. 59 – 61.

[9] Gorshkov A M 2015 *Materials of XIX Int. Symp. students and young scientists in honor academician M A Usov. Problems of geology and mineral resources development* . Some aspects of development of laboratory technique for evaluating of petrophysical properties of ultralow permeability reservoirs of Bazhenov Formation sediments. Vol. **2** pp 66–70.

[10] Luffel D L and Hopkins C W 1993 *SPE Annual Technical Conference and Exhibition.* Matrix Permeability Measurement of Gas Productive Shales. pp. 261–270.

Peculiarities of pyrrhotite mineralization in the Chertovo Koryto deposit (Patom Uplands)

Sinkina E.A.[1,a], Savinova O.V.[1], Nepryakhina O.B.[1], Kolbysheva Yu.V.[1]

[1] National Research Tomsk Polytechnic University, 30 Lenin Ave., Tomsk, 634050, Russia

E-mail: [a]vea1986@mail.ru

Abstract. Pyrrhotite mineralization of black shale gold-ore deposit in the central part of Chertovo Koryto (Patom Uplands) was studied. It was found that pyrrhotite consists of three generations, containing hexagonal and monoclinic modifications in various proportions. The increase in the monoclinic modification from the earlier to later generation was observed.

1. Introduction

Pyrrhotite is widely distributed in the deposits of various genetic types, including hydrothermal gold deposits. It is usually associated with sulfides (pyrite, sphalerite, galena, chalcopyrite, and arsenopyrite), chlorites, carbonates (calcite, ankerite), and quartz and forms several generations. For example, in comparison with such deposits as Ta Nang (Central Vietnam), Kumtor (Kyrgizstan), Sukhoy Log (Russia), where it forms 1-2 generations, pyrrhotite is represented by 3 generations in deposits Awanda (China), Blagodatnoe, Olimpiada (Russia), and its quantity allows relating it to the primary ore minerals [1, 2, 3, 4].

Geological interest to study pyrrhotite is connected with its variable chemical composition. The iron content varies from 46.5 to 50 at.%. The chemical composition of natural pyrrhotites is characterized by intermittent series: monoclinic – 46.45…46.9 at.%, hexagonal – 47.0…48.3 at.%, troilite – 49.8…50.2 at.% [5, 6]. All pyrrhotites can contain isomorphic impurities of Cu, Ni, Co and other elements [7].

Changes in pyrrhotite composition are connected with variable physicochemical conditions of mineral genesis and cause different structural characteristics, mechanical and magnetic properties. In hydrothermal ore deposits pyrrhotite can be found in hexagonal and monoclinic modifications.

The present article describes the study results of pyrrhotite mineralization characteristics in the Chertovo Koryto gold-ore deposit, as well as the chemical composition of pyrrhotites.

2. Research methods

To carry out the research core samples were taken from 4 cross-bedding profiles, including two intersecting profiles and one from each opposite cross-bedding. Wells were drilled at full productivity of the ore body with access to the barren area. The drilled depth varies from 150 to 250 m. Core samples were selected thoroughly at intervals from 10 to 50 cm. Firstly, the core samples were studied macroscopically, and samples containing sulphide mineralization, including pyrrhotite, were selected. Then, the polished sections were made for the mineral-graphic study and implementation of point-counter X-ray fluorescence analysis. The polarizing microscope Zeiss Imager A2m was used to carry out the optical study. The chemical composition was studied at the X-ray fluorescence microscope XGT-7200. 184 point-counter measurements were carried out.

3. Geological structure of deposit

This deposit is located in the north of Patom upland within the Big Patom river basin. The ore deposit was formed in the early Proterozoic carbonaceous terrigenous slates of Mikhailovskoe suite. The suite embraces metasomatic rocks of beresite-propylite formation with vein-disseminated sulfide-quartz mineralization. Sulfides occur rarely in quartz veins and veinlets. Pyrite, arsenopyrite, pirhottite are predominant. Galena, sphalerite, chalcopyrite occur as insignificant impurities. Free gold dominates in quartz [8, 9].

4. Results and discussion

Pyrrhotite is represented by three generations. The generations are separated according to the polishing characteristics, morphology of aggregates, and paragenetic mineral associations.

Pyrrhotite I forms "striates" parallel to the foliation of carbonaceous shales (fig. 1*a*). It has a eugranitic internal structure. "Striates" size varies in the range of 0.1...5 mm. Usually it has a trace of polishing in form of cavities. The concentrations of main mineralizing elements vary in the following range: Fe – 46.61...48.28 at.%, S – 50.69...53.24 at.%. Copper is found in several samples (0.1; 0.12; 0.14, and 0.28 wt.%). The chemical composition of pyrrhotite corresponds to hexagonal pyrrhotite modification (table 1, fig. 2).

Pyrrhotite II has a compact granular texture with thin quartz-carbonate veinlets (fig. 1*b*;. fig. 2). The iron content varies within the range of 46.64...48.16 at.%, the sulfur content – 50.15...53.52 at.%. Nickel is found among impurities (0.01…1.44 wt.%). Pyrrhotite II includes both monoclinic and hexagonal modifications. The number of the first modification predominates (table 1, fig. 2).

Pyrrhotite III occurs as pocket segregations of various sizes in relatively thick (up to several meters) quartz-carbonate veins (fig 1*c*). It is polished better than the first two generations without cavities. It is characterized by eugranitic texture and has a nonstoichiometric composition: Fe – 46.45…48.34 at.%, S – 50.54…53.33 at.%. The mineral composition indicates nickel in the amount of 0.01...0.29 wt.%. Pyrrhotite III is characterized by the predomination of monoclinic modification (table 1, fig. 2).

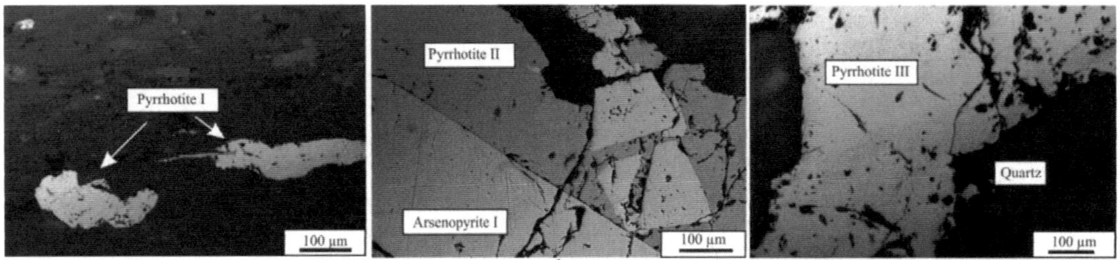

a *b* *c*

Figure 1. Pyrrhotite mineralization in the Chertovo Koryto deposit: a) pyrrhotite I in the form of "stringers" in terrigenous carbonaceous rocks, b) pyrrhotite II in combination with arsenopyrite I in thin quartz-carbonate veinlet, c) pyrrhotite III in quartz-carbonate vein

Table 1. Chemical composition of pyrrhotites in the Chertovo Koryto deposit

Generations	Fe, at.%	Ni, wt.%	Cu, wt.%	Fe, at.% Monoclinic	Fe, at.% Hexagonal
Pyrrhotite I	46.61…48.28	0.01…0.94	traces	46.61…46.97	47.07…48.28
	47.419 (93)	0.284		46.769 (21)	47.609 (72)
Pyrrhotite II	46.64…48.16	0.0…1.44	0	46.64…47.0	47.07…48.16
	47.323 (41)	0.318		46.838 (25)	47.502 (16)
Pyrrhotite III	46.45…48.34	0.0…0.29	traces	46.45…47.0	47.03…48.34
	47.597 (50)	0.1		46.802 (38)	47.325 (12)

Note: Nominator shows the value interval, denominator shows an average value. Number of samples is given in brackets.

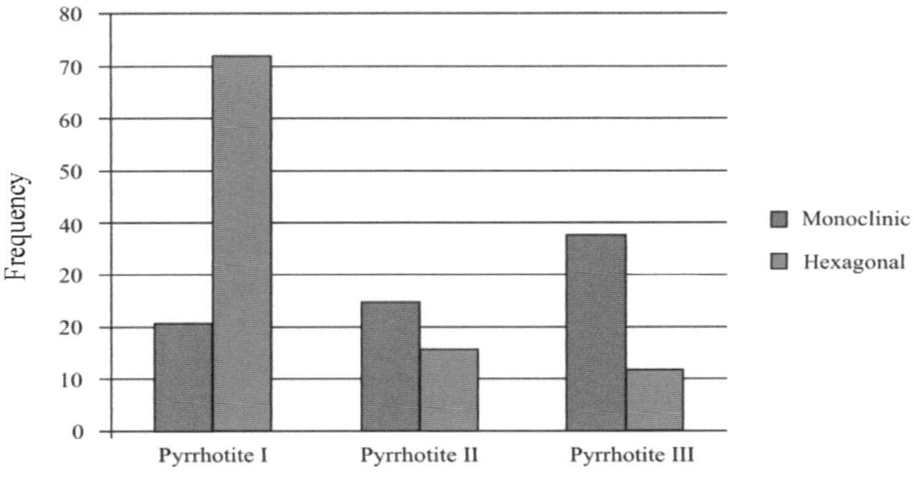

Figure 2. Histogram of frequency of pyrrhotite modifications in various generations (based on the table data)

5. Discussion and conclusion

During the research, it was found that in the Chertovo Koryto deposit pyrrhotite is represented by three generations, distinguished by the polishing, morphology of aggregates, paragenetic mineral associations and a range of impurity elements. Generations of pyrrhotines are distributed equally in the ore body. Pyrrhotite I occurs throughout the ore deposit both in the ore and non-ore intervals. Pyrrhotite II and III occur in the central part of the ore body, where rich concentrations of gold intervals are located. All three pyrrhotites are characterized by similar iron content, but different nickel content. The highest nickel content was found in pyrrhotites of I and II generations in the amount of 0.284 and 0.318 wt.%. The presence of nickel indicates textural deformations in sulfide structure. Hexagonal and monoclinic modifications are found in pyrrhotites of all generations (fig. 2). However, the amount of monoclinic modification increases from early to late generation.

The research was supported by the Russian Foundation for Basic Research (project 16-35-00041\16).

References

[1] Tran Tuan Anh, Gaskov I V, Tran Trong Hoa, Borisenko A S, Izokh A E, Pham Thi Dung, Vu Hoang Li and Nguyen Thi Mai 2015 Ta Nang gold deposit in the black shales of Central Vietnam *Russian Geology and Geophysics* **56/10** 1797-1813 (in Russian)
[2] Rafailovich M S, Mizernaya M A and D'yachkov B A 2011 *Large gold deposits in black shale strata: conditions of formation, features of similarity* 272 (in Russian)
[3] Qing-Feng Ding, Chang-Zhi Wu, M Santosh, Yu Fu, Lian-Hui Dong, Xun Qu, and Lian-Xing Gu 2014 H–O, S and Pb isotope geochemistry of the Awanda gold deposit in southern Tianshan, Central Asian orogenic belt: Implications for fluid regime and metallogeny *Ore Geology* **62** 40-53
[4] Kolmakov Yu V, Sazonov A M, Pinson A Yu, Potekhina E V, Gertner I F and Tishin P A 2006 Petromagnetic and mineralogical heterogeneity and gold mineralization of the Blagodatnoye ore deposit *Bulletin of the Tomsk Polytechnic University* **309/2** 46-53 (in Russian)
[5] Caprenter R H, Desborough G A 1964 Range and solid solution and structure of naturally occurring troilite and pyrrhotite *Amer. Mineralogist* **49**1350-1364
[6] Desborough G A and Carpenter R H 1965 Phase relations of pyrrhotite *Economic Geology* **60** 1431-1450
[7] Ramdor P 1962 *Ore minerals and their intergrowths* 1132 (in Russian)

IOP Conf. Series: Earth and Environmental Science **43** (2016) 012011 doi:10.1088/1755-1315/43/1/012011

[8] Kucherenko I V, Gavrilov R Yu, Martynenko V G and Verhozin A V 2009 Structural-dynamic model of golden-ore deposits formed in non-shale and black-shale substratum. P. 2. Chyortovo Koryto deposit (Patomskoe uplands) *Bulletin of the Tomsk Polytechnic University* **314/1** 23-38(in Russian)
[9] Gavrilov R Yu and Rudmin M A 2015 Methodological approaches in estimating anomalous geochemical field structure *IOP Conf. Ser.: Earth and Environ. Sci.* **24** 012011

Application of Multivariate Statistical Analysis for Delineation of Prospective Geochemical Anomalies in Providenskaya Area (Chukotka, Russia)

M Kirillova[1], V Voroshilov[2], A Yankovich[3], E Yankovich[2], A Maskov[2] and A Chernyshov[4]

[1] JSC "Georegion", 25A Lenin Avenue, Anadyr, Russia
[2] Tomsk Polytechnic University, 30 Lenin Avenue, Tomsk, Russia
[3] LLC "Mangazeya Gold", 5 Kosciusko-Grigorovich Avenue, Chita, Russia
[4] Tomsk State University, 36 Lenin Avenue, Tomsk, Russia
E-mail: v_g_v@tpu.ru

Abstract The secondary geochemical field structure was modelled on the basis of the lithogeochemical dispersion trains of the Providenskaya Area of the Chukotka Peninsula. The factor and cluster analysis were applied to interpret the nature of geochemical anomalies. It was proved that a range of anomalies were prospective for gold-silver, polymetallic, tin, and tungsten deposit allocation.

1. Introduction
The intensity and degree of potential are the main features of ore-associated geochemical anomalies (GA). The intensity is usually assessed by a multiplicative index. The degree of potential is assessed by a complex of favorable geological prerequisites and geochemical characteristics [1]. The methods of a multivariate statistical analysis including the factor and cluster analyses may be used to characterize the GA inner structure and estimate the degree of potential [1], [2].

2. Research technique
The current research is based on the results of dispersion train lithogeochemical sampling in the Providenskaya Area of the Chukotka Autonomous Region (Russia) (Figure 1). 2288 samples of a sandy-sludge alluvial fraction were analyzed. The content of chemical elements was determined in the laboratory of the East-Chukotka Geological Expedition with the use of following methods. A fire assay finish and an atomic absorption method were used for gold and silver, an X-ray spectral one for uranium and thorium; a chemical one for copper, lead, zinc, and tin. The multivariate statistical analysis was carried out by Statistica software of StatSoft company.

3. Research results
The territory under consideration is formed by the variously dislocated and metamorphosed Archean, Rifean, and Paleozoic marine sediments, the volcanogenic continental Mesozoic formations, and the unconsolidated Quaternary sediments (Figure 2).

The Archean is represented by deeply metamorphosed rocks of Penkigney series: various gneisses, marbles, amphibolites, crystalline schists. The Riphean rocks are metamorphosed limestones, porphyrites, crystalline schists, metasiltstones, metasandstones, and rare thin interbeddings of spilite. The Paleozoic sediments include Devonian argillaceous slates and limestones.

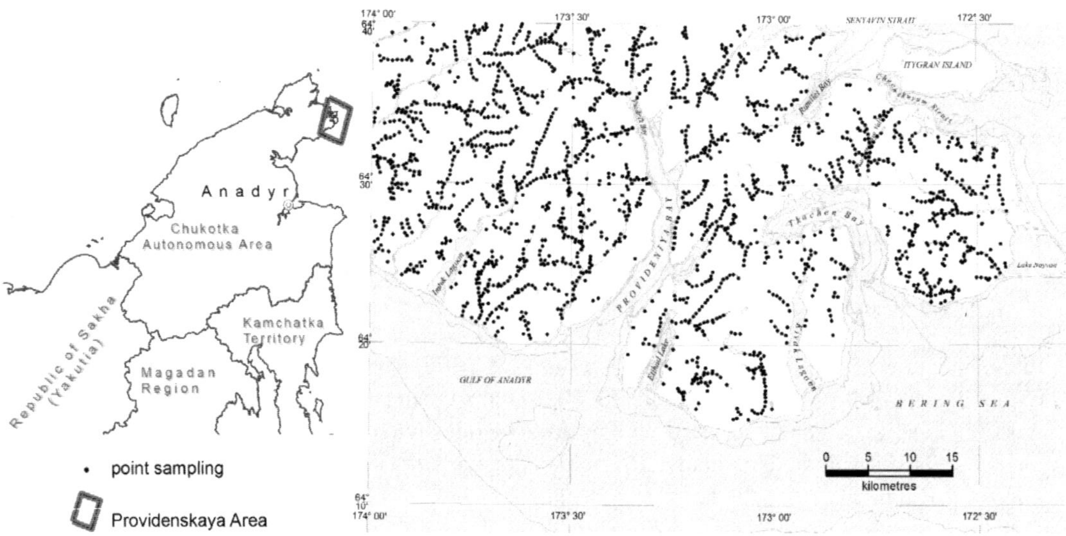

Figure 1. Survey map of the study area

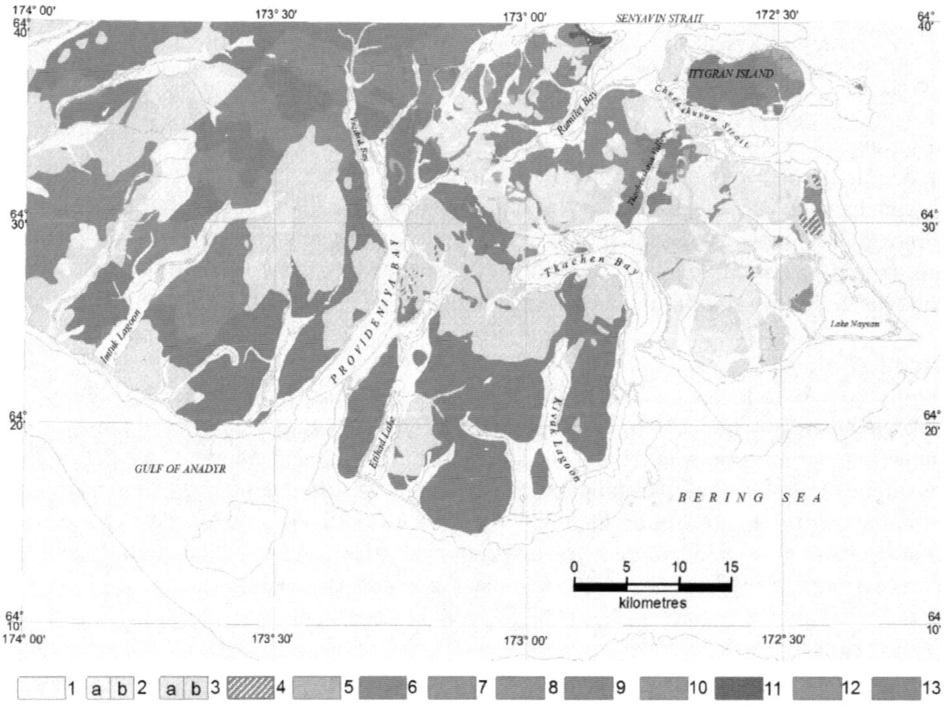

Figure 2. Geological map:

Legend: 1 – Quarternary; Cretaceous: 2 - Upper. a - Coniacian. Providens set, b - Turonian. Leurvaam suite; 3 - Lower. a - Hauterivian. Amgen set, b - Argyt set; 4 - Subvolcanic bodies. Argyt volcanic complex. Andesite; Plutonic bodies: 5 - Erguveem complex. Gabbro; 6 - Taureran complex. Granodiorite; 7 - Leurvaam complex. Granite; 8 - Leurvaam complex. Leucogranite; 9 - Providens complex. Diorite; Devonian: 10 - Ikychuren suite; 11 - Tanatap suite; 12 - Riphean. Kelkhin series; 13 - Archean. Penkigney series.

All the mentioned rocks are covered by the Cretaceous volcanites of felsic, intermediate, and rarely mafic composition that belong to the Okhotsk-Chukotsky Volcanogenic Belt. The Quarternary sediments are presented by marine, glacial, glaciofluvial, lacustrine, alluvial, and deluvial sediments.

The methods of multivariate statistics are widely used for the genetic interpretation of anomalous geochemical fields. Generally, these methods are the factor and cluster analysis [3], [4], [5], [6]. The geochemical fields linked with the ore deposit formation are also clearly distinguished in the multivariate geochemical field structure [7], [8], [9], [10].

The authors studied six factors during the factor analysis that contribute 60% to the total dispersion (Table). The factor values were calculated for each sampling point.

The first factor illustrates 18% of the total dispersion variation. Ti, Sc, V, Mn, Co, and Zr play a major role in the factor feature structure. According to the factor elemental composition and its location, it probably characterizes the least altered host rocks (Figure 3).

Table. Matrix of factor weights.

Substances	Factor 1	Factor 2	Factor 3	Factor 4	Factor 5	Factor 6
Au	-0.01	0.86	-0.07	-0.03	-0.06	-0.02
Ag	-0.03	0.94	0.02	0.04	0.09	-0.05
Pb	-0.09	0.30	0.29	0.29	0.28	-0.06
As	0.00	-0.06	0.14	0.46	0.04	0.04
Cr	0.03	0.00	0.00	0.01	0.00	0.90
V	0.72	0.06	-0.25	0.42	-0.10	0.12
Ni	0.12	0.05	0.00	-0.05	0.02	0.92
Mn	0.70	0.12	0.10	-0.17	0.15	0.21
Ba	0.41	-0.08	0.00	0.36	-0.16	-0.07
Nb	-0.03	-0.05	0.80	0.14	0.02	-0.05
Mo	-0.01	0.05	0.32	0.12	0.24	0.08
Sn	-0.07	-0.06	0.00	0.22	0.75	-0.04
Cu	0.19	0.89	-0.03	-0.02	0.02	0.23
Zn	0.26	0.46	0.21	0.13	0.30	0.04
Sc	0.79	0.04	0.14	0.02	-0.04	0.14
Y	0.03	-0.01	0.80	0.06	-0.04	-0.05
Ga	0.06	-0.02	0.21	0.79	-0.11	-0.11
W	0.35	0.03	-0.21	0.14	0.45	0.11
Co	0.57	0.24	-0.15	0.03	0.06	0.62
Ge	0.01	0.07	-0.09	0.47	0.14	0.11
Bi	0.05	0.18	-0.04	-0.12	0.78	0.07
Ti	0.85	0.03	-0.07	0.15	-0.01	0.04
Be	-0.06	-0.04	0.49	-0.32	0.59	0.02
Zr	0.52	-0.02	0.51	-0.25	0.08	0.08
Li	0.21	0.04	-0.11	0.70	0.15	0.04

The second factor defined by Au, Ag, and Cu is unambiguously treated as the main ore-bearing one. More perspective anomalous zones can be distinguised according to the factor intensity, spatial and associative patterns of monoanomalies. One of such anomalies is situated in the area of Rumilet Bay, where complex ore anomalies have been identified. The anomalies in the area of Endogurov Promontory and further along the Annyuvaam River include monoanomalies of gold, silver, zinc, partially lead and copper. The anomaly in the Tkachenskaya Valley is coherent with the anomalies of gold, silver, and copper. The anomalies in the vicinity of Senyavin Highland include monoanomalies of gold, silver, and copper. All the selected anomalies belong to the ore-associated GA (Figure 3).

Several ore formations can be presumed by the content of chemical elements. They are considered to be a gold-silverore formation, a polymetallic (lead-zinc) with silver one, and a polymetallic skarn one at the current stage of work.

The third factor illustrates 10% of total dispersion variation. This factor predominantly influences yttrium, niobium, zirconium, beryllium, and to a lesser extent, molybdenum and lead. The granitoid Rumilet and Kyndlyagak Massives are characterized by the anomalous values of this factor. They belong to the subalkaline granites of the second phase of the Leurvaam complex (Figure 3). Both massives are spatially coherent with the monoanomalies of niobium and beryllium. The yttrium anomaly is observed in the Kyndlyagak Massive, and the zirconium and molybdenum one in the Rumilet Massive. The outstanding high values of the third factor also occur in the Mainykvyn River headwaters (the Oleniy Stream) and in the Tkachenskaya Valley that might indicate the presence of similar subalkaline granites at these locations.

The set of elements with strong positive bonds can be regarded as a sign of feldspar metasomatism in this factor. The researchers of the region have proposed the hypothesis of a metasomatic origin of the Rumilet Massive granites. The anomalies overlaying the volcanogenic rocks may indicate that the similar metasomatic processes occur in the volcanites as well.

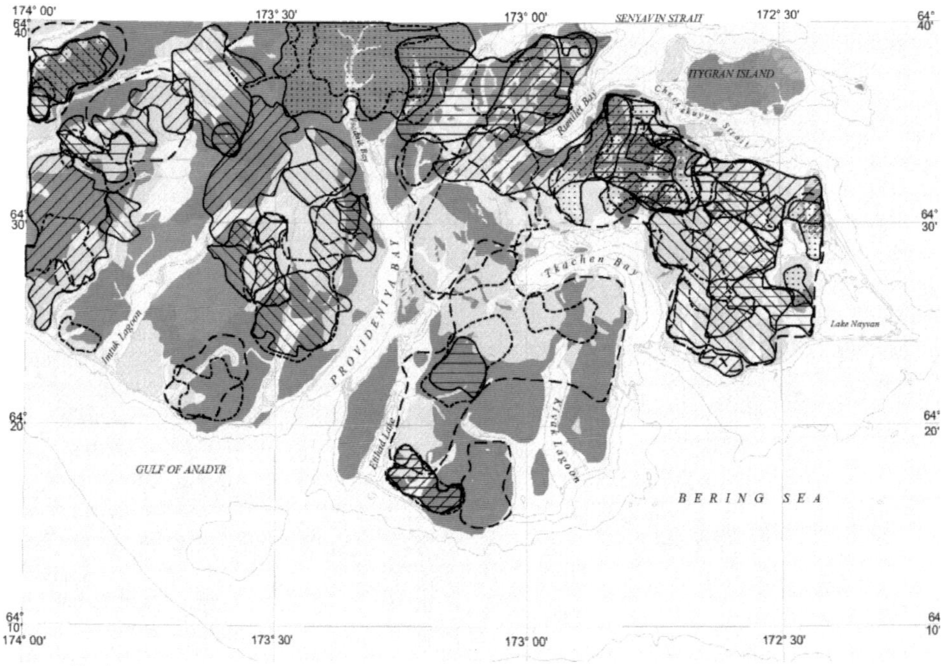

Figure 3. Allocation of factor anomalies in Providenskaya area

The forth factor contains 10% of information. It has a significant positive bond with gallium, lithium, germanium, arsenium, vanadium, barium, lead and a negative one with beryllium and zirconium. Geologically these anomalies are aligned with the zones of volcanite floods (e.g. the largest anomaly on the Chaplin Peninsula). These anomalies mainly coincide with the complex GA associated with ore mineralization, but they apparently are of petrogenic nature.

The fifth factor has strong correlations with bismuth, tin, beryllium, slightly weaker ones with tungsten, zinc, lead, and the weakest ones with molybdenum. The largest anomaly stretches from the Udobnaya River to Rumilet Bay. It is spatially aligned with monoanomalies of bismuth, tin, beryllium, lead, zinc, and partially molybdenum. Three intensive lesser-scale anomalies are outlined on the right bank of the Sineveem River (the Lysaya mountain), in the vicinity of Endogurov Promontory, and alongside the Ulhum River. The anomalies of the fifth factor are almost completely included in the selected GA associated with ore mineralization. Generally, a range of significant elements in this factor is characteristic for tungsten (quartz-forsterite) and/or tin ore formation. This may indicate that the intensity and prospects of GA increases.

The analysis of the sixth factor feature structure shows that this factor load is determined by the nickel, chrome, and cobalt concentration. The factor geological position coincides with the areas of the Paleozoic and Riphean-Archean rocks development. The most intensive and extensive anomaly is situated in Vsadnik Bay headwaters, the Pervomaisky Stream, the Ulev and Udobnaya rivers. This anomaly is formed by the ablation of disintegrated material from Senyavin Highland of the Riphean-Archean Vostochno-Chukotskiy Massive. The anomalies of the sixth factor are largely included in the selected (primarily by a multiplicative factor) geochemical anomalies associated with ore mineralization.

As a result of cluster analysis, five clusters of chemical elements are distinguished: cluster 1 – Ag, Cu, Au; cluster 2 – Pb, Zn, Li, Ga, W, Ge, As, Mo; cluster 3 – Nb, Y, Sn, Bi, Be; cluster 4 – Cr, Ni; cluster 5 – V, Ti, Co, Mn, Sc, Ba, Zr. The multiplicative anomalies were calculated according to the elements grouped in clusters. The gradations of four classes were visually selected, and then the anomalous areas were outlined for each cluster.

The cluster and factor analyses allow carrying out the more unbiased division of geochemical anomalies into petrogenic and ore-associated ones. The combination of the second and fifth factors can be a reliable indicator of the intensity and degree of potential of selected ore-associated anomalies.

4. Conclusion

Complex geochemical fields were outlined in the Providenskaya Area of the Chukotka Peninsula. These fields are generated by metamorphic, magmatic, volcanic, and hydrothermal-metasomatic rock formations. The associations of correlated chemical elements selected by the multivariate statistical methods indicate that various types of ore forming occur in the area. The delineation of complex anomalies provides an opportunity to localize the prospective areas of gold-silver, polymetallic, tin, and tungsten deposits.

References

[1] Voroshilov V G 2009 *Geology of Ore Deposits.* Anomalous structures of geochemical fields of hydrothermal gold deposits: formation mechanism. Methods of geometrization, typical models, and forecasting of ore mineralization. Vol. **51** (**1**) pp. 1-16.

[2] Izenman A J 2008 *Springer New York.* Modern Multivariate Statistical Techniques: Regression, Classification, and Manifold Learning. pp.757.

[3] Cheng Zh Zh, Xie X J, Yao W Sh, Feng J Zh, Zhang Q, Fang J D 2014 *Geochem. Explor.* Multi-element geochemical mapping in Southern China. Vol. **139** pp. 182-191.

[4] Gavrilov R Y, Rudmin M A 2015 *IOP Conference Series: Earth and Environmental Science.* Methodological approaches in estimating anomalous geochemical field structure. Vol. **24** pp. 1-6.

[5] Reis A P, Sousa A J, Cardoso Fonseca E 2003 *J. Geochem. Explor* Application of geostatistical methods in gold geochemical anomalies identification (Montemor-o-Novo, Portugal) Vol.. **77** pp. 45-63.

[6] Salati, S, Moore F 2010 *Environmental Monitoring and Assessment.* Assessment of heavy metal concentration in the Khoshk River water and sediment, Shiraz, Southwest Iran. Vol. **164(1)** pp. 677-689.

[7] Timkin T V, Lavrov D S, Askanakova O Y, Korotchenko T V 2014 *IOP Conference Series: Earth and Environmental Science.* Structure-geochemical zoning of Topolninsk gold-ore field (Gorny Altai). Vol. **21** (**1**) pp. 1-6.

[1] Voroshilov V, Savinova O, Ananev Yu, Abramova R 2014 *IOP Conference Series: Earth and Environmental Science.* Anomaly Geochemical Fields in Siberian Hydrothermal Gold Deposits. Vol. **21 (1)** pp. 1-6.
[1] Wang L, Wang Y, Zhang W, Xu C, An Z 2014*Environmental Earth Sciences.* **7** Multivariate statistical techniques for evaluating and identifying the environmental significance of heavy metal contamination in sediments of the Yangtze River, China. Vol. **1(3)** pp. 1183-1193.
[1] Meng O, Zhang J, Zhang Z, Wu T 2015 *Journal of Geochemical Exploration.* Influence of ore deposits on river sediment compositions in Dan River drainage, China. Vol. **159** pp. 8–19.

Geology, geochemistry and gold-ore potential assessment within Akimov ore-bearing zone (the Altai Territory)

T Timkin, V Voroshilov, O Yanchenko, J Suslov, T Korotchenko
Tomsk, Lenin Avenue, 30, Tomsk Polytechnic University, Russia
E-mail: [2] timkin@tpu.ru

Abstract. The article considers the geological setting of ore-bearing zone. Geochemical zoning of secondary dispersion haloes within Akimov ore-bearing zone was analyzed. Based on analytical data, distribution of gold mineralization trace-elements was examined. Geochemical associations were identified, analyzed and geometrized. The structure of abnormal geochemical field associated with gold-silver mineralization was determined. Three promising areas for exploration activities were identified.

1. Introduction

Akimov ore-bearing zone is located in the north-western part of Gorny Altai (the Altai Mountains). It is an administrative subdivision of Kurinsky and Krasnshchekovsky regions of the Altai Territory (Siberian Federal District of the Russian Federation) (figure 1).

The study area is within the transition zone between the Pre-Altai plain and the Altai Mountains and is comprised of gently-rugged, hillocky foothill and low-hill terrains that rise between 50-120 m at an absolute elevation of 400-430 m. The climate is extreme continental, average precipitation being not more than 450-500 mm. Hydrographic network of the area is poorly-developed. The Charysh river is the main water artery. The river Loktevka runs through the center of the ore-bearing zone, the river Poperechnaya and the river Kukuyka (right affluents of the Charysh river) run in the West and East, respectively.

According to metallogenic zoning, the study area is confined to the western part of North Altai gold belt characterized by intensive Middle Devonian volcanism [1].

Regionally, Akimov ore-bearing zone located within Novofirsovsky area is confined to Charysh block, close to its border with Ore Altai block and within the depression superimposed on Cambrian-Silurian structures and composed of Lower Devonian carbonate-terrigenous deposits (Kamyshenskaia D1km and Baragashskaia D1br formations), volcanic and igneous-sedimentary deposits of Middle and Upper Devonian (Kukuiskaia formation D2-3kk) which are intruded by subvolcanic bodies of diorite porphyry, rhyodacite, and rhyolites of Middle Devonian Kukuisky complex (figure 1) [2]. Spatially, gold-silver mineralization is associated with volcanic deposits of Kukuisky complex.

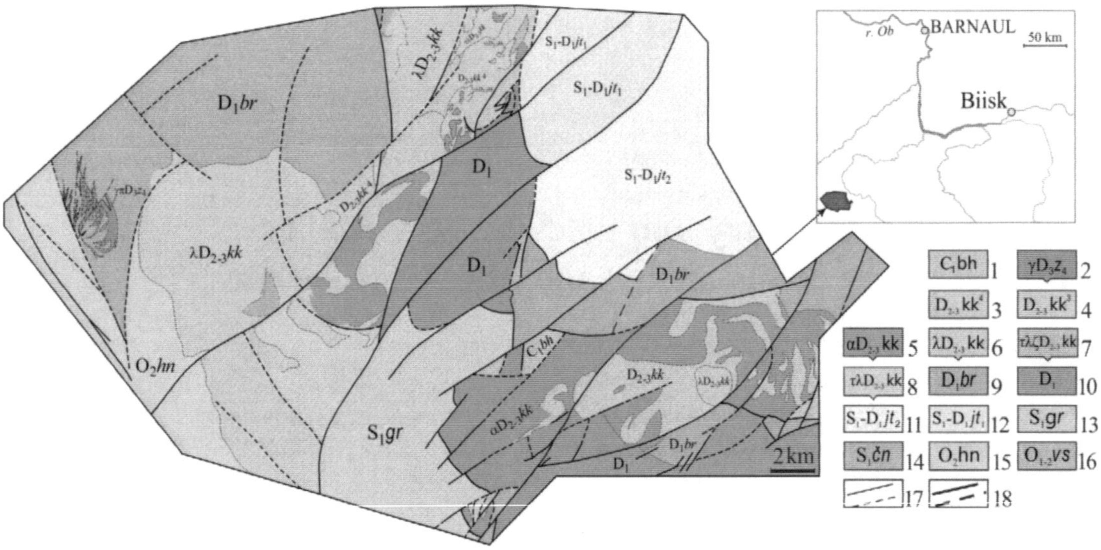

Figure 1. Geological map of the Akimov promising ore area (after Timkin V, 2009)
1 – Bukhtarminskaia Formation: limestones, siltstones, sandstones; 2 – Zmeinogorsky magmatic complex gabbro-granite-leykogranit; 3 – Kukuiskaia Formation, four member: tuff-sandstones and silty sandstone; 4 – third member: siltstones and claystones; 5-8 – Kukuiskii magmatic complex rhyolite-dacite-andesite; 9 – Baragashskaia Formation: sandstones and siltstones; 10 – Undifferentiated Lower Devonian: polymictic sandstones, gritstones, conglomerates and siltstones; 11-12 – arkose and polymictic sandstone, siltstones; 13 – Gromotukhinskaia Formation: siltstones, clayey sandstones, limestones; 14 – Chinetinskaia Formation: clayey sandstones, siltstones, sandstones; 15 – Khankharinskaia Formation: siltstones, calcareous sandstones, gritstones and conglomerate; 16 – Voskresenskaia Formation: calcareous sandstones, siltstones, quartz sandstones; 17 – measured (inferred) geological boundary; 18 – measured (inferred) fault

Structurally, rectilinear elements of the relief that correspond to the fault pattern are basically observed within the study area. NE and NW linear structures are extensionally and quantitatively identified (figure 1). The rings and arcs of the relief are well-defined within the volcanic and metasomatic formations [3].

According to [4], metasomatic formations identified within the zone are classified as metasomatites "deposited throughout progressive stage of magmatic-hydrothermal systems development" and "plutonic metasomatites of the regressive stage" within the linear zones. The vertical zoning with uprise alteration of hydrothermal column of propylites (epidot-calcite and albite-chlorite-calcite facies), argillizated rocks (kaolinite-smectite and kaolinite-quartz facies), and secondary quartzites was identified in the metasomatites. The centers of hydrothermal activity and location of gold-silver mineralization are controlled by the linear metasomatites of progressive stage.

2. Research technique

The present study is based on the lithochemical surveys in scale 1:50 000 along the secondary dispersion haloes and 1:200 000 along dispersion trains carried out by JSC "Gorno-Altaiskaya Ekspeditsiya" in 2006-2009. To estimate a gold-bearing potential of the study area, a total of 2 000 samples were processed (figure 2). The samples were collected at depth between 0.2-0.6 m. Then, the samples were dried and sieved.

All collected samples were sent to the laboratory Stewart Geochemical and Assay, LLC., Moscow, where semiquantative spectral analysis and gold assays (subsample 30g) followed by atomic absorption test were carried out.

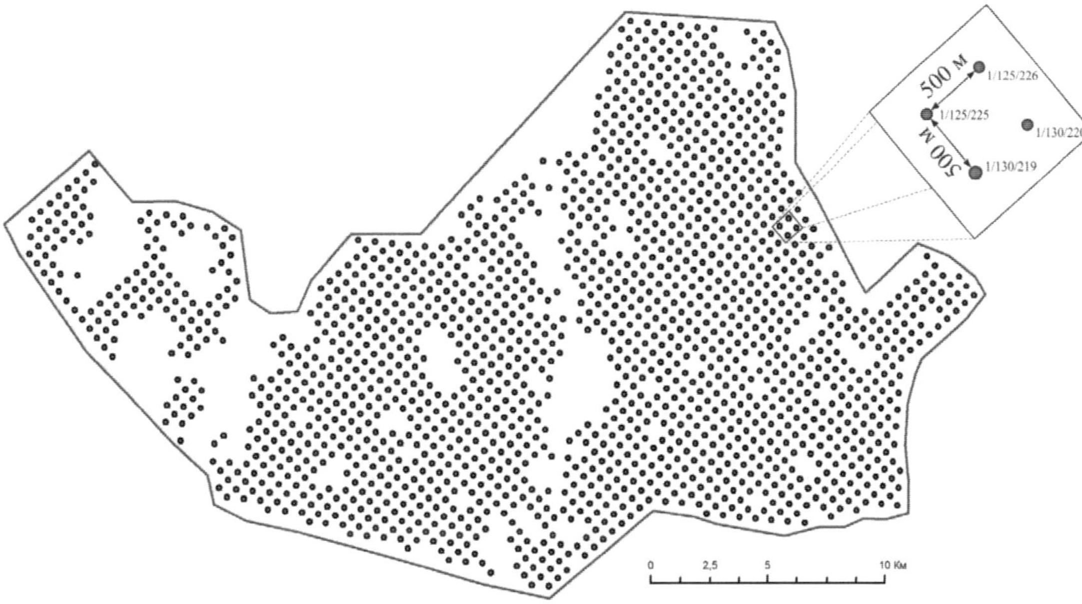

Figure 2. Map showing location of the sample points

Geochemical data were processed using software program Statistica 12, while geometrization was performed by ArcGis and Surfer in accordance with generally accepted methods and techniques [5, 6, 7, 8, 9, 10, 11].

3. Research results

As a result, background and minimum abnormal concentrations of gold and trace-elements in secondary dispersion haloes were identified. In addition, geochemical associations were revealed and association zoning within the study area was outlined. The location of gold-silver mineralization within the abnormal geochemical field was determined. The geochemical criteria for gold-silver mineralization prospecting within Akimov ore-bearing zone were proposed.

The background and minimum abnormal concentrations of elements were calculated according to standard methodology developed by A.P. Solovov [12]. Since element distribution within the study area does not conform to the normal law (A/SA>3 and E/SE>3), the log-normal distribution model was used to calculate background and abnormal values.

In order to identify the resistant associations of elements and determine their location for further analysis of geochemical field structure, factor analysis was used to combine groups of correlated variables [13, 14].

Within the secondary geochemical haloes of the study area, 4 factors were defined (figure 3). They are interpreted on the basis of elemental composition, spatial distribution of factors, and geological environments. Thus, factor 1 (W,Sn,Li) corresponds to Kurino-Akimov volcano-tectonical structure, i.e. paleo caldera.

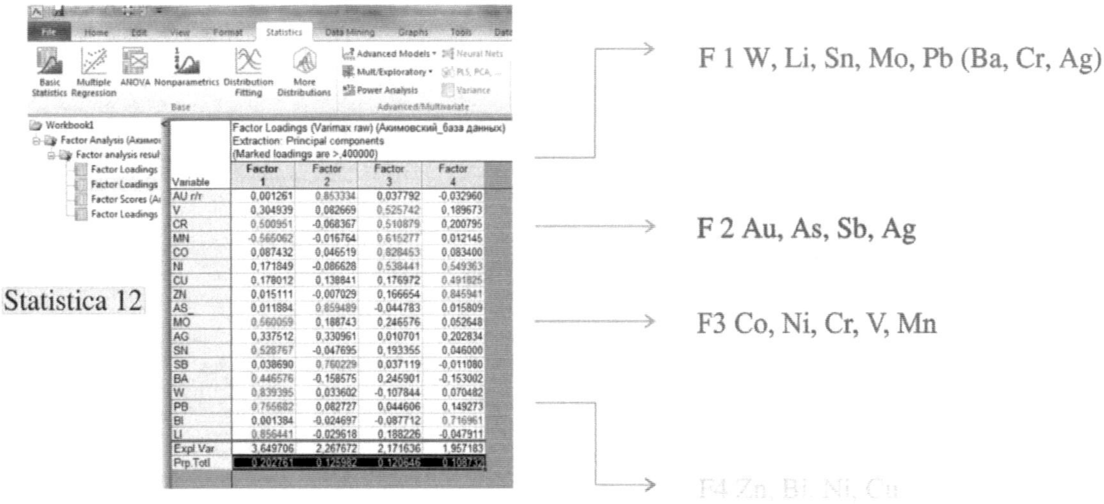

Factor Loadings (Varimax raw) (Акимовский_база данных)
Extraction: Principal components
(Marked loadings are >,400000)

Variable	Factor 1	Factor 2	Factor 3	Factor 4
AU r/т	0,001261	0,853334	0,037792	-0,032960
V	0,304939	0,082669	0,525742	0,189673
CR	0,500951	-0,068367	0,510879	0,200795
MN	-0,565062	-0,016764	0,615277	0,012145
CO	0,087432	0,046519	0,828453	0,083400
NI	0,171849	-0,086628	0,538441	0,549363
CU	0,178012	0,138841	0,176972	0,491825
ZN	0,015111	-0,007029	0,166654	0,845941
AS_	0,011884	0,859489	-0,044783	0,015809
MO	0,560059	0,188743	0,246576	0,052648
AG	0,337512	0,330961	0,010701	0,202834
SN	0,528767	-0,047695	0,193355	0,046000
SB	0,038690	0,760229	0,037119	-0,011080
BA	0,446576	-0,158575	0,245901	-0,153002
W	0,839395	0,033602	-0,107844	0,070482
PB	0,755682	0,082727	0,044606	0,149273
BI	0,001384	-0,024697	-0,087712	0,716961
LI	0,856441	-0,029618	0,188226	-0,047911
Expl.Var	3,649706	2,267672	2,171636	1,957183
Prp.Totl	0,202761	0,125982	0,120646	0,108732

Figure 3. Matrix of factor loadings area

Factor 2 unites elements (Co,Ni,Cr,V) commonly accumulated in the frontal enrichment zones, which is a characteristic feature of all hydrothermal deposit types and can be used for delineation of abnormal geochemical fields of various ranks. In other words, it can be applied for ranking oregenetic geochemical fields. Factor 3 relates to gold-ore mineralization itself (Au,As,Sb), while the elements included in factor 4 (Ag,Mo,Sn,Pb) can be associated with sulphide mineralization within the flanks of ore bodies and supraore space where gold grade decreases.

The analysis of geochemical associations distribution allows identifying 3 abnormal zones of sulphide mineralization (figure 4).

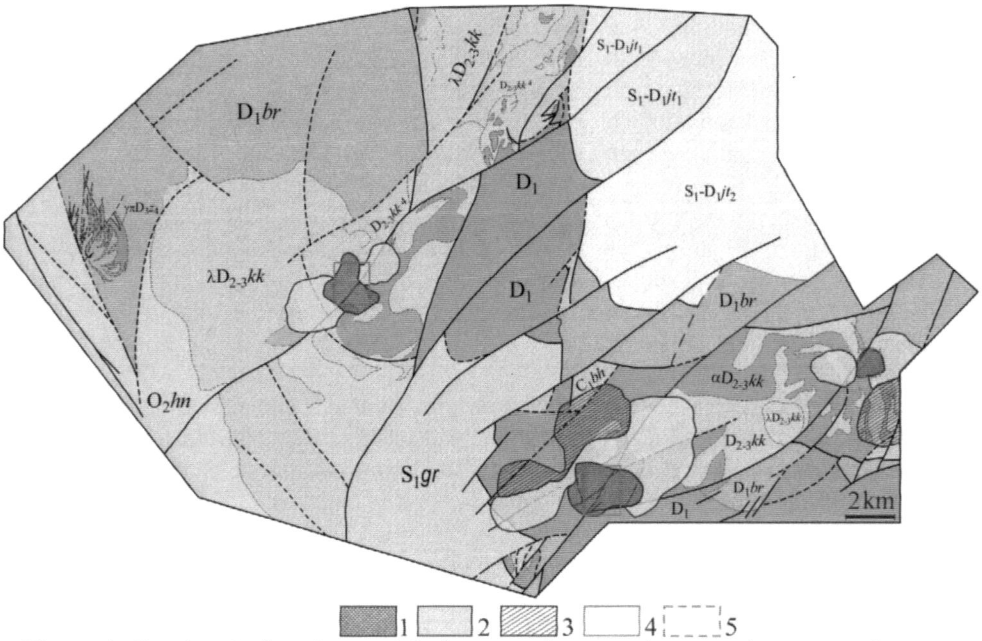

Figure 4. Geochemical zoning model of perspective area; 1 – central concentration zone (Au,As,Sb,Ag); 2 – intermediate concentration zone (Co,Ni,Cr,V,Mn); 3 – high value area (Cu,Zn,Bi,Ni); 4 – promising areas (first stage); 5 – promising areas (second stage)

Based on the character of association correlation, it is possible to conclude about different degree of ordering of abnormal geochemical field structures within the identified zones. As known [2], the degree of ordering reflects the intensity of hydrothermal mineralization. The concentric zonal abnormal geochemical fields with ore elements in their central parts are the most well-ordered structures [4]. Such a geochemical zoning was identified in the southern part of the study area.
Previously, based on the geological survey results, 4 promising zones, 3 of which are well-correlated with the identified geochemical anomalies, were determined. This fact proves the validity of the obtained results.

4. Conclusion

Thus, the highest gold concentrations are confined to the zones of co-occurrence of several diachronous mineral associations. The presence of ore controlling structures contributes to numerous relocation of gold and its concentration within the geochemical barriers.
Based on the analysis of geochemical data including statistical calculations, chemical element mapping, and factor analysis, it has become possible to develop a geochemical model of Akimov promising ore-bearing zone and identify 3 areas for further exploration: 2 promising areas of first stage and 1 promising area of the second stage.

References

[1] Krivchikov V A, Timkin VI 2003 *Tomsk State University Journal*. Stratigraficheskaya posledovatel'nost' i geodinamicheskaya interpretatsiya glubokovodnykh strukturno-veshchestvennykh kompleksov Talitskogo bloka (Gornyy Altay). Vol. **383** pp. 95-98.
[2] Timkin V I 2010. *Prirodnye resursy Gorny Altai.* Novye dannye po geologicheskomu stroeniyu Novofirsovskogo rudnogo uzla (severo-zapadnaya chast' Gornogo Altaya). Vol **2** pp. 30-54.
[3] Voroshilov V G, Ananyev Yu S, Timkin T V 2015 *Regional'nye geokhimicheskie raboty kak osnova dlya otsenki rudonosnosti i neftegazonosnosti territoriy: proceedings of Russian scientific-practical conference*. Kompleksirovanie geokhimicheskikh issledovaniy i geologicheskogo deshifrirovaniya kosmosnimkov – novyy podkhod k vydeleniyu rudonosnykh ploshchadey. pp. 42-43.
[4] Akimov G Yu, Timkin V I, Mal'kov V A 2011 *Ores and metals.* Geologicheskaya pozitsiya, veshchestvennyy sostav i istoriya formirovaniya Cherepanovskogo zoloto-serebryanogo mestorozhdeniya, Rudnyy Altay. Vol. **3-4** pp.7-8.
[5] Voroshilov V G 2009 *Geology of Ore Deposits.* Anomalous structures of geochemical fields of hydrothermal gold deposits: formation mechanism. Methods of geometrization, typical models, and forecasting of ore mineralization. Vol. **51 (1)** pp. 1-16.
[6] Gavrilov R, Rudmin M 2015 *IOP Conference Ser.: Earth Environmental Sci.* Methodological approaches in estimating anomalous geochemical field structure. Vol. **24** pp. 1-6.
[7] Wang E J M, Carranza Liu X, Zhang S, Da Z, Yang L, Sun X, Duan C 2013 *J. Geochem. Expl.* Characterization of primary geochemical haloes for gold exploration at the Huanxiangwa gold deposit, China. Vol. **124** pp. 40-58.
[8] Harraz H Z, Hamdy M M 2015 *J. Afr. Earth Sci.* Zonation of primary haloes of Atud auriferous quartz vein deposit, Central Eastern Desert of Egypt: A potential exploration model targeting for hidden mesothermal gold deposits. Vol. **101** pp. 1-18.
[9] Timkin T, Voroshilov V, Askanakova O, Cherkasova T, Chernyshov A, Korotchenko T 2015 *IOP Conference Ser.: Earth Environmental Sci.* Estimating gold-ore mineralization potential within Topolninsk ore field (Gorny Altai). Vol. **27** pp. 012010.
[10] Timkin T V, Lavrov D S, Askanakova O Y, Korotchenko T V 2014 *IOP Conference Ser.: Earth Environmental Sci.* Structure-geochemical zoning of Topolninsk gold-ore field (Gorny Altai). Vol. **21** pp. 012010.
[11] Cherkasova T, Timkin T and Savinova O 2014 *IOP Conference Ser.: Earth Environmental Sci.* First native silica findings in bismuth from garnet skarns of Ribny Log-2 gold ore target in Topolninsk ore deposit (Gorny Altai) Vol. **24** pp. 012033.

[12] Solovov A P 1987 *Mir Publishers. Moscow.* Geochemical Prospecting for Mineral Deposits. Pp. 288
[13] Filzmoser P, Hron K, Reimann C, Garrett R 2009 *Comput. Geosci.* Robust factor analysis for compositional data. Vol. **35** pp. 1854-1861.
[14] Yousefi M, Kamkar-Rouhani A, Carranza E J M 2012 *J. Geochem. Explor.* Geochemical mineralization probability index (GMPI): a new approach to generate enhanced stream sediment geochemical evidential map for increasing probability of success in mineral potential mapping. Vol. **115** pp. 24-35.

Oil and gas presence perspectives of weathering layer reservoir of Nurol'ka mega-basin according to data of geothermics

T. Luneva[1], G. Lobova[2], A. Fomin[3]

[1, 2] Department of Geophysics, Institute of Natural Resources, National Research Tomsk Polytechnic University, 30 Lenin Ave., Tomsk, 634050, Russia
[3]Head of the laboratory of geochemistry of oil and gas, Institute of oil and gas geology and geophysics of A.A. Trofimuk SB RAS, building 3, Academician Koptyuge Ave., Novosibirsk, 630090, Russia
E-mail: [1]tanyao@sibmail.com, [2]lobovaga@tpu.ru, [3]fominan@ipgg.sbras.ru

Abstract. Oil perspective areas of the Nurol'ka mega-basin (south-east of Western Siberia) through the M formation (Permian-Triassic weathering crust) on the basis of the results application of geothermometry were identified. Accumulating power distribution and quality of the M formation collectors were taken into account. The priority area for research are weathering layer reservoirs and its development that cover the South beads of Kulan-Igayskaya and Tamradskaya basins and its insulation joint were proposed. Glukhov's oil field that is located in this area approve its high prospectivity.

1. Introduction

Expansion of the resource base of hydrocarbons in the oil-field regions of Tomsk (in south-east of Western Siberia (figure 1) is possible to use the deep stratigraphic levels, including reservoirs, formed in the lower Jurassic and in the weathering crust of Nurol'ka mega-basin and its framing structures [1].

The investigated reservoir of the weathering crust (M formation) is related to the pre-Jurassic of petroleum-bearing complex. Direct signs of oil saturation in the weathering crust in Rechnoe, Urmanskaya and other areas confirmed the potential of this stratigraphic level [2].

The zoning area in the resources compaction of primary accumulated of Togur's oils for reservoir of weathering crust were conducted in the given paper. According to the results that were early achieved of geothermal regime reconstruction of oil source - Lower Jurassic Togur' suite and intense release of oil generation were conducted as well [1].

2. Characteristic of oil and gas area perspectives

Dispersed organic matter (DOM) of oil-producing of Togur suites within Nurol'ka mega-basin is a typical lake, sapropelitic-humic type, with the index C_{org} to 5% [3]. Spread of the Togur suite in the study area is limited to reduced forms of topography (figure 2a) and has thickness more than 100 m [2].

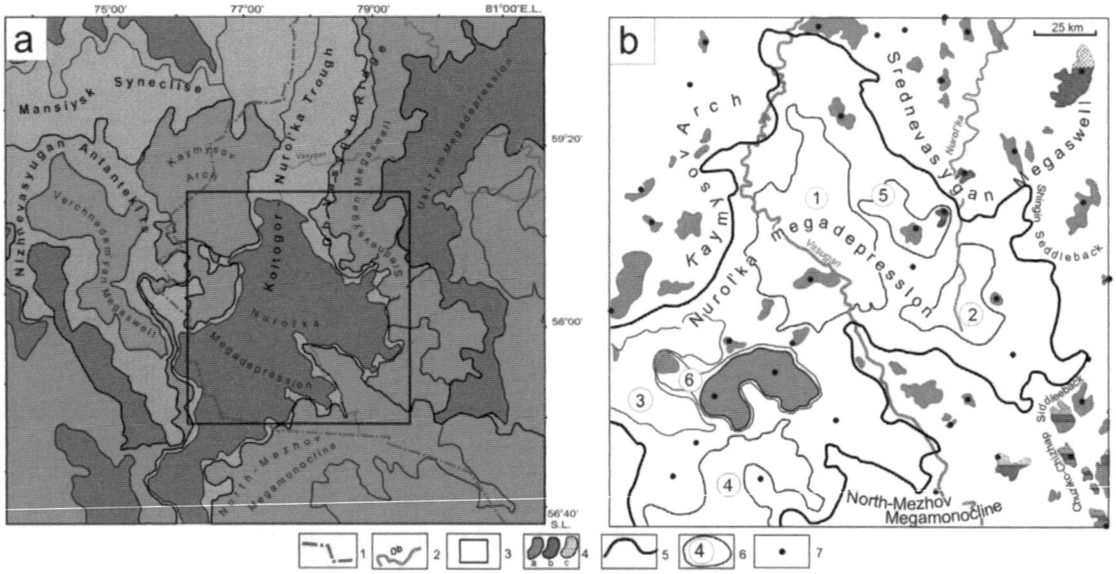

Figure 1. General tectonic scheme (**a**) and scheme of petroleum potential (**b**) of Nurol'ka mega-basin (based on [2]). To figure (**a**): *1* – the south-western border of Tomsk region; *2* – the main rivers; *3* – arial limits of area zoning. To figure (**b**): *4* – oil fields: a) oil, b) condensate, c) gas; *5* – the boundary of Nurol'ka mega-basin; *6* – structure of III order and its number; *7* – wellsite of paleotemperature modeling [1]. Structures of III order: 1 – Kulan-Igay basin; 2 – Tamrad basin; 3 – Axial diflection; 4 – Tamyan diflection; 5 – Festival swell; 6 – Igol-Talov domal upwarping

It is noticed that there is an uneven arial spreading of the sediments of weathering crusts within the research area. The maximum thickness is 120 m in the central and south-western parts of the territory. Pinching-out the sediments of weathering crust is confined to the joint zones of Nurol'ka mega-depression and its framing structures (figure 2a).

The way out on the surface of polyfacies basement rocks (figure 2b) predetermines the formation of a variety of profiles of weathering crusts, which, in its turn, in various degrees have reservoir properties. Granitoid, granodiorite and rhyolitic magma bodies that have been spread here, being exposed to supergene processes and create the preconditions for the reservoir formation with good filtration-capacitive properties (FCP). Thus, the exit zones of clay-siliceous and igneous rocks of acid composition on the surface of the basement are zones of enhanced proliferation of collectors.

Lower Jurassic, Levin local and / or regional Kiterbyut clay rocks are the cap for reservoir of weathering crust. The local Laida and / or Leontiev clay formation of Middle Jurassic plays the role of cap in the case when the Lower Jurassic strata wedge out.

3. Research Methodology

Simulation of the thermal history of Togur suite formation made with the use of computer technology of paleotemperature modeling [4, 5, 6]. Twenty-three representative wells located in the zone of Togur deposition were selected for paleotemperature modeling. Design procedure of paleotemperatures, mapping centers of generation of oil and the distribution map of the calculated values of the resources of the generated Togur oils density are given in [7]. The resource estimate of density is performed in conventional units, that appears to be valid for the subsequent areal zoning [8].

The map of distribution of settlement density of resources of Togur oils (figure 3a) was built for pre-zoning of weathering crust reservoir. This map was built by multiplication the values of the density of the generated resources of Togur oils [1, 7] and values of bark thickness of weathered sediments (figure 2a), taken from the nodes of a regular grid.

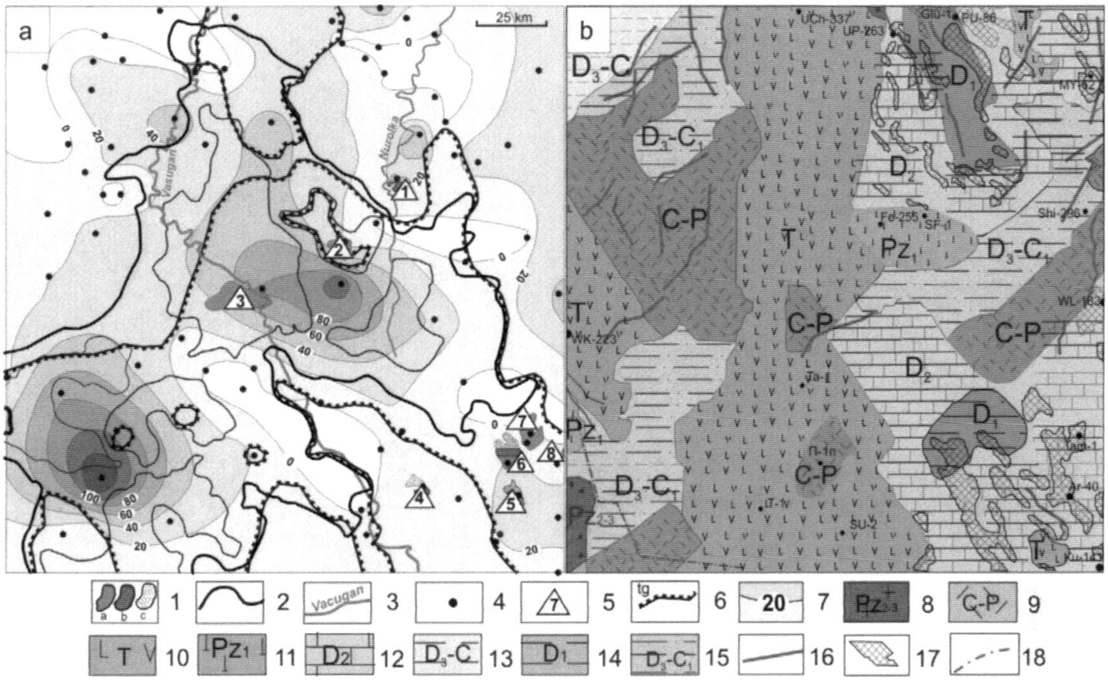

Figure 2. Schematic maps of isopach line weathering crust (**a**) and spread of petrotypes of basement rocks (**b**) in Nurol'ka mega-depression and its framing structures. To figure (**a**): 1 – oil field: a – oil, b – condensate, c – gas; 2 – boundaries of tectonic elements of the I order; 3 – river set; 4 - wells, used to build the maps of isopach line; 5 – conditional number of oil fields; 6 – boundary of Togur suite; 7 – isopach line map of the weathering crust. Oil fields with deposits in the weathering crust: Rechnoe (1), Festival (2), Glukhov (3), Elley-Igay (4), Arch (5), Urman (6), Tambaev (7), South-Tambaev (8). To figure (**b**): facies of complexes of basement rocks showing its age: 8 – Granites, 9 – Liparites 10 – Andesite-basalts, 11 – Ultrabasites, 12 – Carbonates, 13 – Shales, 14 – Clay-siliceous Shales, 15 – clastic breeds, 16 – faults,17 – zone of improved collectors in the basement, 18 –boundary of Tomsk region

Three of the most promising areas in the weathering crust reservoir with resource density of more than 30 conventional units (figure 3b) according to preliminary data were allocated.

An analysis of petrotypes basement rocks [9, 10] was taken into account and were shown areas on the map (figure 4a) with improved filtration-capacitive properties of reservoirs formed by siliceous-carbonate, clay-siliceous rocks and igneous rocks of acid composition, and areas with unfavorable FCP formed by magmatic rocks of basic composition and species of clay-shale formation.

The perspective areas (figure 4b) were indicated, taking into account the density distribution of the generated resources of Togur oils, distribution of collectors petrotypes and distribution of tectonic disturbances density in the reservoir of M formation (figure 2b).

4. Results and Discussion

The most interesting one in relation to oil and gas potential is *section 1*, which stretches to the west and covers the southern side of Kulan-Igay and Tamrad basins, and its junction (figure 4b). The high density resource of Togur oils and improved filtration-capacitive properties of the M formation coincided in this area. The deep well Tamrat 1 parametric (figure 3b, conditional index of well T-1p) is situated within this area while drilling some pre-Jurassic rocks were discovered but the test productivity in this part of the section was not carried out. Glukhov's oil field is located in this region (figure 2a, the conditional oil field number 3) confirmed its high prospectivity.

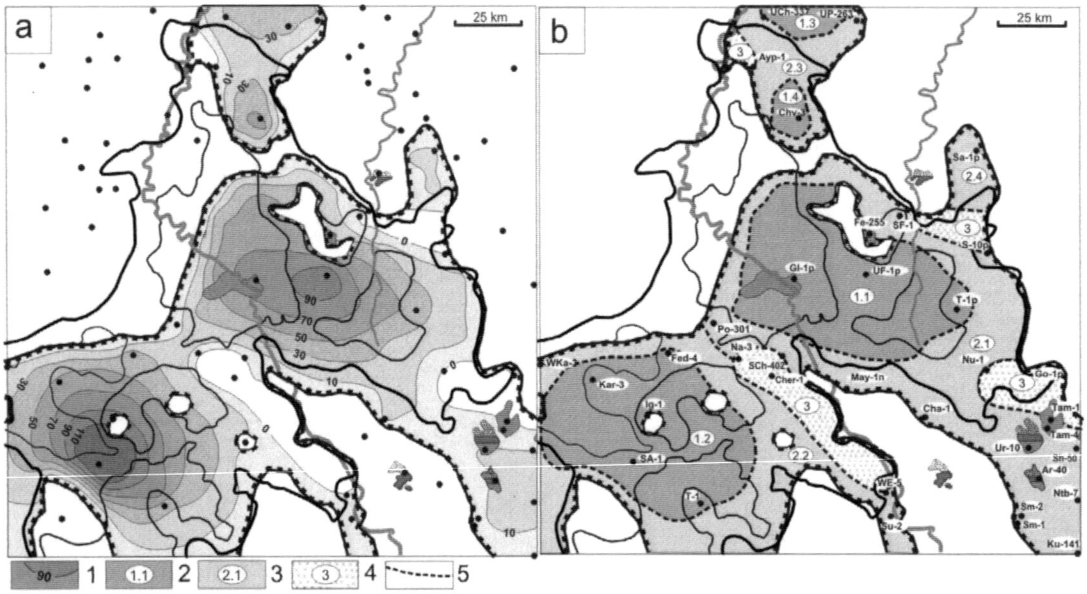

Figure 3. Schematic maps of the calculated density of Togur oil resources in the M layer reservoir according to the power of the weathering crust (**a**) and zoning of Nurol'ka megadepression (**b**). To figure (**a**):1 – contour values of resource density, conditional unit. To figure (**b**): 2-4 – zones (zone number, the value range of resources density, cond.unit): 1.– more than 30, 2. – less than 30, 3. – zones of wedge out the weathering crust within the spread of Togur deposits; 5 – borders of zones. The rest of the symbols are the same as in Fig. 2

The next perspective one – *section 2* (figure 4b.) was indicated in the north-western part of Chuzik-Chizhap mezoseddle. The oil fields, such as Tambaev, Arch, Urman and South-Tambaev (figure 2 a) with different phases of deposits in the M formation (reservoir weathering crust) were opened here and in this case some prospectivity of this area is confirmed.

Potential *section 3* tectonic confined to the southern slope of Srednevasyugan megaswell and its northern edge of Nurol'ka megadepression (figure 4b). The well Salat 1 parametric (figure 3b, conditional index of well is Sa-1p) is situated within this area, while drilling some pre-Jurassic rocks were discovered, but the test productivity in this part of the section was not carried out.

Local section 4 located in the south-western part of Nurol'ka megadepression by ranking is in the fourth place. There are no any data on the direct signs of oil saturation in the cross section on this lot.

Prospective *zone 5* is divided into six sections, with the same density of resources of Togur oils, ranked by area and the presence of tectonic faults. The *section* 5.1 is confined to the northern board of Tamyan trough. The *section* 5.2 is located on the northern hillside of the North-Mezhov megamonoklin. The *section* 5.3 is located in the south downcutting of Nurol'ka megadepression. The *section 5.4* belongs to the northern hillside of the Festival arch. The *section 5.5* is situated in the south-western slope of Srednevasyugan megaswell. The *section 5.6* is located in the south-western side of Kulan-Igay depression.

It should be noted, that the two deposits with the M pools formation were opened outside of spreading Togur suite: Elley-Igay and Rechnoe, (figure 2a.).

5. Conclusion

Thus, carrying out the geothermal modeling with use of geological and geophysical data of the last years allowed to carry out the zone forecast of oil and gas presence perspectives of Nurol'ka megadepression and structures of its frame for the deep-laying structural floor – weathering crust.

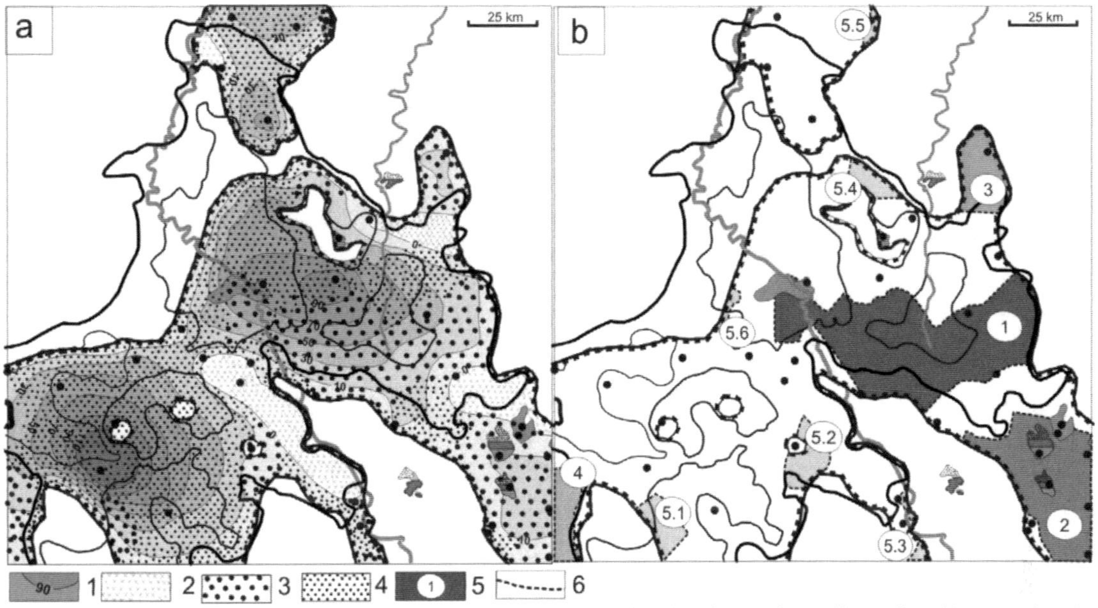

Figure 4. Scheme of zoning ratios (figure 3) and distribution of quality of collectors in the reservoir of the weathering crust (**a**), the scheme selection of priority areas for the discovery of oil deposits in the sediments of the weathering crust (**b**) of Nurol'ka megadepression. To figure (**a**): 1 – isoline of significance of estimated resource densities, condit.unit; 2 – zone of lack weathering crust within the spread of Togur suite; 3 – zone of weathering crust with improved FCP; 4 – zone of weathering crust with unfavorable FCP. To figure (**b**): 5 – prospective area (section), ranking number (the intensity of the fill area (painted) of the lot is proportional to the extent of land prospects); 6 – border areas. The rest of the symbols are the same as in figure 2

The priority area for exploration and development of the weathering crust reservoir of Nurol'ka megadepression – section 1 was indicated and proposed on the basis of performed research, this section (section 1) is situated on the southern edge of Kulan-Igay and Tamrad troughs and also its junction. The coincidence of high density resource of Togur oils and improved filtration-capacitive properties of the M formation (reservoir of weathering crust), makes the area more promising for the production of subsequent exploration work.

References

[1] Lobova G, Osipova E, Isaev V and Terre D 2015 *IOP Conf. Series: Earth and Environmental Science Scientific and Technical Challenges in the Well Drilling Progress*. Petroleum potential of Lower-Jurassic deposits in Nurolsk megadepression. Vol. **24** pp. 012001. http://iopscience.iop.org/1755-1315/24/1

[2] Kontorovich V A 2002 *Novosibirsk: Publishing House of SB RAS Branch "GEO"*. Tektonika i neftegazonosnost mezozoysko-kaynozoyskikh otlozheniy yugo-vostochnykh rayonov Zapadnoy Sibiri. pp. 253.

[3] Isaev V I, Fomin N 2006 *Russian Geology and Geophysics*. Loci of Generation of Bazhenov- and Togur-Type Oils in the Southern Nyurol'ka Megadepression. Vol. **47** pp. 734–745.

[4] Isaev V I, Gulenok R U, Veselov O V, Byichkov A V and Soloveychik Yu G 2002 *Geologiya nefti i gaza*. Kompyuternaya tehnologiya kompleksnoy otsenki neftegazovogo potentsiala osadochnyih basseynov. Vol. **6** pp. 48-54.

[5] Gulenok R Yu, Isaev V I, Kosygin V Yu, Lobova G A and Starostenko V I 2011 *Russian Journal of Pacific Geology.* Estimation of the Oil-and-Gas Potential of Sedimentary Depression in the Far East and West Siberia Based on Gravimetry and Geothermy Data. Vol. **5 (4)** pp. 273–287.
[6] Iskorkina A, Isaev V and Terre D 2015 *IOP Conf. Series: Earth and Environmental Science*. Assessment of Mesozoic-Kainozoic climate impact on oil-source rock potential (West Siberia). Vol. **27** pp. 012023. http://iopscience.iop.org/article/10.1088/1755-1315/27/1
[7] Isaev V I, Lobova G A, Osipova E N 2014 *Russian Geology and Geophysics.* The oil and gas contents of the Lower Jurassic and Achimovka reservoirs of the Nyurol'ka megadepression. Vol. **55** pp. 1418–1428.
[8] Lobova G A, Popov S A, Fomin A N 2013 *Neftyanoe xozyajstvo.* Lokalizaciya prognoznyh resursov nefti yursko-melovyh neftegazonosnyh kompleksov Ust'-Tymskoj megavpadiny. Vol. **2** pp. 36–40.
[9] Syngaevskiy P E 1990 *Geologiya nefti i gaza.* Formatsiya kory vyvetrivaniya v osadochnom tsikle Zapadno-Sibirskom baseyne. Vol. **11-12** pp. 22–30.
[10] Zapivalov N P, Isaev G D 2010 *Tomsk state university journal* . Kriterii otsenki neftegazonosnosti paleozoyskich otlozcheniy Zapadnoy Sibiri. Vol. **341** pp. 226-232.

Adaptive interpretation of gas well deliverability tests

V L Sergeev, Nguyen Thac Hoai Phuong and A B Strelnikova

National Research Tomsk Polytechnic University, 30 Lenina Ave., Tomsk, 634050, Russia
SergeevVL@ignd.tpu.ru, nguyenphuongtpu1512@gmail.com

Abstract. The paper considers topical issues of improving accuracy of data obtained from gas well deliverability tests, decreasing the number of test stages and well test time, and reducing gas emissions. The aim of the research is to develop the method of adaptive interpretation of gas well deliverability tests with resulting IPR curve conducted in gas wells with steady-state filtration, which allows obtaining and taking into account additional a priori data on the formation pressure and flow coefficients, setting the number of test stages adequate for efficient well testing and reducing test time. The present research is based on the previous theoretical and practical findings in the spheres of gas well deliverability tests, systems analysis, system identification, function optimization and linear algebra. To test the method, the authors used the field data of deliverability tests run in the Urengoy gas and condensate field, Tyumen Oblast. The authors suggest the method of adaptive interpretation of gas well deliverability tests with resulting IPR curve, which is based on the law for gas filtration with variables dependent on the number of test stage and account of additional a priori data. The suggested method allows defining the estimates of the formation pressure and flow coefficients, optimal in terms of preassigned measures of quality, and setting the adequate number of test stages in the course of well testing. The case study of IPR curve data processing has indicated that adaptive interpretation provides more accurate estimates on the formation pressure and flow coefficients, as well as reduces the number of test stages.

1. Introduction

Deliverability tests with resulting inflow performance relationship (IPR) curve run in the gas wells with steady-state filtration are one of the most informative and common methods well tests to characterize the behavior of well and the bottomhole conditions. Currently, the data obtained via deliverability tests are interpreted using the methods described in [1-3], which are based on Forchheimer binomial equation for gas filtration:

$$p_{\text{пл}}^2 - p_{\text{з}}^2 = aq + bq^2 , \tag{1}$$

where $p_{\text{пл}}^2, p_{\text{з}}^2$ are formation pressure and bottomhole pressure, respectively; *a* and *b*- flow coefficients dependent on bottomhole zone parameters and bottomhole structure; *q* - flow rate. The coefficients *a* and *b* for IPR curve model (1) should be estimated using least square method, with the formation pressure being known [2-4]. IPR interpretation based on the model (1) and least square method is challengeable as a field method, which is attributed to the following facts: the formation pressure is difficult to determine, estimates should be robust and accurate, the number of test stages (a number of "cycles" characterized by a stabilized flow when the pressure and flow rate are recorded) is reduced.

To ensure that the estimates are accurate and robust, in the work [5] we suggest to interpret the IPR curve using integrated IPR curve models with account of additional a priori data on the formation pressure and flow coefficients. However, the question is how to provide additional a priori data on the formation pressure and flow coefficients and to determine the adequate number of test stages to secure preassigned estimate accuracy.

To overcome the above-mentioned challenges, the method of adaptive interpretation of deliverability tests with resulting IPR curve with variable parameters is suggested and investigated. The method implies that the parameters depend on the number of test stage and additional a priori data on the formation pressure and flow coefficients obtained according to the empirical power law [6] for gas filtration are taken into account:

$$q = \lambda(p_{пл}^2 - p_з^2)^{\gamma} \tag{2}$$

where λ - productivity index; γ - constant factor with theoretical value ranging from 0.5 (turbulent flow) to 1.0 (laminar flow).

It is noteworthy that the empirical law for gas filtration (2) is widely applied in deliverability analysis over the years [7-8].

2. Models and Algorithms for adaptive interpretation of IPR curve

The basis to develop algorithms for gas well deliverability test data interpretation is an integrated system of IPR curve models (1) with variable parameters dependent on the number of the test stage and account of additional a priori data on the formation pressure $\overline{p}_{пл,n}^2$ and flow coefficients $\overline{a}_n, \overline{b}_n$:

$$\begin{cases} y_n^* = p_{пл,n}^2 - a_n q_n - b_n q_n^2 + \xi_n, \\ \overline{p}_{пл,n}^2 = p_{пл,n}^2 + \eta_n, \\ \overline{a}_n = a_n + \nu_n, \overline{b}_n = b_n + \varepsilon_n, n = \overline{1, nk}, \end{cases} \tag{3}$$

where $y_n^* = p_{з,n}^2, q_n$ – values of squared bottomhole pressures and flow rates obtained at test stage number n; nk – the number of test stages appropriate to secure preassigned estimates accuracy for the formation pressure and flow coefficients $p_{пл,n}^2, a_n, b_n$ dependent on number of test stage; $\xi_n, \eta_n, \nu_n, \varepsilon_n$, – random variables, i.e. error in measurements, recovery data, and estimates of flow coefficients, as well as deficiencies of gas filtration models (1),(2) etc.

The additional data on the formation pressure $\overline{p}_{пл,n}^2$ and parameters estimates $\overline{\lambda}_n$ and $\overline{\gamma}_n$ of model (2) can be obtained by solving the following optimization problem:

$$\overline{\boldsymbol{\alpha}}_n = \arg\min_{\boldsymbol{\alpha}} \sum_{n=1}^{nk} r(q_n^* - \alpha_1(\alpha_2 - y_n^*)^{\alpha_3}) \tag{4}$$

where $\arg\min_x f(x)$ is the minimum point x^* of the function $f(x)$ ($f(x^*) = \min_x f(x)$); $\overline{\boldsymbol{\alpha}}_n = (\overline{p}_{пл,n}^2, \overline{\lambda}_n, \overline{\gamma}_n)$ – the vector of estimates; $r(x)$ – the known function. The additional data on flow coefficients $\overline{a}_n, \overline{b}_n$ can be obtained from the system of linear equations

$$z_n = a_n q_n + b_n q_n^2, n = \overline{1, nk} , \tag{5}$$

which is the result of grouping models (1),(2) for depression $p_{пл}^2 - p_з^2$ where $z_n = \sqrt[\overline{\gamma}_n]{q_n / \overline{\lambda}_n}$, q_n – value of flow rate obtained at test stage number n; $\overline{\lambda}_n, \overline{\gamma}_n$ – the optimal estimates obtained by solving problem (4).

The optimal values of squared formation pressure $p_{пл,n}^2$ and flow coefficients a_n, b_n of model (3) represented for convenience as a matrix

$$\begin{cases} \mathbf{y}_n = F_n \boldsymbol{\alpha}_n + \xi_n, \\ \bar{\boldsymbol{\alpha}}_n = \boldsymbol{\alpha}_n + \boldsymbol{\eta}_n, n = \overline{1, nk}, \end{cases} \tag{6}$$

are calculated using the method of adaptive identification by solving optimization problems (7),(8)

$$\boldsymbol{\alpha}_n^*(\boldsymbol{\beta}_n, h_n) = \arg\min_{\boldsymbol{\alpha}_n}(J_0(\boldsymbol{\alpha}_n, h_n) + J_a(\boldsymbol{\alpha}_n, \boldsymbol{\beta}_n)), \tag{7}$$

$$\boldsymbol{\beta}_n^*, h_n^* = \arg\min_{\boldsymbol{\beta}_n, h_n} J_0(\boldsymbol{\alpha}_n^*(\boldsymbol{\beta}_n, h_n)), \tag{8}$$

where $\mathbf{y}_n^* = (y_n^* = p_{з,n}^2, n = \overline{1, nk})$ – the vector of initial data on squared bottomhole pressures; $F_n = (\varphi_n^T = (1, q_n, q_n^2), n = \overline{1, nk})$ – the matrix; $\boldsymbol{\alpha}_n = (\alpha_{1,n} = p_{пл,n}^2, \alpha_{2,n} = a_n, \alpha_{3,n} = b_n)$ – the vector of unknown parameters; $\bar{\boldsymbol{\alpha}}_n = (\bar{p}_{пл,}^2, \alpha_{2,n} = \bar{a}_n, \alpha_{3,n} = \bar{b}_n)$ – the vector of additional a priori data obtained at stage number n; $J_0(\boldsymbol{\alpha}_n) = \sum_{n=1}^{nk} \omega_n(h_n) \cdot \psi_0(y_n^* - \varphi_n^T \boldsymbol{\alpha}_n)$, $J_a(\boldsymbol{\alpha}_n, \boldsymbol{\beta}_n) = \sum_{j=1}^{3} \beta_{j,n} \psi_a(kr_{j,n} \cdot \bar{\alpha}_{j,n} - a_{j,n})$ – measures of IPR curve model quality; $\boldsymbol{\beta}_n = (\beta_{j,n}, j = \overline{1,3})$ – vector of control parameters defining the importance (weight) of additional a priori data $\bar{\alpha}_{j,n}, j = \overline{1,3}$; ψ_0, ψ_a – the known functions; $\omega_n(((n-i)/h_n), i = \overline{1, nk-1}, n = \overline{1, nk})$ – weighting functions with decay parameter h_n to secure adaptive identification and interpretation $(\omega(x_1) < \omega(x_2), x_1 < x_2)$; $kr_{j,n}$ – the adjustment parameter for additional data $\bar{\boldsymbol{\alpha}}_n$.

The solution on the time for deliverability test with resulting IPR curve to be completed can be taken via visual analysis of graph (see figures 2–4) or using the criterion for estimates stabilization, where nk is such a test stage n that

$$\left|(\alpha_{j,n}^*(\beta_{j,n}^*, h_n^*) - \alpha_{j,n-1}^*(\beta_{j,n-1}^*, h_{n-1}^*)) / \alpha_{j,n}^*(\beta_{j,n}^*, h_n^*)\right| \le \varepsilon_j, j = \overline{1,3}, n = 1,2,3,... \tag{9}$$

is a valid inequality, where ε_j is preassigned accuracy.

The algorithm given below represents the method of adaptive interpretation of IPR curve with determination of additional a priori data and flow coefficients:

1. Forming vector $\mathbf{y}_n$ and matrix F_n (6).
2. Defining the vector of additional data $\bar{\boldsymbol{\alpha}}_n = (\bar{p}_{пл,n}^2, \alpha_{2,n} = \bar{a}_n, \alpha_{3,n} = \bar{b}_n)$ by solving problem (4) and system of linear equations (5).
3. Selecting measures of model (6) quality $J_0(\boldsymbol{\alpha}_k, h_k), J_a(\boldsymbol{\alpha}_k, \boldsymbol{\beta}_k)$.
4. Solving problems (7), (8) using the appropriate method of function optimization.
5. Checking condition (9): if the condition is fulfilled, the test is completed; if condition (9) fails to be fulfilled, the next test stage $n+1$ is arranged and one should start new research with step 1 of the algorithm.

3. Results of IPR curve interpretation for gas wells.

The results of a case study of deliverability test with resulting IPR curve run in wells 1 and 2 of the Urengoy gas and condensate field are given in figures 1–4 and tables 1, 2.

For example, figure 1 shows the initial data for IPR curves for wells 1 and 2, with eight and seven test stages, respectively.

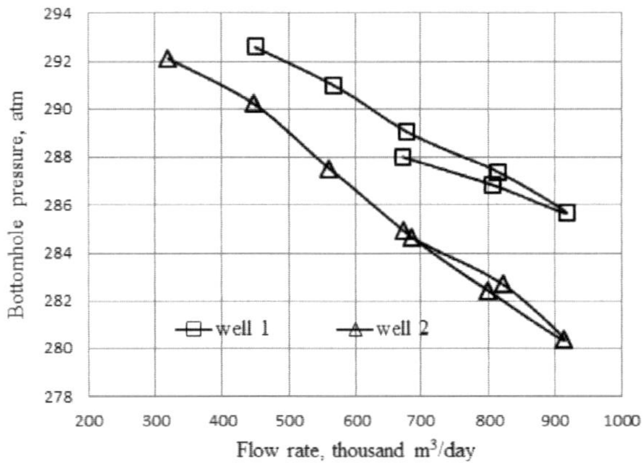

Figure 1. Initial data for IPR curves for wells 1 and 2.

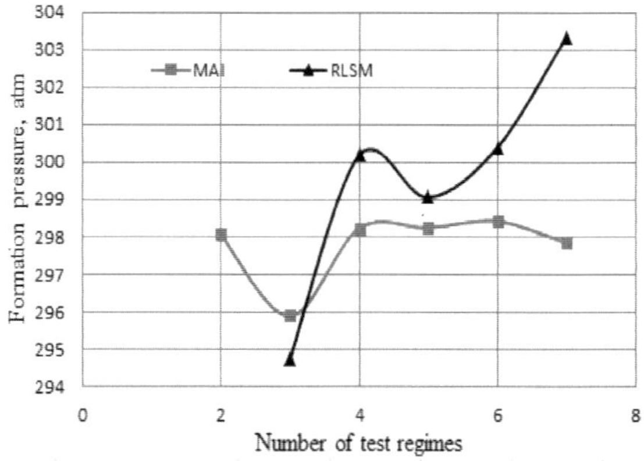

Figure 2. Formation pressure estimates for well 1.

Figures 2–4 show the estimates of formation pressure and flow coefficients of well 1, which are obtained using the following techniques:

1. the method of adaptive interpretation (MAI) (7) with quadratic measures of quality $\psi_0(x) = \psi_a(x) = x^2$ by solving the system of linear equations when $\mathbf{kr}_n = (kr_{j,n}, j = \overline{1,3})$ and $\beta_{j,n} = \beta_n^*, j = \overline{1,3}$ [9-10].

$$(F_n^T W_n(h_n^*) F_n + \beta_n^* \mathbf{I})\, \boldsymbol{\alpha}_n^*(\beta_n^*, h_n^*) = (F_n^T W_n(h_n^*) \mathbf{y}_n + \beta_n^* \mathbf{kr_n} \bar{\boldsymbol{\alpha}}_n), \qquad (10)$$

where the estimates of control parameter β_n^* and decay parameter h_n^* are defined by solving problem (8) using the downhill simplex method [11]; $W(h_n^*) = diag(\exp((n-i)/h_n^*), i = \overline{1, nk-1})$ - diagonal matrix of weighting function values;

2. the regularized least squares method (RLSM) from (10), with $\bar{\boldsymbol{\alpha}}_n = 0$.

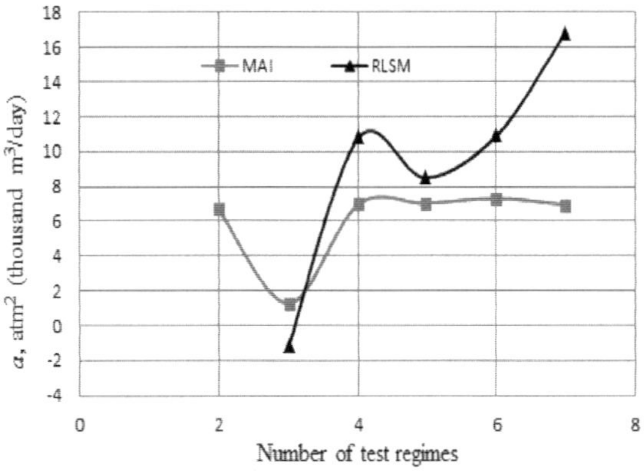

Figure 3. Estimates of flow coefficients a in well 1.

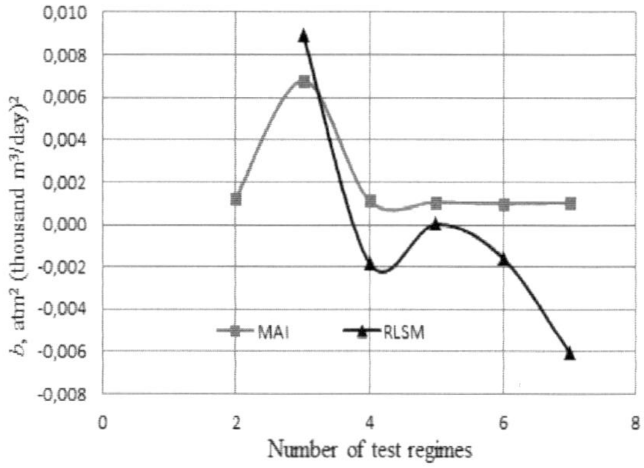

Figure 4. Estimates of flow coefficients b in well 1.

Table 1 shows the estimates of flow coefficients and formation pressure in well 2.

Table 2 gives the estimates of the formation pressure $\bar{\alpha}_{1,n} = \bar{p}^2_{n\pi,n}$, and flow coefficients $\bar{\alpha}_{2,n} = \bar{a}_n, \bar{\alpha}_{3,n} = \bar{b}_n$ of wells 1 and 2, which are used as additional data in (3) and obtained by solving optimization problem (4) using Gauss-Newton method with $r(x)=x^2$ [9-10] and the system of linear equations (5).

Table 1. Flow coefficients and formation pressure estimates.

Number of the test stage (n)	Method	Flow coefficient estimate		Formation pressure estimate
		$a^* = \alpha^*_{2,n}(\beta^*_n, h^*_n)$, atm^2/(thousand m^3/day)	$b^* = \alpha^*_{3,n}(\beta^*_n, h^*_n)$, atm^2/(thousand m^3/day)^2	$p^*_{n\imath} = \alpha^*_{1,n}(\beta^*_n, h^*_n)$, atm
2	MAI	2.25	0.0084	294.82
	RLSM			
3	MAI	1.47	0.0111	294.97
	RLSM	-8.91	0.0227	291.22
4	MAI	1.51	0.0104	294.79
	RLSM	2.29	0.0096	295.10
5	MAI	9.59	0.0022	297.91
	RLSM	8.97	0.0027	297.63
6	MAI	9.27	0.0019	297.49
	RLSM	12.10	-0.0003	298.89
7	MAI	9.07	0.0019	297.36
	RLSM	11.69	-0.0001	298.65
8	MAI	9.12	0.0019	297.30
	RLSM	12.58	-0.0008	299.04

Table 2. Additional data.

Number of the test stage (n)	Well	Formation pressure $\bar{p}_{n\imath,n}$, atm	Flow coefficient		Model parameter (2)	
			$\bar{a}$, atm^2/(thousand m^3/day)	$\bar{b}$, atm^2/(thousand m^3/day)^2	$\bar{\lambda}$, (thousand m^3/day)/atm^2	$\bar{\gamma}$, d e
2	1	295.30	0.06	0.00775	10.95	0.5038
	2	294.20	0.30	0.01093	8.10	0.5176
3	1	295.30	0.001	0.00787	11.19	0.5009
	2	294.40	-0.12	0.01286	8.19	0.5107
4	1	299.00	8.48	-0.00006	0.13	0.9929
	2	294.30	0.30	0.01159	8.56	0.5070
5	1	299.00	8.38	0.00009	0.13	0.9894
	2	295.10	3.37	0.00751	3.13	0.6206
6	1	298.40	7.27	0.00102	0.29	0.9013
	2	296.30	6.86	0.00376	0.81	0.7672
7	1	299.50	9.88	-0.00106	0.12	0.9947
	2	296.50	7.34	0.00325	0.63	0.7960
8	1					
	2	296.90	8.31	0.00247	0.44	0.8323

It is noteworthy that when the coefficient γ of model (2) approaches 1, the flow coefficient b of model (1) approaches 0 (the laminar flow in the well).

As can be seen in figures 2–4 and table 1, the suggested method of adaptive interpretation with account of additional data allows obtaining more accurate estimates of the formation pressure and flow coefficients with less amount of field data compared to the method of least squares. For example, for the adaptive interpretation method three test stages are enough (see figures 2–4 and table 1).

4. Conclusion

To overcome the challenges of interpreting deliverability tests with resulting IPR curve of gas wells, the method of adaptive interpretation with account of additional a priori data has been suggested. This method allows:

1. Obtaining additional a priori data on the formation pressure and flow coefficients.
2. Defining optimal, in terms of preassigned measures of quality, estimates of the formation pressure and flow coefficients within the period of test time.
3. Setting the number of test stages adequate for efficient well testing.

The case study of IPR curve interpretation for two wells of the Urengoy gas and condensate field has indicated that adaptive interpretation provides robust and more accurate estimates of the formation pressure and flow coefficients, as well as allows reducing the number of test stages.

References

[1] Houpert A 1959 *Revue de L'Institut Francais du Petrole*. On the flow of gases in porous media. Vol. **14 (11)** pp. 1468-1684.
[2] Aliev Z S, Gritsenko A I et al 1995 *Moscow Nauka.* Guidelines for Well Testing. pp. 523.
[3] Aliev Z S and Zotova G A 1980 *Moscow Nauka.* Gas reservoir and Well Testing: Handbook. pp. 301.
[4] Akhmedov K S, Gasumov R A and Tolpaev V A 2011 *Technique of data processing of wells hydrodynamic studies Petroleum Engineering.* Vol. **3** pp. 8–11.
[5] Nguyen Th H Ph and Sergeev V L 2015 *Bulletin of the Tomsk Polytechnic University: Georesources Engineering.* . Identification of IPR curve for interpreting gas well deliverability tests. Vol. **326** (**12**) pp. 54–59.
[6] Rawlin E L and Schellhardt M A 1936 *United States Bureau of Mines, Monograph 7.* Back pressure data on natural gas wells and their application to production practices. pp. 210.
[7] Al-Subaie A A, Al-Anazi B D and Al-Anazi A F 2009 Learning from modified isochronal test analysis Middle East gas well case study *Nafta Scientific Journal* **60** pp 405–415
[8] AL-Attar H and Al-Zuhair S 2009 *Journal of Petroleum Science and Engineering.* A general approach for deliverability calculations of gas wells. Vol. **67** pp. 97–104.
[9] Sergeev V L 2011 *TPU Publishing House.* Integrated Systems of Identification. pp. 198.
[10] Polishchuk V I and Sergeev V L 2015 *Journal of Siberian Federal University: Mathematics & Physics.* Adaptive identication method of a signal from stray field magnetic sensor for turbogenerator diagnostics. Vol. **8 (2)** pp. 201–207.
[11] Letova T A and Panteleev A V 2002 *Moscow: Vysshaya Shkola.* Optimization Methods: Examples and Problems. pp. 544.

Adaptive identification and interpretation of pressure transient tests of horizontal wells: challenges and perspectives

V L Sergeev, Dong Van Hoang

[1] Department of Geology and Oil Field Development, National Research Tomsk Polytechnic University, 30 Lenina Ave., Tomsk, 634050, Russia
[2]SergeevVL@ignd.tpu.ru, [3]hoang.tpu@gmail.com

Abstract. The paper deals with a topical issue of defining oil reservoir properties during transient tests of horizontal wells equipped with information-measuring systems and reducing well downtime. The aim is to consider challenges and perspectives of developing models and algorithms for adaptive identification and interpretation of transient tests in horizontal wells with pressure buildup curve analysis. The models and algorithms should allow analyzing flow behavior, defining oil reservoir properties and determining well test completion time, as well as reducing well downtime. The present paper is based on the previous theoretical and practical findings in the spheres of transient well testing, systems analysis, system identification, function optimization and linear algebra. Field data and results of transient well tests with pressure buildup curve analysis have also been considered. The suggested models and algorithms for adaptive interpretation of transient tests conducted in horizontal wells with resulting pressure buildup curve make it possible to analyze flow behavior, as well as define the reservoir properties and determine well test completion time. The algorithms for adaptive interpretation are based on the integrated system of radial flow PBC models with time-dependent variables, account of additional a priori information and estimates of radial flow permeability. Optimization problems are solved with the case study of PBC interpretation for five horizontal wells of the Verkhnechonsk field.

1. Introduction

Today, oil and gas companies intensively drill and exploit horizontal wells, which is attributed to implementation of resource efficient technologies. These wells are equipped with information-measuring systems to conduct transient well tests. Since petroleum reservoirs are structurally complex and the monitoring technologies of field exploration and development need to be improved, transient testing enhancement is a critical issue, which implies development of methods and algorithms to process research data and reduce well downtime.

It is noteworthy that the main source of information on the reservoir properties is transient well testing based on pressure buildup curve (PBC). The obtained data are necessary to manage oil recovery, work out field exploration and development projects, develop geological and technical models of petroleum recovery processes.

Currently, there are three basic approaches to develop methods and algorithms for adaptive identification and interpretation of PBC data:

1. Conventional approach to transient test data interpretation based on PBC analytical models [1-3].

2. Interpretation of transient test data based on hydrodynamic simulation of the reservoir using differential equations containing partial derivatives [4-5].
3. Adaptive identification and interpretation of transient tests in wells equipped with information-measuring systems [6,7].

Conventional approach to transient test data interpretation using PBC analytical models implies flow behavior analysis and solving an inverse problem of identifying filtration properties essential for the reservoir and well using a reservoir simulator corresponding to the definite flow.

One of the methods widely applied in transient test interpretation by both Russian and foreign specialists is the analysis of bottom hole pressure derivative (grapho-analytical method). For example, for the radial flow the slope of the tangent line to the bottomhole pressure derivative $\Delta p_r^{'} = \mathrm{d}p_r / \mathrm{d}\lg(t)$ within the coordinates $\lg(\Delta p_r^{'}) - \lg(t)$ is equal to zero, while for the linear flow it is 0.5 [1–3].

However, with intensive implementation of telemetric systems, which allows monitoring hydrodynamic properties and real-time well operations, the application of conventional methods becomes challengeable.

This makes adaptive identification of transient tests in horizontal wells a promising method of real-time flow behavior analysis [6,7].

The challenge of transient test data interpretation based on hydrodynamic simulation of the reservoir using equations for flow filtration and their finite difference analogues is solving an inverse problem of identification during well test time. Currently, solving problems of real-time identification is a critical issue, since this is one of the limitations for hydrodynamic simulation to be applied in в real-time processing of data obtained via transient tests conducted in horizontal wells [8].

Another limitation essential for conventional methods of transient test data identification and interpretation based on PBC analytical models is their being quite expensive. This is due to the fact that the data are interpreted after well test completion, which results in well downtime and decrease in the amount of hydrocarbon recovery.

Today, obtaining real-time data necessitates implementation of new technologies, which ensure identifying filtration properties and to classify the reserves during well test time, without setting a time limit for test completion in advance. This approach was suggested and is currently developed by the Department of Geology and Oil Field Development, Tomsk Polytechnic University. It based on adaptive identification and interpretation of transient test conducted in the wells equipped with information-measuring systems [6,7,9,10].

The present paper continues the research on adaptive identification and interpretation of transient tests conducted in horizontal wells. It describes new data obtained in the course of the research on real-time flow behavior analysis and identification of reservoir properties during transient well test based on PBC models with time-varying parameters and additional a priori obtained data on the radial flow permeability of the reservoir.

2. Models and Algorithms of Real-Time Flow Behavior Analysis and PBC Interpretation

The method of flow behavior analysis during transient tests in horizontal wells is based on the estimate of radial flow permeability $k_{r,t}^{*}$, obtained at the current moment t:

$$k_{r,t}^{*} = \frac{c_s q \mu B}{L \cdot \alpha_{2,t}^{*}} \tag{1}$$

where $\alpha_{2,t}^{*}$ is the estimate of the system of integrated PBC models for the radial flow with time-dependent parameters $\alpha_{1,t}, \alpha_{2,t}$ and account of a priori data [7,10]:

$$\begin{cases} P_з^*(t) = P_з^*(t_0) + \dfrac{c_s q \mu B}{k_{r,t} L} \ln\left(\dfrac{2,25 k_{xy} t}{m \mu r_{\text{пр}}^2}\right) + \xi_t = \alpha_{1,t} + \alpha_{2,t} \ln(t) + \xi_t, \\ \bar{\boldsymbol{\alpha}}_t = \boldsymbol{\alpha}_t + \boldsymbol{\eta}_t, t \in [t_0, t_{nk}], \end{cases} \quad (2)$$

(1) and (2) contain notations as follows: $P_{з,t}^*, P_з^*(t_0)$ – values of bottomhole pressure at time t and t_0, respectively; $k_{r,t} = \sqrt{k_{z,t} k_{y,t}}$ – radial flow permeability; $k_{z,t}, k_{y,t}$ – vertical and horizontal permeabilities ; q_0 – flow rate at the moment of shut-in t_0, μ, B – oil viscosity and formation volume factor, respectively; $r_{\text{пр}}^2$ – modified well radius; L – effective wellbore length; c_s – constant dependent on the system of units; $\bar{\boldsymbol{\alpha}}_t$ – vector of additional a priori data and parameters estimates; $\boldsymbol{\alpha}_t = (\alpha_{1,t} = f_1(t), \alpha_{2,t} = f_2(t))$ – estimated parameters given as unknown single-valued functions of time $f_1(t), f_2(t)$; t_{nk} – time of transient test completion; $\xi_t, \boldsymbol{\eta}_t$ – random variables, i.e. error in the bottomhole pressure measurement and estimates, PBC model inaccuracy, etc.

Using adaptive identification, the optimal estimates of the parameters $\alpha_{1,t}, \alpha_{2,t}$ (2) with bottomhole pressure $P_з^*(t_n)$ at discrete instants of time $t_n, n = \overline{1, nk}$ with uncertain random variables $\xi_t, \boldsymbol{\eta}_t$ and functions $f_1(t), f_2(t)$ are calculated by solving optimization problems [9,10]:

$$\boldsymbol{\alpha}_n^*(\boldsymbol{\beta}_n, h_n) = \arg\min_{\boldsymbol{\alpha}_n}(J_0(\boldsymbol{\alpha}_n, h_n) + J_a(\boldsymbol{\alpha}_n, \boldsymbol{\beta}_n)), \quad (3)$$

$$\boldsymbol{\beta}_n^*, h_n^* = \arg\min_{\boldsymbol{\beta}_n, h_n} J_0(\boldsymbol{\alpha}_n^*(\boldsymbol{\beta}_n, h_n)) \quad (4)$$

where $\arg\min_x f(x)$ is the minimum point x^* of the function $f(x)$ ($f(x^*) = \min_x f(x)$);

$J_0(\boldsymbol{\alpha}_n) = \sum_{i=1}^{n} \omega(h_n) \cdot \psi_0(\Delta P_з^*(t_i) - \alpha_{1,n} - \alpha_{2,n} \ln(t_i)))$, $J_a(\boldsymbol{\alpha}_n^*, \boldsymbol{\beta}_n) = \sum_{j=1}^{2} \beta_{j,n} \psi_{a,j}(\bar{\alpha}_{j,n} - a_{j,n})$ – empirical measures of PBC model quality; $\boldsymbol{\beta}_n = (\beta_{j,n}, j = \overline{1,3})$ – vector of control parameters defining the importance (weight) of additional a priori data $\bar{\alpha}_{j,n}$; $\psi_0, \psi_{a,j}$ – known loss functions; $w((t_n - t_i)/h_n)$ – weighting functions with decay parameter h_n for adaptive identification and interpretation ($w(x_1) < w(x_2), x_1 < x_2$); $\Delta P_з^*(t_i) = P_з^*(t_i) - P_з^*(t_0)$.

The criterion to define the start of the radial flow regime is such a value of $t_{r,n}^* \in t_r + \tau$, that

$$\left|k_{r,n+\tau}^* - k_{r,n}^*)\right| \le \varepsilon, \tau > 0 \quad (5)$$

is a valid inequality, where $t + \tau$ – time interval, within which the estimates of radial flow permeability are stabilized $k_{r,n}^* = c_s q \mu B / L \cdot \alpha_{2,n}^*(\boldsymbol{\beta}_n^*, h_n^*)$ (1) with account of (3),(4). Stabilization of radial flow permeability estimates (1) means that its derivative is equal to nil:

$$\left|\frac{\partial}{\partial t} k_{r,t}^*\right| \approx 0 \quad (6)$$

Therefore, formula (6), as well as (5), can be used to define time instants of the radial flow.

The solution on the time of test completion t_{nk} can be taken via visual analysis of the dependency graph (3) or using the criterion for estimates stabilization (5).

The algorithm given below represents the method of adaptive interpretation of PBC during transient tests conducted in horizontal wells with flow behavior analysis:

1. Collecting initial data on the bottomhole pressure with the specified volume n=n_0, additional a priori data and evaluation of reservoir properties (for example, see figure 1 and table 1).
2. Solving identification problems:

2.1. Calculating quality measures $J_0(\boldsymbol{\alpha}_k, h_k), J_a(\boldsymbol{\alpha}_k, \boldsymbol{\beta}_k)$ in (3),(4).

2.2. Solving problems (3), (4) using the method of function optimization.

3. Checking condition (5), defining the start of radial flow regime, and estimating radial flow permeability.
4. Research completion if the estimates of radial flow permeability and PBC model parameters are stabilized; otherwise, collecting initial data on the bottomhole pressure with volume $\Delta n \geq 1 (n = n_0 + \Delta n)$ and starting new research with stage 2.

3. Results of PBC interpretation with flow behavior analysis

The results of a case study of PBC interpretation with flow behavior analysis for five horizontal wells of the Verkhnechonsk field are given in figures 1–4 and tables 1, 2. For example, figure 1 shows the data on bottomhole pressure in wells 1 and 2. Figures 2–4 shows estimates of radial flow permeability (1), (3), estimates of radial flow permeability derivative (6) and PBC derivative

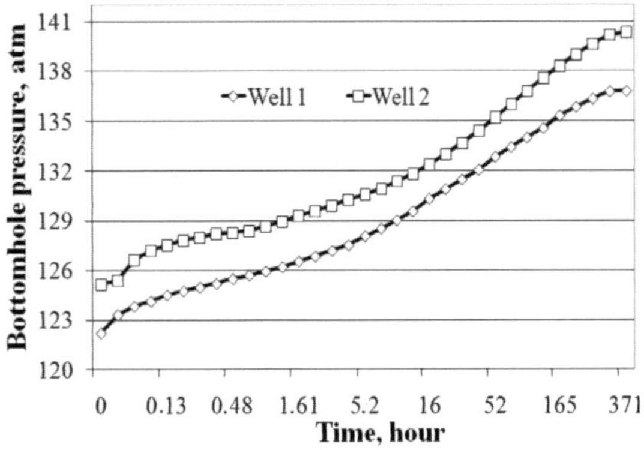

Figure 1. Pressure buildup curves

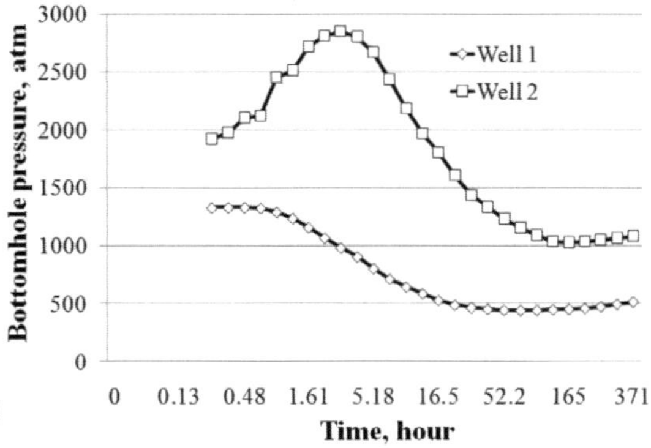

Figure 2. Estimates of radial flow permeability

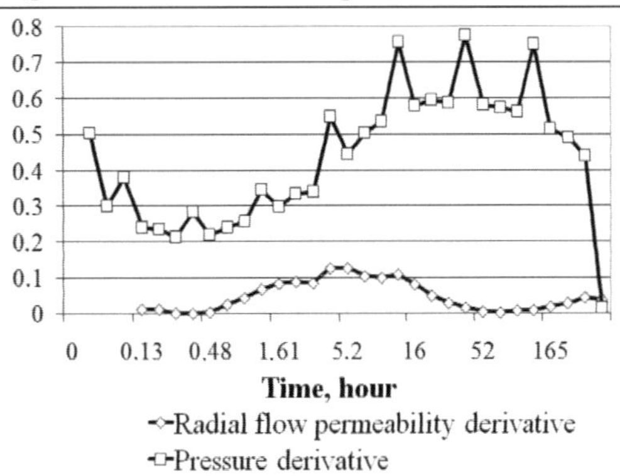

Figure 3. Diagnostic plot identifying flow regime in well 1

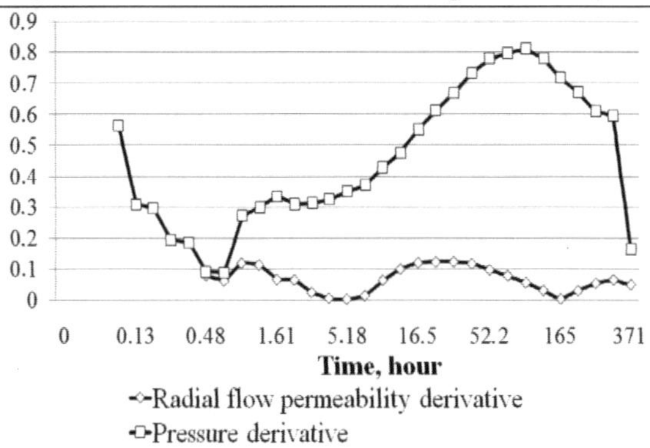

Figure 4. Diagnostic plot identifying flow regime in well 2

The estimates of radial flow permeability obtained using the method of adaptive interpretation (MAI) and quadratic performance indices $\psi_0(x) = \psi_a(x) = x^2$, additional a priori data on the radial flow permeability $\bar{k}_{r,n}$ given in table 1 and by solving the system of linear equations with $\beta_{1,n} = \beta_{2,n} = \beta_n$:

$$(F_n^T W(h_n^*)F_n + \beta_n \mathbf{I})\, \boldsymbol{\alpha}_n^*(\beta_n^*, h_n^*) = (F_n^T W(h_n^*)\Delta P_з^* + \bar{\boldsymbol{\alpha}}_t),$$

where $F_0 = (1, \Delta P_з^*(t_i), i = \overline{1,n})$ is $2 \times n$ matrix; $\mathbf{I}$ – 2×2 diagonal matrix; $W(h_n) = \mathrm{diag}(w(h_n) = \exp((n-i)/h_n^*), i = \overline{1, n-1})$ – diagonal matrix of weighting functions with decay parameter h_n^*; $\bar{\boldsymbol{\alpha}}_t = (0, \bar{\alpha}_{2,t} = c_s q \mu B / L\bar{k}_{r,n})$. The estimates of the control parameter β_n^* и decay parameter h_n^* were defined by solving problem (4) using the downhill simplex method [11].

Figures 2–4 indicate that the criteria (5), (6) allow identifying early-time and late-time radial flow regimes and recommending the time of test completion. For example, late-time radial flow regime will be established in well 1 after about 50 hours, which indicates the appropriate time of transient test completion.

Table 1. Initial and additional a priori data on wells 1 and 2

Initial data and estimates of reservoir and well properties (International System of Units (SI))	Well	
	1	2
1. Dynamic viscosity of oil, cP	3.92	3.92
2. Oil compressibility factor, atm^{-1}	1.78×10^{-4}	4.27×10^{-5}
3. Well radius, m	0.108	0.108
4. Atmospheric pressure, atm	1.033	1.033
5. Temperature, standard conditions (+20^0 C) ^{0}K	293	293
6. System compressibility factor, atm^{-1}	4.32×10^{-4}	4.32×10^{-4}
7. Porosity	0.13	0.13
8. Effective wellbore length , m	4.78	7.2
9. Well yield before shut-in, m^3/day	176	163.2
10. Estimate of radial flow permeability	-	-

Table 2. PBC interpretation results for five wells

Well	Interpretation method	Defined parameters		
		Test time, hour	Radial flow permeability, mD	Reservoir pressure, atm
1	Saphir	371.37	418	138.5
	MAI	52.2	444	137.5
2	Saphir	144.27	3760	141.8
	MAI	40.86	3760	144.5
3	Saphir	179.14	1010	135.1
	MAI	34.39	1089	135.2
4	Saphir	190.97	597	142.2
	MAI	52.83	607	143

5	Saphir	167.8	253	150.2
	MAI	32.75	226	149.4

It is worth noting that the horizontal flow observed between radial flows can be easily identified by the maximum point of the estimate derivative for the radial flow permeability (4), which is proved by the results of data interpretation for other horizontal wells (see table 2).

Table 2 shows that the method of adaptive interpretation with radial flow behavior analysis allows reducing well downtime compared to that caused by the planned tests via Saphir software. For the five wells presented in table 2 there is a fivefold reduction in the test time, from 1053.55 to 213.03 hours. The cost of five wells downtime is $ 1 500 000 if the price is $ 50 per barrel.

4. Conclusion

The present research has described the challenges concerning identification and interpretation of transient tests conducted in horizontal wells equipped with information-measuring systems by analyzing PBC. The perspectives of adaptive identification and interpretation have been explored.

The algorithms have been developed to carry out adaptive identification and interpretation of the data on flow behavior analysis during the time of transient well test with resulting PBC, as well as to define the reservoir properties and determine well test completion time.

The case study of PBC interpretation with flow behavior analysis for five horizontal wells of the Verkhnechonsk field has indicated that the suggested models and optimal algorithms of adaptive identification and interpretation make it possible to significantly reduce well downtime, which results in cost advantage.

References

[1] Ipatov A I, Kremenetskiy M I and Gulyaev D N 2012 *Moscow, Izhevsk: Institute of Computer Science.* Hydrodynamic Modeling of Oil and Gas Wells: Data Support and Technologies pp. 896.

[2] Joshi S D1991 *Oklahoma: PenWell Publ. Co.* Horizontal Well Technology. pp. 381.

[3] Bourdet D, Ayoub J A and Pirard Y M 1984 Use of pressure derivative in well test interpretation SPE **12777** pp. 293 – 302.

[4] Khayrullin M Kh, Khisamov R S, Shamshiev M N and Farkhullin R G 2006 *Moscow. Izhevsk: Regular and Chaotic Dynamics.* Regularization Methods for Intepretation of Transient Well Tests pp. 172.

[5] Azis Kh and Settari E 1982 *Moscow. Nedra.* Mathematical Models for Reservoir Simulation. pp. 407.

[6] Gavrilov K S and Sergeev V L 2012 *Bulletin of the Tomsk Polytechnic University.* Adaptive interpretation of transient well test in the system "reservoir-well" by method of integrated models Vol. **321 5** pp. 72–75.

[7] Romanova E V and Sergeev V L 2013 *Bulletin of the Tomsk Polytechnic University.* Adaptive identification of pressure buildup curve with flow behavior analysis for horizontal wells. Vol. **323 (5)** pp. 20–25.

[8] Iktisanov V A 2001 *Moscow: VNIIOENG.* Determining Reservoir Filtration Properties and Rheological Properties of Dispersed Systems for Oil Field Development. pp. 212.

[9] Sergeev V L 2011 *Tomsk: TPU Publishing House.* Integrated Systems of Identification pp. 198.

[10] Polishchuk V I and Sergeev V L 2015 *Journal of Siberian Federal University: Mathematics & Physics.* Adaptive identication method of a signal from stray field magnetic sensor for turbogenerator diagnostics. Vol. **8 (2)** pp. 201–207.

[11] Letova T A and Panteleev A V 2002 *Moscow: Vysshaya Shkola.* Optimization Methods: Examples and Problems pp. 544.

An assessment of the geotemperature conditions of Bazhenov oil generation (Koltogor mezodepression and its framing structures)

V V Stotskiy[1], V I Isaev[2] and M A Fomin[3]

[1,2]Department of Geophysics, Institute of Natural Resources, National Research Tomsk Polytechnic University, Tomsk
[3]Laboratory #334, Trofimuk Institute of Petroleum Geology and Geophysics of Siberian Branch of Russian Academy of Sciences (IPGG SB RAS), Novosibirsk
E-mail: [1]Stotskiy_VV@sibmail.com, [2]IsaevVI@tpu.ru, [3]FominMA@ipgg.sbras.ru

Abstract. The thermal history of oil source rocks of Bazhenov deposits was reconstructed in 8 cross-sections of representative wells of Koltogor mezodepression and the structures framing it. The tectonic history and geotemperature were reconstructed for well profiles, some located in the depression zone and others in the positive structures. A comparative analysis of the geotemperature conditions accompanying the generation of Bazhenov oils was performed.

1. Introduction
The Bazhenov oil source rock formation (*J_3bg*) is the main source of hydrocarbon deposits formation in Upper Jurassic and Neocomian traps [1]; it is the prominent [2] shale formation in oil and gas regions in South-East Western Siberia. The key factor in the realization of oil generation potential of a source rock is its thermal history [3] defining the starting time and the intensity of oil generation.

In the reviewed paper, the geological time of deposits' entrance into the main oil zone formation was assessed, based on the reconstruction of the paleotectonical and geotemperature modes of the Bazhenov formation in the Koltogor mezodepression and its framing structure (south-east Western Siberia). Also, the conditions under which Bazhenov oils had been generated were determined.

2. Characteristics of the researched territory
The Koltogor mezodepression is a 2nd order negative structure in the south-east Western Siberia (Figure 1). The high oil generation potential of Bazhenov deposits is stipulated by their ubiquitous spread, organic carbon content (Corg up to 12%) and the catagenetic transformation of MC_1^1–MC_1^2 graduations.

The Upper Jurassic oil and gas complex (OGC) consists of Callovian-Volgian deposits of the Vasyugan, Georgiev and Bazhenov formations. The Vasyugan formation was formed in littoral conditions and consists of lower (especially argilliceaous) and upper subformations, in terms of its lithological composition.

The deposits of *Neocomian OGC* in the south-east part of the West Siberian basin were examined with less detail. Neocomian deposits' studies were a research "byproduct" at best.

The eight representative wells have been selected for the analysis (see Table), some of them located in the depression zone while others within the positive structures (Figure 1).

3. Research methodology
The reconstruction of the tectonic and thermal history of Bazhenov formation was done using a software suite for paleotemperature modeling [5]. The method of paleotemeratures calculation is provided in paper [6].

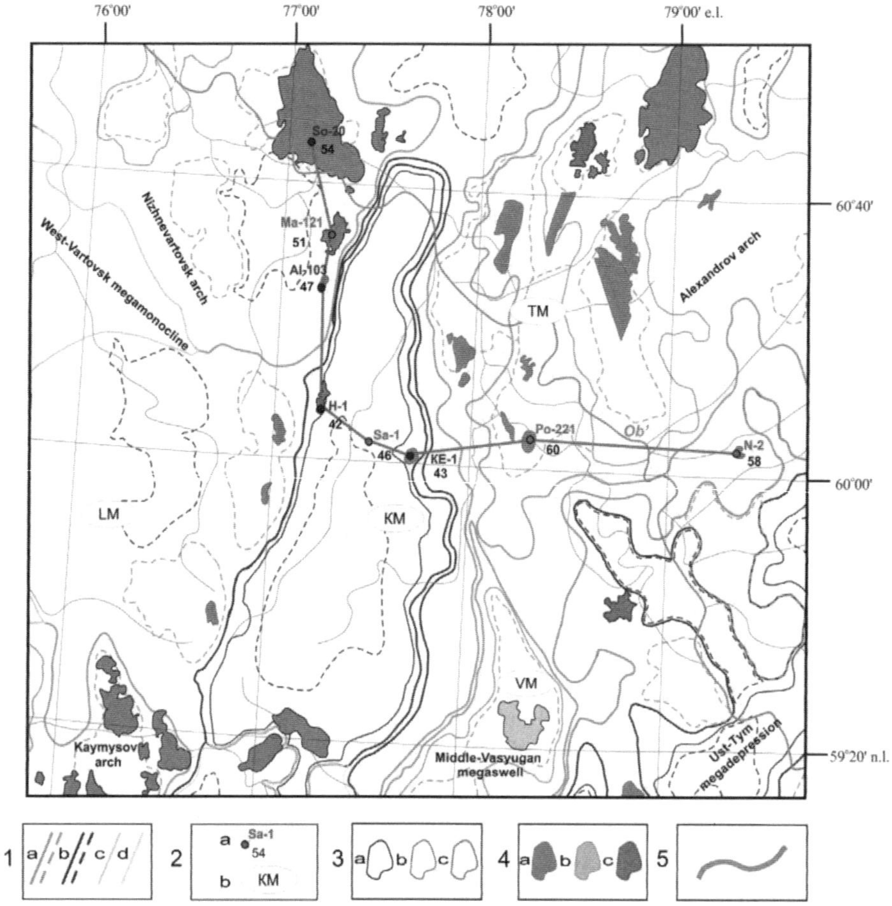

Figure 1. Overview diagram of the oil and gas potential of the Koltogor mezodepression and its framing structure (based on the tectonic foundation [4]): ***1*** – the contours of the tectonic elements: ***a*** – positive; ***b*** – negative; ***c*** – mezosaddle; ***d*** – megamonocline; ***2*** – ***a*** – the research well, its index and the estimated value of the heat flow density, mW/m^2; ***b*** – the index of 2nd order structures; ***3*** – the contours of the fields: ***a*** – production; ***b*** – exploration; ***c*** – ready for production; ***4*** – the type of hydrocarbon field: ***a*** – oil; ***b*** – gas condensate; ***c*** – oil and gas condensate; ***5*** – river system. 2nd order structures: VM – the Vasyugan mezoswell; KM – the Koltogor mezodepression; LM – the Ledyansk mezosaddle; TM – the Traygorod mezoswell.

Table. Characteristics of the research wells (see Figure 1 for well index).

Field	**Sovetskoe**	**Malorechenskoe**	**Alenkinskoe**	**Hvoynoe**
Well	**So-20**	**Ma-121**	**Al-103**	**H-1**
Characteristics	Values			
Downhole, m	2759	2664	2800	3207
Downhole deposits	Paleozoic (Pz)	Paleozoic (Pz)	Paleozoic (Pz)	Paleozoic (Pz)
Top of Bazhenov formation, m	2451	2430	2501	2673
Thickness of Bazhenov formation, m	13	14	15	10

Results of well testing (formation; reservoir; fluid type; flow, m3/day)	Alym; AV_{1-1}; oil; 156.0.	Vasyugan; J_1; gas; 13,2. Vasyugan; J_1; oil; 124.7.	Tara; B_{10}; oil; 20.3.	Vasyugan; J_1 oil; 8.1.
Estimated reservoir temperature (formation; measured depth; reservoir temperature)	-	Vasyugan; 2452 m; 90°C.	-	-
"Estimated" temperature by vitrinite reflectance (formation; sampling depth; (R^0_{vt}); temperature)	Tyumen; 2746 m; (0.76); 115 °C	Tyumen; 2470 m; (0.67); 103°C Tyumen; 2515 m; (0.67); 103°C Tyumen; 2620 m; (0.70); 106°C	Tyumen; 2700 m; (0.67);103°C	Tyumen; 2867 m; (0.64); 99°C Tyumen; 2939 m; (0.64); 99°C
Field	**Saymovskoe**	**Kul'-Eganskoe**	**Poludennoe**	**Nickol'skoe**
Well	**Sa-1**	**KE-1**	**Po-221**	**N-2**
Characteristics	Values			
Downhole, m	4008	3112	2448	2816
Downhole deposits	Paleozoic (Pz)	Paleozoic (Pz)	Paleozoic (Pz)	Paleozoic (Pz)
Top of Bazhenov formation, m	2832	2648	2207	2424
Thickness of Bazhenov formation, m	19	13	19	12
Results of well testing (formation; reservoir; fluid type; flow, m3/day)	Bazhenov; J_0; oil; 0.6.	Vasyugan; J_1; oil; 1.5.	Vasyugan; J_1; oil; 3.2.	Vasyugan; J_1; oil; 7.8.
Estimated reservoir temperature (formation; measured depth; reservoir temperature)	Bazhenov; 2844 m; 98°C.	-	Vasyugan; 2239 m; 94°C.	Vasyugan; 2445 m; 91°C.
"Estimated" temperature by vitrinite reflectance (formation; sampling depth; (R^0_{vt}); temperature)	Bazhenov; 2882 m; (0.67); 103°C Tyumen; 3437 m; (0.85); 124°C	Bazhenov; 2680 m; (0.62); 94°C	Tyumen; 2335 m; (0.70); 107°C	Tyumen; 2652 m; (0.76); 115°C

In the cross-sections of the 8 wells, the paleoreconstructions of the tectonic history and geotemperature for 17 key geological times were done, for the start time and end time of each formation (Figure 2).

The conditional integral index *R* [7] was calculated for assessing the density value of the generated Bazhenov oil which allows for cumulatively taking into account the geotemperature dynamics of the main oil generation zone of the source rocks [8].

4. Results

The thermal history analysis of the Bazhenov formation in well cross-sections (Figure 2) has shown that the deposits entered the main oil formation zone, having "overstepped" the threshold temperature of 85°C due to the dispersed organic matter of the sapropel type. The conditions that are favorable for Bazhenov oil generation persist at present, both in the most loaded and in the elevated parts of the researched territory.

The conclusions can be made by analyzing the time when the Bazhenov formation entered the main oil formation zone. The earliest time of entrance was around 92 million years ago on the territory near wells Sa-1 and Po-221. The Bazhenov formation escaped the main oil formation zone under the influence of secular climate and temperature trend [9] (wells H-1 and KE-1). It can be noted that the thermal history is less favorable for oil generation in well cross-sections on the margins of the Koltogor mezodepression (Figure 3).

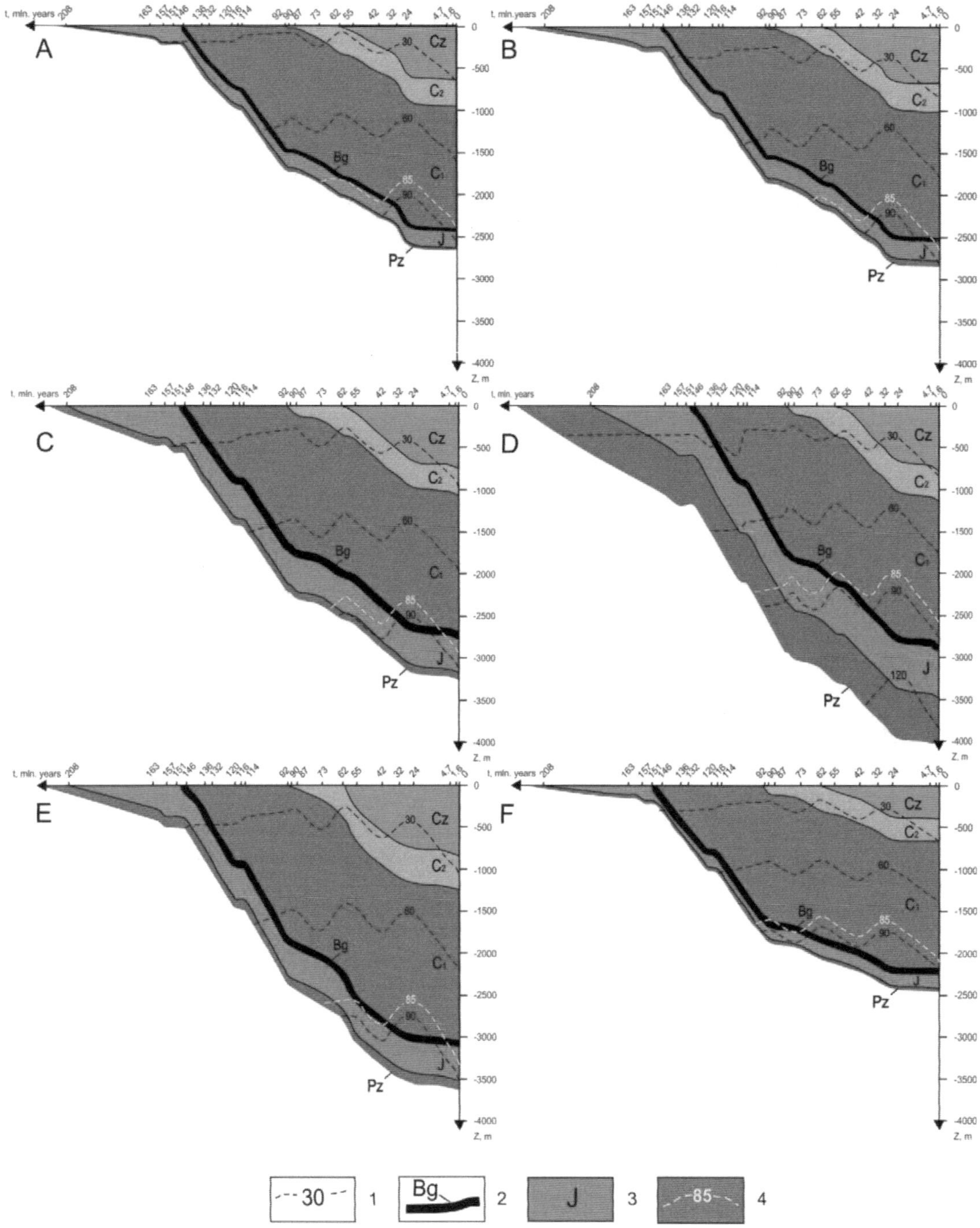

Figure 2. Paleoreconstructions of tectonic and thermal history in well cross-sections (index of wells in Figure 1): A – Ma-121; B – Al-103; C – H-1; D – Sa-1; E – KE-1; F – Po-221. ***1*** – the isotherms; ***2*** – the Bazhenov formation; ***3*** – the stratigraphic confinement of the deposits; ***4*** – the threshold temperature of the main oil formation zone.

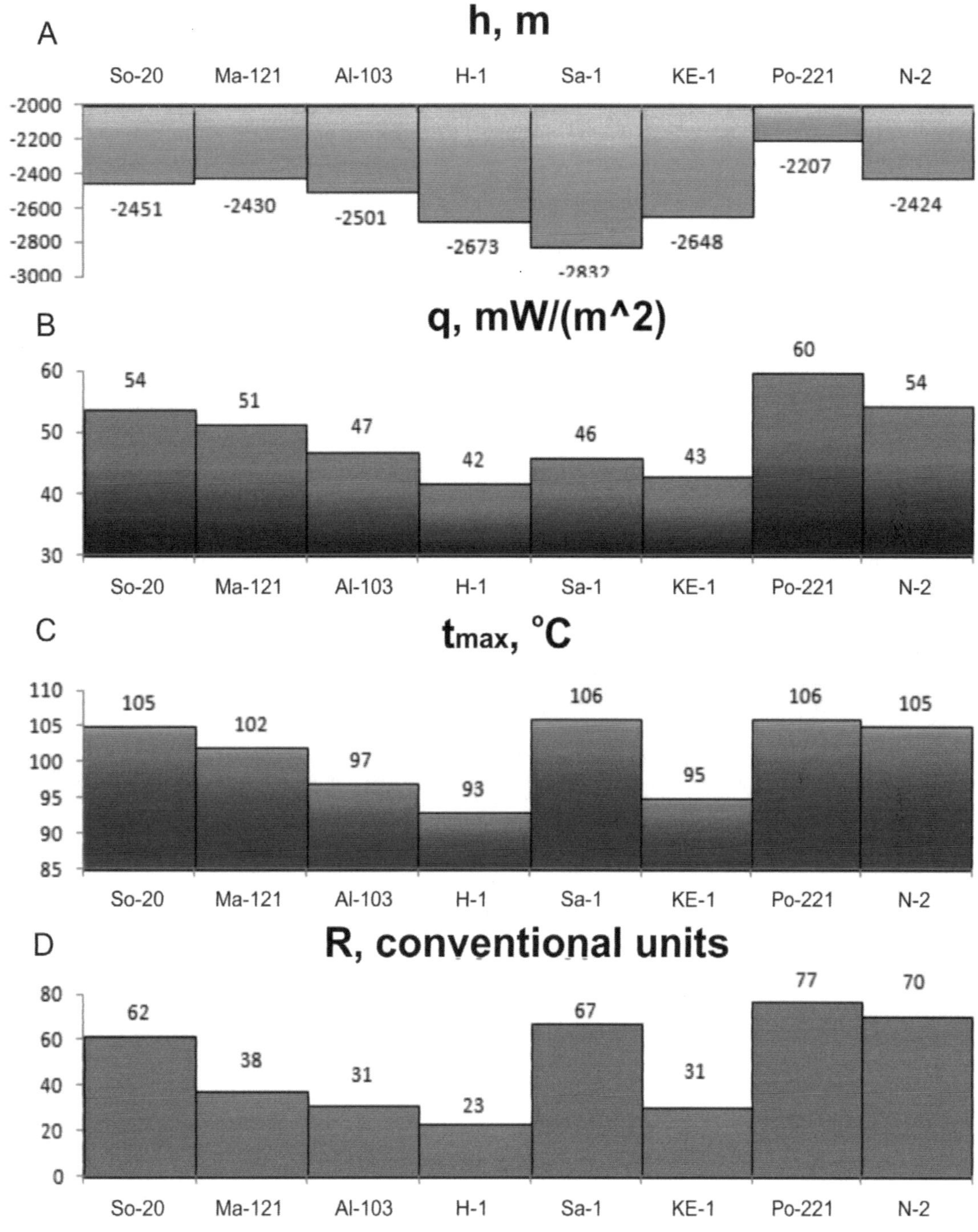

Figure 3. Correlation graphs along the line of the research wells (Figure 1): A – the thickness graphs of the deposits above the Bazhenov formation; B – the values of calculated heat flow density from the base of the sedimentary cover; C – the maximum geotemperature of the Bazhenov formation; D – the estimated resource density of the generated Bazhenov oil.

IOP Conf. Series: Earth and Environmental Science **43** (2016) 012017 doi:10.1088/1755-1315/43/1/012017

5. Conclusion

Oil generation for the researched profile of wells is confirmed by well testing results. Production and non-production oil inflows were mainly obtained from the Upper Jurassic reservoir J_1. In the Saymovskoe field, a non-production oil inflow was obtained from the Bazhenov formation during well testing results of well Sa-1. In the Alenkinskoe oil field in the Neocomian reservoir B_{10}, production oil inflow was obtained from well Al-103. In the Sovetskoe field, a production oil inflow was obtained from the Neocomian reservoir AV_1.

References

[1] Isaev V I and Fomin A N 2006 *Russian Geology and Geophysics.* Loci of Generation of Bazhenov- and Togur-Type Oils in the Southern Nyurol'ka Megadepression. Vol. **47** pp. 734-745

[2] Isaev V I, Lobova G A, Osipova E N, Sungurova O G 2016 *Neftegazovaya geologia. Teoria I praktika.* Rayonirovanie megavpadin Tomskoy oblasti po plotnosti resursov slantsevoy nefti. Vol. **11(1)** http://www.ngtp.ru/rub/4/1_2016.pdf

[3] Burshtejn L M, ZHidkova L V, Kontorovich A E, Melenevskij V N 1997 *Geologiya i geofizika.* Model' katageneza organicheskogo veshchestva (na primere bazhenovskoj svity). Vol. **38(6)** pp. 1070-1078.

[4] Kontorovich V A, Belyaev S Yu, Kontorovich A E, Krasavchikov V O, Kontorovich A A Suprunenko O I 2001 *Geologiya i geofizika.* Tektonicheskoe stroenie i istoriya razvitiya Zapadno-Sibirskoj geosineklizy v mezozoe i kajnozoe. Vol. **42(11-12)** pp. 1832-1845.

[5] Gulenok RYu, Isaev V I, Kosygin V Yu, Lobova G A, Starostenko V I 2011 *Russian Journal of Pacific Geology.* Estimation of the Oil-and-Gas Potential of Sedimentary Depression in the Far East and West Siberia Based on Gravimetry and Geothermy Data. Vol. **5(4)** pp. 273–287.

[6] Lobova G A, Stotskiy V V, Isaev V I 2014 *Neftegazovaya geologia. Teoria I praktika* Vliyanie paleoklimata na geotermicheskij rezhim i realizaciyu neftegeneracionnogo potenciala bazhenovskih otlozhenij yugo-vostoka Zapadnoj Sibiri (Novosibirskaya oblast'). Vol. **9/3** http://www.ngtp.ru/rub/4/31_2014.pdf

[7] Isaev V I, Lobova G A and Osipova E N 2014 *Russian Geology and Geophysics.* The oil and gas contents of the Lower Jurassic and Achimovka reservoirs of the Nyurol'ka megadepression. Vol. **55** pp. 1418-1428.

[8] Tissot B P 2003 *Oil & Gas Science and Technology* . Preliminary Data on the Mechanisms and Kinetics of the Formation of Petroleum in Sediments. Computer Simulation of a Reaction Flowsheet Rev. Vol. **58(2)** pp. 183-202.

[9] Iskorkina A, Isaev V and Terre D 2015 *IOP Conf. Series: Earth and Environmental Science.* Assessment of Mesozoic-Kainozoic climate impact on oil-source rock potential (West Siberia). Vol. **27** pp. 012023 http://iopscience.iop.org/article/10.1088/1755-1315/27/1/012023/pdf

Rheological properties of crude oils in Yaregskoye and Yaraktinskoye oil fields

Manzhai V.N., Clovis Le Grand Monkam Monkam, Terre D.A.

National Research Tomsk Polytechnic University, 30 Lenin Avenue, Tomsk, 634050, Russia

E-mail: mang@ipc.tsc.ru

Abstract. Rotary viscometer tests of crude oil with a high content of resins and asphaltenes (Yaregskoye oil field) and crude oil with high paraffin content (Yaraktinskoye oil field) have been conducted. The typical flow curves for these oil types have been plotted. It has been detected that these oils are non-Newtonian fluids, viscosity of which is dependent on shear rate. Based on Arrhenius-Eyring equation, calculations of viscous flow activation energy and complex structural unit (CSU) sizes have been performed. It has been stated that there is a tenfold reduction in CSU size in asphaltic crude oil with the increase in shear rate in a rotary viscometer, while particle size in paraffinic crude oil does not essentially change under the same hydrodynamic conditions.

1. Introduction
For the extraction and the pipeline transport of crude oils with anomalous properties, detailed information about the features of its rheological behavior at different shear stresses in a predetermined temperature range is required. Non-Newtonian properties are typically discovered in high-viscosity oils with a high content of asphaltenes, resins and paraffins. As the study objects, crude oils with a high content of resins and asphaltenes (Yaregskoye oil field) and crude oils with high paraffin content (Yaraktinskoye oil field) were used. The typical flow curves for the shear stress (τ) versus shear rate (γ) dependence obtained by rotary viscometry method are presented in figures 1, 3.

2. Materials and Methods
Crude oil with high resin and asphaltene content (figure 1) has a pseudoplastic properties (flow index is in the range 0 <n <1) and analytically can be described by Oswald de Ville equations $\tau = K \cdot \gamma^n$ and $\eta_{eff} = K \cdot \gamma^{n-1}$. The effective viscosity of that oil ($\eta_{eff.}$) decreases with increasing shear rate, which is particularly noticeable at low values of γ <50 s^{-1}. There is the largest destruction of supramolecular structures in this range. A further increase in shear stress has little effect on the rheological properties of oil, which becomes essentially a Newtonian fluid with a viscosity depending only on the temperature.

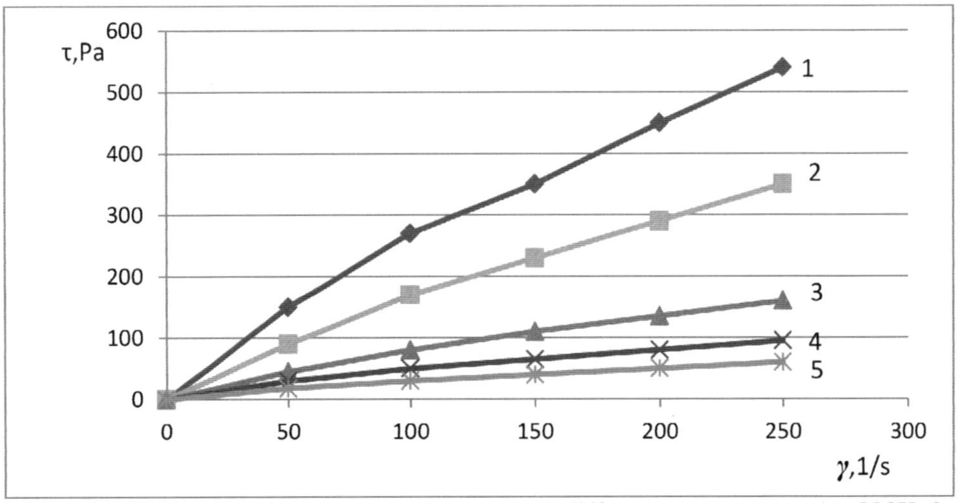

Figure 1. Shear stress (τ) versus shear rate (γ) dependence at different temperatures: 1 – 298K, 2 – 303K, 3 – 313K, 4 – 323K, 5 – 333K, crude oil in Yaregskoye oil field

Asphaltic crude oil in Yaregskoye oil field has thixotropic properties. The hysteresis loop for this oil (figure 2) is obtained by the sequential increase, followed by the reduction of the shear rate during its testing in a rotary viscometer. Thixotropy, depending on the rate of deformation of the material and duration of an external force effect, is the result of mechanical destruction of supramolecular structures (agglomerates and associates) which are present in high-viscosity oils. Restructuration occurs along with the structure destruction. At a certain point the speed of deformation can be equal to the speed of structure formation. Thereby, the equilibrium state is achieved, which is characterized by a definite value of the effective viscosity. As follows from figure 2, the disappearance of the hysteresis loop occurs at a shear rate higher than $\gamma \sim 30\ s^{-1}$, which, in its turn, indicates that oil losses anomalous properties immediately at low shear stress range due to the almost complete destruction of supramolecular structures.

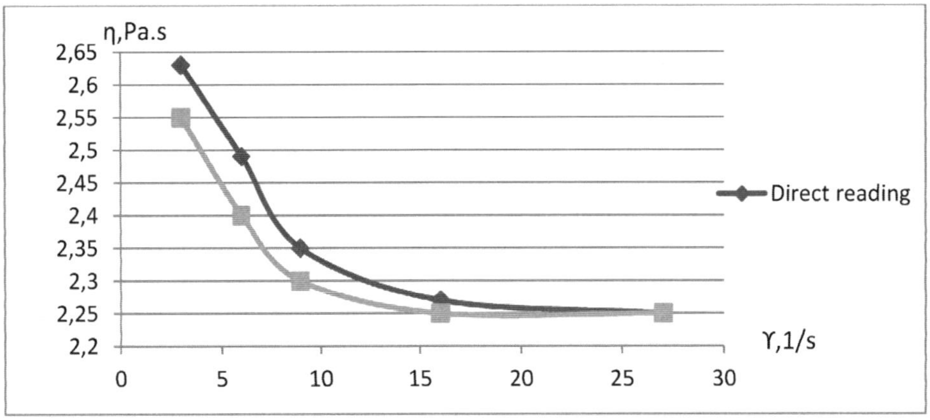

Figure 2. Dynamic viscosity (η) versus shear rate (γ) dependence at the temperature of 298K, crude oil in Yaregskoye oil field

Paraffinic crude oil in Yaraktinskoye oil field is a viscous-plastic fluid (figure 3), which has a yield point (τ_0) at temperature below 263 K. The rheological properties of this oil are described by the equations, which are deduced based on the rheological Bingham - Shvedova model $\tau = \tau_0 + K \cdot \gamma^n$

and $\eta_{eff} = \dfrac{\tau - \tau_0}{\gamma} = K \cdot \gamma^{n-1}$. The yield point distinguished in oils at low temperatures is due to the formation of volumetric massive frame structure from paraffin microcrystals, which is destroyed when the temperature rises because of thermal motion [1], and subsequently, oil becomes a pseudo-plastic fluid. When the temperature is within the yield point, crude oils of Yaraktinskoye oil field also reveal thixotropic properties.

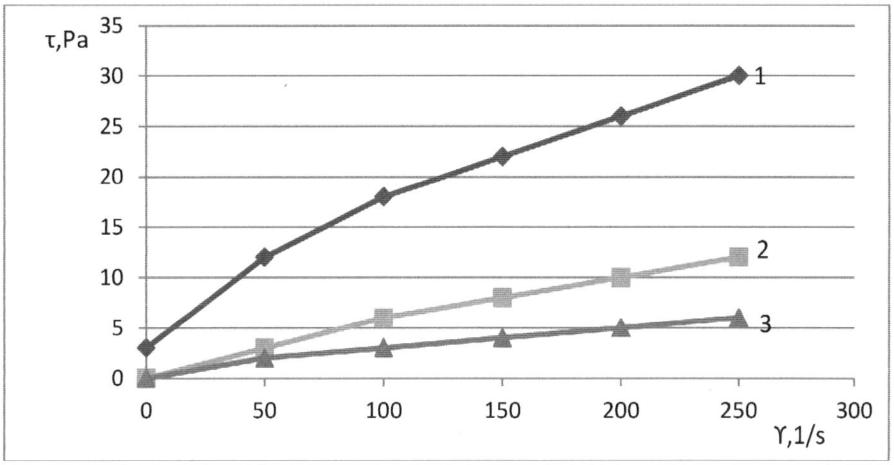

Figure 3. Shear stress (τ) versus shear rate (γ) dependence at different temperatures: 1 – 248K, 2 – 263K, 3 – 283K, crude oil in Yaraktinskoye oil field

3. Results and Discussion

The formal rheological Oswald de Ville and Bingham-Shvedova equations describe the dependence of viscosity on the shear rate only, but does not explicitly consider the temperature effect, which, in turn, makes it difficult to understand the physical nature of the processes occurring in the system at the molecular level. Therefore, under these conditions the search for other ways of analytical interpretation of the rheological properties of oil is reasonable. Using Arrhenius-Eyring equation $\eta = A \cdot e^{\frac{E_a}{RT}}$ for analytical description of the properties of liquids in a wide temperature range can determine the energy and geometric parameters of oil particles for different shear stresses. By analyzing the results of viscometric experiments with the coordinates lnη - 1/T (figures 4, 5), the values of viscous flow activation energy (Ea), pre-exponential factor (A) and average particle size (r), in different hydrodynamic conditions can be identified (table 1, 2).

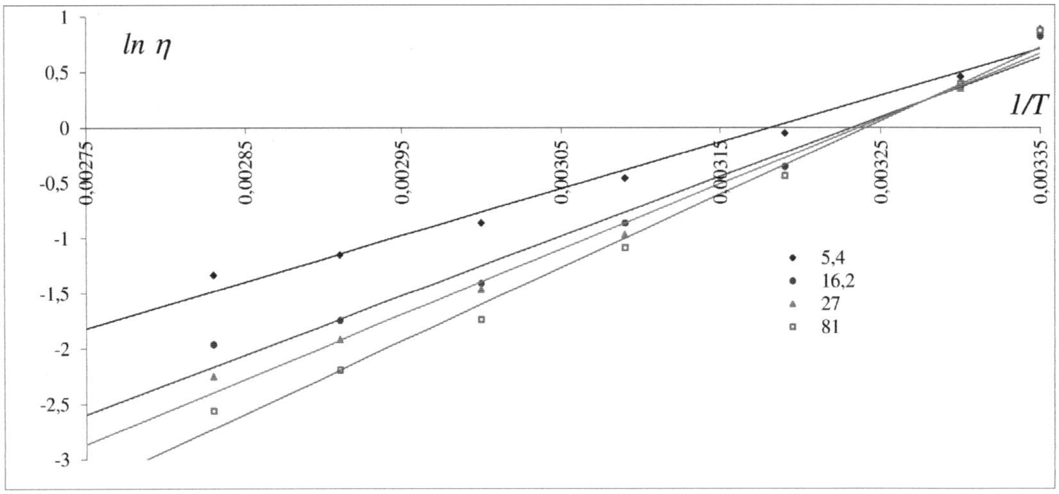

Figure 4. lnη viscosity versus inverse temperature (1/T) dependence, asphaltic crude oil

As follows from figures 4 and 5, the tangents of curve inclination angles are proportional to the activation energy of viscous flow (Ea), and they increase with increasing shear rate in case of Yarega (asphaltic) crude oil, whereas they remain practically constant considering crude oil (paraffinic) of Yaraktinskoye oil field. Average radii of colloidal dispersed associates shown in tables 1 and 2 were calculated according to the formula $r = \left[\dfrac{A}{\tau \cdot (16\pi\rho / 3kT)^{1/2}} \right]^{2/5}$ after experimental determination of the numerical values of the pre-exponential factor (A), which according to Frenkel-Eyring equation is dependent on the geometric dimensions of complex structural units (CSU) varying in structure due to oil composition.

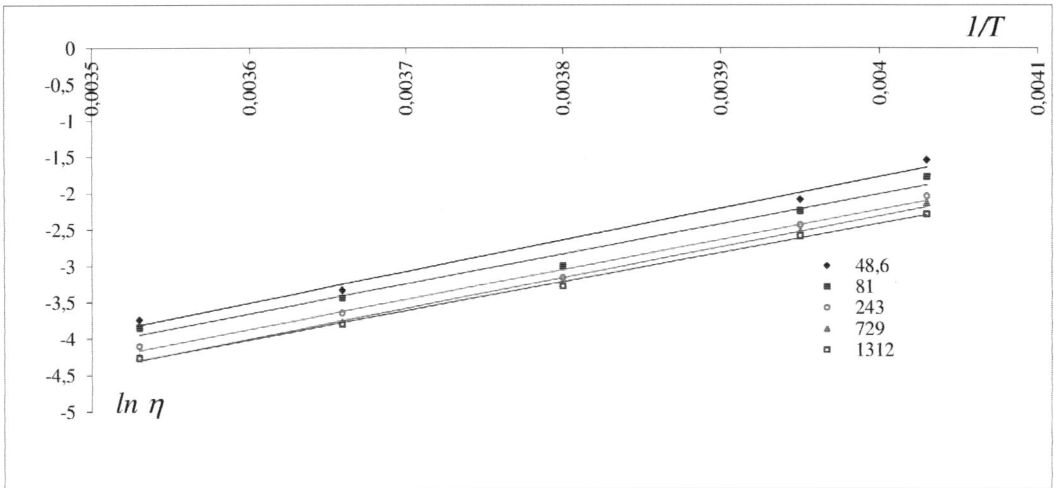

Figure 5. lnη viscosity versus inverse temperature (1/T) dependence, paraffinic crude oil

According to modern concepts, crude oil is not a simple mixture of a large number of organic compounds dispersed to the molecular state, but it is a colloidal dispersion system [2]. The intermolecular forces in this system can form supramolecular associates, composed of many individual molecules which are arranged in ordered structures. The composition and size of CSU depend on relative content of paraffin molecules in the oil or the total amount of resin and asphaltene molecules. The dominance of a certain class of chemical compounds results in particular rheological behavior of crude oil.

The central nucleus of a complex structural unit (CSU) formed in asphaltic crude oil is composed of a large number of high and polycyclic asphaltenes which are bound together in an aggregate due to weak intermolecular interaction. Resin molecules are less connected to each other at the periphery of the nucleus in the solvate-adsorption layer. CSU nucleus formed in paraffinic crude oil contains a great number of high molecular chain alkanes (paraffins), which are tightly packed together. Resin molecules in the solvate-adsorption membrane (periphery of the nucleus) being weakly bound to the nucleus, diffusive scattering of the membrane occurs with a shift of particles due to movement which results in the decrease in CSU size to a constant value characteristic of a competent paraffin nucleus.

Table 1. Activation energy (E_a), pre-exponential value (A) and average particle radius (r) of asphaltic crude oil at various shear rates (γ).

Parameters	Shear rates γ, s^{-1}			
	5.4	16.2	48.6	145.8
E_a, kJ/mol·K	30.6	37.2	45.1	51.3
$A \cdot 10^8$, Pa.s	715	41.4	1.8	0.2
r, nm	47.8	10.9	1.9	1.1

The results in table 1 indicate that asphaltic crude oil associates in Yaregskoye field are highly susceptible to the destructive influence of the shear flow. Their mean radii in low shear rate range decrease from the colloidal particle size (~ 50 nm) to values close to the molecular size (~ 1 nm). Stronger particles (paraffin microcrystals) of oil in Yaraktinskoye field (table 2) almost do not reduce their dimensions in the wide range of shear rates.

Table 2. Activation energy (E_a), pre-exponential value (A) and average particle radius (r) of paraffinic crude oil at various shear rates (γ).

Parameters	Shear rates γ, s^{-1}			
	48.6	81.0	146	243
E_a, kJ/mol·K	36.2	34.3	33.6	34.3
$A \cdot 10^8$, Pa·s	0.47	0.9	1.1	0.73
r, nm	4.5	4.9	4.4	3.1

The work was supported by Russian Foundation for Basic Research Grant 16-43-700866 p_a

References

[1] Mirzadzhanzade A H, Ametov I M, Kovalev A G 1992 *Nedra.* The physics of oil and gas.
[2] Cheng D C, Evans F 1965 *British Journal of Applied Physics.* Phenomenological characterization of the rheological behavior of inelastic reversible thixotropic and antithixotropic fluids. Vol. **16** pp. 1599–1617
[3] Dickie J P, Yen T F 1967 *Anal Chem.* Macrostructures of asphaltic fractions by various instrumental methods. Vol. **39** pp. 1847–1852
[4] Groezin H, Mullins O C 2000 *Energy & Fuels.* Molecular size and structure of asphaltenes from various sources. Vol. **14** pp. 677–684
[5] Mostowfi F, Indo K, Mullins O C 2009 *Energy & Fuels.* Asphaltene nanoaggregates studied by centrifugation. vol. **23** pp. 1194–1200
[6] Mullins O C 2010 *Energy & Fuels.* The modified Yen model. vol. **24** pp. 2179–2207
[7] Sabbah S, Morrow A L, Pomerantz A D 2011 *Energy & Fuels.* Evidence for island structures as the dominant architecture of asphaltenes. Vol. **25** pp. 1597–1604.
[8] Lopes-da-Silva J A, Coutinho A P 2004 *Rheol Acta.* Dynamic rheological analysis of the gelation behavior of waxy crude oils. Vol. **43** pp. 433–441.
[9] Lopes-da-Silva J A, Coutinho A P 2007 *Energy & Fuels.* Analysis of the isothermal structure development in waxy crude oils under quiescent conditions / J. A. Lopes-da-Silva. Vol. **21** pp. 3612–3617.
[10] Syunyaev Z I, Syunyaev R Z, Safieva R Z 1990 *Chemistry.* Oil dispersed systems.

The work was supported by Russian Foundation for Basic Research Grant 16-43-700866 p_a

Designing linings of mutually influencing parallel shallow circular tunnels under seismic effects of earthquake

A S Sammal, S V Antsiferov, P V Deev

Tula State University, Lenin av., 92, Tula, Russia
E-mail: sammal@ mm.tsu.tula.ru

Abstract. The paper deals with seismic design of parallel shallow tunnel linings, which is based on identifying the most unfavorable lining stress states under the effects of long longitudinal and shear seismic waves propagating through the cross section of the tunnel in different directions and combinations. For this purpose, the sum and difference of normal tangential stresses on lining internal outline caused by waves of different types are investigated on the extreme relative to the angle of incidence. The method allows analytic plotting of a curve illustrating structure stresses. The paper gives an example of design calculation.

1. Introduction

The challenges in seismic design of tunnel linings are due to the fact that during earthquakes different types of seismic waves are propagating: compressive-tensile (longitudinal) and shear (transverse) waves. These waves propagate with different velocities and due to numerous refractions and reflections (from both free-field surface and the boundary between rocks with different mechanical properties), they may affect tunnel linings in different combinations and directions. Therefore, designing an underground structure upon the action of only one wave type, longitudinal or shear wave of the specified direction, is a priori inefficient. An original approach to seismic design of underground constructions was developed at Tula State University [1, 2, 7]: according to this approach, the design consists in determining the most unfavorable lining stress states for each standard section under the effect of different seismic waves propagating through the cross section of the structure in different directions and combinations. This approach was applied to design a number of large underground constructions in seismic areas and is prescribed by the effective regulations [2].

The aim of the paper is presenting a new analytical design method for parallel mutually influencing shallow tunnel linings constructed in seismic regions.

2. The design method

Currently, there are several analytical methods of deep underground construction design, which are developed on the basis of the above-mentioned approach. The methods are as follows: multi-layer linings of circular tunnels and vertical shafts; cast-in-place linings with arbitrary cross section shape, which are designed for closed-loop tunnels, including those with the application of grouting; shotcrete linings, including those in combination with anchors; multi-layer linings of mutually influencing parallel circular tunnels of different radii and centered with respect to the right line, with the optimal distances between them being determined. All these method of underground tunnel design upon seismic effects are realized by means of a computer program and specified in technical guidelines [3, 4].

The method proposed is based on the new analytical solutions of the two plane quasi-static problems of the elasticity theory for a semi-infinite linearly deformable medium simulating the rock (soil) mass weakened by a number of arbitrary located circular openings of different radii supported by the rings simulating tunnel linings made of different materials. The design schemes are given in Figure 1 a, b.

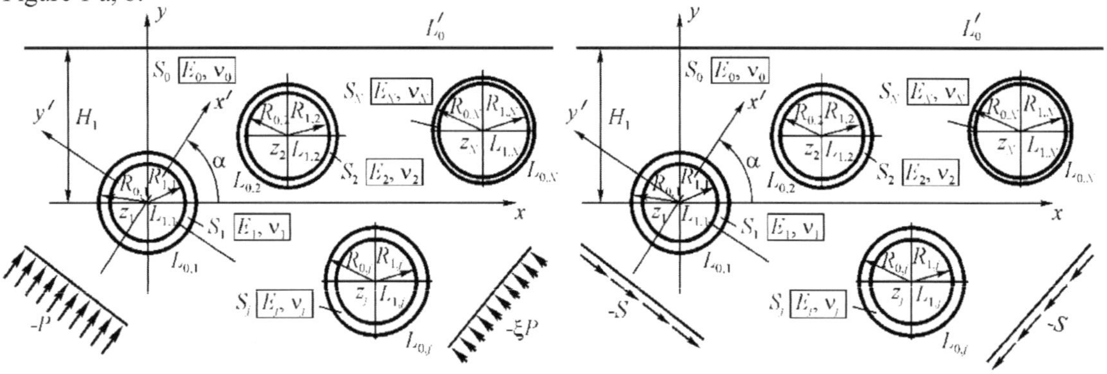

Figure 1. Schemes for designing tunnel lining upon the action of long arbitrarily directed (a) longitudinal waves and (b) shear waves

The problem of compression on the infinity with respect to two axes (fig. 1 a) is solved applying the theory of complex variable analytic functions suggested by I. G. Aramanovich [5], via analytical continuation of the complex potentials characterizing the lower semi-plane beyond the openings in the upper semi-plane across the straight boundary L_0' (the method introduced by Kolosov-Muskhelishvili [6]), and using the apparatus complex series. It allows an original iteration process to be formed, when in every approximation of the process a closed solution for a double-layer ring supporting an opening in the extended plane is obtained. The boundary conditions include additional terms reflecting the straight boundary of the semi-plane, which are represented in Laurent series with unknown coefficients specified on the basis of previous iterations.

The problem of pure shear on the infinity (fig. 1 b) is solved as a particular case of the previous one, where $P = S$, ξ = -1, with the angle $\alpha + \pi/4$ instead of α. Based on the first problem solution, the $\sigma^{(P)}$ stresses are obtained (hereinafter the σ symbol signifies all components of the stress tensor), which are caused in the linings by a long longitudinal wave falling under an arbitrary angle α. Based on the second problem solution, the $\sigma^{(S)}$ stresses attributed to the long shear wave propagating at the angle α are obtained.

Then the sum and the difference of general expressions for $\sigma_\theta^{(P)}$ and $\sigma_\theta^{(S)}$ normal tangential stresses characterizing the lining stress state due to the mutual action of simultaneously passing longitudinal and shear waves (the worst case) are investigated in every point of the internal lining cross-section outline on the extreme relative to the α angle of incidence. With this aim, the following equations are solved:

$$\frac{\partial}{\partial \alpha}\left[\sigma_\theta^{(P)} \pm \sigma_\theta^{(S)}\right] = 0 \tag{1}$$

and for every point such a combination of longitudinal and shear waves and such an angle of their falling are determined at which normal tangential stresses in the given point are maximal by their absolute. This allows analytical construction of a curve showing normal tangential stresses on the internal outline $L_{1,j}$ of each j-lining $(j = 1, 2, ..., N)$. As for the stresses on the external lining outline, the longitudinal forces N and bending moments M in every lining radial section are determined at such a combination and such a direction of waves at which the normal tangential stresses σ_θ on the internal lining outline in the given section have a maximum absolute value. The obtained stresses and forces are assumed to have signs "plus" and "minus" and are summed up with stresses and forces caused by other loads in their most unfavorable combinations. Then the section strength is tested upon compression and tension.

In case the lining is not anchored to the ground and designed with an allowance of crack formation, it is assumed that there are no tensile normal loads on it. Therefore, the effect of the longitudinal wave is not considered and the lining is designed on the basis of two curves of normal tangential stresses constructed using the maximal absolute values of the compressive (negative) and tensile (positive) stresses caused by mutual actions of both shear and longitudinal waves in the compression phase.

The above-described method represented as an algorithm of calculations and realized by means of the computer software allows identifying the most unfavorable stresses in every lining radial section of each lining with due regard to all plausible seismic effects. The method also makes it possible to identify the stress states caused by a long, either longitudinal or shear, wave of the specified direction.

Since the method is based on the assumption that quasi-static problem solutions are appropriate to design dynamic stresses in the shallow tunnel lining, with the length of the seismic wave being more than three times as much as the tunnel diameter, we compared the results obtained analytically from quasi-static model with those of numerical simulation of shear and longitudinal waves diffraction in a single tunnel lining via FLAC program [7]. The results comparison showed a satisfying agreement.

3. Examples of the design

Examples of the design of two identical parallel tunnels under seismic effects are given below. The input data are the following: $H_1 = H_2 = 8$ m; $x_1 = 0$; $x_2 = 9$ m; $R_{0,1} = R_{0,2} = 3.0$ m; $R_{1,1} = R_{1,2} = 2.65$ m; $E_0 = 200$ MPa; $\nu_0 = 0.2$; $E_{1,1} = E_{1,2} = 23000$ MPa; $\nu_{1,1} = \nu_{1,2} = 0.2$ the soil unit weight is $\gamma = 0.017$ MN/m^3, prevailing period of a soil particles oscillations is $T_0 = 0.5$ s, intensity of the Earthquake is 9 point of the MSK-scale. The linings are designed with the cracks allowance.

Diagrams of maximal compressive (negative) and tensile (positive) circumferential stresses (in MPa) which may appear along internal outline of the left tunnel lining due to seismic effects (a) and corresponding stresses in external lining outline (b) are shown in Figure 2 a, b by solid and dotted lines relatively.

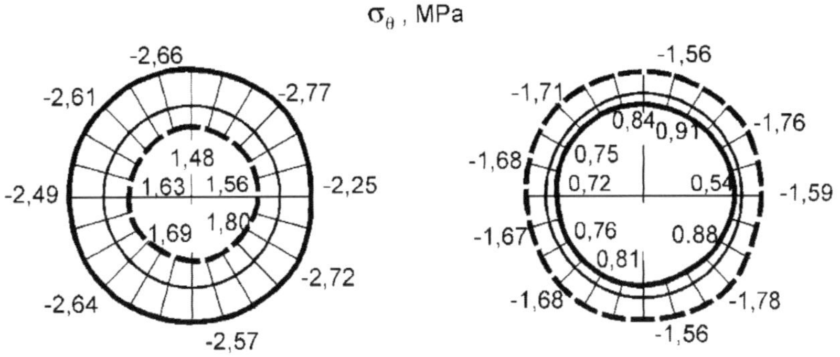

a b

Figure 2. Maximal compressive and tensile stresses along the left tunnel lining internal outline (a) and corresponding them stresses on the external outline (b) (tunnels are situated at the same depth)

The same two tunnels with centers located on a vertical straight line are considered in the next example, here $y_1 = 0$; $y_2 = 9$ m. Diagrams of maximal compressive and tensile stresses (a) and corresponding stresses along external outline (b) are shown in 3 by solid and dotted lines relatively.

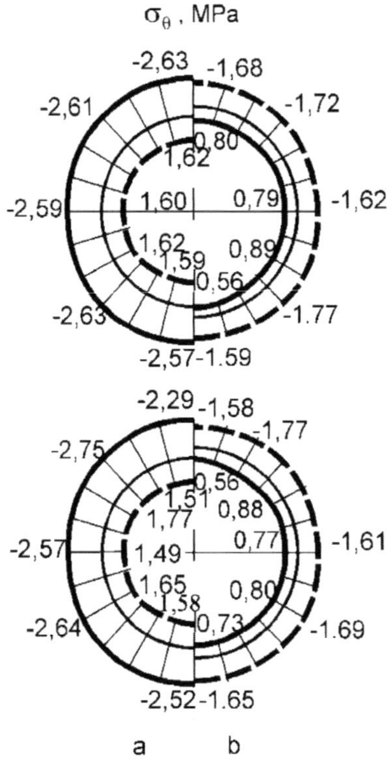

Figure 3. Maximal compressive and tensile stresses along the left tunnel lining internal outline (a) and corresponding them stresses on the external outline (b) (tunnels are situated one under another)

Another example concerns stress state computation performed for a complex of three shallow tunnels. The design scheme is shown in Figure 4.

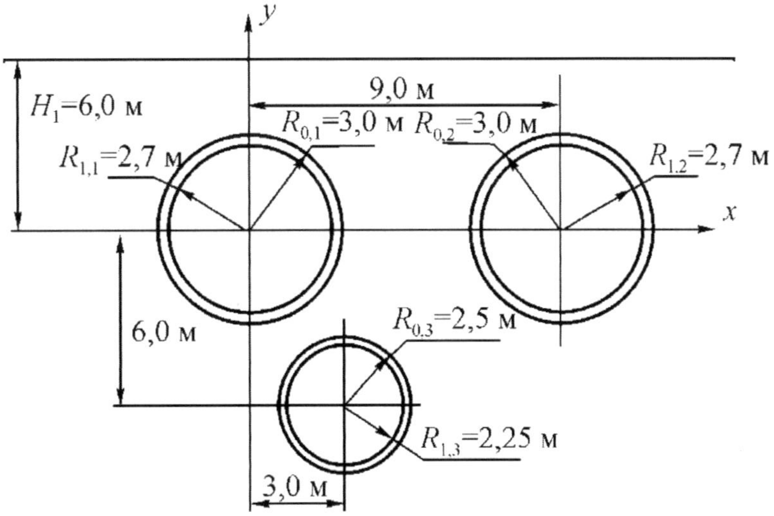

Figure 4. Tunnel size and layout

The input data are the following: E_0 = 700 MPa; ν_0 = 0.27; $E_{1,1}$ =$E_{1,2}$ =$E_{1,3}$ = 23000 MPa; $\nu_{1,1}$ = $\nu_{1,2}$ = $\nu_{1,3}$ = 0.2, γ = 0,017 MN/m^3, prevailing period of a soil particles oscillations is T_0 = 0,5 s, intensity of the Earthquake is 9 point of the MSK-scale. The linings are designed with the allowance of cracks formation.

Diagrams of maximal compressive and tensile stresses $\sigma_\theta^{(in)}$ are shown in Figure 5 by solid and dotted lines correspondingly.

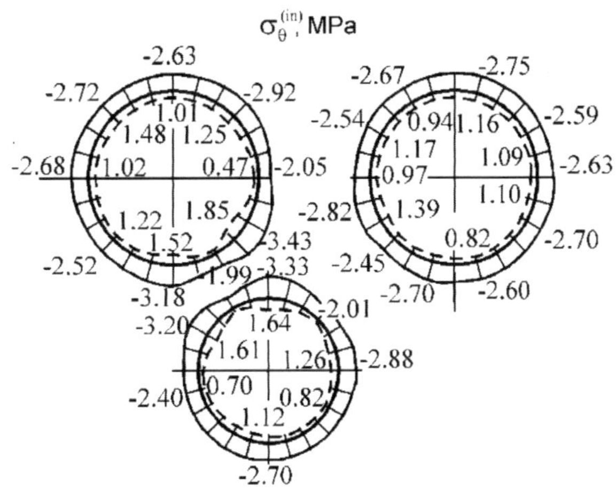

Figure 5. Distribution of maximal stresses $\sigma_\theta^{(in)}$ (in MPa) on the lining cross-section internal outline

Speaking of limitations, the above-described analytical method cannot be used to study Earth surface dynamics and currently the only adequate technique is numerical computer simulation, which was applied in the works [8, 9].

References

[1] Fotieva N N 1980 *Moscow Nedra.* Design of the Underground Structure Support in Seismically

Active Areas pp. 240.
[2] Fotieva N N, Kozlov A N 1992 *Moscow Nedra.* Design of the parallel openings support in seismic areas.
[3] *Moscow, Mintransstroy.* Instruction on account seismic effects at the design of mine transport tunnels VSN 193-811982.
[4] 1996 *Moscow Information Center TIMR*. Guide on designing underground structures in seismic areas pp. 80.
[5] Aramanovich I G 1955 *Dokladi AN USSR*. Distribution of stresses in the elastic semi-plane weakened by the supported circular hole . Vol. 104 (3) pp. 372 – 375.
[6] Muskhelishvili N I 1966 *Moscow Nedra.* Some basic tasks of mathematical theory of elasticity.
[7] Fotieva N N, Sammal A S, Bulychev N S, Lemos J V, Vieira A, Sousa L R 2001 *Conf. on Computer Methods and Advances in Geomechanics.* Designing shallow tunnel linings under seismic effects Proc. of the Int. pp. 1087–1091.
[8] Strokova L A 2009 *Mechanics and Foundation Engineering.* Numerical model of surface subsidence during subway tunneling Soil. Vol. **46** 3 pp. 117–119.
[9] Strokova L A 2015 *IOP Conf. Series: Earth and Environmental Science.* Modeling of tunneling-induced ground surface movement. Scientific and Technical Challenges in the Well Drilling Progress. Vol. **24**. pp.012030 .

Improvement in accuracy of point seismic sources wavefields numerical modeling

A.V. Azarov[1,3] **A.S. Serdyukov**[1,4] **,A.V. Yablokov**[2,5]

[1] Institute of Mining SB RAS, Krasniy ave 54, Novosibirsk, 630091, Russia
[2] Novosibirsk State University, Pirogov st 3, Novosibirsk, 630090, Russia
E-mail: [3] antonazv@mail.ru, [4] aleksanderserdyukov@ya.ru,
[5]yablokov.alexandr93@gmail.com

Abstract. A new scheme of wave fields modelling from point sources in a homogeneous isotropic elastic medium is developed. To set the source the developed technique is based on the use of additional components of the finite-difference grid and the use of all the stress tensor components in the nodes. The results of the scheme employment in the two-dimensional homogeneous isotropic medium are presented; they demonstrate the numerical solutions' improvement for the source being under consideration in the article.

1. Introduction

Numerical modelling of seismic wave propagation is necessary to address a number of seismic survey problems [1]. There are various approaches of modelling. The first one is the application of analytical approaches [2] (e.g. the Haskell-Thompson matrix method) is limited by horizontally layered model of the mediums. The problems of numerical methods for solving elastic equations in the frequency domain are the increased requirements to the memory and the complexity of algorithms paralleling [3]. The finite-difference method in the time domain is universal, easy to use in the processes of implementing and parallelizing [4, 5, 6].

The so-called standard staggered grid scheme is also widely-used [4]. The storage of the various components of stresses and displacements in different nodes of stacks makes it possible to reduce the memory requirements. Nevertheless, this approach has some drawbacks - due to the fact that normal and shearing stresses are set in different nodes of the grid there may appear some problems with the accuracy of the nonisotropic seismic wave sources setting, determined by a seismic moments tensor [7,8].

This work is aimed at improving the accuracy of the numerical solutions with the help of the additional grid nodes and with the increment of all stress tensor components. In this paper we deal with the two-dimensional homogeneous isotropic medium.

The results can be employed if there are sources of seismic wave appearing as a result of cracks' disclosure and closure and the shifts along the cracks in case of the stress relaxation in the rock mass. It is a well-known fact that [9], such displacement discontinuities can be described by introducing the equivalent additives to the stress (seismic moment tensor). Modeling of this type of sources particularly important to deal with microseismic monitoring of geomechanical processes [10].

2. The finite difference scheme at the staggered grids

The modelling of seismic wave propagation is carried out in the framework of the linear elastic theory. In the case of two dimensions the system of equations is:

$$\rho \frac{\partial^2 v_x}{\partial t^2} = \frac{\partial \sigma_{xx}}{\partial x} + \frac{\partial \sigma_{xz}}{\partial z},$$
$$\rho \frac{\partial^2 v_z}{\partial t^2} = \frac{\partial \sigma_{xz}}{\partial x} + \frac{\partial \sigma_{zz}}{\partial z},$$
$$\sigma_{xx} = (\lambda + 2\mu)\frac{\partial v_x}{\partial x} + \lambda \frac{\partial v_z}{\partial z},$$
$$\sigma_{zz} = (\lambda + 2\mu)\frac{\partial v_z}{\partial z} + \lambda \frac{\partial v_x}{\partial x},$$
$$\sigma_{xz} = \mu\left(\frac{\partial v_x}{\partial z} + \frac{\partial v_z}{\partial x}\right).$$

in which v_x, v_z - displacements, $\sigma_{xx}, \sigma_{zz}, \sigma_{xz}$ - stresses, ρ - medium density, λ, μ - the Lamé parameters.

The system of equations can be solved by the finite difference method using the scheme at the staggered grids [4]. Figure 1 shows the pattern of the second order finite-difference scheme at the staggered grids for the two-dimensional case. As it is shown in the figure, the components $v_z, \sigma_{xx}, \sigma_{zz}, \sigma_{xz}$ are arranged in the half-integer grid nodes, and v_x - in the integral grid nodes. Calculation of the displacements components is performed on the entire time steps, the calculation of the stresses is performed on the half-integer time steps.

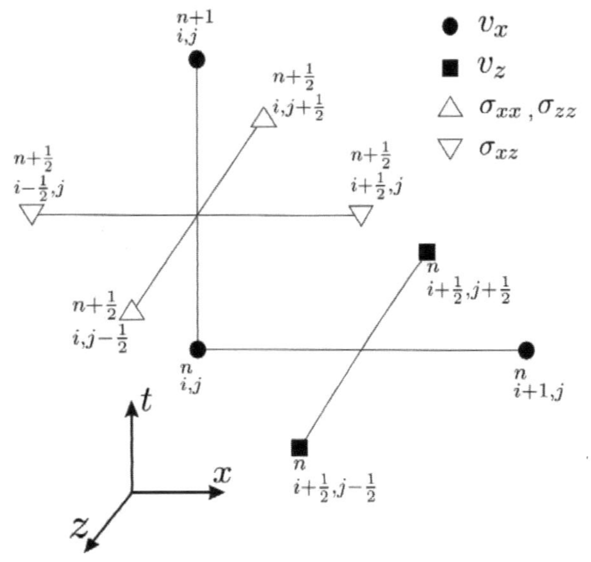

Figure 1. The pattern of the finite difference scheme of the second order at the staggered grids. i, j - space nodes, n - time nodes, v_x, v_z - displacement velocity, σ_{xz} - stresses.

The point sources are perceived as elastic waves sources, determined by a symmetric seismic moment tensor M_{pq} [9]. The components of this tensor correspond to a double pair of forces in the p direction with the moment and with the arm in the q direction. This work [8] shows that in case of the finite difference scheme it is convenient to set the couples of forces by the stress tensor. Thus, the source of the shift type with a fracture plane parallel to xy and with shifts to x directions (see Figure 2), the seismic moment tensor will have only two non-zero components M_{13} and M_{31}. To set the source with the seismic moment tensor M it is necessary to bring the increment to the stress component S_{13} in the corresponding grid node.

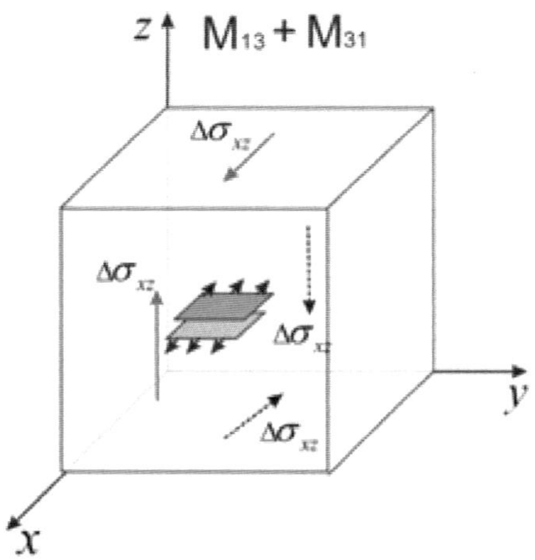

Figure 2. Stresses increments equivalent to the point shifting source.

3. Description of the source setting scheme

The proposed method is based on finding the numerical solution in the form of a sum of the fields from the nodes, in which are caused the increments of the specified stress components are produced. The procedure involves the following steps:

1. The calculation of the fields set U_S^n (U_S^n - a field from the node with number *n*, with the stress increment *S*. For two-dimensional case illustrated in Figure 1 (a), *n* goes from 1 to 9, *S* takes values S_{11}, S_{22} or S_{12}).

2. The construction $U_{sum} = \sum_{n=1}^{N}\left(\sum_S w_S^n U_S^n\right)$, in which w_S^n - the weight with which U_S^n is included in the solution.

3. Construction of the analytical solution for the selected source $U_{ан}$.

4. Construction and minimization of the functional $\|U_{an} - U_{sum}\|^2$.

After the implementation of the above steps, we obtain a set of weights w_S^n which minimizes $\|U_{an} - U_{sum}\|^2$, i.e. we obtain the numerical solution the closest to the analytical solution.

4. Example of the scheme implementation

Let us look at the example of the proposed scheme to increase the accuracy of the numerical solution, in which it will be used by 9 neighboring nodes of the grid (Figure 3).

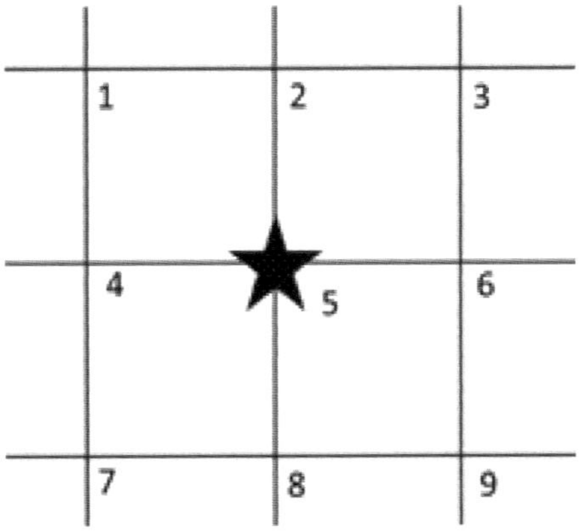

Figure 3. Grid nodes used for the source setting

As an example, we shall take the source, like the separation crack, which is set by a non-zero M_{11}. Let U be the first component of the displacement field. Then, the analytical solution for the first displacement component $U_{ан}$ in the frequency domain has the following form: $U_{ан} = M_{11} * G_{11,1}$, in which Green's function $G_{11}(\mathrm{x,y}) = \frac{1}{i8\rho}(\mathrm{A}-2\mathrm{B}\gamma_1\gamma_1)$, $A = \frac{H_0^{(2)}(\mathrm{qr})}{\alpha^2} + \frac{H_0^{(2)}(\mathrm{kr})}{\beta^2}$, $B = \frac{H_2^{(2)}(\mathrm{qr})}{\alpha^2} - \frac{H_2^{(2)}(\mathrm{kr})}{\beta^2}$, $q = \omega / \alpha$, $k = \omega / \beta$, $r = \sqrt{x^2+y^2}$, $\gamma_1 = x / \sqrt{x^2+y^2}$, $H_n^{(2)}$ - the second order Hankel function, α and β longitudinal and shear waves velocities.

Numerical solution for this source will be sought in the form of a sum:

$$U_{sum} = \sum_{n=1}^{9}\left(\sum_{S=S_{11},S_{12},S_{22}} w_S^n \; U_S^n \right)$$

in which U_S^n - the field obtained numerically from the node number n, wherein the stress S increment is performed; index node runs the grid nodes close to the source 1,2..9 (Figure 1.); index S runs the stresses with the increments (S_{11}, S_{12}, S_{22}); w_S^n -weight coefficients with which field U_S^n is included in U_{sum} .

The aim is to find the weights w_S^n with which the norm $\|U_{an} - U_{sum}\|^2$ takes the minimum value. If we consider L_2 being the norm, then $\|U_{an} - U_{sum}\|_2^2$ is a second degree polynomial from the variables w_S^n, the number of which is equal to $9 \times S$. The polynomial type is:

$$\|U_{an} - U_{sum}\|_2^2 = \sum_{n_1=1}^{9}\sum_{n_2=1}^{9}\sum_{S=S_{11},S_{12},S_{22}}\left[\sum_{i,j}(U_S^{n_1})_{ij}(U_S^{n_2})_{ij}\right] w_S^{n_1}\, w_S^{n_2} -$$

$$-2\sum_{n=1}^{9}\sum_{S=S_{11},S_{12},S_{22}}\left[\sum_{i,j}(U_S^n)_{ij}(U_{an})_{ij}\right] w_S^n + \sum_{i,j}(U_{an})_{ij}$$

in which n_1, n_2 run the nodes 1..9, index S corresponds to the stress components, which get the increments; *i,j* run all the computational grid nodes. To find the minimum point of the resulting polynomial we shall differentiate it in accordance with each variable $\frac{\partial}{\partial w_S^n}\left\|U_{an}-U_{sum}\right\|_2^2=0$. So we obtain $9\times S$ of the linear equations to find $9\times S$ of the unknown w_S^n. We obtain the weights after finding the solutions to the linear equations system. These weights, the medium from the node n with the increment and the stress components with the S index are included in the numerical solution.

As a result, after the implementation of the scheme the following results are obtained. As we are searching U_{sum} through 9 nodes with the increment S_{11}, the following weight coefficients are obtained: at the node 1,3,7,9 – 0.6056; the node 2, 8 – 0.9366; 4, 6 – 0.6510; 5 – 1.9807, i.e. the received values are symmetrical to the central node. In this case, the relative error at the norm L_2 is 0.1333. When the source is set through one node the relative error is 0.1402. When we are searching for U_{sum} through 9 nodes and the sources not only from the increment S_{11}, but also from S_{12} and S_{22} the relative error is reduced to 0.0746. Table 1 shows the relative errors in the various norms and with different methods of searching for the field U_{sum}, which are calculated according to the formula $e_r=\left\|U_{sum}-U_{an}\right\|_2/\left\|U_{an}\right\|_2$.

Table 1. Relative errors at various norms.

Method of the source setting	Norm L_1	Norm L_2	Norm L_∞
Through one node	0.16	0.1402	0.1359
Through 9 nodes (the sum of 9 fields from the sources with the component S_{11} increment)	0.1545	0.1333	0.1424
Through 9 nodes from the sources with the components S_{11}, S_{22}, S_{12} increment	0.0894	0.0746	0.077

The pictures below the diagrams comparing the analytical (red) and the numerical (blue) solutions of (blue) for the given source. In Figure 4, the source is set through one node with the component increment S_{11}. In Figure 5 the source is set through 9 nodes with all the components increment. In the second case the solutions match is the best.

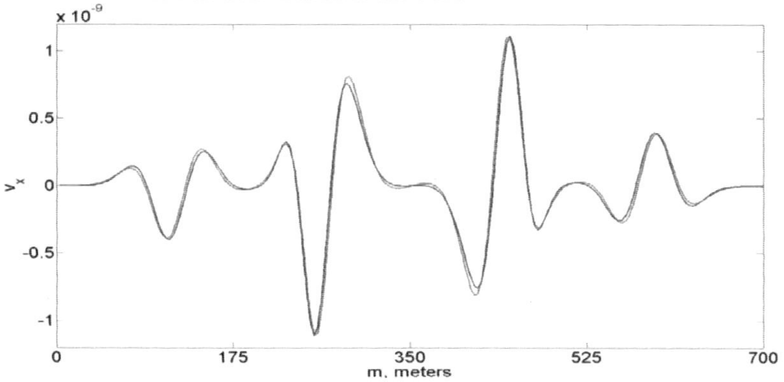

Figure 4. Displacement field v_x from the source. Blue – analytical solution, red – numerical solution determined through the setting of the source in 1 node.

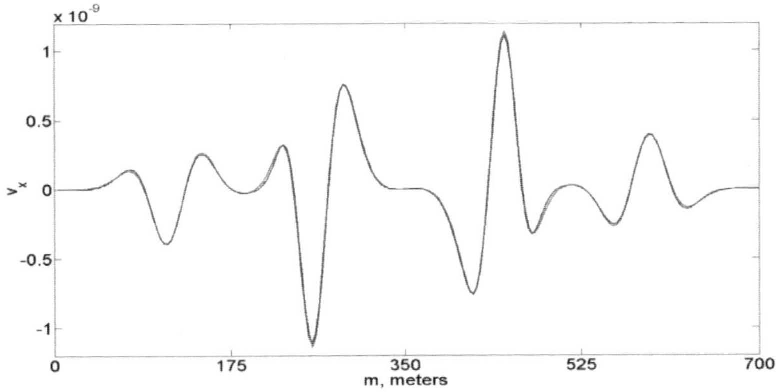

Figure 5. Displacement field v_x from the source. Blue – analytical solution, red – numerical solution determined through the setting of the source in 9 nodes with all 3 stresses increments.

5. Conclusion.

The results show that the proposed method of work helps to increase the accuracy of modelling of the wave fields from the point sources. For the further improvement of the accuracy it is possible to use a larger number of grid nodes to set the sources. Also it is possible to apply the same method for calculation of the weight coefficients and to widen it for the three-dimensional case.

Acknowledgements

The reported study was funded by RFBR according to the research project No. 16-35-00513 мол_а.

Reference list

[1] Boore D M 1972 *Methods in computational physics*. Finite difference methods for seismic wave propagation in heterogeneous materials. Vol. **11** pp. 1-37.

[2] Bouchon M 2003 *Pure and applied Geophysics*. A review of the discrete wavenumber method. pp. 445-465.

[3] Pratt R G 1990 *Geophysics*. Frequency-domain elastic wave modeling by finite differences: A tool for crosshole seismic imaging. Vol. **55.5** pp.626-632.

[4] Virieux J 1986 *Geophysics*. P-SV wave propagation in heterogeneous media: Velocity-stress finite-difference method. Vol. **51. 4** pp.889-901.

[5] Coutant O, Virieux J, Zollo A 1995 *Bulletin of the Seismological Society of America*. Numerical source implementation in a 2D finite difference scheme for wave propagation. Vol. **85.5** pp. 1507-1512

[6] Bohlen T. 2002 *Geosciences*. Parallel 3-D Viscoelastic finite difference seismic modelling. Vol. **28. 8** pp. 887-899.

[7] Aki K, Richards P 1980 Quantitative Seismology: Theory and Methods Vol. **1** pp. 520

[8] Moczo P 2007 *Acta physica slovaca*. The finite-difference and finite-element modeling of seismic wave propagation and earthquake motion. Vol. **57.2**

[9] Saenger E H, Bohlen T 2004 *Geophysics*. Finite-difference modeling of viscoelastic and anisotropic wave propagation using the rotated staggered grid. Vol. 69.2 pp. 583-591

[10] Kurlenya M V, Serdyukov A S, Azarov A V, Nikitin A A 2015 *Journal of Mining Science*. Numerical modeling of wavefields of microseismic events in underground mining. Vol. **51.4** pp. 689-695

IOP Conf. Series: Earth and Environmental Science **43** (2016) 012021 doi:10.1088/1755-1315/43/1/012021

Electrophysical parameters and NMR-characteristics of cryogel

M Shumskayte [1], A Junasheva[2], T Eltsov[1,3], N Golikov[1]

[1]Laboratory of Borehole Geophysics, Trofimuk Institute of Petroleum Geology and Geophysics of Siberian Branch of Russian Academy of Sciences, Koptuga ave., 3, Novosibirsk, Russia
[2]Physics and Technology Faculty, Novosibirsk State Technical University, Karla Marksa ave., 20, Novosibirsk, Russia
[3]Novosibirsk State University, Pirogova str., 2, Novosibirsk, Russia

Abstract. There has recently been great interest in the cryogels based on polyvinyl alcohol. Cryogel usage allows improving existing materials and creating new ones. Cryogel is completely safe, both for humans and for the soil. It protects the soil from drying out and erosion and enhances plant growth, making them more adaptive to the environment. Under the conditions of the Far North, cryogels might be used to strengthen building structures in order to prevent destruction. There is a question of monitoring cryogel distribution in the treated environment. To do this, one needs to know the physical properties of cryogel and composites with ground on its basis. This work aims to study the electrical properties and pore space structure of the composites made from ground and cryogel at various temperatures and various amounts of freezing cycles.

1. Introduction.

Cryogel is a colorless aqueous solution based on a water-soluble polymer, which changes from viscous-elastic polymer to the state characterized by a substantial reversible deformation after passing through the freezing and defrosting cycles. A significant challenge in building in the conditions of the Far North is the construction of foundations and roadways. Because of a large amount of sediments in such areas the foundation of building structures is eroded, with water flows carrying away building material. Measures are needed to strengthen the soil and to create thermal insulation of permafrost to avoid subsidence. For these purposes, it is advisable to apply the new technologies of construction, one of which is the usage of cryogel [1-3] in the Far North in order to prevent the destruction of structures. As a result, there is a question of monitoring cryogel distribution in the treated environment. To map the distribution of cryogel in the soil it is necessary to study the physical properties of cryogel and ground composites. There is also a question of describing the structural features of such ground composites. In this paper, we study the dependence of the electrical resistivity of the ground composites with different amount of cryogel on temperature, with different numbers of freezing/defrosting cycles. One can study the structure of the ground composites with a variety of physical methods, such as NMR-relaxometry, which recommended itself for studying the structure of the pore space of porous media [4-6].

Thus, in this paper we present the results of a study of the physical properties of ground composites with different contents of three types of cryogel, namely, electrical resistivity and NMR-characteristics depending on the number of freezing/defrosting cycles in the temperature range from + 20°C to -17°C. There have been a lot of works aimed at studying physical and chemical properties of cryogels. At the same time, the dependence of electrical resistivity and NMR-characteristics on the temperature in the published literature does not occur.

2. Subject of research.

IOP Conf. Series: Earth and Environmental Science **43** (2016) 012021 doi:10.1088/1755-1315/43/1/012021

We study the sand that is used in the construction of the Yamal-Nenets Autonomous District (YaNAO) and cryogel of different chemical composition (Table 1). The soil used to create composites has a density of 1.657 g/cm^3, mineral density 2.65 g/cm^3 and porosity of 37.47%. Preparation process consisted of two phases: weighing and mixing. The sand and cryogel were blended in the right proportions until being smooth. The maximum proportion of the cryogel was 35%. It is necessary to completely fill the pore space of the ground used. In addition, composites with 25% and 15% content of cryogel were prepared. The prepared samples were stored at room temperature in a closed container.

Table 1

Cryogel number	Description
1	10% polyvinyl alcohol solution (density 1.022 g/cm^3)
2	10% polyvinyl alcohol solution with boric acid (density 1.083 g/cm^3)
3	10% polyvinyl alcohol solution and borax (density 1.064 g/cm^3)

3. Measurement of the electrical properties.

Resistivity measurements were performed using LCR Meter E7-8. It is known that this device is highly reliable and suites for harsh operating conditions. The main disadvantage is that the measurements are carried out only on one frequency of 1000 Hz. E7-8 meter measures the parameters of capacitors, inductors, and resistors. The measuring principle is based on the method of bridge balancing with balancing phase change detectors. First class accuracy electronic scales were used for weighing samples. The error of the analytical balance is ± 0.5e during the initial calibration and ± 1.0e during operation. Three textolite 2-electrode cells with different inner diameters d_1 = 1.96 cm, d_2 = 1.86 cm, d_3 = 1.96 cm and a height of 2.96 cm were made. The cells contained a built-in electronic temperature sensor (DS18B20). It enables to determine the temperature in the range from -55°C to +125°C. The measurement accuracy is ± 0.5°C. The samples of ground composites were placed in the measurement cell and in the freezer where measurements were conducted. After reaching the minimum temperature, the cells were taken out and the resistivities of the samples during the natural thawing were measured. The electrical resistivity was calculated using a standard formula:

— where - electrical resistivity; R- resistivity; S - cross sectional area; L - length of the sample [7].

It follows from the measurements that ground composites based on cryogels №3 have the highest resistivity. During freezing the electrical resistivity of the sample reaches a maximum in the temperature range from -7 to -10 °C. After reaching a maximum with temperature drop, resistivity drops 2-3 times when reaching the minimum temperature of the freezer (- 17°C). During thawing, the maximum electrical resistivity of ground composites is observed in the temperature range from -2°C to 0°C, after this electrical resistivity decreases to initial values, which were obtained at room temperature. The results of the measurements of pure cryogel № 1 are presented in Figure 1.

When pure cryogel is freezed/defrosted, a hysteresis is observed in the dependence of electric resistivity from temperature. With an increase in the number of freezing cycles the form of the curve is saved, and the values of electrical resistivity change only in a small range.

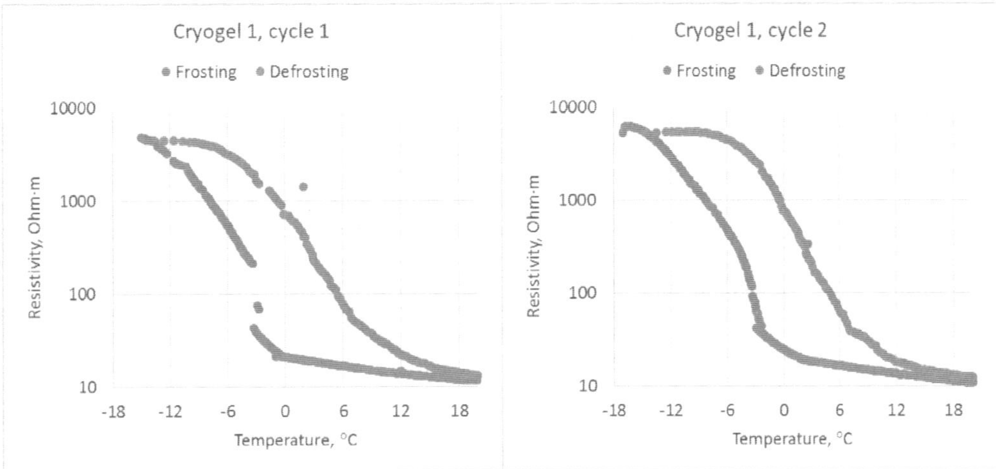

Figure 1. Dependence of the electrical resistivity of cryogel №1 from temperature during freezing, defrosting. The first cycle is on the left, the second cycle is on the right.

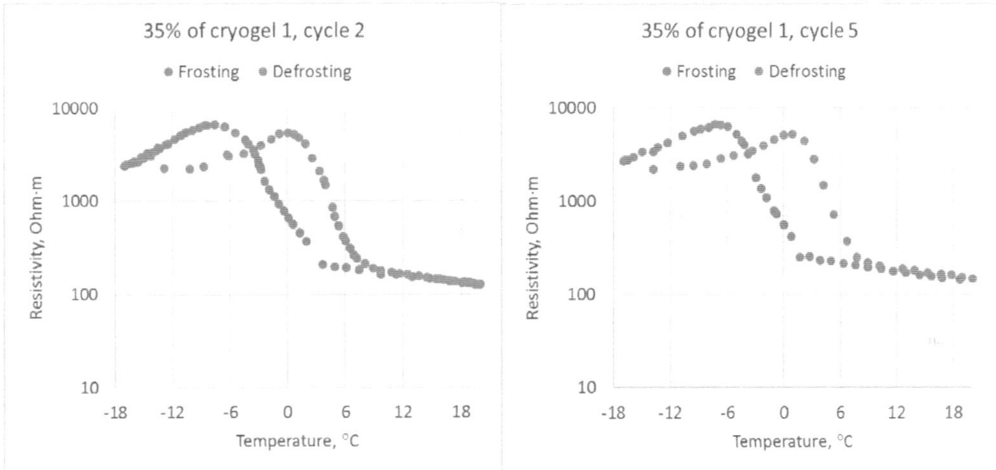

Figure 2. Dependence of the electrical resistivity of the soil composite containing 35% of cryogel №1 from temperature. The second cycle is on the left, the fifth cycle is on the right.

The samples with 15%-content of cryogel №3 have the highest electrical resistivity; the dependence of resistivity on temperature remains despite the number of cycles. Such results seem not to have been published before, and the dependence of the resistivity of the ground composites from temperature differs from a similar dependence in case of pure cryogel (see. Figures 1, 2 and 3). The measurement results show that at temperatures below -15°C the values of electrical resistivity differ by several times from those at room temperature. At the same time, the values obtained are different from the electrical resistivity of frozen sand at low temperatures (over 15000 ohmm [4]), allowing one to regard it as an identifying feature. Figures 2 and 3 show that when the samples experience one freezing/defrosting cycle, the subsequent freezing cycles do not seriously affect the observed electrical resistivity values.

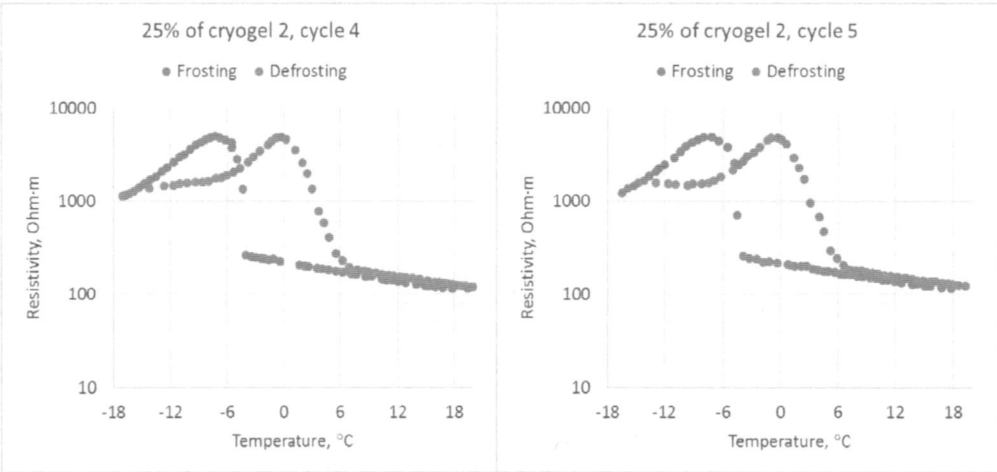

Figure 3. Dependence of the electrical resistivity of the soil composite containing 25% of cryogel №2 from temperature. The forth cycle is on the left, the fifth cycle is on the right.

4. Study of NMR characteristics.

NMR measurements were made using the relaxometer «MST-05» at a frequency range of 2.2-2.3 MHz. The NMR method is known as a reliable method for studying the pore structure of a porous medium. This method does not have a mechanical impact on a sample [8, 9].

The laboratory NMR measurements of three cryogels and four ground composites based on cryogel show that the cryogel №3 has shorter transverse relaxation time (T_2) than the cryogels №1 and №2 (Fig. 4). All three cryogels have a high hydrogen content – more than 65%.

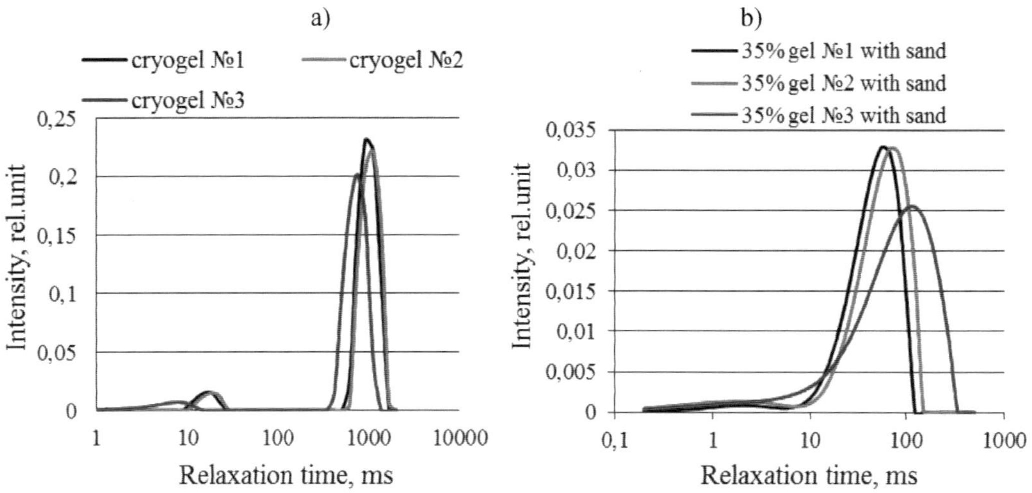

Figure 4. Spectra of the transverse relaxation times of three cryogels (a) and of their mixture with sand (b).

The Spectra show that the cryogel №3 has the lowest transverse relaxation times, i.e. it is the most viscous. All the NMR spectra are narrow, it confirms the uniformity of the samples' composition. The reverse situation is observed for the sample consisting of 35% of cryogel and 65% of sand: a mixture of sand with the cryogel №3 has the highest transverse relaxation times and the highest value of porosity index (φ), a mixture of cryogel №1 with sand has the lowest transverse relaxation times. Probable, this is due to different interaction of the cryogels with sand. The cryogel №1, as the least viscous, mixes well with sand and forms a porous sample with good reservoir properties. The cryogel

№3 does not provide a homogeneous mixture with sand, partially remaining in its original form, which leads to an increase in transverse relaxation times as opposed to the cryogel №1.

In addition, we carried out the measurements of NMR spectra of cryogels and ground compositions, depending on temperature. The measurements were performed in the following way. A sample was placed in a recipient with 20 ml volume, and then in a foam container with a build-in temperature sensor (DS18B20), which was placed in the freezer with a minimum temperature -8°C. Then NMR measurements were carried out with a measurement interval of 3°C. Then the container was removed from the freezer and the NMR measurements were carried out during natural defrosting of the ground composite sample with the same interval of 3°C.

During frosting we observe a decrease in transverse relaxation times at the temperature of -2°C for ground composite and at the temperature of +5°C for cryogel. The relaxation times remain constant at these temperatures, which indicates the invariability of pore structure and state of cryogel, which is a part of a porous medium. During defrosting both samples are characterized by a constant increase in transverse relaxation times. A hysteresis is observed in both cases. (Fig. 5).

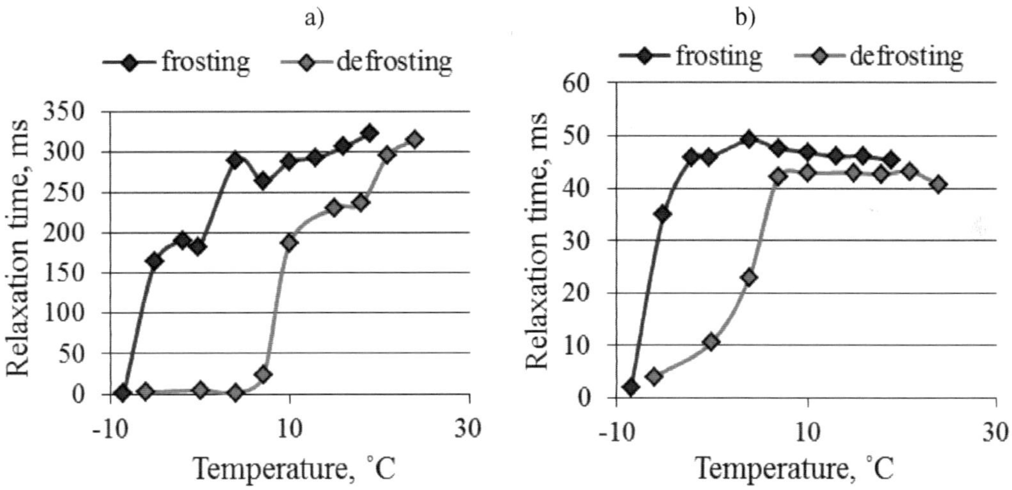

Figure 5. Temperature dependence of transverse relaxation time for the cryogel (a) and ground composite with 35% cryogel (b).

As noted previously, during frosting we observe a typical decrease in transverse relaxation times for any cryogel content in the sand. While frosting, the porosity index decreases at temperatures below 0°C, which indicates a decrease in the mobile fluid portion. The porosity index at positive temperatures slightly changes with decreasing temperature. During defrosting the porosity index gradually increases, which indicates an increase in the mobile fluid portion contained in ground composite (Fig. 6). The 3rd and 4th frosting cycles are characterized by the smallest changes of both T_2 and φ. During defrosting, each cycle is described by an increase in typical transverse relaxation times and porosity index for all the samples. All the cycles except the first are characterized by the approaching of a constant value φ and T_2 at temperatures above 0°C.

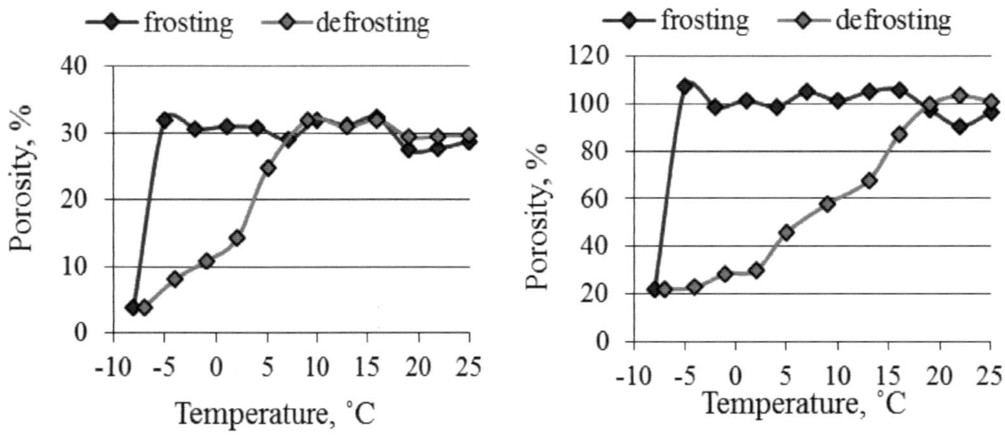

Figure 6. Temperature dependence of the porosity index for the ground composite with the 35% cryogel №3 (a) and cryogel №3 (b).

It should be pointed out that the NMR spectra of the study samples change slightly during frosting/defrosting at all the observed changes of NMR parameters (φ and T_2) except for pure cryogel. When frosting, the transverse relaxation time spectra of ground composites based on cryogel begin to move towards low T_2 values at sub-zero temperatures. At the same time, the NMR spectra of ground composites move towards high relaxation times only to temperatures about +4°C, after which the change hardly occurs. The pure cryogel spectra shift towards high relaxation times when the temperature rises. The cryogel №1 has the highest transverse relaxation times (≈1000 ms), whereas the cryogel №3 has the lowest ones (≈750 ms). During frosting, the cryogels №1, 2 also have similar relaxation times, and the cryogel №3 is characterized by high transverse relaxation times. Probably it is related to the change of its internal structure caused by temperature (Fig. 7).

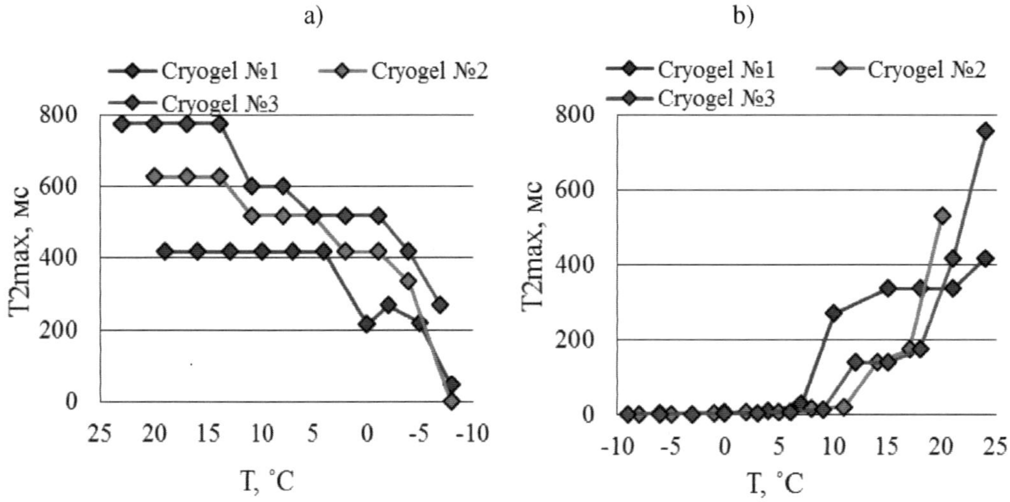

Figure 7. Temperature dependence of the transverse relaxation times corresponding to the maximum spectra amplitude when frosting (a) and defrosting (b) the cryogels.

The transverse relaxation times of all three cryogels gradually decrease with decreasing temperature (Fig. 7a). The cryogel №3 is characterized by the highest value of T_2 at both the maximum and minimum temperature. During defrosting a significant increase in transverse relaxation times is observed only at temperatures above +5°C. The cryogel №3 has highest T_2 value. Such considerable relaxation time changes are not observed in case of ground composites. The T_2 values change significantly only at temperatures below -2°C during frosting, until reaching the temperature of 0°C during defrosting.

5. Summary.

We made the temperature-dependent measurements of the electrical properties and NMR-characteristics of ground composites. The measurement results show that at temperatures below -15°C the values of electrical resistivity differ by several times from its electrical resistivity values at room temperature. After the second cycle of freezing/defrosting, the type of resistivity-temperature curve does not change, and the values vary slightly despite an increase in the number of freezing/defrosting cycles. After the 3 cycles NMR-characteristics vary slightly with an increase in the number of freezing/defrosting cycles. At the same time the values of electrical resistivity obtained are different from the electrical resistivity of frozen sand at low temperatures (over 15000 ohmm [10]), allowing one to regard it as an identifying feature. Temperature effect on the transverse relaxation times and porosity index of cryogels and ground composites based on cryogel was studied. It is found that both the relaxation time and porosity index decrease during frosting and increase during defrosting. It allows detecting a change in the mobile fluid portion contained in ground composite. NMR spectra during frosting/defrosting show alteration of a pore structure formed by cryogel and sand under the influence of temperature.

References

[1] Altunina L K, Gan-EHrdehneh T, Rozhdestvenskij E A, Svarovskaya L I, Filatov D A, and Fufaeva M S 2014 *Pedology* **Vol. 5** pp. 563-570 [In Russian].
[2] Manzhai V N, Fufaeva M S 2011 *Oil and gas* **Vol.** 6 pp. 102-107
[3] Lozinsky V I, Damshkaln L G, Kurochkin I N, Kurochkin I I 2012 *Colloid journal* 3 **Vol.** pp. 74 343-352
[4] Aksel'rod S M, Nerettin V G 1991 *Nuclear magnetic resonance in Petroleum Geology and Geophysics* (Moscow: Nedra) p. 192 [In Russian]
[5] Dolomansky Y K, Murav'ev L A 2010 *Ural geophysical messenger* **Vol.** 1(16) pp. 33-39
[6] Donaldson E, Tiab D 2009 *Petrophysics: theory and practice of study of reservoir rock properties and reservoir motion* (Moscow: «Premium Inzhiniring») p. 840 [In Russian]
[7] Kaufman A A, Anderson B I 2010 *Principles of Electric Methods in Surface and Borehole Geophysics* **Vol.** 44 pp. 1-439
[8] Dgafarov I S, Syngaevsky P E, Hafizov S F 2002 *Application of nuclear magnetic resonance method for characteristic of the consist and distribution of the reservoir fluids* (Moscow: Chemistry) p. 439 [In Russia]
[9] Coates G R, Prammer M G, and Xiao L 2001 *NMR Logging, Principles&applications* (Houston: Hulliburton Energy Services Publishing) p. 342
[10] Bogolyubov, A N 1973 The study of the dependence of the electrical resistivity of frozen rocks in the bedrock from cryogenic temperatures and textures p. 59 [In Russian]

Hydrogeological condition patterns of Kuznetsk Basin coalbed methane fields for estimating hydrodynamic calculations

Gridasov Alexander, Kuzevanov Konstantin and Bogdanova Anna
National Research Tomsk Polytechnic University, Tomsk, Russia
[1]gridasov@tpu.ru, [2]kki@tpu.ru, [3]bogdanovaag@tpu.ru

Abstract. The paper describes the natural regional conditions and key formation factors of groundwater dynamics. Characteristic hydrogeological structure elements for Kuznetsk Basin coal were identified providing parameter-oriented hydrodynamic calculations and hydrogeological models in predicting coalbed methane mining impact on the regional groundwater.

1. Introduction

A number of potential productive coalbed methane (CBM) areas have been isolated in the Kuznetsk Basin. These areas are concentrated in south Kuznetsk Basin within Erunakovsk, Tersinsk, Tutuyassk and Tom-Usinsk mining areas. Basically, these potential mining areas are confined to the brachysyncline flanks, due to relatively coalbed accessibility within the structures not subjected to weathering. Methane is accompanying - gas in coalbeds as it is generated throughout all coalification stages of organic sediments and being sorbed by parent material under proper thermodynamic conditions.

Predominately, CBM is a hazard factor in deep coal mining. Drainage and coal - seam uncovering disturb "in place" conditions and furthers methane desorption which subsequently migrates into low-pressure area, i.e. mining areas. In this case, coal producers have been dashing via underground mine ventilation. Nowadays, due to many mining techniques CBM extraction is possible before underground pit construction. This mining method is based on the identified methane properties- desorption and migration after releasing hydrostatic pressure. In this case, CBM extraction involves exposed gas seams, hydraulic fracturing to improve permeability and then in-mine pumping of formation water. As a result groundwater pressure decrease generates coal methane desorption, either dissolved or free base, migrating to the borehole. In view of this specific feature, CBM by borehole mining is inevitably accompanied by the extraction of formation water to the surface. This defines the significant hydrogeological aspects of coal-methane mining. At the stages from pilot commercial mining to extensive production, it is important to forecast the influence of new production activities on underground hydrosphere and adjacent spheres of subsurface resources.

2. Physicogeographical conditions

Hydrodynamic conditions of Kuznetsk Basin formed as a result of the interaction of various natural and anthropogenic factors. The physicogeographical conditions of studied region and influence factors on the groundwater dynamics, as well as CBM mining areas are discussed.

Kuznetsk coal basin, an intermontane basin, is located in central Eurasia: in the N-W Altai-Sayan folded zone, near West Siberian Plain border. The infrastructure for CBM by borehole mining is being developed within the central southern basin province.

Climate

The Kuzbass climate is continental, being determined by its geographical position. The climate is characterized by long and cold winter and short and hot summer. Average annual temperature is 0°C, varying between −50°C in January to +35°C in July. Total annual precipitation is about 750 mm of average normal 500 mm, whereas groundwater recharge is 100 mm/yr. Two-meter snow cover reduces ground freezing depth to 0.2-0.4 m. In general, the excessively humid climate provides favorable conditions for groundwater recharge.

Terrain

The Salair and Kuznetsk Alatau mountain systems, extend meridionally along Kuznetsk Basin border, preventing Arctic and Atlantic air mass circulations, which, in its turn, determines the climate pattern of this region. Geomorphological conditions of the Kuznetsk Basin are different, including 5 landscape zones. The studied region is situated in dissected low-mountain topography, where elevation levels between watershed divide and water course is 300 meters and the sea level varies from 180 m to 580 m. Wide watershed areas include dissected relief. The river valleys are colateral to the mountain bordering of basin indicating their neo-tectonic origin. The relief and the landscape features form favorable conditions for groundwater recharge.

3. Geological structure and lithological description

Regional geological structure and tectonic conditions determine distinctive features of the hydrogeological stratification, as they govern the subsurface filtration heterogeneity. Studying hydrodynamic conditions of CBM fields, it is very important to consider the middle and upper structural levels of the basin deposits, which are comprised of coal formation and overlying rocks.

Upper Permian (P_2) coal formations of the Kolchugino series, composed of the middle structural level are prospective for CBM mining. Kolchugino series is subdivided into Ilyinsk (P_2il) and Erunakovsk (P_2er) subseries and suites. Total coal formation thickness is more than 2000 m, while overall thickness of coal layers is about 320 m.

Lithological composition of Kolchugino series includes sandstone, aleurolite, argillite and coal. Aleurolite is more abundant, up to 60% and sandstone up to 24% within the cross-section. Lithological variations are changeable, whereas composition and thickness vary spatially lateral. Sandstone layers are rather continuous, the thickness of which reaches 50 m. Coal beds are irregular geometrical shapes: pinching-out, branching or merging.

Coal-bearing deposit structures are various. Sandstone thicknesses are characterized by cross bedding and / or swaley-cross bedding. Aleurolites exhibit cross and/or or horizontal bedding, while argillites exhibit poor horizontal bedding including layers of insignificant thickness, confined to surface coal beds. Coal beds exhibit distinct horizontal bedding and cleavage, i.e. non-tectonic fracturing. This property determines high coal bed permeability comparable to adjacent rocks.

Kolchugino series deposits are characterized by layered structure, with alternating lithotypes of different permeability. The thickness is composed of numerous small amplitude discontinuous fractures.

Upper structural level deposits do not include coal beds suitable for methane mining. Nevertheless, the authors consider these deposits to be of interest in terms of hydrodynamic structure elements. Upper structural level basin deposits include Triassic (T), Jurassic (J) and Quaternary (Q) deposits.

All bedding rocks are lithified and involve regional diagenetic or tectonic fracture system. Fracture system intensity and openness are irregular vertically and horizontally and are defined by rock stress and strain state. Highly-developed fracture openness is found in the upper cross-section at a depth of 100-150 m. This factor significantly influences the formation of regional hydrodynamic conditions.

4. Hydrogeologic characteristics

As mentioned above, the middle and upper structural levels of the Kuznetsk Basin deposits are considered being practical in studying the hydrodynamic conditions of the CBM fields. Based on lithologic-stratigraphical and hydrodynamic features the following four geological units were identified:

Aquifer system of Quaternary alluvial deposits – phreatic continuous aquifer, confined to river deposits the thickness of which varies from 0 to 10 meters. Groundwater is pore-stratal with free surface or weak artesian pressure.

Aquifer system of Jurassic deposits – located within limited areas of some syncline structures thickness of which is from 0 to 800 m. Jurassic rocks have high active porosity due to less compaction and diagenetic changes comparable to Perm deposits, resulting in rather high water inflow. According to movement, groundwater is either fissure and fissure-stratal, generally unconfined.

Aquifer system of Triassic deposits is within the Kyrgay-Ostashkinsky syncline, the thickness of which is about 570 m. Water inflow system is irregular, where groundwater is fissure and fissure-stratal, unconfined.
Aquifer system of Permian deposits is everywhere, the thickness of which is more than 2000 m. Water inflow system is irregular, where groundwater is fissure and fissure-stratal. The most permeable zones are in cross-section of prevailing sandstone and less permeable aleurolite and argillite. Coal bed permeability is significantly higher than the permeability of adjacent rocks. Graded deposit interbedding with different permeability determines the filtration heterogeneity of coal-bearing deposits. The reservoir pressure increases with depth.

Table 1. Vertical hydrodynamic zoning.

Water exchange zone	Thickness, m	Depth, m	Open rock porosity, %	Filtration coefficient Ff, m/day	Mineralisation, gr/l
Active	10-150	< 150	5-15	0,1-1,0	<1
Moderate	more than 2000	-	2-7	0.001-0,4	0.7-8

Vertical hydrodynamic zoning (table 1) within studied area is governed by various degrees of rock fracturing and fracture poenness at different depths. All bedding rocks are disrupted by regional fracture system of various genesis. Fracture openness and, consequently, rock permeability decrease with the depth [6]. In this case, there are two zones with various intensive groundwater circulation within this region cross-section: active water exchange zone and slow water exchange zone. Active water exchange zone is characterized of lateral hydrodynamic heterogeneity. Maximum rock permeability was revealed within the branched zone system fracture openness which coincides with the river network structure.

5. Data and procedures

Interpretation of CBM field hydrogeological conditions is based on field research materials [6, 7], analysis of previously published data, key reference literature [1, 2, 5, 12], scientific publications [3, 4, 13] and production reports, as well as similar target researches [9, 10, 11, 12]. The exploration results of coal, coalbed methane and groundwater in Erunakovsk, Tutuyassk and Tom-Usinsk mining areas. Hydrodynamic sampling data is the target of this research. The applied methods are based on the key principles of groundwater dynamics [8].
Natural and technogenic condition patterns for hydrodynamic modeling includes determining time mode, spatial structure and areal filtration characteristics within initial-boundary conditions. Based on the analysis results of physiogeographic features, geological structure and hydrogeological characteristic of the studied area, the main hydrodynamic structure elements were determined. The following determined parameters are necessary for developing geofiltration models of CBM fields and analytical calculations.
Decomposition of the filtration area including a set of standard elements was conducted in the following sequence:

1. detecting basic elements in filtration area.
2. determining initial-boundary conditions for each element.
3. evaluating generalized element characteristics, determining range and average values for each parameter.

6. Results and discussion

Hydrogeological conditions of the studied area are governed by the cumulative effects of natural and anthropogenic factors. Increasing development of anthropogenic activities further the significance of the technogenesis itself , which, in its turn, could change this or that nature component. Climatic, landscape and geological factors determine groundwater recharge and discharge regimes, forming the hydrodynamic conditions of considered territory. The climate and relief contribute to the active groundwater recharge which are subjected to intensive seasonal fluctuations. The fluctuation amplitude of seasonal groundwater table level decreases with the depth. Evaluating hydrodynamic processes in the moderate water exchange zone, groundwater recharge value could be constant. The infiltration in this area is about 15% of atmospheric precipitation [1]. During a year these aquifer systems embrace about 100 mm precipitation layer throughout the recharge area.

Table 2. Filtration area parameters.

Water exchange zone	Element	Filtration coefficient Ff, from-to (average), m/day		
		a Q $_{1-4}$	***J*** $_{1-2}$	***P*** $_{1-2}$
Active	Valleys	10-70 (30)	1-50 (20)	0.3-1 (0.5)
	Slopes		0.1-1 (0.5)	0.05-0.3 (0.1)
	Watersheds		0.01-0.1 (0.05)	0.01-0.1 (0.05)
Moderate	Exploited layer			0.001-0.04 (0.01)
	Adjacent rock thickness		0.01 lateral 0.001 in cross-section	

Geological factors determine the filtration flow structure and intensive water exchange. The geological structure features and regional tectonic conditions of the region exhibit distinct vertical hydrodynamic zonation section (tables 1, 3) and lateral filtration heterogeneity which could be traced in active water exchange zone (table 2).The regularities are: decreasing section permeability with the depth (table 1). In the active water exchange zone most permeable rocks are confined to the valleys of surface stream flows, comparable to watershed areas which are characterized by significantly less permeable rocks [7].

Coal-bearing beds at considerable depths are characterized by sharp filtration heterogeneity, which is conditioned by different permeability variations of the layers (Ff = 0.001-0.04 m/day) composing the section thickness. The rock permeability in the moderate water exchange zone is vertically less than in subhorizontal direction confined to the surfaces of the lithologic boundaries. Thus, vertically, section permeability is governed by low permeable layers (from 0.001 m/day), and subhorizontally - permeable ones (up to 0.04 m/day) [7].

So, the section of coal-bearing deposits including graded interbeddings of argillite, aleurolite, sandstones and coal beds could be depicted as single-layer thickness with vertical and horizontal inhomogeneous permeability.

Table 3. Hydrodynamic structure elements of coalbed methane deposits.

Water exchange zone	Element	Boundary conditions	Notes
Active	Watershed	II, Q = 0	Suitable for establishing of filtration area boundaries
	Stream flow	III, H = const	
	Terrain	II, Q = const	Infiltration recharge
	Bottom border	II, Q = 0	Conventional impenetrable
Moderate	Borehole	I, H = const, Q = f(t)	Pumping parameters
	Upper boundary	I, H = const	Conventional defined, beyond pumping influence zone.
	Bottom boundary	II, Q = 0	
	Model perimeter	I, H = const	

Groundwater regime in the active water exchange zone is subjected to seasonal fluctuations, which are connected with the change of recharge and discharge conditions during a year. This can be clearly identified by the influence of river network and watersheds and infiltration stream flow (table 3). Modeling active water exchange zone is possible as a single-layer thickness exhibiting zonal permeability increase from watersheds to valleys. Perimeter of the modeled area could be confined to the 2nd and 3^{d} boundary types.

Moderate water exchange zone, where CBM beds are located, do not exhibit noticeable influence of the recharging hydrodynamic boundaries. Due to the remoteness of the recharge and discharge areas decrase the influence of the lateral boundaries. The influence of hydrodynamic interaction decreases with the depth due to the significant thickness of low permeable deposits, separating the coal bed from the active water exchange zone and surface water. In this case, the developed thickness could be modeled as unlimited and limited in the layer cross-section.

7. Conclusion

The analysis of physiogeographic characteristics, geological structure and hydrogeological conditions of the central southern Kuzbass allows formulating the principles of hydrodynamic generalization of CBM production fields and exhibits the following results:

1. determined major regime-forming factors and conditions of their interaction;
2. identified initial-boundary conditions for task-solving;
3. identified regularities of the heterogeneous structure in filtration area; proposed interpretation methods of this heterogeneity in analytic models; and
4. enumerated hydrodynamic structure elements and their characteristic properties.

Proposed concept of hydrodynamic conditions in CBM field is important in the development of numerical hydrogeological models for solving prospecting tasks to determine the hydrodynamic behavior of coal methane wells and assess their possible influence on other hydrosphere targets. Generalized parameters of filtration area and elements of hydrodynamic structure (tables 2 and 3) could be used at the formation stage of conceptual CBM field model. These parameters should be determined in terms of model calibration based on factual observations of the groundwater regime on a modeled object.

References

[1] Hydrogeological map of the USSR. Scale 1:200 000 1965 Sheet N-45-XXII Explanatory letter Novokuznetsk p 103

*[2]*Hydrogeological map of the USSR 1972 Kemerov Oblast and Altai territory **Vol.** 17 Moscow Nedra

[3] Hydrogeology of Erunakovo area in Kuznetsk Coal Basin Dissertation 1967 Tomsk

[4] *Gidrogeologija Osinovskogo rajona Kuzbassa i uslovija otrabotki uglej pod obvodnjonnymi Jurskimi otlozhenijami* Dissertation 1965 ed. Plevako G Tomsk Polytechnic University Tomsk

[5] State geological map of the Russian Federation 2007 Scale 1:1 000 000 Altai-Sayan series. Sheet N-45 - Novokuznetsk. Explanatory letter

[6] Gridasov A 2014 *Problems of Geology and Mineral Resources: Proceedings of the 18th International Symposium after academician M.A. Usov* Rezul'taty gidrodinamicheskih issledovanij strukturnoj skvazhiny № SN-15 na Chaltokskom uchastke rabot po poisku ugol'nogo metana (Kuzbass) **Vol.** 1 pp. 442-444

[7] Gridasov A 2015 *Problems of Geology and Mineral Resources: Proceedings of the 19th International Symposium after academician M.A. Usov* Gidrogeologicheskie uslovija Podobassko-Tutujasskoj depressii v svjazi s perspektivoj dobychi metana iz ugol'nyh plastov (Juzhnyj Kuzbass) pp. 372-374.

[8] Lenchenko N 2004 *Groundwater Dynamics* Moscow

[9] Nikitenkov A, Dutova E and Pokrovsky D 2016 *IOP Conference Series: Earth and Environmental Science* Performance assessment of groundwater drainage basin in small-sized mountain rivers based on DTM **Vol.** 33

[10] Pokrovsky V, Dutova E, Kuzevanov K, Pokrovsky D, and Nalivaiko N 2015 *IOP Conference Series: Earth and Environmental Science* Hydrogeological Conditions Changes of Tomsk, Russia **Vol.** 27

[11] Popov V, Kozina M, Levak Yu and Shvagrukova, E 2016 *IOP Conference Series: Earth and Environmental Science* Ecologization of water-land property matters on the territory of the Tom lower course **Vol.** 33

[12] Purgina D, Strokova L and Kuzevanov K 2016 *IOP Conference Series: Earth and Environmental Science* Modeling of changing hydrogeological conditions during construction of pier foundations on the Kama river bank **Vol.** 33

[13] Rogov G and Popov V 1985 *Hydrogeology and catagenesis of the Kuznetsk Basin rocks* [in Russian] Tomsk University Tomsk

[14] Shvartsev S, Khryukin V, Domrocheva E, Kuzevanov K, Rasskazov N, Popova T, Lepokurova O, and Shvachko E 2006 *Geology and Geophysics* Hydrogeology of the Erunakovo area of the Kuznetsk Basin in the context of the problem of coal methane formation and mining **Vol.** 47 (7) pp. 881-891

Classifications of bog peat sensitivity to anthropogenic impact (Western Siberia)

V V Kramarenko[1], I A Matveenko[2], A.N. Nikitenkov[3], V Y Molokov[4], A P Khoroshko [5]

[1] Associate professor, Tomsk Polytechnic University, 30 Lenin Av., Tomsk, 634050, Russia

[2]Associate Professor, Tomsk Polytechnic University, 30 Lenin Av., Tomsk, 634050, Russia

[3] Senior Teacher, Tomsk Polytechnic University, 30 Lenin Av., Tomsk, 634050, Russia

[4] Post-graduate. *Tomsk Polytechnic University, 30 Lenin Avenue, Tomsk, 634050, Russia*

[5] Student, Tomsk Polytechnic University, 30 Lenin Av., Tomsk, 634050, Russia

e-mail: [1]kramarenko-v-v@mail.ru; [3]mia2046@yandex.ru; [4]vik3011347@yandex.ru;

Abstract. The article deals with strength parameters of peats in Western Siberia, evaluates their transformations under the anthropogenic mechanical impact, presents peat classification in terms of sensitivity allowing the forecast of strength loss when destructing their structure in the process of building roads, pipelines. Sensitivity classification also permits predicting roadability for construction design.

1. Introduction.

In the territory of Western Siberia soft soils, such as peat, silt, sapropel, quick clay, are widely distributed. They are rather sensitive to anthropogenic impact when developing wetland areas. Reliable impact forecast design would prevent serious problems and permit calculating properly a bearing capacity of structure base. Sensitivity indicator (S_t) is of particular interest as it characterizes peat properties, the structure of which has been destructed as a result of site stripping in oil field development, partial peat reclamation while building, chiselling and trimming polluted top layers with washing under hydraulic pressure, land remediation, transport passage (off-roaders, tractors, grubbers, dozers, etc.).

The purpose of the article is to study transformations of peat strength properties, their evaluation in remolded (c_{ud}) and undisturbed states (c_u), classification in terms of peat sensitivity to mechanical effects in peat lands as well as unnamed peat massifs in the oil-gas field sites of Tomsk Oblast and Khanty-Mansiysk Autonomous District. The tasks include studying engineering-geological conditions of the territory; processing the results of peat vane field test; investigation of changes in strength characteristics of peat formation – before and after structure deformation; revealing the relationships between depth and strength characteristics, peat classifications in terms of strength properties and sensitivity.

The study of peat shear strength characteristics and peaty soil was carried out in the works by Amaryan L.S., Yevgeniev I.Ye., Kazarnovsky V.D., Shaposhnikov M.A. as well as Kanmuri, H., Landva, A.O, Marachi, N. D., Mesri, G. Ogino, T., Oikawa, H., Tsushima, M., Mitachi, T , Miyakawa, I, Radforth, J. , Sodha V.G., Helenelund K.V. [1-13]. Anthropogenic impact of oil pollution and its influence on soil physical properties are considered in the work [14]. Among the widely used classifications of strength properties according to the vane test defining the building type of soft soil in groundwork base are those by Shaposhnikov M.A., Yevgeniev I.Ye., Kazarnovsky V.D., et al [3, 11, 15], roadability classifications [1, 16], soft soil sensitivity classifications [17-21].

2. Methods

Within the problems being solved, the thickness of boggy deposits has been stated, the parameters of peat reaction has been studied at peat deposits Klukvennoye (upland and lowland deposits), Plotnikovskoye, Vasyugan (site 5, Gavrilovka), Chelbak-2, Tagan, Ust-Kandinskoye and others, the density (ρ), the density of the solids (ρ_s), the density of the dry soil (ρ_d), water content (w), the parameters of the state, such as the void ratio (e), porosity, saturated limit, ash content (D_{as}) and the degree of decomposition were obtained by the gravimetric method (D_{dp}). Work was carried out in accordance with the methods provided in the existing regulatory standards [22].

In boggy sites peat was tested by vane shear test apparatus in 0.5 m through the whole peat formation thickness according to GOST 20276-2012. The foreign analogue to the method of rotary cutter to determine soil undrained resistance is vane shear test [22]. Field tests for determining undrained shear strength of soil with natural (c_u) and remolded structure (c_{ud}), as well as sensitivity (S_t) were carried out using vane shear test, according to [23, analogues 24].

Vane shear test allows determining the peat natural strength in formation after destruction of its structural bonds by rotary blades of the vane tester. The natural maximum peat strength (c_u) is determined by dependence curve peak of test rotational moment on soil deformation, i.e. angle of vane rotation in the formation. After destruction of natural soil texture and its structural bonds by the vane tester the steady-state strength (c_{ud}) is left, which is conditioned by water-colloid bonds of particles. This strength characterizes residual shear strength with which soil can resist to load after its structural bond destruction by construction machinery.

The relationship of maximum natural and steady-state values, i.e. the so called sensitivity index ($S_t=c_u/c_{ud}$) and difference in values of maximum and steady-state strengths ($S_t'=c_u-c_{ud}$), characterizes relative and absolute strength of structural bonds (chemical, ion, covalent, etc.), composition properties, compression, and anthropogenic impact rate, as well as, indirectly, gives data on lithification rate of soft soils and their structures.

3. Results and discussion

The results obtained have shown that the value of shear strength in natural state changes from 4.2 to 44.6 kPa, in remolded state from 0.01 to 16.1 kPa (fig. 1). The minimal values of S_t are of greatest interest as they define mechanical cohesion of peat particles – the less the value, the better the cohesion. The minimal difference of it is found in fuscum, magellanicum peat, and cottongrass-sphagnum, which indicates high cohesion of peat particles with sponge texture even in remolded state. Being more decomposed, small size particles of other peats loose most of their bonds after their natural structure destruction, long fibers of sphagnum moss often remain undisturbed. The results of correlation analysis show that there are the closest significant relationships between S_t and porosity coefficient e (correlation coefficient r=-60) that demonstrates low cohesiveness of peat particles with low porosity (e <12–13) containing high amount of decomposed elements: sedge, woody-sedge, wood. The minimal values of S_t are observed for sphagnum peat with porosity coefficients ranging from 16–17 to 20, but at higher porosity coefficients the values of S_t grow again, as c_u and c_{ud} decrease, and the number of contacts among the peat particles is likely to decrease when destructing the peat structure. The changes of S_t with the increase of ρ_s are closely connected with variations of ρ_s in peats of different botanical composition. The increase in S_t after 1.70 g/cm^3 is connected with the decrease in c_{ud} with growth of D_{as}.

According to the strength properties classification [3], the study peats are referred to I and II construction types, in remolded state they are mainly referred to II–III types. In site stripping, tree and bush roots serving as reinforcement for improved roadability are removed. For example, after peat structure destruction, cross-country capability of peat deposits Klukvennoye and Chelbak-2 are referred to heavy ones in terms of the Amaryan classification [1], whereas that of rest deposits – to particular ones. Those aspects are necessary to take into account at wetland operation performance to avoid heavy traffic and long transport stops in such sites.

As has been noted, soil classification is performed separately in terms of the following indicators c_u and S_t. When soft soils lose their strength in the remolded state, the classification including strength and sensibility can be used to group them. It may be the classification models combining c_u and S_t or c_u and

S_t' [1, 3, 11, 15-21] in different ways. They are intended for site investigation of linear structure construction.

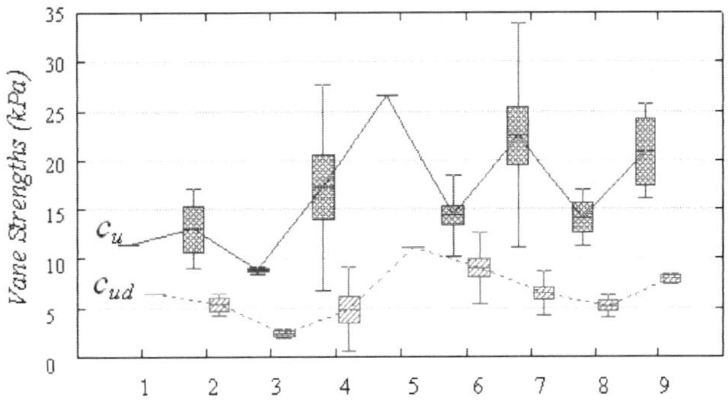

Fig. 1. Distribution of shear strength of different peat types: 1 – cottongrass-sphagnum, 2 - magellanicum, 3 – sphagnum-ridge, 4 – sedge, 5 – grass, 6 – fuscum, 7 –woody-sedge , 8 – sedge-hypnaceous, 9 - hypnaceous.
"Boxes" denote standard errors, dotted line – standard deviations, lines – average values of c_u and c_{ud}

Let us consider the examples of peat classifications using the results of vane test (fig. 2) obtained from peat deposits in landscaping development of Mamontov oil field. In terms of their inflow regime and location the bog are referred to moss type confined by flow runnels and topographic lows. The bog deposits include low-ash peats of different decomposition degree. As a result of geological survey, the peat deposit thickness is established: least decomposed peat 0.3-1.5 m, medium decomposed peat 0.2-3.8 m, and fibrous one 1.2-2.2 m. The least decomposed peat has organic matter content 94.8%, degree of decomposition – 15.8 %, water content – 9.1 u.f. (unit fraction), deformation modulus – 0.11 MPa, shear strength from 3 to 9 kPa; medium decomposed peat contains 92.2% of organic matter, degree of decomposition – 32.5 %, average water content – 6.8 u.f., density 1.05-1.08 g/cm^3, density of the solids – 1.47 g/cm^3, deformation modulus – 0.23 MPa, shear strength from 7.3 to 7.5 kPa; fibrous peat makes up 8.6% of organic matter content, degree of decomposition – 51.3 %, average water content – 4.2 u. f., density 1.20-1.28 g/cm^3, density of the solids – 1.81 g/cm^3, deformation modulus – 0.25 MPa, shear strength from 11.1 to 11.5 kPa. With depth water content decreases, whereas degree of decomposition, ash content, density, deformation modulus and shear strength increase.

The soil strength in both remolded and undisturbed states increases with depth (fig. 2), the most sensitive to structure destruction are peats of up to 2-3 meters deep, the index values become stabilized (S_t=2÷3) deeper than 4 meters. The close relationships of strength properties and test depth are reflected in the correlation equations:

$S_t = 3.77\text{-}1.36log10(h)$ (r=-60);
$c_u = 2.07+2.10h-0.087h^2$ (r=93);
$c_{ud} = 0.28+0.83h-0.02h^2$($r$=90).

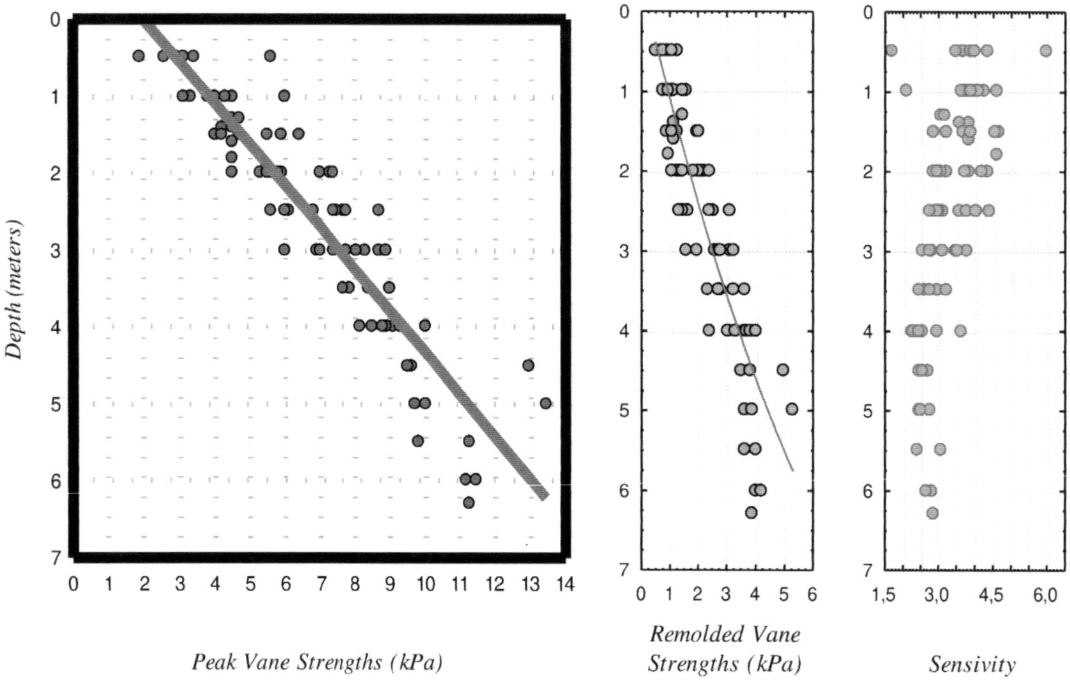

Fig. 2. Undrained shear strength profile of peat

In authors' opinion, to evaluate sensitivity to peat structure destruction in an integrated manner, the most conclusive and simplest classifications are those uniting indexes c_u and c_{ud} (tables 1-3). As the first example, the classification of soft soils in terms of road construction types is used from [26]. It should be reminded that 1 type includes peat that is not squeezed out under stress from loading, 2 type – squeezed out only under rapid stress, 3 type – always squeezed, 1,2 type requires additional consolidation tests to verify it. Hence, in the proposed classification of the same type are designated by the letters A (type 1), B (1, type 2), C (2 types), and D (3 types). According to table 1 and fig. 2 in the section of up to the depth of 2 m 2D type prevails, lower – 2C type up to 3-4 m, some points from the depth of 4.5-6.3 m can be referred to 1, 2 C types. Consequently, in the upper part of the bog soils with remolded structure would be squeezed out at any speed of filling.

Table 1. Peat types in terms of sensitivity according to road construction classification [26]

№ sample	Peat type Peak strength	Remolded strength Undrained shear strength, кPa	A (1) >20	B (1,2) 20-10	C (2) 10-3	D (3) <3
1	1	>20	1A	1B	1C	1D
2	1, 2	20-10	-	1.2B	1.2C	1.2D
3	2	10-3	-	-	2C	2D
4	3	<3	-	-	-	3D

To classify peat in terms of its sensitivity, let us consider table 2 using another classification for pipeline construction [15]. According to Industry-Specific Construction Standards (ISCS), 51-3-85 peat types are determined depending on maximum shear strength c_u: 1 type – c_u=10 kPa, bogs are entirely filled with dense peat up to its mineral bed, which is a reliable foundation for pipelines; 2 type – c_u=5-10 kPa, bogs are filled with peats of stable structure with water layers and peat formation thickness of more than 2 pipeline diameters between them up to its mineral bed; such peat can be used as a bearing stratum for pipelines; 3 type – c_u =5kPa, bogs are filled with fibrous peat or water with organic remains, with water layers and peat formation thickness of not less than 2 pipeline diameters between them up to its mineral bed; such peat cannot be used as a pipeline foundation.

Table 2. Peat types in terms of sensitivity according to the classification for pipeline construcion [25]

№ sample	Peat type Peak strength	Remolded strength Undrained shear strength, кPa	A (1) >10	B (2) 5-10	C (3) <5
1	1	>10	1A	1B	1C
2	2	5-10	-	2B	2C
3	3	<5	-	-	3C

According to table 2 and fig. 2 up to the depth of 2 m in the study areas 3C type prevails, lower there is 2C type, some points from the depth of 4.5-6.3 m can be referred to 1C type. Thus, independently on their initial strength, all peats are included in C category and, hence, such conditions are not acceptable for pipeline construction.

Table 3. Peat types in terms of radability [1]

№ sample	Peat type Peak strength	Remolded strength Undrained shear strength, кPa	A (1) >18	B(2) 12-18	C(3) 8-12	D(4) <8
1	(easy conditions)	>18	1A	1B	1C	1D
2	(medium conditions)	12-18	-	2B	2C	2D
3	(heavy conditions)	8-12	-	-	3C	3D
4	(particular conditions)	<8	-	-	-	4D

In terms of roadability (table 3) the peats under consideration can be taken from "heavy conditions" to "particular" ones – type 4D. Such bogs are suitable for passage of floating track machines or vehicles with arch tires, as well as passage in winter with the exception of warm winters.

The presented classifications (tables 1-3) and fig. 1 allow evaluating the behavior of different peats in a construction base. In terms of strength properties and sensitivity the most favorable peats are sphagnum – fuscum and magellanicum, other types are rather sensitive to structure destruction. On the other hand, the classifications do not consider deformation properties, whereas sphagnum peat compressibility can excess 80% of bed thickness.

Conclusion

As a result of the research performed the peat strength properties of some peat formations have been characterized, the dependencies of peat strength properties parameters on the depth of occurrence have been revealed. They can be used to evaluate peats as a construction foundation. Based on the suggested classifications the most appropriate sites for transport and construction have been revealed. The suggested approach to evaluation of peat soil sensitivity by means of well-known and reliable classifications allows researchers and engineers to apply peat strength properties more carefully and consider soft soils attentively when prospecting, constructing, and using facilities.

References

[1] Amaryan L S 1969 *Strength and deformation of peat soils* (Moscow, Nedra) (in Russain)
[2] Delft Geotechnics, Geotechnical classification of peats (in Dutch) Technical report 1994 No. CO-341820/29
[3] Evgenev I E and Kazarnovsky V D 1976 Automobile road base on soft soils (Moscow, Transport) (in Russian)
[4] Kanmuri H, Kato M, Suzuki O and Hirose E 1998 Shear strength of K0 consolidated undisturbed peat *Proc of the Int Symposium on Problematic Soils* (Japan, Sendai) 25-28

[5] Kramarenko V V and Savichev O G 2014 Engineering-geologic survey: research methods of peat soil (Tomsk, Tomsk Polytechnic University Press) (in Russain)
[6] Landva A O 1980 Vane testing in peat *Canadian Geotechnical Journal* **17** **1** 1-19
[7] Marachi N D, Dayton D J and Dare CT 1983 Geotechnical Properties of Peat in San Joaquin Delta Testing of Peats and Organic Soils, ASTM, Special Technical Publication **820** 207-217
[8] Mesri G and Ajlouni A M 2007 Engineering Properties of Fibrous Peats *J Geotechn and Geoenvironmental Eng ASCE* **133** **(7)** 850–864
[9] Radforth J I 1969 Preliminary engineering investigation Muskeg Engineering Handbook (The muskeg subcommittee of NRC) 127-149
[10] Shaposhnikov M A 1974 Principles of developing engineering-construction classification of undrained bogs *Types of bogs in the USSR and principles of their classification* (Leningrad, Nauka) 228-234
[11] Sodha VG 1966 Some aspects of the behaviour of peat in shear MEngSc thesis (Dublin, University College)
[12] Helenelund K V 1967 Vane tests and tension tests on fibrous peats *Proc. Getech. Conf* Oslo **1** 199–203
[13] Brakorenko N N and Korotchenko T V 2016 Impact of petroleum products on soil composition and physical-chemical properties *IOP Conference Series: Earth And Environmental Science* **33** http://iopscience.iop.org/article/10.1088/1755-1315/33/1/012028
[14] VCN 51-3-85 Standards of design of infield steel pipelines 1985 (Moscow)
[15] SP 36.13330.2012 Main (Trunk) pipelines
[16] Bjerrum L 1954 Geotechnical Properties of Norwegian Marine Clays *Geoiechnique* **IV** 49
[17] GOST 25100-2011 Soil. Classification
[18] Methodological guidelines of road base design on soft soils 1968 (Moscow, Orgtransstroy)
[19] Rosenqvist I Th 1953 Considerations of the sensitivity of Norwegian quick clays *Géotechnique* **3(5)** 195-200
[20] Skempton AW and Northey R D 1952. The sensitivity of clays *Géotechnique* **3** 30-53
[21] GOST 30416 2012. Laboratory testing. General requirements
[22] GOST 20276-2012. Field methods for determining the strength and strain characteristics
[23] ASTM STP 1014 on Vane Shear Strength Testing in Soils
[24] Guidelines on automobile road base design on soft soils 2004 (Moscow, Mintrans Rossii, FDA)

IOP Conf. Series: Earth and Environmental Science **43** (2016) 012024 doi:10.1088/1755-1315/43/1/012024

Triggering of great earthquakes: calculation and analysis of combined tidal effect of the Moon and Sun

Chubarov D.L.[1], Kochnev V.A.[2], Terre D.A.[1]

[1]National Research Tomsk Polytechnic University, 30 Lenin Avenue, Tomsk, Russia

[2]Institute of computational modeling Siberian Branch of the Russian Academy of Sciences, 50 Academgorodok, Krasnoyarsk, Russia

E-mail: convert008@mail.ru

Abstract. The largest part of solid minerals (with the exception of those which are at the earth's surface) is being extracted world-wide by surface and underground mining techniques, with adits, mines and other mine workings being used. A considerable amount of mineral deposits (including oil reservoirs) is located either close to a fault-line or immediately within the zone of high seismic activity. To prevent economic and environmental damage under the effect of an earthquake, thorough seismic monitoring of the area must be performed, as well as the study of all possible mechanisms of an earthquake occurrence.

In analysing the trigger effect of moon- and sun-induced tidal forces on seismic activity, six great earthquakes which occurred close to equatorial latitude over the last 15 years have been considered. Based on the positions of the Sun and Moon during the day relative to the point mass, the maps of horizontal, vertical components and vector of gravitational forces per unit mass have been plotted. The developed technique can be applicable to a set of methods to study integration and stress unloading mechanisms at the boundaries of block structures.

1. Introduction

Great earthquakes are known to have devastating consequences. The earthquake which occurred near the eastern coast of Honshu, the largest and most populous island of Japan, on 11 March 2011 provides an example of such disastrous effect. It caused a major radiation accident rated level 7 on International Nuclear and Radiological Event scale. This event prompted a worldwide response in many environmental organizations due to its huge impact on people's health and environment.

It should be noted that mineral deposits can also be located in seismic zones and their development may involve factors which can damage both environment and health of people. For example, if Sakhalin oil platforms are exposed to an earthquake followed by a tsunami, they can be damaged, which, in its turn, can lead to an ecological disaster similar to the Gulf of Mexico oil spill in 2010.

In this respect, the study of all possible mechanisms of earthquake occurrence is of great importance. The effect of tidal forces induced by the Sun and Moon can be regarded as one of these mechanisms. One of the possible tidal effects on the Earth's shells is considered in [1]. The purpose of the present research is to detect the trigger effect of tidal forces in great earthquake occurrence. Some data from [2] concern numerical modeling of trigger effect induced in geological environment and seismic waves that are generated as a result of it.

2. Physico-mathematical Modeling

The following is the data from [3]. Horizontal and vertical components of tidal force vector are calculated, with angle θ taken into account.

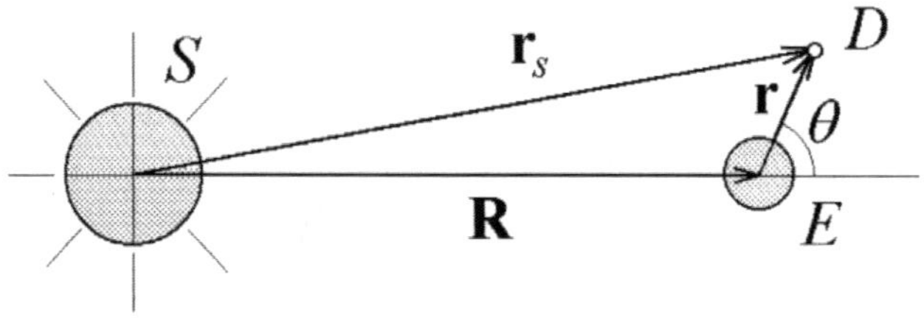

Figure 1. Variables r and θ used to define a point location near the Earth surface [3]

$$F_{hor} = -3G\frac{mM_{sun}}{R^3} r\cos\theta\sin\theta =$$

$$= -\frac{3}{2}G\frac{mM_{sun}}{R^3} r\sin 2\theta \qquad (1)$$

$$F_{vert} = G\frac{mM_{sun}}{R^3} r(3\cos^2\theta - 1) =$$

$$= \frac{3}{2}G\frac{mM_{sun}}{R^3} r(\cos 2\theta + \frac{1}{3}) \qquad (2)$$

As follows from the formulas (1, 2), maximum values of horizontal component are observed when θ =45° and 135°, whereas in case of θ =0 °, 90°, 180°, 270 ° they are zero. The corresponding extrema of a vertical component are offset by 45°.

Let us consider the case illustrated in figure 2. Equatorial latitude being considered, the value of horizontal component of tidal force induced by the Sun and Moon and affecting a point will be calculated using the formula (1). In the first case, (Sun effect) θ_1 angle will be applied, while in the second case (Moon effect) – θ_2 angle. The total value of horizontal component will be identified by summing these two quantities. In the same way, the vertical component will be calculated using the formula (2).

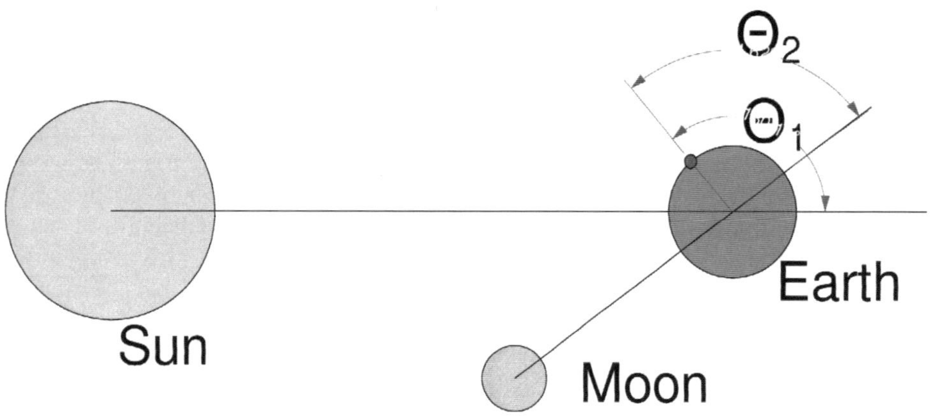

Figure 2. Example of relative positions of Moon and Sun with respect to a point location on the Earth surface

The value of vector sum tidal forces was determined by the formula $F = \sqrt{F_{hor}^2 + F_{vert}^2}$. The obtained values are presented in figure 3.

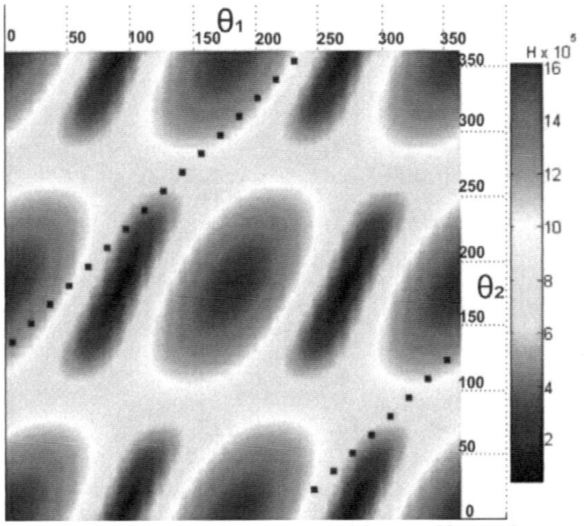

Figure 3. Values of vector sum tidal forces (the combined effect of the Moon and Sun). Axis: x – $\theta_1$values, y – $\theta_2$values.

Legend:

■ Daily variations (measurement interval – 1 hour)

3. Correlation of Tidal Forces with Earthquakes

The effect of tidal forces on great earthquakes which occurred near the equatorial latitude will be further analyzed. Over 10 events were selected. Table 1 summarizes the reference data on those which were used in calculations.

Table. Earthquakes selected for calculations

N	Location	Ms	Date	Time (UTC)			Latitude	Longitude
				h	M	s		
1	Indonesia	8.8	26.12.2004	0	58	48	3.39	95.84
		7.3	26.12.2004	4	21	26	6.8	92.86
		6.1	26.12.2004	19	19	55	3.08	93.49
2	Japan	9.1	11.03.2011	05	46	23	38.32	142.37

The technique to determine the effect of tidal forces on the earthquake epicenter immediately at the time of its occurrence consists in the following.

Using the known latitude and known earth angular velocity, the time correction for an earthquake epicenter (accurate to second's time difference compared to Greenwich) was determined. Based on this correction, the exact local time of earthquake occurrence was calculated, and this value was referred to, when calculating θ_1 angle, which defines Sun's location relative to the given point at the shock moment by the formula (3).

$$\theta_1 = t \cdot v_s \quad (3)$$

t – local time of the earthquake, v_s – Earth's angular velocity (15 degrees/hour).

The elongation angle between the Moon and the Sun was determined (figure 4). To define the relation of α angle and that of α' elongation angle the formula is given in [4] and is presented as follows:

$$\cos\alpha' = \cos\alpha - \frac{a}{r}\sin^2\alpha \quad (4)$$

Subsequently, based on the given angle which defines Moon location relative to the epicenter point at the moment of the shock, θ_2 angle was calculated. Finally, considering θ_1 and θ_2 angles, it is possible to calculate the value of horizontal and (vertical) component of tidal force as well as the value of its vector sum.

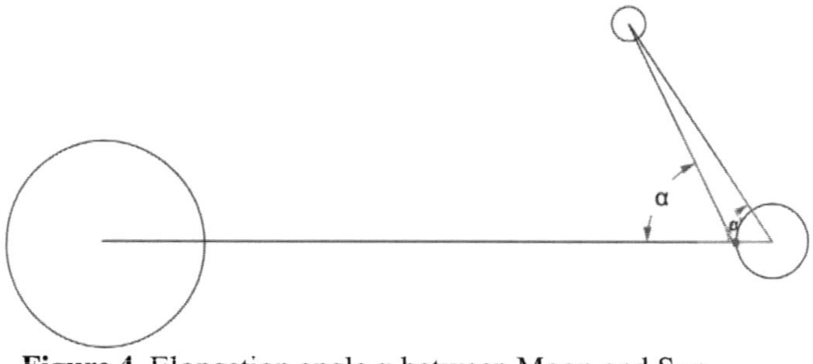

Figure 4. Elongation angle α between Moon and Sun

To understand the degree of tidal force effect on an earthquake it is necessary to observe the changes in values of vector sum tidal forces over some period before the event. It should be considered that at every moment not only θ_1 and θ_2 angles but also distance to the Sun and the Moon change, as well as the tilt of the Earth's axis relative to these two bodies. The changes in tidal forces at earthquake epicenter points are shown figure 5.

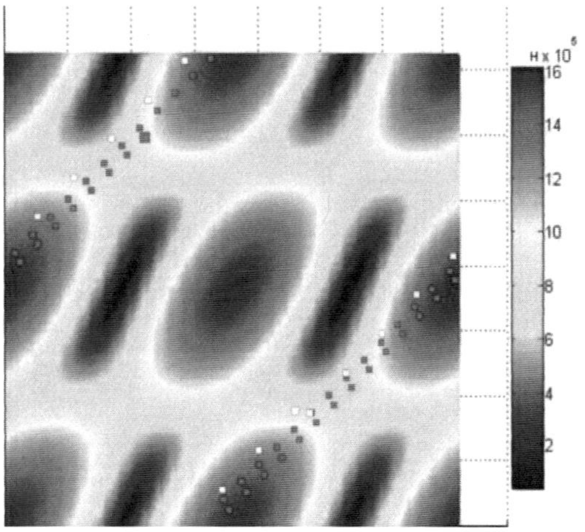

□	Tidal force effect on epicenter point during the day before the shock, Japan, 11.03.2011 (the shock is marked as an increased squire); the interval – 2 hours
■	Tidal force effect on epicenter point during the day before the shock, Indonesia, 26.12.2004 (the shock is plotted as an increased square); the interval – 1 hour

Figure 5. Variation of vector sum tidal forces (calculated for epicenter points of earthquakes in Indonesia and Japan)

4. Conclusion

The investigations carried out by both Russian and foreign researchers [5] show that the effect of tidal forces in triggering earthquakes is generally not great. However, any researcher can hardly ever deny that the tides affect the stress field processes in the crust. Moreover, it is quite significant to study any possible mechanism which triggers earthquakes in the regions regarded as zones of the highest seismic activity (Japan, Indonesia). Based on the positions of the Sun and Moon during the day relative to the body under study which is referred to as a point mass, the maps of horizontal, vertical components and magnitudes of gravitational forces per unit mass have been plotted. Furthermore, the analysis of tidal force effect on epicenter points of several great earthquakes which occurred near the equatorial latitudes has been made.

The present research can be applied to a set of methods when studying integration and stress unloading mechanisms which act at the block structure boundaries, as well as for more precise understanding and numerical description of earthquake generation.

References

[1] Kochnev V A 2013 *Geophysical journal.* Kinematikogravtatsionnaya model gedynamo. Vol. **35** **(4)** pp. 3–15.

[2] Nemirovich-Danchenko M M 1998 *Physical Medium.* A model for the brittle hypoelastic medium: application to computation of deformations and failure in rock. Vol. **1** **(2)** pp. 101-108.

[3] Butikov E A 2002 *American Journal of Physics.* Dynamical Picture of the Oceanic Tides. Vol. **70** **(10)** pp. 1001–1011.

[4] Melchior P 1974 *Geophysical Surveys.* Earth tides. Vol. **1** pp. 275 – 303.

[5] Rebetsky Yu L 2007 *Devyatye geofizicheskie chteniya Fedynskogo.* Mekhanizm generatsii tektonicheskikh napry azheniy obuslovlenny gravitatsionnymi silami, plasticheskim techeniem i verticalnymi peremeshcheniyami pp. 80–81.

Geothermal heat in a heat pump use

A Pavlova[1,2], J Hansen[2], H Obermeyer[2], I Pavlova[3]

[1]Karlsruhe Institute of Technology (KIT), Kaiserstraße 12, 76131 Karlsruhe, Germany
[2] Society for sensing and locating, Südbeckenstr. 20 a, D-76189 Karlsruhe, Germany
[3] Tomsk Polytechnic University, 30 Lenin Ave., Tomsk, 634050, Russia
E-mail: pavlovaiv@tpu.ru

Abstract The considered innovative technology proposes to use alternative energy sources for the process efficiency in low-height construction. The world economy depends on price rises for energy sources and the danger of environmental pollution increases. Geothermal energy is the basic resource saving and environmentally safe renewable heat source that is characterized by inexhaustibility, permanent all the-year-round use, universal prevalence of resources and the ability to replace considerable volumes of traditional energy carriers. The expediency and power efficiency to apply a heat pump with the use of geothermal heat is proved for low-height construction.
Keywords: geothermal heat, power saving, renewable energy sources, heat pump

1. Introduction

The project dimensions a geothermal heat pump for the single family house located at 8 Mozart street, Karlsruhe, 76133.

The large amount of fossil energy sources are known to have serious impact on the environment. Such impact is the pollutant increase in the atmosphere, for example sulphur dioxide and nitrogen oxide, that occurs in the process of combustion. Also the oil and gas restricted reserves make it difficult to use the fossil energy sources for energy supply. The energy production must be aimed at renewable energy sources to solve these problems in future.

It is considered that geothermal energy is located below the earth's surface in the form of warmth. Such energy belongs to the renewable energy, i.e. the energy from permanent sources that are inexhaustible by human reckoning. It is possible to use the geothermal energy to heat a building at low temperatures of the subsurface layers (subsurface geothermics). There are many advantages of geothermics private use along with the environmental condition improvement: space saving, availability, flexible use, cost-effective energy. The heat pump technology is applied in private heating, along with the geothermal energy use.

The heat pump is not a combustion engine in contrast to the conventional heat producers. It uses free environmental energy (from water, air or geothermal energy) and converts the lower temperature heat into higher temperature one. There are many kinds of heat pumps. Two the most suitable types to heat the given house are presented in the project: a groundwater-source heat pump and heat pump with geothermal probes.

2. Heat pump types

The heat pump technology works on the following principle. The stored geothermal heat as useful heat is transferred from the lower temperature reservoir to the higher temperature heating system. The heat always flows from the higher temperature to lower one according to the second law of thermodynamics. A heat pump should bring the received energy (using high-quality energy) at the required temperature level. Such natural environments as soil, water, and air can be used as a heat energy reservoir.

One of the heat pump types is a groundwater-source heat pump. Ground water heat is used as a regenerative energy source in the groundwater-source heat pump. This heat pump needs two wells: a transport and absorbing well (Figure 1). The transport well is used to pump ground water to the pump heat exchanger. The thermally changed ground water drains again in the same groundwater bed through the absorbing well. The groundwater-source heat pump is suitable for plots of land with the small groundwater depth and sufficient amount of high-quality ground water.

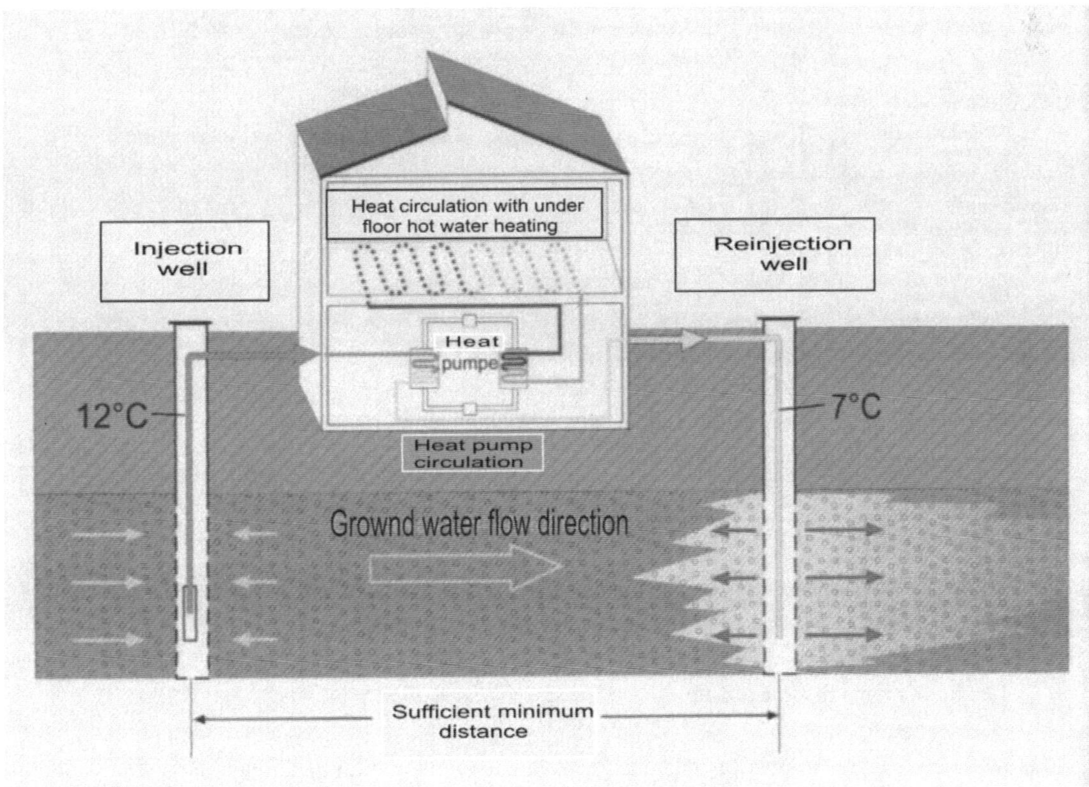

Figure 1. Groundwater-source heat pump

The second heat pump type, which is considered in the dimensioning, is the heat pump that takes the required energy out the earth by means of a geothermal probe and transfers it to the pump heat exchanger through heat-transfer liquid (brine water) (Figure 2). The geothermal probe is a closed pipe system with circulating heat-transfer liquid that is placed vertically in the subsoil. This heat pump can be designed very space-saving and, therefore, be suitable for small plots of land.

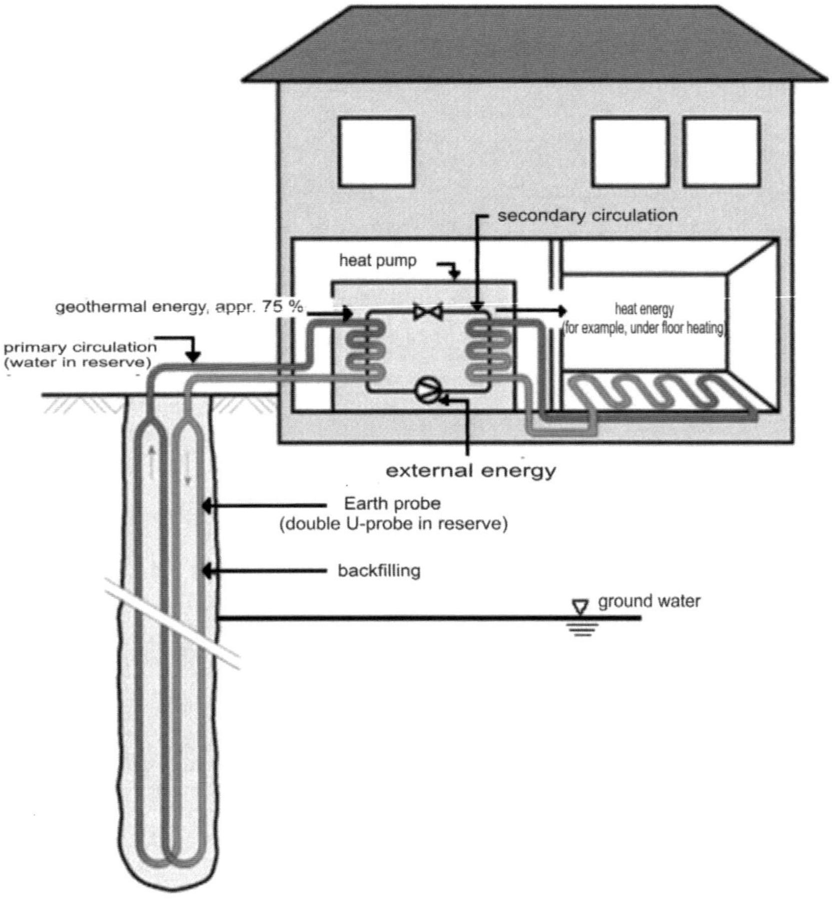

Figure 2. Brine-water heat pump

3. The heat pump dimensioning

- **Heat consumption**

The heat pump dimensioning has been carried out for the single family house located at 8 Mozart street, Karlsruhe. The first stage of the dimensioning was to calculate the house heat consumption.

The considered house consists of three heated floors (including a usable roof floor) and an unheated cellar. The ceiling height is 3.07 m for the ground floor, 3 m for the first floor, and 2.5 m for the usable roof floor. The overall heated area is 418 m^2 and overall volume is 1210 m^3. The more exact data are presented in the following tables.

Table 1. The ground floor area and volume

Room	**Area, m²**	**Volume, m³**
Room 1	15.75	48.35
Room 2	11.90	36.53
Room 3	22.00	67.54
Room 4	8.75	26.86
Room 5	27.00	82.89
Room 6	15.60	47.89
Room 7	16.00	49.12
Room 8	4.62	14.18
Room 9	3.08	9.45
Room 10	10.59	32.51
Floorboard	19.98	61.33
Bathroom	5.47	16.79
WC	2.82	8.65
Overall:	163.56	502.12

Table 2. The first floor area and volume

Room	**Area, m²**	**Volume, m³**
Sitting Room	12.87	38.61
Utility Room	13.65	40.95
Study	12.61	37.83
Kitchen	8.32	24.96
Living Area	27.00	81.00
Dining Room	13.44	40.32
Bedroom	19.10	57.30
Floorboard	25.86	77.58
Bathroom 1	5.60	16.80
Bathroom 2	4.25	12.75
WC	1.00	3.00
Overall:	143.70	431.10

IOP Conf. Series: Earth and Environmental Science **43** (2016) 012025 doi:10.1088/1755-1315/43/1/012025

Table 3. The usable roof floor area and volume

Room	**Area, m²**	**Volume, m³**
Study	9.12	22.80
Bedroom 1	22.5	56.25
Bedroom 2	8.96	22.40
Bedroom 3	14.00	35.00
Living Area	17.88	44.70
Room 1	13.05	32.62
Room 2	20.66	51.65
Bathroom	3.20	8.00
WC	1.60	4.00
Overall:	110.97	277.42

The house standard heating load has been calculated according to the EU norm of German Institute for Standardisation EN 12831. According to the norm, the standard heating load is the sum of transmission heat loss, ventilation heat loss and the additional required heat supply for all heated rooms. The house standard heating load is 22.733 W according to the calculation. The data are shown in the following table.

Table 4. The house standard heating load

Floor	**Transmission heat loss, F_T, W**	**Ventilation heat loss, F_V, W**	**Additional required heat supply, F_{RH}, W**	**Standard heating load, F_{HL}, W**
Ground floor	4,974.65	3,042.73	1,472.04	9,489.43
First floor	3,193.86	3,031.17	1,293.3	7,518.33
Usable Roof Floor	3,070.42	1,478.59	998.73	5,725.23
Overall:	11,238.94	7,729.98	3,764.07	22,733

The additional hot water consumption is 1,250 W for five people. The overall house heat consumption is 23,983 W, which is the sum of the standard heating load and hot water consumption.

Another house standard heating load has been calculated according to the oil fuel consumption and is 34,914.4 W.

The heat pump dimensioning has been carried out for both load values.

- **Groundwater-source heat pump**

This heat pump directly uses groundwater as a heat source. This is the most effective heat source because the groundwater annual average temperatures are the highest. The average groundwater temperature is 13.9 °C in this case.

It is necessary to check whether the plot of land is located in the groundwater protection area because the groundwater-source heat pump construction and operation is prohibited in some water protection areas. The following map created by the Baden-Württemberg State Office for the Environment, Measurements and Nature Protection shows the water conservation areas in and around Karlsruhe. The house in 8 Mozart street is located in the Weststadt (the district in the west of Karlsruhe) outside the water protection area (Figure 3).

Figure 3. Water protection areas in Karlsruhe and surroundings

The ground water suitable chemical properties are an important condition to use ground water for the groundwater-source heat pump. The ground water should contain no sediments such as iron, manganese, and some other chemical elements. The limiting concentration should not be over the limit values for these elements. The data of the annual ground water catalogue (2000 - 2014) created by the Baden-Württemberg State Office for the Environment, Measurements and Nature Protection have been taken from the well closest to the house to study the ground water chemical composition. It can be clearly seen that some elements exceed the limit values, nevertheless, the excess is not critical. The data are shown in the following table.

Table 5. Ground water chemical properties

	Fe [mg/l]	**Mn [mg/l]**	**pH**	**O_2 [mg/l]**	**LF [mS/m]**	**NO_3 [mg/l]**	**SO_4 [mg/l]**	**Cl [mg/l]**
Average value for 14 years	0.342	0.054	7.207	1.183	66.383	7.003	81.362	34.025
Limit values for special steel heat exchanger	<0.2	<0.1	>7.5	<2	1-50	<100	>1	<300

The drinking water supply is the top priority among the other purposes of ground water use. Consequently, the area hydrogeological situation is always very important while heat pump dimensioning. The use of the upper groundwater bed near surface with free water level is quite smooth. The distance between the groundwater bed and ground surface is approximately 5 m for the given house (Figure 4).

The upper groundwater bed thickness under the plot of land is approximately 30 m. The impermeable horizon starts between the first and second groundwater beds at approximately 35 m. One may drill not lower than 35 m in order not to touch the second groundwater bed.

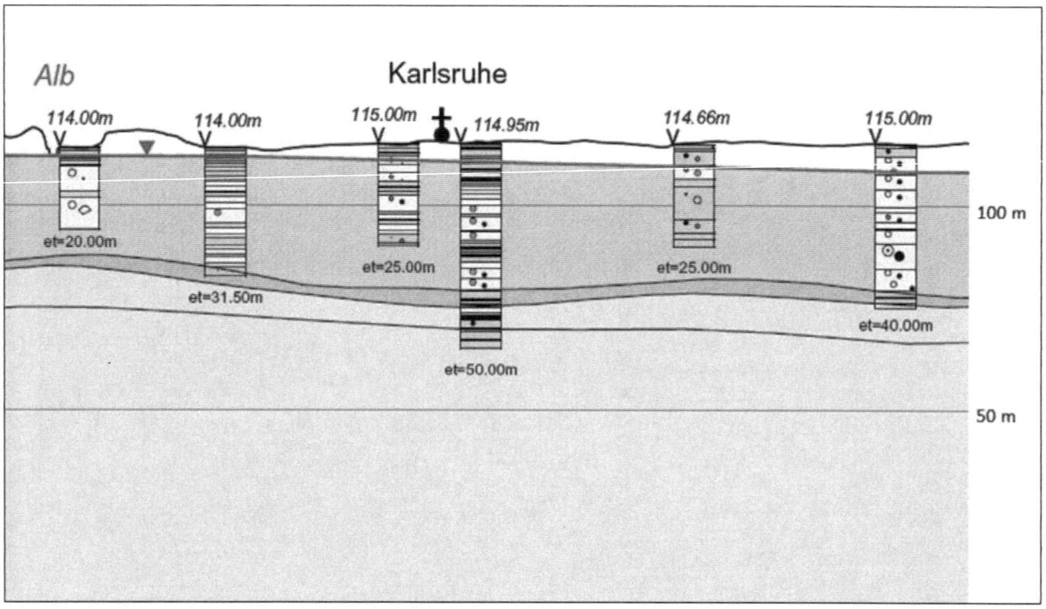

Figure 4. Geologic profile of Karlsruhe

Two heat pumps have been dimensioned for two different heating load values. The Heliotherm HP28S 40W-WEB heat pump with the heat supply of 38,050.0 W has been taken for a heating load of 23,983.0 W. The Heliotherm HP32S45W-WEB heat pump with the heat supply of 42,560.0 W has been taken for a heating load of 34,914.4 W.

The rated consumption is 5-6 m³/hr for the required heating load. Table 6 shows the calculated data.

Table 6. Rated consumption

Heatig load, W	**Pump heat supply, W**	**Evaporator capacity, W**	**Rated consumption, m³/hr**
23,983.0	38,050.0	31,971.7	5.4
34,914.4	42,560.0	35,750.4	6.1

The well capacity and productivity have been calculated depending on the rated consumption (Table 7).

Table 7. Well productivity and capacity

Minimum diameters, m	Maximum penetration, m	Operational range, m	Allowed entry speed, m/s	Filter length, m	Well productivity, m^3/h	Well capacity, m^3/h
0.05	0.03	4.67	0.003	3	12.14	11.93

The minimum drilling depth is approximately 10 m (5 m of water level + minimum 3 m of filter length). The transport well should be deeper than the absorbing one.

The minimal distance between the transport and absorbing well should be 15 m. It is possible to keep the maximum distance between the wells of approximately 40 m at the considered plot of land. The wells should be drilled perpendicularly to the ground water flow direction. Figure 5 shows the approximate well location at the plot of land.

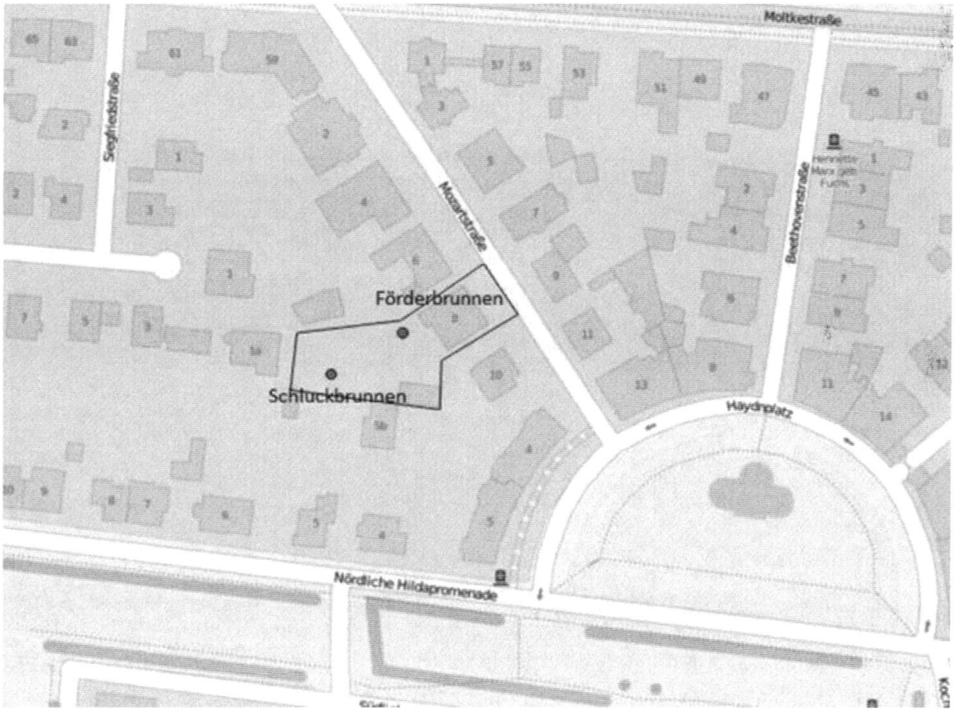

Figure 5. Well location

- **Geothermal probe**

The other heat pump type that has been dimensioned is a brine-water heat pump with geothermal probes. The ground is a heat source for this pump.

Two cases to supply the building with the possible required heat have been considered like by the groundwater-source heat pump dimensioning. The GeoTherm VWS 300/3 heat pump with the heat supply of 30,000 W has been taken for a heating load of 23,983.0 W. A heating load of 34,914.4 W is used for the GeoTherm VWS 460/3 heat pump with the heat supply of 45,700 W. The evaporator capacity has been calculated for both heat pumps (Table 8).

Table 8. Evaporator capacity

Heating load, W	Pump heat supply, W	Evaporator capacity, W
23,983.0	30,000	23,333.3
34,914.4	45,700	35,544.4

The drillings with the geothermal probes from 50 to 100 m in length should be carried out through three beds (Figure 6):

- the higher sandy gravel groundwater bed;
- the intermediate bed from clay and sandy clay;
- the second groundwater bed from sand and gravel.

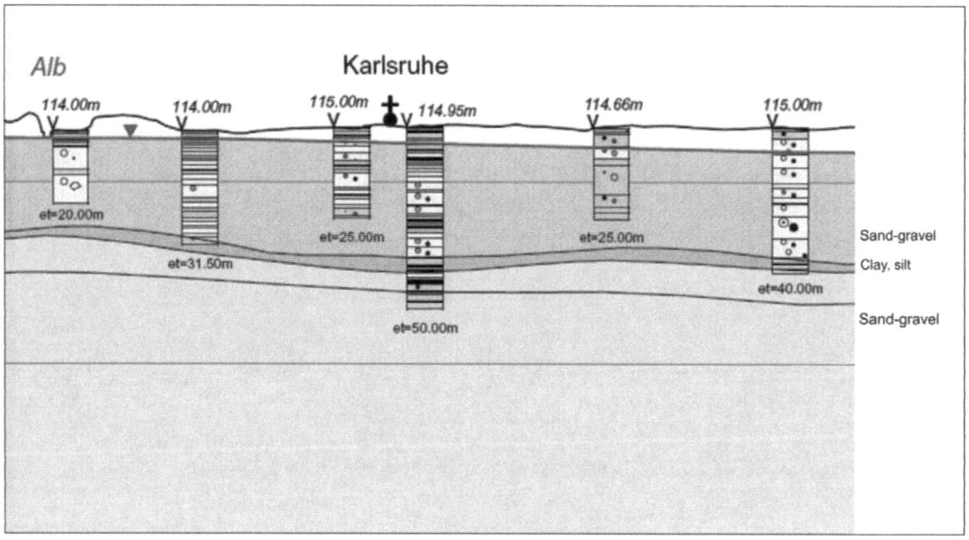

Figure 6. Geologic profile of Karlsruhe

The geothermal probe length has been calculated for the lowest and highest heat generation of the beds. It has also been calculated for two evaporator capacities (Table 9).

Table 9. Geothermal probe length

Evaporator capacity, W	Geothermal probe length, m, lowest heat generation	Geothermal probe length, m, highest heat generation
23,333.3	534.9	428.1
35,544.4	814.9	652.1

It can be clearly seen that the geothermal probe length is from 428.1 m to 534.9 m for the lower evaporator capacity. Two calculations have been carried out to find the number of the geothermal probes: with probe minimum length 50 m and with maximum one 100 m (Table 10).

Table 10. Number of probes, the first case

Geothermal probe length, m	Number of probes, 50 m	Number of probes, 100 m
534.9	11	6
428.1	9	4

The maximum number of probes from 50 to 100 m in length is 11 and the minimum number is four for a heating load of 23.983 W.

The same calculation has been carried out for the higher evaporator capacity (Table 11).

Table 11. Number of probes, the second case

Geothermal probe length, m	Number of probes, 50 m	Number of probes, 100 m
814.9	16	8
652.1	13	7

The maximum number of probes is 16 and the minimum one is seven for a heating load of 34,914.4 W. The calculations show that the minimum overall length of probes is approximately 428 m and the maximum one is approximately 815 m for the given house.

The heat pump with geothermal probes is not cheaper than the groundwater-source heat pump due to the drilling costs.

4. Conclusion

Two heat pump types have been dimensioned in the project: the groundwater-source heat pump and heat pump with geothermal probes.

The first stage of both dimensionings has been to calculate the house standard heating load. Two values of the heating load have been calculated as a result of different calculation methods and used in further dimensioning. The suitable heat pumps have been found for both values in heat pump prospectuses.

The first dimensioning has been carried out for the groundwater-source heat pump. The following aspects have been considered by dimensioning:

- The house location (water protection areas);
- The ground water quality (chemical properties);
- The hydrogeological situation (the distance between the groundwater bed and ground surface, the groundwater bed thickness and its hydrogeological properties).

The hydrogeological research carried out by the Baden-Württemberg State Office for the Environment, Measurements and Nature Protection has been used as well as the hydrogeological maps and profiles.

The well location shown in the report has been provided by means of QGIS.

The second dimensioning has been carried out for the brine-water heat pump. Several cases with different evaporator capacity, heat generation, and geothermal probe length have been considered. Moreover, the minimum and maximum number of probes have been calculated.

The result of the dimensioning suggests that the groundwater-source heat pump is best suited for the given house.

References

[1] The house and plot of land plan.

[2] SGD-Personenkreis Geothermie der Ad-hoc-Arbeitsgruppe Geologie. Fachbericht zu Bisher Bekannten Auswirkungen Geothermischer Vorhaben in Den Bundesländern

[3] The textbook on geothermal heat and groundwater-source heat pumps. The Baden-Württemberg State Office for the Environment, Measurements and Nature Protection, the first edition, April.

[4] 2014 Heat pumps for heating and hot water production, the project engineering and installation book, Dimplex.

[5] Design documentation of air-to-water and brine-water heat pumps. Wolf.

[6] 2011 Heat pump plants, the small guide for beginners, Johannes Wegesin, the first edition.

[7] 2003 *The German Institute for Standardization* EN 2831.

[8] 2011 *Virtual Desktop Infrastructure* 4640

[9] Bleicher A[a], Gross M[ab] 2016 *Renewable and Sustainable Energy Reviews*. Geothermal heat pumps and the vagaries of subterranean geology: Energy independence at a household level as a real world experiment Vol. **64**, pp. 279-288

[10] Lund J.W. Sanner B. Rybach L. Curtis R. Hellström G. 2014 *A world overview (2004) GHC Bulletin*. Geothermal (ground-source) heat pumps. pp. 1-10.

[11] Weber J, Ganz B, Schellschmidt R, Sanner B, Schulz R 2015 *July Congress Proceedings of World Geothermal*. Geothermal energy use in Germany.

Treatment of petroleum-contaminated water resources: modern techniques

Pogharnitskaya O.V.[1,2], **Konovalov V.V.**[1,3], **Dmitrieva N.V.**[4], **Belozerova D.S.**[1], **Strelnikova A.B.**[1]

[1] National Research Tomsk Polytechnic University, 30 Lenina Ave., Tomsk, 634050, Russia
[4] Novosibirsk State University, 2 Pirogova Str., Novosibirsk, 630090, Russia

E-mail: [2]pov@tpu.ru, [3]sltan@ngs.ru

Abstract. The article deals with the issue of petroleum-contaminated water resources. The authors have analyzed the dynamics of oil spills, including the world's largest ones, and claimed the issue to be global. The modern methods of mitigating oil spill effects have been studied, as well as the modern techniques of water resource treatment. The particular attention is paid to peat sorbent production, which is considered a promising trend of petroleum-contaminated water treatment.
Key words: water resources, oil, petroleum contamination, peat, water resources treatment

Introduction

Over the past decades, negative environmental impact attributed to oil spills has substantially increased. The major damage has been caused to water resources.

In practice, oil spills and leaks inevitably occur at different stages of recovery, processing, and transportation, with oil pipeline accidents being particularly dangerous. Therefore, it is quite difficult to develop the whole range of preventive measures to reduce the environmental impact.

Having penetrated into the ground, oil results in complicated processes, which makes the treatment of petroleum contaminated soil and water a challenging task. Currently, water resource contamination caused by oil spills is a topical issue on both national and global scale.

Another urgent question to be solved today is water resource treatment, which involves wastes recovery and recycling. It is also important to choose the optimal methods to mitigate the negative environmental impact.

Materials and methods

Today, different economic sectors search for the methods to liquidate oil spill under various plausible scenarios.

Laboratory experiments, land survey, and practical experience made it possible to describe oil properties and develop methods of oil spill liquidation. The research was mainly focused on the moderate climate regions, however, the Arctic scenarios, which imply low temperatures and presence of sea ice, were investigated as well.

According to the data by the Division of Mineral Resource Management, the USA, there were 858 fires and explosions on offshore platforms over the period of 2001–2010, which means approximately one accident per four days. Annually more than 1.5 billion tons of oil, which accounts for about 40% of all recovered oil, is transported by tankers. This figure is significantly higher than that of oil recovery in offshore fields, the latter being within the range of 30–34% over the past years. In accordance with statistics, the amount of oil spilled during transportation is 23-26 times as high as that during offshore recovery [1].

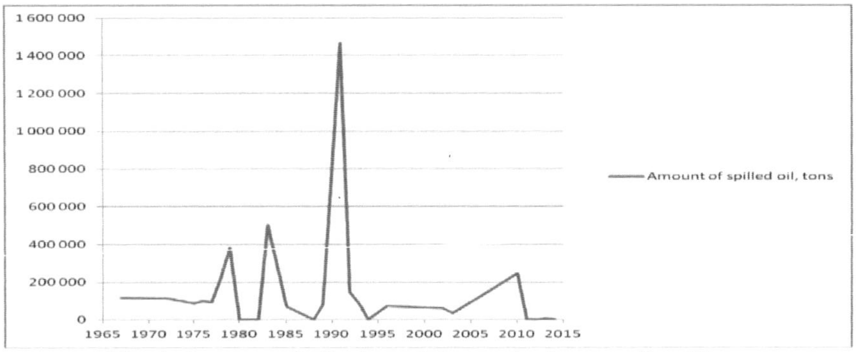

Fig. 1. Amount of oil released due to the largest oil spills

Figure 1 illustrates the data on oil released in the result of the largest oil spills, without reference to minor ones. Statistical data on offshore oil spills indicate that the share of major oil spills (more than 700 thous. tons) is relatively small and has decreased over the studied period. Most accidents occurred before 2000: 1970s – 245 spills (54%), 1980s – 94 spills (21%), 1990s – 77 (17%), 2000s – 35 (8%) [2].

Table 1 shows figures on the largest oil spills (the data are based on the materials published by the International Tanker Owners Pollution Federation (ITOPF) and elaborated by the authors of the present article).

Table 1. World's largest oil spills

Tanker / Platform	Year	Location	Responsible party	Amount of spilled oil, tons	Damage caused
The Gulf War oil spill	1991	The Persian Gulf	Iraq	1 091 405	$540 million
Ixtoc I oil spill, Mexico's semi-submersible drilling rig Sedco 135-F	1980	The Gulf of Mexico, about 100 km northwest of Ciudad del Carmen	Mexican state-owned petroleum company PEMEX	467 000	$42 million
Tanker Atlantic Empress	1979	Tobago, West India	South Gulf Shipping Co. Ltd., Greece	287 000	not available
Oil spill on the Colva river, the pipeline Kharyaga-Usinsk	1994	The Republic of Komi	AO Komineft, Russia	264 000	not available
Tanker ABT Summer	1991	700 nautical miles off the Angolan coast	Liberia	260 000	not available

IOP Conf. Series: Earth and Environmental Science **43** (2016) 012026 doi:10.1088/1755-1315/43/1/012026

Platform Deepwater Horizon, Macondo field, BP	2010	The Gulf of Mexico	British Petroleum, Great Britain	245 566	More than $40 billion
Tanker Castillo de Bellver	1983	Saldanha Bay, South Africa	Empresa Nacional Elcano De La Marina Mercante Madrid	252 000	not available
Platform Nowruz	1983	The Persian Gulf	Iran	250 000	not available
Tanker Amoco Cadiz	1978	5 km (3.1 mi) from the coast of Brittany, France	Amoco Corporation	223 000	$85.2 million
Tanker Odyssey	1988	700 nautical miles off the coast of Nova Scotia, Canada	Polembros Shipping Ltd. Great Britain	132 000	not available
Supertanker Torrey Canyon	1967	The Isles of Scilly, Great Britain	Barracuda Tanker Corporation, British Petroleum, Great Britain	119 000	not available
Supertanker Sea Star	1972	The Gulf of Oman	South Korea	115 000	not available
Tanker MT Haven	1991	Genoa, Italy	Cyprus	114 000	not available
Tanker Irenes Serenade	1980	Navarino Bay, Greece	Greece	100 000	not available
Supertanker Urquiola	1976	La Coruna, Spain	Bilbao, Spain	100 000	not available
Tanker Hawaiian Patriot	1977	300 miles off Honolulu, Hawaii	Liberia	95 000	not available
Tanker Independenta	1979	The Bosphorus, Turkey	Romania	95 000	not available
Tanker Jakob Maersk	1975	Porto, Portugal	Denmark	88 000	not available
Tanker MV Braer	1993	The Shetland Islands, Great Britain	Braer Corporation, Canadian Ultramar Ltd	85 000	not available
Khark 5	1989	120 nautical miles off Morocco	Iran	80 000	not available
Tanker Sea Empress	1996	Milford Haven waterway, Great Britain	Oriental Ocean Shipping, Spain	72 000	not available
Tanker Katina P	1992	Maputo Bay, Mozambique	Malta	72 000	not available
Tanker Nova	1985	Kharg Island, Iran	Iran	70 000	not available
Tanker Prestige	2002	The Spanish coast	Mare Shipping Inc., Liberia	60 000	€4 billion

Tanker Exxon Valdez	1989	Prince William Sound, Alaska, the USA	Exxon Corporation	37 000	$ 3.5 billion
Tanker Tasman Spirit	2003	Near the city of Karachi, Pakistan	Spain	35 000	not available

As shown in the table above, there were numerous offshore oil spills over the past 50 years, which had negative impacts on the marine ecosystem.

For instance, the Exxon Valdez oil spill, which occurred in Prince William Sound in 1989, caused great damage to the particularly sensitive ecosystems of Alaska. This is due to the fact that the coastal waters of Alaska are cold and fail to provide habitat for the bacteria capable of hydrocarbon degradation. Warm waters, on the contrary, are rich habitat for the above-mentioned bacteria and oil degradation is regulated by the environment, however, this fact does not permit the parties to waive their responsibilities for the damage caused.

Having integrated the data obtained from various information sources, one can state that the amount of oil that annually enters the world ocean ranges between 5 and 100 million tons, with oil spill accidents accounting for only 12–15% of this volume. The sources of hydrocarbon input to the world ocean are given in table 2 [3].

Table 2. Sources of oil input to the world ocean

Source	million tons per year
Oil transporting and shipping (except for accidental oil spills)	1.83
Accidental oil spills	0.3
Discharges with rivers (including urban municipal sewage)	1.9
Discharges of waste water from coastal industries	0.8
Atmospheric fallout	0.6
Natural seeps	0.6
Offshore production	0.08
Total:	6.11

As for Russia, petroleum transported by sea is mainly export, and the major sources of petroleum input to the marine environment are ship-related:

- discharges of tank-washing residues, oil-contaminated ballast water, and bilge water;
- discharges in the harbors;
- accidental discharges.

An example of accidental discharges is the Russian tanker Nakhodka oil spill: the tanker broke up en route to Kamchatka in 1997, which resulted in contamination of some 200 km of the coastal area. Another Russian oil tanker, the Volgoneft-139, broke apart in Kerch Strait, with 1.2 thous. tons of oil residue released [4].

According to Greenpeace Russia, the Russian Federation, where the oil pipeline stretches for 400 thousand km, is among top countries in terms of the frequency of pipeline accidents. Annually at least 5 million tons of petroleum is released into the environment, which is five times as much as the amount of oil spilled in the Gulf of Mexico in 2010 [5].

Petroleum contaminated water resources are an urgent issue for West Siberia, since many companies involved in oil recovery, transportation, and processing are located in this region. Having entered the water, the oil forms a thing film on the surface. This film prevents oxygen ingress into the water, which has a negative impact on fish and plants [6].

More often than not, oil spills on soil lead to groundwater contamination [7].

Due to the above-mentioned reasons, there should be enhanced techniques for the treatment of petroleum contaminated water and soil. This question is particularly topical under the circumstances of intense petroleum production and necessitates new approaches.

Contaminated soil is usually recovered via technological and biological methods. Technical methods are as follows: mulching, raking and removal of the contaminated layer, burning out. However, it takes a long time for the ecosystem to recover, about 10-25 years and even more.

Biological methods are more promising and include various agricultural engineering techniques, such as liming, use of sorbents and fertilizers [8].

It is supposed that natural and synthetic sorbents, such as peat, polymers, sand, and containment booms are only auxiliary means, which prevent oil spread. Peat sorbents are also widely used today and tend to be more efficient than many organic sorbents, including those exported from other countries.

Siberian Research Institute of Agriculture and Peat of Russian Academy of Agricultural Sciences, Tomsk, received a patent on removal of petroleum from water surface. One gram of peat is enough to adsorb from 8–11 to 15 grams of oil. Peat sorbent developed at the research institute to liquidate oil spill effects, i.e. hydrocarbon contaminants and oil slick. The sorbent is produced from the upper peat with low decomposition rate (5–10%), which is modified to obtain hydrophobic properties. The sorbent porosity and sorption capacity are significantly higher than those of hydrocarbons.

The peat sorbent has many advantages in comparison with other sorbents, such as activated carbon, straw and bulrush chaff, buchwheat, which are characterized by lower capacity for oil (0.5–2.7 grams of oil per a gram of sorbent) and hydrophilicity causing release of trapped oil and preventing sorbents from floating.

In addition, synthetic sorbents fail to decompose under natural conditions and should be recycled, unlike peat sorbent, which is environmentally friendly. Moreover, synthetic sorbents absorb both oil and water, but peat absorbs only oil spill from the surface.

The issue of water deposits removal is especially topical for oil producing regions as there are many oil leaks caused by pipeline failures. When in water, hydrocarbons have a negative impact on fish, shell fish, etc.

The scientists from the Institute of Biology, Tomsk State University, in cooperation with NTO Priborservis have developed a technique which allows lifting up oil from the bottom. The technique is based on flotation effect: the place where petroleum is accumulated get under pneumo-mechanic impact, which makes oil "stick" to the contact of two phases, air and water, and rises to the surface. The method has already been efficiently implemented in the Republic of Komi, where it helped to raise 157 tons of oil from the bottom of Lake Schuchye [9].

Conclusion

Russia is rich in peat as the significant share of global peat reserves is located within the territory of the country, and particularly in West Siberia. Tomsk Oblast is ranked second in terms of peat reserves (first is Tyumen Oblast). The estimated peat reserves account for 30 881 412 thous. tons, which makes up 18 % of all national reserves. More and more people and organizations get interested in using peat, however, peat recovery in Tomsk Oblast is at its initial stage since there is no appropriate infrastructure, regulations are inapplicable to monitor peat recovery, reserves are poorly estimated, there is no actual data on peat bogs, etc.

To overcome these challenges, there should be innovations implemented in the industry. It is necessary to consider in detail the questions as follows: peat fields exploration and development, environmental assessment of peat bogs, the rate of fire hazard, innovative products which can be obtained from peat, the efficiency of peat in agriculture and treatment of petroleum-contaminated resources, etc.

Oil spills inevitably lead to negative environmental and economic effects, which causes biological misbalance of water resources. To cope with the consequences, it is important to adequately response to the oil spills, implementing modern techniques and technologies, following world's best practices contingency management, for instance, in case of fire [10–12].

References

[1] Bogoyavlensky V I 2014 Oil and gas exploration in the Arctic and the World' ocean: emergency response (in Russian) *Arctic: Environment and Economy* **4 (16)** 48

[2] Oil Tanker Spill Statistics 2015 (London, United Kingdom: ITOPF) 4

[3] Vladimirov V A 2014 Oil spills: causes, scales, effects (in Russian) *Civil safety strategy: issues and research* **1** 217

[4] Supreme Commercial Court took jurisdiction over revision of an award on 507 billion rubles for oil spill in the Kerch Strait 2013 (in Russian) [Electronic resource] URL: *http://portnews.ru/news/156514/* reference date 19.09.15

[5] Accidents comparable to The Gulf of Mexico oil spill tale place in Russia several times a year 2012 (in Russian) [Electronic resource] URL: *http://news2.ru/story/348542/* reference date 15.09.15

[6] Savichev O G, Tokarenko O G, Pasechnik E.Yu, Nalivaiko N.G., Ivanova EA., Nadeina L.V. 2015 Microbiological composition of river waters in the Ob' basin (West Siberia) and its connection with hydrochemical indices *IOP Conf. Series: Earth and Environmental Science 27 80292-8*

[7] Brakorenko N N, T.V. Korotchenko 2016 Impact of petroleum products on soil composition and physical-chemical properties *IOP Conf. Series: Earth and Environmental Science 33 012028*

[8] Kablov V F, Ioschenko Yu P 2004 Oil spills and the issue of oil removal *Fundamental Research* **6** 64–65

[9] The technique developed at the Institue of Biology, TSU, allows lifting up oil from water resource bottoms and saving… 2015 (in Russian) [Electronic resource] URL: http://www.tsu.ru/news/vernut_ozero_k_zhizni/?sphrase_id=13110 reference date 19.09.15

[10] Vazim A A, Romanyuk V B, Akhmadeev K N, Matveenko I A 2015 Associated petroleum gas utilization in Tomsk Oblast: energy efficiency and tax advantages *IOP Conf. Series: Earth and Environmental Science* **27** 012078

[11] Tsibulnikova M R, Kupriyanova O S, Strelnikova A B 2015 Economic assessment of environmental impact in the course of oil field development and production *IOP Conf. Series: Earth and Environmental Science* **27** 012076

[12] Matyugina E G, Glyzina T S, Kolbysheva Yu V, Vusovich O V 2016 Compensation for oil pollution damage *IOP Conf. Series: Earth and Environmental Science* **27** 012070

Changes in chemical composition of drilling waste water in taiga zone of Western Siberia (the Russian Federation) on the basis of thermodynamic approach

O.G. Savichev, I.A. Matveenko, D.V. Savchenko

Tomsk Polytechnic University, 30 Lenin Avenue, Tomsk, 634050, Russia
E-mail: OSavichev@mail.ru, mia@tpu.ru

Abstract. Condition and process change modeling of chemical composition of liquid drilling wastes has been performed in waste pits located in the boggy areas of Western Siberia. Dissolved substance migration and waste water saturation indexes were calculated in relation to a number of minerals and organic-mineral compounds. It was shown that reclamation of disturbed land can be made with the use of substrate produced by mixing liquid drilling wastes, high-bog (mostly at a technical stage) and low-bog (for forest reclamation) peats.

1. Introduction

Western Siberia (the Russian Federation) is known for production of large amount of oil and gas. It results in emission of a great deal of waste water and drilling wastes into the environment that defines the relevance of research in both ecosystem conditions – sites of oil and gas industrial facilities, and processes of storage, refining, utilization, and treatment of waste water and refinery wastes. The key feature of the regional environment is its extremely high bogginess achieving 30 % or more in some areas [1, 2]. Besides, at present in taiga zone of Western Siberia rather a high rate of bog formation – up to 2.62 mm/year vertically and ten thousands of hectares horizontally is observed [3, 4]. Bogs are distinguished by elevated concentration of organic matter and some transformation products including very toxic ones, which define general conditions of bog ecosystem functions as extremely specific and unfavorable for many species of plants and animals [5, 1, 6]. One might assume that waste waters and drilling wastes (taking into account their chemical composition) can be used in forest reclamation of boggy areas, that conditions the goal of the research, namely, forecast evaluation of chemical composition changes in waste waters which are accumulated in reservoirs («waste pits») in the course of drilling operations. In particular, the conditions of waste water self-purification in waste pits formed at exploration well drilling. The basic assumptions: 1) waste water and drilling wastes inflow the waste pits on a tight schedule; 2) drilling operations and infrastructure building are to be performed within a year, after which disturbed land is remediated. Usually, reclamation reduces to site surface leveling (technical stage), fertilizer treatment, and seeding (biological stage).

The general purpose of reclamation is natural condition recovery. However, in case of site reclamation located in oligotrophic bogs such actions lead to development of a new, quite sustainable body, the properties of which are close to those of eutrophic bogs. One more question raised in the course of reclamation in Western Siberia consists in expedience of wetland restoration that replaces forest ecosystems rapidly. In this case industrial activity often contributes to, but not prevents, bog formation. Actually, this fact is not described in the nature protection strategy of the region.

2. Initial data and research methods

As a research object we chose a boggy area of the Cluch River catchment (an element of «the Kluch River – the Bakchar River – the Chaya River – the Ob River – Kara Sea» river system) within the South taiga subzone of Western Siberia. A brief hydrometeorological characteristic of the river is presented in Table 1, more detail information including the technique of water balance calculation, – in works [7, 8].

Table 1. Climatic and hydrologic characteristics of the Cluch River and its catchment area (an element of «the Kluch – the Bakchar – the Chaya – the Ob» river system) [8]

Indicator	I	II	III	IV	V	VI	VII	VIII	IX	X	XI	XII	Total
T_a, °C	-20	-18	-10	1	9	15	18	14	9	0	-11	-18	-1
T_w, °C	-28	-24	-12	1	10	16	18	12	6	-3	-16	-26	-4
v_w, m/s	3.1	3.3	3.6	3.7	3.7	3.2	2.5	2.6	3	3.7	3.7	3.3	3.3
P_{sa}, mm	18	14	15	26	45	67	74	84	46	42	31	23	485
P_{la} mm	1	0	2	6	42	67	74	84	46	26	7	1	356
O_{sn}, mm	0	0	0	90	22	0	0	0	0	17	0	0	129
P_{la} + Osn, mm	1	0	2	96	64	67	74	84	46	43	7	1	485
E_{ws}, mm	3	5	11	21	53	64	75	62	75	38	6	4	417
E_{sn}, mm	3	5	11	0	0	0	0	0	0	0	6	4	29
E_{sw}, mm	0	0	0	58	101	147	161	126	83	0	0	0	676
Y, mm	0	0	0	11	37	12	3	1	1	2	1	0	68

Note: T_a – average monthly air temperature; T_w – calculated temperature of water surface; vw – wind velocity; P_{sa} – total amount of precipitation; P_{la} – liquid precipitation (rain); O_{sn} – water yield of snow cover; P_{la} + O_{sn} – actual total catchment moisture; E_{ws} – total evaporation of catchment surface; Esn – evaporation from snow cover (a part of total evaporation from catchment surface); E_{sw} – evaporation from water surface; Y – total (surface and underground) runoff.

It was assumed that waste pit (a reservoir for drilling waste and drilling waste water storage) is located in the site, which possesses the following typical parameters: pit volume – 800 m^3 (average volume of waste pit per one well), volume of liquid flowing into the pit per one time (with regard for moisture 70 %) – 560 m^3; average depth – 2.5 m [9, 10]. Chemical composition of waste, rain (Table 2), and bog waters was taken in terms of the results of previous research [11-14].

The research technique includes:

1) calculation of annual liquid balance in simulated waste pit using data on precipitation and evaporation from water surface (at assumed negative temperature of water surface the evaporation from snow was taken into account);

2) calculation of annual hydrochemical balance using the formula (1);

$$C_x \cdot (V_{dr} + P_{la} + O_{sn} - E_{sn} - E_{sw}) = C_{dr} \cdot V_{dr} + C_{la} \cdot P_{la} + C_{sn} \cdot O_{sn}, \quad (1)$$

where V_{dr} is the waste pit volume; P_{la} is liquid precipitation (rain); O_{sn} is the water yield from snow cover; E_{sn} is evaporation from snow surface; E_{sw} is evaporation from water surface; matter concentration in: C_{dr} – in waste pit water; C_{la} – liquid precipitation; C_{sn} – melt water; the balance is defined for the substances from Table 2;

3) evaluation of change in waste pit water concentration by means its single mixing with high and low bog peat with the assumption that matter balance is described by the equation (2):

$$M \cdot k_a \cdot C_{x,s} + V_l \cdot C_{x,s} = V_l \cdot C_{x,0}, \quad (2)$$

where M is the mass of sorbent; V_l is the liquid volume; k_a is a coefficient; $C_{x,0}$ and $C_{x,s}$ are the matter concentration in the initial liquid and after mixing liquid and sorbent [15]; the value k_a is determined

by matching the data on chemical composition of uncontaminated water in oligotrophic and eutrophic bogs [12] with the assumption that the impact of contamination source is limited by the site of up to 500 m in length; average depth of peat deposit is taken as 1.95 m in size, the peat moisture is 90.1 %;
4) calculation of migration conditions and saturation index L of waste pit water in relation to a number of minerals; negative magnitudes of index L indicate the potential undersaturation, whereas positive ones – solution oversaturation in relation to a number of minerals that can form or be present in the research medium [16];
5) simulation of chemical composition change towards minimization of Gibbs energy in «water – rock» system; thermodynamic calculations were performed by means of software Solution+ based on the method of constant [17].

Table 2. Average chemical composition of waste pit water and precipitation

Indicator	Waste pits [Bleuten et al., 1999; Savichev, Bernatonis, Bernatonis, 2011]	Snow [Savichev, Ivanov, 2010]	Rain [Savichev, Ivanov, 2010]
pH	8.00	6.10	6.50
Sum of principle ions	7211.80	23.10	50.90
Ca^{2+}	248.40	3.10	5.40
Mg^{2+}	31.80	0.70	2.00
Na^+	2124.80	1.90	2.30
K^+	202.20	0.70	0.90
HCO_3^-	338.30	12.10	29.10
SO_4^{2-}	388.10	2.30	8.00
Cl^-	3874.40	2.40	3.20
NO_3^-	4.96	0.98	0.01
NO_2^-	0.026	0.030	0.960
NH_4^+	1.60	0.74	1.23
PO_4^{3-}	0.68	0.14	0.72
Si	7.32	0.28	0.15
COD	364.69	4.00	4.00
Fe	19.277	0.090	0.110
Cu	0.0072	0.0046	0.0096
Zn	0.0336	0.0125	0.1390

Note: COD – chemical oxygen demand

When performing the thermodynamic calculations we took into account the presence of fulvic acids (PA) (3) and humic acids (HA) in the solution (4), concentration of which was determined depending on chemical oxygen demand COD:

$$PA=0.3122 \cdot COD,\ R^2=0.57, \quad (3)$$

$$HA=0.0832 \cdot COD,\ R^2=0.59, \quad (4)$$

where R^2 is the square of correlation ratio. In case of thermodynamic calculations just for waste pit water we assumed that fulvic and humic acids flow into the waste pit mostly with precipitations, but in case of peat – with bog water.

3. Results and discussion

If we assume the absence of additional water inflow in waste pit and correlation of evaporation from water surface and waste pit water, the decrease in total liquid volume of waste, drilling waste water, and precipitation is to be observed during a year (Fig. 1). In particular, in case of simulated waste pit the liquid volume decreases from 560 m^3 to 490 m^3 during a year due to evaporation (despite precipitations). Due to solution concentration it is possible to increase concentration of some substances (Fig. 1, Table 3). This results in changes of conditions in «water – rock» system including fall-outs of some slightly soluble compounds and, as a consequence, decrease in concentrations of soluble substances (Table 4).

Additional decrease in a number of substances in waste pit water can be potentially achieved by means of its mixing with peat, mostly with high-bog one (Table 3). It does not preclude mixing with low-bog peat, but changes the technique of the substrate obtained. Taking into account higher content of uncomplexed ions of Na^+ and Cl^- in waste pit water after its mixing with low-bog peat (consequently – lower concentration in the substrate), it is reasonable to apply the obtained substrate for forest reclamation of disturbed boggy areas. The use of high-bog peat for sorption would presumably allow the efficient decrease of substance concentration in liquid waste, but the substrate based on it is better to be applied at the technical stage of land reclamation due to elevated concentrations in the absorption complex.

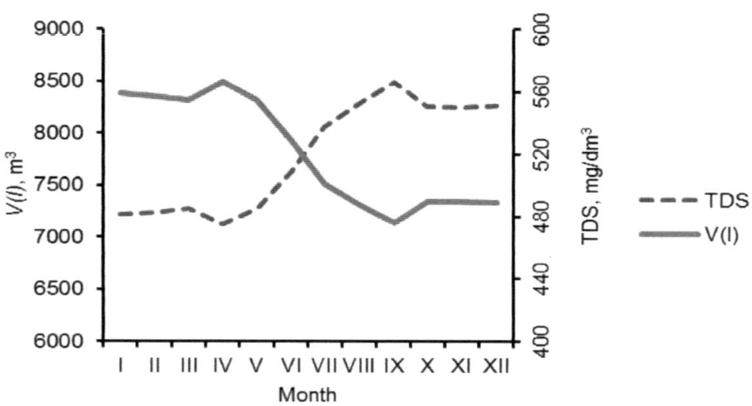

Fig. 1. Change of liquid volume in waste pit during a year (V(l)) and sums of dissolved substances in it (TDS)

Table 3. Calculated concentrations of uncomplexed ions in waste pit water, mg/dm^3

Ion	Alternate design					
	I	II	III	IV	V	VI
Ca^{2+}	228.7	94.9	256.1	109.1	253.6	232.9
Mg^{2+}	29.4	12.2	33.3	14.1	33.5	31.8
Na^+	2105.6	811.6	844.9	928.6	1490.0	569.4
K^+	201.6	77.2	230.8	88.5	167.3	118.0
HCO_3^-	313.8	321.8	374.0	373.6	374.0	356.6
CO_3^{2-}	0.6	0.6	<0.1	<0.1	<0.1	<0.1
SO_4^{2-}	347.5	148.2	394.8	170.4	313.7	336.3
Cl^-	1740.6	1479.9	1694.2	1693.2	1692.5	622.4
Сумма	4967.9	2946.4	3828.3	3377.6	4324.5	2267.5

Note: I – calculation of migration conditions without taking into account waste pit water dissolving with precipitations (Table 2); II – simulation of chemical composition change towards minimization of Gibbs energy in «water – rock» system» without taking into account waste pit water dissolving with precipitations (based on design I); III – calculation of migration conditions taking into account waste pit water dissolving with precipitations; IV – simulation of chemical composition change towards minimization of Gibbs energy in «water – rock» system» taking into account waste pit water dissolving with precipitations (based on design III); V – concentration after waste pit water mixing with low-bog peat (based on design III); VI – concentration after waste pit water mixing with high-bog peat (based on design III)

Judging by the saturation index calculation (Table 4), it should be noted that more efficient decrease in concentration of some metals (with the exception of Na and K) might be achieved by additional introduction of humic acids in waste pit due to slightly dissolved compound formation.

Table 4. Magnitudes of waste pit water saturation index L in relation to some minerals and organo-mineral complexes

River point	Alternate design					
	I	II	III	IV	V	VI
$CaCO_{3\,\text{(calcite)}} = Ca^{2+}+CO_3^{2-}$	0.13	-0.09	-1.18	-1.17	-8.53	-8.45
$CaCO_{3\,\text{(calcite)}}+CO_2+H_2O = Ca^{2+}+2\cdot HCO_3^-$	0.69	0.47	-0.29	-0.62	0.10	-0.71
$CaMg(CO_3)_{2\,\text{(dolomite)}} = Ca^{2+}+Mg^{2+}+2\cdot CO_3^{2-}$	0.39	-0.05	-2.23	-2.21	-16.91	-16.71
$CaMg(CO_3)_{2\,\text{(dolomite)}} + 2\cdot CO_2 + 2\cdot H_2O = Ca^{2+}+Mg^{2+} +4\cdot HCO_3^-$	1.51	1.07	-0.45	-1.11	0.35	-1.22
Ca(ГК) = Ca^{2+}+ГК	1.28	0.99	1.37	1.03	1.35	1.43
Mg(ГК) = Mg^{2+}+ГК	1.11	0.81	1.20	0.86	1.19	1.32
$SiO_{2\,\text{(quartz)}} + 2\cdot H_2O=H_4SiO_4^0$	0.49	0.07	0.56	0.14	0.55	0.54
$CaAl_2Si_2O_{8\,\text{(anorthite)}} + 3\cdot H_2O + 2\cdot CO_2 = Al_2Si_2O_7\cdot 2\cdot H_2O_{\text{(kaolinite)}} + Ca^{2+} + 2\cdot HCO_3^-$	-256.2	-256.4	-258.5	-258.8	-257.6	-259.3
$2\cdot NaAlSi_3O_{8\,\text{(albite)}} + 11\cdot H_2O + 2\cdot CO_2 = 2\cdot Na^+ + 2\cdot HCO_3^- + Al_2Si_2O_7\cdot 2\cdot H_2O_{\text{(kaolinite)}} +4\cdot H_4SiO_4^0$	-4.36	-6.71	-7.18	-8.74	-5.84	-8.47
$3\cdot KAlSi_3O_{8\,\text{(microcline)}} + 2\cdot H^+ + 12\cdot H_2O = KAl_3Si_3O_{10}OH_{2\,\text{(muscovite)}} + 2\cdot K^+ + 6\cdot H_4SiO_4^0$	-16.88	-20.13	-18.68	-22.01	-18.12	-20.29
$2\cdot KAl_3Si_3O_{10}OH_{2\,\text{(muscovite)}} + 2\cdot H^+ + 3\cdot H_2O = 3\cdot Al_2Si_2O_7\cdot 2\cdot H_2O_{\text{(kaolinite)}} +2\cdot K^+$	4.63	3.84	2.41	1.59	2.98	0.86
$CaAl_2Si_2O_{8\,\text{(anortite)}} + 2\cdot H^+ + 6\cdot H_2O = Al_2O_3\cdot 3\cdot H_2O_{\text{(gibsite)}} + 2\cdot H_4SiO_4 + Ca^{2+}$	-3.36	-4.47	-5.49	-6.66	-4.65	-6.43
$CaAl_2Si_2O_{8\,\text{(anortite)}} + 2\cdot H^+ + H_2O = Al_2Si_2O_7\cdot 2\cdot H_2O_{\text{(kaolinite)}} + Ca^{2+}$	-19.47	-19.90	-23.94	-24.59	-22.28	-25.70
$NaAlSi_3O_{8\,\text{(albite)}} + 7\cdot H_2O + H^+ = Al_2O_3\cdot 3\cdot H_2O_{\text{(gibsite)}} + Na^+ + 3\cdot H_4SiO_4^0$	-200.8	-202.4	-202.1	-203.3	-201.5	-202.8

4. Conclusions

Oil and gas production in Western Siberia has a sufficient impact on the environment [18, 19] that often can be estimated as regional and long-term. In most cases this impact may be decreased due to improvements in locations and reclamation of waste pits in boggy areas as well as more efficient use of natural conditions. Besides, it is possible to reduce the costs for nature protection measures without decreasing their performance. In particular, disturbed land reclamation can be performed using the substrate obtained by means of mixing drilling waste water, high-bog (mostly at the technical stage) and low-bog (for forest reclamation) peats.

The research has been supported by the grant issued in accordance with VIU_VAF_144_2014 and RFBR r_a 16-48-700230.

References

[1] Bazanov V A, Berezin A Ye, Savichev O G and Skugarev A A 2009 The phytoindication method for mapping peatlands in the taiga zone of the West-Siberian Plain *International Journal of Environmental Studies* **4** 473–484

[2] Zemtsov V A and Savichev O G 2015 Resources, regime and quality of surface waters in the Ob River basin: history, current state and problems of research *International Journal of Environmental Studies* **3** (Published online: 17 Mar 2015) DOI:10.1080/00207233.2015.1019299

[3] Neishtadt M I 1972 Bogs of Ob-Irtysh interstream area In: Zhukov, M.V. (Eds) *Natural conditions of Ob-Irtysh interstream area development* (Institute of Geography Publishing House) 322–346
[4] Pologova N N and Lapshina Ye D 2002 Accumulation of hydrogen in peat deposits of Large Vasyugan bog In: Kabanov M V (Eds.) *Large Vasyugan bog. Modern condition and development processes* (Tomsk, Atmosphere Optics Institute RAS SB Publishing House)174–179
[5] Liss O L, Abramova L I, Avetov N A, Berezina N A, Inisheva L I, Kurnishkova T V, Sluka Z A, Tolpysheva T Yu and Shvedchikova N K 2001 Bog ecosystems of Western Siberia and their nature protection significance (Tula: Grif i C°)
[6] Eckstein Y, Savichev O G and Pasechnik E Yu 2015 Two decades of trends in ground water chemical composition in The Great Vasyugan Mire, Western Siberia, Russia *Environmental Earth Sciences* **1** 3-15 DOI 10.1007/s12665-014-3908-z
[7] Savichev O G, Skugarev A A, Bazanov V A and Khanzhevskaya Yu A 2011 Water balance of boggy catchment areas of Western Siberia (by the example of the small river Kluch, Tomsk Oblast) *Geoinformatika* **3** 39 – 46
[8] Savichev O G, Bernatonis P V and Bernatonis V K 2012 Hydrological foundation of peat bog development (by the example of the River Kluch catchment, Western Siberia) *Bulletin of Tomsk Polytechnic University* **320:1** 155 – 162
[9] Regulation of engineering design of gathering, transportation, refining facilities of oil, gas, ans water at oil fields. VNTP 3-85 1986 (Moscow, Minneftprom SSSR)
[10] Instruction on nature protection at well construction on land at polycomponent hydrocarbon field including sulphur containing. RD 51-1-96 1998 (Moscow, IRC Gazprom)
[11] Bleuten W, Lapshina E, Ivens W, Shinkarenko V and Wiersma E 1999 Ecosystem recovery and natural degradation of spilled crude oil in peat bog ecosystems of westsiberia *International Peat Journal* **9** 73–82
[12] Savichev O G 2009 Chemical composition of bog water over the territory of Tomsk Oblast (Western Siberia) and its interaction with mineral and organic-mineral compounds *Bulletin of Tomsk Polytechnic University* **314:1** 72-77
[13] Savichev O G and Ivanov A O 2010 Precipitations in the Middle Ob River basin and their influence on hydrochemical river flow *Izvestia RAN. Series of Geography* **1** 63-70
[14] Savichev O G, Bernatonis P V and Bernatonis V K 2013 Hydrochemical conditions of drilling waste storage and utilization in peat-bog geosystems of Siberia *Tomsk State University Journal* **375** 183–186
[15] Zhukov A I, Mongait I L and Rodziller I D 1977 Methods of industrial waste water treatment (Moscow, Stroyizdat)
[16] Grenthe I and Puigdomenech I 1997 Symbols, standards, and Conventions *Modelling in aquatic chemistry* (Paris, Nuclear energy agency) 35 – 68
[17] Savichev O G, Kolokolova O V and Zhukovskaya E A 2003 Composition and equilibrium of bottom sediments of the River Tom with river water *Geoekologia* **2** 108-119
[18] Adam A M 2005 Evaluation of ecological conditions of the Western Siberia territory for the purpose of ecological safety in the context of sustainable nature use In: *Okhrana Pripody* (Ed. by A.Ye. Berezin) (Tomsk, NTL Publishing House) 3–12
[19] Savichev O G, Kolesnichenko LG and Saifulina E V 2011 The ecologo-geochemical state of water bodies in the Taz-Yenisei interfluve. *Geography and Nature Resources* 32. 4 333-336, DOI: 10.1134/S1875372811040056

Basin Model of Total Dissolved Salts Transformation in Water of a Small River (the Kirgizka River, Tomsk, Russia)

O.G. Savichev, I.A. Matveenko, I.V. Titov

Tomsk Polytechnic University, 30 Lenin Avenue, Tomsk, 634050, Russia
E-mail: OSavichev@mail.ru, mia@tpu.ru

Abstract. The basin model of total dissolved salt transformation in river water has been developed. It was tested in the Kirgizka River, the right tributary of the Tom River (Russia, Western Siberia, Tomsk). It was shown that the river system has the capacity of self-purification and is characterized by rather stable salt composition. It is explained by the fact that the growth in dissolved salt concentration in river water is limited to some extent by, firstly, dilution of more mineralized groundwaters drained by rivers, and, secondly, relatively low solubility of some compounds.

1. Introduction

The effective wastewater treatment is a necessary condition for sustainable development of natural-manmade systems of all levels. However, it is not only quantity and type of treatment facilities that are of great significance, but also standard development for treated wastewater discharge into water bodies within river basins or large water sites. This development includes both administrative and management measures as well as techniques of accumulation and processing of initial information and calculation of permissible concentrations (PC) of pollutants in treated wastewater, at which discharge impact on water body would not result in sufficient deterioration of water quality. Common techniques of PC determination have been developed, described in detail in literature, and included in the list of current regulations in the sphere of environment protection [1–4]. Yet, the technique of basin calculation has not been sufficiently developed, because of which, firstly, it is still a nontrivial problem [2, 5]. Secondly, PC basin calculation is performed relatively rare and, in most cases, the effect is calculated for a single source.

Similar situation is observed in the watershed of the Bolshaya Kirgizka River, a component of "the Bolshaya Kirgizka River – the Tom River – the Ob River – the Kara Sea" system. This area, included into the Northern Industrial hub (NIH) of Tomsk agricultural agglomeration (fig. 1), is characterized by sufficiently high density of residential and industrial sites, which are the sources of great amount of wastewater and pollutants [6–8]. Taking it into account, the authors have developed and tested pollutant transformation model in terms of total dissolved salts (TDS) discharged into the Bolshaya Kirgizka River and its tributaries from lumped sources (discharge of treated and untreated effluents) using average long-term data. Selection of average long-term values is explained by the need for hydrochemical balance calculation based on the average long-term values of hydrological and hydrochemical factors with model parameter optimization not determined directly by experiment, but selected by minimizing the sum of squared differences of calculated and measured concentrations [9].

2. Initial data and research methods

The Bolshaya Kirgizka River flows into the Tom River 58 km from its mouth; the river length – 85 km, catchment area – 898 km^2; major tributaries – the Malaya Kirgizka, Omutnaya, Chernaya, Kamenka, Topkaya Rivers (fig. 1).

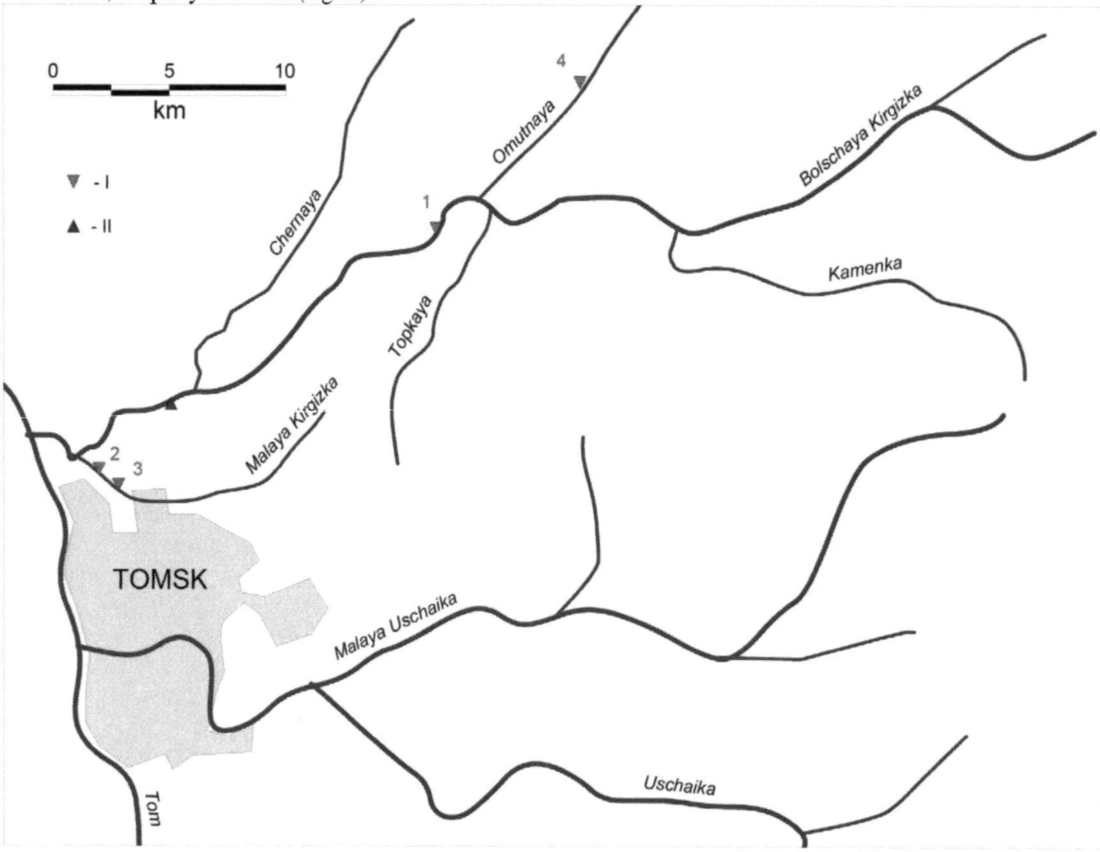

Fig. 1. The scheme of the river network of the study area; symbols: I – wastewater inflow (the number of inflow corresponds to table 1); II – point of hydrological observation on the Bolshaya Kirgizka River

The basic inflows of wastewater of lumped sources come from treatment facilities of Tuganskaya poultry plant (discharge into the Bolshaya Kirgizka River), Tomsk poultry plant (discharge into the Omutnaya River) and JSC "Tomsk plant of chipboard production" (discharge into the Malaya Kirgizka). Wastewater is treated, but the treatment effectiveness is not sufficient enough for a number of substances: maximum – for suspended matter and easily oxidized organic substances in terms of 5 day biochemical oxygen demand (BOD_5); minimum – for nitrogen and phosphorus compounds (in many cases post-treatment is required to remove organic decomposition products; concentration of substances resistant to oxidation in terms of chemical oxygen demand value (COD) and sum of total dissolved salts (TDS) after treatment virtually do not change or change within the margin of error for average value determination [10]. Being the first stage of research, the current paper deals with the changes in sum of total dissolved salts as a basic hydrochemical indicator for assessment of general geochemical condition of the river ecosystem.

The initial data on composition of waste and river water used for calculation of total dissolved salt transformation in water of the Bolshaya Kirgizka River are presented in tables 1 and 2. TDS values of the Malaya Kirgizka and Omutnaya Rivers sources are taken according to table 2 for the upstream points of wastewater discharge, for the sources of the Bolshaya Kirgizka, Chernaya, and Kamenka Rivers – using average values of TDS (467.7 mg/dm^3) calculated for downstream points of the Bolshaya Kirgizka, Malaya Kirgizka, and Omutnaya Rivers from the water discharge. Wastewater discharge is taken as an average value without taking into account daily and hourly irregularity, m^3/sec: wastewater treatment facilities (WTF) of Tuganskaya poultry plant – 0.0110; WTF of Tomsk poultry plant – 0.0151; WTF of JSC "Tomsk plant of chipboard production" – 0.0139; discharge of untreated storm and melt waters – 0.0016. In addition, water abstraction from the Bolshaya Kizgizka River is taken into account in two points in the amount of 0.0048 and 0.0265 m^3/sec.

Table 1. The total dissolved salts in the Bolshaya Kizgizka, Malaya Kirgizka, and Omutnaya Rivers (according to [7, 10])

The number of outlet in fig. 1	Water users	Wastewater receiver	Wastewater description	A, mg/dm^3	N
1	Tuganskaya poultry plant	The Bolshaya Kirgizka River	Before treatment	1189.2	29
			After treatment	1023.4	29
2	"Tomsk plant of chipboard production"	The Malaya Kirgizka River	Before treatment	460.5	8
			After treatment	369.1	8
3	Thermal Power Plant -1	The Malaya Kirgizka River	Sewage system without treatment	473.4	6
4	Tomsk poultry plant	The Omutnaya River	Before treatment	1382.9	17
			After treatment	835.9	18

Note: A – arithmetic average; N – the number of samples

Table 2. The total dissolved salts in the Bolshaya Kizgizka, Malaya Kirgizka, and Omutnaya Rivers (according to [7, 10])

River	Point (the number of outlet in fig. 1)	Point	A, mg/dm^3	N
The Bolshaya Kirgizka River	Rassvet settlement, WTF of Tuganskaya poultry plant (1)	above the outlet	447.1	15
		below outlet	524.6	13
	Tomsk	inlet	265.7	16
The Malaya Kirgizka River	Tomsk, WTF of JSC "Tomsk plant of chipboard production" (2)	above the outlet	516.2	15
		below outlet	454.7	14
The Omutnaya River	Molodezhny settlement, WTF of Tomsk poultry plant (4)	above the outlet	439.8	10
		below outlet	492.4	10
	Tomsk city	inlet	358.7	2

Note: A – arithmetic average; N – the number of samples; WTF – wastewater treatment facilities

Hydrological parameters are taken according to the observation data of the Bolshaya Kizgizka River at Kuzovlevo settlement (fig. 1): mean water discharge Q=5.17 m^3/sec; mean unit discharge M=6.27 L/(s·km^2); mean slope of water surface S=1.35 m/km; roughness code of free channel $n_{M.OR}$=0.16. the calculation of hydrological parameters of the Bolshaya Kizgizka River and its tributaries in other points is carried out using the unit of water flow 6.27 L/(s·km^2), catchment area, and dependencies of water discharge, discharge cross section area, flow velocity, and mean depth calculated at the point of Kuzovlevo settlement. In this case, it is sure to be an oversimplified but acceptable assumption based on similarity of morphometric parameters of water flows.

The TDS transformation is calculated by means of the model according to [1, 5]:

$$C_p = C_1 \cdot \exp(-k_C \cdot t_m) + \sum_{i=1}^{Nw} \frac{(C_{w,i} - C_1)}{n_{C,i}} \cdot \exp(-k_C \cdot t_i), \qquad (1)$$

where C_p and C_1 – substance concentration in the control point and sources of the Bolshaya Kizgizka River; C_w – substance concentration in wastewater; Nw – the number of pollution sources; k_C – self-purification factor of river water in natural condition and under the impact of i-th outlet or inlet; t_m and t_i – the time of water movement from the source of the Bolshaya Kizgizka River and pollution sources (junctions); n_C – wastewater dilution factor (or water inflow) calculated by Frolov-Rodziller method [3]. The calculation method applicable to the set of sources is described in detail in [5]. The same model is applied to calculate TDS transformations of water inflow. In this case, the TDS value in the junction is presented in the form of source together with wastewater outlets.

3. Results and discussion

The calculation performed has shown that, firstly, the model (1) is an adequate description of average long-term changes in TDS of the Bolshaya Kizgizka River and its tributaries (fig. 2). Secondly, changes in total dissolved salts in river water along the river length are best described by k_C=0.093 day^{-1}. Therefore, TDS value should be considered as nonconservative. Thirdly, the river system is determined as rather resistant to dissolved salt discharge with wastewater. One of the explanations of this phenomenon is a combination of, on the one hand, dilution of more mineralized groundwater drained with rivers and precipitations, but, on the other hand, interactions in the "water –rock – organic matter" system.

According to Kolubayeva [8], the total dissolved salts in groundwaters of the area ranges from 507 mg/dm^3 on average (loose deposit water) to 527 mg/dm^3 (rock waters), whereas that of precipitation, the sum of principle ions is typical for this area: rain water in the Tom River basin within Kemerovo Oblast (to the south from Tomsk) – 73.6 mg/dm^3; melt water in the territory of Tomsk Oblast – 31.6 mg/dm^3 [7].

The effect of the second factor is shown and explained in detail for groundwaters in some works [11–13]. But, the same explanation is applicable for average long-term values of river water. In particular, in [14] it was stated that the most intensive interaction of river water and sediment is observed spatially – in the upstream river flow, but in time – in the flood periods. In this case, the following peculiarities are identified:

1) the state close to equilibrium is mostly observed only in interaction of water with calcite, clay minerals, quartz, metal compounds with humic acids;

2) as for primary aluminosilicates, river waters are undersaturated everywhere and during the whole year, since equilibrium state is limited by calcium extraction from solution in combinations with carbonic and humic acids;

3) the growth of dissolved salt concentrations in the river water – wastewater receiver is limited to a definite extent by relatively low solubility of some principle ion compounds that prevents from increasing total dissolved salts or even results in its decrease [14].

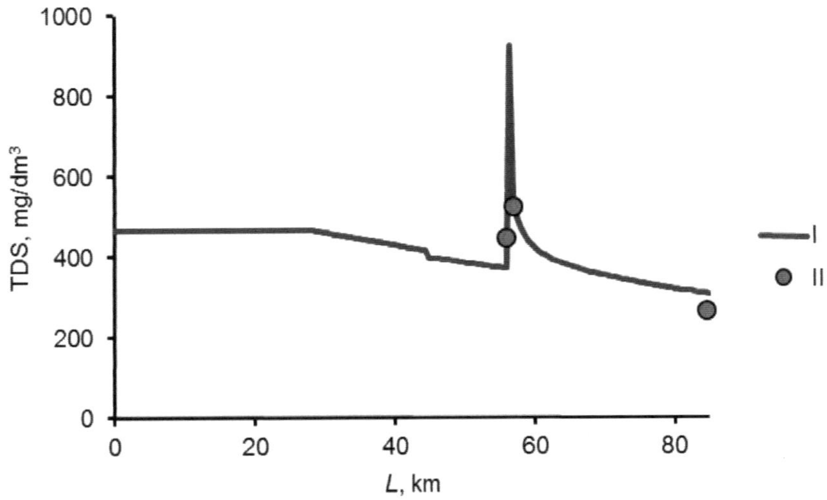

Fig. 2. Changes in predicted (I) and mean measured values of TDS (II) along the length (L) of the Bolshaya Kizgizka River (from the source)

On the whole, the analysis of modeling results (fig. 1) confirms these conclusions and reflects the significant river system capability of self-regulation.

4. Conclusions

In the course of study, the model of total dissolved salt transformation has been developed and tested in the Bolshaya Kizgizka River under the impact of wastewater discharge and the effect of tributaries (the Malaya Kirgizka, Omutnaya, Chernaya, Kamenka, Topkaya Rivers). It is stated that the river system has a significant capacity of self-purification (from inorganic salts) and is characterized by sufficiently stable salt composition. It is explained by the fact that the growth of dissolved salt concentrations in river water is limited to some extent by, firstly, dilution of more mineralized groundwater drained by the rivers and, secondly, relatively low solubility of some compounds. One more important conclusion supported by the modeling results consists in the fact that total dissolved salts should be considered as nonconservative hydrochemical factor that, according to [14], is a complex function of a number of physical, chemical, and biochemical processes in "river water – sediment – river load" system at the surface of catchment basin and active groundwater exchange zone.

The research has been supported by the grant issued in accordance with VIU_VAF_144_2014 and RFBR r_a 16-48-700230.

References

[1] Zhukov A I, Mongait I L and Rodziller I D 1977 Methods of industrial wastewater treatment (Moscow, Stroyizdat)

[2] Loucks D P and Van Beek E 2005 Water resources systems planning and management. An Introduction to Methods, Models and Applications (Turin, UNESCO Publishing)

[3] Methods of developing norms for permissible discharge of matter and microorganisms into water bodies from water users 2008 (Moscow, Ministry of Natural Resources, Russian Federation)

[4] Benedini M and Tsakiris G. 2013 Water quality modelling for river and streams (Dordrecht, Springer)

[5] Savichev O G and MatveenkoI A 2013 Evalution of chemical composition changes of surface

IOP Conf. Series: Earth and Environmental Science **43** (2016) 012028 doi:10.1088/1755-1315/43/1/012028

water in Boguchan Reseroir (Siberia, Russia) *Hydrological Sci J* **58** **3** 706 – 715.
[6] Ecology of the North of the Industrial Hub, Tomsk city. Problems and solutions 1994 (Tomsk, Tomsk State University Press)
[7] Savichev O G 2010 Vodnye resursy Tomskoy Oblasti (Tomsk, TP Univesity Press)
[8] Kolebayeva Yu V 2015 *Hydrochemistry of the north-eastern part of Kolyvan-Tomsk folded area A*bstract of dissertation for Cand. of Geol.-Mineral. Science degree (Tomsk, Tomsk Polytechnic Un-ty Press) 22
[9] Nash J E and Sutcliffe J V 1970 River flow forecasting through conceptual models. P. I. A discussion of principles *J of Hydrology* **10** **(3)** 282–290
[10] L'gotin V A, Savichev O G and Nigorozhenko V Ya 2006 The state of surface water bodies, water-management systems and facilities in the territory of Tomsk Oblast in 2000-2005 (Tomsk, Tomskgeomonitoring)
[11] Garrels R M and Christ C L 1965 Solution, minerals and equilibria (San Francisco, Freeman, Cooper)
[12] Grenthe I and Puigdomenech I 1997 Symbols, standards and conventions Modelling in aquatic chemistry (Paris, Nuclear energy agency) 35–68
[13] Shvartsev S L 2008 Geochemistry of fresh Groundwater in the Main Landscape Zones of the *Earth Geochemistry International* **13** 1285–1398
[14] Savichev O G 2009 Influence of the interactions within the water - earth materials system on the formation of river water composition in the Ob basin *Geography and Natural Resources* **30, 2** 161-166

Determination of water content in clay and organic soil using microwave oven

V V Kramarenko, A N Nikitenkov, I A Matveenko, V Yu Molokov and Ye S Vasilenko

Tomsk Polytechnic University, 30 Lenin Avenue, Tomsk, 634050, Russia
E-mail: kramarenko-v-v@mail.ru

Abstract. The article deals with the techniques of soil water content determination using microwave radiation. Its practical application would allow solving the problems of resource efficiency in geotechnical survey due to reduction of energy and resource intensity of laboratory analysis as well as its acceleration by means of decreasing labour intensity and, as a result, cost reduction. The article presents a detail analysis of approaches to soil water content determination and soil drying, considers its features and application. The study in soil of different composition, typical for Western Siberia including organic and organic-mineral ones, is a peculiarity of the given article, which makes it rather topical. The article compares and analyzes the results of the investigation into soil water content, which are obtained via conventional techniques and the original one developed by the authors, consisting in microwave drying. The authors also give recommendation on microwave technique application to dry soil.

1. Introduction

When developing modern economy the most urgent task is to increase the resource efficiency in all branches of industry. Particularly, in building industry one of the most significant stages is engineering-geological survey. One of the main trends in solving this problem is time reduction of soil tests and increase in field and laboratory work efficiency due to rapid dry of disperse soils when determining water content, preparing samples for grading test, water permeability test, natural slope angle determination, solid particles density and other tests. In international practice, household microwave ovens are widely used to dry soil, which reduces the time of soil water content determination, test and facility cost. Therefore, a comparative analysis of water content determination of clay and organic soils in microwave and drying ovens is of particular interest.

The purpose of the article is to study the process of drying clay, organic, and organic-mineral soils in microwave oven to determine their water content. The tasks include: review of domestic and foreign literature dealing with methods of soil water content tests; identification and description of specific features of clay and organic soil drying, laboratory test to determine the identification indicators of soil composition and physical properties, determination and comparison of their water content when drying in microwave and convection ovens, developing brief recommendations on using microwave oven to dry finely dispersed soils rapidly.

As it was evident from the literature and standard review, there are many ways of soil water content test [1-18]. In our case, according to the existing specifications, the drying technique in convection oven up to constant moisture is required. According to GOST 5180 [2] soil samples of 15-50 g (10-20 g for hydroscopic content, 10-15 g for water content at plastic limit) are dried at temperature 105±2 °C during 5 hours, after that soil samples are weighed in 2 hours before mass difference at two series weighing of not less than 0.02 g. According to GOST 11305 [3] peat samples of 5–10 g are placed in the oven heated up to 105–110°C and dried during 2.5–4.0 hours, after weighing the portions are dried during 30 min, if losses are not more than 0.01 g, test ends. Using rapid technique [3] the peat portions of 5–6 g are placed in the oven heated up to 165–170 °C, after which they are dried at 145–150 °C during 30 min, but at water content (water content is a relationship of water weight to the initial weight of wet peat) more 55 % – 45 min. According to ASTM D2216 [17] drying of dispersed and rocky soil is performed at 110±5°C. To decrease gypsum dehydration of saline soil or organic loss in peats, they are dried at 60° C or in a desiccator.

According to ASTM D 4643 [18], to determine water content a portion of 100-200 g (for soil containing up to 10 % of fraction more than 2 mm) are placed in the oven for 3 min at defrost mode, then after a minute it is weighed up to the moment when the weight loss is less than 0,1 % at reweighting. The initial time of clay soil drying can be increased up to 12 min. Water content is calculated as a relationship of water weight to the dry soil weight with accuracy up to 0.1 %.

Clay and organic soils have a number of features developed while drying. Organic soils can lose organic matter and increase in mass after removal of free water. As for clay soils [19], the processes taking place at their firing are divided into several stages depending on the temperature. The 1-st stage of water removal includes three phases: water movement inside the clay soil, vapor development, and vapor transfer from the surface. Oversanded clay is less sensitive to drying than clays with more content of clay particles. The process of water removal is reversible and, after water saturation, clay has the same plasticity as before drying. Some researchers have indicated some loss of swelling property.

Drying depends on mineral composition of clays: montmorillonite gives off and preserves water slowly and heavily. The value of relative drying shrinkage is within the range 2-8 %, dispersed clays show more shrinkage, oversanded clays – less one. At temperatures from 100 to 400 °C the 2-nd stage is observed: in the course of water removal montmorillonite losses 5-6 % of chemically bound water (2/3 of the whole water content), kaolin - 13 %. Within this period the crystal lattice of clay minerals weakens and there occurs a preparation for the third stage – dehydration process taking place at temperatures from 400 to 700 °C [19]. As T.G. Makeeva and Yu.M. Yegorov stated [20], *"microwave field have an effect on soil at the atomic-molecular level generating dielectric heating. In this case there is a migration of moisture and phase transfer of bound water in soil depending on the structure and texture of soil and molecular-structure features of water as well as microwave field parameters and polarization at the interface of bound water and transition water. Energy absorption of electromagnetic waves from absorbed water of dispersed system is a limiting stage and corresponds to the water-binding energy equal to the value of layered charge by radiation frequency"* [20].

According to Prokhina A.V. [21], X-ray diffraction analysis shows that no significant changes are observed at clay microwave treatment. Grouping in terms of the particle sizes indicates that in samples of green clay the prevailing fraction is particles of 5-10 mkm radius, but after microwave treatment the particles of radius less than 0.01 mkm prevail. Uniform sample heating in microwave oven is an advantage of the technique [22]. In drying ovens at temperature 100–160°C the loss of physical and weakly bound water is observed. Sample drying occurs from the edge to the center, which is the main disadvantage, as due to low heat conductivity insufficient energy penetrates to the inner part of sample to evaporate water.

As a result of brief review, it should be noted that despite complexity and ambiguity of clay and organic soil drying in microwave oven, free water is largely removed from them and, as it has been mentioned, ovens are rather helpful for soil laboratory tests [26, 27].

2.Materials and methods

In this article the authors study typical for Western Siberia peats, clays, and peaty soils (table1, figure 1) from different sites: Bely Yar settlement (№1-7), Dvurechensk oil field (№ 8, 17, 18), Kalinovoye field (№ 13-15), Danenbergovsk field (№ 9,16,17), and from sites of Tomsk city (№ 10-12). According to the methods [1-4, 18, 23-25], the following values were determined: water content (w), organic content (I_r), and water content at plastic and liquid limits (w_L and w_p) of clay and peaty soils.

In laboratory tests the following equipment was used: drying oven DO–0.25–100, air-dry chamber ADC–80–Kasimov, microwave ovens LG of 900 Watt power and Wellton WMO–1700GW of variable power, balance of the accuracy 0.01. To compare the results of drying in convection and microwave ovens every soil sample was divided in two parts and, according to the methods [2, 18] 4-8 portions of different weight and/or different water content were selected from every part. Drying was performed up to the moment of soil mass difference with container at two series weighings not more than 0.02 g [2]. The time of sample drying in convection oven met the specifications [2, 18], the test time in microwave drying was defined by water content and sample weight.

Table 1. Results of water content determination in clay and peaty soils

№ sample	Soil type	Mass of charge, g	Water content, % when drying in oven microwave			convection		
			average	minimum	maximum	average	minimum	maximum
1.	Clay loam (Bely Yar settlement)	50	26.3	25.5	27.3	25	24.5	25.7
2.	Clay loam (Bely Yar settlement)	15	25.7	24.5	26.8	25.3	24.3	26.3
3.	Clay loam (Bely Yar settlement)	100	42	41	42	41	40	41
4.	Clay loam (Bely Yar settlement)	9-10	15.5	14.8	15.9	16.7	16.5	16.9
5.	Clay loam (Bely Yar settlement)	18-22	32	32	33	33	33	33
6.	Clay loam (Bely Yar settlement)	18-22	21.8	21.2	22.3	23.8	22.4	25.8
7.	Clay loam (Bely Yar settlement)	18-22	19.8	19.7	19.9	19.5	19.2	19.8
8.	Clay loam (Dvurechinsk field)	18-22	35	35	36	35	33	37
9.	Low peat clay loam (Danenbergovsk field)	18-22	30	29	31	29	27	30
10.	Clay sand (Tomsk, Solyanaya st.)	18-22	24.4	23.6	25.2	24.3	23.5	24.6
11.	Clay sand (Tomsk, Solyanaya st.)	30-40	20	19.6	20,3	20	19.7	20.4
12.	Clay sand (Tomsk, Ivanovskogo st.)	40-50	14.3	14.1	14.7	14.1	13.8	14.6
13.	Clay (Kalinovoye field)	18-22	32	32	33	33	33	33
14.	Clay (Kalinovoye field)	18-22	24.7	24.1	25.1	25.6	25.2	26
15.	Clay (Kalinovoye field)	18-22	17.3	16.9	17.8	16.2	15.8	17
16.	Low peat clay (Danenbergovsk field)	18-22	58	56	60	60	58	61
17.	Least decomposed moss peat (Dvurechinsk field)	15–20	2527	2253	2794	2816	2390	3003
18.	Least decomposed moss peat (Dvurechinsk field)	5–10	2513	2200	2754	2587	2250	3063
19.	Least decomposed moss peat (Dvurechinsk field)	100-200	3188	3125	3255	2994	2412	3327

Tests were performed using several schemes: the same soil of different water content and in portions of different weight (table 1, samples № 3-7, 13-15); at different water content in the same portions (№ 5-10 and № 13-16); at the same water content (№ 1, 2 and № 17, 18) at water content of plastic limit by compression (№ 4).

3. Results and discussion

In the course of drying tests it was observed that from the very beginning of test intensive evaporation is typical for almost all studied samples (figure 1). The given stage took from 30 (most samples) to 80% of test time, which was defined by both sample water content and its initial weight (figure 1).

Clay soils gave positive results: in most cases the spread of water content values met the specification requirements for the results of simultaneous determinations of the parameter at microwave and convection drying (table1). The permissible ranges were exceeded for the portions of 15 g (№ 2), which was not evidently enough to obtain reliable results for clay of natural moisture. At the same time sample № 4 of 9-10 g (after compression) had an acceptable spread of values.

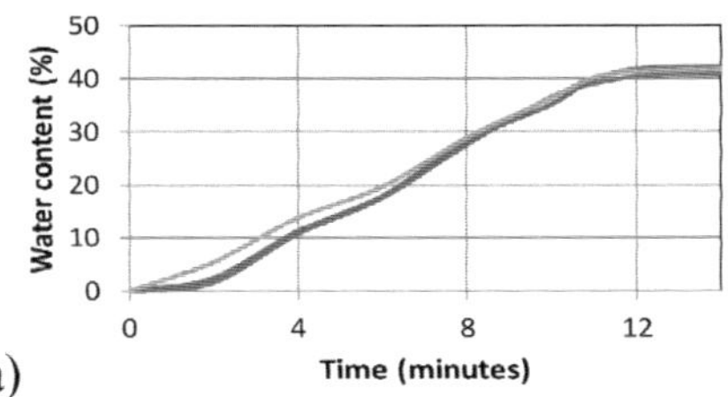

a)

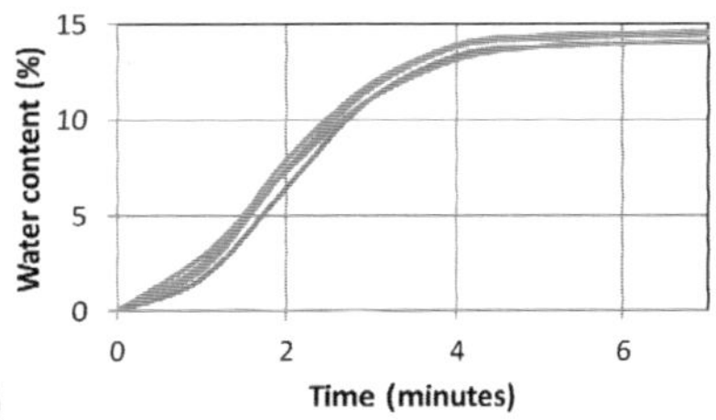

b)

c) 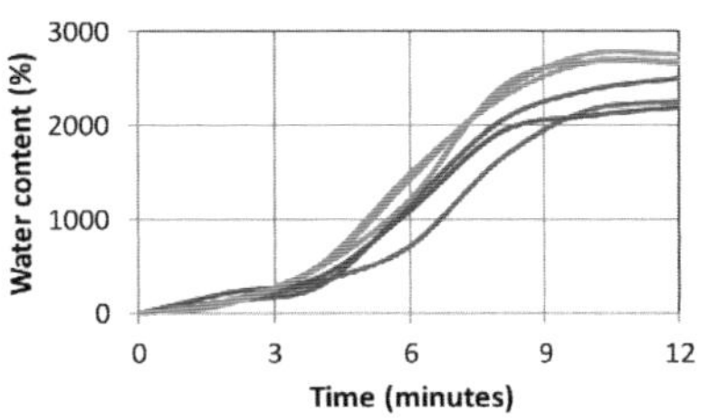

d) 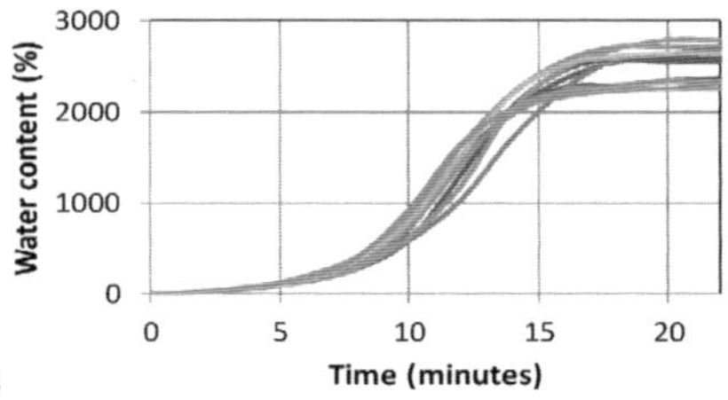

Fig. 1. Dependence of soil water content on drying time: a) clay loam and b) clay c) least decomposed moss peat (5-10 g); d) least decomposed moss peat (15-20 g)

Right after beginning of drying, a dense dry crust formed on the clay soil samples. Further evaporation through it took place with strong heating of the sample. Hence, it was necessary to mix soil after the first weighing to give it a shape providing more evaporation area.

The study of peat has shown that spread of water content values while drying in microwave oven is higher in most cases than at drying in convection oven (table1, figure 1). In both cases it does not meet the specification requirements for the results of simultaneous parameter determination (at moisture more than 100% the permissible difference is 5% [2], water content – 1 % [3]). Maximal water content (1500–3000 % and more) is registered for the samples of low decomposed moss peat. The spread of its values and drying time increase as the water content grows. With the increase in the weight of portions from 5-10 to 15-20 g, the spread of values decreases (figure 1, c, d; table 1). The increase in sample number up to 8–10 portions did not improve the results as well (figure 1, c). For the sample weight of 100–200 g of moss peat from the other site (№ 19) significant variations of simultaneous determination were also observed.

Consequently, it is necessary to distinguish good reproducibility of the results obtained by both methods as well as the fact that spread of values obtained by microwave technique is comparable with that of conventional technique of convection heating. While mineral soil has a narrow spread of water content values [26], as it has been demonstrated by working at undrained peats [27], it is not simple to obtain the same results for peats. Therefore, it is appropriate to determine the parameters of physical properties using one portion [4], increasing the number of selected samples up to 15–20, taking into account their different types. In all test there was no increase in soil weight at control weighing.

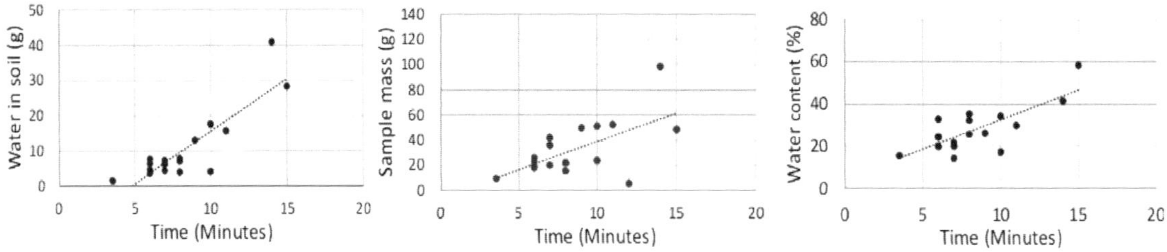

a) b) c)

Fig. 2. Dependence of water in soil (a), sample mass (b) and moisture (c) on soil drying time

To analyze the effect of such experimental factors as water content and sample weight on the time of drying, they were compared side-by-side by means of dependency graphs showing the determination coefficient (figure 2). It was revealed that the maximum determination coefficient R^2 =0.75 was obtained for dependency of water weight in a sample on the drying time (figure 2, a). The determination coefficients R^2 were considerably less (0.58 and 0.38, respectively) and characterized the dependencies of drying time on water content and portion weight of the initial soil samples (figure 2, b, c). Hence, it is obvious that in case of clay soil samples the drying time depends mostly on mass of containing water. It is defined by the fact that drying is regulated by, first of all, absorbed microwave energy used for heating liquid volume in a sample and, to a lesser extent, depends on sample structural features and its initials parameters that are characterized by its initial weight and natural moisture. From the practical point of view it means that optimal weight of sample portion may be considered the one at which the obtained results would have good repeatability with each other and the results obtained by conventional methods.

The tests allowed developing some recommendations to improve the technique of water content determination using microwave oven for cohesive soils.

1. Application of microwave oven eliminates the use of metal containers, therefore, it is recommended to use porcelain ones. Samples should also contain low amount of conducting minerals.
2. The optimal weight of portions is recommended to be in the range from 20 to 50 g, which corresponds to the existing specifications. To determine water content at plastic limit, the portion weight of 5-10 g would be enough. The optimal portion weight is defined by the initial soil water saturation and ranges from 15 to 50 g for medium-, highly decomposed peats and organic mineral soils. For low decomposed water-saturated peats the portions of more than 100 g are recommended to be used, the number of samples being not more than 2–3. At multiple water content determination it is reasonable to use rapid method according to [4].
3. The time recommended for drying to determine water content is 10–15 min and more – for clay with high water content (40-50%); 10 min – for peaty soil; 5 min – to determine hygroscopic water content, clay sand water content, and water content at plastic limit of all clay soil types. The intervals between control weighing are 1–2 min. The drying time of peat portions of more than 100 g is 30–40 min, weight 30–50 g – 10–15 min, weight 5–10 g – 3–5 min. The intervals between control weighing of portions of more than 50 g is about 3–5 min and 1–2 min of portions more than 5–10 g and low peaty soil.
4. Increase in evaporation area of test samples would allow intensifying the soil drying process by means of their thin layer rolling and covering the channel surface, using pallet knife as well as their mixing at control weighing.

4.Conclusions

Hence, the research results have proved that household microwave ovens are an effective means to rapidly determine water content in clay, sandy, and organic soils [26, 27], since tests are carried out in shorter time and the results obtained are more accurate than they are when using drying ovens. An attractive feature of this technique is that sample weight does not increase in control weighing at microwave drying (figure 1), which increases the accuracy of soil water content determination. The authors anticipate that the suggested recommendations would allow wide application of low-cost technique in both field and laboratory conditions, as well as its possible inclusion in the updated version of GOST 5180 as an additional method.

References

[1] GOST 23061–2012 Soils. Methods of radio isotopic measurements of density and water content
[2] GOST 5180–2015 Soils. Methods of laboratory determination of physical characteristics
[3] GOST 11305–2013 Peat. Method of water content determination
[4] GOST 19723–74 Peat. Method of water content determination in deposits
[5] ASTM D 1558 – 10 Standard Test Method for Moisture Content Penetration Resistance Relationships of Fine-Grained Soils
[6] ASTM D 4944–11. Standard Test Method for Field Determination of Water (Moisture) Content of Soil by the Calcium Carbide Gas Pressure Tester
[7] ASTM D425–88(2008) Standard Test Method for Centrifuge Moisture Equivalent of Soils
[8] ASTM D6938–10 Standard Test Method for In-Place Density and Water Content of Soil and Soil-Aggregate by Nuclear Methods (Shallow Depth)
[9] Ernest S, Berney I V, James D Kyzar and Lawrence O Oyelami 2011 Device Comparison for Determining Field Soil Moisture Content *Geotechnical and Structures Laboratory U.S. Army Engineer Research and Development Center* 3909 Halls Ferry Road Vicksburg, MS 39180-6199

[10] Ernest S, Berney I V, Mariely Mejias-Santiago and James D Kyzar 2013 Non-Nuclear Alternatives to Monitoring Moisture-Density Response in Soils *Geotechnical and Structures Laboratory U.S. Army Engineer Research and Development Center* 3909 Halls Ferry Road Vicksburg, MS 39180-6199 117
[11] Mendoza M J and Orozco M 1999 Fast and accurate techniques for determination of water content in soils *Geotechnical Testing Journal* **22 (4)** 301-307
[12] Gee G W and Dodson M E 1981 *Soil Science Society of America Journal.* Soil water content by microwave drying: a routine procedure. Vol. **45 (6)** pp. 1234-1237.
[13] Carter M and S Bentley 1986 *Canadian geotechnical journal.* Practical guidelines for microwave drying of soils. Vol.**23 (4)** pp. 598-601.
[14] Bentley S P and Carter M 1987 *Highways and transportation.* Microwave drying time chart. Vol. **34 (3)** pp. 36-37.
[15] Gilbert P A 1988 *Technical Report - US Army Engineer Waterways Experiment Station (GL-88-21).* Computer controlled microwave oven system for rapid water content determination pp. 123.
[16] Vaitekunas D, Raghavan G S V and Van de Voort F R 1989 *Canadian Agricultural Engineering* .Drying characteristics of soil in a microwave environment. Vol. **31(2)** pp. 117-123.
[17] ASTM D2216–10 Standard Test Methods for Laboratory Determination of Water (Moisture) Content of Soil and Rock by Mass
[18] ASTMD 4643–08. Standard Test Method for Determination of Water (Moisture) Content of Soil by Microwave Oven Heating
[19] Stanevich V T 2008 Building ceramics (Pavlodar, Kereku) 96
[20] Makeeva T G and Yegorov Yu M 2011 *Proceedings of the I Russian working session "Clay, clay minerals, and layered materials" devoted to the 90-th anniversary of B.B. Zvyagin* Experimental observation of first-order transition of bound water similar to the second-order to mono- and polymineral clay soil. (Moscow, ODGI RAS) pp. 161.
[21] Prokhina A V, Shapovalov N A and Latypova M M 2011 *Sovremennye Naukoyemkiye tekhnologii.* Modification of clay mineral surface with high content of montmorillonite in electromagnetic field of high frequency. Vol. **1** pp. 135-136.
URL: www.rae.ru/snt/?section=content&op=show_article&article_id=6744
[22] Davletbakov R R 2014 *Proceedings of All-Russia Scientific-Methodological Conf*. Influence of particle dispersion and microwave radiation on strength of siliceous ceramics [Electronic resource] Universitetskiy complex kak regionalny center obrazovaniya, nauki i kultury In: (Orenburg, Orenburg State University) pp. 1331.
[23] GOST 27784–88. Soils. Methods of ash determination in peat and peaty soil horizons
[24] GOST 11306–2013 Peat and products of its processing. Methods of ash determination
[25] Kramarenko V V, Nikitenkov A N and Molokov V Y 2005 *Modern problems of science and education.* On the applicability of microwave method for determining the moisture content of sandy soils. Vol. **1** pp. 1-12. URL: *http://www.science-education.ru/121-18451* (in Russian)
[26] Kramarenko V V, Nikitenkov A N, Molokov V Yu, Shramok A V and G P Pozdeeva 2016 *IOP Conf. Ser.: Earth Environ. Sci.* Application of microwave method for moisture determination of organic and organic-mineral soils. Vol. **33** pp. 012040.
http://iopscience.iop.org/article/10.1088/1755-1315/33/1/012040

Geochemical groundwater peculiarities of Paleogene sediments in S-E Western Siberia artesian basin

A A Balobanenko[1], V L'gotin[2], E M Dutova[3], D S Pokrovsky[4], A N Nikitenkov[5], I V Raduk[6]
[1,2]Tomskgeomonitoring, Tomsk
[3,5,6]Tomsk Polytechnic University, Tomsk, Lenin Avenue, 30, Russia
[4]Tomsk University of Architecture and Civil Engineering, 634003, Tomsk, 2 Solyanaya square, Russia

E-mail: [1]SiJay@sibmail.com, [2]tgm@tgm.ru, [3]dutova@sibmail.com, [4]dsp@sibmail.com, [5]nik@tpu.ru, [6]radyuk_ilya@mail.ru

Abstract. The geochemical peculiarities of groundwater in Paleogene deposits in southeastern part of Western Siberia artesian basin are considered in the paper. Landscape, climate, geo-structural and hydrogeological conditions define the water composition and quality peculiarities in this region. It has been established that ion-saline composition, mineralization and water quality changes arre governed by the horizontal zonal distribution. Groundwater of taiga landscapes generally is in equilibrium with kaolinite and quartz, mainly involving Ca- and Mg–montmorillonite, illite, carbonate minerals, sometimes barite. Groundwater in woodland grass and grassland, together with previously mentioned minerals, is usually in equilibrium with barite, colestine, and particularly, fluorite and gypsum. As a result, all relevant elements are removed from the groundwater and their accumulation level is restricted.

1. Introduction

The most important and priority issue today is to supply a population with good quality water. Numerous scientists are paying more and more attention to this problem. Health and environmental effects, sanitary and epidemiological safety, environmental comfort and safety are reliant on the quality of the water [1-4]. This issue is more acute for arid and semi-arid regions where natural waters are usually unconditional due to the existing level of some components. A prime example is southern Siberia.

Paleogene aquifer system is widely distributed throughout and is the major supply of drinking water. Urban and rural water supply is from these groundwaters and provides the central water supply of Tomsk, Tumen and Barnaul, whereas small water catchment areas and wells supply water for small populated areas.

Hydrogeological, hydrogeochemical conditions of West Siberian Platform sedimentary cover have discussed by different scientists. In spite of this fact, such issues the peculiarities of fresh groundwater distribution and its quality are still underinvestigated. Updated methodical basis and software mapping of the hydrogeochemical conditions, as well as analysis of hydrogeochemical processes allows furthers a possibly new approach in solving the above-mentioned issues. This research is a continuation of previous research investigations [3-5]. The research target is the analysis of groundwater chemical composition changes in Paleogene deposits, mechanisms of water quality by studying its equilibrium with water bearing rock minerals and secondary minerals.

2. Materials and methods

Research base is factual data of JST "Tomskgeomonitoring" which includes water chemical composition from wells of water catchment areas of populated places. Groundwater chemical composition analysis was based on modern certificated analytic methods in the laboratories of JST "Tomskgeomonitoring", TPU and several regional geological organizations.
Methods of mathematic statistics and GIS technologies were used for information generalization and processing via programs EXCEL, Statistica, ArcGIS, as well as physico-chemical calculations by HydroGeo. The use of the software HydroGeo for solving various problems has been shown by us earlier [6 - 8]

3. Results and Discussion

3.1. Chemical composition of groundwater

The generalized chemical composition data are represented in table 1. Regionally, groundwater embraces various chemical compositions and is characterized by signifcant mineralization alterations.
Variety of landscape and climate conditions (fig.1), zonal changes of wetness conditions and infiltration charge, aquifer lithofacies peculiarities determine the main features of Paleogene deposit groundwater formation. This, in its turn, reflects the horizontal zonal changes of Paleogene waters ion-saline composition and salinity (fig. 2, 3). Fresh water is located within most of the territory. TDS regularly increases from east-northward and eastward to west-southward (fig.1) and westward from moderate wetting and overwetting charge zones to zones with difficult water exchange conditions and underwetting. Groundwater chemical composition (fig.3) and its quality also alters directionally.
It is estimated that 80 % of the above-described territory embraces fresh water with TDS of up tol g/l, while water with mineralization of more than 1 g/l are within 20% of this territory. Groundwater with mineralization of 5 g/l was found only within 7 % of the territory.

Table 1. Paleogene deposits groundwater chemical composition.

	Components	TLV	Amount	Mean	Min	Max	% higher TLV
Generalized components	TDS, mg/l	1000	4168	851.7	56.7	4993.5	42.9
	pH	6-9	3854	7.4	6.0	10.9	-
	Eh, мВ	-	94	23.57	-187.00	210.00	-
	Total hardness, mg-eq/l	7	4113	6.9	0.3	59.9	45.3
	Permanganate oxidation, mgO$_2$/l	5	1820	2.67	0.01	39.0	26.1
	Oil-products, mg/l	0.1	481	0.026	0.0005	1.77	5.6
	Anionic surfactant, mg/l	0.5	24	0.03	0.009	0.09	-
	Detergent, mg/l	0.5	73	0.023	0.008	0.2	-
	Phenolic index	0.25	340	0.003	0.0001	0.062	-
	Phenols, mg/l	0.001	52	0.0016	0.0001	0.012	3.8
Macrocomponents	Hydrocarbonate (HCO_3^-), mg/l	-	4166	449.3	20.0	1281.0	-
	Carbonate (CO_3^{2-}), mg/l	-	202	11.1	0.2	150.0	-
	Chlorides (Cl^-), mg/l	350	4087	75.9	0.14	2906.0	31.5
	Sulfates (SO_4^{2-}), mg/l	500	3491	71.9	0.01	1819.0	7.4
	Calcium (Ca^{2+}), mg/l	-	4168	60.7	0.15	502.4	-
	Magnesium (Mg^{2+}), mg/l	50	4156	38.4	0.28	372.0	38.2
	Sodium (Na^+), mg/l	200	4162	125.1	0.2	1612.0	49.5
	K^+, mg/l	-	359	1.8	0.25	59.88	-
	Ammonium (NH^{4+}), mg/l	1.5*	2555	0.32	0.001	12.61	4.6
	Nitrate (NO_3^-), mg/l	45	2169	1.1	0.001	44.0	-
	Nitrites (NO_2^-), mg/l	3	1507	0.019	0.00005	2.97	0,2
[illegible]	Aluminium (Al), mg/l	0.5	180	0.039	0.0001	0.056	-

	Barium (Ba), mg/l	0.7	293	0.1	0.0001	0.6	-
	Beryllium (Be), mg/l	0.0002	99	0.00011	0.00002	0.2	8.1
	Boracium (B), mg/l	0.5	132	0.13	0.002	1.42	6.8
	Bromide (Br^-), mg/l	0.2	130	0.11	0.002	6.5	13.1
	Components	TLV	Amount	Mean	Min	Max	% higher TLV
	Protoxidic iron form (Fe^{2+}), mg/l	-	147	0,.68	0.04	51.12	86.2
	Oxide iron form (Fe^{3+}), mg/l	-	199	0.2	0.001	27.2	63.9
	Iodine (I), mg/l	-	147	0.17	0.003	5.0	-
	Cadmium (Cd), mg/l	0,001	159	0.00027	0.00001	0.00	8.8
	Silicium (Si), mg/l	10	473	6.98	0.2	35.9	24.3
	Lithium (Li), mg/l	0.03	143	0.0046	0.00003	0.047	2.1
	Manganese (Mn), mg/l	0.1	475	0.16	0.00001	2.45	74.3
	Copper (Cu), mg/l	1	969	0.01	0.00003	0.92	-
	Molybdenum (Mo), mg/l	0.25	259	0.003	0.00003	0.2	-
	Arsenium (As), mg/l	0.05	199	0.0038	0.0001	0.04	0.5
	Nickel (Ni), mg/l	0.1	231	0.0069	0.00063	0.08	-
	Polyphosphates (PO_4), mg/l	3.5	382	0.096	0.001	3.49	-
	Hydrargyrum (Hg), mg/l	0.0005	135	0.000081	0.00001	0.0006	0.7
	Plumbum (Pb), mg/l	0.03	427	0.0021	0.0001	0.25	0.7
	Selenium (Se), mg/l	0.01	49	0.00051	0.0001	0.005	-
	Strontium (Sr), mg/l	7	439	0.33	0.00004	5.9	-
	Fluoride (F^-), mg/l	1.5	780	0.25	0.00048	1.4	-
	Chromium (Cr), mg/l	0.05	262	0.0066	0.0007	0.043	-
	Zink (Zn), mg/l	5	765	0.012	0.0002	4.43	-

N o t e * Normalized by ammonia (by nitrogen).

Water chemical composition changes from hydrocarbonate calcium and sodium to chlorite sodium, pH increases, however, iron and organic substance content decreases (fig.2).

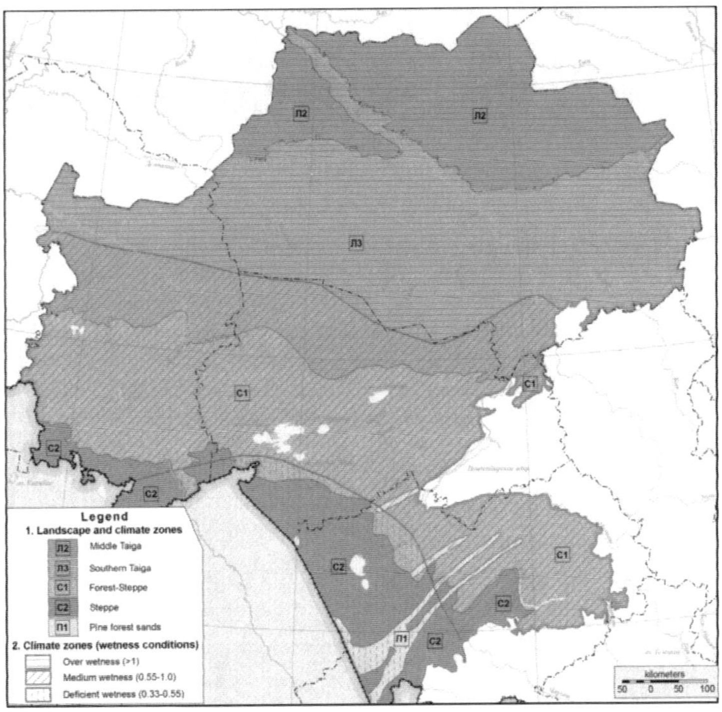

Figure 1. Landscape and climate zones.

Hydrocarbonate calcium and magnesium-calcium, fresh water (TDS up to 450 mg/l), often acid and neutral (pH less than 7) with increased oxidability and iron and manganese content are widely distributed in central taiga conditions. It has been estimated, there are wells with off-spec water with iron is 95% , manganese - 74%, and oxidability - 56%.

Hydrocarbonate calcium and magnesium-calcium or sodium-calcium, fresh water (TDS 400-800 mg/l), neutral and weakly alkaline (pH 7-7.7) water with relatively increased Fe and Mg content, sometimes organic substances and total hardness are widely distributed in south taiga and sub-boreal taiga conditions. It has been estimated, there are wells with off-spec water where iron is 93 % , 37 % - manganese, 10% - oxidability, whereas total hardness – 7%.

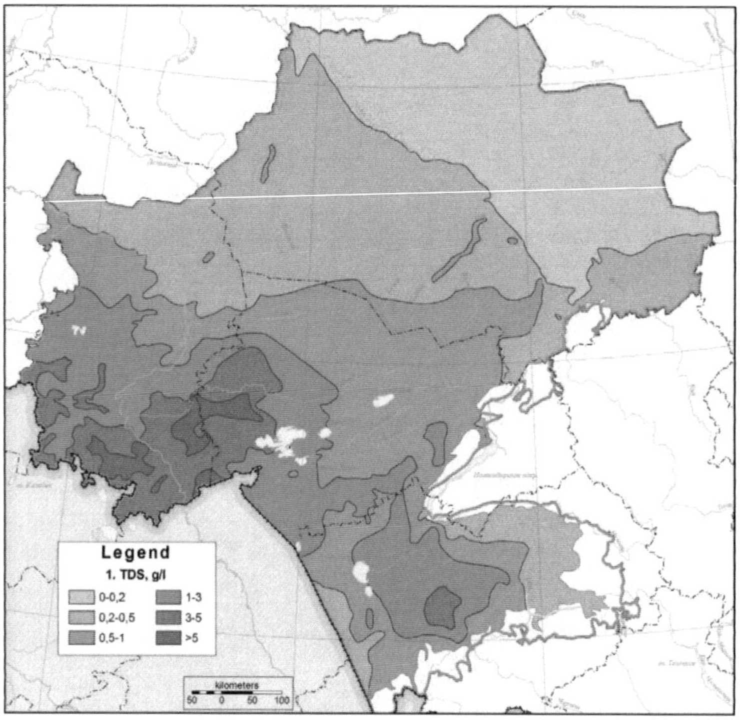

Figure 2. Paleogene deposits groundwater mineralization.

TDS growth is the result of total macrocomponent decrease (Cl, SO_4, HCO_3, Na, Mg, Ca).

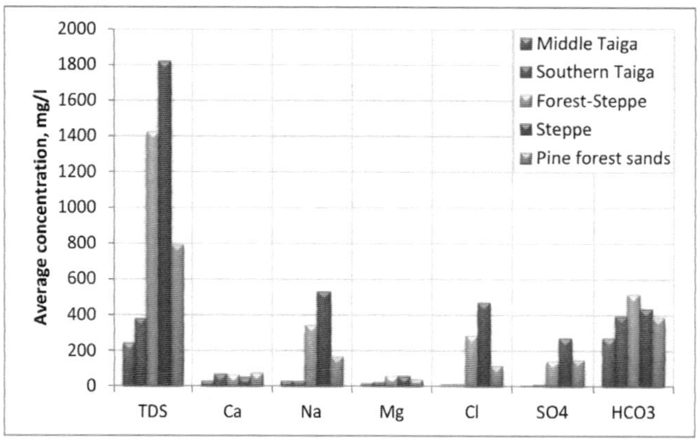

Figure 3. Mean content of macrocomponents in groundwaters of different landscape zones.

Both saltish and fresh water are typical for woodland grass zones. The saltish waters are widely distributed in typically grassland landscape with deficient wetness conditions and continental salt accumulation processes. Significantly high concentrations of sulfates (540—610 mg/l and over 1000 mg/l in Altai region), chlorides (over 350 mg/l) and high hardness are noted in mineralized Cl-SO_4, SO_4-HCO_3, SO_4 and SO_4-Cl water types. Wells with water of over 7 mg-eq/l composes 50% of Paleogene deposits in Kulundin artesian basin. Ground water in grassland and woodland grass zones comparable to taiga zones differ by high microcomponent concentrations.

3.2 Physico-chemical calculations of complexable and mineral-forming groundwater capacity

The complexable formation calculation results show that the role of compounds in microcomponents migration is insignificant. For example, tenth percent of Na migrates as complex compounds. Complex calcium and manganese compounds are more prevalent in migration, i.e. Mg migration - from 3,3 to 4,6 % and Ca migration – from 3,3 to 4,7 % of total content. Complex compounds of these elements are represented by hydrocarbonate forms.

Among the considered microcomponents the migration prevalence involves native ions ions , mainly ferrous iron. Ferrous complex compounds are only from 5.7 to 9.1% of total content. They are mainly carbonate and hydrocarbonate complex compounds.

Native ions also play an important role in the migration of manganese and zinc, i.e. from 66 to 79% and from 72 to 87%, respectively of total manganese and zinc content. The predminate complex compounds are carbonate and bicarbonate.

Comparable to micronutrients, Fe, Cu and Pb migration involves mainly complex compounds, i.e. only 0.03 - 0.24; 1.38 - 6.18 and 3.29 - 7.07%, respectively. Iron oxide migration involves prevailing carbonate and hydroxide, Pb - carbonate, bicarbonate and hydroxide, Cu - carbonate and bicarbonate compounds. In the groundwater of woodland grass and grasslands the water migration - chemical element relation transfers to complex compounds, mainly bicarbonate, carbonate, and even, sulfate (table 3).

Table 3. Macrocomponents (Na, Mg, Ca) and microcomponents (Mn, Fe, Cu, Zn, Pb) migration types in regional groundwater.

Landscapes	**Migration compounds**	
	Major	**Secondary**
Central taiga	Na^+, Mg^{2+}, Ca^{2+}, Mn^{2+}, Fe^{2+}, $(FeOH)^{2+}$, CuФК, $Cu(ФК)_2^{2-}$ Zn^{2+}, $(PbHCO_3)^+$, $(PbOH)^+$, $PbCO_3$	$(MgHCO_3)^+$, $(CaHCO_3)^+$, $(MnHCO_3)^+$, $MnCO_3$, $(FeHCO_3)^+$, $Fe(HCO_3)_2$, $FeCO_3$, $Fe_2(CO_3)_3$, Cu^{2+}, $(CuHCO_3)^+$, $CuCO_3$, $(ZnHCO_3)^+$, $ZnCO_3$, Pb^{2+}, PbФК
South taiga	Na^+, Mg^{2+}, Ca^{2+}, Mn^{2+}, Fe^{2+}, $(FeOH)^{2+}$, $CuCO_3$, Zn^{2+}, $(PbHCO_3)^+$, $(PbOH)^+$, $PbCO_3$	$(MgHCO_3)^+$, $(CaHCO_3)^+$, $(MnHCO_3)^+$, $MnCO_3$, $(FeHCO_3)^+$, $Fe(HCO_3)_2$, $FeCO_3$, $Fe_2(CO_3)_3$, Cu^{2+}, $(CuHCO_3)^+$, CuФК, $Cu(ФК)_2^{2-}$, $CuOH^+$, $(ZnHCO_3)^+$, $ZnCO_3$, Pb^{2+}
Woodland grass	Na^+, Mg^{2+}, Ca^{2+}, Mn^{2+}, $MnCO_3$, Fe^{2+}, $(FeOH)^{2+}$, $Fe_2(CO_3)_3$, $CuCO_3$, Zn^{2+}, $ZnCO_3$, $(PbOH)^+$, $PbCO_3$, $(PbHCO_3)^+$	$(MgHCO_3)^+$, $MgSO_4$, $(CaHCO_3)^+$, $CaCO_3$, $CaSO_4$, $(MnHCO_3)^+$, $MnSO_4$, $(FeHCO_3)^+$, $Fe(HCO_3)_2$, $FeCO_3$, Cu^{2+}, $(CuHCO_3)^+$, $CuOH^+$, $(ZnHCO_3)^+$, $ZnSO_4$, Pb^{2+}

Note* major- 25 %; secondary – from 1 % to 25% of total content.

The results of water mineral - forming capacity are represented in table 4. Results analysis showed that groundwater formed in central taiga conditions generally are equilibrium to hematite, magnesite, kaolinite, quartz and sometimes with siderite. Water in south taiga territories comparable to the groundwater in the northern territories are equilibrium to a wide spectrum of carbonate and clay minerals. Besides previously mentioned minerals, groundwater of Quarternary and Paleogene deposits is equilibrium or practically equilibrium to goethite, calcium montmorillonite, siderite, calcite and aragonite.

Groundwater of woodland grass and grassland is equilibrium to clay minerals (kaolinite, hydromica, Ca- and Mg-montmorillonite, and sometimes Na-montmorillonite), carbonates (often, calcite, aragonite, dolomite, siderite, strontianite, cerussite sometimes malachite), quartz and sulfites (almost always barite) (table 4).

Table 4. Saturation index (L) of groundwater to carbonate and sulfate minerals.

Landscape	Equilibrium (L > 0)	Practically equilibrium (0> L> -5)	Moderately unsaturated (-5> L> -10)	Unsaturated (-10 >L >-15)
Central taiga		$CaCO_3$, $CaCO_3$ aragonite, $CaMg(CO_3)_2$, $FeCO_3$, $MnCO_3$, $SrCO_3$, $BaSO_4$	$ZnCO_3$, $Cu_2CO_3(OH)_2$, $CaSO_4$, $CaSO_4{\cdot}2(H_2O)$, $SrSO_4$	$BaCO_3$, $MgCO_3$, $PbSO_4$
South taiga	$CaCO_3$, $CaCO_3$ aragonite, $CaMg(CO_3)_2$	$FeCO_3$, $MnCO_3$, $PbCO_3$, $SrCO_3$, $Cu_2CO_3(OH)_2$, $BaSO_4$	$BaCO_3$, $MgCO_3$, $ZnCO_3$, $CaSO_4$, $CaSO_4{\cdot}2(H_2O)$, $SrSO_4$	$PbSO_4$
Woodland grass	$CaCO_3$, $CaCO_3$ aragonite, $CaMg(CO_3)_2$, $PbCO_3$, $SrCO_3$, $FeCO_3$, $BaSO_4$, $SrSO_4$	$MnCO_3$, $ZnCO_3$, $Cu_2CO_3(OH)_2$, $CaSO_4$, $CaSO_4{\cdot}2(H_2O)$, $PbSO_4$	$BaCO_3$, $MgCO_3$	

4. Conclusion

West Siberian territory is characterized by a variety of landscapes, climate and geomorphological conditions, which, in its turn, provides suitable conditions for different chemical groundwater types in hypegenesis zone of moderate climate and flat structures. This groundwater is equilibrium to different minerals.

From north to south, due to the influence of different horizontal landscape alterations, groundwater TDS increases, as well as chemical elements, furthering migration type evolution as complex compounds and expanding the secondary mineral equilibrium phase.

There are no problems concerning the drinking water quality in taiga conditions. Non-conventional water via appropriate simple water treatment methods could be easily converted be to the sanitary standards and regulations. In the grasslands of Western Siberia, with highly populated areas, the water of almost all water-bearing rocks, except for Quaternary alluvial deposits, often have components which exceeding total hardness and salinity, bromine, lithium, and sometimes sulfate, chloride, boron, strontium and barium content. Drinking water in taiga lands as groundwater contains high contents of iron, magnesium and organic compounds.

References

[1] Banks D, Parnachev V, Frengstad B, Holden W, Karnachuk O, Vedernikov A 2004 *Applied Geochemistry* The evolution of alkaline, saline ground- and surface waters in the southern siberian steppes **Vol.** 19 (12) pp. 1905-1926

[2] Parnachev V, Berezovsky A, Banks D, Garbe-Schönberg D 1999 *Hydrogeology Journal*Hydrochemical evolution of na-so4-cl groundwaters in a cold, semi-arid region of southern Siberia **Vol.** 7 (6) 0546-0560

[3] Poliyenko A, Dutova E 2000 *4th Korea-Russia International Symposium on Science and Technology art* Urinary stones investigation and influence of the water factor to their formation in a human organism proceedings - KORUS 2000: 865968 278-283

[4] Dutova E, Vologdina I, Pokrovsky D, Nalivaiko N, Kuzevanov K and Pokrovsky V 2016 *IOP Conference Series: Earth and Environmental Science* Hydrogenous mineral neoformations in Tomsk water intake facility from underground sources **Vol. 33** (1)

[5] Pokrovsky V, Dutova E, Kuzevanov K, Pokrovsky D, Nalivaiko N 2015 *IOP Conference Series: Earth and Environmental Science* Hydrogeological conditions changes of Tomsk, Russia **Vol.** 27 (1)

[6] Dutova E, Bukaty M, Nevol'ko A, Pokrovsky D, Shvartsev S 2006 *Geologiya i Geofizika* Hydrogenic concentration of gold in alluvial placers of the Egor'evsk area (Salair) **Vol.** 47 (3) 364-376

[7] Shvartsev S, Dutova E 2001 *Geology of Ore Deposits* Hydrochemistry and mobilization of gold in the hypergenesis zone (Kuznetsk Alatau, Russia) **Vol.** 43 (3) pp. 224-233

Groundwater chemical and microbiological composition in Aptian-Cenomanian deposits (Kaimisovsky oil-gas bearing province)

N G Nalivaiko[1] · E M Dutova[2], T S Spiridonov[3], D S Pokrovsky[4] and V D Pokrovskiy[5]
[1,2,3,5]Tomsk Polytechnic University, 30 Lenin Avenue, Tomsk, 634050, Russia
[4]Tomsk State University of Architecture and Building, 2 Solyanaya sq., Tomsk, 634003, Russia
E-mail: [1] nng@tpu.ru, [2]dutova@sibmail.com, [3]tim3899@yandex.ru, [5]vdp@sibmail.com

Abstract. This paper reveals the investigation results of the groundwater chemical and microbiological composition in Aptian-Cenomanian deposits, Kaimisovsky oil-gas bearing province. The mineral-forming behavior of the groundwater was evaluated. It was determined that the diversity of microbial communities could be corrosive- harmful for the groundwater used in reservoir pressure maintenance systems. According to the research results it is necessary to study the groundwater microorganisms in Aptian-Cenomanian deposits and their influence on groundwater itself.

1. Introduction

Water-flooding is the major method in reservoir pressure maintenance during oil field development. Geochemical processes interacting with layer-injection water co-operation result in emerging produced water, saturated inorganic salts, salt precipitation in the wells, and sometimes further suppressing corrosive processes [1-5].

Kaimisovsky oil-gas bearing area is located within the southern West-Siberia mega-province, territory of 140 thousand km^2. This area includes Kaimisovsky and Mezhovky arch, Nurol and partially Ugansky depressions, Upper-Demyanosky megaswell and others. Within this province there are about 50 oil fields confined to anticline and / or combined reservoirs. Kaimisovsky oil-bearing province embraces approximately 2.3% of the hydrocarbon reserves in West Siberian mega-province [6]. Aptian-Cenomanian aquifer system within Kaimisovsky oil-bearing province was intensively developed for water production and further usage in the reservoir pressure maintenance system over a number of years (fig. 1).

The authors studied the hyrogeochemical and microbiological processes, as well as sediments formed from fresh groundwater [7,8]. This research is a continuation of previous research investigations. The research target object is saline groundwater of Aptian-Cenomanian aquifer system within Kaimisovsky oil-bearing province itself.

2. Materials and methods

Water samples for microbiological analysis are collected in sterilized glass bottles in accordance with aseptic regulations. A wide spectrum of microorganisms affecting both the water chemical composition in investigated aquifer system and oil quality and technological petroleum production equipment service were identified. Simultaneously, the chemical composition of water, as well as water as an environmental niche for microorganism was also investigated.

Figure 1. Overview map of Kaimisovsky oil-gas bearing province.

Identification and quantitative estimation of microorganisms are determined on the basis of traditional methods. The aerobiotic and anaerobic bacteria involved in the carbon, nitrogen and sulfur cycle were detected and calculated in the water samples from production wells. The studied spectrum of physiological bacteria group included: saprophile, oligotrophic, oil oxidizing microorganisms, autotrophic, thionic, ammonifying, dentrification, nitrite, sulfur reducing, as well as oxidizing hydrocarbons of methane group, aromatics, saturated and unsaturated bacteria. The traditional methods based on elective habitat are used in investigating the groundwater microflora [11]. Investigated water plates on solid medium (Petri dish) and on fluid medium (test tubes). The plate number of grown colonies was calculated. Limiting dilution culture was used for plating in vitro. The number of bacteria is determined by visual environment change, typical for each physiological bacteria group. Growth intensity was evaluated in scores and conditional units. MacCready tables, based on statistic methods, were used in the quantitative estimation of growths. Saprophile are cultivated on meat-and-peptone agar at temperature 20–22 °C. Grown colonies were calculated two days after seeding (plating). Oil oxidizing microorganisms are used in the metabolism of petroleum and its derivatives, which were revealed in agarized Muntz medium with oil as the only source of carbon and energy. The seeding was cultivated at 20– 22 °C during two weeks. At the end of inoculation the number of colony growths was calculated. The quality control involved the seeding from one and the same water tube in solid Muntz medium without oil. Seeding was performed four-times. Bacteria oxidizing individual oil hydrocarbons were revealed in fluid Muntz medium. Seeding was conducted in aseptic serum vials placed in disiccator at the atmosphere comparable to hydrocarbon atmosphere three times. Inoculation period was 2 weeks, after which the growth of the microbes in the tubes was evaluated according to the forming films based on conditional unit in Mogilevski. Nitrite bacteria were revealed in the medium by Vinogradsk method of limiting dilution. Seeding inoculation period was for two weeks at 22 °C. The appearance of nitrites and nitrates in the medium are evidence of the desired bacteria group. Dentrification bacteria were revealed in the medium by Giltaya method of limiting dilution during 2-week inoculation at 22 °C. The occurrence of bacteria in the medium is the result of pH changes, color of cultured liquid, nitrogen and /or ammonia which could be evident in the smell and foam. Oligotrophic bacteria were revealed in agarized distilled water without organic substances. Water plates (0.1 or 0.05ml) on Petri dish, whereas seeding was cultivated at room temperature during 2-3 weeks. Colony growths were calculated by increasing through colonometer. Sulfur-reducing bacteria were cultivated in the Postgate medium with calcium lactate as source of organic substances. Seeding was conducted by limiting dilution. Intensive sulphate reduction could be observed due to the black iron sulphide sediment in the medium formed as a result of hydrogen sulphide reaction (microbic sulphate reduction) and existing iron in the medium. Psychrophile (20 °C) and thermophile (45 °C) types of sulphate-reduction were revealed.

3. Results and Discussion

3.1. Chemical composition of waters

Aptian-Cenomanian aquifer system embraces horizontal hydrogeochemical zonation, typical for this region. This involves increase of mineralization, content of macro- and micro-components north-westward. Mineralization varies from 18.03 (Pervo-Maiskoe field) to 21.18 gr/l (Katilginskoe field); however, throughout most of the territory mineralization includes 20gr/l. Predominate water type is chlorite nitrate. Basically, aquifer water exhibits weakly acid reaction (pH ranges within 7.0-7.6), rather low content and/or no sulphate -ions. Water in Aptian-Cenomanian aquifer system embraces a wide spectrum of micro-components (table 1).

Table 1. Groundwater chemical component composition content in Aptian-Cenomanian deposit within studied areas (based on averaged data).

Components	Deposits				
	Pervo-Maiskoe	Krapivinskoe	Dvurechenskoe	Zapadno-Moiseevskoe	Lineinoe
M,mg/l	18.03	19.21	19.20	19.37	20.9
pH	7.1	7.3	7.6	7.5	7.3
Ca^{2+}	514	720	680	635	662
Mg^{2+}	136	403	329	357	117
Cl^-	10956	11998	11491	11780	12723
$Fe_{общ}$	2.5	1.5	1.5	1.5	6/3
NH_4^+	29.0	27.8	23.7	25.0	23.0
HCO_3^-	165	190	171	185	159
SO_4^{2-}	2.0	45.2	39.0	38.8	29.0
K^+	34	39	36	33	38.0
Na^+	6469	5150	7425	6778	7299
Br^-	54.3	43.7	51.6	52.1	55.8

The following conclusions are based on the investigated results: Aptian-Cenomanian aquifer groundwater is in non-equilibrium to the majority of sulphates and chlorites, but in equilibrium to carbonates (calcite, magnesium, dolomite and siderite) under the thermobaric conditions governing this complex. Carbonate minerals are also found in the sediments incrustating petroleum production equipment, especially on electrical driven centrifugal pumps.

To study the mineral-forming behavior of groundwater, physico-chemical calculations were conducted by applying software tool HydroGeo [9, 10]. Experience in the use of the last shown by us in many works. It should be noted that obtained calculations of positive equilibrium index values prove the fact that these systems could form this or that mineral, while negative values- reveal their deformation (table 2).

Table 2. Groundwater saturation degree (L) in carbonates, sulphates and chloride minerals.

Deposit	Equilibrium (L > 0)	Sub-equilibrium (0 > L > -5)	Moderately undersaturated (-5 > L> -10)	Undersaturated (-10 >L >-15)
Pervo-Maiskoe	$CaCO_3$к, $MgCO_3$, $CaMg(CO_3)_2$, $FeCO_3$	$CaCo_3$a, $SrCO_3$, $SrSO_4$	NaCl, $CaSO_4$	KCl
Krapivinskoe	$CaCO_3$к, $MgCO_3$, $CaMg(CO_3)_2$, $FeCO_3$	$CaCo_3$a, $SrCO_3$, $SrSO_4$	NaCl,$CaSO_4$	KCl
Dvurechenskoe	$CaCO_3$к, $MgCO_3$, $CaMg(CO_3)_2$, $FeCO_3$	$CaCo_3$a, $SrCO_3$, $SrSO_4$	NaCl	$MgCO_3$
Zapadno-Moiseevskoe	$CaCO_3$к, $MgCO_3$, $CaMg(CO_3)_2$, $FeCO_3$	$CaCo_3$a, $SrCO_3$	NaCl, $CaSO_4$	NaCl, $CaSO_4$
Lineinoe	$CaCO_3$к, $MgCO_3$, $CaMg(CO_3)_2$	$CaCo_3$a, $FeCO_3$	$CaSO_4$	KCl

3.2. Microbiological composition of waters

Microbiologic sampling of Aptian-Cenomanian aquifer deposit groundwater within the oil field territory of Kaimisovsky oil-bearing province was conducted for some years. These investigations showed that Aptian-Cenomanian aquifer deposit is a unique groundwater ecosystem, characterized by diverse physiological bacterial groups (table 3).

Table 3. Groundwater microflora of Aptian-Cenomanian aquifer in oil fields of Kaimisovsky oil-bearing province.

Physiological bacterial groups	Deposits				
	Pervo-Maiskoe	Krapivinskoe	Dvurechenskoe	Zapadno-Moiseevskoe	Lineinoe
Oligotrophile, kl/ml	159520/38	84880/79.6	450/6.7	650000/51.8	440000/46.3
Saprophile, kl/ml	256760/61	14500/13.6	5270/78.0	600000/47.8	510000/53.6
Oligotrophic index	0.2	5.0	0.0	1.0	0.0
Ammonifying, kl/ml	1000/0.23	100/0.09	100/1,5	1000/0.1	100/0.01
Nitrifying, kl/ml	0/0	0/0	0/0	0/0	0/0
Denitrifying, kl/ml	1000/0.23	0/0	200/1.5	100/0.01	1000/0.1
Hydrocarbon-oxidizing c.u.	8360/2	6050/5.7	720/10.7	3970/0.31	600/0.06
Pentanol-oxidizing, c.u.	0	300	170	300	120
Haxanlo-oxidizing, c.u.	0	310	330	250	150
Heptanon-oxidizing, c.u.	150	0	110	0	0
Octane-oxidizing, c.u.	0	150	100	0	0
Nonan-oxidizing , c.u.	0	300	180	300	200
Decane-oxidizing, c.u.	250	150	110	0	230
Beneze-oxidizing, c.u.	120	150	0	300	150
Toluol-oxidizing, c.u.	150	200	60	270	0
Phenol-oxidizing, c.u.	300	0	0	0	0
Naphtalene-oxidizing, c.u.	320	0	0	0	120
Thione-authophytic, kl/ml	1000/0.23	10/0.1	10/0.14	100/0.01	100/0.1
Sulphate-reducing, kl/ml	0/0	0/0	100/1.5	0/0	0/0

Note: left- number of cells in 1ml of water; right- number of microbes in one specific physiological group, % comparable to total microbe number in water from the deposit

The initial segment in the microbic trophic chain, degrading organic substances, including oil, is aerobiotic and anaerobic organotrophic microorganisms. Aerobiotic organotrophic microflora, such as saprophiles, oligotrophiles, ammonifying and oil-hydrocarbon oxidizing bacteria are predominate in the investigated groundwater. Saprophiles are permanent and predominate (except in very rare cases). These microorganisms are used as carbon and nitrogenous nutrition in both protein and organic compounds (including oil). During the oxidation of organic substances saprophiles form different intermediate metabolism products, especially organic acids: oxalic, citric, vinic, glutamic, fulvic and other acids which are directly and/or indirectly attack metallic equipment.

Saprophile content is one to tens of thousands kl/ml in the formation waters of some of Tomsk Oblast oil fields. In the case of deep groundwater not confined to oil fields, the number of saprophiles is cells, rarely tens in 1ml of water. Based on the investigation results of groundwater from Aptian-Cenomanian aquifer system within most of the oilfields the number of psychrophile saprophytes is hundreds kl/ml. Maximum psychrophile saprophyte content was determined in Zapadno-Moiseevskoe field groundwater (600000 kl/ml/) and minimum- Dvurechenskoe (5270 kl/ml). This fact proves that habitat factors in Aptian-Cenomanian aquifer system groundwater are favorable for life sustaining saprophile bacteria. According to the morphological forms of saprophile microflora these bacteria are predominantly sporeless rod-like bacteria. Coccus-form bacteria are rare. It is well-known that globular (spherical) bacteria are prevailing in the mature stage of organic matter degradation within the water ecosystem. Rod-like bacteria abundance indicates the increased number of hard degradable organic substances.

Oligotrophiles are widely spread, except within Dvurechenskoe oil field, where quantity distribution of saprophile bacteria is more significant. These microorganisms can exist in ecosystems with low organic substance content (0.1 – 1 mg/l). The number of oligotrophiles is a direct indicator of the mineralization level of organic substances. The microflora content in Aptian-Cenomanian aquifer system groundwater embraces less than 50% oligotrophile bacteria of all microbes, except in Krapivinskoe oil field. The quantitative margin of oligotrophiles in different oil fields ranges from thousand to hundred thousands kl/ml. Maximum number of these bacteria can be found in the groundwater within Pervo-Maiskoe oil field. The oligotrophile- saprophile ratio is the so-called oligotrophic index. If more than 1, active processes of organic substance decomposition are in progress in the ecosystem. If less than 1 mineralization of organic substances is not in progress and, as a result, ecosystems lose the capacity of self-purification. According to obtained results, microbic decomposition of organic substances occurs only in the groundwater of Krapivinskoe and Zapadno-Moiseevskoe oil fields.

Number of hydrocarbon-oxidizing bacteria in groundwater of studied oil fields is insignificant and is only from a basis point to 10% of total amount. Maximum number of bacteria in Pervo-Maiskoe oil field groundwater is 8360 kl/ml, while minimum- in Lineinoe oil field groundwater. These hydrocarbon-oxidizing bacteria have oxidation capacity towards a wide spectrum of individual hydrocarbons, i.e. not only methane group, but also aromatic group (beneze-, toluol-, phenol-, naphtalene-) (table 3). In spite of the cosmopolitan distribution, the intensive development of hydrocarbon-oxidizing bacteria is rather insignificant- 200-300 c.u. (maximum 500 c.u.). Such a diversity of hydrocarbon-oxidizing organisms is indicative of dissolved hydrocarbons in the groundwater which have migrated from the underlying oil-bearing formation of Jurassic sediments as gaseous emanation.

Aptian-Cenomanian groundwater in each of investigated oil fields embraces its own specific spectrum of hydrocarbon -oxidizing bacteria, being formed consistent to the ecosystem conditions. The role of hydrocarbon -oxidizing bacteria in investigated microbal communities involves oxidation of hydrocarbons and further formation of partially-oxidized intermediate products -spirit, aldehyde and others. Consistent composition of hydrocarbon -oxidizing bacteria includes predominately rod-like bacteria, possessing active oxidation activity of oil and its derivatives: Pseudomonas, Arthrobacter, Rhodococcus, Corynebacterium and Mycobacterium. Oil-oxidizing bacteria decomposing oil and gaseous hydrocarbons terminate the cycle by freeing carbon from the stable compounds to atmosphere as carbon dioxide.

Ammonifying bacteria – the initial chain in the nitrogen cycle within the hydrosphere, and found in all investigated oil field groundwater. However, this group is less in number comparable to the above-described ones and is only ten and hundred units kl/ml. Microbes in the next nitrogen chain- nitrifying bacteria within investigated groundwater were not detected. Thus, nitrogen cycle in Aptian-Cenomanian groundwater becomes "disconnected". This could be associated with insignificant ammonifying bacteria productivity revealed in the investigated groundwater.

Due to oxygen consumption during hydrocarbon oxidation, oxygen deficiency emerges, promoting the development of facultative-anaerobic and anaerobic microorganisms. i.e. denitrifying and sulphate-reducing bacteria.

Denitrifying bacteria were revealed throughout all oil fields except in Krapivinskoe oil field. The number of which is tens to hundreds kl/ml. These bacteria reduce nitrates to molecular nitrogen with conjugated oxidation of organic acids, sugar and spirit to gas and water. Liberated energy (during oxidation) is used by the microorganisms for life sustainability. Denitrification could occur in aerobiotic and anaerobic conditions; however, it is more intensive in conditions without oxygen. It has been established that nitrogen source for these bacteria is nitrogen- containing heterocyclic oil components. Denitrifying bacteria are within the identical ecological niche (habitat) as sulphate-reducing bacteria, consuming both free and bound oxygen and forming favorable ecosystem (habitat) for sulphate-reducing bacteria.

Sulphate-reducing bacteria have been found only in the groundwater of Aptian-Cenomanian sediments within Dvurechenskoe oil field. Thionic bacteria were revealed in Aptian-Cenomanian groundwater within the territories of all the oil fields, the number of which is from one to tens thousand kl/ml. Maximum number of these bacteria was revealed in Pervo-Maiskoe oil field groundwater. Thionic bacteria attack forming favorable conditions for sulphate-reducing bacteria. Moreover, thionic bacteria, as other aerobiotic bacteria, consuming oxygen form favorable conditions for sulphate-reducing bacteria. However, during metabolism sulphur, pyrite and life sustaining products of sulphate-reducing bacteria oxidize into sulphides, while hydrogen sulphide- into sulphates, forming energy sources for sulphate-reducing bacteria.

4. Conclusions

Aptian-Cenomanian aquifer system within Kaimisovsky oil-bearing province has been being intensively developed for water production and further usage in the reservoir pressure maintenance system. Research results proved the significance of studying the chemical microbiological composition of above-mentioned groundwater. It is important if the water-flooded layer excludes salt-deposition and further suppressing corrosive processes. Obtained results showed that not only sulphate-reducing bacteria, but also organotrophic microflora found in groundwater, throughout its life sustaining activities could further corrosive active metabolites. In this case, this fact should be considered in oil field development.

References

[1] Frank Y, Banks D, Avakian M, Antsiferov D, Kadychagov P and Karnachuk O 2016 *Geomicrobiology Journal* Firmicutes is an important component of microbial communities in water-injected and pristine oil reservoirs, Western Siberia, Russia **Vol.** 33 (5) pp. 387-400

[2] Kashavtsev V and Mishenko I 2004 *Moscow. Orbit.* Scaling in oil extraction p 432

[3] Nalivaiko N, Dutova E and Spiridonov T 2016 *IOP Conference Series: Earth and Environmental Science.* Groundwater microflora of the Aptian-Cenomanian deposits at the Igolsko-Talovoe field in Tomsk Region **Vol.** 33 (1)

[4] Nazarov A 2004 *Idea-Press.* Oil and gas hydrogeochemistry of the South-Eastern part of the West Siberian oil and gas province p 288

[5] Trifonov N, Nazarov A and Alexeev S 2016 *IOP Conference Series: Earth and Environmental Science.* Groundwater compatibility with formation water and pay zone rocks in Pervo-Maiskoe oil-gas-condensate field to maintain formation pressure **Vol.** 33 (1)

[6] Kontorovich A, Nesterov I, Salmanov F 1975 *Nedra Press. Moscow.* Geology of oil and gas in West Siberia p 521

[7] Pokrovsky V, Dutova E, Kuzevanov K, Pokrovsky D and Nalivayko N 2015 *IOP Conf. Seri.: Earth Environ. Sci.* Hydrogeological conditions changes of Tomsk, Russia **Vol.** 27 pp. 12031

[8] Dutova E, Vologdina, Pokrovsky D, Nalivaiko N, Kuzevanov K and Pokrovsky V 2016 *IOP Conference Series: Earth and Environmental Science.* Hydrogenous mineral neoformations in Tomsk water intake facility from underground sources **Vol.** 33 (1)

[9] Dutova E, Bukaty M, Nevol'ko A, Pokrovsky D and Shvartsev S 2006 *Geologia and Geophysics.* Hydrogenic concentration of gold in alluvial placers of the Egor'evsk area (Salair) **Vol.** 47 (3) pp. 364-376

[10] Shvartsev S and Dutova E 2001 *Geology of Ore Deposits.* Hydrochemistry and mobilization of Gold in the hypergenesis zone (Kuznetsk Alatau, Russia) **Vol.** 43 (3) pp. 224-233

[11] Netrusov A, Egorova M, Zakharchuk L 2005 *Moscow. ACADEMA. Practical guide on microbiology* p 608

Hydrogeochemical characteristics of water intakes from groundwater sources in Seversk

A I Karmalov[1], E M Dutova[2], I V Vologdina[3], D S Pokrovsky[4], V D Pokrovskiy[5] and K K Kuzevanov[6]
[1] JSC "Seversky Vodokanal", Seversk, Russia
[2,5,6]Tomsk Polytechnic University, Tomsk, Russia
[3]Tomsk State University,Tomsk, Russia
[4]Tomsk State University of Architecture and Building,Tomsk, Russia
E-mail: [1] karmalov@svk.sels.ru, [2], [3]irvalvo@inbox.ru, [4]dsp@sibmail.com, [5]pokrovskiy.v@gmail.com, [6]kuzevanovkk@mail.ru

Abstract. The article describes the hydrogeochemical environment behavior analysis of groundwater intake which, in its turn, provides the utility and drinking water supply for Seversk. The reasons for temporary changes of the hydrogeochemical aquifer indicators in the producing areas have been highlighted. The main factor could be upset hydrodynamic conditions during long-term operation. Changed hydrogeochemical indicators have been revealed not only during the technological water treatment process but also during water transportation to consumers. Chemical composition water changes are related to secondary mineral and sludge formation on technological equipment. Precipitation is a polymineral mixture predominantly a ferrous phase, whereas phosphate and carbonate phases are secondary. Clay minerals are also found.

1. Introduction

One of the most common types of human activities involves water supply, which, in its turn, results in significant. The study of the chemical and mineral precipitation composition, its structure and formation mechanisms is of both theoretical and practical interest. The authors have been studying precipitation formations within water supply systems of the water intakes from Tomsk and Tomsk Oblast groundwater sources [1-5]. Based on investigation results, the precipitation composition and structure were identified, as well as the precipitation property- water aeration regime and **tile line** relationship. This research is focused on studying the variability of hydrogeochemical environments and accumulation of secondary solid mineral neocrystallizations in Seversk water intake facilities.

2. Research target characteristics

The utility and drinking water in Seversk, the second largest city in Tomsk Oblast, is supplied from two groundwater source intakes, located on the right Tom River bank and embracing aquifers of the West Siberian Megabasin (fig.1). Water intake facility No.1, operating since 1965, includes 32 wells within Paleogene aquifer. Water intake facility No.2 operating since 1971 , includes 22 wells within Paleogene and Upper Cretaceous aquifers. All the wells and water intake facilities are equipped with wire-wrapped screens and submersible pumps. Their depth varies from 84 to 129 meters while their productivity is from 40 to 120m^3/day. Total productivity is about 50 thousand m^3/day.

Operating regulations for analyzing and predicting hydrogeological condition changes within the areas of water intake facilities are stipulated by the monitoring system for groundwater level and quality. Analytical work was performed in accredited laboratories of OJSC Seversk Vodokanal and OJSC Tomskgeomonitoring.

Figure 1. Layout of Tomsk and Seversk water intake facilities: 1 – lines of water intake production wells in Tomsk; 2 – water intake facilities in Seversk.

3. Groundwater chemical composition

Chemically, water in production wells is exclusively hydrocarbonate in composition with different calcium and magnesium ratios and could be classified as fresh, neutral and weakly alkaline, moderately hard water (Table 1).

This groundwater is typical of taiga and subtaiga landscapes in geochemical ferromanganese province of Western Siberia. Water values exceeding maximum allowable concentrations for drinking and household water include iron, manganese and silicon, whereas iron and manganese are found in practically 100% of all investigated cases. Chemical composition variation indicators,changing patterns and their magnitude depend on spatial location and lithologic and facies rock characteristics, as well as the temporal sampling coordinates.

Table 1. Groundwater chemical composition and technological water cycle in water intake facilities (based on water intake facility service data of 2012-2014).

Indicators	Water intake facility No.1			Water intake facility No.2		
	Water in production well	*Technological water	**Water supplied to networks	Production well waters	*Technological water	**Water supplied to networks
Dry residues, mg/dm^3	244…895 355	319…320 1	330	200…306 247	257…264 7	252
TDS, mg/dm^3	375…1066 520		485	318…474 377	377-364 -13	375
pH	7.1…7.6 7.3	7.0…7.5 0.5	7.6	6.9…7.3 7.1	7.0…7.4 0.4	7.5
Hardness, mg-eq/dm3	4.1…6.9 5.1	4.9…4.85 -0.5	5.3	3.5…6.1 4.2	4.2…4.0 -0.2	4.0
HCO_3^-, mg/dm^3	262…386 330	351…330 -21	311	228…366 260	271…271 0	245
Cl^-, mg/dm^3	2…253 29	23.5…25.5 2	19.3	2.0…5.1 2.3	0.2…2.4 2.2	4.4
SO_4^{2-}, mg/dm^3	2…36 8	19.5…5.86 -13.64	6.1	2.0…14.2 5.3	16.6…2.8 -13.8	5.5
Ca^{2+}, mg/dm^3	60…242 85	78.2…82.2 4	77	50…81 63	62…65 3	62
Mg^{2+}, mg/dm^3	4…34 16	12.6…9.45 -3.15	13	7…28 13	13.7…8.7 -5.0	11

Fetotal, mg/dm^3	3.0…6.7 / 4.8	4.74…0.1 / -4.73	0.1	1.1…4.5 / 2.4	2.4…0.1 / -2.3	0.11
Mn, mg/dm^3	0.2 0.6 / 0. 4	0.29…0.2 / -0.09	0.26	0.02 0.39 / 0.24	0.23…0.20 / 0.03	0.29
Si, mg/dm^3	10.3…14.8 / 12.0	12…11.7 / -0.3	9.6	11…14 / 12	13-14 / 1	10
Oxidability, mg O_2/dm^3	1.4…4.7 / 2.4	2.04…0.88 / -1.16	1.1	0.8…2.3 / 1.5	1.3…0.7 / -0.5	1.5
Number of values	25	10	4	21	10	4

Note. *numerator- range and denominator – mean value of the indicator.
**numerator - value indicator range of water supplied to and in water treatment cycle, denominator – change of indicator values

4. Hydrogeochemical characteristic changes during field development

Operating water intake facilities furthered depression cone formation in developing Paleogene and overlying Neogene-Quaternary aquifers. In both aquifers the depression cones are separated by groundwater divisions, spatially coinciding with orographic water division. The cone configurations and dimensions have not practically changed in recent years, while their level surface ranges from 0.5 to 1.0 m depending on water intake rate. In highly intensive production wells cone depression surface level decreases to 10…23 m.

It is obvious that hydrodynamic condition changes lead to variations of extracted water quality. The main mechanisms of these changes are the following: (1) subsurface aeration occuring within forming depression cones and aeration water flows from overlying aquifers and (2) as a result of elastic deformation processes occuring in bedded formations during hydro-technical utilities and the active areal water flows from underlying aquifers which could be rather intensive in lithologic windows [5]. Spatially, both mechanism changes of hydrogeochemical conditions embrace a wide areal distribution.

In terms of continuous water intake facilities operation and underinvestigated Tomsk water intake facility [2] these processes result in the chemical groundwater composition within the aquifers. Year-to-year observation data indicated a rather pronounced increase in mineral content, especially chloride ion, as well as iron content decrease in the waters of developed aquifers. Such changes are apparently associated with intensive water flows from deep aquifers. Iron content decrease is associated with solution-solid transformation during water flows from overlying oxygen-enriched Neogene-Quaternary aquifers.

According to data analysis, water supplied to the water treatment process cycle comparable to production well water is characterized by a low variability of indicators due to equalization (table 1). The quality of production well water is enhanced to drinking standards via water treatment technology based on natural processes. In terms of its physical nature this could be an analog to groundwater discharge at the surface. This water treatment technology includes deferrization of simplified aeration, further penetration into a contact chamber and filtration through granular medium, as well as hypochlorite sodium purification. Water treatment reduces the indicator values (table 1). Iron content in after-treatment water does not exceed maximum allowable concentrations. Water of required hygienic standards is supplied to consumers via distribution networks being about 200 km. Water composition and quality indicators also change.

5. Changes of hydrogeochemical indicators and secondary mineral formation

Water chemical composition changes are accompanied by secondary mineral formation and precipitation on water intake equipment. Precipitation accumulates near the wellbore in pre-filtering zone and on well equipment (pumps, water gauges, pipes) (fig. 2), on filters of water intake facilities and distribution networks, plumbing equipment and household appliances.

Two morphological types of mineral neocrystallization are formed on the equipment of water intake facilities: structureless ochreous (iron) masses and mineral aggregate having rigid-structured framing. There are no significant genetic differences between these precipitation types: both are formed due to physical-chemical environmental condition changes, whereas their morphological peculiarities are predominantly associated with the hydrodynamic depositional environment [4].

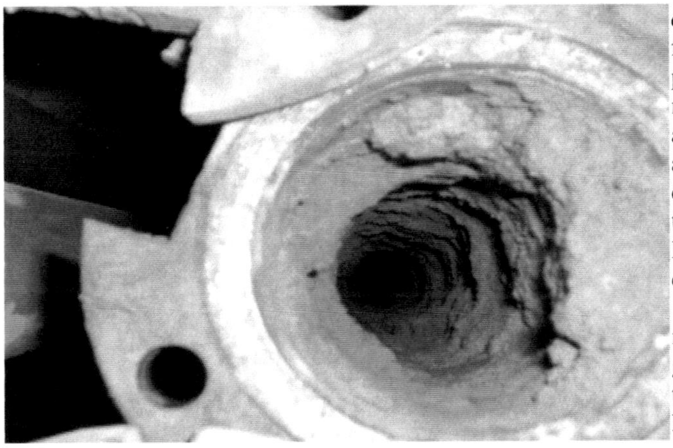

Figure 2. Precipitate in an ascending pipe in well No. 14.

Ochreous (iron) masses accumulate in dynamic water flow conditions on the filters and production well ascending pipes, on submersible pumps and measurement equipment. They also accumulate in granular medium filters after water purification. They have different consistency: near the surface – thick creamy liquid and and internal layers- dense, paste-like; harden during drying up and easily triturable into dust.

Second-type mineral neocrystallization accumulates on the granular medium filters in water treatment facilities, having concentrically layered structure with alternating beds of different thicknesses and color. Based on these characteristic features they could be referred to as oolites [4]. After air drying, oolite structure is preserved while aggregates can be decomposed only under mechanical effect. Dispersion particles remain on the grain surface forming an increasing thick film due to filter flushing.

Oolite composition and structure were analyzed on the VEGA I LMU scanning electron microscope equipped with INCA Energy dispersive spectrometer in the Analytical Center for Geochemistry of Natural Systems, TSU (by A.S. Kulkov -analyst, assistant, Department of Petrography, TSU). The concentrically layered precipitation film structure formed around the grain medium was detected on the electron microscope. The precipitation film thickness depends on filtration operating time and could vary from mm fractions to 1 mm [4]. In our case, separate film layer thickness ranges from 10 to 45 μm (fig. 3).

Precipitation is mainly composed of iron compounds (up to 75.34%), less of silicon (up to 14.52%), calcium (up to 7.71%) and phosphorus compounds (up to 7.89).

Mineralogical identification of investigated material has not been performed yet. However, according to the obtained data, one and the same aquifer is the groundwater source for both Tomsk and Seversk water intake facilities. The ferrous phase includes ferrous hydrite, goethite and hematite and carbonate phase – calcite. Iron hydrophosphates, such as rockbridgeite, dufrenite and vivianite precipitations were identified in Tomsk water intake facility [4].

Besides the main components and mineral neocrystallization, impurity elements are also found and their concentrations in the primary solution depend on both natural factors and the environmental system conditions.

Figure 3. Concentrically layered oolite structure formed around grain medium and points of analysis BSE image.

The chemical precipitation composition was studied via energy-dispersive X-ray spectroscopy (EDX) (table 2).
Impurity elements could be removed from water in several ways. Apart from iron mineral formation, the isomorphic or sorption mechanisms of microelement-capture forming mineral neocrystallization is significant. Although many undersaturated microelements in groundwater are not removed during the water migration, for example, iron minerals, they are either sorbed or precipitation of colloidal oxides and iron hyrooxides, manganese oxides, clay minerals and carbonates [4].

Table 2. Chemical composition of the newly formed oolite around grain medium (%).

Oxides	Content (mass fraction, %)			
	Points of analysis			
	1	2	3	4
Si_2O	11.62	12.96	10.11	14.52
FeO*	75.34	75.24	74.30	75.18
CaO	7.13	7.04	7.71	6.64
P_2O_5	5.91	4.76	7.89	3.66
Total	100.00	100.00	100.00	100.00

Note: FeO* = FeO + Fe_2O_3.

The microelemental composition of precipitation was analyzed on ICP-MS at LLC Chemical-Analytical Center Plasma. The analysis results showed that precipitation formed at different water treatment stages, mainly, in cooling tower and in granular filters, differ in their composition (table 3). The precipitation from the cooling tower has a more ferrous composition (34.6%) comparable to the precipitation from the filters (13.6%). The difference involves the manganese content which is more significant. There are also microelement content differences, which indicate the complex mechanisms of chemical element precipitation during water treatment.

Table 3. Elemental and microelemental composition of precipitation (mass fraction, %)

Sampling No. *	Elements								
	Fe	Mn	Cu	Ni	Cr	Ba	Zn	Co	Sr
1	13.6	26.0	0.0017	0.0005	0.00098	0.55	0.029	0.00057	0.070
2	34.6	0.72	<0.0001	0.0005	0.00028	0.12	0.018	0.00004	0.024

*Sampling location: 1 – deferrization filters in water intake facility No.1 in Seversk; 2 – precipitation from a cooling tower in water intake facility No.1 (2) in Seversk.

6. Conclusion

Changes of hydrogeochemical conditions during groundwater field development is accompanied by secondary mineral formation and precipitation both in reservoir conditions and during operating water intake facilities, as well as at subsequent stages of water treatment and transportation.
Precipitation from water supply systems are a mixture of finely dispersed and poorly crystallized minerals that are predominantly composed of nanometer-sized ferrous oxide and hydroxide particles, which could be of some interest for researchers involved in nanotechnology development. Based on precipitation from water supply systems, technologies for producing nanomaterials could be developed, which, in its turn, could generate composite materials with specific catalytic and oxidation-reduction properties [6,7].
According to research findings, precipitation formed in service water systems reveal similar composition and structure comparable to those of natural formations [8,9,10]. Analyzing the chemical and mineral composition of these precipitation, their structure patterns and formation mechanisms is of theoretical and practical importance.
Thus, further detailed studies of natural and man-caused precipitation can significantly extend our knowledge of mineral formation in hypergenesis areas of both modern and ancient environments.

References

[1] Pokrovsky D, Dutova E, Vologdina I and Taylashev A 2002 *Izvestiya Tomsk Polytechnic Universit*y Natural and Manmade Mineral Formation on Deferrization Filters of a Water Intake in Tomsk Academic Town **Vol.** 305 (6) pp. 319–329

[2] Dutova E, Vologdina I, Pokrovsky D and Zamorovskaya L 2008 Change of the Hydrogeo chemical Environment when Developing the Aquifer Academichesky *Izvestiya Tomsk Polytechnic Universit*y **Vol.** 312 (1) pp. 59-63.

[3] Pokrovsky D, Dutova E and Vologdina I 2010 *Izvestiya Higher Educational Establishments. Construction* Hydrogeochemical Environments and Mineral Neocrystallizations from Subsurface Sources **Vol.** 11-12 pp. 54-61

[4] Pokrovsky D, Dutova E, Rogov G 2002 *edited by Pokrovsky Tomsk NTL Publishing House* Mineral Neocrystallizations at Water Intake Facilities of Tomsk Region p 176

[5] Dutova E, Vologdina I, Pokrovsky D, Nalivaiko N, Kuzevanov K and Pokrovsky V 2016 *IOP Conference Series: Earth and Environmental Science.* Hydrogenous mineral neoformations in Tomsk water intake facility from underground sources. **Vol.** 33 (1), art. no. 10 088

[6] Ryltsova I, Nestroynaya O, Lebedeva O, Vorontsova O, Kosova N and Kurzina I 2014 *Russian Journal of Inorganic Chemistry* Synthesizing and studying new layered double magnesium-cobalt-iron hydroxides with a hydrotalcite structu*re* **Vol.** 59 (12) pp. 1652

[7] Novoselova L 2016 *Powder Technology* Hematite nanopowder obtained from waste: iron-removal sludge **Vol.** 287 pp. 364-372

[8] Rudmin M, Mazurov A and Ruban A 2015 *IOP Conference Series: Earth and Environmental Scien*ce Facies and sedimentation model of iron-ore sequence in Bakchar deposit **Vol.** 24 (1) art. no. 012028

[9] Lyutoev V, Lysyuk A and Silaev V 2010 *Vestnik Institute of Geology of the Komi Science Center UB RAS* Micro-nanomineralogic mysteries of natural pigments **Vol.** 9-1 (189) pp. 18 – 20

[10] Banks D, Pumar A and Watson I 2009 *Quarterly Journal of Engineering Geology and Hydrogeology* The operational performance of Scottish minewater-based ground source heat pump systems **Vol.** 42 (3) pp. 347-357

Assessment of fresh groundwater vulnerability to contamination caused by production operations in oil and gas fields of Nizhnevartovsk Region (Western Siberia)

Yu V Bespalova[1,3], V M Matusevich[1], V K Popov[2] and K I Kuzevanov[2]

[1]Tyumen Industrial University, Volodarsky St. 56, Tyumen, 625000, Russia
[2]National Research Tomsk Polytechnic University, 30 Lenin Avenue, Tomsk, 634050, Russia

E-mail: [3]bespalova_y@mail.ru, [2]pvk@tpu.ru, [2]kki@tpu.ru

Abstract. Groundwater quality is possible to provide under the condition of aquifers effective protection from surface contamination and bottom pollution prevention in the areas of intense oil-gas production operations. Qualitative and quantitative assessment of groundwater vulnerability is based on regional characteristics of lithological composition and thickness of impermeable deposits in unsaturated zone and overlying deposits. The correlation of head and water table levels, absorption capacity of soils and clay rocks are also considered. An integrated approach to natural vulnerability analysis of groundwater has been presented. It suggests a combination of three most common methods. Based on performed calculations, the map has been made to plot fresh groundwater vulnerability of Atlym-Novomikhailovsk aquifer system (Nizhnevaryovsk Region, Khanty-Mansiysk Autonomus Okrug).

1. Introduction

Assessment of natural groundwater vulnerability to contamination is a vital issue in hydrogeology. At present, the technogenic impact on groundwater can be seen most directly at regional scale. This problem is considered in the research with respect to natural vulnerability of aquifer system used for domestic water supply. Attenuation capacity has been assessed.

Groundwater vulnerability is defined as a degree to which main aquifer is being capped with low-permeable deposits capable to prevent surface contaminant seepage into groundwater. Groundwater vulnerability depends on many factors which fall into three groups: natural, technogenic and physico-chemical [1].

To prevent deterioration of fresh drinking groundwater quality and occurrence of new geochemical types due to anthropogenic impact, it is necessary to analyze natural vulnerability of groundwater in the study area.

2. Materials and methods

2.1. Study area

The target of the present research is groundwater of Atlym-Novomikhailovsk aquifer system composed of alternating sandy-clayey deposits which are 100 – 140 m thick. Water head varies from 108 to 154 m. Static groundwater level is at the depth of 2 – 8 m. The actual elevation ranges within

37.5 – 53 m. The overlying rocks contain Quaternary lacustriner-alluvial deposits. The water-saturated bottom section forms an aquifer. Groundwater occurs at the depth of 5 – 10 m. The thickness of Quaternary aquifer ranges within 22 – 28 m. This description has been made by authors of article on the basis of a hydrogeological section of Nizhnevartovsk Region [2].

There is a slight difference in groundwater table and head water level in Ob valley and its tributary basin, and it hardly exceeds two meters in watershed.

The deep-lying fossiliferous layer of permafrost has been traced. Sporadic permafrost occurs in the central part of the study area. Permafrost embraces sand and sandy-clayey deposits of Atlym and Novomikhailovsk suites. The depth to the top ranges within 60 - 105 m in watershed; while thickness varies from 45 to 100 m.

The unsaturated zone is composed of non-homogeneous layers containing different permeable, low-permeable and impermeable deposits which vary in composition and thickness. In flood-plain clayey deposits are hardly ever present.

In watershed, unsaturated zone and underlying Quaternary deposits are mainly clayey. Total thickness of sand and clay layers is defined by the ratio 2:1. Total thickness of overlying deposits is 40 - 70 m [3].

2.2. Sampling and analytical procedures

Based on the most frequent techniques, calculations of predicted time (T_1, T_2, T_3) of contaminant filtration from the surface to overlying bed of main aquifer have been performed. The basics of hydrodynamic calculations are described in standards [1, 4] and published literature [5, 6].

Apart from basic geological parameters (filtration coefficient, thickness of overlying rocks, porosity of water-resistant rocks), the first method considers difference in levels of groundwater and underlying bed.

—, (1)

where T_1 – the predicted filtration time calculated using the first method; $\Delta H = H_1 - H_2$ the head gradient, H_1 – the head of overlying bed; H_2 – the head of main aquifer; n – the porosity of water-resistant rocks; k –the weighted mean of filtration coefficient which is defined by the formula:

—, (2)

where m_i - the i-layer thickness, which overlies main aquifer; k_i – the i-layer filtration coefficient; k - the weighted mean of filtration coefficient.

The second method takes account of groundwater recharge rate and depth of groundwater runoff (h). In the study area groundwater recharge rate (M_e) is 4.02 $dm^3/sec \times km^2$ [7], consequently, depth of groundwater runoff is 126.774 mm/year or 0.127 m/year. With the reference to the given groundwater recharge rate, intensity of infiltration recharge (E) is 0.0003 m/day.

Contamination travel time is calculated by the formula:

— (3)

where T_2 – the predicted filtration time calculated using the second method; n_0 – the absolute porosity of overlying rocks (mean value 0.15), m_0 – the thickness of overlying rocks, k_0 – the coefficient of vertical filtration.

When using the third method, the intensity of infiltration recharge is the main parameter [5]. In this case the thickness of overlying deposits is indirectly considered:

—, (4)

where T_3 – the predicted filtration time calculated using the third method; W – the intensity of infiltration recharge at the groundwater surface. It accounts for 20 % of total precipitation, which, in its turn, makes 94 mm/year or 0.00026 m/day taking into account that average annual rainfall intensity is 471 mm/year; k_0 - coefficient of vertical filtration, which is defined as 1/20 average filtration coefficient in Quaternary deposits in watershed; μ – absolute porosity of rocks in Quaternary aquifer which is 0.1.

Based on the above-mentioned parameters, the time of vertical filtration through semipermeable deposits to domestic water producing aquifer was calculated, with each well tests considered individually.

3. Results and discussions

Time required for a contamination to migrate through overlying rocks to overlying bed of Atlym-Novomikhailovsk aquifer has been calculated using thirty well test series. Average values of contamination seepage time are given in table below according to data series from wells located along cross-sectional line.

Table. Results of average filtration time calculations in case of neutral pollutant (T average).

№ well	T_1	T_2	T_3	$T_{average}$	Rank
551	13910	10714	12500	12375	4
N12-A,B,V	20376	12857	12000	15078	5
254	27670	15750	15750	19723	7
535	17158	12000	11200	13453	4
NZ-79	11178	10125	10800	10701	3
7-857	8003	8020	7952	7992	1
7729	8672	8571	8000	8414	2
7669	20840	12500	12500	15280	5
9 bN	13181	10500	9800	11160	3
27aN	9152	9750	10400	9767	2
6E	13766	10285	9600	11217	3
7-231	8862	9429	11000	9764	3
6	8939	8750	7000	8230	2
7166	8862	9500	11100	9821	2
Krp28	8862	9510	11150	9841	2
120e	8862	9440	11253	9852	2
7394	8862	9443	11000	9768	2
156	11178	11571	10800	11183	3
7675	21024	12900	12000	15308	5
nz-267	15649	12900	10900	13150	4
98-3	14955	11143	10400	12166	4
3	14955	11130	10500	12195	4
356	21024	15000	12200	16075	6
733	14955	11143	10400	12166	4
nz-230	25452	12900	12300	16884	6
nz-118	23954	14750	11800	16835	6
409	16875	13250	13250	14458	2
4k-3	20759	14250	14250	16420	6
429-3	14198	10928	12750	12625	4
383	10002	10071	9400	9824	2

Based on averaged calculated travel times of contaminants percolating unsaturated zone and Quaternary water-saturated deposits, the vulnerability map of Atlym-Novomikhailovsk aquifer system was made (figure 1).

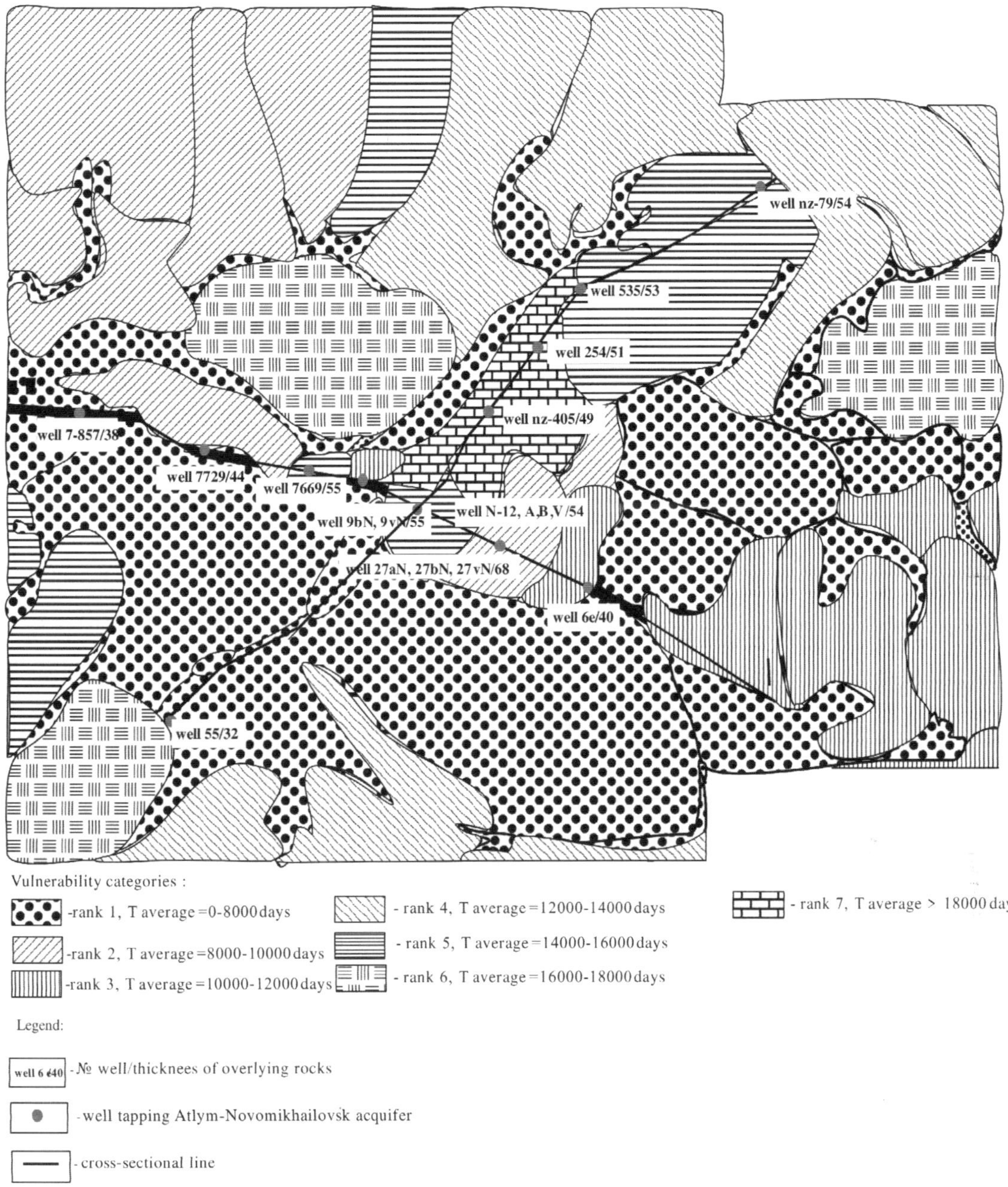

Figure. Natural vulnerability map of Atlym-Novomikhailovsk aquifer system (scale 1:500000).

It has been proposed to rate natural vulnerability of groundwater in Nizhnevartovsk oil-gas bearing area on 7-point scale: 1 - vulnerable, 2 – essentially vulnerable, 3 - weakly vulnerable, 4 – relatively invulnerable, 5 – sufficiently invulnerable, 6 – invulnerable, 7 – effectively invulnerable.
The scale takes account of estimated time for safe exploitation of an aquifer (before contamination due to vertical filtration), which, as a matter of actual practice, makes 25-50 years.
The most widespread are vulnerable groundwater (rank 1), with infiltration time being 0 – 8000 days. This estimate is a characteristic of boggy areas where highly sandy rocks are abundant in floodplains. Contamination travel time in these areas is 7992 days (well 7-857). Vulnerable groundwater occurs in

Ob valley and its tributary basin which are significantly deep. In these areas estimated time of safe groundwater exploitation is 22 years.
Boggy deposits are common in all geomorphological landscape types. Lithologically the deposits are composed of peats of different decomposition degree, clay loams, clay sands, with extensive inclusions of vegetable detritus and subordinate sand interbeds. Deposit thickness varies from 1.5 – 3 m to 4 – 5 m.
Talagaika suite deposits formed by alluvium laid down in Ob, Vaha, Vatinsk Egan paleo valleys are pervasive except for watershed zones. The sediments are fine-grained quartz sands interlayered with thin loams and clay sands. Graded bedding is observed in sands with gravel and pebble. Thickness of the deposits ranges within 8 – 22 m.
Essentially vulnerable groundwater (rank 2) is confined to the first terrace above the floodplain, which is located in the western and central parts of the study area. Though total thickness of overlying deposits is significant (35 - 68 m), clayey deposits are not more than 8 – 11 m thick which, in its turn, is not effective enough to protect groundwater from vertical contamination over the period of safe groundwater exploitation which is 25 years.
Channel facies composed of very-fine and fine-grained quartz sand is predominant formation in the cross-section. The upper layers are dust sands interlayered with subordinate loam and clay-sand; inclusions of vegetable detritus can be observed as well.
In the central and eastern parts of the study area there are local zones of weakly vulnerable groundwater (rank 3), which is assumed to be safe for exploitation over the period of 33 years. The thickness of low-permeable deposits is 12 – 14 m. The third terrace above the floodplain deposits are composed of fine-grained sands in the top and silts and silt clays – in the bottom. Based on different lithological composition, two deposit series have been distinguished in the cross-section of the second terrace above the floodplain. The bottom series is sand deposits while the upper – clayey deposits. The bottom series contains mainly sands and loams, whereas upper series contains silts, loam clays and clays.
The zones indicated as relatively invulnerable (rank 4) are mainly located along the watershed lines in the northern part of the study area. In these areas conventional contamination travel time increases to 14000 days, this is equal to safe exploitation time corresponding to 38 years. In these zones total thickness of overlying rocks is 54 m on average, with 15 m being composed of clayey deposits.
The zones of sufficiently invulnerable groundwater (rank 5) are less frequent, where groundwater is assumed to be safe over 44 years. These zones are observed in the central and eastern parts of the study area. They contain low-permeable deposits which are 20 m thick. This area mainly corresponds to deposits of second- and third terraces above the floodplain where the upper rock series is silts, loam clays and clays which function as barriers preventing migration of pollutants from the surface.
The zones of invulnerable groundwater (rank 6) which is safe for exploitation over 46 years are restricted to watershed lines and spatially correspond to areas of second- and third terraces above the floodplain. Low-permeable deposits differ in thickness from adjacent formations insignificantly.
The zones of effectively invulnerable groundwater (rank 7) are confined to permafrost which functions as a cryogenic acquiclude (northern part of the study area). In this case the aquifer is safe over the whole exploitation period.
Due to permafrost, clay-sand and loam deposits which occur in overlying beds, groundwater is least contaminated. However, indications of groundwater contamination, which is assumed to be irreversible, are obvious to the north of West-Siberian megabasin. This is due to the fact that groundwater contamination is directly proportional to technogenic impact and inversely proportional to its vulnerability. Under the conditions of areal distribution of potential contamination sources vulnerability analysis is the most vital issue.

4. Conclusion

Based on the analysis of concurrent calculation techniques, it has been indicated that according to the first method [1] maximum travel time is $T_{average}$ (25452 days), whereas other methods to estimate this time [4, 5] yield a conservative value which is approximately 15750 days.
Thus, a substantial difference in quantitative estimates of groundwater vulnerability has been detected. When only one method being applied, assessment of groundwater vulnerability can be controversial.

More objective assessment of vulnerability can be provided when an integrated approach is being used. This methodology suggests using not only averaged values of safe exploitation times obtained as a result of concurrent analytical calculations but also methods of numerical modeling [9, 10, 11, 12, 13] taking into account rock sorption properties [8] when solving problems of geomigration forecasting.

Areal zoning of a particular oil-gas bearing area according to natural vulnerability enables to optimize location of observation stations which supplement the existing groundwater monitoring network in areas unprotected against contamination.

Acknowledgements

The authors would like to thank L.A. Kovjatkina both for submitting supplementary materials and critics.

The investigations were supported by The Russian Foundation for Basic Research (Project No 16-35-50181)

References

[1] Zabuzov A A 1985 Regulations for groundwater protection (Moscow) (in Russian) p. 21
[2] Vashurina M V 2011 The study of fresh groundwater in Vartovsk petroleum province (Tuymen) (in Russian) p. 138.
[3] Bespalova Y V 2015 Towards multifactor approach to vulnerability analysis of aquifer systems in developing West Siberian petroleum provinces *Higher Educational Institutions News. Oil and Gas* **Vol.** 3 pp. 6-13.
[4] Lapshin N N and Oradovskaya A E 1983 Guidelines for hydrogeological computation to detect protective sanitary zone (II and III) boundaries of undergroundwater sources for domestic water supply (Moscow, VNII VODGEO) (in Russian) p. 102.
[5] Shestakov V M and Opzdnyakov S P 2003 Geohydrology (Moscow) (in Russian) p. 176.
[6] Fitter C W 2001 Applied hydrogeology (New York) p. 691
[7] Palkin S S 2001 An assessment of security the population Khanty-Mansiysk Autonomous Okrug with resources of underground waters for economic and drinking water supply (the II stage). Report on accomplishment of thematic works with Department on oil, gas and mineral resources of Khanty-Mansiysk Autonomous Okrug. State register No. 50-98-38/57. Hydrogeoecological research and production and GIDEK engineering firm (Moscow) (in Russian).
[8] Krainov S R and Shvets V M 1980 Essentials of groundwater geochemistry (Moscow) (in Russian) p. 285.
[9] Kuzevanov K I and Savichev O G, Reshetko M V 2012 Mathematical modeling of natural processes (Tomsk) (in Russian) p. 144.
[10] Wang H F and Anderson M P 1982 Introduction to groundwater modeling-Finite difference and finite element methods (San Francosco).
[11] Young H L et all 1986 Nothern midwest regional aquifer system stady. In Regional aquifer-system analyses program of the U.S. Geological Survey Circular 1002.
[12] Barlow P and Moench A F 2011 WTAQ version 2 – a computer program for analysis of aquifer tests in confined and water-table aquifers with alterna tive representations of drainage from the unsaturated zone U.S. Geological Survey.Techniques and Methods 3-B9.
[13] Tomlin C D and Berry J K 1979 A Mathematical Structure for Cartographic Modelling in Environmental Analysis *In Proceedings of the **39th** Symposium of the American Conference on Surveying and Mapping* pp. 269-283.

The nonlinear effects based on peat chronology data in paleoclimatic reconstructions

O G Savichev, M V Reshetko and Yu A Moiseeva

Tomsk Polytechnic University, 30 Lenin Avenue, Tomsk, 634050, Russia
E-mail: OSavichev@mail ru, mrechetko@tpu ru, julchiky@mail ru

Abstract The simplified mathematical model of peat deposits growth is suggested based on assuming the swamp is a thermodynamics system Its validation was performed according to the data of the swamps in Siberia It is shown that the peat deposits growth is nonlinear related to climate change Therefore, for the appropriate methods applying in paleoclimatic reconstructions is expedient to establish the optimal values of functioning particular bio- and ecosystem in natural conditions that determine the most probable existence of life forms of biological species

1. Introduction

Paleoclimatic reconstructions play an important role in the study of regularities and mechanisms of climate and environmental changes and their prediction and development of regional socio-economic programs, before the adoption of global strategies for the development of the various fields of the world economy The used methods have different disadvantages [1, 2, 3] As a result, some methods are usually used simultaneously, with used as dendrochronology [4, 5], palynology [6, 7] and algorithms of complex analysis of marshy processes [4, 8] the reference methods (used for "calibration")

Although the above methods have positive features (first of all, the relative simplicity), when applying in paleoclimatic reconstructions, it is necessary to take into account the nonlinear relation of the environment components, and, especially, nonlinear reactions of vegetation to climate change It is caused by the influence of several factors from astronomy to “atmosphere-ocean” system Taking this into account, some aspects of the analysis of the relations between climate characteristics and the intensity of growth of peat deposits are described below In this work the peat chronology is understood as the detrmination and the restoration of natural events, so it is the same as dendrochronology

2. Materials and methods

Methods of research consist of construction of a simplified mathematical model of peat accumulation based on assuming that the swamp is a thermodynamic system (arising and existing under excess moisture) Testing this model is carried out according to the published data on the depth of the peat deposit taking from a Siberia bogs [9-14] and on materials of the Roshydromet observation over the temperature of atmospheric surface layers [15-17] The choice of research objects is due to exceptionally high bogginess of the region, which reaches over 70 % in the number of river catchments [17, 19]

2.1. Subject of research
As the objects of study, we have chosen Vasyugan swamp and a number of swamps in the Salym-Yugansk interfluve, near the township Igarka, Lake Samotlor, the village Sytomino, the catchment of the rivers Kas and Shuchya Vasyugan swamp is located in the interfluve of the Ob and the Irtysh rivers, mainly in the southern taiga subzone and sub taiga and situated in the Tomsk, Omsk and Novosibirsk oblasts [18, 19] Areas of headwaters of the rivers Shegar, Chaya, Parabel, Vasyugan, Om, Tara are located within Vasyugan swamp Some researchers also include parts of the catchment of the rivers in Vasyugan swamp, such as the Bolshoi Yugan and the Demyanka (of Khanty-Mansi Autonomous Area and Tyumen oblast) The rivers Om, Tara and Demyanko – the tributaries of the Irtysh River; the Shegarka, Chaya, Parabel, Vasyugan and Bolshoi Yugan rivers – the tributaries of the Ob River The Kas River is a left tributary of the Yenisei River Its catchment is located on the eastern border of the West Siberian Plain, in subzone of the southern taiga (the Krasnoyarsk Territory) The swamps on the watershed of the rivers Salym and Bolshoi Yugan, near the village Sytomino and Lake Samotlor are located in the middle taiga subzone in Khanty-Mansi autonomous area The swamp in the catchment the Shchuchya River (the tributary of the Ob) is located in the southern part of the tundra zone (within the Yamal-Nenets Autonomous district) and the swamp near village Igarka is located in the forest-tundra zone in the watershed the Yenisei River (in the Krasnoyarsk Territory)

3. Results and discussion
Suppose that the condition of the swamp ecosystem can be described by the equation according to the first law of thermodynamics:

$$d\Theta = dU + A, \tag{1}$$

where $d\Theta$ is the thermal energy added to the system; dU is the change of inner energy of the system; A is the work performed by the system

$$m \cdot \phi \cdot k_{sa} \cdot dT_a = k_U \cdot k_{sa} \cdot d(m \cdot T_a) + m \cdot g \cdot \delta \cdot dh_p, \tag{2}$$

$$\delta = \begin{cases} 0, & H - Y - E \le 0 \\ 1, & H - Y - E > 0 \end{cases}, \tag{3}$$

where m and h_p is the mass and the average vertical size of the system; k_{sa} is the ratio of air temperature T_a and the systems T_s ($T_a = k_{sa}\ T_s$); H, Y, E is the layers of atmospheric moistening, runoff and evaporation; ϕ is the specific heat of swamp ecosystem (conditional); g is the acceleration of gravity Assuming that $m = k_{p,1} \cdot h_p^{\ b}$ and the average air temperature during the period equals $T = k_{p,2} \cdot (T_0 + T_t)$, then we obtain the following equation at time t:

$$\delta \cdot \int dh_p = \int k_{p,3}\, dT_a - k_{p,4} \cdot T_a \int \frac{dh_p}{h_p}, \tag{4}$$

$$h_{p,t} = h_{p,0} + k_{p,3} \cdot (T_{a,t} - T_{a,0}) - k_{p,4} \cdot b \cdot k_{p,2} \cdot (T_{a,t} + T_{a,0}) \cdot \ln\left(\frac{h_{p,t}}{h_{p,0}}\right), \tag{5}$$

where $k_{p,1}$, $k_{p,2}$, $k_{p,3}$, $k_{p,4}$ is the empirical coefficients; δ is the average value of the delta-function of water balance in a specified period; $T_{a,t}$ and $T_{a,0}$ are the average annual air temperature at the end and beginning of the period; $h_{p,t}$ and $h_{p,0}$ is the depth of peat deposit at the end and beginning of the period

The equation (5) takes the form (figure 1) for the West Siberia Plain according to [9-14]:

$$\begin{aligned} h_{p,t} - 0.10 &= (-12.36 \pm 3.44) - (1.81 \pm 0.51) \cdot T_{a,t} - \\ &- (0.73 \pm 0.16) \cdot T_{a,t} \cdot \ln\left(\frac{h_{p,t}}{0.10}\right) + (5.08 \pm 0.91) \cdot \ln\left(\frac{h_{p,t}}{0.10}\right) \end{aligned} \tag{6}$$

the squared correlation relation R^2=0,74

IOP Conf. Series: Earth and Environmental Science **43** (2016) 012034 doi:10.1088/1755-1315/43/1/012034

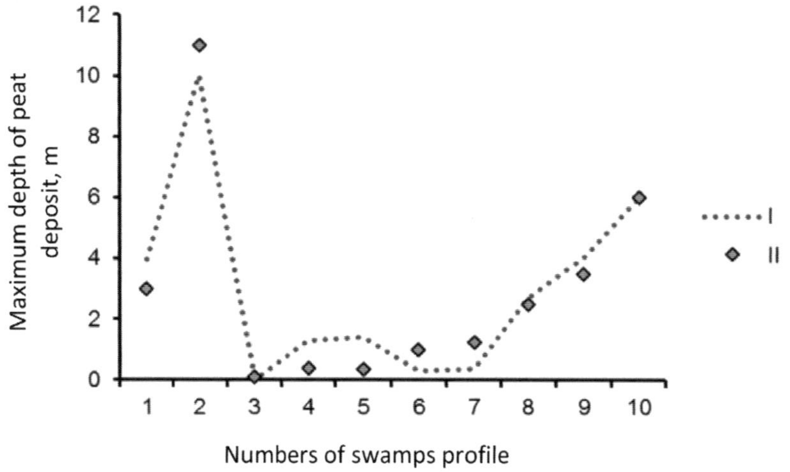

Figure 1. Calculated (I) and measured (II) values of the maximum depth of peat deposits; swamps profile: 1 – the catchment of the Kas River [14]; 2 – Vasyuganskoye swamp, the interfluves of the rivers Parabel, Tara, Om [12]; 3, 6 – the catchment of the Shuchya River; 4, 5, 7 – area of the township Igarka; 8 – Lake Samotlor's area; 9 – area of the village Sytomino; 10 – the Salym-Yuganskaya interfluve [11]

According to the analysis of the dependence (6) it follows that, firstly, in the taiga zone in Western Siberia the calculated maximum depth of peat deposit is equal to 7 7 m, which is close to the average maximum depth of peat bogs in the region, that corresponds to a period of constant temperature –3°C Secondly, when the average annual air temperature is over –7°C, the marked growth of peat deposits (over 0 1 m) is observed The peat deposit is not formed when the air temperatures are about or lower –7°C, but perhaps it is preserved due to peat formation in the previous warmer period Thirdly, there is a nonlinear dependence between growth of peat deposits and maximum increment air temperature corresponding to not maximum, but optimal one for the given ecosystem (figure 2)

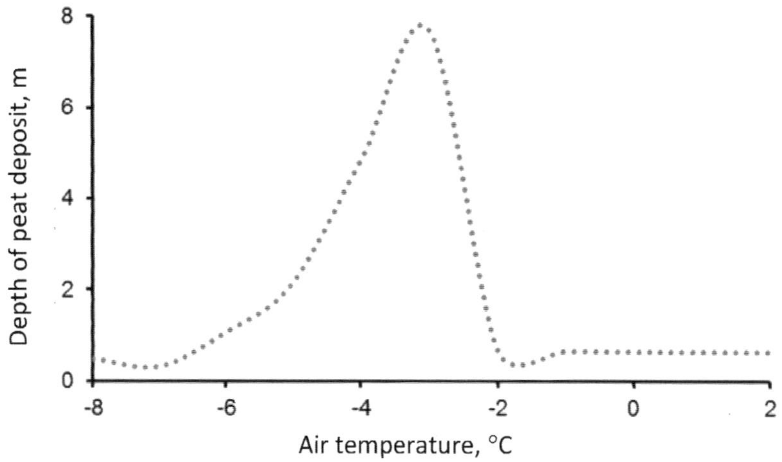

Figure 2. The relation between the temperature of surface layers of the atmospheric air and the depth of peat deposits, calculated according to the equation (6)

Fourth, if we assume minor change in internal energy of the system with changes in its mass, it is possible to simplify the equation (5) to linear relation, which has the following form for the taiga zone of Western Siberia:

$$h_{p-\max,t} = (6.97 \pm 1.42) + (0.87 \pm 0.26) \cdot T_{a,y} \tag{7}$$

where $h_{p\text{-}max,t}$ is the maximum depth of peat deposits; R^2=0,58

However, the use of equation (7) is limited by the range of temperatures from –8°C to –2 °C, after which the non-linearity of the relation is very difficult to be ignored (figure 3)

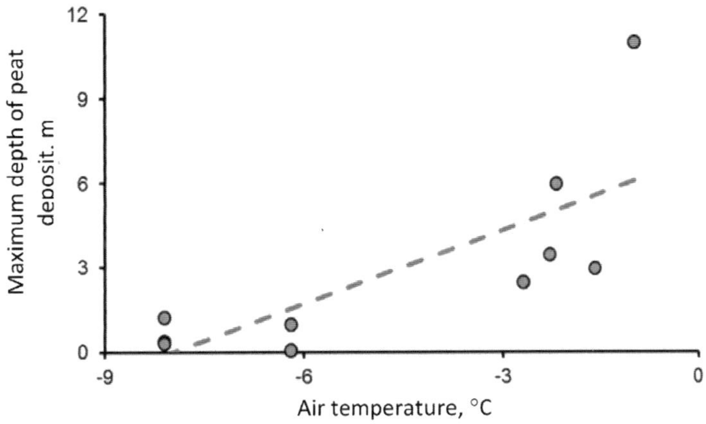

Figure 3. The relation between the temperature of surface layers of the atmospheric air and a maximum depth of peat deposits according to [11, 12, 14]

4. Conclusions

As a result of the study it was shown that peat deposits growth is nonlinear related to climate change Therefore, for the appropriate methods to be applied in paleoclimatic reconstructions it is expedient to establish the optimal values of functioning particular bio- and ecosystem in natural conditions that determine the most probable existence of life forms of biological species

The research was performed with the financial support of the Government of the Russian Federation (project No 14 Z50 31 0012), and Tomsk Polytechnic University (project ВИУ_VAF_144_2014)

References

[1] Chernavskaya M M , Zolotokrylin A N , Krenke A N , Lyahov M E , Brazdil R 1995 *Moscow. Nauka.* Izmenchivost klimata Evropy v istoricheskom proshlom. pp. 218.

[2] Vagner G A 2006 Nauchnye metody datirovaniya v geologii, arheologii i istorii, Moscow. pp. 575.

[3] Velichko A A, Arhipov S A, Ahmetev M A, Volkova V S, Zhidovinov N Ya, Korotkij A M, Kulkova I A, Nechaev V P and Fradkina A F 1998 Istoriya landshaftov i klimata Severnoj Evrazii v kajnozoe *Globalnye izmeneniya prirodnoj sredy* ed. by N N Dobrecov and V I Kovalenko Novosibirsk, *Izd-vo SO RAN, NIC OIGGM.* pp. 261-268.

[4] Bostonalieva Zh K 2015 *PhD dissertation, Berlin, Freie Universität Berlin.* Late Quaternary Climate Change in Western Eurasia A spatio-temporal review of climate proxies. pp. 142

[5] Rui-bo Zhang, Yu-jiang Yuan, Wen-shou Wei, Xiao-hua Gou, Shu-long Yu, Hua-ming Shang, Feng Chen, Tong-wen Zhang and Li Qin 2015 *Dendrochronologia.* Dendroclimatic reconstruction of autumn–winter mean minimum temperature in the eastern Tibetan Plateau since 1600 AD. Vol. **33** pp. 1–7.

[6] Premke-Kraus M 2008 Holocene environmental history of the Kara Sea (Siberian Arctic, Russia) inferred from marine palynological records, Bremerhaven, Alfred Wegener Institute for Polar and Marine Research (AWI) Bremerhaven. pp. 146.
[7] Gajewski K 2015 *Global and Planetary Change.* Holocene paleoclimate data from the Arctic: testing models of global climate change. Vol. **128** pp. 14–23.
[8] Antipina T G, Panova N K and Korona O M 2014 *Russian Journal of Ecology.* The Holocene Dynamics of Vegetation and Environmental Conditions on the Eastern Slope of the Northern Urals. Vol. **45 (5)** pp. 351–358.
[9] Nejshtadt M I 1972 *Akademiya nauk SSSR, Institut geografii AN SSSR.* Bolota Ob'-Irtyshskogo mezhdurech'ya. Prirodnye usloviya osvoeniya mezhdurech'ya Ob' – Irtysh, ed. by Nejshtadt, G D Rihter, Moscow. pp. 322–346.
[10] Lvov Yu A 1991 Bolotnye resursy, Prirodnye resursy Tomskoj oblasti. red I M Gadzhiev, A A Zemcov – Novosibirsk: Nauka, Sibirskoe otdelenie pp. 67–83
[11] Liss O L, Abramova L I and Avetov N A 2001 *Grif i Ko.* Bolotnye sistemy Zapadnoy Sibiri i ikh prirodookhrannoe znachenie Tula pp. 584.
[12] Pologova N N and Lapshina E D 2002 *Sovremennoe sostoyanie i protsessy razvitiya. Tomsk. Institute of Atmospheric Optics SB RAS* . Nakoplenie ugleroda v torfyanykh zalezhakh Bolshogo Vasuganskogo bolota, Bolshoe Vasuganskoe boloto. pp. 174–179
[13] Bezrukova E V, Vershinin K E, Orlova L A, Letunova P P, Krapivina S M, Chepinoga V V, Verhozina A V, Dudareva N V and Abzaeva A A 2003 *Geologiya i geofizika.* Rastitel'nost' i klimat vysokogornoj zony Vostochnogo Sayana v pozdnem golocene. Vol. 44 **(4)** pp. 380–384.
[14] Yamskih G Yu, Karpenko L V and Grenaderova A V 2004 *Geografiya i prirodnye resursy.* Rekonstrukciya sukcessij rastitel'nosti i paleogidrologicheskogo rezhima bolot (na primere torfyanika v doline reki Kas). Vol. **4** pp. 128–134.
[15] Stroitel'naya klimatologiya SNiP 23-01-99* 2003 Moscow, *Gosstroj GUP CPP* 162
[16] Nauchno-prikladnoj spravochnik po klimatu SSSR Seria 3 1993 Mnogoletnie dannye Chasti 1–6 Vypusk 20 Tomskaya, Novosibirskaya, Kemerovskaya oblasti, Altajskij kraj, SPb, *Gidrometeoizdat*, 718
[17] Nauchno-prikladnoj spravochnik po klimatu SSSR Seria 3 1998 Mnogoletnie dannye Chasti 1–6 Vyp 17 Tyumenskaya and Omskaya oblasti, SPb: *Gidrometeoizdat*, 702
[18] Bazanov V A, Berezin A E, Savichev O G and Skugarev A A 2009 *International Journal of Environmental Studies.* The phytoindication method for mapping peatlands in the taiga zone of the West-Siberian Plain. Vol. **66 (4)** pp. 473-484
[19] Eckstein Yoram, Savichev Oleg G and Pasechnik Elena Yu 2015 *Environmental Earth Sciences.* Two decades of trends in ground water chemical composition in The Great Vasyugan Mire, Western Siberia, Russia. Vol. **1** pp. 3-15. DOI 10 1007/s12665-014-3908-z

Monitoring-based analysis of agriculture in Iraq

O Tokareva[1], O Pasko[2], A Alshaibi[3] and M Mochalov[4]

[1] Associate professor of computer engineering department, candidate of technical science, Tomsk Polytechnic University, 30 Lenina ave., 634050,Tomsk, Russia,
[2] Professor of Tomsk Polytechnic University, doctor of agricultural science, 30 Lenina ave., 634050, Tomsk, Russia,
[3] Master student of computing engineering department, Tomsk Polytechnic University, 30 Lenina ave., 634050,Tomsk, Russia,
[4] Manager of innovation programs and projects division, research department, Tomsk State University, 36 Lenina ave., 634050, Tomsk, Russia,

e-mail: ost@tpu.ru, oap@tpu.ru, Ahmedalshaiby88@gmail.com, mvm063@yandex.ru

Abstract. The paper deals with change in area and structure of Iraq agricultural lands. It revealed the main reasons for the change: crisis (war, sanctions, etc.); economic (swamp and lake drainage, melioration, etc.); weather condition. Land-use intensification as a reason for reduction of agricultural land areas was not proved. The area of cultivated lands proved to correlate significantly with the level of precipitation, wheat productivity –with the average temperature in Iraq.

1. Introduction

The agriculture lands are one of the main factors of production in all countries [1]. They have special legal status and are subjected to special protection aimed at maintaining the area, improving soil fertility [2] and supporting adequate cadastral value [3]. However, since 1985, there has been reduction of agricultural land areas in many countries regardless their natural, demographic and economic conditions [4].

The experts identify two reasons: crisis, which is conditioned by political and social-economic reforms, as well as wars, and intensification, which is connected with the development of technologies aimed at increasing crop fertility.

The aim of this work is to study the causes of change in the agriculture land areas in the Republic of Iraq. Thus, to achieve the goal the following tasks were completed:

- The change in agricultural land areas over the period of 35 years was studied;
- The main reasons for changes were revealed;
- The correlation of area, fertility, gross product of the key agricultural crops with the meteorological conditions was defined.

2. Objects and methods of research

Agricultural lands of Iraq were chosen as an object of study, since this country is agrarian, and characterized by different climatic characteristics, social-political conditions, and features of agriculture (ameliorative, bogar) [5]. The following characteristics were used as the criteria of their condition: crop yield, gross product, total area and areas for different crops (rice, watermelon, tomatoes, cucumber, grapes, dates, chickpeas, barley, wheat, corn, clover, etc.).

Iraq is located between 29°15' N. lat. and 38°15' N. lat., 38°45' and 48°45', has a total area of 438,320 km^2, is bordered by Iran in the East, Turkey in the North, Syria and Jordan in the West, Saudi Arabia and Kuwait in the South and the Persian Gulf in the South-East. The main rivers are the Tigris and the Euphrates, which are widely used for irrigation of crops (figure 1). They flow through the great Mesopotamian plain, surrounded by mountains rising 3,550 m above the sea level on the North and East, and desert areas in the South and West occupying more than 40% of the territory [6].

a б

Figure 1. Agricultural lands in the valley of the Tigris river on the north of Iraq (a); plantation of the date palms (photo of A Alshaibi) (b)

The climate of Iraq is mainly continental, subtropical in the North and mediterranean in the North-East. The average temperature in winter is + 16°C during daytime and 2°C at night; in summer, the daytime temperature is 43°C and 26°C at the night. Rainfalls are very unevenly distributed over the country, as well as over the years (figure 2) and seasons: in most regions – mainly in winter months, in the mountains - from November to April; from < 100 mm in the South and South-West to 1000 mm per year in the North and North-East. Significant differences in the amount and distribution of rainfalls increase the risk in rain-fed agriculture [6]. Values of precipitations and annual temperatures in the period from 1980 to 2011 are shown in the figure 1.

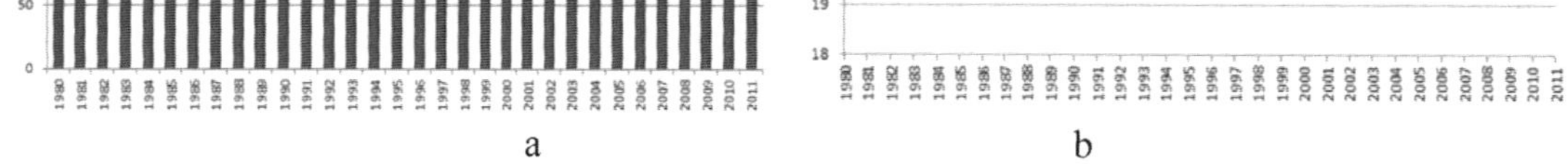

a b

Figure 2. Values of precipitations (a, mm) and annual temperatures (b, °C)

Agricultural lands occupy about 20% of total land surface in Iraq. About half of them are located in the North-Eastern plains and mountain valleys with sufficient rainfalls. The other part of the lands occupies the valleys of the Tigris and Euphrates rivers. Due to scarce rainfalls, the rivers supply water to irrigate fields with crops wheat and barley in winter and rice, corn, dates, cotton, vegetables and fruits in summer (figure 2). Alfalfa and other legumes are grown for cattle [7]. The list of crops with low harvest area consists of cotton (seeds), okra, clover, fresh vegetables, legumes, apples, potatoes, beans, onions, grapes, peas, pumpkin, and sesame seeds.

Cultivation of drought-resistant technical fodder and grain crops on bogar (using moisture of spring rains in the areas inconvenient in terms of irrigation) is widespread In Iraq. Changes in water availability, installation of drainage systems and reclamation of saline lands, the introduction of new varieties of crops, scientific advances in agronomic practices and other agricultural technologies also influence the efficiency of agricultural production.

To solve the tasks, the abstract-logical, monographic and correlation methods were used.

3. Results and discussion

Analysis of statistical data showed that the area of cultivated land in Iraq experiences significant changes [7, 8]; for example, the cultivated area reduced fourfold during the period 1993-2000 (figure 3a). The ratio of cultivated areas for summer crops also changed (figure 3b). The period 1990-2010 is characterized by the following changes: the reduction in the areas of land occupied by rice crops and the growth of the areas occupied by corn and figs (the main export crop).

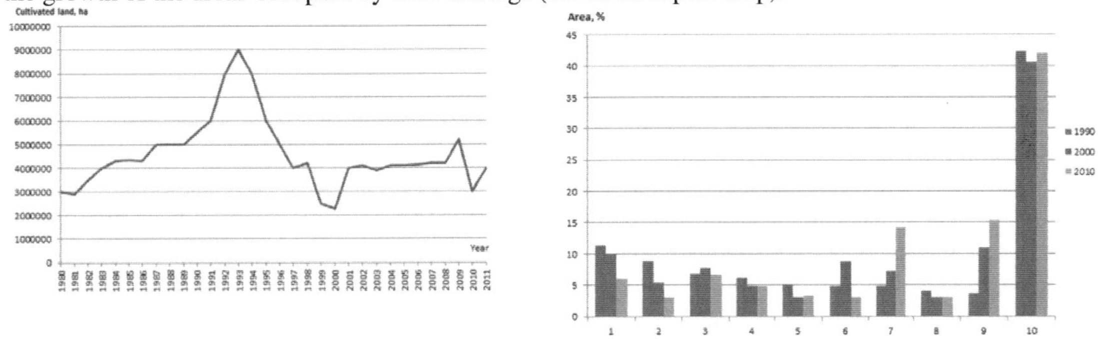

a b

Figure 3. Change of areas of cultivated lands (a) and percent ratio of summer crops areas in 1990, 2000 and 2010 (b). Descriptions: 1 – rice, 2 – watermelon, 3 tomatoes, 4 cucumbers, 5 – grape, 6 – nut, 7 – corn, 8 – clover for silage, 9 – dates, 10 – other cultures

It is notable that a steady rise of cropland area began in the early 80s, then, there was a sharp increase in cultivated areas in the early 90s followed by a sharp fall after 2003. In our opinion the reason for these changes is the implementation of large-scale governmental ameliorative projects aimed at draining wetlands and lakes and using new areas for agriculture. These projects were supported by income from petroleum export (the early 80s). In response to Iraq's invasion of Kuwait in 1990 the United States and other countries launched a military operation, which resulted in destruction of dams, power stations, irrigational facilities and pump stations. The economic sanctions against Iraq made the situation worse, destroying the life support system of the country. To prevent hunger, the population used actively all available land resources for growing crops. After the Iraq war (2003) the labor resources migrated to the cities, agricultural production decreased almost twice, and the cultivated areas decreased [9].

Moreover, winters were dry from 2004 to 2010 [9], which along with the destroyed ameliorative systems had negative impact on the effectiveness of agricultural production. We studied the correlation of agricultural land areas, general product and productivity of rice barley and wheat with temperature and rainfalls. The authors revealed that the areas of cultivated lands have significant correlation with the level of precipitation (significance level is 0.05, correlation coefficient is 0.45), which may be explained by the use of naturally wet areas for rice growing; the increase in rice production is also registered in the years rich in rainfalls. The significant correlation of wheat productivity with the average temperature in Iraq was identified (significance level is 0.05, correlation coefficient is 0.42). The same correlation for rice and barley was not identified. Figure 4 shows a chart of wheat productivity change over 30 years [7].

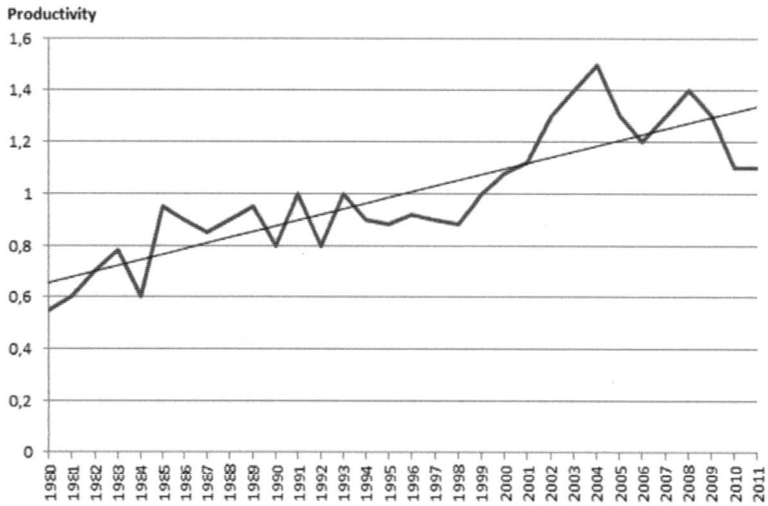

Figure 4. Productivity of wheat (tons/year)

4. Conclusion

The research resulted in the following conclusion:

1. Significant changes in areas and structure of agricultural lands in Iraq are conditioned by crisis (wars, sanctions etc.); economics – implementation of large-scale governmental projects (drainage of the wetlands and lakes, amelioration etc.); and weather conditions. Land-use intensification as a reason for reduction of agricultural land areas was not proved.

2. The area of cultivated lands has great correlation with level of precipitation (significance level 0.05, correlation coefficient 0.45) which may be explained by the use of additional naturally wet areas for rice growing. The wheat productivity, a drought-resistant plant, is correlated with the average temperature in Iraq (significance level 0.05, correlation coefficient 0.42).

References

[1] Baumane V and Pasko O 2014 Comparison of land reform of Latvia and Russia in conditions of transition period. *Baltic Surveying* **1** 40-44 Date Views 1.05.2015: *http://www.lif.llu.lv/print_pdf.php?article=269*

[2] Pasko O 2013 Vergleichende Analyse des effizienten Flaechenverbrauchs in der Landwirtschaft in Deutschland und Russland *Economic development and perspectives of cooperation between the USA, Europe, Russia and CIS* states 1 (New York: CIBUNET Publishing) 4560

[3] Pomelov A, Pasko O and Baranova A. 2015 Comparative analysis of land management in the world //*IOP Conf. Series: Earth and Environmental Science* **27**

[4] Kovyazin V, Pasko O, Romanchikov A., Belyaev V. 2014 Taxation Indices of Forest Stand as the Basis for Cadastral Valuation of Forestlands. *IOP Conf. Series: IOP Conference Series: Earth and Enviromental Science* **21**, 012026 11(5).

[5] Chapin Metz H 1988 Iraq: A Country Study. *Washington: GPO for the Library of Congress* Date Views 25.02.2016: *http://countrystudies.us/iraq/58.htm*

[6] *Agriculture in Iraq: Resources, Potentials, Constraints, and Research Needs and Priorities* Date Views 25.02.2016: *http://www.ars.usda.gov/SP2UserFiles/Place/50600000/Products-Reprints/2002/1107.pdf*

[7] Soppe R and Dr Raad O Saleh (Editors). Iraq Salinity Project. *Technical Report 8* Date Views 25.02.2016:

https://www.google.ru/url?sa=t&rct=j&q=&esrc=s&source=web&cd=1&ved=0ahUKEwjI6sDz5dbMAhVBjCwKHQS4D8kQFggeMAA&url=https%3A%2F%2Fapps.icarda.org%2FwsInternet%2FwsInternet.asmx%2FDownloadFileToLocal%3FfilePath%3DIraq_Projects%2FIraq_Salinity%2F08_CompB_Review_Agricultural_Production.pdf%26fileName%3D08_CompB_Review_Agricultural_Production.pdf&usg=AFQjCNFR_jfHWA1Lww-3eTwq8dnUqI94qw&sig2=2sqUn4FQ-oKxR_UItQ9xkA&bvm=bv.122129774,d.bGg

[8] *Land availability* Date Views 1.05.2015: *www.fao.org*

[9] *Iraqi Agriculture in Crisis* Date Views 25.02.2016: *http://www.al-monitor.com/pulse/originals/2013/01/iraq-farmers-abandon.html*

The issues of weed infestation with environmentally hazardous plants and methods of their control

V L Bogdanov [1], T S Posternak, O A Pasko [2], V F Kovyazin[3]
[1]St. Petersburg State University, St. Petersburg, Russia
[2] National Research Tomsk Polytechnic University, Tomsk, Russia
[3]National University of the mineral resource "Mountain", St. .Petersburg, Russia

Abstract

The authors analyze expansion of segetal and ruderal vegetation on agricultural lands in Leningrad and Tomsk oblasts, typical for the European and Asian parts of Russia. The spreading conditions, composition of species, biological features and ecological requirements of the most aggressive species are identified. Some effective ways of weed control are suggested.

1. Introduction

Agricultural land areas in Russia and other countries of the former Soviet Union have reduced for recent years [1, 2], as well as the agricultural land use efficiency [3]. Soil fertility decline, water logging and weed infestation of the fields are conditioned by the shortage of funds, necessary for modern farming techniques, [4]. The relevance of research is determined by the urgent need for agricultural producers to develop strategic directions for weed control, based on long-term prognosis. The weeds include "wild plants growing on farmlands and reducing the size and quality of the products" [5]. They are classified into segetal (field weed, plow land weed), plants of natural lands, ruderal (weeds, coarse grasses, and wasteland plants) and plants of special areas [6, 7]. Weeds are the companions of cultural and compete with them for minerals and moisture, shading them, slowing growth and development. Species composition of weeds agrocoenosis includes: a) a complex of species naturally and steadily growing in the fields, b) species originating from ruderal habitats due to lower levels of agricultural technology, c) adventitious species classified as harmful in other regions. Weed dispersal speed depends on the biological characteristics: plasticity, competitiveness, productivity, etc.

The objectives of the study were to analyze the species composition, identify the most aggressive weeds in European and Asian parts of Russia using Leningrad and Tomsk oblasts as an example, and test effective ways of weed control.

2. Objects and methods of research

The objects of the study were weeds common to the European and Asian parts of Russia. Long-term data were used for analysis, with minimum period of five years. Statistical, analytical and cartographic methods were used in the research.

3. Results and discussions

298 weed species from 38 families are currently growing in Leningrad oblast; 25 representatives of Asteraceae, Leguminosae, Buckwheat, Brassicaceae, Poaceae families dominate [8]. More than 50 years of observations reveals the stability of the floristic structure of the spectrum. The fields of Tomsk oblast are the habitat for 22 species of weeds. Many of them are cosmopolitans and common to the crops all over Russia, including Leningrad oblast (quinoa, field sow thistle, couch grass and mayweed, etc.) (figure 1). The most aggressive are field sow thistle and pink, couch grass, quinoa and duck wheat (representatives of Asteraceae,

Chenopodiaceae and Polygonaceae families) [9]. Their leadership is conditioned by high seed productivity, long-term organic rest of seeds and several life forms. The roots of the field sow thistle, sow-thistle pink and quinoa are able to penetrate the soil to the depth of four to five meters and reach moistened layers. Thlaspi arvense, Fumaria officinalis, Pinweed and Hemp nettle are probably the least typical for the region (figure 1). High moisture in Tomsk region (hydrothermal coefficient is 1.4-1.2 [10]) facilitates aggressive spread of segetal and ruderal vegetation. Studies of spring and winter crops fields, which were carried out in 2010-2015, revealed their 100% infestation [11] (figure 2).

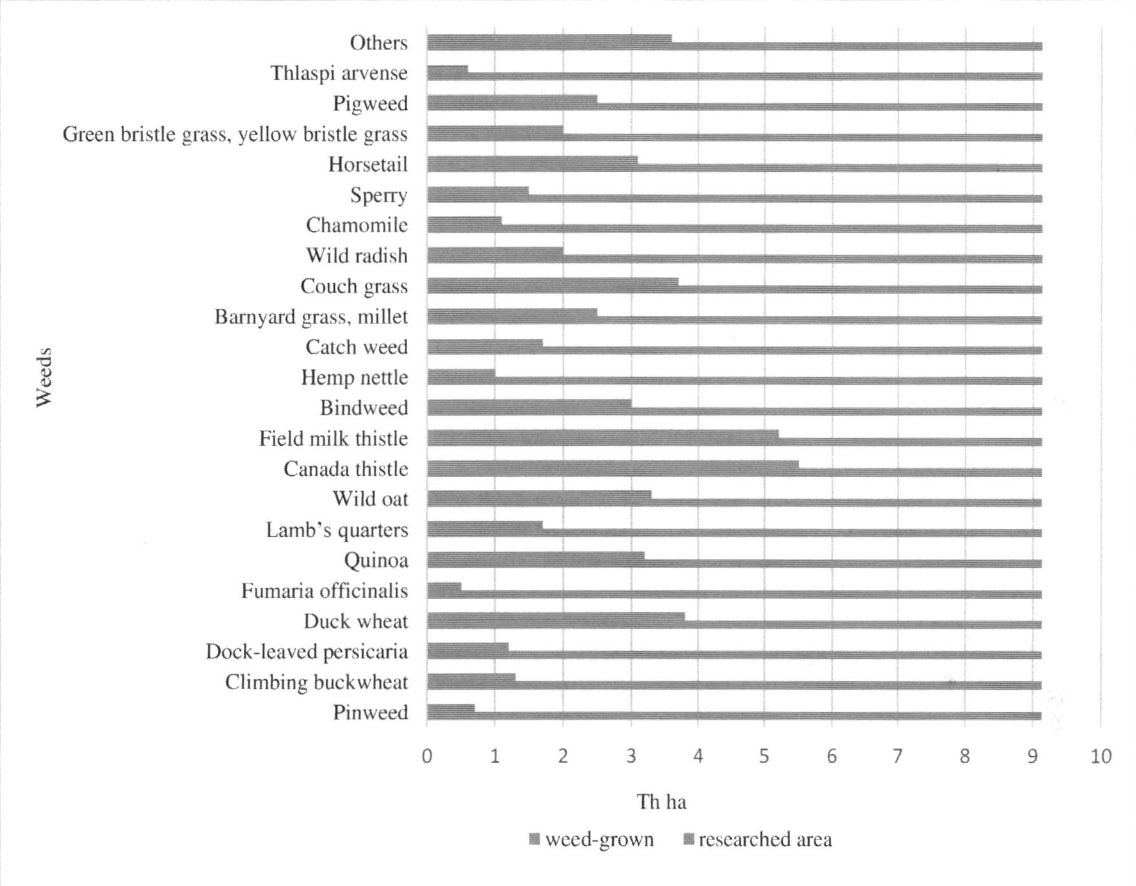

Figure 1. Correlation between weeds in spring wheat culture of Tomsk oblast in 2015 [data 8]

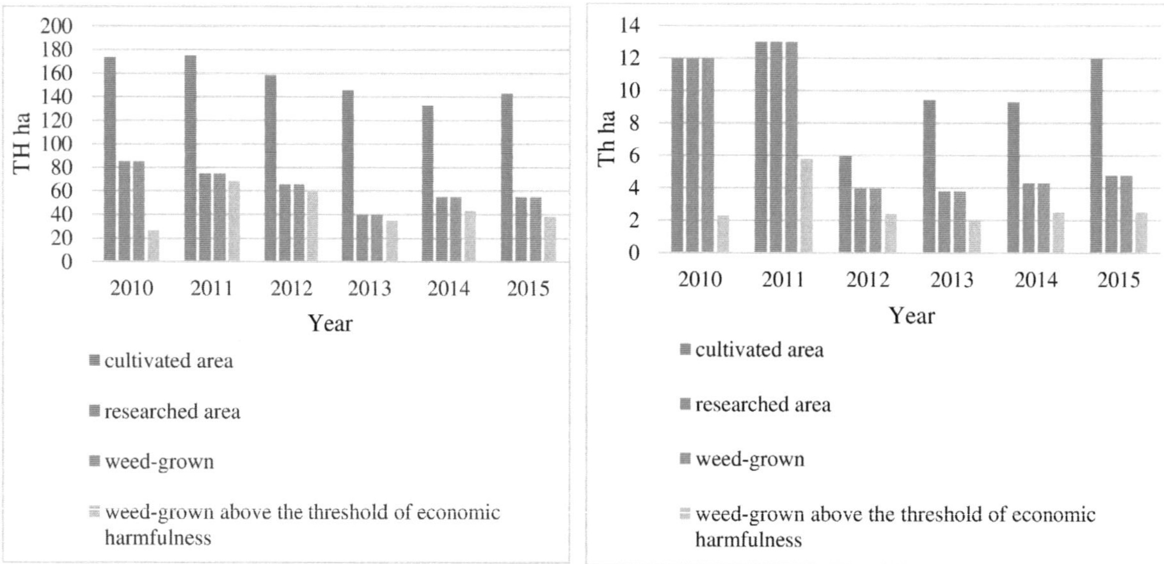

Figure 2. Dynamics of weediness in Tomsk oblast for 2010-2015
A - Spring crops, B – winter crops [data 10]

There are some ways of weed infestation: abandoned and infested areas, sowing of untreated seeds, and fresh manure application. High level of weed infestation in Tomsk region results from the lack of agro-technical works, lack of mineral nutrition for crops and favorable agro-climatic conditions for regional mass distribution of cosmopolitan weeds.

A giant Sosnovskyi cow parsnip plant (Heracleum Sosnovskyi Manden) (hereinafter cow parsnip) is the major environmental threat for the European part of Russia. It was brought from the Caucasus to the northern regions of Russia in the mid 40-ies of the last century. A type of cow parsnip resistant to harsh climatic conditions was fancied as a forage crop. It allowed obtaining about 500-600 kg / ha of green mass over a growing season [12, 13]. With the dramatic decline in agriculture, it stopped being mowed. The plant began to spread on derelict lands, garden plots, slopes of drainage canals, roadsides, forming arrays ranging from a few square meters to several hectares. The plant is 1.5-3.0 m high, the thickness of the base is 4-6 cm. Root system is strong, well developed, penetrates the soil to 70 cm and deeper. Having high reproductive capacity, one plant produces 5 - 20 thousand seeds per year, which are able to germinate for 5-6 years and are spread by wind, surface water, birds, and vehicles. The average life of the plant is 8 years, with the reproductive stage starting in the 2nd year [14].

There are no natural enemies of cow parsnip due to the presence of furocoumarins, so it is little damaged by herbivores. The plant is hazardous for human health, since skin-to-furocoumarins contact under the influence of ultraviolet sunlight causes burn-like dermatitis (the1st, 2nd and 3rd degree - depending on contact time). There are cases of toxicological poisoning, accompanied by disorder of the nervous system and the heart muscle work [15].

Strong competitiveness in the ecosystem and sustainability to the adverse factors of environment allow Sosnovskyi cow parsnip to be more aggressive and impervious compared to related species. [16] It captures the sunniest and most fertile areas, displaces natural vegetation, and forms a new monospecific successional system. According to our estimates, cow parsnip is able to capture about 5-7% of new territories annually, which will cause significant economic losses due to infestation and non-use of lands.

The area, covered with cow parsnip, and the rate of its spread are so great, that it is almost impossible to make an up-to-date map of cow parsnip infestation. We made an attempt to analyze and visualize the open access information in the issue for the European part of Russia. The administrative region was taken as an information unit. The areas "damaged" by uncontrolled spread of the species are marked on the map (figure 3).

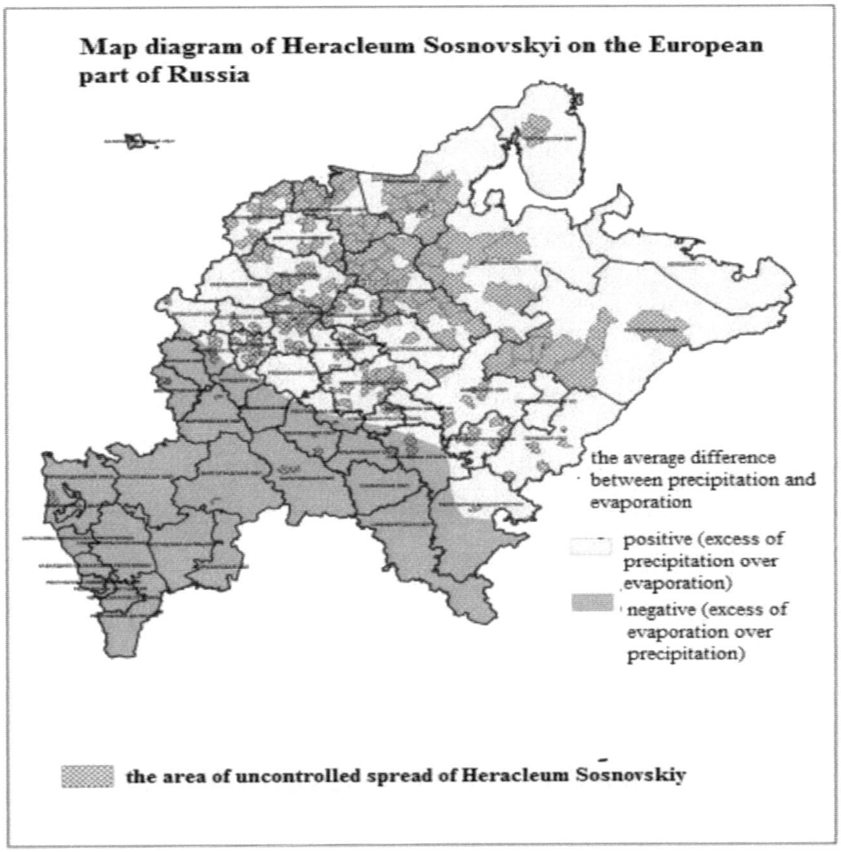

Figure 3. Map diagram of Heracleum Sosnovskyi spread on the European part Russia

The secondary area of cow parsnip distribution in the European part of Russia is almost identical to the zones of excessive (Northwestern Federal District) and adequate (the major part of Central Federal District) moistening [17]. It can be obvious from the map (figure 3), where the areas with positive and negative values of the difference between the average annual precipitation and evaporation, which characterizes moisture content of the area, are shown in different colors. The only factor impeding further spread of cow parsnip southward is insufficient humidity of climate, which previously prevented its cultivation. However, the cow parsnip has recently appeared in the areas associated with unstable and insufficient moisture (Belgorod, Voronezh, Penza, Saratov, Ulyanovsk regions). The obtained data are incomplete; nevertheless, it allows evaluating the scale of the disaster and can help to forecast the areas of further spread of the species. The most infested areas are Leningrad, Moscow, Ivanovo, Kirov, Novgorod, Pskov, Vologda, Arkhangelsk, Tula, Yaroslavl regions. The southern part of the Komi Republic is threatened by cow parsnip infestation: the plant communities intensively spread the agricultural lands and settlements of the region.

In 2013, more than 83 thousand hectares of land in Leningrad oblast were reported to be cow parsnip infested. The analysis of current situation and the forecast showed that unless the problem is taken under control, 25-30% of the land in natural landscapes and up to 15-20% of agricultural lands may be infested in 10 years. In this regard, the studies related to developing effective low-cost measures to control this dangerous type of weed and restoration of natural ecosystems become extremely important. Environmental conditions in Leningrad oblast are favorable for cow parsnip [17], which allows it to spread on the whole territory.

A field experiment was used to study the effect of agricultural activities on the growth of cow parsnip. We carried out plowing, disking, harrowing of soil infested with cow parsnip, followed by planting of vegetable crops, as well as burning of old vegetation. Before agronomic tillage, the abundance of cow parsnip in phytocenosis was 5 - 12 plants per 10 m2. Next autumn plowing was followed by disking. In spring only individual plants of cow parsnip were observed on the plowed field. At the end of May the farming techniques were repeated and dill was planted. In September, cow parsnip was reported to occur very rare and grow in single specimens. Thus, we assume that the agricultural techniques ensure effective elimination of cow parsnip. It should be noted that the agricultural methods are to be carried out regularly for several years, since large supply of seeds in the soil allows the plant to restore the population quickly. The burning of old vegetation in ecosystem of old-growth reclaimed meadows did not have negative impact on the spread of cow parsnip.

In order to study the effect of herbicides on the reproductive capacity and destruction of cow parsnip was field experiment was conducted in Vyborg district, Leningrad oblast. The mixture of herbicides (glyphosate, banvel) at doses of 1 - 8 l / m2 of the soil surface was tested. The plants were treated in the stem extension stage, on June 9. The most effective dose for the destruction of cow parsnip was reported to be a mixture of glyphosate and banvel BP 8 l / m2 in the ratio 3: 1 respectively. Three weeks after the treatment the stems and leaves of the cow parsnip wilted, and the roots of the 1st, 2nd and 3rd order blackened. However, the seeds, that were present in the soil, were not damaged and after a while they gave shoots. The ecosystem started to restore gradually one month after the chemical treatment. Barbarea vulgaris, galeopsis common, pigweed white, and foal foot were the first herbs that occurred in the area.

One more field experiment was conducted to study the impact of mulching with black film on the growth and development of the cow parsnip. The field experiment took place in Vyborg district, Leningrad oblast. In early May the meadow land infested with cow parsnip (on the average 2 plants per 1 m2; rosette phase) was covered with black film, which was removed after 3.5 months. 100% of cow parsnip plants died, moreover, the seeds taken from the soil surface after mulching, did not germinate.

Thus, we assume that the method of cow parsnip mulching on the early stages of its development is a highly efficient way to control it. Further soil cultivation and sowing of perennial grasses allowed restoring the aboriginal ecosystem.

References

[1] Pasko O 2013 *Agricultural science*. The Use of agricultural lands in Tomsk region. Vol. **6** pp. 9-12

[2] Baumane V, Pasko O 2014 *Baltic Surveying*. Comparison of land reform of Latvia and Russia in conditions of transition period. Vol. **1** pp. 40-44

[3] Pasko O 2013 *Economic development and perspectives of cooperation between the USA, Europe, Russia and CIS* states 1. Vergleichende Analyse des effizienten Flaechenverbrauchs in der Landwirtschaft in Deutschland und Russland (New York: CIBUNET Publishing) pp. 45-60

[4] Pomelov A, Pasko O, Baranova A 2015 *IOP Conference Series: Earth and Environmental Science*. Comparative analysis of land management in the world. Vol. **27**

[5] GOST 16265-89 "Agriculture. Terms and definitions", weeds *GOST 16265-89* 2

[6] Berezina N, Afanaseva N B 2009 *Plant ecology* (Moscow: Academy) pp. 400

[7] Kurdyukova O, Hemp N 2014 *Plant protection news*. Seed production of various types of weed plants. Vol. **1** pp. 32-36

[8] Mysnik E 2014 *The dissertation on comp. of a scientific degree of cand. of biol. Sciences* Features of formation of the species composition of weeds in agroecosystems in the North-West region of the Russian Federation S-Petersburg.

[9] 2015 *A summary statement of the basic survey of infestation of crops by 1 December 2015 "Rosselkhoztsentr" in the Tomsk region* (Tomsk)

[10] 1975 *Agroclimatic resources of the Tomsk region* (Gidrometeoizdat) pp. 146

[11] 2014 *Report on the implementation of the list and volume of works stipulated by the state task* Federal state budgetary institution "Agrochemical service Station "Tomsk" (Tomsk)

[12] Bogdanov V L, Shmeleva I V, Nikolaev R V 2012 *Conf "Geographical Sciences in ensuring sustainable development strategies in the context of globalization"* (Minsk: Belarus Publishing center of BSU). Ecological problems of land use in rural areas in the Russian Federation.

[13] Bogdanov V, Nikolaev R, Shmelev I 2011 *Regional ecology.* Invasion of environmentally harmful plants Hogweed Sosnowski (Heracleum Sosnowskyi Manden) on the territory of the European part of Russia. Vol. **1-2 (131)** pp. 43-52

[14] Orlov B, Gelashvili D, Ibrahimov A 1990 *Poisonous animals and plants of the USSR* (Moscow: "Vysshaya SHKOLA")

[15] Moiseev K A, Aleksandrova M I 1968 *Hogweed Sosnowski.* Institute of biology of the Komi branch of the USSR Academy of Sciences Syktyvkar.

[16] Laman N, Prokhorov V, Maslovskaya O *Giant Hogweed is a dangerous invasive species to natural systems and population of Belarus* (Minsk: Institute of experimental botany of them. V. F. Kuprevich of NAS of Belarus) 40

[17] *Agroclimatic resources of the Leningrad region* 1975 (Gidrometeoizdat) pp. 118

Geotechnical properties of gullying in Tomsk Oblast

A V Leonova, K M Lomakina, S A Dmitrieva, A V Baranova

National Research Tomsk Polytechnic University, 2, building 5, Lenin Avenue, Tomsk, tel. 3822-606-385,
[1]8-952-880-3608 avleonova@tpu.ru
[2] 8-952-887-6254 lokris@mail.ru
[3] 8-913-100-1005 dmitrieva_sa93@mail.ru
[4] 8-923-402-10-54 baranskikh@yandex.ru

Abstract. The article deals with causes, conditions and factors of gullying in Tomsk Oblast, the case of site Grodnenskaya. A nuclear-power station was planned to be constructed in Closed Administrative Territorial Unit Seversk, which led to geotechnical investigation of the area to study the geological processes developing there. The investigation included reconnaissance traverse of the area, geological and geomorphic descriptions, and sampling. The area and percentage of the territory affected by the geological processes were estimated. As a result, it was proved that the geological processes of the area are conditioned by surface water activity leading to gullying and river erosion, and by both ground water and surface water activity resulting in bog formation. It appears that gullying is caused not only by natural factors but also by technogenic ones.

1. Introduction

Geological processes developing in any areas are unfavorable factor that impedes construction work, destroys buildings and facilities and can lead to economic losses and environmental damage. To reduce the possible risks, it is necessary to provide a valid prediction of hazardous geological processes development, which is impossible without understanding the causes and factors of the process.

A nuclear-power station was planned to be constructed in Closed Administrative Territorial Unit Seversk, Tomsk Oblast. It caused a geotechnical investigation to be conducted on the site, which included the study of geological processes.

2. Materials and Methods

To study geotechnical conditions of the site, the area was traversed, samples were taken, and the geology and geomorphology of the area were described. The area and percentage of the territory affected by the geological processes were studied and estimated. The dynamics of the processes was studied by calculating the number and length increment of active and newly developed processes. Besides, laboratory testing was conducted to obtain the information on physical properties of the soils and rocks that build up the studied area [1].

3. Results and Discussion

The investigation of the site allows us to make the conclusion that there are two genetic groups of the processes conditioned by the natural factors:

- Processes trigged by geological activity of surface water – gullying and river erosion (Grodnenskaya site);
- Processes trigged by surface water and groundwater activity – bog formation (Orlovskaya site).

These processes are known to depend on climate features (high moisture or excessive moisture is necessary), surface topography (flat areas and sink-holes for water accumulation and bog formation, and slopes for gully formation), specific properties of superficial deposits (loose, unconsolidated rocks), groundwater flows and anthropogenic activity.

Gullying is the most developed geological process on Grodnenskaya site, which occupies almost 40% of the site [2] making the areas unsuitable for construction (fig.1).

There are a lot of branches and ravine tributaries in the gully sides, which form a developed gully network limited to the slopes of interfluve area Mostovka-Chernaya.

Figure 1. The gully on the studied site.

The gully network can be classified as highly eroded [3]. A developed gully network with both dry and wet gullies spreads from the top of the site down toward the river valleys. Most of the gullies are grass-covered and forested; however, there are rare cracks and failures of the sides, which means that the erosion still continues.

Gully formation and expansion is known to start from the top with a flow forming a narrow rill several meters wide. As long as the flow transports more and more soil particles, the crack develops into a ravine several dozens of meters wide. The depth of the gully depends on its topography and a base level of erosion.

The gullies of the studied area are of different size: from 2 to 43 m wide, with 0.5-7.0 m deep. They are at different stages of development: from water galls and developing gullies to grass-covered and forested ravines. The length of the gully network is 1.5 km and more. The side slopes are usually covered with grass and subdued. The longitudinal section has the form of a smooth curve. The cross-section is of U-type. The slopes of gullies and ravines are covered with birch, aspen, abies, and small spruce (up to 40 cm high). There are a lot of windthrown trees. The sides are covered with thick grass: sedge, horsetail, fern, and etc. There occur small hills of sedge 7 cm high and of a diameter up to 25 cm, they are covered with moss. The side slopes are prone to slumping, cracks and paps (fig. 2). As a rule, a temporal flow cuts the gully bottom. The observed fresh cut, 40-105 cm wide and 15-30 cm deep, continues to develop the gully. Sand and sand clay are exposed in the cut (fig. 3).

The basic factors contributing to the gullying on the site are as follows:

- Soil properties: the soil consists of silt loam, sand clay and sand.

- Temporal water flows that make gullying possible even with gentle slope of the area.
- Man-made factors: road construction and operation, turf cutting on the areas with gentle slope, which leads to sink-holes development.

Figure 2. Slumping.

Figure 3. Sand clay in the gully side.

Obviously the processes have been taking place for several decades, which is proved by almost unnoticed ravines in the forest to the right of the road to Krasniy Yar along the abandoned roads. It should be noted that the active gullies occur in flat areas with gentle slope. Gully formation in such areas is conditioned by man-made factors such as plant destruction and deforestation, destruction of rock integrity. Gully formation on the area is an intensified process, and it can be hardly expected to stabilize soon due to natural causes.

Having studied the conditions of gully development, we cannot confidently determine the nature of gully network formation on the studied site. This geological process may be conditioned by modern tectonic movements resulting in change of the local base level. Temporal water flows intensify erosion moving to the base-level. Having done it, they stabilize for a while. This supposition is proved by the forms of erosion – the edges and sides of the gullies are covered with vegetation, while the bottoms are eroded by fresh cuts.

4. Conclusions

To come to the final conclusion on the nature of this geological process, the site needs further observation and study.

However, it can be asserted that the unfavorable geological processes are strongly developed on the area. The processes are conditioned by both natural and man-made factors. The processes being in progress, it is necessary to take protective and preventive measures in case of the area development.

References:

[1] Leonova A V 2009 *Proceedings of the 13th International Scientific Symposium of Students, Postgraduates and Young Scientists in Honor of Academician M.A. Usov.* Factors of gully formation on the site of Seversk nuclear power station, Tomsk oblast Problems of geology and subsurface development. pp. 206-207.

[2] Leonova A V 2009 *International Conference Engineering protection of construction sites in terms of unfavourable geological processes.* Factors and conditions of gullying pp. 56-57.

[3] Ivanova I P, Trzhtsinskiy Yu B 2001 St. *Petersburg: Nauka.* Engineering geodynamics.

[4] Strokova L A 2015 *IOP Conference Series: Earth and Environmental Science.* Modeling of tunneling-induced ground surface movement Vol. **24(1)** pp. 012030.

[5] Pokrovskiy V, Pokrovskiy D, Dutova E, Nikitenkov A. 2014 *International Multidisciplinary Scientific GeoConference Surveying Geology and Mining Ecology Management.* The research

underflooding processes of architecture monuments on the territory of tomsk with using gis-technology, SGEM. Vol. **2** pp. 935-941.
[6] Belousova A P 2011 *Water Resources.* Risk assessment of underflooding of areas by groundwater during floods. Vol. **38(1)** pp. 39-46.
[7] Zhukov M, Tishaieva A 2010 *9th International Conference on Geoinformatics: Theoretical and Applied Aspects.* Compositional model of data distribution in problems of underflooding detection by means of remote sensing Geoinformatics.

Mercury content in soils on the territory of Mezhdurechensk

A N Nicolaenko, N A Osipova, E G Yazikov, I A Matveenko

Tomsk Polytechnic University, 30 Lenin Ave., Tomsk, 634050, Russia

E-mail: saiiiek@mail.ru

Abstract. The geochemical features of mercury content and distribution in the zone of coal producers have been studied (Mezhdurechensk town). Mercury content in soil (30 samples) was determined by atomic absorption method using mercury analyzer PA-915+ with pyrolytic device. Mercury content in soil samples changed from 0.12 to 0.17 mg/kg, the average value being 0.057 mg/kg. Within the town territory five zones with mercury elevated concentrations in soil were distinguished. 25-year observation period showed a 2.8 time decrease in average mercury content in soil. The major contribution to soil pollution in the urban territory was made by the two factors: local and regional. The mercury content in soil is affected by the emissions from boilers operating on coal as well as coal dust from the open pits near the town.

1. Introduction

Mercury and its compounds are one of the most hazardous ecotoxicants or even "superecotoxicants" [1]. The relevance of the given issue is supported by its consideration at the international level [2]. The study in mercury transformations and migration in connection with different technogenic processes is of great significance for understanding toxicants' geochemical cycle in the environment. Along with other media soil is not only deposit environment of mercury and its compounds, but also its source in other ecosystem constituents including water and plants, which defines its further way through trophic chains [3-5]. Under intensive local pollution soil can be a source of secondary mercury emission in the environment [6]. One of the key anthropogenic sources of mercury in the environment is coal producers. At present, coal producers in Kemerovo Oblast are largest enterprises in coal mining of Russia. These enterprises of different capacity surround Mezhdurechensk, which is located 60 kilometers east from Novokuznetsk city and 302 kilometers south-east from the administrative center, Kemerovo city. Mercury is released into the environment in the process of coal mining and combustion as a part of coal dust and combustion products.

The purpose of the given paper is to determine mercury content in soil of Mezhdurechensk town, to reveal technogenic pollution at its initial stage, when it does not pose a serious threat yet, but is of great importance for monitoring and forecasting. There is an opinion that constant exposure to mercury with concentration lower than the maximum permissible concentration (MPC) may also be hazardous for human health [7, 8]. The paper presents the data on spatial distribution of mercury in the territory of Mezhdurechensk as one of the towns located in the vicinity and zone of largest coal producers.

2. Methods and material

2.1. Study area and sampling

Soil samples were taken in May, 2015 from the entire territory of the town in the densely populated areas as well as in vicinity of schools and kindergartens (fig. 1), according to GOST 17.4.2.01-81, GOST 17.4.3.01-83, GOST 17.4.1.02-83, GOST 17.4.4.02-84, GOST 17.4.3.02-85 and methodical guidelines. To select samples the areal monitoring network was used at the scale 1:80000. The total number is 30. Samples were selected by depth traversed technique, at the depth of 5 – 10 cm. The weight of one sample was 1.5 – 2 kg. Then the samples were dried at room temperature 18 – 19^{0}C, large foreign particles were removed. The samples were manually grinded and sieved through the meshes of 2.5 mm and 1 mm in diameter. 100-150 g of every sample were grinded up to 0.074 mm at the grinding mill.

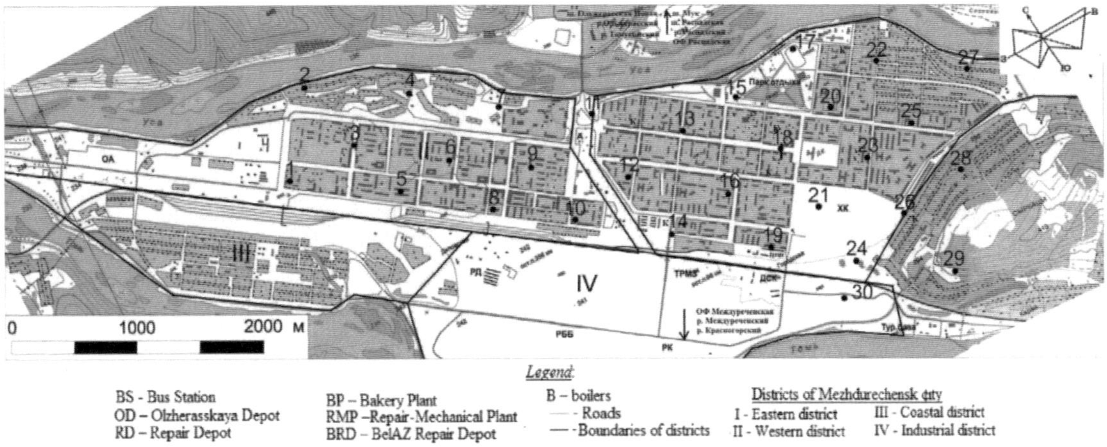

Figure 1. Schematic map of sampling.

2.2. Analysis

Mercury examination was carried out by atomic adsorption analysis using mercury analyzer RA-915+ with PYRO-915 attachment according to the method of PND F 16.1:2.23-2000 M 03-05-2005. As a background for mercury determination in soil the State Standard Sampling (SSS) of soddy-podzolic sandy-loam soil 2500-83 with mercury content 0.29 mg/kg was used. The portions of pre-grinded and dried at room temperature samples were 75.0±0.1 mg. The error range of measurement was 28 % depending on weight fraction of mercury in the samples with the confidence coefficient 0.95 and two parallel measurements.

3. Results and discussion

The data analysis has shown that according to the aerial survey of Mezhdurechensk town site the mercury content in soil samples changes from 0.12 to 0.17 mg/kg, with the average value 0. 057 mg/kg (figure 2).

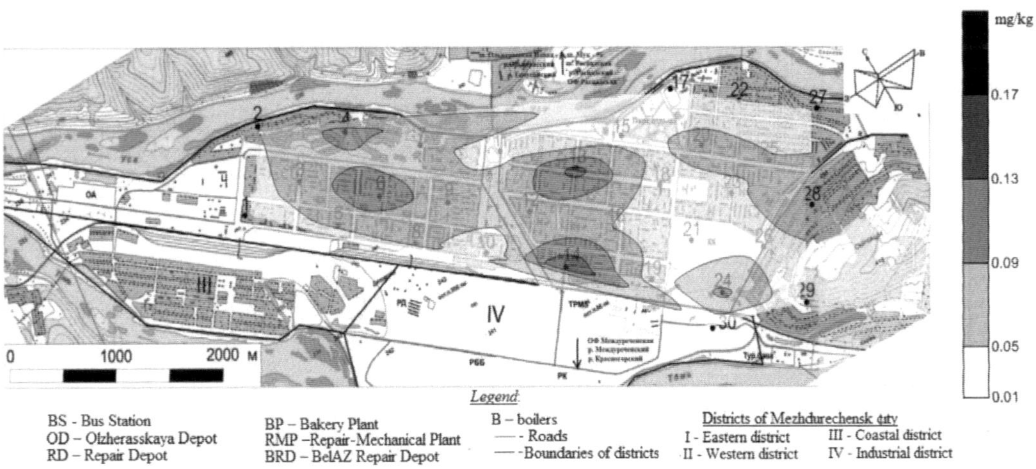

Figure 2. Mercury content in soil of Mezhdurechensk town.

Under these circumstances five zones with elevated mercury content were distinguished, three of which are located in the Eastern region, but two – in the Western one (fig. 2). Mercury content in the first zone (point № 4) is 0.104 mg/kg, in the second one (point № 6) is 0.116 mg/kg, in the third one (point № 13) is 0.148 mg/kg, in the forth one (point № 14) is 0.173 mg/kg, in the fifth one (point № 24) is 0.102 mg/kg. The first zone is located in the country part of the town where detached houses with stove heating are located, using coal as a fuel. The second, third, and fifth zones are located in vicinity of boilers that heat these zones. The fact can explain development of mercury elevated concentration zones in this part of the town. The forth zone is included in the industrial site in vicinity of the transport repair factory.

Thermal industry is still one of the major sources of mercury in the environment, which is proved by the example of Western Siberian cities and towns [9-11]. The unique properties of mercury (low melting point (-38.9°C) and high vapour pressure (boiling point 356.66°C) demonstrate the fact that it can exist only in the form of elementary mercury vapour Hg^0 at coal burning points. Regarding these properties an important feature of mercury is its volatility. Mercury emissions from coal combustion depend on the volume of burning fuel, combustion conditions, and mercury content in coal. According to [12], the average mercury content in Kuzbass coal changes from 0.01 to 0.1 g/t. However, in some coal fields, in the south-eastern part of Kuznetsk Basin particularly (pits Olzherassk, Mezhdurechensk, Tomusinsk), mercury content is several times higher: 0.3, 0.5, 0.6 g/t, respectively.

The comparison of mercury content in soils of some areas in the south of Western Siberia (figure 3) has revealed that at present the average mercury content in Mezhdurechensk town is 1.63 times lower than in Kemerovo Oblast and nearly equal to the average mercury content in Altai Territory [13]. It also should be noted that in comparison with the conducted research (1991, [14]) mercury content decreased 2.8 times.

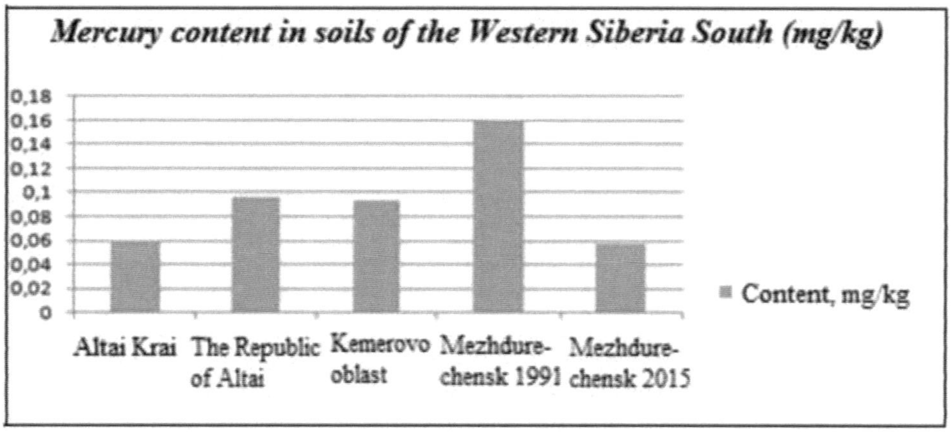

Figure 3. Mercury content in soil of some areas in the south of Western Siberia (mg/kg).

4. Conclusion

As a result of research performed it was stated that in the territory of Mezhdurechensk town five zones are distinguished with elevated mercury content in soil, three of which are located in the Eastern region, but two – in the Western one. The mercury content in the first zone – 0.104 mg/kg, in the second – 0.116 mg/kg, in the third – 0.148 mg/kg, in the forth – 0.173 mg/kg, in the fifth – 0.102 mg/kg. Besides, it was established that the major contribution to soil pollution in the territory of Mezhdurechensk town is made by the two factors: regional and local. The regional factor includes fallouts from coal mine development and explosion due to pollutants' transport. The local factor comprises boilers and enterprises located in the territory of Mezhdurechensk town as well as vehicles exhausts.

References

[1] Maistrenko V N, Khamitov R Z and Budnikov G K 1996 Ecological-analytical monitoring of supertoxicants (Moscow, Khimiya) 319 p (in Russian)
[2] Global and regional mercury cycles: sources, fluxes and mass balances Proceedings of the NATO advanced research workshop on regional and global mercury cycles: fluxes and mass balances. Novosibirsk, Russia, July 10—14, 1995 / Eds. W. Bayens, R. Ebinghaus, O. Vasiliev (Dordrecht—Boston—London, Kluwer Academic Publishers) 1996, p 563
[3] Mercury in the Environment 1970 *US Geol. Prof. Paper* (Washington) **Vol.** 713 p. 68
[4] Anoshin G N, Malikova I N, Kovalev S I, Androsova N V, Sukhorukov F V, Tsibulchik V M and Shcherbov B L 1995 Mercury in the environment in the south of Western Siberia *Khimiya v interesakh ustoychivogo razvitiya* 3 **Vol.** 1-2 pp. 69—111 (in Russian)
[5] Osipova N, Stepanova K and Matveenko I 2015 Evaluation of metal content in perch of the Ob River basin *IOP Conference Series: Earth and Environmental Science* **Vol.** 27 [012041, 5 p.]
[6] Ebinghaus R, Tripathi R M, Walischlager D and Lindberg S E 1999 Natural and anthropogenic mercury sources and their impact on the air-surface exchange of mercury on regional and global scales In: Mercury contaminated sites / Eds. R. Ebinghaus, R.R. Turner, L.D. De Lacerda, O. Vasiliev, W. Salomons. Springer-Verlag (Berlin, Heidelberg) pp. 3–50
[7] Glazovskaya M A 1988 Geochemistry of natural and technogenic landscapes of the USSR (Moscow, Vysshaya Shkola) p 325 (In Russian)
[8] Ivanov V V 1997 Mercury In: Ecological geochemistry of elements (Moscow, Ecology) **Vol.** 5 pp. 497–562
[9] Talovskaya A V, Filimonenko E A, Osipova N A, Lyapina E E and Yazikov E G 2014 Toxic elements (As, Se, Cd, Hg, Pb) and their mineral and technogenic formations in the snow cover in the vicinity of the industrial enterprises of Tomsk *IOP Conference Series: Earth and Environmental Science* **Vol.** 21 1 doi:10.1088/1755-1315/21/1/012042
[10] Filimonenko E A, Lyapina E E, Talovskaya A V and Parygina I A 2014 Eco-geochemical peculiarities of mercury content in solid residue of snow in the industrial enterprises impacted areas of Tomsk *Proceedings of SPIE The International Society for Optical Engineering* Article number 929231 20th International Symposium on Atmospheric and Ocean Optics: Atmospheric Physics; Novosibirsk 23 June 2014 **9292** Code 110711
[11] Osipova N A, Filimonenko E A, Talovskaya A V and Yazikov E G 2015 Geochemical Approach to Human Health Risk Assessment of Inhaled Trace Elements in the Vicinity of Industrial Enterprises in Tomsk, Russia [Electronic resource] In: *Human and Ecological Risk Assessment: An International Journal* **Vol.** 21 6 pp. 1664-1685
[12] ASAP Assessment of mercury input in the environment in the territory of the RF. Action plan of the Arctic Council for prevention of Arctic pollution, Federal service for Ecological, Technologic, and Atomic Inspection in cooperation with Dutch Agency for Environmental Protection 2005 Copenhagen 312 [Electronic resource] http://acap.arctic-council.org/projects.cfm
[13] Malikova I N, Anoshin G N and Badmayeva Zh O 2011 Active forms of mercury in soils of natural and natural-technogenic landscapes *Geology and geophysics* **Vol.** 52 3 pp. 409–425 (in Russian)
[14] Yazikov E G 2006 Ecogeochemistry of urban territories in the south of Western Siberia Dissertation for Doctor of Geol-Mineral Science Degree 25.00.36 (Tomsk, Tomsk Polytechnic University) 423 (in Russian)

Mineralogical and geochemical features of human body ash residue of spatially-localized technogenic system (Norilsk city)

L P Rikhvanov, M A Deriglazova, N V Baranovskaya and I A Matveenko

[1]Department of Geoecology and Geochemistry, Department of Foreign Languages Tomsk Polytechnic University, 30 Leninastreet, Tomsk, 634050, Russia
E-mail: rikhvanov@tpu.ru

Abstract. The paper presents the information on the element and mineral composition of the human body ash residue of Norilsk residents. The data are obtained by such methods as X-ray analysis, inductively coupled plasma mass spectrometry, and electron microscopy. This research has identified the content of 47 elements, 17 of which have their own mineral phases. Besides, the estimation of average content of chemical elements in terms of sex and correlation analysis of the data was conducted in this research. The final results indicate the influence of Norilsk industry on the element and mineral composition of ash residue. For example, the accumulation of such elements as Zr, Al, Ca, Y, some rare earth, and radioactive elements has been observed in the human body ash residue of Norilsk residents. The presence of Ag, Au, Pt microphases and numerous compounds of Cu, Ni etc. was detected among all mineral phases.

1. Introduction

Nowadays, study in element and mineral composition of human body is one of the priorities in modern geochemistry, medical geology and ecology. To develop these scientific areas, the Department of Geoecology and Geochemistry of Tomsk Polytechnic University is studying such material as human body ash residue (HBAR), which has been sampled in different Russian cities (Novosibirsk, Novokuznetsk, Yekaterinburg, St. Petersburg, and Rostov-on-Don). A long-term study of this material shows that the human body ash residue of every studied city has specific geochemical features reflected in high or low accumulation of certain chemical elements [1, 2].

In previous research of HBAR samples from the cities mentioned above it was required to take into account multifactorial industrial impact. In fact, there are several industrial engineering, chemical and metallurgical enterprises in large cities. Hence, it is interesting to study samples of human body ash residue of Norilsk residents.

Norilsk city is one of the Russian and world's most polluted cities. However, in contrast to other areas, the industry of Norilsk is presented not by the complex of different plants, but by the monoproduction, forming the technogenesis of the environment. Thus, this city represents spatially-localized technogenic system, as the local industrial enterprise uses ore, sand and coal, which are extracted in vicinity of Norilsk. Besides, the city is located in the north of the country, far from other industrial centres, which excludes the additional contribution of pollutants from other territories. One of the world's largest enterprises, which extracts, enriches and processes Cu-Ni ores, is the only industrial plant in Norilsk. The plant uses ores of Norilsk deposit, which are rich in such elements as Cu, Ni, Co, Se, Ag, Au, Cd, Te, Re, Bi, Zn, Pb, Cr, and Sn, as well as elements of the platinum group

(Ru, Rh, Pd, Ir, Pt etc). The main products of the local industry are Cu, Ni, Co, Se, S, and platinum concentrates.

Medical research of Norilsk and the surrounding area residents has shown that diseases of the respiratory and digestive systems prevail in the morbidity rate. It was also noted in the literature, that local residents have the elevated content of Cu and Ni in the blood plasma comparing with the control group [3].

2. Materials and methods

The material of the research is 22 samples of the human body ash residue, which were sampled in the Norilsk crematorium from the material unclaimed by the relatives with the official permission of administration. The ratio of men to women in the Norilsk samples is 13:9. The results of element composition of Norilsk residents' HBAR samples were compared with the previous results from the other cities (Novosibirsk, Novokuznetsk, Yekaterinburg, St. Petersburg, and Rostov-on-Don). Since ash residue of human body is a material burnt at high temperature (900-1100°C), it leads to the loss of some volatile elements. Therefore, only the comparative analysis of element content from different cities was performed.

Mass spectrometry with inductively coupled plasma was used to study the element content of HBAR. This method includes dissolution of sample at the preparatory stage. Thus, it is impossible to extract some elements thoroughly, that may lead to errors in determination of concentrations. Electron microscopy and X-ray analysis were used to study the mineralogical content of human body ash residue of Norilsk residents. Scanning electron microscope Hitachi S -3400N was applied to determine microphases, which content was studied by microanalysis (detection limit is 0.1-0.2%). The identification of mineral phases was conducted by Bruker D2 Phaser diffractometer (the detection limit is 0.1-0.5%).

3. Geochemical features of human body ash residue of Norilsk residents

The content of 47 chemical elements was detected in the human body ash residue of Norilsk residents using ICP-ms. Based on these results, the coefficients of concentration (relative to the average values of 6 cities) were calculated to determine the geochemical features of HBAR of this city (figure 1). The diagram shows that the ash residue of Norilsk residents is characterized by high accumulation of such elements as Al, Ca, Y, Zr, Ce, Pr, Nd, Cd, Tb, Dy, Ho, Yb, Lu, Th, and U in comparison with other cities.

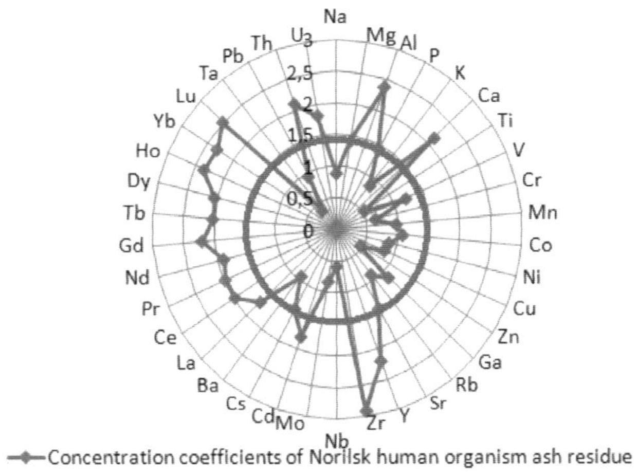

Figure 1. Concentration coefficients of human organism ash residue, Norilskcity.

The specificity of element accumulation is unlikely to reflect the content of Cu-Ni ores which are used in local industry, because the spectrum of concentrated elements in ash residue of Norilsk residents differs much from those concentrated in the ore.

The comparison of different cities accumulation specifics revealed the similarity in elements accumulation between ash residue of 2 Russian cities – Norilsk and Novokuznetsk. The simplest Jaccard similarity coefficient showed the 25% similarity in the accumulation of these cities that is likely to demonstrate the metallurgical features of the largest Russian industrial centres of ferrous and non-ferrous metallurgy.

The comparative analysis of element accumulation in terms of sex showed that most elements concentrate equally in the male and female organism of Norilsk residents (figure 2). However, such elements as Zr, Rb, Y, Sn, Dy, Ho, Tm, Yb, Lu, Hf, Th, U have a tendency to accumulate in male organism; but Mo, Gd, Pb, Bi tend to concentrate in female organism.

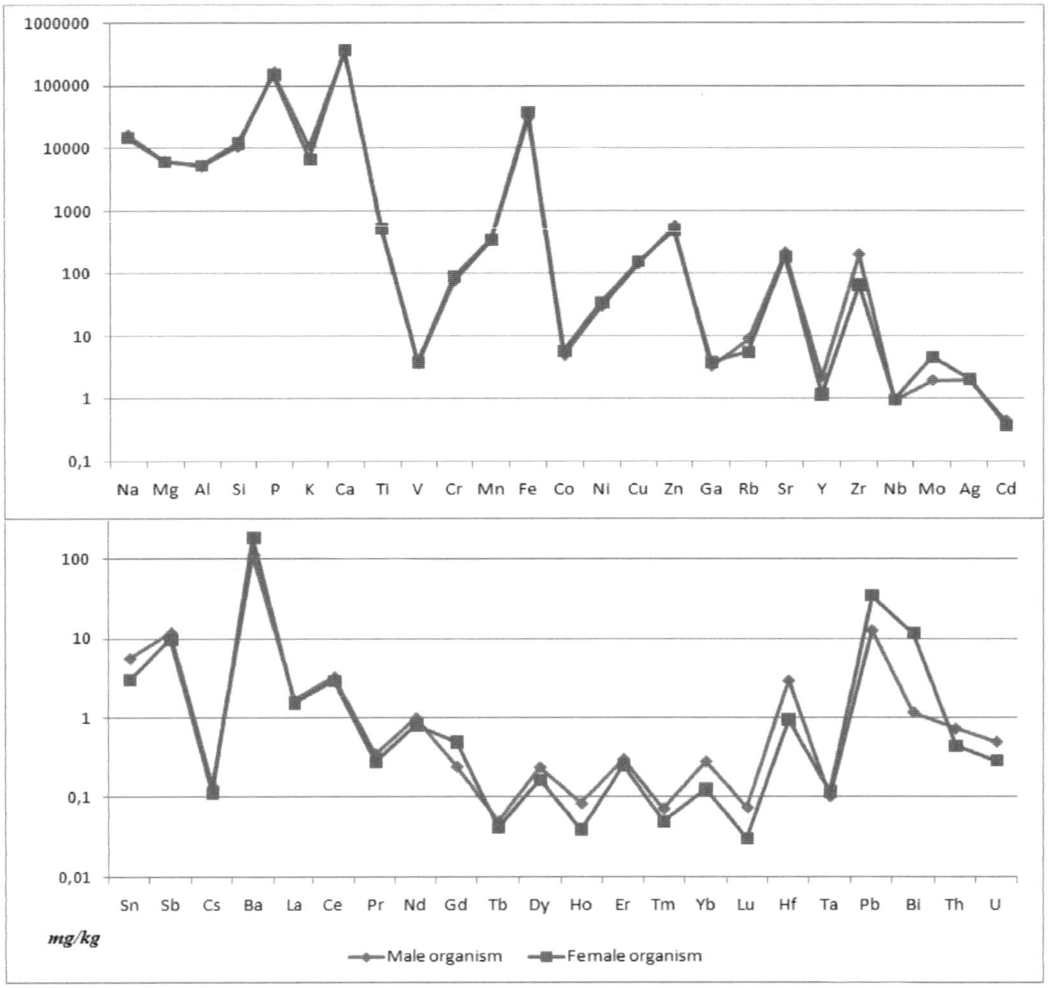

Figure 2. Estimation of the average content of elements in the human body ash residue of Norilsk residents in terms of sex.

Correlation analysis showed close relations of many elements. The highest positive pair correlation coefficient ($r = 0.99$) belongs to the pair of Zr-Hf (critical coefficient $r = 0.404$, $p = 0.05$). Besides, 19 pairs are characterised by significant relations with correlation coefficient from 0.90 to 0.98; these include: Y-Tb, Y-Dy, Zr- Tm, Zr-Yb, Zr-Lu, Y-U, Zr-Th, U-Th, Nd-Pr, Dy-Yb, Dy-U, Tm-Yb, Tm-Lu, Tm-Hf, Yb-Lu, Yb-Hf, Yb-Th, Hf-Lu and Hf-Th. The strongest negative correlation coefficient ($\bar{r} = 0.95$) belongs to the pair Ca-Nd. Wherein, such element as Ca is associated with almost all elements with negative relations except P.

Results of cluster analysis (Figure 3) allowed determining several groups of elements, which are likely to have different source and mechanism of accumulation. The pair Ca-P was chosen as significant relation, because this is one of the strongest relations in a human body, which constitutes the main mineral of human bones –hydroxyapatite. Thus, the following groups of tightly related elements in the human body ash residue of Norilsk residents may be distinguished:

- Ho, Yb, U, Hf, Zr;
- Nd, Pr, La, Ce, Dy, Th, Y;
- Ta, Ga, Fe;
- Sn, Cr, Mn, Ti;
- Cu, V, Nb, Si, Al;
- Rb, K, Sb, Co, Mg;
- Zn, Ba, Na;
- Pb, Mo.

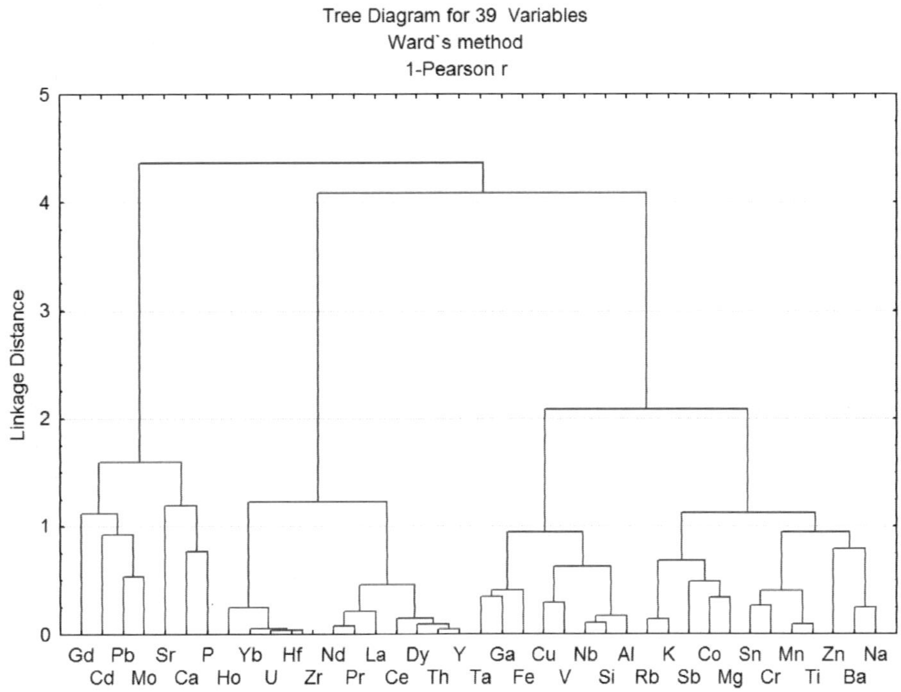

Figure 3. The diagram of cluster analysis.

4. Mineralogical features of the human organism ash residue of Norilsk residents

The previous mineralogical research of the ash residue samples from different Russian cities has shown that material consists of more than 99.5% of hydroxyapatite ($Ca_5(PO_4)_3(OH)$), that is the main mineral of bone tissue. The X-ray analysis of HBAR samples from Norilsk identified not only hydroxyapatite in the mineral content, but also hydroxyapatite with the significant impurity of the silicate ($Ca_5(PO_4)_3(SiO_4)(OH)$). Electron microscopy analysis of the Norilsk samples also showed the impurities of Na, K, Mg, Al, S, Si, and Cl in the content of hydroxyapatite. Besides, the aggregates of hydroxyapatite were found in the samples, in which Na, K and Mg replace Ca more than 70-90%.

Microphases of different composition with less than 0.5% of the sample volume were also found in the studied material. In general, 27 mineral phases of different composition were found in the human body ash residue of Norilsk residents including mineral phases of Cu, Ni, and Pt, that are likely to reflect the impact of local industry. Thus, the specificity of mineralogical content of Norilsk material is a large number of mineral phases characterized by a complex composition which is not typical for the samples from other cities. Some micromineral phases of Zn, Cu, Zr, and Pt are shown in figures 4, 5, 6, and 7.

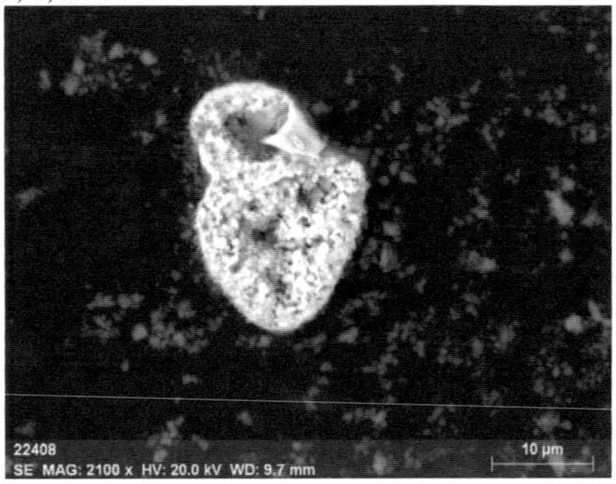

Figure 4. Micromineral phase of Zn in the HOAR of Norilsk residents.

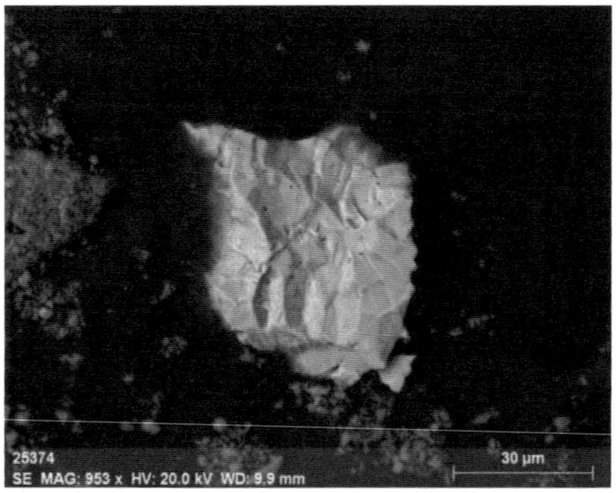

Figure 5. Micromineral phase of Cu in the HOAR of Norilsk residents.

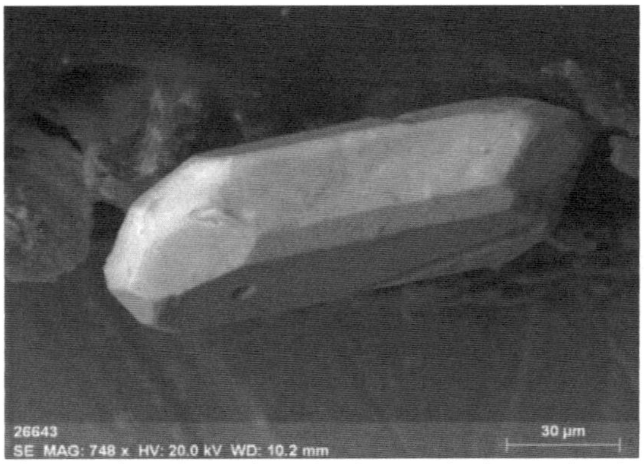

Figure 6. Micromineral phase of zircon in the HOAR of Norilsk residents.

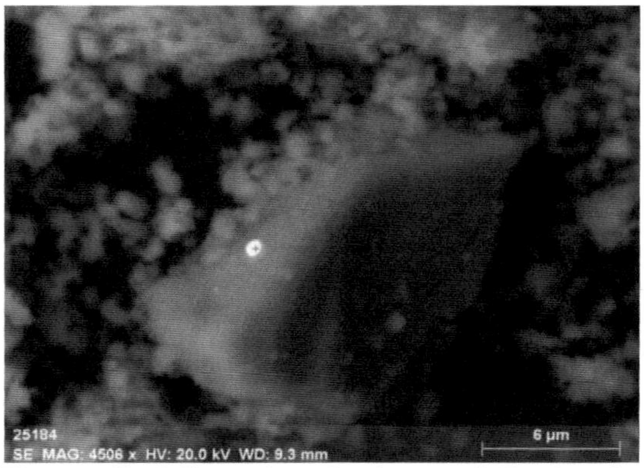

Figure 7. Micromineral phase of Pt in the hydroxyapatite aggregate in the HOAR of Norilsk residents.

5. Conclusion

The content of 47 chemical elements was studied in the human body ash residue of Norilsk residents, as well as modes of occurrence were determined for some of them. Based on the results of the research, some conclusions about the specificity of the element accumulation and distribution in the HBAR of Norilsk residents can be made. Probably, the spectrum of accumulated elements (Al, Ca, Y, Zr, Ce, Pr, Nd, Cd, Tb, Dy, Ho, Yb, Lu, Th and U) and mineral phases (Cu, Pt, Ceetc) indicate two main sources of elements: ore used in the local industry and emissions of the industrial plant. Besides, the accumulation of radioactive and rare-earth elements and the presence of such minerals as zircon and monazite indicate the possible use of zircon-ilmenite sands or zirconium crucibles in the production. Nevertheless, the source of the studied elements in the organisms of Norilsk residents has not been determined yet and requires additional study of the local water and industrial materials.

IOP Conf. Series: Earth and Environmental Science **43** (2016) 012039 doi:10.1088/1755-1315/43/1/012039

References

[1] Ignatova T, Baranovskaya N, Rikhvanov L and Sudyko A 2010 *Bulletin of the Tomsk Polytechnic University.* Regional biogeochemical features of human body ash residue. Vol. **317** pp. 178-183.

[2] Rikhvanov L, Baranovskaya N, Deriglazova M and Strelnikova A 2014 *Procedia Chemistry.* Mineralogical and Geochemical Characteristics of the Human Body Ash Residue. Vol. **10** pp. 454-459.

[3] Kolpakova A 1999 *Human Ecology Journal.* Influence of anthropogenous pollution on the heavy metals content in blood of the Tymir autonomous district inhabitants. Vol. **2** pp.15-17.

The level of air pollution in the impact zone of coal-fired power plant (Karaganda City) using the data of geochemical snow survey (Republic of Kazakhstan)

T E Adil'bayeva, A V Talovskaya, Ye G Yazikov, I A Matveenko

Tomsk Polytechnic University, 30 Lenin Avenue, Tomsk, 634050, Russia
E-mail: mega.adilbaeva@mail.ru, talovskaj@yandex.ru, yazikoveg@tpu.ru, mia@yandex.ru

Abstract. Coal-fired power plants emissions impact the air quality and human health. Of great significance is assessment of solid airborne particles emissions from those plants and distance of their transportation. The article presents the results of air pollution assessment in the zone of coal-fired power plant (Karaganda City) using snow survey. Based on the mass of solid airborne particles deposited in snow, time of their deposition on snow at the distance from 0.5 to 4.5 km a value of dust load has been determined. It is stated that very high level of pollution is observed at the distance from 0.5 to 1 km. there is a trend in decrease of dust burden value with the distance from the stacks of coal-fired power plant that may be conditioned by the particle size and washing out smaller ash particles by ice pellets forming at freezing water vapour in stacks of the coal-fired power plant. Study in composition of solid airborne particles deposited in snow has shown that they mainly contain particulates of underburnt coal, Al-Si-rich spheres, Fe-rich spheres, and coal dust. The content of the particles in samples decreases with the distance from the stacks of the coal-fired power plant.

1.Introduction

The increasing global and local atmospheric pollution is associated with development of power industry to a significant extent. It is known from the domestic and foreign literature [1-14] that construction and operation of fuel-power facilities has led to pollution of all environmental components. In this case, pollution of biosphere is a multi-component, dynamic and complex phenomenon connected with the large number of simultaneous environmental damage: soil-geological, hydrological, geochemical, climatic, and biogeochemical. At high temperature of coal combustion in boilers of fuel-power complexes significant amount of impurities is released into atmosphere as a part of solid particles and gaseous compounds. In all forms they are capable of affecting living organisms including human being [12-14]. Urban atmospheric pollution with thermal power plants emissions is especially harmful to children's health due to intensive metabolic processes in children's organisms, homeostasis immaturity, and immune instability [11, 18]. At present, there is a problem of determining indicators to assess the share of a certain emission source in the general pollution level of the environment. Snow geochemical survey is effectively used by many researchers to assess the conditions of urban environment. Snow is considered to be an extremely accessible and reliable medium for studying air pollution due to its high sorption capacity [3, 14]. Geochemical information on snow allows revealing the dynamics of pollution in winter season, and one sample of the entire snow depth provides the information on pollution rate within the whole period from snow cover formation to sampling time. Mass determination of solid airborne particles deposited in snow is a basis for establishing dust load and permits revealing polluted sites within the period of steady snow cover [5, 12, 18-19]. The study in composition of solid airborne particles deposited in snow contributes to determining mineral and non-mineral technogenic formations as parts of industrial emissions and distinguishing the source of pollution in the enterprise impact zone [12-13]. Investigation of composition is also of crucial significance from the standpoint of human health risk assessment of solid airborne particles. In the work [5] it is shown that long-term impact of technogenic mineral phases composing solid airborne particles is a cause for numerous diseases.

The purpose of the article is 1) to determine the level of snow dust pollution 2) to reveal types of mineral and non-mineral phases in solid airborne particles deposited in snow in the zone of coal-fired power plant impact (Karaganda).

2. Research methods

2.1 Study area

Karaganda City is the largest city in the province and the forth one in Kazakhstan in terms of population (after Almaty, Astana, and Shymkent). During a year in Karaganda the south-west winds prevail, their share within a long term period is 65%. The opposite winds blow in 14% cases, but the frequency of other wind directions does not exceed 7-8%. Karaganda city is administratively divided into two regions: one named after Kazybek bi and Oktyabrskiy. The study area of Oktyabrskiy region is 22.4 thous. hectares in square, which amounts 41.3% of general city square, by 2014 the population of the city was 221.5 thousand people. In the northern part of the city there is one of the largest energy-producing enterprises in Karaganda Oblast – coal-fired power plant. The plant is a major heat and electrical energy producer in the city heat- and energy supply system. The general rated output power of the plant is 592 MW, available one – 418.1 MW. In terms of thermal power – the rated one is equal to 1634 Gcal/h, available is about 1010.3 Gcal/h. The plant operates on Ekibastuz coal, uses fuel oil as a start-up fuel. Ekibastuz coal basin is located in Pavlodar Oblast, Kazakhstan Republic, Ekibastuz city. The disadvantage of this coal consists in the fact that it is high ash (more than 40%) and contains relatively high amount of impurities that decreases thermal effectiveness of coal combustion increasing its transportation cost [14]. Upon the beginning of heating season, due to coal combustion a significant amount of fine dust with technogenic particulates is released into atmosphere.

2.2 Sampling and sample preparation

In January, 2015 snow was sampled in the zone of coal-fired power plant (Karaganda City). In addition, sampling was performed in the residential area located 2.5 km from the coal-fired power plant to the South-West. Sampling was performed using vector network taking into account the principle wind direction (south-west): in the north-east, east, south-east, south, south-west, north-west, and north directions at the distances 0.5; 0.7 km;1.6 km; 2.2; 3.2 and 4,5 km from the stacks. The total amount of samples was 42.

The two areas located 80 km from Karaganda city in the south-east direction were chosen as background sites. In the background sites 4 samples were taken.

Planning sampling points, sampling, and sample preparation were carried out according to the recommendations of RD 52.04.186 № 2932-83 using the methodical guidelines of Institute of Mineralogy, Geochemistry and Crystal Chemistry of Rare Elements, supervision manual over air pollution [14]. According to those specifications, it is necessary to take into account the principle wind direction and stack height when planning the sample procedure, i.e. to select sampling points at the distance calculated as a multiplication of stack height (in our case it is 100 m) by the value from 10 to 40. Snow samples were selected all its depth long with the exception of a 5-sm layer over the ground. The weight of each sample was 15-16 kg. In each sampling the sides and depth of pits were measured the time from stable snow cover formation to snow sampling was fixed (in days). The snow samples were melted in the laboratory condition at room temperature. Snow-melt water was filtered through the paper "Blue ribbon" filter. The solid precipitation obtained after filtration was dried and sieved isolating the fractions of less than 1 mm, then snow solid precipitation was weighed. Solid precipitation consisted of solid airborne particles deposited in snow. After that, the snow solid precipitation was analyzed. In general, 46 snow samples were selected.

2.3.Research methods

The samples of snow solid precipitation were studied in the laboratory of optical diagnostics, International research center "Uranium Geology", Department of Geoecology and Geochemistry, Tomsk Polytechnic University. Microscopic study of samples was conducted by means of binocular stereoscopic microscope LeicaZN 4D with video and scanning electron microscope (SEM) Hitachi S-3400N with EDS Bruker XFlash 4010. Analysis of composition of snow solid precipitation samples with the subsequent percentage calculation of all mineral and technogenic compounds was carried out in accordance with the patent design [Patent 2229737].

Dust load (P_n, mg·(m^2·day)$^{-1}$) was calculated based on the weight of snow solid precipitation (P_0, mg), square of the snow pit (S, m^2) and the number of days from snow-up day to sampling day (t):

$$P_n = P_o / S \times t \quad (1)$$

In Russia, there is a dust load gradation for detection of pollution rate and morbidity level: less than 250 – low, 251-450 – average, hazardous, 451-850 – high, hazardous, more than 850 – very high, highly hazardous [12, 15].

3. Results and discussion

It was stated that in the zone of Karaganda coal-fired power plant the dust load value changes from 84 to 1145 mg·(m^2·day)$^{-1}$, at the background 47 mg·(m^2·day)$^{-1}$. According to the standard classification [15] the obtained value changes from low to very high pollution rate and from non-hazardous to highly hazardous morbidity rate of population, respectively.

The increased value of dust load is observed at the distance of 0.5 and 0.7 km from the coal-fired power plant in all directions that amounts 864 mg·(m^2·day)$^{-1}$on average, which exceeds the background 18 times and corresponds to very high pollution degree and highly hazardous morbidity rate.

The analysis has shown that the maximum value of dust load is found in the north-east part from the coal-fired power plant, exceeding the background from 3 to 31 times, which can be explained by pollutant transportation in the direction of prevailing wind.

The data obtained for coal-fired plant of Karaganda city are comparable with the calculations [17] of dependencies for fallout intensity of solid airborne particles with the distance from the source in the direction of principle wind. Based on the results of observations the decrease in dust loads up to 9 times occurs with the distance from the coal-fired power plant at the distance 1.6; 2.2; 3.2 and 4.5 km.

The least values of dust load were observed in the south direction from the coal-fired power plant. This value corresponds to low pollution degree and non-hazardous morbidity rate.

In work [17] based on the results of long-term observation over the dust pollution level in the zone of fossil fuel power plant (Tomsk city) operating on coal (~ 40 %) and natural gas (~ 60 %) as fuels, tha value of dust pollution changed from 44 to 115 mg·(m^2·day)$^{-1}$. Besides, it was noted that maximum dust fallouts on snow was observed at the distance of 1.0 km. According to the research, it may be explained by the process of washing out ash small particles with ice pallets formed in water vapour freezing in the thermal plant stack plumes, according to the research [16]. Due to this phenomenon a large part of dust emissions may be deposited at close distances from the plant in winter despite the significant height of stacks.

This phenomenon is likely to be typical for coal-fired power plant in Karaganda city as well that may also explain high values of dust pollution at the distance of 0.5 and 0.7 km from the coal-fired power plant. The increased dust load values may be connected not only with emissions from the studied coal-fired power plant, but also with intake of solid particles due to wind transport from the open coal storage located in the site of the coal-fired power plant or due to the dust formed in coal unloading.

The increased dust load values in the study area are conditioned by the content of different mineral and technogenic particles in solid airborne particles deposited in snow (table1).

Table 1. The value of dust load mg·(m²·day)⁻¹. Content (%) of mineral and technogenic particulates in snow solid precipitation in the North-East direction from the coal-fired power plant, Karaganda city

Distance from TPP, km	**Dust load, mg·(m²·day)⁻¹**	**Pollution level***	**Morbidity level***	**Mineral particles, %**		**Technogenic particulates, %**				**Plant remains, %**
				quartz	feldspar	underburnt coal	coal dust	Fe-rich spheres	Al-Si-rich spheres	
0.5	1445	very high	highly hazardous	3	5	32	13	15	20	3
0.7	1186	very high	highly hazardous	9	5	32	13	13	17	3
1.6	398	average	hazardous	10	5	28	12	13	12	5
2.2	269	average	hazardous	10	5	18	10	12	12	5
3.2	227	average	hazardous	22	9	15	10	9	9	5
4.5	165	low	not hazardous	25	9	10.	9	7	6	5

Note: * - according to the data [15]

The binocular microscopic analysis of sample composition has shown that snow solid precipitation at the distance of 0.5 and 0.7 km from the coal-fired power plant contains mostly particles typical for emissions of coal-fired power plant, i.e. underburnt coal (8-32 %), Al-Si-rich spheres and Fe-rich spheres (4-20%), coal dust (6-13 %). The least amount of technogenic particles was found in the south direction from the coal-fired power plant (3-9 %). In all samples the mineral part of snow solid precipitation includes mainly quartz (3 - 55 %) as well as feldspar particles (5 - 12 %). The prevalence of technogenic particulates (4-32%) over the mineral ones (3-55%) was revealed in the samples in all directions at the distances 0.5 km; 0.7 km. With the distance from the coal-fired power plant at the distances 1.6; 2.2 km; 3.2 km; 4.5 km the content of mineral particles in samples prevails over that of technogenic particulates (5-18%).

4.Conclusion

Based on the research results obtained it has been determined that high values of dust load are revealed at the distance 0.5 km and 0.7 km from the coal-fired power plant. With the distance from the coal-fired power plant dust load value tends to decrease in all directions. Composition of snow solid precipitation includes different technogenic (coal dust, alumosilicate and metallic microspherules, underburnt coal particles) and mineral (quartz, feldspar) particles. The source of the given particles is suggested to be high-temperature coal combustion at TPP, open coal storage, handling operations. Besides, the possibility of remote pollutant transportation from the plants of Karaganda and its satellites should also be taken into account regarding the principle wind direction.

References

[1] Miloš Miler and Mateja Gosar 2009 *Materials and Geoenvironment.* Characterisation of solid airborne particles in urban snow deposits from Ljubljana by means of SEM/EDS. Vol. **3** pp. 266–282.

[2] Adil Elik 2010 *Journal of Environmental Analytical Chemistry.* Monitoring of Heavy Metals in Urban Snow as Indicator of Atm osphere Pollution International. Vol. **1** pp. 37-45.

[3] Geophys J Res 1996 Aerosols and atmospheric optics. Vol. **14** pp. 19185–19369.

[4] Bowen H J M 1966 *London. New York, Academic Press.* Trace elements in biochemistry pp. 241.

[5] Carpi A 1997 *Water Air Soil Pollut* .Mercury from combustion sources: A review of the chemical species emitted and their transport in the atmosphere Vol. **3–4** pp. 241–254.

[6] Seinfeld JH 1998 *Atmospheric chemistry and physics.* From Air Pollution to Climate Change John Wiley& Sons Inc. pp. 1326.

[7] Magiera T, Jabnska M, Strzyszcz Z and Rachwal M 2011 *Atmospheric Environment* Morphological and mineralogical forms of technogenic magnetic particles in industrial dusts. Vol. **45** pp. 4281–4290.

[8] Senior C L, Helble J J and Sarofim A F 2000 *Technol.* Emissions of mercury, trace elements, and fine particles from stationary combustion sources II Fuel Proces. Vol **65—66** pp. 263—288.

[9] Dai S, Ren D, Chou C L, Li S and Jeang Y 1995 Mineralogy and geo- Kluwer Academic Publishers. pp. 24–50.

[10] Osipova N A, Filimonenko K A, Talovskaya A V and Yazikov E G 2015 *Human and Ecological Risk Assessment.* Geochemical approach to human health risk assessment of inhaled trace elements in the vicinity of industrial enterprises in Tomsk, Russia. Vol. **21** pp. 1664–1685. doi: 10.1080/10807039.2014.972912

[11] Raputa V F, Khodzher T V, Gorshkov A G and Koutzenogii K P 1998 *Journal of Aerosol Science.* Aerosol falls on snow cover on the outskirts of Siberian towns. Vol. **29 (2)** pp. 807–808. doi:10.1016/S0021-8502(98)90586-9

[12] Talovskaya A V, Filimonenko E A, Osipova N A, Lyapina E E and Yazikov E G 2014 *IOP C. Ser. Earth. Env.* Toxic elements (As, Se, Cd, Hg, Pb) and their mineral and technogenic formations in the snow cover in the vicinity of the industrial enterprises of Tomsk. Vol. **21** pp. 012042.

[13] Saet Yu E, Revich B A, Yanin E P, Smirnova R S, Basharkevich I L, Onishchenko T L, Pavlova L N, Trefilova N Ya, Achkasov A I and Sarkisyan S Sh 1990 *Moscow. Nedra.* Geochemistry of the environment . pp. 335.

[14] Belyaev S P, Beschasnov S P, Khomushku G M, Morshina T I and Shilina A I 1997 *Meteorology and hydrology.* Some regularities of environmental pollution with coal combustion product by the example of Kyzyl city. Vol. **12** pp. 54–63.

[15] Talovskaya A V, Raputa V F, Filimonenko E A and Yazikov E G 2013 *Optics of atmosphere and ocean.* Experimental and numerical research of long-term dust-aerosol pollution in vicinity of thermal power plant (by the example of Tomsk HPP-2). Vol. **8** pp. 642-646.

[16] Vasilenko V N, Nazarov I M, Fridman Sh M 1985 *Leningrad. Gidrometeoizdat.* Monitoring of snow cover pollution pp. 181.

[17] 1990 *Moscow. IMGRE.* Methodical recommendations on assessment of air pollution level in the settlements in terms of the pollutant content in snow cover and soil. pp. 66.

Resource conservation program in terms of Vostokgazprom environmental policy

M R Tsibulnikova[1,2] **S V Nadyumov** (Надымов)[1] **A M Adam**[2] **T V Korotchenko**[1]
1 Tomsk Polytechnic University, 30 Lenina street, Tomsk, 634050, Russia;
e-mail: tsibulnikova2011@yandex.ru
2 Tomsk State University,30 Lenina street, Tomsk, 634050, Russia;
e-mail: adam@green.tsu.ru

Abstract. The article examines a number of key areas of environmental policy of Vostokgazprom. The Associated Petroleum Gas program is an important step within the resource conservation and environmental protection framework. In addition, the company undertakes the extensive work on emergency response programs, and carries out continuous protection of the subsurface and control over environmental safety in all production sites. Vostokgazprom continuously modernizes the basic industrial facilities and invests money in new projects. The study analyzes the steps being taken by the company within the energy saving policy that leads to significant costs cut.

Key words: environmental policy, oil and gas production, resource efficiency, energy conservation.

1. Introduction

Worldwide, energy and resource conservation is considered the first priority in sustainable development as it is intended to balance the economy and the sustainability of environment [1]. In many countries, the development and implementation of resource efficiency programs [2] which are primary based on introducing new technologies aimed at reducing consumption of fuel and energy resources [3] are an integral part of environmental policy. Resource consumption optimization is regarded as a tool for enhancing competitive capacity [4], while reduction of fuel and energy is supposed to avert global climate changes [5].

Today, natural resources play a significant role in the Russian economy. Production of natural resources contributes 50% of GDP (gross domestic product) of Russia. About 70% of foreign exchange earnings of federal and constituent entity's budgets of the Russian Federation directly or indirectly come from the natural resource sector. Due to these facts, the issue of energy and resource conservation occupies a central place in petroleum companies' policies [6].

The petroleum industry recognizes that energy efficiency and resource conservation programs, including well-balanced environmental policies, can make a major contribution to optimal use of natural resources and reduction of environmental impact [7, 8].

2. Materials and Methods

The present study is focused on the analysis of a number of programs implemented within the environmental policy of "Vostokgazprom". Precisely, the performance indicators of resource conservation programs and environmental policy itself are thoroughly examined.

3. Results and Discussion

Environmental protection implemented in accordance with the Environmental Policy of Gazprom and unified HSE program is among Vostokgazprom's prioritized activities. The resource conservation program of Vostokgazprom is focused on ensuring the ecological balance, developing new technologies that would save resources and provide consumers with consistent source of energy.

The environmental protection policy of the company encompasses the following targets:

– development and improvement of the environmental management system of the company in accordance with the current environmental legislation;

– implementation of a number of environmental protection measures both at the stage of field infrastructure design and construction, and stage of field operation;

– facilitation of efficient control over implementation of technological and administrative decisions.

Design and engineering solutions which are thoroughly examined by the company departments and government agencies allow reducing environmental impact. Reclamation of disturbed land and forests makes it possible to restore fertility and quality of soils and territories that are not planned to be used during facilities operation and maintenance.

At the first stage of Vostokgazprom production facilities construction, environmental safety was secured due to carefully considered and balanced engineering decisions provided by the field infrastructure development projects. Precisely, efficient natural gas processing technologies, as well as, up-to-date methods of industrial solid and liquid waste disposal, were implemented. In addition, emergency prevention decisions and activities were elaborated.

Assuring permanent control over the quality of production facilities construction, used materials and component parts, test procedures and maintenance activities is considered essential constitute of technological, and as a result, environmental safety.

When developing hydrocarbon fields, the company takes steps on reducing negative environmental impacts associated with hydrocarbon extraction, processing and transportation. To be more precise, production facilities are renovated, corrosion protection practice is revised, gas-field production facilities and pipeline systems are inspected on the schedule basis. In addition, technologies of natural gas gathering and processing are constantly being improved.

In order to enhance production and transportation facilities safety, the following steps were taken: implementation of up-to-date well development and associated gas utilization technologies, as well as, regular non-destructive testing, control over production facilities and piping system tightness testing, automation of the most essential technological processes, and increase in competence and consciousness of the company's employees in addressing the environment-related issues [9].

The industrial wastes are disposed in accordance with the nature conservation laws. The wastewater is treated and purified in the fields via biological treatment facilities. The peculiar design of wastewater facilities (the soil absorption system in Severo-Vasyuganskoye gas condensate field and the root zone waste water treatment system in Severo-Ostaninskoye oil field) secures efficient treatment of wastewater at minimum cost. To dispose and neutralize industrial solid waste, Vostokgazprom introduced onsite waste landfills appropriate for burial disposal techniques [10].

As the current study reveals, drilling cuttings removed from the exploration and production boreholes are classified as low-hazard and hazardous industrial waste (4 and 5 classes of hazard) in accordance with the waste management policy of Vostokgazprom. At the same time, these drilling cuttings are also composed of mineral substances found in the ground and required for plant nutrition. It is these characteristics of drilling cuttings that allowed engineers of Vostokgazprom to prove the possibility of using drilling cuttings as a by-product, i.e. fertilizer, in reclamation of the forest lands disturbed during the construction activities.

Recently, development of a new framework for financing and implementing the Associated Petroleum Gas program has become an essential step towards resource conservation and environmental safety within the environmental policy of Vostokgazprom.

Besides, implementation of emergency response programs developed for all hazardous production sites of the company is also an important constituent of the industrial and environmental safety.

The company timely carries out the inventory of the sources that create negative environmental impacts. Due to this, draft standards for allowable emissions of pollutants, discharge of pollutants into surface water sources, and waste generation and limits on waste disposal have been developed and are regularly revised.

Field activities that occur during exploration and production stages also include protection of the subsurface and control over environmental safety. In general, based on monitoring and control results, it can be stated that industrial environmental impact is within the allowable limits set by the design documentation and specifications.

Energy conservation is the prioritized activity of Vostokgazprom. The company is one of the largest consumers of energy resources among the enterprises of Tomsk oblast.

In 2013, the company spent 31940 thous. rubles on energy resources, 26404 thous. rubles on electricity, 5536 thous. rubles on thermal energy. Due to the certain activity of the company, the extracted gas is not only transported but also used for on-site needs. Therefore, the company does not bear any expenses related to gas consumption.

In 2013, in order to reduce energy cost, the furnace smoke exhaust motor was equipped with adjustable speed drive (Myldzhinskoye field). From 2007 to 2011, only two of four smoke exhausts were in service, each consuming 45 kW. The operating time of the smoke exhaust was 24 hours, with the average duration of the heating season being 201 days.

In addition, to secure continuous energy supply, outdated panels of electric equipment control centers were replaced by modern ones manufactured by «Siemens». This program, initiated in 2005, was planned to continue until 2016. According to this program, 250 panels of electric equipment control centers, 27 transforming substations and distributing points were either replaced or renovated.

In 2012, within the joint program on energy supply reliability enhancement and reduction of risks of power cuts in Myldzhinskoye, Severo-Ostaninskoye, Kazanskoye, and Severo-Vasyuganskoye oil and gas condensate fields (OGCF), outdated isolating switches, short-circuit throwing switches, and oil switches were replaced by modern vacuum switches and SF breakers. In the course of Ostaninsk fields development, power stations PS-35/6kW "Severo-Ostaninskaya" and PS-35/10kW "Ostaninskaya" were commissioned in 2015. Besides, transforming substation is planned to be constructed in Ostaninsk field itself. It would supply power to Mirnoye, Pidzhinskoye, and Ostaninskoye fields. Today, energy supply to these fields is secured by diesel-engine power plant located in Mironoye field. Assuring continuous energy supply will definitely increase reliability of the field facilities operation [11].

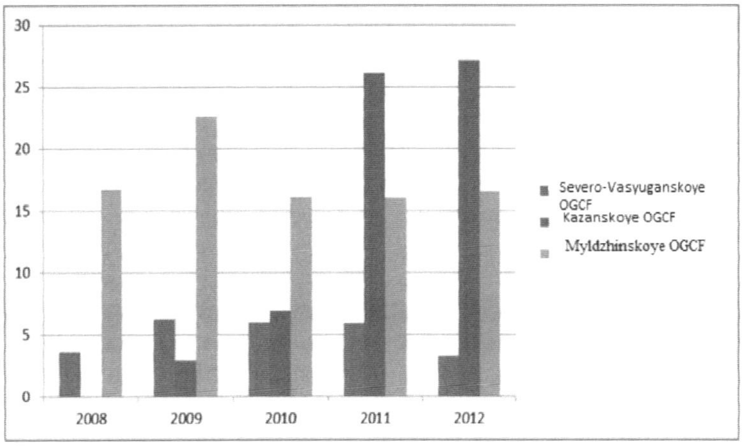

Figure 1 – Dynamics of energy consumption, mln kW/h.

Figure 1 illustrates the energy consumption analysis carried out for each of the above-mentioned fields. It shows that the annual increase in energy consumption is primarily stipulated by industrial and technological needs, while energy use for on-site needs, including supply of power to the contractors and compensation of industrial losses, remains constant. The volume of energy consumption in Kazanskoye oil and gas condensate field (OGCF) stands out against the general background.

Since the moment of the field commissioning in 2009, energy consumption increased 8.8. times and amounted to 23.074 mln kWh in 2012. In general, this fact identifies the steady level of energy consumption.

In August 2011, "Vostokgazprom" was issued "Power Certificate of Industrial Consumer of Fuel-Energy Resources" valid until 2016.

For further implementation of energy saving activity to reduce energy consumption and based on the Federal Law 261-FZ "On Energy Saving and Energy Efficiency Increase and Amending Certain Legislative Acts of the Russian Federation" adopted on 23 November 2009, Vostokgazprom developed the third program on energy saving for 2011-2015 with a vision to 2020. It is planned to save about 360 mln. kWh of electric power and 870 thous. Gcal of thermal energy.

The following results were achieved with the support of this program:

- 52009 kWh of electric power in kind and 144.8 thous. rubles in cash were saved. Active harmonic filters and adjustable speed drive were installed in furnace smoke exhaust motor in Mirnoye gas condensate field;
- 16.175 Gcal\year of thermal energy in kind and 114.7 thous, rubles in cash were saved. Thermal energy supply of the field facilities operated by Vostokgazprom is secured due to on-site energy generation: wasteheat exchangers have been placed behind gas turbines of power stations and gas-compressor units, which makes it possible to save money on heating buildings. To ensure more efficient saving, the defects in enclosing structures were laminated, and thermostatic valves were installed;
- 0.208 mln. m^3/year of gas in kind and 2. 17 mln. rubles in cash were saved for on-site needs. In order to reduce gas consumption in the furnace rooms due to full use of generating unit thermal capacity, the efficiency of exhaust heat boilers was increased;
- 187.12 Hcal of thermal heat was saved for hot-water supply, as well as 2265 $м^3$ of cold water. The savings per year from installation of touchless faucets were 1.44 mln. rubles.

Totally, the company consumed energy resources of 43.05 mln. rubles. The total savings with the support of the energy saving program were 39.58 mln. rubles.

In general, the company spent 31.940 mln. rubles on developing and implementing energy saving program, with 29.582 mln. rubles being saved.

In addition to resource conservation policy, Vostokgazprom continuously modernizes the basic industrial facilities and invests money in new projects.

4. Conclusion

As it was mentioned, resource conservation is an integral part of successful development of an enterprise. And, it is of particular importance for petroleum engineering since petroleum sector is the most resource consuming one. The activities of petroleum companies are directly dependent on resources consumption.

The current study analyzes the activities of the petroleum company in the sphere of resource conservation. The resource conservation program resulted in saving resources with full cost of 43.05 mln. rubles, which, in its turn, positively affected the profit of the company. Program implementation costs amounted to 31.94 mln. rubles, i.e. 39%, 61% being of the company's profit. The continuous improvement of industrial process rests on developing resource conservation technologies.

References

[1] Calvin, K.,Rose, S.,Wise, M.,McJeon, H.,Clarke, L.,Edmonds, J. Global climate, energy, and economic implications of international energy offsets programs. Climatic Change. V. 133, Issue 4, 2015, p. 583-596.

[2] Berg, S.V. Energy efficiency in developing countries: Roles for sector regulators. Energy for Sustainable Development. V. 29, 2015, p. 72-79.

[3] Galli, A.M., Fisher, D.R. Hybrid arrangements as a form of ecological modernization: The case of the US energy efficiency conservation block grants Sustainability (Switzerland) V. 8,

IOP Conf. Series: Earth and Environmental Science **43** (2016) 012041 doi:10.1088/1755-1315/43/1/012041

Issue 1, 2016, p. 1-19
[4] Motie, S., Keynia, F., Ranjbar, M.R., Maleki, A. Generation expansion planning by considering energy-efficiency programs in a competitive environment. International Journal of Electrical Power and Energy Systems, V.80, 2016, p. 109-118.
[5] Farrokhifar, M. Optimal operation of energy storage devices with RESs to improve efficiency of distribution grids; Technical and economical assessment. International Journal of Electrical Power and Energy Systems. V.74, 2016, p 153-161
[6] Ventsyulis, L.S., Skorik, Yu.I. Energo- i resursosberezhenie — odno iz osnovnykh napravleniy povysheniya ekologicheskoy bezopasnosti toplivno-energeticheskogo kompleksa Rossii. Regional'naya ekologiya. 1-2 (31), 2011, p. 65-68.
[7] M. R. Tsibulnikova, O. S. Kupriyanova, A. B. Strelnikova. Economic assessment of environmental impact in the course of oil field development and production // IOP Conference Series: Earth and Environmental Science. — 2015. — Vol. 27: Problems of Geology and Subsurface Development. [Electronic resource] / http://dx.doi.org/10.1088/1755-1315/27/1/012076
[8] Vazim A., Tretyakov K., Khloptsov Dm., Lemeshko E. Energy efficiency, low-carbon energy and economic growth in the Arctic countries - the exporters of hydrocarbons // III International Scientific Conference 'Information Technologies in Science, Management, Social Sphere and Medicine' (ITSMSSM 2016), held in Tomsk, Russian Federation, May 2016. dx.doi.org/10.2991/itsmssm-16.2016.37
[9] Minigulov, R.M., Laptev, A.G.,Taraskin, M.M. (2009)Vnedrenie nauchno-tekhnicheskikh razrabotok pri dobyche i podgotovke prirodnogo gaza *Bulletin of Kazan State Power Engineering University* **3**, pp.8-13.
[10] Vostokgazprom. Environmental Protection. [Electronic Resource]. – http://vostokgazprom.gazprom.ru/ecology.
[11] Drozdovskiy, A.A. (2013) Engineering report №IK-67/13 on energy inspection of Tomskgazprom. 574 p.

Evaluating perspectives of economy "decarbonation" – environmental aspect

Matugina E.G.[1,4], Pogharnitskaya O. V.[1,5], Dmitrieva N.V.[2], Grinkevich L.S.[3], Selenchuk J.O.[1], Strelnikova A.B.[1]

[1] National Research Tomsk Polytechnic University, 30 Lenina Ave., Tomsk, 634050, Russia
[2] Novosibirsk State University, 2 Pirogova Str., 2, Novosibirsk, 630090, Russia
[3] National Research Tomsk State University, 36 Lenina Ave., Tomsk, 634050, Russia

E-mail: [4]emk512542@mail.ru, 5pov@tpu.ru

Abstract. The present paper deals with evaluation of economy "decarbonation" perspectives in post-Soviet states. The dynamics of this process over the period of 1990–2014 has been studied and the countries have been ranked respectively. Decoupling of GDP growth and CO2 emissions has been identified. The authors suggest a classification of tools to manage the process.

Introduction

Despite natural resource depletion, growing society's needs, and environmental challenges, current economic systems are based on intensive use of conventional energy sources. This necessitates a reexamination of prevailing management approaches and implementation of new management modes. The promising solution is "decarbonation" of national economies, which implies implementation and development of green economy policies by the end of the 21st century. By 2050, minimum 50% to 70% of electricity will be supplied from renewable energy sources [1]. The decreased technogenic load upon the environment makes decarbonation a promising trend in economic development, which implies intensive development of civil society based on alternative economic approaches.

1. Materials and Methods

The previous research has indicated that the dynamics of greenhouse gas emission depends on the following factors: 1) GDP growth; 2) changes in production patterns, opening up new industries; 3) dynamics of prices for energy sources; 4) implementation of enhanced technologies and innovations; 5) introduction of measures directly or indirectly reducing the amount of emissions.

The 90s of the 20th century turned out to be a hard time for the post-Soviet countries. Breaking of the ties existing within the former economic complex, development of independent national policies, and integration into the global market caused GDP decrease by 18–72%. Country ranking by GDP (PPP) per capita growth (absolute) over the period from 1990 to 2014 is shown in table 1.

Table 1. Dynamics in country ranking by GDP (PPP) per capita growth (absolute) over the period from 1990 to 2014 (dollars, sorted in ascending order)*

Country	1990		2014		GDP (PPP) growth per capita over 25 years	Country's rank by GDP (PPP) growth per capita
	GDP (PPP) per capita	Rank	GDP (PPP) per capita	Rank		
Kazakhstan	3073	5	24020	4	20947	1
Lithuania	6 660	3	27051	1	20391	2
Estonia	7 160	2	26999	2	19839	3
Russia	5709	4	24805	3	19096	4
Latvia	7 720	1	23707	5	15987	5
Azerbaijan	1669	10	17618	7	15949	6
Belarus	2328	8	18161	6	15833	7
Turkmenistan	2191	9	14165	8	11974	8
Armenia	1144	12	7374	11	6230	9
Ukraine	2641	6	8668	9	6027	10
Georgia	2500	7	7653	10	5153	11
Uzbekistan	547	15	5609	12	5062	12
Moldova	1611	11	4979	13	3368	13
Kyrgyzstan	699	14	3361	14	2662	14
Tajikistan	708	13	2688	15	1980	15

**Data provided by International Monetary Fund*

The countries which have improved their rating in terms of GDP based on PPP over 25 years are as follows: Kazakhstan (moved from 5th place in 1990 to 4th place in 2014), Lithuania (moved from 3d to 1st), Estonia (remained at 2nd place), Russia (moved from 4th to 3d), Azerbaijan (moved from 10th to 7th), Belarus (from 8th to 6th). Moldova, Kyrgyzstan, and Tajikistan are at the bottom of the list: Kyrgyzstan remained at the same place, while Tajikistan and Moldova had lower rating in 2014 compare to 1990.

Table 2 presents data on carbon dioxide emissions per capita in post-Soviet countries, as well as country ranking in terms of the emissions. The countries which have significantly decreased their rating over the 25-year period are western post-Soviet states, except for Belarus and Georgia. Russia improved its ranking in 2014 compared to that at the beginning of the period.

Table 2. Dynamics in country ranking by carbon dioxide emissions per capita (absolute) over the period from 1990 to 2014 (sorted in ascending order)

Country	1990		2014		Reduction of CO_2 emissions per capita over 25 years, tons	Country's rank by reduction of CO_2 emissions per capita
	CO_2 emissions per capita, tons	Rank	CO_2 emissions per capita, tons	Rank		
Ukraine	15.10	12	5.50	10	-9.60	1
Estonia	23.58	15	15.14	15	-8.44	2
Lithuania	9.45	9	4.11	8	-5.34	3

Moldova	7.32	6	2.24	5	-5.08	4
Georgia	6.32	5	1.51	4	-4.81	5
Armenia	5.85	3	1.49	3	-4.36	6
Kyrgyzstan	5.37	2	1.31	2	-4.06	7
Azerbaijan	7.38	7	3.33	6	-4.05	8
Latvia	7.64	8	3.69	7	-3.95	9
Russia	16.10	14	12.40	12	-3.70	10
Belarus	10.40	10	7.28	11	-3.12	11
Tajikistan	2.30	1	0.43	1	-1.87	12
Uzbekistan	5.90	4	4.20	9	-1.70	13
Kazakhstan	15.50	13	14.20	14	-1.30	14
Turkmenistan	12.44	11	13.30	13	0.86	15

The dynamics of carbon dioxide emissions per capita in post-Soviet states is illustrated in figure 1.

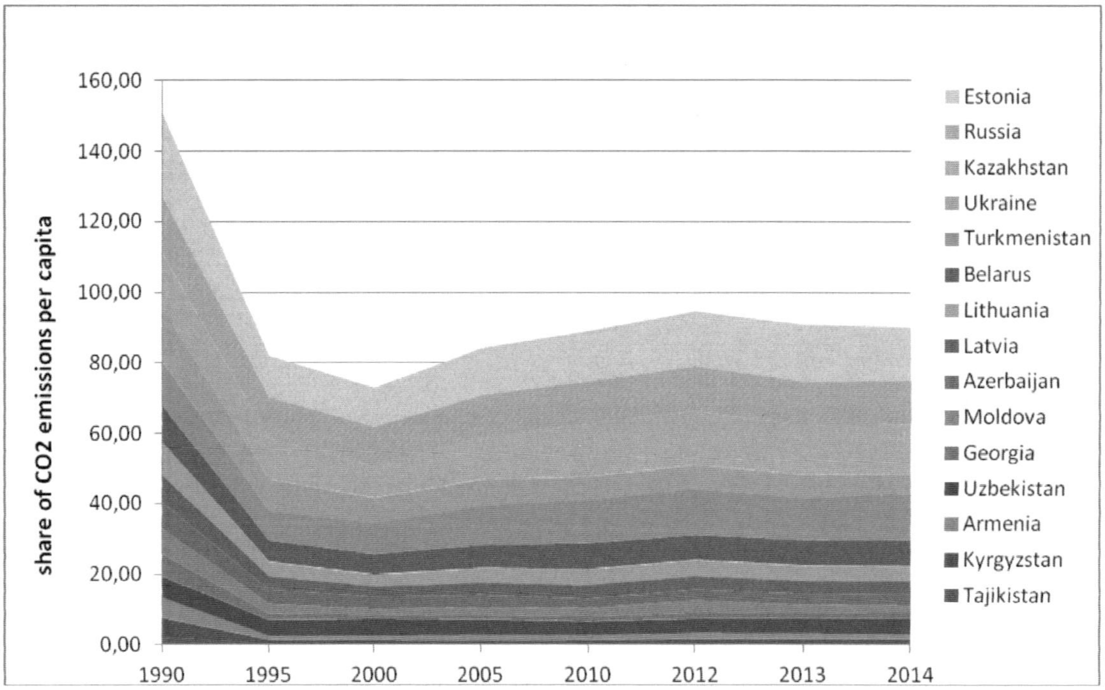

Fig. 1 Dynamics of CO_2 emissions, tons per capita

Over the period 1990–2014 carbon dioxide emissions reduced in all countries by 40 % on average (except for Turkmenistan, where the amount of emissions increased by 6%). Top five countries in terms of emission reduction are Armenia, Georgia, Kyrgyzstan, Moldova, and Tajikistan (75% on average).

The same trend can be observed when the countries are ranked in terms of CO2 emissions per 1,000 dollars of GDP (see figure 2).

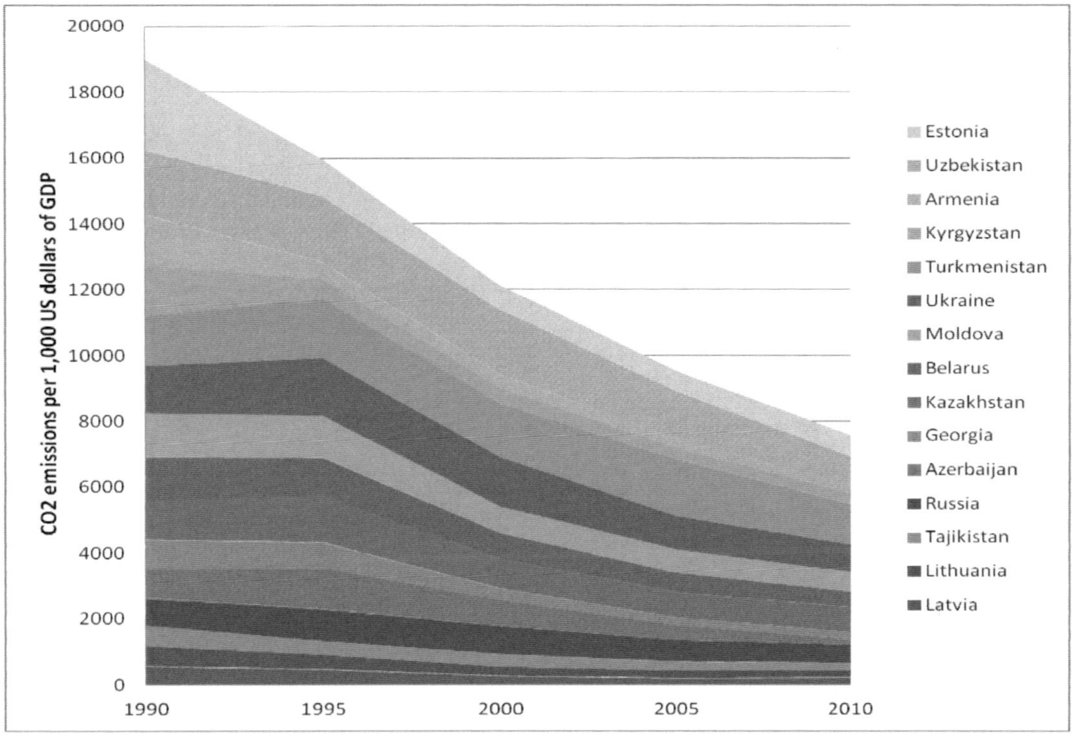

Fig. 2 Dynamics of CO2 emissions, tons per 1,000 US dollars

Table 3. Dynamics in country ranking by carbon dioxide emissions (absolute) (kg of CO2 per 1,000 dollars of GDP) over the period from 1990 to 2014 (sorted in ascending order)

Country	1990		2014		Reduction of CO_2 emissions per 1,000 dollars of GDP, kg over 25 years	Country's rank by reduction of CO_2 emissions per 1,000 dollars of GDP
	CO_2 emissions per 1,000 dollars of GDP, kg	Rank	CO_2 emissions per 1,000 dollars of GDP, kg	Rank		
Estonia	2743.23	15	572.14	11	-2171.09	1
Armenia	1566.63	13	192.34	4	-1374.29	2
Uzbekistan	1952.22	14	749.82	14	-1202.4	3
Kyrgyzstan	1547.04	12	399.8	7	-1147.24	4
Moldova	1346.22	9	461.43	9	-884.79	5
Belarus	1294.86	8	413.02	8	-881.84	6
Ukraine	1435.03	10	690.71	13	-744.32	7
Georgia	896.16	6	200.72	6	-695.44	8
Azerbaijan	874.27	5	197.53	5	-676.74	9
Kazakhstan	1204.55	7	591.75	12	-612.8	10
Turkmenistan	1489.68	11	900.47	15	-589.21	11
Tajikistan	632.05	3	171.26	3	-460.79	12
Lithuania	605.31	2	165.78	1	-439.53	13

Latvia	573.85	1	167.43	2	-406.42	14
Russia	829.09	4	519.16	10	-309.93	15

Over the 25-year period, carbon dioxide emissions reduced in all post-Soviet states. The top countries are Baltic states (75% reduction) and also Armenia and Kyrgyzstan. It is noteworthy that some of these countries occupy top positions in ranking by ecological effectiveness as well: Estonia, Latvia, and Lithuania are ranked 1st, 2nd, and 3d, respectively (Russia is ranked 5th).

One of the ways to decrease the amount of emissions is reducing dependence on conventional energy sources. The share of renewable energy is significant in Georgia (28.3%), Kyrgyzstan (39.4%), and Tajikistan (57.5%), while in Russia it makes up only 9%, and even less in Kazakhstan (1%).

Energy consumption within different economic sectors in Russia is shown in table 4 [2].

Table 4. Fossil fuel energy consumption within different economic sectors in Russia

Industrial sector	Unit of measurement	2012	2013	2014
Use of motor vehicles	reference fuel, kg per unit	2 127.2	2 065.7	2 022.6
Fertilizer production	reference fuel, kg per ton	494.3	487.0	470.1
Pulp and paper production	reference fuel, kg per ton	1 109.8	1 068.8	1 062.2
Ferrous metallurgy	reference fuel, kg per ton	650.6	657.4	647.8
Petrochemistry	reference fuel, kg per ton	1 027.7	1 000.1	956.1

According to the data, there is a decrease in fossil fuel energy consumption over the time period considered, moreover, it is possible to identify the economic sectors with enhanced manufacturing, which proves the potential for energy saving performance (see table 5).

Table 5. Potential for energy saving performance within different economic sectors in Russia

Industrial sector	**Unit of measurement**	**2015**	**2016**	**2017**	**2018**	**2019**	**2020**
Use of motor vehicles	reference fuel, kg per unit	1 986.3	1 948.2	1 893.8	1 849.7	1 811.0	1 771.7
Fertilizer production	reference fuel, kg per ton	457.9	441.7	427.1	408.1	393.5	381.0
Pulp and paper production	reference fuel, kg per ton	998.6	957.5	946.8	924.5	901.9	875.6
Ferrous metallurgy	reference fuel, kg per ton	650.2	640.2	624.9	601.0	579.5	558.1
Petrochemistry	reference fuel, kg per ton	953.9	951.8	948.1	898.7	873.9	857.6

According to the World Bank and the Center for Energy efficiency forecasts, Russia is capable to reduce energy consumption by 45% if the measures for energy efficiency are implemented. By 2020, the annual saving may be up to 250–270 million tons of reference fuel [3].

Conclusion

It is possible to conclude that economy "decarbonation" is attributed to current economic conditions. The trend is essential not only for alternative production, but also for present-day manufacturing based on conventional energy sources and proved to have potential for energy saving performance. The tools of economy "decarbonation" may be grouped as follows:

1. economic tools, which change the structure of the national economy, modify its technological characteristics, and imply the following processes [4–13]:

- enhancement of current manufacturing practices and implementation of new ones that allows reducing carbon intensity;
- development of alternative energy sources;
- forest conservation and management to improve carbon sequestering;

2. institutional tools, which imply revision of current rules and regulations (development of national strategy of carbon intensity reduction, modification of national energy strategy, etc.) and implementation of low carbon fuel standards for national enterprises; this can only be done by accepting the problem, on the one hand, and encouraging economic entities, on the other hand;
3. management tools, which introduce new approaches to business management and enhance the systems of economic interaction, personnel training, and forming personnel reserve, allow implementing innovations at all production stages, from "making" product to placing it at the market, etc.

References

[1] Tarlavsky V 2016 Economy Decarbonation (in Russian) [Electronic resource] URL: http://www.eg-online.ru/article/287725, free, reference date 06.02.2016
[2] National report on energy saving and improvement of energy efficiency in the RF in 2014 2015 Ministry of Energy of the Russian Federation (Moscow) 24
[3] World Bank & International Finance Corporation, Energy efficiency in Russia: Untapped Reserves 2008 [Electronic resource] URL: http://www.ifc.org/ifcext/rsefp.nsf/AttachmentsByTitle/FINAL_EE_report_Engl.pdf, free, reference date 01.02.2016
[4] Safonov G V 2015 Low-carbon development: global challenges and Russia's perspectives *Proceedings of the International Conference "Towards Green Economy and Sustainable Development of Altai krai: perspectives, mechanisms, key destinations"* (Barnaul) 34–44
[5] Kokorin A O, Fyodorov A V, Senova O N and Chuprov V A 2012 Measures to Reduce Greenhouse Gas Emissions and Non-government Organizations' Priorities in the RF (Moscow, WWF in Russia)
[6] State Programme of the RF "Energy Efficiency and Energy Sector Development" [Electronic resource] URL: http://minenergo.gov.ru, free, reference date 06.02.2016
[7] Thomson A M, Calvin K V, Smith S J, Kyle G P, Volke A, Patel P, Delgado Arias S, Bond-Lamberty B, Wise M A, Clarke L E and Edmonds J A 2011 A pathway for stabilization of radiative forcing by 2100 *Climatic Change* **109** 77–94 (DOI 10.1007/s10584-011-0151-4)
[8] Kriegler E, Mouratiadou I, Luderer G, Bauer N, Calvin K, DeCian E, Brecha R, Chen W, Cherp A, Edmonds J, Jiang K, Pachauri S, Sferra F, Tavoni M and Edenhofer O 2013 Roadmaps towards sustainable energy futures and climate protection: A synthesis of results from the RoSE project. [Electronic resource] URL: http://www.rose-project.org/Content/Public/RoSE_REPORT_310513_ES.pdf, free, reference date 01.02.2016
[9] The emissions gap report 2013 United Nations Environment Programme (UNEP, Nairobi)
[10] Anandarajah G and Dessens O 2013 Modelling of global energy scenarios under CO2 emissions pathways with TIAM-UCL, report for the CCC
[11] Makasheva Yu S, Makasheva N P, Remnyakov V V, Burykhin B S and Shenderova I V 2015 Petroleum staff reluctance and adjustment to innovative changes *IOP Conf. Series: Earth and Environmental Science* **27** 012069
[12] Romanyuk V B, Baitova Y S, Safronova E V, Pozdeeva G P 2015 Regulatory framework of pricing and estimate standards in the petroleum sector *IOP Conf. Series: Earth and Environmental Science* **27** 012073
[13] Sharf I, Borzenkova D, Grinkevich L 2015 Tax incentives as the tool for stimulating hard to recover oil reserves development. *IOP Conf. Series: Earth and Environmental Science* **27** 012079

Ecological policy in oil-gas complexes, HSE MS implementation in oil and gas company

Kochetkova O.P.[1], Glyzina T.S.[2], Vazim A.A.[3], Tugutova S.S[4]

[1] Institute of Natural Resources, National Research Tomsk Polytechnic University, 30 Lenin Av., Tomsk, 634050, Russia
kochetkova@tpu.ru
[2] Institute of Natural Resources, National Research Tomsk Polytechnic University, 30 Lenin Av., Tomsk, 634050, Russia
gts@tpu.ru
[3] Institute of Natural Resources, National Research Tomsk Polytechnic University, 30 Lenin Av., Tomsk, 634050, Russia
vazim@tpu.ru
[4] Institute of Natural Resources, National Research Tomsk Polytechnic University, 30 Lenin Av., Tomsk, 634050, Russia
Sarana@mail.ru

Abstract. The paper considers the following issues: HSE MS international standard implementation in oil and gas industry, taking into account international practices; implementation of standards in oil and gas companies; policy in the field of environmental protection and occupational health and safety; achievement of planned indicators and targets in environmental protection and occupational health and safety.

Key words: ecological policy, audit, certification, GOST, occupational health and safety

1. Introduction

The research of economic mechanisms providing environmental safety in oil and gas industry is a topical scientific problem, which allows making scientifically reasoned decisions. These decisions contribute into development of environmentally determined economy, which is characterized by a minimal negative impact on environment and low resource intensity.

The object of the research is a complex of methods and economic management instruments for HSE MS international standard implementation in oil and gas industry, taking into account international practices.

Currently, international certification of quality systems presents a logic classification of standards. It reflects various requirements for manufacturers and suppliers, which are focused on the interests of an ultimate customer. This standard matrix is defined by the abbreviation «ISO» - International Organization for Standardization. ISO has its efficient analogue in Russia; it is the system of standards «GOST R ISO», which appears to be a unique instrument regulating market relations in the conditions of positive competition.

The presence of certified system of environmental management system at the enterprise has become an indispensable condition for business operations on world-wide basis. This certificate provides advantages at conclusion of transactions, participating in competitions. The company becomes more competitive with this certificate, and it cultivates company's image.

2. HSE MS implementation in oil and gas companies

HSE MS implementation in oil and gas companies is necessary for a number of reasons.

Firstly, it helps to achieve and demonstrate more efficient occupational health and safety and environmental performance.

Secondly, the presence of ISO 14001 and OHSAS 18001certificates gives to oil-gas complexes indisputable benefits.

International standards require determining the ecological policy, which is suitable for the organization.

This policy reflects the responsibilities of the top management in keeping legal and other applied requirements, pollution prevention and continual improvement.

Ecological aspects, dangers and company's risks are identified and analyzed.

Ecological aspects are the activities of the organization, its products or services, which may interact with environment.

Danger is a source, situation or action, which are able to damage a human or cause his health deterioration.

Environmental changes, negative or positive, which follow either in part or in whole from ecological aspects are defined as environmental impact. Correlation between ecological aspects and impact means that the aspect is the cause of impact.

As far as the organization has many ecological aspects and dangers, HSE specialists conduct risk assessment in order to assess efficiency of management measures in respect to the particular risk.

There are several types of major risks in oil-gas complexes: the loss of resources, personal injury, direct fire and thermal influence [9].

Being identified, legal and other requirements are divided into 2 groups. In this case, each group is divided into sections.

The first group (legal and other requirements in environment protection) contains the following sections: atmosphere air protection; water-resources conservation, soil conservation, forest conservation, conservation of mineral resources, environmental audit, damage prevention and control of environmental accidents, other impacts.

The second group (legal and other requirements in occupational health and safety) contains the following sections: documents regulating the responsibilities of the employer in OHS-system implementation and providing safe working conditions at the enterprise; documents regulating the responsibilities of the employer in control of OHS-system operation and providing safe working conditions at the enterprise; documents regulating the responsibilities of the employer in taking measures in OHS at impact of harmful and dangerous production factors on personnel health; documents regulating the responsibilities of the employee in fulfilling OHS requirements.

In oil- gas complexes operation and activity management is performed, which is connected with identified dangers and ecological aspects. It is necessary to take measures on de-risking and risk management in the field of environmental protection and occupational health safety. So, the activities connected with revealed significant ecological aspects and dangers and leading to deviations from ecological policy, ecological aims and objectives, must be performed with the methods that allow monitoring or decreasing their harmful impact.

Indexes of monitoring are connected with significant ecological aspects, dangers, planned indicators and targets.

MS noncompliance involves any failure in requirements of:

- International standard ISO 14001:2004;
- International standard OHSAS 18011:2007;
- Business procedures of MS oil and gas play;
- Regulatory legal acts and others in HSE.

Noncompliance is revealed as a result of monitoring and measurements, manufacturing inspection in the field of HSE, internal audit, external inspections, assessment of legal and other requirements correspondence in HSE, analysis of interested party references, analysis performed by top management.

The procedure of noncompliance management includes the following activities:

- Noncompliance recording;
- Report about revealed noncompliance to authorities;
- Correction implementation (prompt elimination/moderation of noncompliance sequences);
- In-depth analysis of noncompliance causes;
- Development of corrective/preventive actions for eliminating noncompliance causes and preventing their repeat, assessment of potential material and financial costs associated with their implementation;
- Fulfillment of corrective/preventive actions;
- Performance control and analysis of efficiency for corrective/preventive actions taken.

The results of commission activities in determining emergency causes are recorded officially. The commission, having investigated emergency causes, arranges distribution of materials obtained within 3 days after the end of investigation [5].

The results of commission activities in determining technological incident causes are recorded officially according to the sample specified by Association. The investigation report is signed by all the members of the commission [5].

3. Audit

Audit is the process of revealing HSE noncompliance to the requirements of international standards. Also, it is a systematic, independent and documented process of collecting and unbiased assessment of audit evidence in order to determine the level of criteria's fulfillment established in the organization [6].

In oil-gas complexes internal audit of MS is performed in accordance with the requirements of «Internal audit procedure of HSE MS in oil-gas complexes». The procedure determines the requirements and sequence of actions in internal MS audit performance in oil-gas complexes, and also it establishes forms for documents and records in internal audit [8].

Audit is performed in order to:

- Assess MS compliance to the requirements of international standards ISO 14001:2004, OHSAS 18001:2007 as well as to the requirements of MS documents in oil-gas complexes and its interrelated documents [6], [7].
- Assess MS efficiency with regard to Policy implementation in the field of environmental protection and occupational health and safety, and achievement of planned indicators and targets in environmental protection and occupational health and safety.
- Inspect efficiency of corrective /preventive actions aimed at MS improvement.

Internal audit of the first level is the internal audit in the Central Administrative Office and oil-gas complex, its arrangement and performance is carried out by auditors of Central Administrative Office.

Internal audit of the second level is the internal audit in oil-gas complex, its arrangement and performance is carried out by the department of environmental protection and occupational health and safety.

Table 1. MS internal audit check-list

MS internal audit check-list **Standard/MS Document** OHSAS 18001:2007 and ISO 14001:2004		
Department	***Audit Questions***	***Audit Findings***
Production properties of oil and gas, Oil & Gas Production Department № Console operator in oil and gas production	1. What for HSE MS is needed for the company? 2. When has HSE MS implementation been started in LLC «TN»? 3. What actions have been taken for it? 4. Does LLC «TN» have a certificate of international standard compliance? What? 5. How staff participates in EMS implementation and its support in Association?	To achieve and demonstrate efficient OHS and environmental performance. It is achieved by control over activities, products and services and their impact on environment according to its ecological policy and targets. HSE MS implementation. Ecological policy establishment. Assignment of executives in HSE MS Working team in HSE MS is formed. Regulation on HSE MS and company standards are adopted. Ecological aspects, dangers and risks are indentified. The analysis is conducted and significant non-conformance is revealed. The documents are prepared: Environmental Aspect Register. Consolidated Register of significant ecological aspects of Association; Register of Association dangers; Consolidated Register of dangers with prohibited or significant Association risks. Register of legal and other requirements in the field of HSE MS is formed. Staff training is conducted periodically in issues of OHS, E and IS as well as instructing, internal audits and assessment

		of HSE MS conformance Yes, it has. ISO 14001 and OHSAS 18001 In accordance with duty regulations

4. Conclusion

In order to provide economical conversion of the country from a resource-based type to the efficient and highly technological production, it is necessary to solve two problems. The first problem is formation of permitting and monitoring environmental infrastructure (licensing and certification of environmental activities, independent environmental assessment, ecological risk insurance, environmental audit, complex statistical monitoring of environment pollution). The second one assumes changes in environment legislation aimed at forming economic incentives for the use of modern international standards.

References

[1] Groshilin S M 2012 *A textbook. M.: GBOU VPO Pervyj MGMU im. I.M. Sechenova.* Methodological and legal principles of vital activity safety. pp. 195
[2] Accredited certification authority SERCONS [web-site] URL: http://www.serconsrus.ru (21st April, 2016)
[3] International Organization for Standards [web-site] URL: http://www.worldico.org (21st April, 2016)
[4] MES of Russia Order d.d. February, 28, 2003, № 105 "About establishing requirements for preventing emergencies at potentially hazardous facilities and vital infrastructure". In Rus.
[5] Federal act d.d. July, 27, 2010, N 225-FZ (ed. d.d. November, 4, 2014) "About compulsory liability insurance of an owner of a hazardous facility against damage resulting from an accident at a hazardous facility". In Rus.
[6] ISO 14001:2004 EMS. Requirements and application guide. In Rus.
[7] OHSAS 18001:2007 MS of occupational health and safety. Requirements.
[8] Vazim A A, Romanyuk V B, Akhmadeev K N, Matveenko I A 2015 *IOP Conf. Ser.: Earth and Environ.* Associated petroleum gas utilization in Tomsk Oblast: energy efficiency and tax advantages. Vol. 27 pp. 1-5. http://dx.doi.org/10.1088/1755-1315/27/1/012078

IOP Conf. Series: Earth and Environmental Science **43** (2016) 012044 doi:10.1088/1755-1315/43/1/012044

Stress-strain state of pipeline depending on complicated environment

P V Burkov [1,2], M A Filimonenko [2], S P Burkova [2]

[1] National Research Tomsk Polytechnic University, 30, Lenin Ave., Tomsk, Russia
[2] Tomsk State University of Architecture and Building, 2, Solyanaya Sq., Tomsk, Russia
E-mail: [1]burkovpv@mail.ru

Abstract. The paper presents the 3D model and FEM analysis of the stress-strain state of the soil–pipe interaction system. The analysis shows that the geological environment has shown a strong effect on the pipelines. This stress-strain analysis is carried out using the ANSYS finite element program.

1. Introduction

The stress-strain state of the soil–pipe interaction system is influenced by a number of factors, such as humidity, chemical composition and geological structure of soils, water saturation, thermo physical soil properties, air temperature, precipitations, snow accumulation and snow melting, waterlogging, land configuration, etc. The operational characteristics also have an effect on the stress-strain state of the pipelines [1]. Pipelines ballasting is a method of pipelines securing by means of placing weights or concrete when laying pipes in swampy or wet lands. Weights (saddle shaped, articulate, with flexible elements and others) are laid on pipes when constructing them by a pipe layer, a swamp excavator, an amphibious crane or a helicopter. For concrete ballasting, a continuous uniform pipe coating is provided (external concreting) often on a stationary base or concreting the space between the pipe and external casing (pipe-in-pipe system).

The majority of pipelines in Siberia and even in Russia are located mostly on terrains with sharply pronounced seasonal fluctuations. This condition evokes negative processes that, finally, result in the buckling of pipelines.

The finite element method (FEM) is one of the main methods used to evaluate the stress-strain and other states of the pipeline transportation systems. The ANSYS finite element program is the most suitable for this evaluation and has such advantages as a wide spectrum of soluble problems; the high accuracy and degree of approximation to real situations; convenient interface; many types of finite elements; material models; and the embedded algorithmical programming language that facilitates the automation of certain simulation procedures [2–11].

This work is focused on the investigations of the stress-strain state of the pipeline under the complicated conditions of geological environment and internal stress. The ANSYS finite element program and Autodesk Inventor software are used to determine the conditions for the pipelines ballasting or the use of float-type systems intended for pipelines laid in bogs.

2. Results and discussion

The stress-strain state of the soil–pipe interaction system is affected by a variety of forces. The effect from the gravitational force, the internal working pressure, and the forces of seasonal and perennial soil bulging affecting the underground pipeline [1] is considered below.

Figure 1 presents the schematic effect from the soil–pipe interaction system represented by the force vector distributed on the lower generator line of the pipeline system subjected to the gravitational forces. The analysis of the stress-strain state of the soil–pipe interaction system is carried out for the underground pipelines of West Siberia having the following characteristics: 426 mm pipe diameter; 8 mm wall thickness; the type 09G2S[1] low-alloy steel; 206 MPa Young's modulus; 0.3 Poisson number. The following loading conditions are considered: 5 MPa internal pressure; 0.8 MPa stress due to geological forces; 0.4 MPa stress due to gravitational forces.

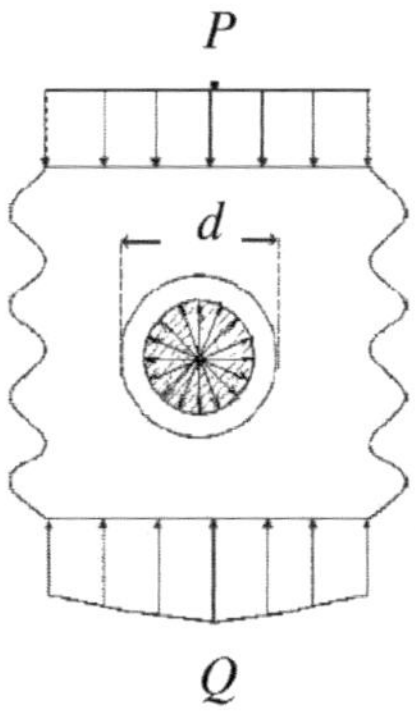

Figure 1. Sectional view model of the soil–pipe interaction system:
P – gravitational forces; *d* – pipe diameter; *Q* – force vector.

Figure 2a,b,c illustrates the finite element models of the stress-strain state of the pipeline subjected to different loads.

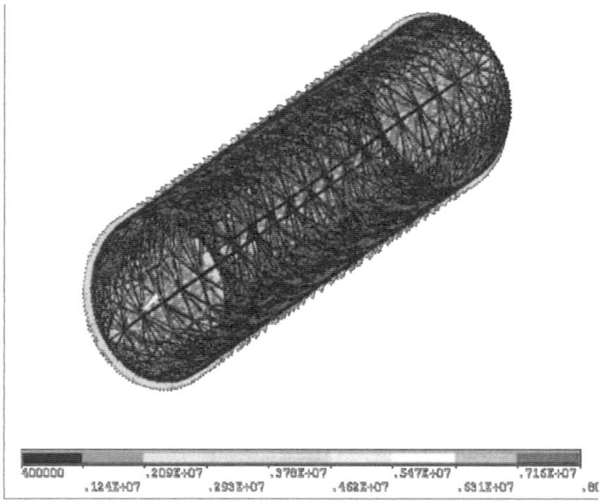

Figure 2 a. FEM of pipeline initial state under gravitational and geological environment forces

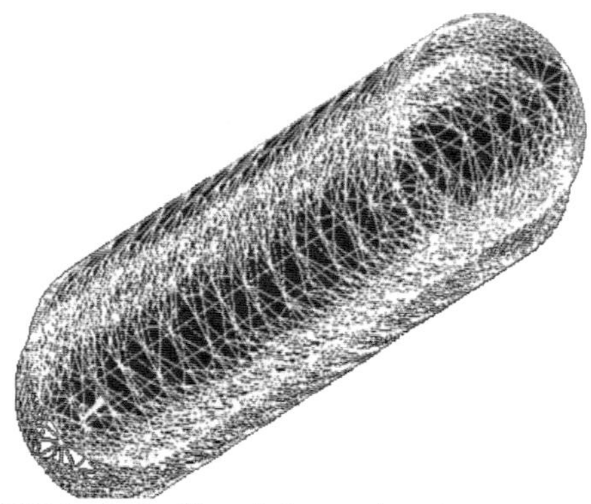

Figure 2 b. FEM of pipeline deformation due to gravitational forces

[1] Steel composition: 0.09% carbon; 2% manganese; 1% silicon.

IOP Conf. Series: Earth and Environmental Science **43** (2016) 012044 doi:10.1088/1755-1315/43/1/012044

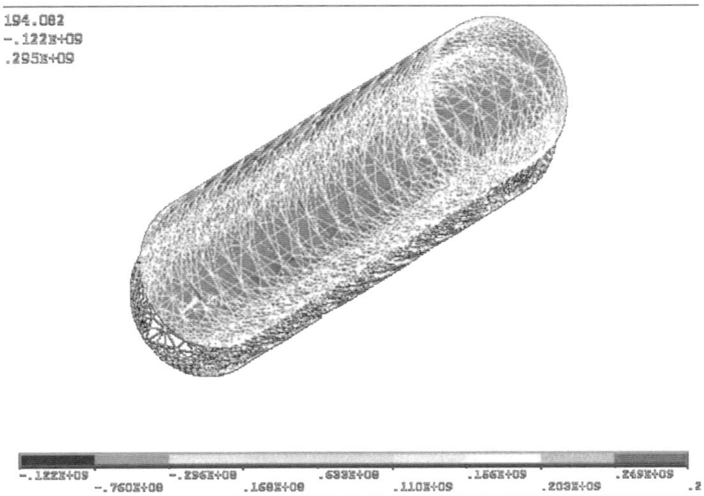

Figure 2 c. FEM of pipeline stress state due to gravitational forces

The graphical representation of the pipeline FE models is shown in figure 3.

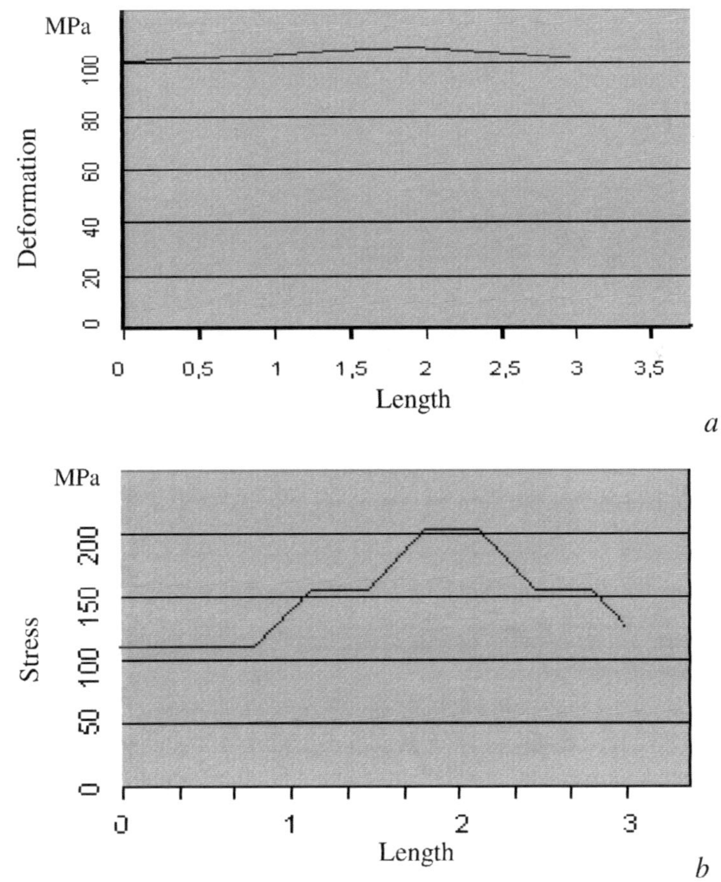

Figure 3. Stress-strain state of the pipeline: *a* – deformation; *b* – stress.

The analysis of research results shows that the pipeline displacements due to the geological environment forces can be rather essential even during a short-term operation. The practical experience shows that during 4.5–6 years of operation, the pipeline deformation due to the geological environment comes to one meter and more provided that the range of the pipeline laying depth is 0.8–1 m at different route sections. Under these conditions, the main pipelines require rather expensive diagnostics and repairs. Otherwise, further operation of the main pipelines will lead to the accidental events. The FEM analysis allows detecting the points of the pipeline strains induced by the geological environment forces.

The stress-strain state of any bearing element of the pipeline linear section is defined by the loads applied. The underground pipelines laid in bogs are affected by the dead load of the environmental buoyancy forces and the internal pressure load of the delivered product similar to the dead load of the pipeline section in case of its rigid embedding in soil. This pipeline section is laid in terrains with the most cold five-day period raging from –46 to –49 °C. Thus, the northern-type pipe is considered herein, the original data of which are given in table below.

Table. Original data of the northern-type pipe

External diameter, m	0.325
Wall thickness, m	0.008
Pipeline section length, m	48
Density of water, kg/m^3	1000
Density of oil, kg/m3	850
Stress limit of steel, N/mm^2	502
Yield stress of steel, N/mm^2	353
Weight of 1 m pipe, kg/m	62.14
Buoyancy forces of water, N	39004
Dead load of pipeline, N	29231
Working pressure, MPa	7.4

Three-dimensional modeling is used to analyze the stress-strain state of the pipeline laid in the complicated conditions of bogs of the type III. The 3D model of the stress-strain state of the pipeline is presented in figure 4.

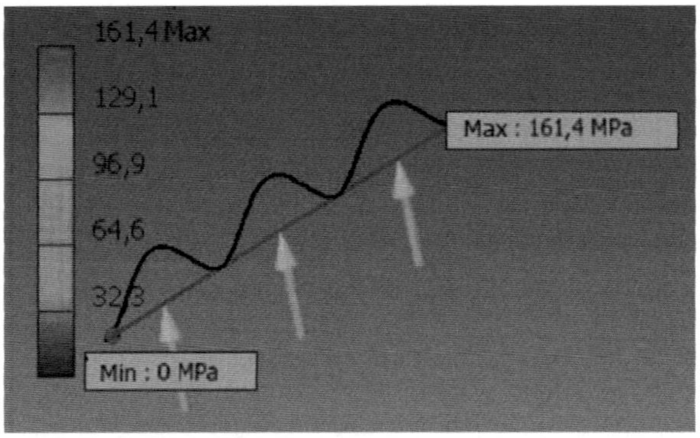

a

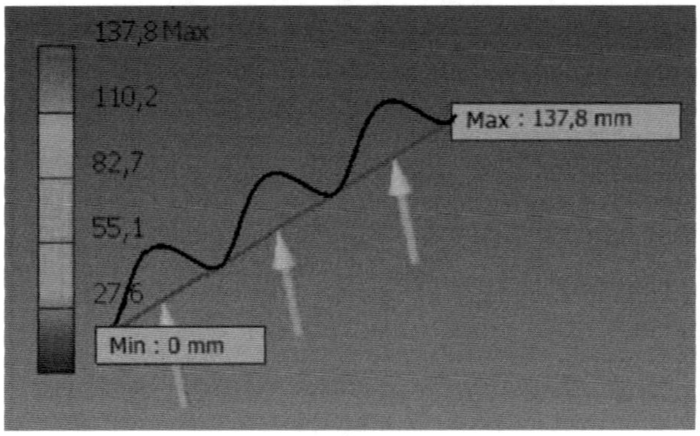

b

Figure 4. 3D models of stress-strain state: *a* – von Mises stress; *b* – strain.

Information obtained after the stress-strain state analysis of the linear section of the main pipeline allows detecting the pre-emergency situation areas including pipelines still having no defects and, taking measures for their elimination, increase the system reliability.

The graphical representation of the obtained stress-strain state are presented in figure 5.

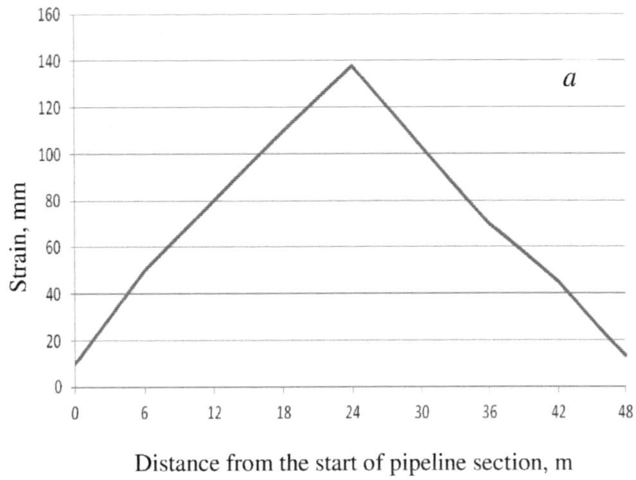

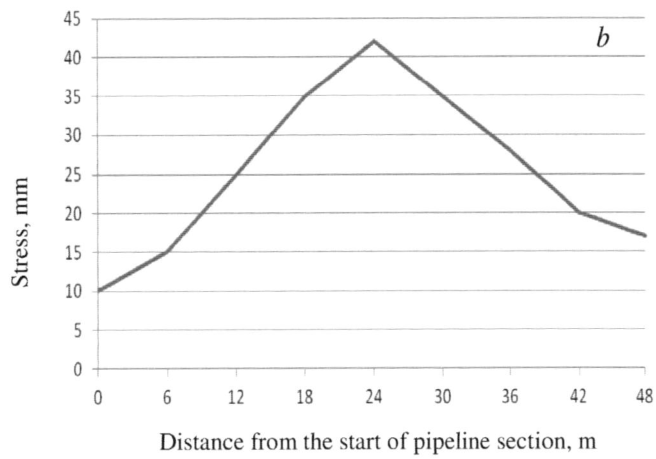

Figure 5. Stress-strain state of the pipeline: *a* – strain distribution; *b* – stress distribution.

The obtained results show that the maximum deformations are observed in the center of the pipeline section in which the transversal deviation from the normal is 137.8 mm, while the maximum stress is 42.6 MPa and observed inside the pipe.

3. Conclusions

The computation results showed that the equivalent stresses of the pipeline were 161.4 MPa. Therefore, in the described conditions, the pipeline embedded in soil from the both sides of the crossing, requires no balancing, since its deformations were insignificant.

- Stresses occurred at the pipeline buckling achieved values close to the yield stress of steel, thereby resulting in degradation of its reliability.
- The stress value varied throughout the pipe length. The pipeline sections adjacent to the solid ground were characterized by a higher stress level.
- The modified conditions of the soil–pipe interaction system complicated the deformation processes in the pipeline. Therefore, more detailed investigations of the stress-strain state are required with the account for mechanical-and-physical properties of soils.

IOP Conf. Series: Earth and Environmental Science **43** (2016) 012044 doi:10.1088/1755-1315/43/1/012044

4. References

[1] Shevchuk V V, Ivanik E M, Lavrenyuk V I and Lavrenyuk N V 2008 *Geophysical Journal.* 1 pp. 74-80.

[2] Burkov P V, Kalmykova K G, Burkova S P and Do T T 2014 *Proc. IOP Conf.* (*Bristol: IOP Publishing Ltd, Series: Earth and Environmental Science* vol **21** pp 5-7.

[3] Antropova N A, Krets V G, Luk'Yanov V G, Baranova AV 2015 *IOP Conf. (Bristol: IOP Publishing Ltd, Series: Earth and Environmental Science.* Reliability assessment of tunneling flow charts. Vol **24** pp. 012019.

[4] Butov V G, Nikulchikov V K, Nikulchikov A V, Yashchuk A A s 2016 *Albena, Bulgaria: SGEM Conference Proceeding.* Study of the stress-strain behavior and strength characteristics of sealer devices for oil-trunk pipelines 16th, International Multidisciplinary Scientific GeoConferences SGEM. (28June -7 Jule, ISSN 1314-2704) .

[5] Chigarev M A 2003 *ANSYS for engineers Moscow: Mashinostroenie-1.* pp. 320 (in Russian)

[6] Shammazov A M, Mugallimov F M, Nefedova N F 2000 *Nedra-Business-Center.* Main Oil-Pipelines Underwater Crossings. pp. 237 ISBN 5-8365-0049-5.

[7] Basov K A 2002. *Computer Press.* ANSYS in Examples and Problems. pp. 224.

[8] Chigarev A V, Kravchuk A S, Smalyuk A F 2004 *Mashinostroyeniye 1.* Ansys for Engineers: Reference Book. pp. 512.

[9] Methodological Instructions on Examination and identification of Steel Tubes for Oil and Gas Pipelines. Industrial Guideline 39-1.11-014-2000.

[10] Zakharkin F I, Fomin V A 2008 *Gas industry.* The definition of stress deformed state of the main pipe line sections, which are in non-design state. Vol. **12** pp. 40-43.

[11] Dimov D A 2008 *Gas industry.* Ground deformation capacity and the calculation of underground main gas pipe lines. Vol. **2** pp. 82-85.

Chemical composition of the small mammal reproductive system as an indicator of enterprise technogenic impact on the environment

N Baranovskaya [1], A Belyanovskaya[2], V. Bezel[3], S. Mukhacheva[4], M Anufrieva[5]

[1, 2,5] Tomsk Polytechnic University, Russia, 634050, Tomsk, Lenin Avenue, 30.
[3,4]Institute of Plant and Animal Ecology UB RAS, Russia, 620144, Yekaterinburg, 8 Marta Street 102
E-mail: [1]natalya.baranovs@mail.ru, [2]toobiovet@mail.ru, [3]bezel@ipae.uran.ru, [4]msv@ipae.uran.ru, [5]anufrieva@tpu.ru

Abstract. In this paper we consider the indicative role of chemical composition of the small mammal (specifically, the bank vole, or Myodes glareolus) reproductive system with the purpose of studying the impact of a large-scale nonferrous metal-processing enterprise on living organisms through the expample of Middle Ural copper-smelting plant OJSC. We have analysed the chemical composition of the placenta-embryo system in the areas which are 2 km and 30 km away from the plant.

1. Introduction

Under the conditions of urban environment technogenic impact the living organism is constantly exposed to multiple effects of diverse factors. Polluting the adjacent areas with waste products, nonferrous metal-processing enterprises cause environmental changes and acceleration of natural landscape degradation [8]. Contamination by heavy metals and other chemicals is one of the negative impacts on the environment and living organism in particular. From this perspective, modern industrial cities tend to be extreme habitats [1, 7, 10]. The barrier systems of the living organism are intended for its defense from negative environmental impact. Each of the organ systems functions as a barrier in its own way. In this paper we consider functioning of the placental barrier included in the reproductive system. The placental barrier plays an essential part in the formation of the embryo as a future organism. Chemical analysis of organs and tissues, the function of which is to reduce migration ability of chemicals by accumulating them, enables us to draw conclusions about the state of the environment and its technogenic transformation.

2. Research materials and methods

2.1 Research subject

The research subject is the biological material (placenta-embryo) of small mammals, namely, the bank vole (Myodes glareolus), which represent rodents of Myodes genus. The animals are characterized by a relatively short lifecycle, high level of metabolism, and high sensitivity to ambient conditions [9]. Small mammals are a traditional and very informative subject of bioindicative research which gives the possibility to effectively study biosystem reaction to pollutant impact at the level of populations and cenosis [6]. The reproductive system is a marker and biological indicator of the environment state

[5], as adverse environmental factors have a significant impact on the state of the mother-placenta-foetus system, in which the placental tissue holds a central position. On the one hand, the placenta implements relations between the mother and the foetus, on the other hand, it plays an exceptional part in protection of the foetus from negative impacts [3].

2.2 Sampling and research methods

The research is based on the materials collected by staff members of the Institute of plant and animal ecology UB RAS with the article author's participation when they were studying small mammals in the operating area of the large-scale copper-smelting plant (the town of Revda, Middle Ural) from 2005 to 2013. Based on the content of heavy metals in the natural deposit environment (soil, leaf-litter, and snow cover) we distinguished background, or reference (conditionally 'clean', 30 km off the plant) and impact (heavily and moderately polluted, 2 km) areas. The animals were caught during a snowless period in both areas simultaneously. For the purpose of the analysis we only used embryos and placentas at late stages of gestation (18-20 days). For sample assaying we used the method of instrumental neutron activation analysis, the analyst was A.F. Sudyko, senior researcher. Compared to the others, this method has a number of advantages, including capability to detect chemicals content in a wide range (from n.1 to n.10–6%). In addition to this, chemical preparation of samples is not used, which excludes errors caused by adding or removing elements together with reagents [11].

3. Geoecological characteristics of the object

Middle Ural copper-smelting plant OJSC is located in Sverdlovsk Oblast, the town of Revda near Pervouralsk. Revda is situated 47 km west of Yekaterinburg, in that part of Middle Ural which geographically demarcates Europe from Asia. The town is located on the banks of a pond and the river Revda in the place where it flows into the river Chusovaya. Middle Ural copper-smelting plant OJSC is the largest plant, which smelts copper from fresh raw materials, in Ural, it produces up to 100 000 tonnes of blister copper a year. The plant uses zinc-copper ore supplied by the Gaiskoye copper-pyrite deposit.

Main sources of hazardous substances emissions from engineering buildings in nickel industry are burning and smelting furnaces and converters. Unorganized sources of atmosphere pollution, such as unsheltered warehouses, slag disposal areas and end products shipping, are also present.

Main pollutants are inorganic dust, sulphur dioxide, copper, cupric oxide, lead and its compounds, arsenic, zinc, zinc oxide, cadmium, and iron [2].

4. Results and discussion

4.1 Chemical analysis of the placenta-embryo system

Results of biological material research conducted in the operating area of Revda copper-smelting plant are presented as diagrams in Figures 1 and 2.

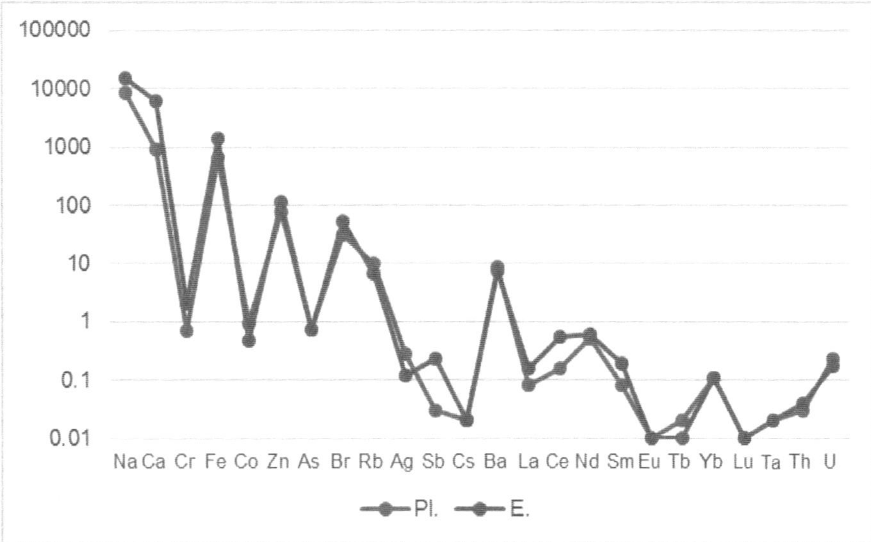

Figure 1. Chemical substance content in placenta (Pl.) and embryos (E.) of small mammals, taken in the area 2 km away from Middle Ural copper-smelting plant OJSC (impact zone), mg/kg of ignition residue

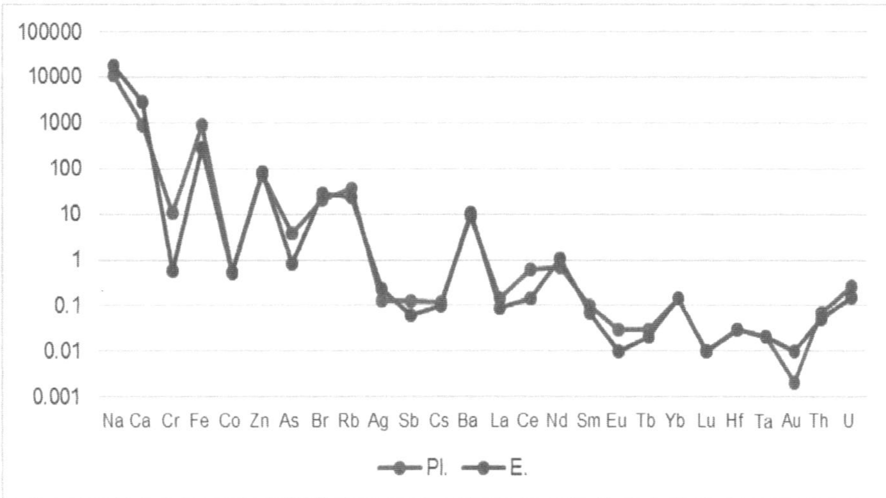

Figure 2. Chemical substance content in placenta (Pl.) and embryos (E.) of small mammals, taken in the area 30 km away from Middle Ural copper-smelting plant OJSC (background zone), mg/kg of ignition residue

In the impact zone the following groups of elements are mainly accumulated in the tissues of the embryo: essential and necessary elements: Ca, Na, Fe, and Zn; accompanying elements: Cr, Sb, and Br; rare-earth elements: La, Ce, Nd, and Sm; radioactive elements: Th. In the reference zone chemical substance content in the embryo tissues changes in the following way: content of such elements as Na, Ca, Zn, Nd, and Sm in the embryo still exceeds the content of the same elements in the placenta, and the content of Ag and Yb additionally increases. Such metals as Na, Ca, Zn, Nd, and Sm overcome the placental barrier in the area that is close to the plant as well as in the remote one. Na and Ca are vital for new organism development, and Zn content can be related to its biological function as well as a specific character of the enterprise emissions. The placenta barrier does not protect against rare-earth elements (Sm, Nd, and Yb) and they enter into the embryo.

4.1 Chemical analysis of the embryo systems of small mammals selected in the areas in question

The content of chemical elements in the embryo samples of the reference and impact zones enables us to draw a conclusion about the state of the small mammals' natural habitat. To visualize the analytical material we constructed a scatter diagram showing distribution of chemical elements in small mammals' embryos.

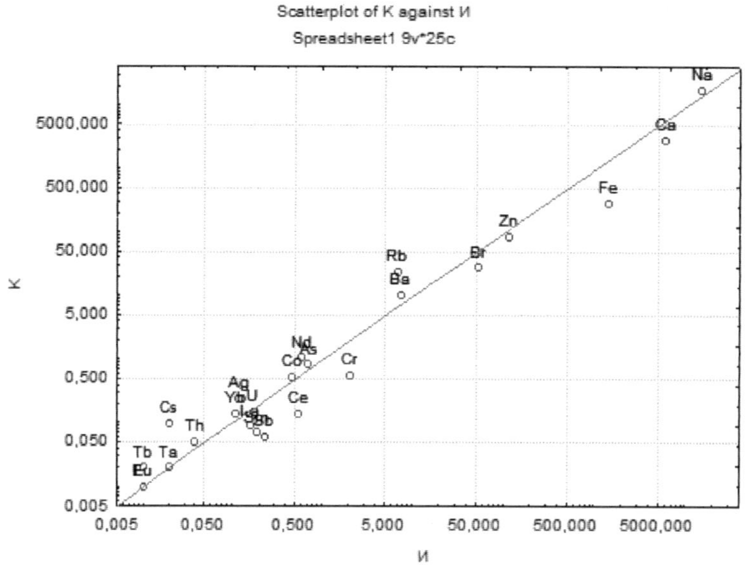

Figure 3. Distribution of chemical elements in small mammals' embryos in the impact and background areas, mg/kg of ignition residue

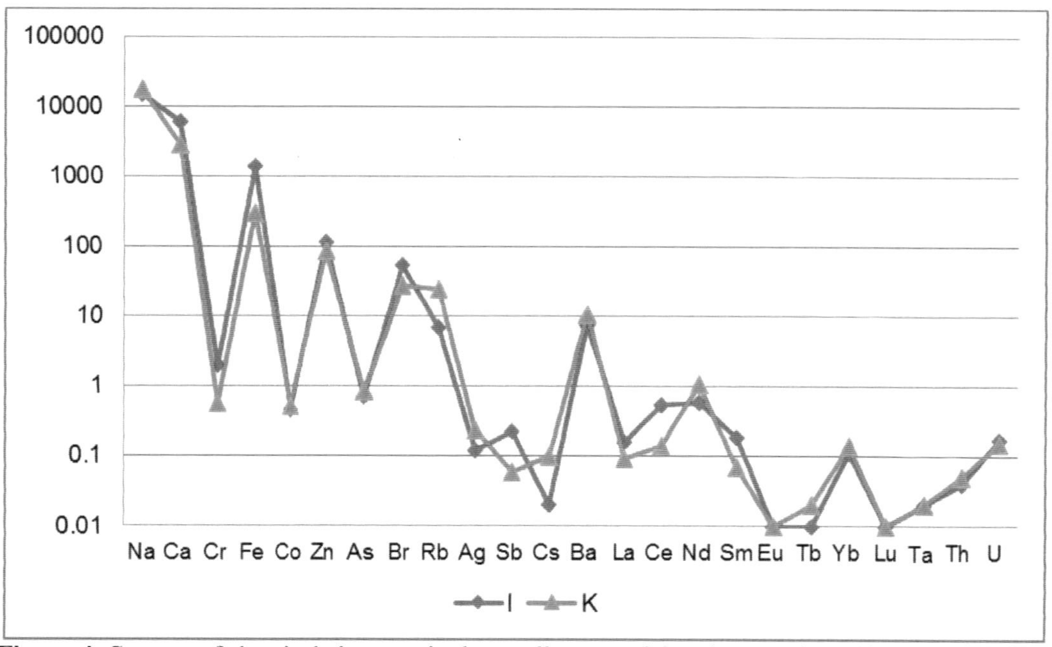

Figure 4. Content of chemical elements in the small mammals' embryos selected in the areas 2 km

(impact zone) and 30 km (background zone) away from the enterprise, mg/kg of ignition residue

In the area, which is 2 km away from the plant (impact zone), such chemical elements as Fe, Zn, Cr, La, and Ce concentrate in a greater degree than in the districts located 30 km away. Fe, Zn, and Cr are typical pollutant metals emitted by nonferrous metal enterprises. The content of these metals in the samples selected in impact and background zones ranges from 1396±75 to 292±50 mg/kg, from 114±5 to 86±7 mg/kg, and from 2±0.4 to 0.6±0.06 mg/kg for Fe, Zn and Cr respectively. La and Ce refer to rare-earth elements (REE), which flames and smokes of industrial emissions contain. REE can enter living organisms together with drinking water and atmospheric dust, the content changes from 0.2±0.02 to 0.9±0.01 mg/kg for La, and from 0.6±0.3 to 0.1±0.02 mg/kg for Ce.

Embryos, selected in areas which are more distant from the plant (30 km), accumulate a larger amount of Na as compared to those selected in areas closer to it, Na concentration varies from 15120±785 mg/kg in the impact zone, to 17313±807 in the reference one. Na is a vital (essential) element, without which healthy functioning of the future living organism is impossible.

4. Conclusion

1. Chemical elements are classified into three groups by their content in the samples of embryos and placentas selected in the studied areas as well as by their content in the human body regardless of the location. A number of researchers stated that such elements as Na, Ca, and Fe are more concentrated in the human body than the others, which is also confirmed by the research of the small mammals. In both cases high concentration of the macroelements (Na, Ca, and Fe) is observed. The barrier mechanism to protect a foetus from environmental impacts better operates in the background area, which is connected with its remoteness from the source of the impacts. In both zones under study the placental barrier is overcome by Na, Ca, and by the rare-earth elements - samarium and neodymium - as well.

2. Embryos selected in the areas located in the vicinity of the source of environment pollution (2 km) accumulate a larger amount of nonferrous industry pollutants than those selected in the distant areas (30 km). It is estimated that the above mentioned elements enter embryos due to malfunction of the placental barrier caused by technogenic impact increase.

References

[1] Avtsyn A P, Zhavoronkov A A, Rish M A, Strochkova L S 1991 *Human microelementoses: aetiology, classification, organopathology* (Moscow) pp. 496.

[2] Babajan A A, Poplaukhin A S, Zavyalov M M 1986 *Basic trends of technological advance in copper-smelting industry Nonferrous industry* . Vol **5** pp. 10–13.

[3] Bezel V S 1987 *Population ecotoxicology of mammals* (Moscow) pp. 129.

[4] Medvedev N V 2004 *Ecotoxicological analysis of natural population of birds and mammals in Karelia under conditions of increasing technogenic pollution* (Petrozavodsk). pp. 261.

[5] Vorobeichik E L 1997 *In addition to methods of measurement of leaf litter thickness for diagnostics of technogenic ecosystem breach Ecology.* Vol **4** pp. 265–269.

[6] Dobrynina N A 2007 *Bioenergenic chemistry Student guide* (Moscow)

[7] Mukhacheva S V, Bezel V S 1995 *Levels of toxic elements and functional structure of small mammals' populations under conditions of technogenic pollution Ecology.* Vol **3** pp. 237 – 240.

[8] Onishchenko G G, Epifanova T M 1993 *Topical questions of sanitary and epidemiological safety of children Proc.* (Moscow). pp. 3-6.

[9] INFOMETGEO LLC 2004 *Nonferrous industry in Russia Geography* . Vol. 38 pp. 10-12.

[10] Opekunova M G 2004 *Bioindication of pollution* (Saint-Petersburg). pp. 354.

[11] Revich B A 1994 *Microelement composition of bio-substrate of population in some industrial cities of the former USSR Activation analysis in environmental protection Proc.* (Dubna) pp. 486-518.

IOP Conf. Series: Earth and Environmental Science **43** (2016) 012045 doi:10.1088/1755-1315/43/1/012045

[12] Rikhvanov L P, Yazikov E G, Sukhikh Yu I 2006 *Ecogeochemical peculiarities of the natural environment and morbidity rate in Tomsk region* (Tomsk) pp. 216.

The change of surface wind in the north Western Siberia since 1966

M V Reshetko, A V Chernikova and E V Vasina

Tomsk Polytechnic University, 30 Lenin Avenue, Tomsk, 634050, Russia
E-mail: mrechetko@tpu.ru

Abstract. The statistical analysis of near-surface wind characteristics has been performed in the territory of the North of Western Siberia. The conclusion on wind speed decrease of different intensity for different directions and increase of calmness frequency is made. The maximum changes in wind speed are observed in winter and summer seasons, in transitional seasons these values are lower. The most intensive changes in wind direction frequency occur in summer; the less intensive ones are typical for winter.

1. Introduction

Modern climate changes in different parts of the globe are connected with changes of atmospheric processes durability that lead to changes in meteorological characteristics. Taking into account the climate changes it is necessary to highlight the relevance of assessing modern and expected climate-related changes in water balance, and, hence, water resources and cycle. One of the major meteorological variables for different research spheres is a speed and direction of near-surface wind. As a result of changes in atmospheric circulation, the air mass behavior coming to the territory of the Lower Ob River also changes. Therefore, it defines the peculiarities of hydrologic behavior of water bodies and environmental flow. Based on these facts the results of statistical analysis of near-surface wind are considered in the territory of the North of Western Siberia. The goal of the research is to reveal the trends in changes of near-surface wind speed and direction.

2. Materials and methods

2.1. Subject of research

The study area is characterized by long winters (7–8 months), mean monthly temperatures of which are -23–29 °C, whereas minimal ones reach -60 °C. Harsh climate is complicated by strong winds [1, 2]. In summer cyclones intensify. Summers are cold due to frequent north winds and cloudy weather. The mean temperature in July changes from 10 to 14 °C, the maximal temperature sometimes reaches 30 °C. During summer season light frosts and snow are possible. On average, summer lasts about 40–60 days. The mean annual air temperature in the territory is -5–7 °C, in some regions it is -10°C. The research area is characterized by excessive moisture and is considered the most hyperhumid area of Russia [1, 2].

To study the wind behavior the following weather stations were chosen: Berezovoye (63.93 N 65.05 E) and Tolka (63.98 N 82.08 E) located on approximately the same latitude and Marresal (69.72 N 66.80 E) and Salekhard (66.50 N 66.68 E) located on approximately the same longitude. The altitude of these weather stations is within the range of 15-31 meters above the sea level.

Marresal weather station is located on the coast of the Kara Sea, where winds are stronger with wind mean speed being 8-9 m/s in winter and 5-6 m/s in summer. At Salekhard station mean wind speed is 3.5-7 m/s [1, 2]. At Marresal weather station winds of south quarter prevail in winter, north and north-west wind directions – in summer, the number of calmnesses is 4% in winter, and 17% in summer. At Salekhard weather station south winds prevail in winter period, but north-east ones are in summer, the amount of calm winds in winter is 16%, but in summer 8%. At Marresal weather station mean annual wind speed is higher in winter period, but at Salekhard the maximums are observed in summer period.

At Tolka weather station the maximum wind speed is observed in April, May (3m/s), the minimum one – in January, February (2.2 m/s). At Berezovoye weather station the maximum wind speed is observed in May, June (4.3 m/s), whereas the minimum one is in February, January (3.3 m/s). The prevailing wind direction at Tolka weather station is south-east quarter in winter season, and those of north-west and north directions in summer. At Berezovoye station the prevailing wind direction is winds of south-west quarter in winter season, but those of north-east and north directions in summer season. At both weather stations the wind speed is higher in transition seasons than in warm and cold seasons. At all weather stations, except Marresal station located on the coast, calm wind conditions are more often in winter than in summer [1, 2].

2.2. Materials and methods

The research materials were taken from special database of current observations (every 3 hours) of wind direction and speed at the stations Berezovoye, Tolka, Salekhard within the period from 1966 till 2014 and Marresal weather station within the period from 1977 till 2014 [1]. At each station the wind direction frequency from 8 points of compass was calculated for each month and mean wind speed for each month and a year in every direction over the research period.

Afterwards, the statistical analysis of the data on time series of surface wind was performed. For long-term data record the test of statistical inhomogeneity was conducted. The research methods included verification of null hypothesis of observation series homogeneity using Student's, Abbe's [3] and Fisher's tests; randomness and presence of trend was tested by means of Pitman's and reversal criteria [4]. The conclusion on series homogeneity or assignable change was made at the significance level 0.05 when estimated module values exceeded the corresponding critical or extended the allowed limits.

3. Results and discussion

Based on the research the statistical inhomogeneity of more than half of data series has been established. For Tolka 77% of mean annual wind speed series are inhomogeneous, but for Berezovoye all series are inhomogeneous. Besides, for Tolka weather station 63% of direction frequency series and mean wind speed in regard to every direction are inhomogeneous, but for Berezovoye they are 53% and 73%, respectively. Within the research period Berezovoye weather station is known to be relocated 1.5 km to the south-west in 1983 that can be a cause for climatic data inhomogeneity. For Marresal weather station 67% of mean annual wind speed series and 25% of direction frequencies are inhomogeneous; for Salekhard station 92% and 67% are respectively inhomogeneous.

The statistically significant negative trend of mean monthly and annual wind speed is established (table 1), which corresponds to the previous research, for instance [5]. One of the reasons for changes of near-surface wind characteristics is change in large-scale atmospheric circulation, namely, increase in durability of anticyclonic weather over Western Siberia [6].

It should be noted that the most changes in wind speed are observed in winter and summer periods. In transition seasons the values of trends are lower. In addition, positive trends of wind speed are recorded, for example, at Marresal weather station which is located on the coast. Besides, absence of statistically significant changes in wind speed at Tolka station in the second half of winter is to be taken into account.

Table 1. The values of statistically significant changes of mean wind speed (m/s over 10 years).

	v, m/s over 10years			
month	Berezovoye	Tolka	Marresal	Salekhard
I	-0.40	-	-0.14	-0.37
II	-0.29	-	-0.05	-0.33
III	-0.25	-	-0.02	-0.27
IV	-0,21	-0.08	**+0.06**	-0.31
V	-0.23	-0.15	-0.26	-0.43
VI	-0.33	-0.25	-0.15	-0.44
VII	-0.31	-0.13	-0.14	-0.34
VIII	-0.31	-0.15	-0.05	-0.25
IX	-0.3	-0.08	-0.13	-0.26
X	-0.35	-0.13	-0.01	-0.27
XI	-0.4	-0.17	**+0.12**	-0.31
XII	-0.38	-0.13	**+0.08**	-0.30
year	-0.31	-0.13	-0.06	-0.31

The research has revealed the trends in long-term series of mean speed and frequency of wind directions from 8 points of compass every month. The increase in frequency was mainly established for north-east, south-east, south-west, and north-west winds. For Berezovoye and Marresal stations the increase in calmness frequency in all months was proved, but for Tolka station – in June. As for Berezovoye and Salekhard stations the negative trend in mean speed was statistically proved in all months (table 1) and for a number of annual values. On the whole, during a year at Salekhard station the increase in frequency of south-west, south-east, and north-west winds was revealed (figure 1). At Marresal station positive trends in frequency of north and south winds as well as calm conditions were stated; the frequency in south-west and south-east wind decreased.

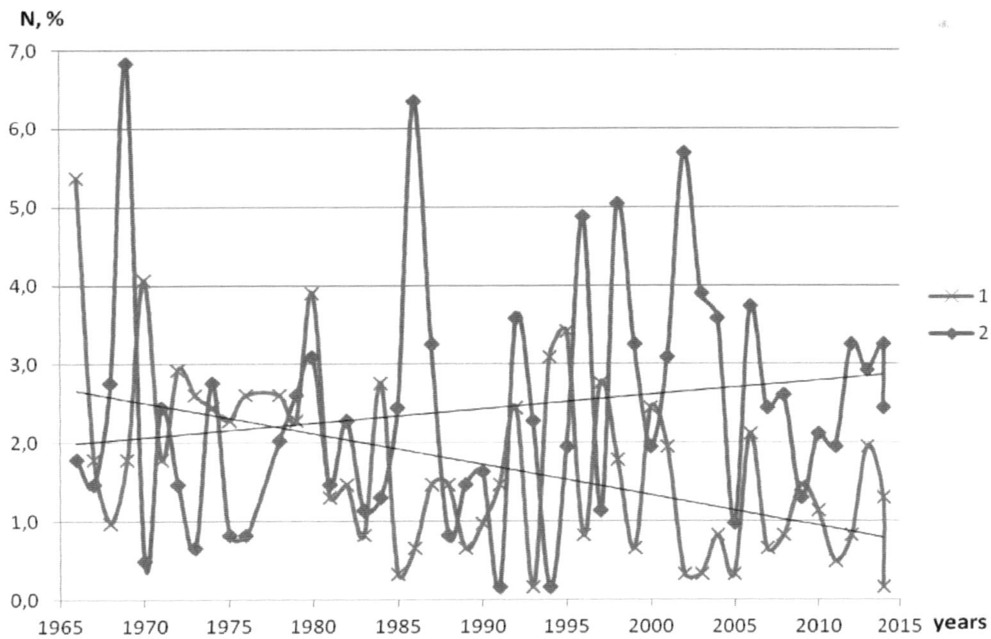

Figure 1. Long-term changes in frequency (N) of east (1) and north-west (2) wind of Salekhard weather station in March with linear trend, 1966-2014

As a result of research in near-surface circulation the changes in wind direction frequency and speed have been considered in each direction for the seasons of typical annual circulation.

Considering Tolka weather station the authors distinguished 4 types of seasons typical for near-surface circulation – winter (November, December, January, February), summer (June, July, August) and transition seasons – spring (April and May) and autumn seasons (September, October). Similar to autumn season the wind rose is observed in March. In winter the frequency of south-west and north-east directions increases and wind speed decreases 0.1 m/s on average in south (figure 2), south-west and south-east directions (south quarter), as well as east direction.

As for Berezovoye station the following types of seasons typical for near-surface circulation are to be distinguished – winter (from October till February), summer (May - July) and transitional seasons – autumn (August, September) and spring in March. Similar to autumn season the wind rose is observed in April. In October wind rose is close to autumn season, but it has its peculiarities. In summer the frequency of south-east and south-west directions increases. At the same time, the frequency of north-west and west directions does not change. The frequency of calmness increases. The decrease in wind speed occurs mostly in south-west and south directions as well as in north-east and east within the range from 0.25 to 0.35 m/s on average.

For Salekhard station – winter (November, December, January, February), summer (May, June, July, August) and transitional seasons: autumn (September, October) and spring (but March and April are different). In winters the frequency of south-east, south-west and north-west directions increases, but the speed of south-west quarter and north and north-east directions decreases 0.3 m/s on average. In summer the frequency of south-east, south-west and north-west directions increases, wind speed decreasing in all direction except the winds of south quarter (S-W, S, S-E).

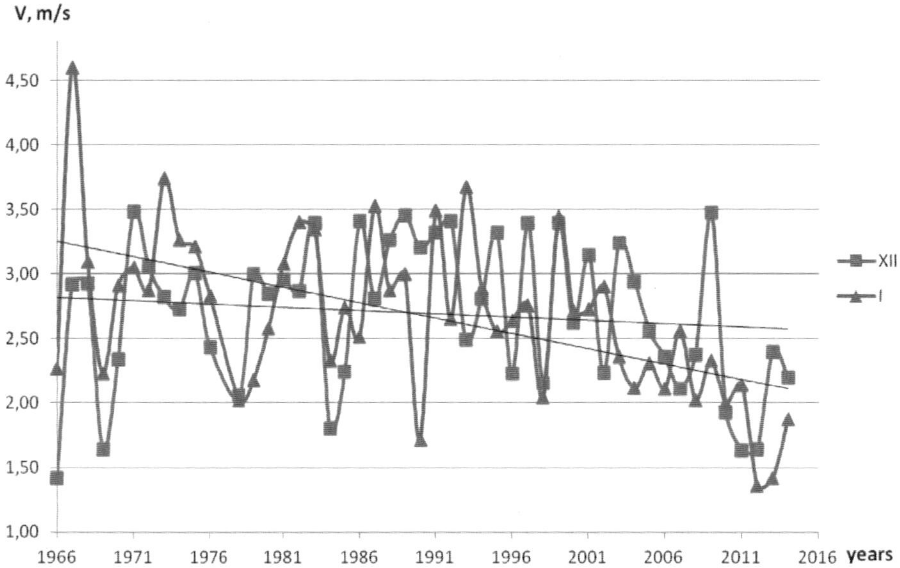

Figure 2. Long-term changes in mean wind speed of south direction at Tolka weather station in January (I) and December (XII), 1966-2014

At Marresal weather station 5 types of seasons typical for near-surface circulation are distinguished – winter (from November till February), summer (May - August) and transitional seasons – the first one occurs with north direction increase (March - May) and separately (October) and (September). In winter the frequency of north, east winds increases, whereas that of south-west, north-west and west decreases. The frequency of calmness increases in all months, not only in winter. In this case there is a slowdown of wind speed in south-west direction in January and east one in December. There is also a 0.1 m/s increase of wind speed in north direction on average. In summer a decrease of north-east, east and south-east direction as well as south and south-west is observed, but there is an increase of north one. The frequency of calmness grows as well. The speed mostly decreases for winds of west direction.

The most intensive changes in wind direction frequency take place in summer period, the less intensive ones are observed in the first half of winter period. Most often the mean monthly near-surface speed of south quarter wind decreases (S-E, S, S-W). Less often the changes occur for the winds of north quarter (N-W, N, N-E).

4. Conclusions

In the course of research it is stated that the mean annual and monthly wind speed slows down and calmness frequency increases. The maximal changes in wind speed are observed in winter and summer periods, whereas in transitional seasons the values are lower. Over the year most often monthly near-surface wind speed of south quarter directions decreases. Less often similar changes occur for the north quarter winds.

The most intensive changes in frequency of wind directions are registered in summer period, less intensive ones are typical for the first half of winter period. It should be noted that there is an increase in frequency of wind direction not typical for the given season. Besides, decrease in speed of winds typical for the given season is observed nearly at all weather stations. The growth of wind frequency in the north quarter in winter and in the south quarter in summer may result in more temperature variability, changes in heat and moisture advection. Thus, the analysis of long-term transformations in near-surface wind characteristics allows the conclusion on slowing down of wind speed, different in intensity in directions of different points of compass and increasing calmness frequency at weather stations in the north of Western Siberia.

References

[1] FGBU VNIIGMI MCD [Electronic resource] access: http://meteo.ru/data (in Russian)

[2] National Atlas of Russia, Ecology and Nature vol 2 [Electronic resource] URL: http://национальныйатлас.рф/ (in Russian)

[3] Rapp J Ch D 1996 Schönwiese Atlas der Niederschlags- und Temperaturtrends in Deutschland 1891-1990 *Frankfurter Geowissenschaftliche Arbeiten Serie B Meteorologie und Geophysik* (Frankfurt a. M.) **5** 255

[4] Bendat J and Piersol A 1989 *Applied analysis of random data* (Moscow: Mir) p 540 (in Russian)

[5] McVicar T R, Roderick M I, Ling Tao Li 2012 *J. of Hydrology*. Global review and synthesis of trends in observed terrestrial near-surface wind speeds: Implications for evaporation. Vol. **416-417** pp. 182–205.

[6] Gorbatenko V P, Ippolitov I I and Podnebesnykh N V 2007 *Russian Meteorology and Hydrology*. Atmospheric Circulation Over Western Siberia in 1976-2004 32. Vol. **5** pp. 301-306

IOP Conf. Series: Earth and Environmental Science **43** (2016) 012047 doi:10.1088/1755-1315/43/1/012047

Biotesting of modeled drilling mud as an indicator of environmental risk

A S Mushunina, S V Azarova, E G Yazikov, I A Parygina

Department of Geoecology and Geochemistry, Institute of Natural Resources, National Research Tomsk Polytechnic University, 30 Lenin Av., Tomsk, 634050, Russia

E-mail: sashenbka@yandex.ru, svetazara@tpu.ru

Abstract.

In the system of environmental monitoring, biomonitoring plays an important and independent role. The essence of the method determining the toxicants' effect consists in the specially selected organisms under standard conditions and recording different behavioral, physiological and biochemical factors. The paper studies the biological influence of modeled drilling mud components. The research results have shown the toxic effect of investigated muds. The obtained data confirm that biotesting can be applied for drilling fluids certification by the complex assessment of hazardous substances.

1. Introduction

The biotesting methods of environmental management have found wide application in the sphere of detection of pollutants hazard. Such methods are known to be widely implemented abroad by environmental monitoring and protection organizations. One of the advantages of biotesting is detection of toxic substances, most of which are not specified by the operating standards, but can lead to the toxic or mutagenic effects. Today, a wide range of test objects have been studied: from microorganisms to animals. Each of them has its own characteristics, advantages and disadvantages, knowledge of which helps to choose an indicator with appropriate "sensitivity" to definite pollutants. For example, since 1985 in the USA there has been an obligatory toxicological biotesting for drilling fluids with the use of mysids as a test object [1]. In Russia, water biotesting using daphnids and infusoria is included in the obligatory list of the National State Standard 17.1.2.04-77 for fishery waters [2].

Ready and waste drilling muds are chemical products, which are an integral part of drilling process in oil industry. For their preparation, a wide range of organic and non-organic substances, chemicals and additive compounds are used [3, 4, 5, 6]. In the USA more than 1900 different components are produced by about 100 enterprises. The used materials are sent to a sludge reservoir (a pit) after operating cycle in the well. According to some authors' data [7, 8, 9], production activity connected with the drilling muds, which are the part of drilling waste, affects negatively some environmental components. Transfer of drilling muds, drilling flush fluids and other chemical products into the environment is dangerous in spite of its self-purification capacity. Drilling cuttings and waste water contact the drilling mud and become toxic. The stored drilling mud components are toxic by themselves. Besides, it is impossible to prevent completely the potential accidents.

The research goal is to assess the hazard of drilling mud components for the environment using biotesting.

The relevance and practical application of the research is defined by the opportunity to assess and to predict the biological impact of each component of certain chemical elements using modeled drilling muds. Having analysed the results obtained by means of Chlorella vulgarisbeijer weed, it is possible to simulate the effect of water-soluble substances in studied materials. The use of Drosophila melanogaster fly allows studying the effect of these components (including solids). In addition, the obtained results can be used for development of substances list in the National State Standard 17.1.2.04-77.

Toxicity detection of modeled drilling mud components is necessary due to the emission of one major ingredient used in drilling.

2. Methods and materials

For the first time Chlorella vulgarisbeijer weed and Drosophila melanogaster fly are used as test objects to study the biological effect of drilling muds. The toxicity was experimentally estimated in the analytical laboratories (Tomsk Polytechnic University) following the requirements of the guidance. According to "Toxicity detection of drink, natural and waste waters, soil-water extracts, sewage sludge production and consumer waste for optical density measurement of Chlorella vulgarisbeijer weed" [10], a series of experiments were conducted (8 samples) to determine toxicity of drilling mud water extracts in the Fundamental Research Laboratory of Hydrogeochemistry (analyst: Vorobieva D.A.). The analysis includes measurement of optical density of weed suspension. The special multiple-cell propagator allows timely revealing the changes in cell number in the reference and experimental samples in acute toxicological condition. A criterion is a growth inhibition or promotion in values of weed optical density [11].

To detect content of the toxic substances sample biotesting was conducted using Drosophila melanogaster fly as a test object in the Department of Geoecology and Geochemistry according to the recommendations applied for the extractive industry waste [11], medicines, and urban dust. The results were statistically processed with the fitting criterion – X2. During biotesting 5 samples were studied at 1% concentration of basic material in the medium. For the experiment, yellow (y) and singed (sn) Drosophila lines were taken. The yellow ones had yellow bodies and straight bristles, the singed ones had grey bodies and scorched bristles. Two females and one male were placed in the test-tubes with the prepared medium for two days. Then, the development of new generation was controlled. The changes in external features and ratio of males indicated the impact of studied muds. The reference and experimental groups were formed simultaneously and identically. The lethal dose (LD_{50}) was determined for every sample.

3. Results and discussion

The research conducted in 2000 by P.S. Chubik in Tomsk Polytechnic University proved the urgency for development of environmental friendly drilling fluid. In his paper it was mentioned that the ecotoxicity assessment of drilling flush fluids should be made at all stages of their "life cycle" to obtain reliable data, which can be achieved only by instrumental biotesting [8, 12].

The authors examine the "first" life cycle of drilling muds. These are modeled drilling muds produced in the chemical laboratory of non-production cycle.

Actually, there is no universal standard composition of drilling muds. Their application depends on location. Besides, they can drastically differ in terms of a well depth and different ore structure. Currently, no special universal methods of hazard assessment of drilling muds or their compounds for the environment have been developed. The characteristics of mud cellulose esters-based components are estimated as non-hazadous.

The reagents of Russian origin (polyanionic cellulose (PAC)) and those of foreign origin (carboxymethylcellulose (CMC 85/1000)) were selected for analysis. It is generally accepted that the polyanionic cellulose reagents are not toxic.

Polymer-containing drilling muds started to be applied in the first half of the 1970s, and at the moment, they are the most commonly and widely used reagents in drilling. So, for biotesting, reagents most of which are polysaccharides were chosen [12]. Production of water-soluble cellulose ethers reaches about 380 thousands of tons per year, with 180 thousand tons of carboximetyl cellulose (CMC). The use of polymeric reagents (CMC) in mud composition has made it possible to improve the wellbore wall conditions, limit the oil content in the mud, and enhance the well cementing quality.

CMC is an imported reagent, which is an interaction product of cellulose with monochloroacetic acid. Dry sodium salt CMC has a weak corrosive action. The main property of the reagent is a gelation [13-18].

Polyanionic cellulose (PAC) is made from the same basic substances as a typical CMC. There are various modifications of the production technologies, which allow obtaining PAC with different degrees of substitution. PAC is characterized by increased bioresistance to the polymineral exposure.

"Polypac R" contributes to formation of a thin, dense, firm and tight filter cake, reducing the area of filtrate invasion into the reservoir, preventing the invasion of solids; it also reduces the risk of differential sticking. The reagent is resistant to bacterial exposure; it does not require any bactericide that complicates its further processing by the bacterial recycling.

PAC (BB, HV) is a multi-purpose reagent used as a thickener, clay inhibitor, lubricating staff, and agents to reduce the loss of drilling fluid. The substitution degree is not less than 90; Brookfield viscosity ranges from 120 to 4500. The country of origin is Russia. The type of reagent is a surface-active agent. In terms of salt resistance the reagent is 10% NaCl salt resistant. Regarding heat resistance the reagent is resistant. In terms of application the reagent is a clay rock inhibitor.

CMC. The polymerization degree of up to 1200 is used to increase viscosity of CMC drilling mud with the polymerization degree of up to 85 / 1000-85 / 1200 in drilling deep wells. The country of origin is Russia. The reagent type is a surface-active agent. In terms of salt resistance the reagent is limited 3 - 10% NaCl salt resistant. In terms of heat resistance the reagent is estimated as resistant. As for application the reagent is used as a filtrate reducer.

"Polypac R" is polyanionic cellulose of high quality, water-soluble polymer used for control of water-based muds filtration, manufacturer is MI-SWACO. The country of origin is the USA. "Polypac R" is used to reduce water loss of fresh, salted, potassium chloride and salt-saturated muds as well as sea water muds. It promotes the formation of a thin, dense, firm, and tight filter cake, reducing the area of filtrate invasion into the reservoir, preventing the invasion of solids; it also reduces the risk of differential sticking. The reagent is resistant to bacterial exposure; it does not require the use of bactericides. "Polypac R" is cost-effective and efficient.

4. Using *Chlorella vulgarisbeijer* weed as a test object

Methods of toxicological analysis using Chlorella vulgarisbeijer weed were approved for the state environmental monitoring purposes. It can be applied to certify drilling muds in combination with other test objects for the comparative evaluation of drilling mud toxicity and hazard class definition.

Using the Kichemasov's method [10], the quality of the water extract of the studied components was determined in terms of their toxicity. For this purpose, the relative difference (in %) of the optical density value for each dilution in comparison with the reference one was calculated. As a result, in all the studied samples the toxicity is indicated (table 1).

Table 1. Detection of water sample quality in drilling mud components

№	Drilling mud component	Value of toxic multiplicity of dilutions	Frequency r*, %	Result of toxicological analysis (quality of water extract)
1	CMC	7	10	Moderately toxic
2	Polypac R	4.4	3	

3	Polypac R+ NaCl	8	21	
4	PAC HV	24	12	Toxic
5	CMC +NaCl	9	5	
6	PAC BB+ NaCl	22	7	
7	PAC HV+ NaCl	9.6	15	
8	PAC BB	59	23	Highly toxic

Based on the research results, it can be concluded that the toxicity degree of studied components is increasing with the addition of salts in spite of the fact that salts are generally not toxic. Presumably, the interaction of additional components with the drilling mud base (cellulose esters) may cause increased toxic effects.

4. The use of Drosophila melanogaster fly as a test object

For a long time, biotesting on Drosophila flies has been used in the medicine. Drosophila melanogaster is the most studied object, which helps to identify the effect of a substance on the living organisms. Different parameters of Drosophila melanogaster are widely used for hazard assessment of chemicals [19, 20].

Using fruit flies in the experiments with the drilling mud components, the following biological factors are estimated: sex ratio and morphoses in relation to sample concentration in the medium. The values of these factors in the experiment groups are compared with the values in the control groups. When processing the data, the statistical index X2 is used to determine the conformity degree of experimental data to expected ones. The obtained results are presented in table 2.

Table 2. Quantitative description of studied factors.

№	Drilling mud component	Sex ratio, %		Morphoses, %
		Males	Females	
1	CMC	43	57	7.5
2	Polipac R	28	72	14
3	PAC BB	48	52	39
4	Duo-vis	17	82	25
5	Control	50	50	1

Obtained X2 values in the components: Polypac R, Duo-vis exceed the critical value of 3.84 ($p = 0.05$; at the number of freedom degree equal to 1) for Polypac R. Hence, the suggestion that male/female ratio in the reference medium and in the medium with sample addition is equal (1:1), is incorrect for Duo-vis. So, it can be concluded that the studied compound samples have influenced the Drosophila flies sex ratio. In the components of PAC-BB, CMC X2 value does not exceed 3.84; it shows that there are no samples with toxic effect for this factor. The "morphoses presence" factor was calculated in a similar way. The drilling mud components Polypac R, PAC-BB, Duo-vis belong to the toxic substances. Therefore, the use of Drosophila melanogaster fly as a test object allows referring Polypac R, PAC-BB and Duo-vis to the substances with toxic effect. The summarized data on the studied test objects are shown in table 3.

Table 3. Summarized data on the studied test objects.

№	Drilling mud components	Test objects	
		1	2
1	CMC		n.d
2	Polypac R		
3	Polypac R+ NaCl		n.d
4	PAC HV		n.d

5	CMC +NaCl		*n.d*
6	PAC BB+ NaCl		*n.d*
7	PAC HV+ NaCl		*n.d.*
8	PAC BB		

Note: 1 - *Chlorella vulgarisbeijer*, 2 - *Drosophila melanogaster*

– presence of biological effect

– absence of biological effect

n.d. – not detected

The analysis performed shows that toxicity was detected in all studied samples using different test objects.

Conclusions

The research findings confirm the presence of the toxicity in drilling mud components. Chlorella vulgarisbeijer weed and Drosophilia melanogaster fly can be reliable environmental risk indicators of drilling mud components. Biotesting can be recommended for certification of drilling muds for the purpose of state environmental control.

Reference:

[1] Abdullin R A 1990 *Moscow: Izd VNIIEgazprom*. New technological tools and processes providing the drilling costs reduction and environmental protection Drilling of gas and gas-condensate wells. Pp. 52.

[2] Standard 17.1.2.04-77 Protection of nature. Hydrosphere. Indicators of condition and taxation rules of fishery water bodies

[3] Heinze T, Liebert T 2001 *Progr. Polym. Sci.* Vol. **26** pp.1689.

[4] Balser K, Hoppe L, Eicher T, Wendel M and Astheimer AJ 1986, in: Ullmann's Encyclopedia of Industrial Chemistry, 5th ed W Gerhartz, YS Yamamoto, FT Champbell, R Pfefferkorn, JF Rounsaville Eds VCH Weinheim New York pp. 419.

[5] Chowdhury J K 1924, Biochem. Z 148 76.

[6] Ogawa T J 1976 *Jpn. Soc. Starch Sci.* Vol. **23** pp. 49.

[7] Novikov V S 1977 *Overview Drilling Series*. The results of industrial tests of potassium solution. Vol. **6** pp. 32-36.

[8] Chubik P S 2000 Methodological fundamentals of the quality optimization of drilling flush fluids: the dissertation of the doctor tehn. sciences: 05.15.14: protected 04.05.00 pp. 370.

[9] Kaufov M A, Khandokhov T X and Kerefova M K 2013 *Fundamental Researches Magazine*. Morphoses of Drosophila melanogaster by irradiation with alternating magnetic field of different frequency. Vol **10 (10)** pp. 2219-2221.

[10] Kichemasov A N 2012 Methods Toxicity detection of drink, natural and waste waters, soil-water extracts, sewage sludge, production and consumer waste for optical density measurement of Chlorella vulgarisbeijer weed FR.1.39.2007.03223 (Moscow: Publishing House Akvaros) pp. 42.

[11] Azarova S V, Yazikov E G and Ilyinskikh N N 2004 *Tomsk, Tomsk Polytechnic University news*. Assessment of environmental risks of mining enterprises waste in the Republic of Khakassia using the biotesting method. Vol. **4** pp. 55-59.

[12] Drovnikov P G 1992 *Moscow: Publishing house VNIIOENG*. Methods for neutralization of drilling wastewater by the construction of deep wells Construction of oil and gas wells by land and by sea. Vol. **5** pp. 18-20.

[13] Stigsson, V, Kloow G and Germgard U 2001 Paper Asia pp. 16.

[14] Thewlis B H 1969 21 21 Stärke.

[15] Hsieh, P T, Chen C S and Yang C T Hua Hsueh 1978 on February 29, Chem. Abstr. 92 (1980) 24714).

[16] Stojanovic Z K, Jerevic S and Jovanovic 2000 Starch / Stärke 52 413.
[17] Kisung Kwon P, Joong Hyuck Auh J W, Kwan Hwa Park S, Chan H P and Cheul Jong Ko K D 1997 Starch / Stärke 69 499.
[18] Bhattacharya D, Singhal R and Kulkarni P R 1995 Carbohydr. Polym 27,167.
[19] Bochkov N P 1989 Human heredity and the environment mutagens Medicine. Vol. **8** pp. 163-167.
[20] Kozak M F 2007 Drosophila - the model of genetics: study guide (Astrakhan: Astrahan University Press) pp. 8.

Geochemical features of the elemental composition of meadowsweet (*Filipendula ulmaria (L).Maxim*) in Kemerovo Oblast

V A Zhdanov, I S Sobolev, N V Baranovskaya, E A Kolesnikova, E V Chernenkaya, and A R Yalaltdinova

National Research Tomsk Polytechnic University, 30 Lenin Av., Tomsk, 634050, Russia
E-mail: [2]hyfrvadim@gmail.com, [3]geolsob@yandex.ru, [4]yalaltdinova.aly@mail.ru

Abstract. Biogeochemical sampling of the aboveground part of meadowsweet (*Fillipendula Ulmaria (L).Maxim*) allowed us to study ecological and geochemical features of 10 regions in Kemerovo Oblast, including both natural and man-made landscapes. The content of 55 elements in the plant is determined by ICP-MS. Statistical analysis of the results allowed us to establish the effect of the soil mineral composition and the mining region specificity on the elemental composition of meadowsweet, to reveal significant positive correlations of the elements and to establish a statistically significant difference in the studied areas on the basis of the content of some elements. Sample reference to one of the clusters, followed by an assessment of their geochemical features is determined by the K-average method.

1. Introduction

Biota is the most dynamic component of any landscape. Due to the numerous studies, the significant amount of both geo-botanical and chemical data has been accumulated, which pointed to close relationship between the plant and the environment. The application of biogeochemical methods in searching for deposits of nickel, cobalt, copper, chromium, lead, molybdenum, selenium, and other elements was described by A.E. Fersman, D.P. Malyuga, A.N. Perelman, H.L. Cannon [1-4 et al.]. On the other hand, the accumulation of information on the plants' elemental composition reflects clearly not only the natural component properties, but also man-made factors [5-8 et al.]. Thus, the goal of our research is to prove that the elemental composition of the widespread plants can be used to detect ecological and geochemical features of a region.

2. Materials and methods

Biogeochemical sampling was conducted for this purpose in 10 regions of Kemerovo Oblast in summer of 2015 in accordance with GOST 27262-87 and GOST 24027.0-80 standards. Meadowsweet (*Filipendula ulmaria (L). Maxim*) is characterized by high tolerance to the environment and occupies a dominant position in tall herbaceous layer of lowland meadows in Western Siberia. Samples of its overground parts were collected in blossom and early fruiting periods within the various natural and man-made landscapes. Air-dried raw material was crushed and sifted through 1 mm sieve. The total number of samples was 50.

The elemental composition study was carried out using mass spectrometry with inductively coupled plasma (ICP-MS) based on the standards in the accredited laboratory of the "Plasma" analytical center, Tomsk city (accreditation certificate ROSS RU.0001.516895).

3. Results and discussion

The content of 55 elements in the dry matter of meadowsweet aboveground parts was determined by ICP-MS. It was revealed that qualitative composition of the samples collected in different places was almost identical, while the quantitative content of the chemical elements differed significantly.

Concentrations of Sc, As, Cd, Eu, Er, Lu, Ta, Tl, Bi were below the sensitivity threshold of analysis in certain samples of plant material. Excess of Clarke concentrations in plant dry matter relative to the data obtained by B. Markert [9] was observed for Ca, As, Sc, Ba, Sb (table 1). High degree of concentration variabilities was established for Sm, Yb, Lu, Hf, Ta, Th, Rb, Eu, Sb.

Table 1. Clarke concentrations and variation coefficients in plant dry matter.

		Variation coefficient, %		
		Uniform (< 39)	Non-uniform (40 – 79)	Extremely non-uniform (> 80)
Clarke concentration	< 0.5	Cr	Na, Fe, Zn, Ag, Cs, La, Ce, Nd, Tb	Sm, Yb, Lu, Hf, Ta, Th, U
	0.5 – 1	Co	Sr	Rb, Eu
	1 – 2	Ca	As, Ba	Sb
	> 2		Sc	

Rank correlation assessment shows a large number of mainly positive significant correlations between the elements. The strongest of them ($R\alpha = 0.05 \geq 0.95$) are included in the lithophils group. It reflects the influence of the soil mineral composition and perhaps mining specialization of the region (figure 1). According to principal components analysis the same components, including Cs, are under the load of the first factor, constituting 32% from the general explained dispersion.

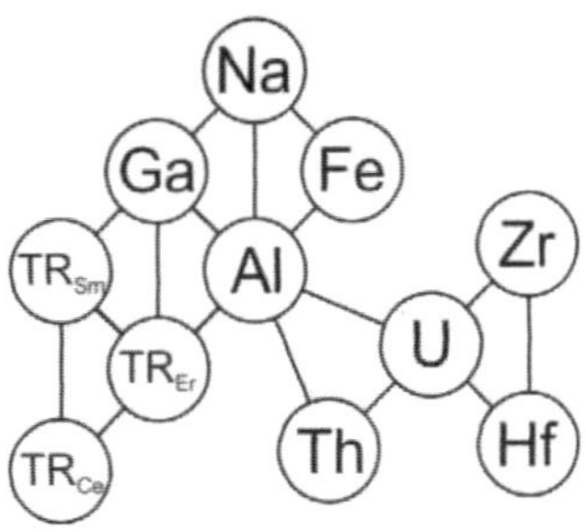

Figure 1. Graph of associations in plant dry matter.

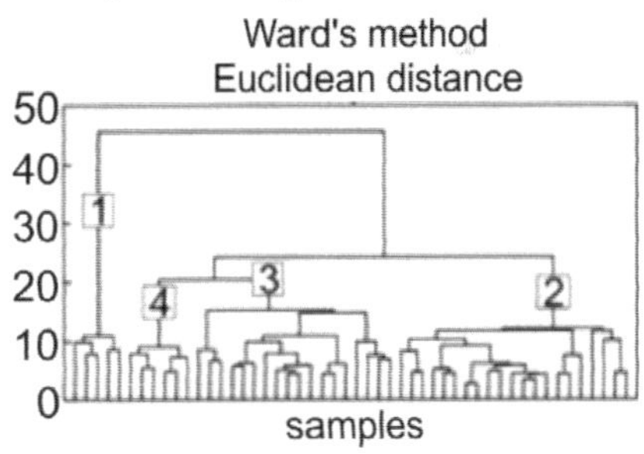

Figure 2. Dendrogram of plant dry matter correlation matrix: 1, 2, 3, 4 - number of clusters.

According to the results of nonparametric Kruskal–Wallis one-way analysis of variance, statistically significant difference between the studied areas is found by the content of some elements (table 2). Concentration coefficients presented in the tables are calculated using the average values.

Cluster analysis using standardized values of the component concentrations reveals a significant effect of local factors on the generalized biogeochemical characteristics of the area. Four clusters, including samples with close element distributions are allocated based on the dendrogram structure features (figure 2). Sample reference to one of the clusters, followed by an assessment of their geochemical features, is determined by the K-average method (table 3, figure 3).

Samples of the second and third clusters are likely to present "near background" element concentrations with some geochemical specificity of the regions and plant sample points. Samples from Tisul region are the most number-representative in the second cluster. The third cluster, where samples of Kemerovo Oblast dominate, is characterized by the relative concentration (CC ≥ 1.5) of rare and rare-earth lithophilic elements.

Table 2. Concentration coefficients of elements in *Filipendula ulmaria* (underground pert) with statistically significant biogeochemical difference in the regions of Kemerovo Oblast.

Region (number of samples)	Coefficient of concentration (CC)															
	Sc	Ti	V	Co	Ni	As	Ag	Hf	Ga	Pb	Bi	Th	U	TR(Ce)	TR(Sm)	TR(Er)
Ch (6)	1.2	1.3	0.9	1.0	0.8	0.9	1.2	0.9	0.7	1.0	1.4	1.0	1.0	0.9	0.9	0.9
G (7)	**0.1**	0.6	0.7	1.2	**0.5**	**0.5**	**0.2**	1.0	**1.5**	**1.8**	0.9	1.0	1.0	0.8	0.7	0.9
I (5)	1.1	**3.2**	**1.6**	1.2	**1.7**	**1.9**	1.0	1.0	**1.8**	1.0	1.0	1.3	1.3	1.4	1.3	1.4
K (5)	**0.4**	0.7	0.9	1.2	**1.8**	0.9	0.6	**1.7**	1.4	**1.6**	0.9	1.3	1.2	**2.1**	**1.8**	**2.4**
L (4)	0.6	0.7	**1.7**	1.3	**0.4**	1.0	**1.5**	**2.1**	**3.0**	1.4	**2.3**	**1.7**	**2.0**	1.2	1.2	1.3
M (7)	1.4	1.4	1.4	0.8	1.4	**1.5**	1.1	0.8	0.7	0.6	0.9	0.7	1.0	1.2	1.2	1.2
T (12)	1.1	1.2	1.0	0.8	0.7	1.3	1.0	0.7	0.6	0.7	1.0	**0.5**	0.8	0.6	0.6	0.6
Ya (1)	1.1	**4.7**	0.7	**0.5**	**1.5**	1.2	1.0	**0.5**	0.6	1.4	1.1	**0.4**	0.6	0.7	0.6	**0.3**
B (2)	**0.5**	1.0	1.1	1.1	0.9	0.9	**0.4**	**3.8**	**1.8**	**2.0**	1.1	**3.1**	**3.2**	**2.1**	**2.2**	**2.5**
P (1)	**0.2**	0.6	1.4	1.2	1.0	**0.2**	**0.5**	**4.2**	**2.2**	1.2	**1.5**	**3.8**	**2.7**	**2.4**	**2.6**	**3.3**

Note: regions: Ch - Chebula, G – Gurevsk, I – Izhmorsky, K – Kemerovo, L – Leninsk-Kuznetsk, M – Mariinsk, T – Tisul, Ya – Yaya, B – Belovo, P – Promyishlennovsky; 0.5 ≤ CC ≥ 1.5

Table 3. Geochemical characterization of meadowsweet (*Fillipendula Ulmaria*) in different clusters.

Number of cluster	Ordered by the value of the concentration coefficient series of chemical elements: chemical element (coefficient of concentration)	Region (number of samples)
1.	U (5.0), TR(Er) (4.6), Hf (4.3), TR(Sm) (4.3), Zr (4.1), Th (4.1), Cs (3.9), Nb (3.7), Ta (3.6), Al (3.5), Y (3.4), TR(Ce) (3.3), Na (2.9), Fe (2.7), Ga (2.5), V (2.1), Tl (1.8), Sb (1.7), Pb (1.6), Si (1.6), Bi (1.5), Sn (1.5), Ti (1.5), Mo (1.4), Cd (1.3), Co (1.3), Rb (1.3), Li (1.2), Ba (1.2), Zn (1.1), Ni (1.1), Cr (1.1), Mn (1.1), P (1.1), Sc (1.1), Cu (1.0), W (1.0), Ag (1.0), Ca (1.0), Sr (1.0), K (0.9), As (0.8), Mg (0.8), B (0.7)	I (2), B (1), L (1), P (1)
2.	Sb (1.0); K (1.0). W (1,0), Bi (1.0), Rb (1.0), Cd (1.0), Mg (1.0), As (1.0), Sr (1.0), P (1.0), Ag (1.0), Cu (1.0), Cr (0.9), Ca (0.9), Co (0.9), Ba (0.9), Mo (0.9), B (0.9), Fe (0.9), Sn (0.9), V (0.8), Pb (0.8), Cs (0.8), Ta (0.8), Ti (0.8), U (0.8), Mn (0.8), Na (0.8), Zr (0.7), Si (0.7), Zn (0.7), Ga (0.7), Al (0.7), Sc (0.7), Tl (0.7), Nb (0.7), Ni (0.7), Hf (0.6), Li (0.6), TR(Sm) (0.6), TR(Er) (0.6), TR(Ce) (0.6), Y (0.5), Th (0.5)	I (1), M (1), Ch (3), T (7), B (1), G (4), L (1)
3.	W (2.2), TR(Er) (1.7), Ta (1.7), Y (1.6), Hf (1.6), Cs (1.5), TR(Ce) (1.4), Al (1.4), TR(Sm) (1.4), Zr (1.4), Bi (1.4), Nb (1.4), Fe (1.3), Ga (1.3), U (1.3), Th (1.3), Mn (1.3), V (1.2), Na (1.2), Co (1.2), Sb (1.2), Ti (1.2), Li (1.1), Sn (1.1), B (1.1), Si (1.1), Pb (1.1), Ca (1.1), Mo (1.1), Cr (1.1), Tl (1.0), Sr (1.0), Mg (1.0), Zn (1.0), Cd (1.0), Cu (1.0), P (1.0), As (1.0), Ba (0.9), Ag (0.9), K (0.9), Ni (0.9), Sc (0.9), Rb (0.8)	I (1), M (3), Ch (3), T (3), K (5), G (2), L (2)
4.	Ni (2.9), Ba (1.8), Cd (1.8), Ti (1.8), Zn (1.6), Cu (1.4), As (1.4), Mn (1.3), Rb (1.3), Sc (1.1), P (1.1), K (1.1), Cr (1.1), Ag (1.0), Bi (1.0), Mg (1.0), Li (0.9), Ca (0.9), Sr (0.9), B (0.9), Si (0.9), V (0.9), W (0.8), Sn (0.8), Mo (0.8), Y (0.8), Fe (0.8), Ga (0.7), Hf (0.7), Zr (0.7), Cs (0.7), TR(Ce) (0.7), TR(Sm) (0.7), Co (0.6), Na (0.6), U (0.6), Al (0.6), Pb (0.6), Nb (0.5), Tl (0.5), TR(Er) (0.5), Th (0.5), Ta (0.4), Sb (0.4)	Ya (1), I (1), M (3), T (2), G (1)

Note: regions: Ch - Chebula, G – Gurevsk, I – Izhmorsky, K – Kemerovo, L – Leninsk-Kuznetsk, M – Mariinsk, T – Tisul, Ya – Yaya, B – Belovo, P – Promyishlennovsky

Samples of the first cluster are distinguished by quite contrast relative accumulation of mainly lithophilic elements. At the same time, a clear indication to natural radioactive, rare earth and rare elements is of particular interest in this cluster.

The fourth cluster samples are distinguished by a mixed chalco-sidero-lithophilic association of concentrating elements.

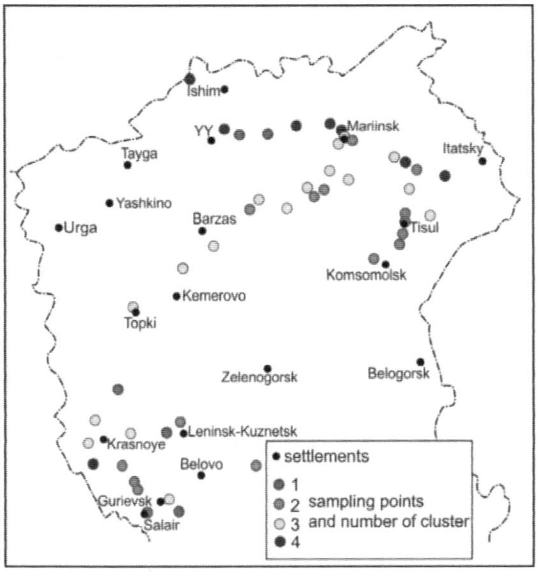

Figure 3. Geochemical features of meadowsweet (*Fillipendula Ulmaria L.*) in different clusters of Kemerovo Oblast.

4. Conclusions

Thus, the regional biogeochemical differentiation of the study areas of Kemerovo Oblast should be carried out taking into account the natural and man-made features of the sampling sites. The dependent number of samples with abnormal deviation at the accumulation levels of element series introduces a significant distortion in the generalized biogeochemical characteristics of the territories.

Acknowledgment

The research is supported by the Russian Science Foundation (grant № 15-17-10011).

References

[1] Fersman A E 1937 *Geochemistry* Leningrad: ONTI-Himteoret.

[2] Malyuga D P 1963 *Moscow: SA USSR Press.* Biogeochemical methods of ore deposit prospecting.

[3] Perelman A I 1975 *Moscow: High School.* Landscape geochemistry.

[4] Cannon H L 1960 *Science.* Botanical prospecting for ore deposits. Vol. **132** pp. 591–598.

[5] Hegland S J, Dunne J and Nielsen A 2010 *Biol. Cons.* How to monitor ecological communities cost-efficiently: The example of plant–pollinator networks. Vol. **143** pp 2092–101.

[6] Reimann C, Englmaier P and Fabian K 2015 *Sci. Total Environ.* Biogeochemical plant–soil interaction: Variable element composition in leaves of four plant species collected along a south–north transect at the southern tip of Norway. Vol. **506–507** pp. 480–95.

[7] Favas P J C, Pratas J and Mitra S 2016 *Sci. Total Environ.* Biogeochemistry of uranium in the soil-plant and water-plant systems in an old uranium mine. Vol. **568** pp. 350–68.

[8] Favas P J C, Pratas J and Prasad M N V 2012 *Sci. Total Environ.* Accumulation of arsenic by aquatic plants in large-scale field conditions: Opportunities for phytoremediation and bioindication. Vol. **433** pp. 390–7.

[9] Markert B 1991 *Water Soil Air Poll.* Establishing of "Reference plant" for inorganic characterization of different plant species by chemical fingerprinting. Vol. **64** pp. 533–8.

The investigation of dangerous geological processes resulting in land subsidence while designing the main gas pipeline in South Yakutia

L A Strokova, A V Ermolaeva, V V Golubeva

National ResearchTomsk Polytechnic University, 30 Lenin Ave., Tomsk, 634050, Russia
E-mail: [2]sla@tpu.ru, alyona7@inbox.ru, golubeva@tpu.ru

Abstract. The number of gas main accidents has increased recently due to dangerous geological processes in underdeveloped areas located in difficult geological conditions. The paper analyses land subsidence caused by karst and thermokarst processes in the right of way, reveals the assessment criteria for geological hazards and creates zoning schemes considering the levels of karst and thermorkarst hazards.

1. Introduction

According to Gazprom VNIIGAZ LLC, currently there are risks of accidents through dangerous geological processes. While investigating the influence of natural factors on the continuity of the Unified Gas Supply System (UGSS) it is important to take into account the experience of gas transmission network operations under dangerous natural processes. One of the factors of potential accidents is UGSS expansion in underdeveloped areas where there are no statistics for long-term observations of dangerous geological processes. Subjective hazards are mistakes made during the gas pipeline design and construction [1].

The main loads influencing gas mains are the following:

- Gas pipeline weight;
- Coating weight;
- Backfill stress;
- Voltage (due to elastic bending);
- Internal pressure;
- Piped product weight;
- Thermal stresses;
- Special loads.

Special loads on gas mains are those which are caused by ground deformations. Such loads should be calculated on the basis of analyzing ground conditions and their potential change while constructing and operating the pipe [2]. Subsidence processes are one of the main reasons for special loads on gas mains. They influence the surface relief substantially by forming large areas of subsidence. Such processes while impacting on the gas main create additional mechanical stresses thus increasing the influence of other defects in the pipe wall. Due to these sources, stresses beyond design limits are the most common and dangerous as it is often impossible to take them into account [3]. However, it is possible to determine the stresses beyond limits during the construction and operation stage by calculating stress-deformed state of the gas main in the areas of subsidence.

Currently the investment project "The Power of Siberia" is one of Gazprom's largest projects. The implementation of this program will allow increasing the level of social and economic development of the Republic of Sakha (Yakutia), the Amur Region, as well as the export potential within Asia-Pacific Region [4].

The paper will consider the potential subsidence hazards in the section "Chayanda-Lensk" (160 km long) of the gas main "The Power of Siberia", which is being built in South Yakutia. In this section both karst and thermokarst may cause land subsidence, which creates mechanical stresses in the pipe and leads to the loss in its reliability.

As mentioned before, there are no time-lapse data concerning dangerous geological processes. Therefore, it is necessary to develop the methods assessing the impact of these processes on the pipe despite the absence of long-term observations.

2. Experimental design

The following stages of fulfilling the task can be singled out. During the first stage the analysis of available information on complex engineering testing was carried out and the relevant map material was selected. The data were processed by means of such software programs as GIS MapInfo Professional, AutoCAD. During the next stage karst and thermokarst processes were ranked according to hazard levels. The overall procedure is presented in Fig.1.

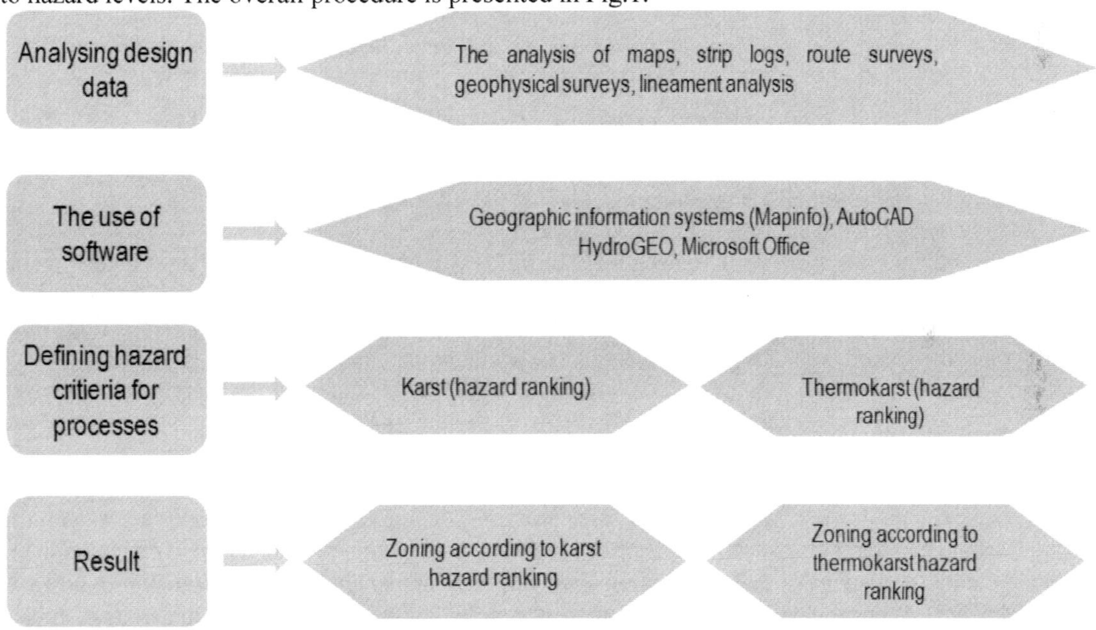

Figure 1. The Overall Procedure

GIS "MapInfo Professional" organized information with layers. The following tables of information (layers) were constructed: wells, the state geological map, the map of quaternary deposits, hydrography, geophysical research, karst development segments (Fig.2). The factual evidence on the wells comprises more than 1500 elements. Thus, the database of geological conditions in the construction area was created.

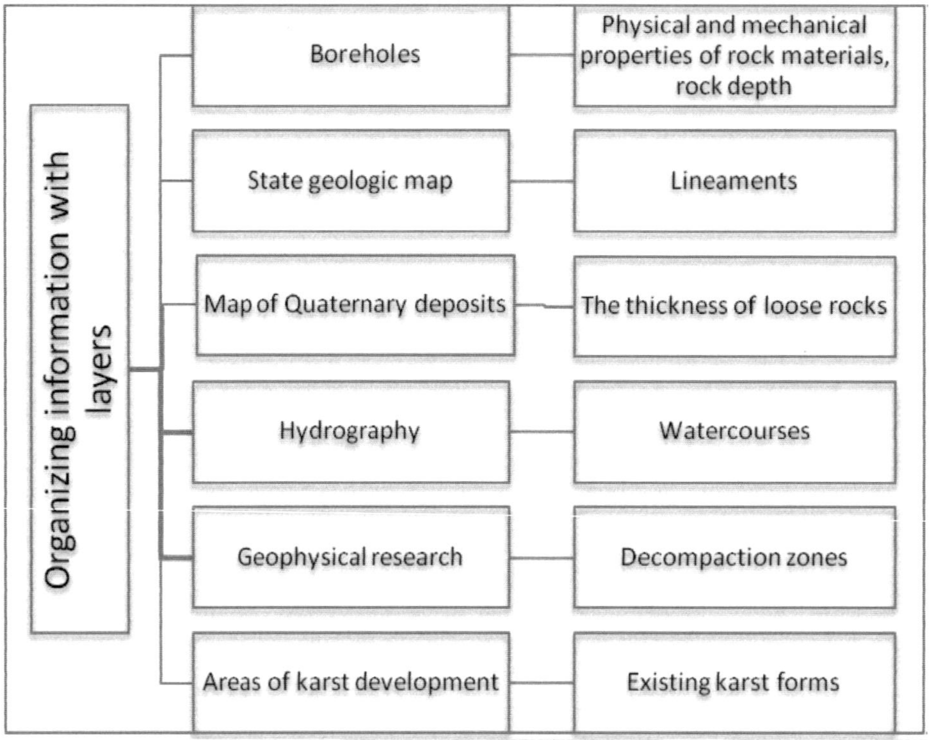

Figure 2. Data Structure in GIS «MapInfo Professional»

We suggest assessing hazards by employing the integral criterion of potential karst development. The following groups of indicators should be mentioned: structural and tectonic, hydrogeological, geological, geomorphological and geophysical groups. Among the indicators characterizing karst hazards are the number of crossing lineaments, groundwater depth, indicator of groundwater aggressiveness, existence of karsting rocks, existence/ absence of aquiclude, the remoteness of river networks, existence of karst landforms, existence of loosening zones. Thus, we suggest ranking the main gas pipeline route depending on its proneness to karst processes according to such criteria as:

- Dangerous area: (the number of crossing lineaments >2 crosses/km, groundwater depth > 3 m, indicator of ground aggressiveness <-1, presence of karsting rocks, absence of aquiclude, presence of karst landforms, presence of loosening zones according to geophysical research);
- Potentially dangerous area: (the number of crossing lineaments >1 crosses/km, groundwater depth > 5 m, indicator of ground aggressiveness $0 > A > -1$, presence of karsting rocks, presence of aquiclude < 1 m, presence of karst landforms);
- Safe area (the number of crossing lineaments <1 crosses/km, groundwater depth > 10 m, indicator of ground aggressiveness $A > 0$, presence of sufficiently powerful aquiclude).

This resulted in the zoning according to karst hazard ranking in the right-of-way (Fig.3).

While assessing thermokarst hazards it is necessary to take into consideration the construction and operation experience of the first gas main in cryolythic zone – VSTO-1. In our case the main factor increasing the thermokarst hazard is the heat transfer change on the ground surface when seasonal deicing depth exceeds the depth of seasonal ice bedding or icy permafrost due to human impact – tree-cutting due to pipe operation [5,6,7].

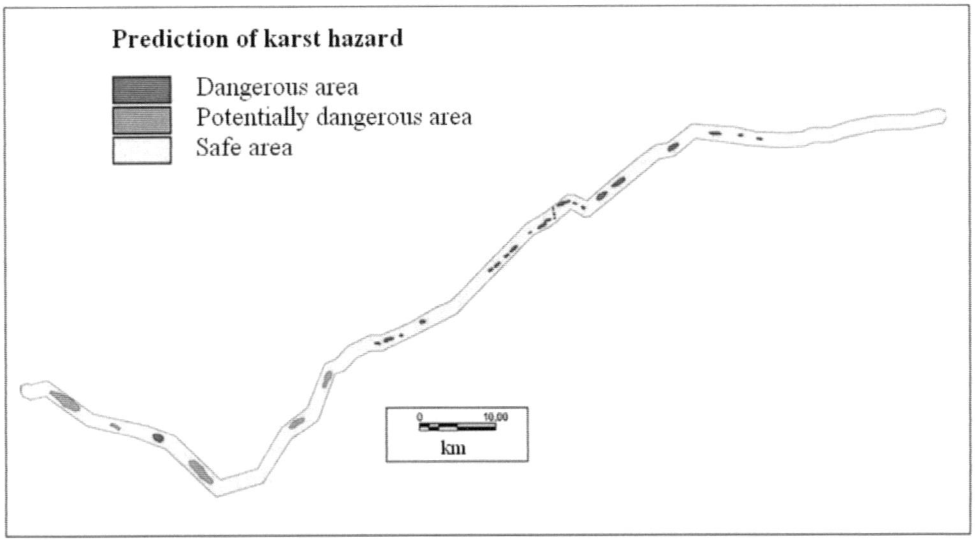

Figure 3. Zoning according to Karst Hazard Ranking

There are also hazards in areas with embedded subterranean ice and polygonal wedge subterranean ice. To define dangerous areas we propose the following algorithm: the analysis of geologic-lithological columns for detecting the areas with ice bodies based on the ranking (table 1).

Table 1. Indicators

Area	Indicators
Dangerous area	The presence of monomineralic ice (depth more than 0.1 m) in geological column
Potentially dangerous area	The presence of ice content in rocks (unit fraction > 0.3) in geological column
Safe area	The absence of ice content rocks (unit fraction > 0.3) and monomineralic ice in geological column

After selection in GIS "MapInfo Professional", the areas prone to thermokarst hazards were identified according to the indicators mentioned in table 1. Potentially dangerous areas were also identified along the main gas pipeline route (Fig.4).

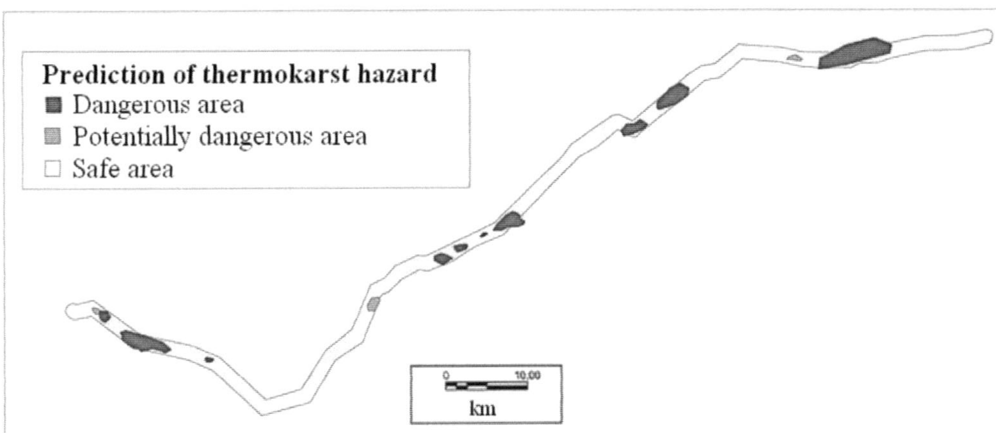

Figure 4. Zoning according to Thermokarst Hazard Ranking

Thus zoning according to karst and thermokarst hazards was made.

3. Summary

We propose to classify dangerous areas according to the degree of karst and thermorkarst hazards by mapping geological information on the base material in GIS "MapInfo Professional" and executing query in the existing database. The data obtained (the length of the area with the maximum degree of danger and the vesicle depth) are the basis for assessing the endurance capability and consistency of the gas main. The proposed procedure of assessing the impact of karst processes on the gas main may be applied while designing and operating in order to take into account engineering measures in project documentation on the operational stage including the operation with geotechnical monitoring.

References

[1] McAllister E W 2009 *Gulf Professional Publishing.* A manual of quick, accurate solutions to everyday pipeline engineering problems pp. 762.
[2] Singh R 2013 *Elsevier Inc.* Arctic pipeline planning: design, construction, and equipment. pp. 120.
[3] Antaki G 2003 *CRC Press.* Piping and pipeline engineering: design, construction, maintenance, integrity, and repair. Pp. 564.
[4] Smith P 2007 *Gulf Publishing Company.* The Fundamentals of Piping Design: Drafting and Design Methods for Process Applications. pp. 240.
[5] Kent Muhlbauer W 2004 *Elsevier Inc.* Pipeline Risk Management Manual. pp. 442.
[6] Boca Raton 2004 *CRC Press.* Water Resources Engineering in karst. pp. 340.
[7] White W B 2002 *J. Engineering Geology.* Karst hydrology: recent developments and open questions. Vol. **65** pp. 85-105.
[8] Worthington S R H, Ford D C 2009 *J. Ground Water.* Self-organized permeability in carbonate aquifers. Vol. **43(3)** pp. 326-336.
[9] Bayasan R M, Golubin S I, Pustovoit G P, Proshina T V, Korotchenko A G 2008 *Heat pipes, Heat Pumps, Refrigerators, Power Sources. VII Minsk International Seminar (Minsk Belarus).* Optimization of engineering solutions for thermal stabilization of saline permafrost soils at bases of structures by means of two-phase heat pipes
[10] Tolmachev V, Leonenko M 2011 *Karst management.* Experience in collapse risk Assessment of building on Covered Landscapes in Russia (Berlin: Springer) chapter. Vol. **4** pp. 75-102.
[11] Klimchouk A, Cucchi F, Calaforra J M, Aksem S, Finocchiaro F, Forti P 1996 *Int. J. Speleol.* Vol. **25 (3-4)** pp. 37-48.
[12] Chilingarian G V, Mazzullo S J, Rieke H H 1996 *A Geologic-Engineering Analysis. Part II. Amsterdam, Lausanne, New York, Oxford, Shannon, Tokyo: Elsevier.* Carbonate Reservoir Characterization. Vol. **16 (2)** pp. 994.
[13] Schorr M 2011 *Desalination, Trends and Technologies* (InTech) pp. 12.

Novorossiysk agglomeration landscapes and cement production: geochemical impact assessment

A V Alekseenko, M A Pashkevich
[1] Department of Geoecology, Saint Petersburg Mining University, 21st line V.O., 2, Saint Petersburg, 199106, Russia
E-mail: [2] al.vl.alekseenko@gmail.com, [3] mpash@spmi.ru

Abstract. The article deals with assessing the environmental impact of marl mining and cement production in Novorossiysk city (Krasnodar krai, Russia). The existing methods of studying the environmental effects caused by the cement industry have been reviewed. Soil and aquatic vegetation sampling has been carried out and the gross concentration of metals in the samples has been defined. The research has been conducted in the certified and accredited laboratory using emission spectral analysis. The external control has been carried out via X-ray fluorescence analysis. Based on the collected data, main chemical pollutants in soil cover and water area near the cement plant have been identified. The contaminants released by urban enterprises and motor vehicle emissions, as well as fugitive dust from dumps and the cement factory, lead to multi-element lithogeochemical anomaly at geochemical barriers in soils. Accumulation of pollutants in soil depends on the type of land use and the area relief. The most contaminated aquatic landscapes have been identified in the inner bay. According to this information, the technical proposals can be prepared for environmental safety management in strongly polluted city areas, as well as for the reclamation design in the areas currently experiencing the negative impact of cement production.

1. Introduction

Novorossiysk city is located in the foothills of the North-West Caucasus on the Black Sea coast. The Novorossiysk industrial agglomeration consists of the city over 300,000 people, the eldest Russian cement factory "Novoroscement", the largest regional Novorossiysk seaport, as well as other enterprises and transport infrastructure objects. The cement factory was put into commission in 1882, and is currently producing 4 million tons per year. Open-pit mining of marl deposit in the agglomeration is a source of raw material for Portland cement manufacturing. According to the report "State and Protection of the Environment of the Russian Federation in 2011", the cement plant is recognized as a major air pollution source in the city.

The research is devoted to monitoring the environmental impact of cement factory, i.e. soil and water pollution. Dust emission is the key environmental pollution driver of a cement industry [2, 7, 17, 24, 25]. An impact of technogenic precipitations on geochemical landscapes is manifested in changing chemical element concentrations in soil [5, 11, 22], plant cover and agricultural lands [8, 13, 23], which can also affect human health in the course of environmental migration [20, 21]. Compared to those of other geochemical landscapes, urban soils experience the strongest technogenic load [1, 4, 6, 14–16]. In the studied landscapes, polluting element anomalies are formed and redistributed depending on the geomorphological conditions. The subsequent surface run-off leads to contamination of the water area [3, 9, 10, 19].

2. Materials and methods

The *transeluvial* (steep slopes), *transaccumulative* (low-angle gentle slopes), and *trans-superaquatic* (overwater slopes) geochemical landscapes are located in different geomorphological conditions in the Novorossiysk agglomeration (figure 1). These landscapes were ranged by the district building types: *single-storeyed, 2–5 storeyed* and *5–10 storeyed*; the other studied areas were *industrial enterprise sites* (including cement factory, seaports, and railway station) and *wastelands*. As a result of intensive water contamination with toxic substances and increased turbidity, meso- and macro-benthos have almost completely disappeared from aquatic landscapes of Tsemes (Novorossiysk) bay. Landscapes with predominating brown algae *Cystoseira* are characterised by the greatest biomass. These landscapes are located along the both shores of the bay at the depth of 1.5–3.0 m.

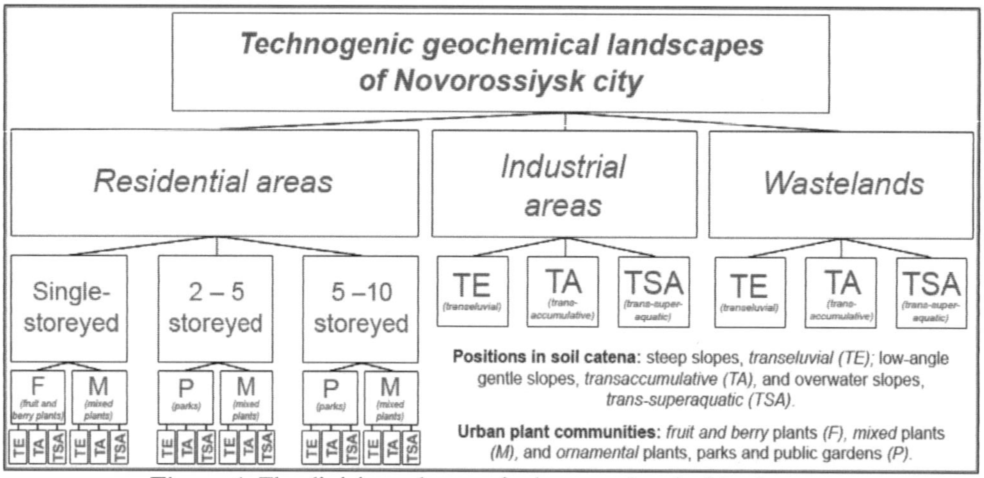

Figure 1. The division scheme of urban geochemical landscapes

Monitoring soil sampling (over 100 samples) was carried out twice, in 2014 and 2015. The samples were taken from the uppermost 10-cm soil horizons with a maximal intensity of geochemical processes in the landscape [12]. The brown algae used for aquatic landscape monitoring are the most sensitive to the changing environmental chemical content. The *Cystoseira barbata* algae used in the research are the most common in coastal aquatic landscapes coast of Novorossiysk. In 2009, 87 algae samples were collected in order to assess the current state of coastal landscapes of Tsemes bay, with the entire coastline being covered (figure 2). All spectral emission analyses of soil samples and seaweed ashes were carried out in the certified and accredited "Kavkazgeolsyomka" laboratory; and the external X-ray fluorescence control analysis was conducted in the Common Use Centre Laboratory of Mining University.

Figure 2. The schematic map of sampling points in Tsemes bay (based on the Yandex satellite image)

3. Results and discussion

Soil pollution was assessed separately in geochemical landscapes of residential areas, wastelands, and industrial sites. The joint effect of cement factory emissions, relief features and building height led to certain dispersion in elements' mean concentrations in soils of landscapes with various geomorphological conditions. The comparison between the *maximal* average concentrations and the *minimal* ones in the *residential area* shows that for 8 out of 22 studied elements the former is more than 1.5 times as much as the latter: *Pb* (4.5 times), *Sr* (1.7), *Ag* (2.2), *Cu* (1.7), *Zn* (2.5), *Ga* (1.5), *Sn* (1.6), and *Yb* (1.5). In soils of *wasteland* landscapes, under different morphological conditions, the excess of maximal average contents over minimal ones by more than 150 % was recorded for four elements only: *Zn* (1.7 times), *Pb*, *Co*, and *Ag* (1.6). We note that the maximal average contents of these elements in soils of *wastelands* are significantly lower than those in soils of *residential areas*. The distribution of the majority of chemical elements' average contents is more uniform in soils of *industrial areas*, including the main pollution source – the cement factory. The largest exceed of maximal average contents over minimal ones was recorded for two elements only: *Sn* (2.9 times) and *Pb* (1.7). The maximal average content of these elements in soils of *industrial areas* is higher than that in soils of *wastelands*, but lower than in *residential areas*.

Thus, *residential areas* mostly tend to be affected by the technogenic impact, resulting in the intensive soil pollution. The *maximal* for the entire city *average polluting element contents* were established in these areas. The greatest number of *unevenly distributed average element contents* was recorded in soils of *residential areas* under the influence of geomorphological and anthropogenic factors in combination. The concentrations of *Ag*, *Sn*, and *Mn* are maximal in soils of transeluvial landscapes in *residential areas* as compared with the whole considered area. At the same time, the minimal average contents in the uppermost parts of the slopes were defined for 10 elements: *Ag, Cu, Zn, Mo, Ti, Cr, Ga, Y, Yb,* and *Sc.* There has been a six-fold increase in the number of elements with the maximal mean contents, i. e. the number has changed for 18 *(Cu, Zn, Pb, Mo, Ba, Co, Ni, Ti, V, Cr, Ga, Li, Sr, Y, Yb, Sc, Zr,* and *Nb)* in soils of *transaccumulative* and *trans-superaquatic* landscapes of residential areas located below. This pollution is associated with relief characteristics [18]: the average contents in soils of upper slope parts were mainly due to the impact of pollutants precipitating after the sorption by cement industrial emissions.

The major part of defined element concentrations were lower in the ash of *Cystoseira* growing in the zones under the influence of biogenic terrestrial landscapes than in those influenced by technogenic landscapes. The highest average contents were noted at the observation points in the areas of urban wastewater treatment plant discharge (Aleksino settlement) and breakwater area, facing the greatest impact of precipitating cement plant emissions. The lowest contents of all the elements considered, except for *Mo*, are generally characteristic for *Cystoseira* of the western biogenic coast (table 1).

Table 1. The average element content in the ash of *Cystoseira barbata* in Novorossiysk bay, ppm

Part of Tsemes bay	Chemical element											
	Cu	Zn	Pb	Co	Ni	V	Cr	Sn	Mo	Mn	Ti	Ba
Western Biogenic	30.0	30.0	7.30	6.0	18.3	2.0	1.0	3.0	3.7	233	367	300
Myskhako	44.4	53.8	11.3	8.9	24.4	2.6	1.3	6.8	9.7	563	531	425
Aleksino	47.5	178	17.1	8.4	30.6	3.1	2.8	8.9	9.4	625	575	363
Central	51.8	57.3	13.8	8.2	22.7	2.6	1.2	6.0	6.2	464	636	455
Eastern Biogenic	41.7	48.3	10.3	6.5	22.5	3.0	1.0	7.0	5.3	583	617	467

The eastern biogenic coast is characterised by a slightly higher content of elements in comparison with the western one. The concentrations of five elements – *Cu* (41.7 ppm), *Zn* (48.3), *Pb* (10.3), *Co* (6.5), and *Cr* (1.0) are lower than those of *Cystoseira* sampled in the technogenic landscape influence zone. In this case, the concentrations of *Ni, V, Sn, Mo, Mn,* and *Ti* are close to the values typical for the coastal strip of Novorossiysk. Thus, the effect of pollutants precipitating from the air remains even on the coast with forest landscapes in the absence of direct technogenic impact. It should be noted that the cement plant is located on the eastern shore of the bay, and the increased metal content in algae may be related to the influence of atmospheric emissions.

Analysing the average element contents in *Cystoseira* ash samples of different years (table 2), we identified certain regularities of changes. The peaks of *Cu, Cr, Ni,* and *Mo* contents were observed in 2000 and 2009; *Mn, Ti,* and *Zn* – in 2009 only; *Pb* – in 2000 only. A slight increase in the average *V* content was noted in *Cystoseira* ash in 2004, and it remained at the same level in the samples of 2009. The increased content in the 2004 samples was also characteristic of *Sn*.

Table 2. The average Pb, Cr, Ni, and V content (with a 95% confidence level) in the ash of *Cystoseira barbata* in Novorossiysk bay, ppm

Year of sampling	Chemical element			
	Pb	Cr	Ni	V
1999	42.2±10.4	51.1±12.5	36.5±5.4	14.7±3.3
2000	46.1±13.0	58.5±7.6	45.9±5.7	11.5±1.1
2001	19.4±4.2	19.3±3.0	21.6±3.0	7.3±1.5
2002	12.0±2.0	13.5±3.0	19.1±2.0	6.3±1.1
2004	11.5±1.8	6.0±1.8	15.1±3.0	7.0±2.0
2009	10.2±3.0	7.5±2.4	22.8±6.0	7.0±3.0

Thus, based on the received data, it is possible to assume the persisting decrease tendency in the average contents of most studied elements in *Cystoseira* ash throughout 1999–2004, and elevated levels by 2009. Monitoring researches have shown that during the years 2001–2004 there were processes in the bay leading to a decrease of most investigated element contents in *Cystoseira* ash.

4. Conclusion

The urban soil pollution under the impact of mining and processing industry is determined by the local factors of chemical element migration and accumulation: geomorphological structures of urban area and types of land use. Monitoring researches in Novorossiysk city revealed that the change of these factors had caused a significant redistribution of chemical elements and their maximal accumulation in soils of residential areas. Pollutants are brought into the soil directly from the source of contamination, as well as due to the migration from landscapes located hypsometrically above.

The biogeochemical investigation allowed revealing the water areas of Tsemes bay that can be described as "relatively pure"– the western and eastern outer bay parts in the zones influenced by biogenic terrestrial landscapes. The inner bay part, experiencing the greatest impact of fugitive dust plant emissions, as well as the neighbouring Aleksino settlement, where urban wastewater treatment plants are located, may be recognised as the most polluted.

The pollutant migration factors should be accounted to determine the reasons for the landscapes to be particularly contaminated. The data obtained characterise the ecological and geochemical state of Novorossiysk urban environment, and allow developing the measures for improvement.

Acknowledgements
The analyses were performed using the equipment of the Common Use Centre of Saint Petersburg Mining University. This research was partly funded by the joint Russian-German program "DAAD—Mikhail Lomonosov".

References
[1] Alekseenko V and Alekseenko A 2014 *J Geochem Explor.* The abundances of chemical elements in urban soils. Vol. **147** pp. 245–249.
[2] Allué J, Garcés A M, Bech J, Barceló J and Poschenrieder C 2013 *J Soils Sediments.* Fractionation of chromium in tannery sludge-amended soil and its availability to fenugreek plants. Vol. **14 (4)** pp. 697–702.
[3] Balykin D N, Puzanov A V and Balykin S N 2013 *Geogr Nat Resour.* Ecologo-geochemical assessment of the Vasyugan River valley (Tomsk Oblast). Vol. **34 (2)** pp. 166–171.
[4] Bech J, Reverter F, Tume P, Sanchez P, Longan L, Puente A and Oliver T 2011 *J Geochem Explor.* Pedogeochemical mapping of Al, Ba, Pb, Ti and V in surface soils of Barcelona Province (Catalonia, NE Spain): Relationships with soil physicochemical properties. Vol. **109 (1-3)** pp. 26–37.
[5] Bech J, Roca N, Barceló J, Duran P, Tume P and Poschenrieder C 2012 *J Geochem Explor.* Soil and plant contamination by lead mining in Bellmunt (Western Mediterranean Area). Vol. **113** pp. 94–99.
[6] Bech J, Tume P, Sokolovska M, Reverter F, Sanchez P, Longan L, et al 2008 *J Geochem Explor.* Pedogeochemical mapping of Cr, Ni, and Cu in soils of the Barcelona Province (Catalonia, Spain): Relationships with soil physico-chemical characteristics. Vol. **96 (2-3)** pp. 106–116.
[7] Bech J, Tume P, Longan L, Reverter F, Bech J, Tume L and Tempio M 2008 *Environ Monit Assess.* Concentration of Cd, Cu, Pb, Zn, Al, and Fe in soils of Manresa, NE Spain. Vol. **145 (1-3)** pp. 257–266.
[8] Bezel' V S and Zhuikova T V 2007 *Russ J Ecol.* Chemical pollution: Transfer of chemical elements to the aboveground phytomass of herbaceous plants. Vol. **38 (4)** pp. 238–246.
[9] Buzmakov S A, Voronov G A and Zaytsev A A 2013 *J World Appl Sci.* The characteristics of the state of protected areas of Perm Krai. Vol. **22 (7)** pp. 956–963.
[10] Chalov S R, Shkol'nyi D I, Promakhova E V, Leman V N and Romanchenko A O 2015 *Geogr*

Nat Resour. Formation of the sediment yield in areas of mining of placer deposits. Vol. **36 (2)** pp. 124–131.

[11] Il'in V B, Syso A I, Baidina N L, Konarbaeva G A and Cherevko A S 2003 *Eurasian Soil Sci.* Background concentrations of heavy metals in soils of southern Western Siberia. Vol. **36 (5)** pp. 494–500.

[12] Kasimov N S and Perel'man A I 1992 *Eurasian Soil Sci.* The geochemistry of soils. Vol. **24** 59–76.

[13] Kim J, Yalaltdinova A, Sirina N and Baranovskaya N 2015 *J Sci Food Agric.* Integration of life cycle assessment and regional emission information in agricultural systems. Vol. **95 (12)** pp. 2544–2553.

[14] Kosheleva N E, Kasimov N S and Vlasov D V 2015 *Eurasian Soil Sci.* Factors of the accumulation of heavy metals and metalloids at geochemical barriers in urban soils. Vol. **48 (5)** pp. 476–492.

[15] Minkina T, Mandzhieva S, Chapligin V, Motuzova G, Sushkova S, Fedorov Y, Kolesnikov S and Bauer T 2014 *J Environ Eng Manag.* Accumulation and distribution of heavy metals in plants within the technogenesis zone. Vol. **13 (5)** pp. 1307–1315.

[16] Motuzova G V, Minkina T M, Karpova E A, Barsova N U and Mandzhieva S S 2014 *J Geochem Explor.* Soil contamination with heavy metals as a potential and real risk to the environment. Vol. **144 (PB)** pp. 241–246.

[17] Pashkevich M A and Petrova T A 2015 *Int J Ecol Dev.* The investigation of toxic waste burial polygon impact on the atmospheric air using lichenoindication method (Russia). Vol. **30 (3)** pp. 98–105.

[18] Perel'man A I 1986 *Appl Geochemistry.* Geochemical barriers: theory and practical applications. Vol. **1 (6)** pp. 669–680.

[19] Puzanov A V, Baboshkina S V and Gorbachev I V 2015 *Water Resour.* Concentration and distribution of major macro- and microelements in surface waters in the Altai. Vol. **42 (3)** pp. 340–351.

[20] Rikhanov L P, Baranovskaya N V, Ignatova T N, Sudyko A F, Sandimirova G P and Pakhomova N N 2011 *Geochem Int.* Trace elements in human organs and tissues and their significance for environmental monitoring. Vol. **49 (7)** pp. 738–742.

[21] Rikhvanov L P, Baranovskaya N V and Sudyko A F 2013 *Gornyi Zhurnal.* Chemical elements in human body as the basis for realization of ideas of medical geology. Vol. **3** pp. 37–42.

[22] Samonova O A, Aseyeva E N and Kasimov N S 2014 *J Geochem Explor.* Metals in soils of erosional systems in forest zone in the central part of European Russia. Vol. **144 (PB)** pp. 247–259.

[23] Syso A I, Kolpashchikov L A, Ermolov Y V, Cherevko A S and Siromlya T I 2014 *Contemp Probl Ecol.* Elemental chemical composition of soils and plants in Western Taimyr. Vol. **7 (6)** pp. 636–642.

[24] Van Oss H G and Padovani A C 2002 *J Ind Ecol.* Cement manufacture and the environment: Part I: Chemistry and technology. Vol. **6 (1)** pp. 89–105.

[25] Van Oss H G and Padovani A C 2003 *J Ind Ecol.* Cement manufacture and the Environment: Part II: Environmental challenges and opportunities. Vol. **7 (1)** pp. 93–126.

Identification and assessment of geohazards affecting pipelines and urban areas

L A Strokova[1,a] **E A Teterin**[1,b]
[1]Tomsk Polytechnic University, 30 Lenina Ave., Tomsk, 634050, Russia

E-mail: [a]sla@tpu.ru, [b]badscrool@gmail.com

Abstract. The paper addresses methods and criteria of risk assessment associated with land subsidence threatening pipelines, buildings, and constructions. Currently, there are some practical issues relating to geohazards that should be taken into account while constructing a pipeline. The article provides comparison data on the effects of Spitak earthquake and the natural disaster in Neftegorsk in terms of geohazards impact on the pipeline systems. The suggested risk assessment procedure embraces a wide range of aspects: from soil properties to economic and management issues.

1. Introduction

The Russian gas pipeline system crosses the areas with different climatic, geological, tectonic, and hydrogeological conditions that may have negative impact on the pipeline operation and maintenance. Geological or geotechnical factors can lead to the pipeline failure. Regarding South Yakutia, there are a lot of dangerous geological and geocryological processes, such as erosion, swamping, karst, thermokarst, thermal erosion, aufeis formation, bloating, screes, slope displacement caused by thawing of soil, etc. [1].

Geohazards present a separate group of possible threats for the pipeline. Geohazard risk assessment involves evaluating failure frequency of pipelines subjected to geohazards. A geological process is considered «safe» or «dangerous» in terms of its impact on the stability of a buried pipeline. It means that if a dangerous geological process does not cause a threat to the pipeline construction and safety operation, it is regarded as "safe". However, if a geological process threatens the pipeline integrity and can cause some damage to the construction, it is classified as "dangerous". At the same time, other occasion can break integrity of pipeline, which considered as damaging event. Thus, geohazards risk assessment for pipelines has its particularities.

2. Geohazards and their consequences regarding to a pipeline

The pipeline is considered to be in «safe» condition if there is no factor threatening its integrity or the pipeline can resist all possible stresses. However, some elements of the pipeline system are less reliable than a pipe, thus, they should be evaluated separately (figure 1).

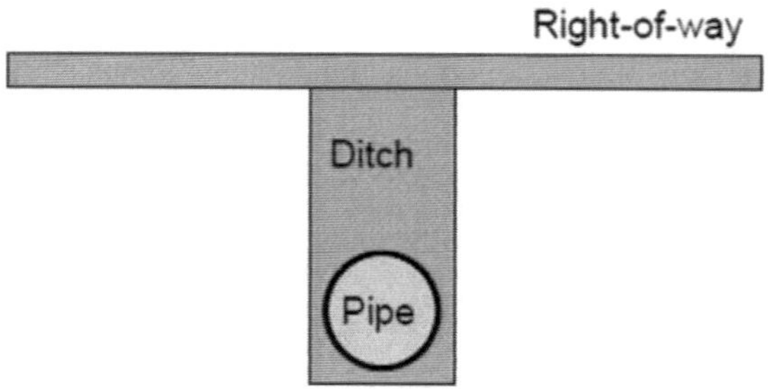

Figure 1. Pipeline elements.

Land subsidence has the most significant influence on the pipeline system. The earth's surface subsidence may occur slowly or suddenly due to different causes: oil or gas production, karsting rocks, mining activities, destruction of permafrost, etc. Land subsidence involves a downward displacement of soil that can be accompanied by horizontal movement of the soil. Both land displacements can cause negative effect on the pipeline and generate a compressive stress resulting in pipeline failure and disturbance of the pipeline stability. The deformation of land surface can occur in the form the bending, tilting, or sinkholes. Table 1 shows the list of geohazards and their impacts on a pipeline system [2].

Table 1. List of hazards and their impacts on a pipeline.

Disaster	Effects	Elements affected (directly only)		
		Pipe	Ditch	ROW
Thawing of permafrost under the pipe	It can result in deformation of pipes due to bending in pipes in areas of cavities	Yes	No	No
Thawing of permafrost around the ditch	It can induce pipe destabilization and displacement due to soil erosion along the ditch	No	Yes	No
Thawing of permafrost above the pipe	It can lead to drainage violation caused by ground subsidence	No	No	Yes
Earthquake shocks	It increases dynamic strain on pipe	Yes	No	No
Soil liquefaction	It decreases soil strength	Yes	Yes	Yes
Soil erosion	It leads to erosion over the pipe	Yes	Yes	No
Sinkhole development (karst)	It leads to pipe deformation	Yes	No	No

3. Risk assessment procedure

The table above presents only a part of possible geohazards that can induce a pipeline failure. However, the consequences may lead to significant economic loss and environmental problems. To avoid the negative consequences, it is necessary to evaluate all the risks in order to predict and prevent possible failures. Risk assessment is the determination of quantitative or qualitative estimate of risk related to a certain situation defined as a threat. The procedure implies two specific stages. The first stage is to determine risk frequency; the second one is to identify the influence of the risk, i.e. severity of consequences [3]. Generally, risk management consists of the following steps:

- Risk management planning;
- Risk identification;
- Qualitative and quantitative risk analysis;
- Risk response plan;
- Risk monitoring and control.

Basic principles of response to any threats are shown in figure 2. To avoid any accident associated with hazardous geological processes, it is necessary to assess the possibility of their occurrence and the scale of their negative influence. Then, the probability level is correlated to the scale for each risk category. A certain color corresponds to a specific level of threat. This method allows estimating and predicting consequences of the hazard. In addition, it identifies the value of each situation and provides the solution. The disadvantage of this method is that probability level and scale value are estimated subjectively.

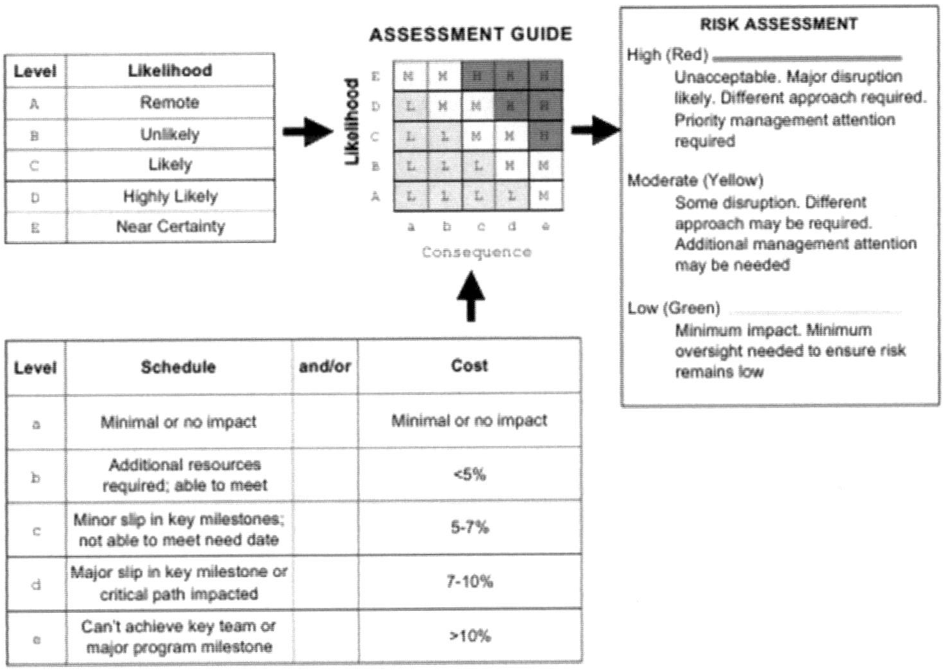

Figure 2. Risk evaluation procedure.

4. Integrated approach for risk assessment of geohazards

The method and criteria mentioned above may not be enough to predict land subsidence with required accuracy. In this case, it is necessary to use additional parameters such as rock properties, tectonic fractures, morphometric characteristics of the relief, physical and mechanical properties of soils, groundwater characteristic, etc. One of the methods implies taking into consideration geological indicators that are often used in karst research. Rock stratum assessment in terms of engineering-geological conditions implies studying composition, condition, structure and texture of sediments and overlying strata of soluble rocks and their physical properties. Usually, the list consists of 21 parameters divided into 6 group [3]: structural-tectonic, geological, hydrogeological, geomorphological, engineering-geological and geocryological (table 2).

Table 2. List of criteria to assess the risk of land subsidence

Group	Parameters
Structural and tectonic	Lineaments density Ll, km/km^2
	Number of lineament intersections Ml, $number/km^2$
	Modularity Bl, km^2
	Distance from lineaments Rl, m
Geological	The of quaternary loose layer thickness
	Overburden thickness
	Composition of bedrock
Hydrogeological	Groundwater level HQ, m
	Fissure-karst water level Hk, m
	Salinity of groundwater M, g/dm^3
	Hydrochemical facies of groundwater
Geomorphological	Slope of a terrain , degree
	Distance from river network U, m
	Watershed slope tg α, u. f.
Geotechnical	Density ρ, g/cm^3
	Total deformation modulus E_0, MPa
	Angle of internal friction of soil φ, degree
	Specific cohesion of the soil c, kPa
Geocryological	Soil temperature t, °C
	Ice content i_{tot}, u. f.
	Total humidity W_{tot}t, u. f.

One of the most important factor in the development of subsidence is a geological structure of the area, which involves texture and structure of soluble rocks; content and composition of insoluble residue; composition, structure, thickness and nature of bedding layers; genesis and age of karst rocks. Karst is dangerous due to possible underground cavities that are sometimes difficult to identify by engineering geological hazard assessment Currently, geological assessment of subsidence risk is quite an efficient way to determine potential threat and to prevent negative impact.

5. Quantitative risk assessment calculations

Quantitative risk evaluation is becoming worldwide the main constituent in decision-making related to risk management activities. Differentiated estimations of economic risks caused by landslides, sinkholes and subsidences resulted from karst development, karst-suffusion and suffusion, liquefaction and mining should be carried out in the form of total and specific (normalized to unit of area) values by the following formulas (1), (2):

$$R_e(H) = P(H) \cdot P_s(H) \cdot V_e(H) \cdot D_e \quad (1)$$

$$R_{s\,e}(H) = R_e(H) S_0^{-1} \quad (2)$$

where $R_e(H)$ and $R_{s\,e}(H)$ - total ($€/year$) and specific ($€/(m^2 \cdot year)$ respectevely, $€/(ha \cdot year)$ or $€/(km^2 \cdot year)$ is risk of loss caused by hazard H of certain nature and intensity; $P(H)$ – is frequency of hazard occurrence (H) within a certain area, which is numerically equal to its statistical probability ($cases/year$); $P_s\,(H) = S_0/S_t$ - geometric probability of hitting the object in hazardous area; S_0 - section of the object (m^2, ha, km^2); S_t – section, where the hazard H can occur (m^2, ha, km^2); D_e - cost of the object before its destruction (€).

$V_e(H)$ - economic exposure of the evaluated object to the hazard H.Equation (3) is used for some area (the formula is different for buildings and structures) [5].

$$V_{t\,e}\,(H) = D_{td} D_{te}^{-1} \quad (3)$$

where D_{td} – cost of the destroyed territory, D_{te} – overall cost of evaluated territory.

For better risk assessment, it is reasonable to introduce such parameters as average speed of a geological process and average amplitude of the deformations into the formulas presented above. Differentiated evaluation of loss for buildings and structures should also involve the geological risks of adjacent areas.

Geological risk assessment should be carried out at all stages of geotechnical survey and should result in geological risk maps, reports and conclusion.

6. Vulnerability evaluation

A vulnerability assessment is the process of identifying, quantifying, and ranking the vulnerabilities of a system. It may be conducted in the physical, social, economic or environmental fields.

In the simplest case, it may be defined by the following formula (4):

$$V(H) = n_i \cdot n^{-1} \quad (4)$$

where V(H) is the object vulnerability in any sphere of consideration of possible losses (fractions of unit, f.u.); n_i – is the number of destroyed and damaged elements by hazard H or the cost of these elements in object; n – is the total number of elements or the total cost of these elements before disaster.

The equation allows identifying vulnerability of a particular area. For example, the 8-9 magnitude earthquake in Spitak occurred in the area with 1million residents and 120000 houses. It resulted in destruction of 58 settlements and about 61000 houses. According to the equation [6]:

- Physical vulnerability of settlements is $58 \cdot (365)^{-1}$ = 0.159;
- Vulnerability of dwelling houses is $61 \cdot (120)^{-1}$ = 0.51;
- The population vulnerability is equal to 0.025.

The disaster in Neftegorsk was of the same power. However, the vulnerability of 3-5-storey houses was 1 and that of the population was 0.58 [4].

Today, experimental statistical tables of vulnerability of buildings and their occupants are developed to be applied to earthquakes, karst collapses, subsidence, and underflooding [5-8]. Until recently, the vulnerability evaluation remains a poorly developed element in the general scheme of natural risk analysis.

7. Conclusion

The methods described above and the assessment criteria can be used to predict and monitor land subsidence and protect the pipelines and other urbanized facilities from their negative impact. The impact of natural factors on the stability of objects in karst areas should be studied. Such study contributes to the development of petroleum industry in geologically complicated areas and investments in high-tech industry. To ensure effective prediction and monitoring of geological hazards and to manage preventive activities it is reasonable to apply the complex approach that takes into account all the aspects of possible hazards.

References

[1] Strokova L A, Dutova E M, Ermolaeva AV, Alimova I N and Strelnikova AB 2015 Karst hazard assessment in the design of the main gas pipeline (South Yakutia) / *IOP Conference Series: Earth and Environmental Science* **27** 012032 (http://iopscience.iop.org/1755-1315/27/1/012032)

[2] Strokova L A 2015 Modeling of tunneling-induced ground surface movement *IOP Conference Series: Earth and Environmental Science* vol 24 (Tomsk: Scientific and Technical Challenges in the Well Drilling Progress) pp 1–5

[3] Rizkalla M., Read R S 2013 Overview of pipeline geohazard assessment approaches and strategies // Paper No. IPG2013-1950, V001T02A007. ASME 2013 International Pipeline Geotechnical Conference. Bogotá, Colombia, July 24–26, 2013.

[4] Ragozin A L 2009 Geological risks, formation and assessment in urbanized territories in Russia. In Culshaw M G, Reeves H J, Jefferson I and Spink T W *(ed.) Engineering Geology for Tomorrow's Cities* **22** p 282

[5] Meng Y, Dai J, Jia D and Lei M 2013 Typical methods for forecasting karst collapse in China. In Land L, Doctor DH, Stephenson JB, (ed.) *Sinkholes and the Engineering and Environmental Impacts of Karst: Proc. of the Thirteenth Multidisciplinary Conference (Carlsbad, New Mexico, 6–10 May 2013) NCKRI Symposium 2. Carlsbad (NM):* National Cave and Karst Research Institute pp 239–245

[6] Thinh H P and Strokova L A 2015 Prediction maps of land subsidence caused by groundwater exploitation in Hanoi, Vietnam *Resource-Efficient Technologies: electronic scientific journal* **1** pp 80–89

[7] Fotieva N, Bulychev N, Sammal' A S, Antziferov S, Deev P V, 2007 Influence of soil grouting on the shallow tunnel linings stress state in urban areas *The 4th Dimension of Metropolises: Proc. 33rd ITA-AITES World Tunnel Congress* (Underground Space) pp 439-444

[8] Fotieva N, Sammal A, Deev P, Bulychev N 2005 Design of shallow tunnel linings under seismic effects of earthquakes *Proc. 16th Int. Conf. Soil Mechanics and Geotechnical Engineering: Geotechnology in Harmony with the Global Environment, September 12-16 2005* Osaka, Japan vol **3** 1607-1610

Impact of environmental factors on the demographic characteristics in Tomsk Oblast (Russia, 1980-2015)

E Pugacheva[1], A Mezhibor[2] and T Makarenko[3]

Department of Geoecology and Geochemistry, Institute of Natural Resources, National Research Tomsk Polytechnic University, 30 Lenin Av., Tomsk, 634050, Russia;
E-mail: [1]pugachevatomsk@yandex.ru , [2]amezhibor@gmail.com, [3]t.vl.makarenko@yandex.ru

Abstract. The research represents the analysis of essential demographic indexes in Tomsk Oblast (Russia): birth-rate, death-rate, natural increase (1980-2015), migration increase (1997-2014), and child mortality (1990-2015). Environmental factors were determined as influencing the health and as a consequence, having the impact on the demographic characteristics of the studied region.

1. Introduction

Tomsk Oblast is situated in the southeast of the Western Siberian lowland in the Middle Ob River. It takes the territory of 314,400 square km (1.9% of the Russian Federation) [1]. The climate in Tomsk Oblast is continental due to its geographic location in the temperate latitude: 55-56° N. Average-year air temperature ranges from -0.5° C in Tomsk to -3.5° C in the northeast of the region.

On January 1st, 2016 the constant population base accounted 1076,800 people; the density of population was 3.43 people per 1 square km [2]. The region includes 6 cities, 1 urban-type settlement, and 694 villages. Distribution of the population in the region territory is irregular. The essential part of the population lives in the southeast of the region because remote places have undeveloped traffic system and weak economical-geographical characteristics. By the ratio of urban and village population Tomsk Oblast relates to the highly urbanized territories with 71.6% of urban population. The industrial production is predominately concentrated in two cities – Tomsk and Seversk. The northern areas of the region are the places of the location of large oil-producing facilities.

2. Results and discussion

The analysis of the indexes of natural population migration in Tomsk Oblast over the last 35 years allowed detecting the following essential attributes of the demographic situation.

1. The birth-rate index constantly decreased from 1983 to 1995, at the same time the mortality increased (figure 1) [2-4]. The natural increase of the population from 1980 to 1993 was positive. The maximum value of the natural increase index was marked in 1983 (11.4 for 1000 people).

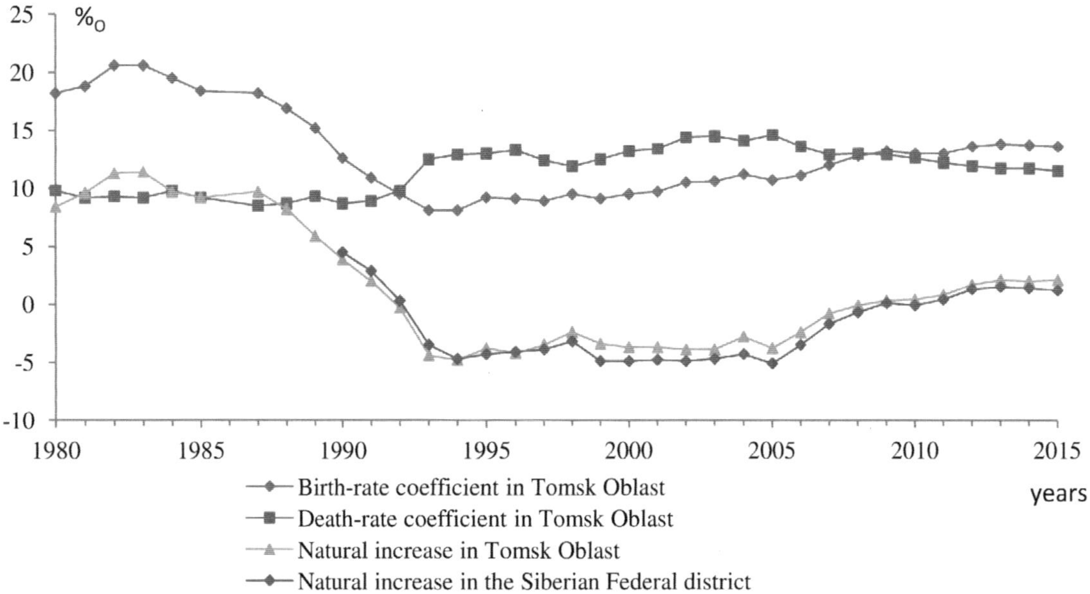

Figure 1. Dynamic of the indexes of natural population change (per 1000 people) from 1980 to 2015. The graphs were made using the data from [2-4].

In 1992 the increase of the population in Tomsk Oblast sharply stopped, as well as in the whole Russia, the birth-rate index corresponded to the death-rate. This situation was named "Russian cross". In the graphs of the natural movement of the population the death curve crosses the birth cross. After 1992 the dead amount gradually began to exceed the birth number. In the opinion of demographers, this situation was formed after the collapse of the USSR. Due to the incompleteness of the modernization processes a challenging social-economic situation has formed.

Since 2006 the death-rate in Tomsk Oblast began to decrease and in 2009 the index of natural increase was positive. In 2009 the situation of the "Russian cross" repeated. Since 2010 the birth-rate gradually began to exceed the death-rate. In 2015 the birth-rate was 13.6 births per 1000 people, the death-rate was 11.5, that was 15% lower than the birth-rate. Such a trend in the natural population increase from 1994 to 2015 was also observed for the Siberian Federal District (SFD) [2, 5]. At the same time the whole Central Russia and many regions of the Russian Federation had natural decrease of the population in 2014.

Such an index of demographic well-being as child mortality for the last 24 years has approximately corresponded to the data of the SFD with the tendency to decrease (figure 2). The main reasons of the child death under 1 year are often perinatal condition (54.4%), congenital anomalies (17.8%), indeterminate diagnosis (11.9%), accidents and poisoning (8.9) [5, 6]. The period from 2007 to 2015 in the Tomsk Oblast has been notable as the most indicative in the child mortality decrease. In the beginning of 2016 the region had first place in the SFD (the lowest level of the mortality – 4.7 per 1000 of living born children) and fifth place in Russia.

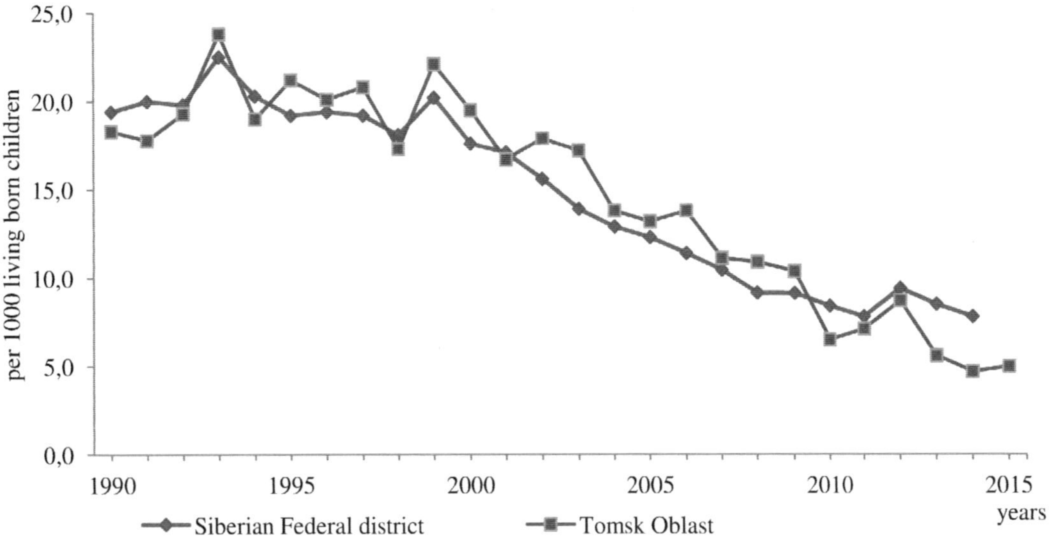

Figure 2. Child mortality from 1990 to 2015 (per 1000 living born children). The graphs were made using the data from [6].

One of the determining factors influencing the demographic indexes is people health. In the dynamic of the morbidity index a growth tendency of the first-diagnosed diseases has been found (figure 3).

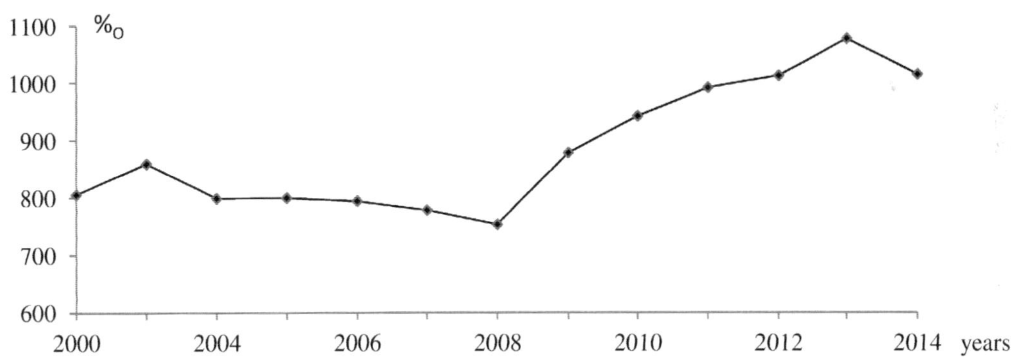

Figure 3. Morbidity of the population (registered with the first-diagnosed diseases, per 1000 people). The graphs were made using the data from [4].

The primary diseases incidence from 2000 to 2013 increased from 805.3 to 1014.7%o with the first-diagnosed diseases [4]. In the structure of the primary morbidity of the population of the Tomsk Oblast the largest part is due to respiratory diseases, accidents and poisonings, diseases of the genitourinary system, infectious and parasitic diseases, and eye and it's appendages diseases.

For the last 10 years among the reasons of the population death the diseases of circulation organs, neoplasms, diseases of the digestive system, respiratory diseases, and infectious and parasitic diseases have had the leading place (table).

Table. Death-rate by the main classes of the death reasons (number of the dead per 100,000 people) [4].

Death reasons	2005	2006	2007	2008	2009	2010	2011	2012	2013	2014
neoplasms	204.8	213	216.5	222.7	219.4	214.3	219.9	212.4	205.7	205.1
diseases of circulation organs	677.4	615.1	596.2	617.5	619.7	612.7	573.9	528.5	519.6	515.5
respiratory diseases	70.1	60	57.1	53	57.9	45.5	57.4	54.9	64.8	67
diseases of the digestive system	74.1	73	65.2	66	66.6	66.1	58.7	67	62.5	79.5
infectious and parasitic diseases	23.4	18	17	16.1	17.6	14	13.1	13.7	15.3	19.1

In the structure of the mortality of employable population of the region the leading places have the same death reasons. But the first place is accidents and poisonings (29.5%), second one – diseases of circulation organs (28.9%), third place – neoplasms (16%).

One more important demographic index is the migration processes of the population (figure 4).

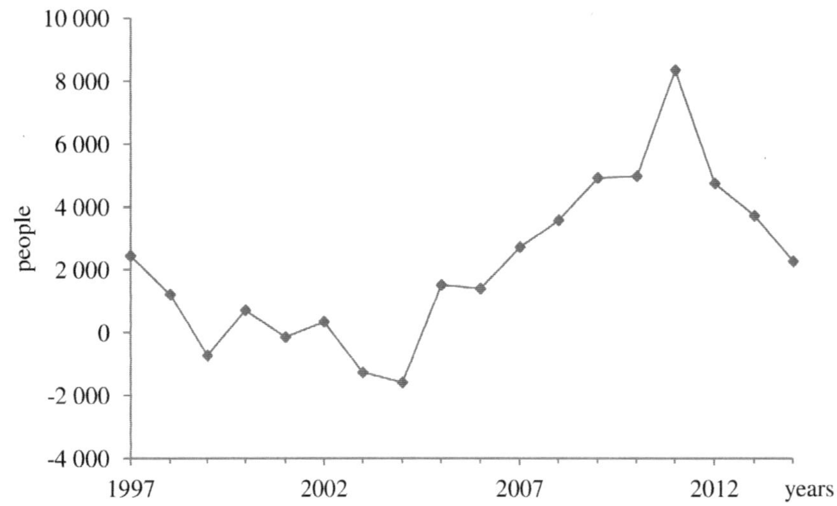

Figure 4. Migration increase of the population of Tomsk Oblast from 1997 to 2014. The graph was made using the data from [3, 4, 6]

From 2005 to 2011 an influx of foreign and Russian immigrants has grown in Tomsk Oblast in comparison with the period from 1997 to 2004. This fact was probably caused by the increase of the status value of Tomsk universities and student arrival, and by the attractiveness of the essential social-economic situation in the region [7].

2. Demographic characteristics are first of all determined by the condition of the population heath, which is defined by environmental factors in a greater extent.

Since 2011 a steady reduction of emissions in the atmosphere has been observed in Tomsk Oblast [1]. This tendency is connected with the decrease in the oil and gas extraction and with the realization of programs on the utilization of oil-associated gas. Since 2010 the provision of the population by safe water has been about 60-65%. But along with this situation, in 2014 524 water pipes were used for the domestic water consumption, and 81.3% from them did not satisfied sanitary requirements due to the absence of the protective sanitary zones and necessary complex of water-treatment equipment [4]. In 2014 367.20 billions m^3 of sewage water were discharged in water ponds, which is 3.71 billions m^3 less than in 2013. Since 2005 this indicator decreased by 170.6 billions m^3.

The radiation situation in Tomsk Oblast was generally formed due to the atmospheric deposition of radionuclides caused by the former nuclear tests in polygons of other regions of Russia, and by the radionuclide pollution because of the Siberian Chemical Combine (SCC) functioning, radioactive waste storage, and accidents [1, 8]. For the last 15 years the radiation doze rate in the settlements of Tomsk Oblast was within the limits of the natural radiation background – from 6 to 13 $\mu R \cdot hour^{-1}$ with the mean values of 7-10 $\mu R \cdot hour^{-1}$. The radiation situation connected with the trace of the accident in 1993 in SCC now is normalized. The radionuclide concentration in food, drinking water, soil, and atmospheric air is within the standards.

All types of waste cause negative environmental impact in different extent. From 1990 to 2001 the volume of waste has been about 1,500 000 tons a year, from 2002 to 2009 it is decreased to 641,000 tons a year. In 2014 771,000 tons of wastes were formed from industry and 330,400 tons were domestic wastes.

By the condition of the atmosphere, surface and ground water the region is determined as a zone of high environmental hazard. Environmental factors can have indirect impact on some diseases, first of all on child pathologies, child mortality, and also on respiratory system, circulation and digestion system, which take a leading place in the structure of the main population diseases.

Natural-environmental factors of Tomsk region determine sanitary-epidemiological situation by the tick-borne encephalitis and opisthorchiasis diseases [9, 10]. Since 2001 a constant tendency in the decreasing amount of patients with the tick-borne encephalitis has been observed. In 2001 504 people were revealed, in 2007 – 307 people, in 2008 – 263 people, in 2012 – 131 people, in 2013 – 143 people. Due to the location of Tomsk Oblast in the water area of Ob-Irtysh basin and due to the active consumption of river fish by the population, one of the most socially important problems for the region is the opisthorchiasis disease. The accounted 400-600 cases per 100,000 people for the last 10 years.

In the structure of the death reasons from infectious and parasitic diseases for the last 10 years the death-rate coefficient has been declining, with the negligible rise in 2013 and 2014 (Table).

3. Social regional programs, maintained by the country, are directed to the heath protection, healthcare delivery, and, as a consequence, to the demographic indexes. For example, from January 1st, 2007 the program "mother (family) capital" began functioning [11]. The main objective of the program is the stimulation of the birth rate and the improvement of the demographic situation in the country. The opening of the "Regional Perinatal Center" in 2010 contributed to the decline of the child deaths (under 1 year). The Center is equipped by unique medical facilities satisfied to modern requirements [12]. Since 2012 the program "Population health survey" has contributed to the preventive examination and early definition of diseases of the adult population; the application of additional reproductive technologies (extracorporal fertilization) has been provided under the base program of the compulsory health insurance. The system of the specialized medical help includes the application of new complicated and (or) unique methods of therapy, and also resource-intensive methods of therapy with the proved effectiveness, including cell technologies, robotic technics, information technologies and genetic engineering, developed on the basis of medical advances and related fields of science and technology. Arrangements on the development of high-technology medical help were taken as a basis for the state program "The development of the heath care protection of Tomsk Oblast" for years 2015-2020.

The strategic objectives of the social-economic development of the region supposes the decrease of the morbidity and mortality, increase of longevity of the population, improvement of the quality of medical services with the maintaining free medical service in the region [13].

3. Conclusion

Thus, the main demographic characteristics of Tomsk Oblast since 90-s and until 2008 have reflected pronounced features of the demographic crisis. After 2008 a negligible, but stable excess of the natural increase has been fixed. Since 2000 the child mortality declined. These tendencies have been determined to a large extent by the regional social-economic reasons, but they have been aggravated by the local natural and environmental conditions. Taking into account the large territory of Tomsk Oblast and different social-environmental-economic situation in every municipal-administrative unit, the further analysis of demographic characteristics is necessary to carry out on the basis of indexes of ranking and zoning.

References

[1] Trapeznikov S Ya (Ed.) 2015 *State report "About the condition and protection of the environment in Tomsk Oblast in 2014"* (Tomsk: Deltaplan) pp. 156 (in Russian).

[2] *Official statistics, Federal Agency of State statistics* URL: http://tmsk.gks.ru (date of access: 10.01.2016).

[3] Pugacheva E E 2008 Main demographic characteristics of the population in Tomsk Oblast *Proceedings of All-Russian Inter-Institutional Scientific-Practical Conference Demographic situation in central regions of Russia and realization of the conception demographic politic of the Russian Federation* (Voronezh: Voronesh State University) pp 125-127 (in Russian).

[4] *Data of Tomskstat: Regional office of the Federal State Statistics Service of the Tomsk region, 2016* (Tomsk) pp. 56 (in Russian).

[5] Pugacheva E E, Makarenko T V 2016 Analysis of demographic indexes of Tomsk region (1980-2014) *Proceedings of International Scientific-Practical Conference LXIX Herzen readings Geography: development of science and education* (Saint-Petersburg: Herzen State Pedagogical University of Russia) pp. 146-150 (in Russian).

[6] EMISS. State statistics. URL: https://www.fedstat.ru/indicator/data.do?id=31166 (date of access: 12.01.2016).

[7] Embrecht R V 2011 *Bulletin of Tomsk State University.* Role of educational migration in the demographic development of Tomsk region (2000-2009). Vol. **348** pp. 76-79 (in Russian).

[8] Gauthier-Lafaye F, Pourcelot L, Eikenberg J, Beer H, Le Roux G, Rhikvanov L P, Stille P, Renaud Ph, Mezhibor A 2008 *Journal of Environmental Radioactivity.*Radioisotope contaminations from releases of the Tomsk–Seversk nuclear facility (Siberia, Russia). Vol. **99** pp 680–693.

[9] Udintseva I N, Poltoratskaya T N, Shikhin A V, Poponina A M, Zhukova N G, Lukashova L V, Malysheva L A 2010 *Bulletin of Siberian Medicine.* Tick-borne viral encephalitis in the Tomsk Region for the last decade. Vol. **4** pp. 156-162 (in Russian).

[10] Sanitary-epidemiological situation *Government of Federal Service on the Control in the Sphere of Protection of Consumer Rights and People Well-being in Tomsk Region.* URL: http://70.rospotrebnadzor.ru/epidemiologic_situation/ (date of access 10.01.2016)

[11] Federal law, December 29th 2006 N 256-ФЗ "About additional measures of the state support of families with children" *GARANT.* URL: http://base.garant.ru/12151286/ (date of access 10.01.2016)

[12] *Department of the health protection of Tomsk region.* URL: http://zdrav.tomsk.ru/ (date of access 08.01.2016)

[13] Nikulina I E, Khomenko I V 2015 *Procedia - Social and Behavioral Sciences.* Interdependence of demographic and economic development of regions. Vol. **166** pp. 142-146.

REE, Uranium (U) and Thorium (Th) contents in Betula pendula leaf growing around Komsomolsk gold concentration plant tailing (Kemerovo region, Western Siberia, Russia)

D V Yusupov, Yu A Karpenko

Department of Geoecology and Geochemistry, Institute of Natural Resources, National Research Tomsk Polytechnic University, 30 Lenin Av., Tomsk, 634050, Russia

yusupovd@mail.ru , schaman734@gmail.com

Abstract. The article deals with the research findings of peculiarities of REE, Uranium and Thorium distribution in the territory surrounding the tailing of former Komsomolsk gold concentration plant according to the data from *Betula pendula* leaf testing. In the leaf element composition the slight deficiency of MREE and substantial excess of HREE are presented. In the nearest impacted area around the tailing, La, Yb, U and Th content, and Th/U ratio are lower than in the distant buffer area. It is shown, that value of Th/U ratio and REE can be an indicator for geochemical transformations of technogenic landscapes in mining districts. The results of the research can be used for biomonitoring of the territory around the tailing.

1. Introduction

For applied problem solving in geochemical researches, rare earth (REE) and radioactive (U, Th) elements are often used. The study of their distribution in natural and technogenic environments gives an opportunity for detecting of different classification features, determining of regularities, assessment of substance income sources and its differentiation degree [1-4]. For example, Th/U ratio, REE in plant and soil matters can reflect geochemical compound of underlying rocks in the conditionally background areas, they also allow detecting of the nature imbalance of these elements in technogenic landscapes [5-10].

The mining industry wastes affect negatively the environment and humans. The non-natural geological features – tailings – have the most negative impact, where the significant amount of waste with untreated mineral components is stored.

As a result of hypergenesis processes (decomposition, oxidation, solution, hydrolysis and others) primary ore minerals transform into secondary mineral forms. Heavy metals containing there migrate actively into the environment, pollute its components: atmosphere, surface water, ground waters, soil, flora, exceeding background concentrations and maximum permissible concentrations [11, 12].

The research purpuse is to study concentration distribution of rare earth elements, Uranium and Thorium in *Betula pendula* leaf as bioindicators in tailing areas of former Komsomolsk gold concentration plant for their usage in monitoring of polluted areas.

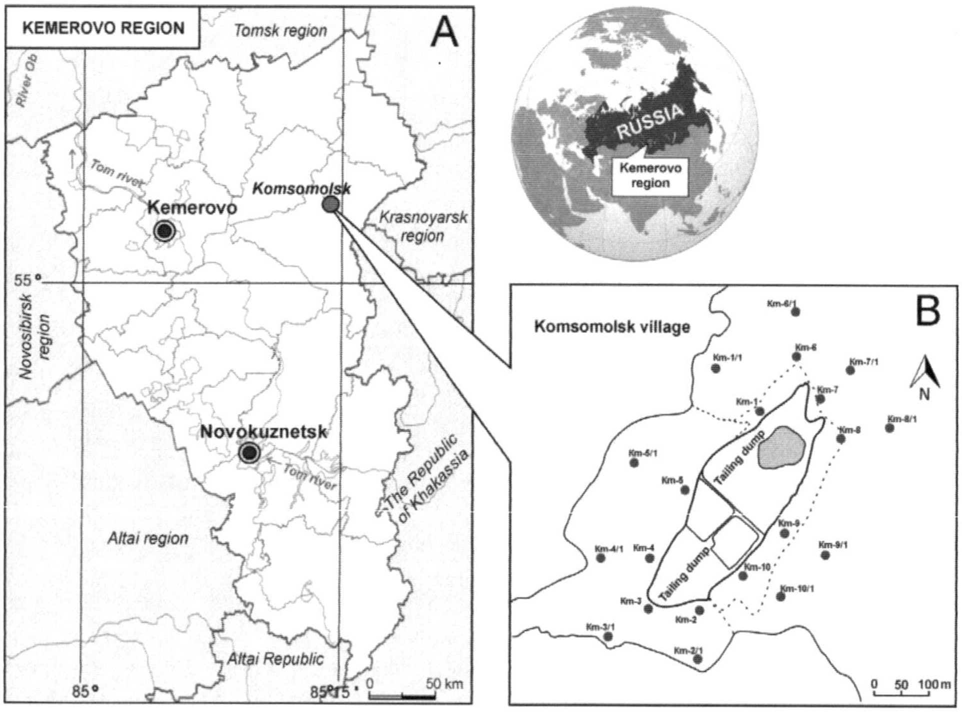

Figure 1. Map of the study area position **(A)**, location of Komsomolsk tailing and *Betula pendula* leaf testing scheme (Km, •) **(B)** (latitude = 55°38'6.34", altitude = 88°11'52.51").

2. Materials and methods

The research area is the tailing of former Komsomolsk gold concentration plant in the Komsomolsk village of Kemerovo region (Western Siberia, Russia) (fig. 1, A). At the plant aurum-arsenic pyrit-quarz ores, as well as gold-containing wastes of Kadamzhaisk antimony plant (Kyrgyzstan) and Berikul gold-processing plant were recycled using cyanation method [13].

The tailing is situated in natural basin in 250-500 meter from residential area of the village. It takes 147 thous. m^2 and its volume is 810 thous. m^2, the tailing contains about 1,1 million m^3 of waste and it is a shrinkable technogenic lake (fig. 2, a). From the south-western side the tailing is bounded by filling dam. Oxidizing wastes (fig. 2, b) contain (first %) sulphidic minerals: pyrite, sphalerite, galenite, pyrrhotine, and arsenic pyrite [13].

The subject of study was *Betula pendula* leaf. It is a widespread species and wood edificator in Siberian region. Being a mesophyte, a birch-tree tolerates dry conditions well; it is light-demanding, but undemanding to soil capability. It often grows on the disforest places or burned forest areas. The birch is also often used in safeguarding foresting under the conditions of increased technogenic load.

Leaf testing was conducted in the second ten-days of July 2015 using radial network from the outer boundary of the tailing with a 150-200 meter pitch along 10 lines (fig. 1, B). The leaves were sampled from the approximately evenaged trees using middle sample method at a height of 1,5-2 m from the ground level. Totally, 20 samples were collected. The baseline sample was collected in the village Makaraksky by the side of the river Kiya in 10 km south-westwards from the tailing. For sample packing the strong envelopes "Sterit" were used. Sample preparation for quantitative element analysis includes drying by the environmental temperature, mechanical grinding, weighted portion taking and decomposition in concentrated nitric acid by standard methods.

a b

Figure 2. Photographs of the northern part (a) and of the southern part (b) of the tailing landscape.

The total content detection of chemical elements in the samples of dry solids of leaf was conducted using mass-spectrometry with inductively coupled plasma in the accredited chemical-analytical centre "Plasma" (Tomsk). For the control of analysis quality the standard sample of *Betula pendula* leaf content is used (SSS 8923-2007). The estimation error has not exceeded 10%. The experimental data were processed with the use of different software (MS Office 2003, Statistica 8.0, Surfer 10, Corel Draw X6).

3. Results and discussion

The REE content of *Betula pendula* leaf around Komsomolsk tailing is shown in fig. 3. REE distribution has serrate composition, deficiency of medium rare earths (MREE - Sm) is slightly expressed, and heavy rare earths (HREE – Ho, Er) are in sufficient excess. In comparison with the data [14], REE content (except Er) in *Betula pendula* leaf is 3-10 times lower.

Geochemical peculiarities of rare earth and radioactive elements distribution in the territory of Komsomolsk tailing according to the data of *Betula pendula* leaf testing are shown in the diagrams of indicator ratios of Th/U and La/Yb (fig. 4). By consideration of La and Yb distributions in the leaf samples, we separate two groups of samples (fig. 4, a). The samples from the first group collected in the impacted area at the tailing border, by comparison with the samples from the second group collected in the buffer area far from the tailing are characterized by the lower contents of La and Yb, but by higher values of La/Yb ratio (from 16 till 122, average - 42).

The similar regularity appears by the consideration of Th and U distributions (fig. 4, b). It is found, that Th/U ratio has lower values (1,1 – 1,2) in the leaf samples collected in the impacted area at the tailing border than in the samples collected in the buffer area at a distance of 150-200 m from the outer tailing border (2,3 – 3,7). The Th/U ratio for the average content of these elements in the selection is 2,2. The low values of Th/U ratio indicate that the natural balance of radioactive elements is upset towards the increased U content determined by the technogenic transformation of the environment. High values of Th/U reflect probably the natural factor of the environment – the impact of igneous granitoidal ore complex developed in the territory on the U and Th content in flora. These values of Th/U ratio are comparable with the values obtained by the other researchers for soils (4,0), grassland vegetation ash (3,1) and *Betula pendula* leaf ash (2,7) in the background areas of the southern part of Western Siberia [15].

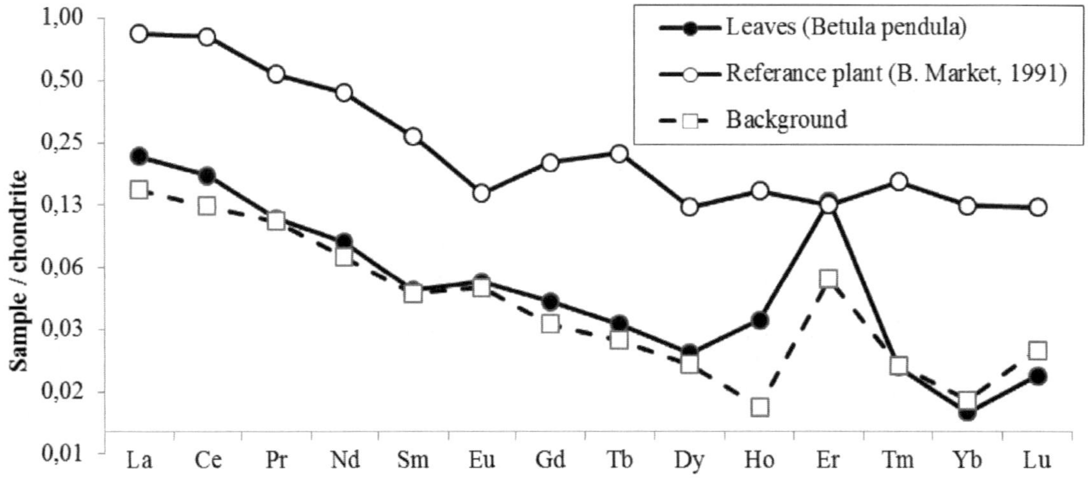

Figure 3. The rare earth element content of leaves (*Betula pendula*) around the Komsomolsk gold concentration plant in comparison with the baseline data and published data[14].

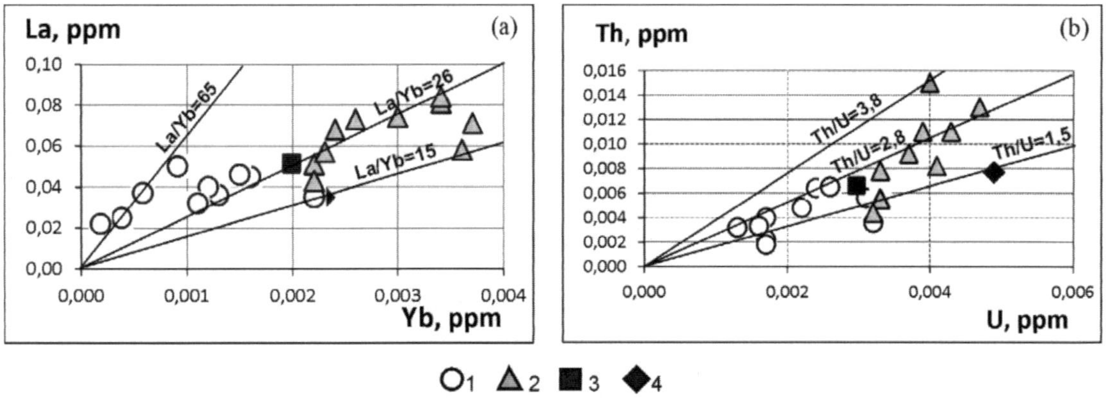

Figure 4. La/Yb (a) and Th/U (b) ratios diagrams for leaves (*Betula pendula*). Legend: 1 – samples taken at the border of the tailings; 2 – samples taken at a distance of 150-200 meters from the boundary of the tailings; 3 – the average content in the samples (n = 20); 4 – background value.

The peculiarities of spatial structure of geochemical fields around Komsomolsk tailing according to the indicator ratios of La/Sm, La/Er, La/Yb and Th/U are presented in the figure 5. Halos with contours 9-12 of La/Sm ratio cover most of the area around the tailing, maximal ratio is 12,5; it is shown in the central part (fig. 5, 1). Halos with maximal values 60-85 of La/Er ratio takes the north-western and western part of the tailing (fig. 5, 2). In the northern part of the tailing the halo with maximal value of La/Yb ratio – 122 is detected (fig. 5, 3). When the plant was in operation, there was a sludge pond near this area, but now it is a shrinkable lake (fig. 2, a), which is used by local children for swimming. In the southern part of the tailing the halo with minimal La/Yb ratio – 16 is detected. The oxidation zone of sulphide-containing wastes with the removal of sulfur, ferrum oxide and hydroxide formation is developed here (fig. 2, b). The halos with minimal Th/U ratio are detected in the basin in the impacted area, where the tailing bed is situated; the halos with maximal Th/U ratio are detected in the buffer area, on watersheds with natural landscape (fig. 5, 4).

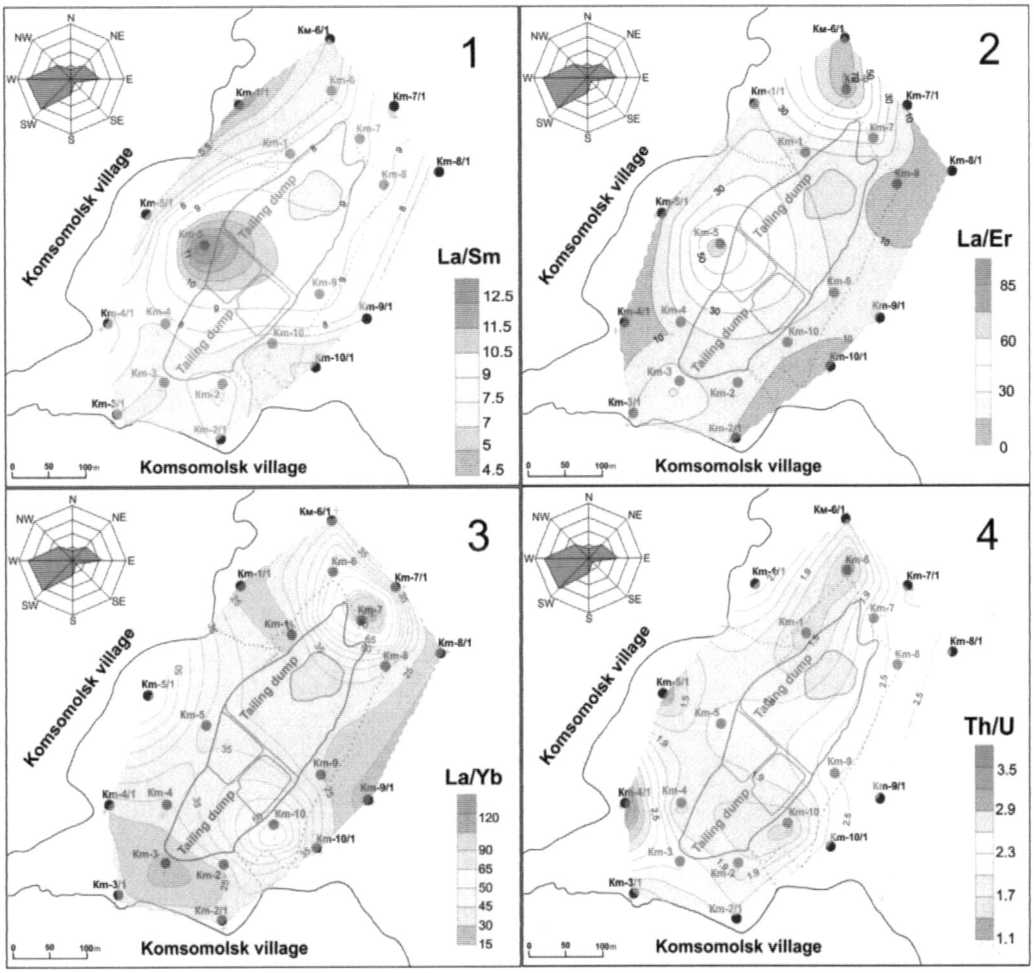

Figure 5. The maps of La/Sm (1), La/Er (2), La/Yb (3) and Th/U (4) ratios ("the nearest neighbor" method) showing the halos of geochemical transformation of landscape around the tailing of former Komsomolsk gold concentration plant

4. Conclusion

For the foregoing information, we can draw the following conclusions:

1. For the REE distribution in the territory of Komsomolsk tailing according to the *Betula pendula* leaf testing data the slight deficiency of MREE (Sm) and sufficient excess HREE (Ho, Er) are observable.

2. The distribution regularity of La and Yb and of U and Th in the *Betula pendula* leaf samples depending on the distance from the tailing is discovered. On the tailing border, in the impacted area the lower contents of La, Yb, U and Th, as well as low value of Th/U ratio were indicated. The halo with minimal value of La/Yb ratio maps the oxidation zone of sulphide-containing wastes. With distance from the tailing border, in the buffer area the La, Yb, U, Th contents and the value of Th/U ratio in growth samples increase.

3. The halos with maximal values of La/Sm ratio and La/Er ratio cover most of the area around the tailing bed.

4. The research findings confirm the conclusion, that values of Th/U ratio and REE can be indicators of the environment transformation degree in the limits of nature landscapes, and they can be used in biomonitoring.

Acknowledgment
This work was financially supported by the Russian Science Foundation grant №15-17-1001.

References

[1] Bargagli R 2005 *Moscow Geos Publ.* Trace elements in terrestrial plants: an ecological approach to biomonitoring and biorecovery. pp. 457.
[2] Markert B, Zhang D L 1991 Natural background concentrations of rare-earth elements in a forest ecosystem Science of The Total Environment. Vol. **103** pp. 27–35.
[3] Baumann N, Arnold T, Haferburg G 2013 *Germany Environ Sci Pollut Res.* Uranium contents in plants and mushrooms grown on a uranium-contaminated site near Ronneburg in Eastern Thuringia. Vol. **21** pp. 6921–6929.
[4] Reimann C, Arnoldussen A, Boyd R, Finne T E, Koller F, Nordgulen Q, Englmaier P 2007 *anthropogenic and geogenic concentration gradients Science of the Total Environment.* Element contents in leaves of four plant species (birch, mountain ash, fern and spruce) along. Vol. **377** pp. 416–433.
[5] Dunn C E 2011 Biogeochemistry in mineral exploration. URL: http:// www.books.google.com (date of access 03.05.2016)
[6] Chiarenzelli J, Asplerb L, Dunnc C, Cousensd B, Ozarkoe D, Powisf K 2001 *Applied Geochemistry* . Multi-element and rare earth element composition of lichens, mosses, and vascular plants from the Central Barrenlands, Nunavut, Canada. Vol. **16** pp. 245–270
[7] Gjengedal E, Martinsen T, Steinnes E 2015 Background levels of some major, trace, and rare earth elements in indigenous plant species growing in Norway and the influence of soil acidification, soil parent material, and seasonal variation on these levels Environ Monit Assess. pp.187 1–28.
[8] Kabata-Pendias A 2011 *CRC Press BocaRaton USA*. Trace Elements in Soils and Plants. pp. 505.
[9] Purvis O W, Longden J, Shaw G, Chimonides P D J, Jeffries T E, Jones G C, Mikhailova I N, Williamson B J 2006 *Journal of Environmental Radioactivity.* Biogeochemical signatures in the lichen Hypogymnia physodes in the mid Urals. Vol. **90** pp. 151–162.
[10] Purvis O W, Chimonides P J, Jones G C, Mikhailova I N, Spiro B, Weiss D J, Williamson B J 2004 *Proc. R. Soc. Lond.* Lichen biomonitoring near Karabash Smelter Town, Ural Mountains, Russia, one of the most polluted areas in the world. Vol. **271** pp. 221–226.
[11] Dolgopolova A, Weiss D J, Seltmann R, Dulski P 2006 *Journal of Hazardous Materials.* Dust dispersal and Pb enrichment at the rare-metal Orlovka–Spokoinoe mining and ore processing site: Insights from REE patterns and elemental ratios. Vol. **132** pp. 90–97.
[12] Lottermoser B 2007 *Berlin Heidelberg: Springer-Verlag.* Mine wastes: characterization, treatment and environmental impacts
[13] Bortnikova S B, Gaskova O L, Bessonova E P 2006 *Novosibirsk Geo Publ.* Geochemistry of technogenic systems. pp. 169
[14] Markert B 1992 Establishing of 'reference plant' for inorganic characterization of different plant species by chemical fingerprinting Water, Air, and Soil Pollution. Vol. **64** pp. 533–538.
[15] Rikhvanov L P, Arbuzov S I, Baranovskaya N V et al. 2007 *Bulletin of the Tomsk Polytechnic University.* Radioactive elements in the Environment. Vol. **311 (1)** pp.128–136.

Pyritization effect on well logging parameters in Jurassic reservoirs within S-E Western Siberian oil fields

A Janishevskii, T Ten and A Ezhova

Department of Mineral Exploration, Institute of Natural Resources, Tomsk Polytechnic University, Tomsk, Russia

E-mail: eav@tpu.ru

Abstract. Authigenic sulfide mineralization in hydrocarbon-saturated reservoirs distorts the electrical and density properties of rocks. The correlation between volumetric density, electro-conductive minerals and open porosity in 300 samples were determined. This fact made it possible to develop a nomograph in evaluating oil saturated reservoirs and could be applied in well geophysical survey data interpretation.

1. Introduction

Determining the reservoir fluid content (saturation) during exploration and hydrocarbon, reserve estimation is based on the interpretation of production well logging data. Formations with low resistivity (R_{por} = 3.5-5.5 ohm.m.) are identified as water-saturated ones [4, 7, 9]. Nevertheless, according to induction logging data (Fig.1),there are cases of cumulative oil inflow from Upper Jurassic reservoirs, namely, from water-saturated intervals within Tomsk Oblast oil fields.

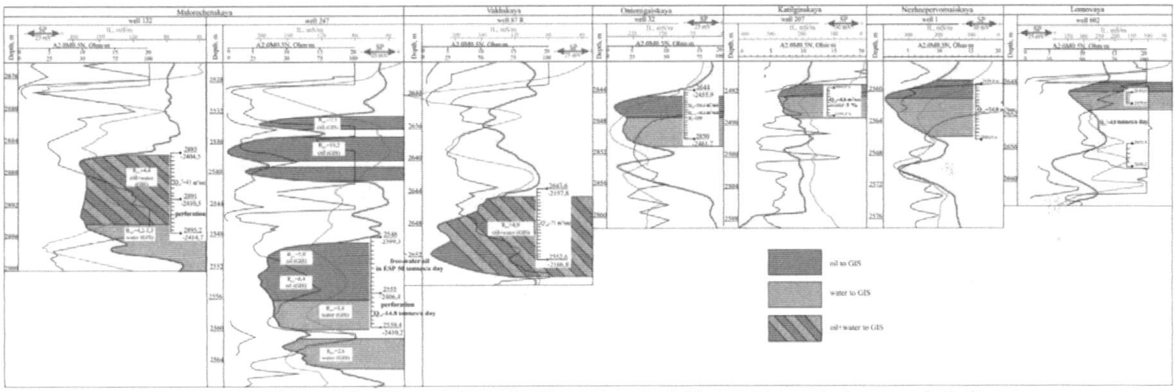

Figure 1. Examples of oil – saturated Upper Jurassic low-resistivity

Similar reservoirs were also exposed in Shirotni Prioby(Tumen Oblast) oil fields within NezhneVartovsk, Surgut and Nojabrsko-Pursk oil-gas bearing areas. This fact was established in the 90s of the last century by engineers of the Petrophysics Laboratory, Siberian Research Institute of Petroleum Industry (F.J. Borkun and others) in collaboration with the Department of Geology and Oilfield Development, TPU (A.V. Ezhova and others) [3]. These anomaly field geophysical characteristics of oil-saturated formations being determined as water-saturated due to low electric resistance values were the result of semi-conductor mineral occurrences in the rocks, such as sulphides, titanium and ferrum oxides. This mineral quantitative correlation indicated a well-defined predominance of sulphides. Titanium and ferrum oxides were only 1-2% and found in water-bearing horizons or water-saturated sections within hydrocarbon reservoirs, while sulphides were pervasive, including from 4 to 14 %.

Sulphides (FeS_2 compounds) are crystalline formations (pyrite and white iron pyrite) and powdered black matter (melnicovite). Pyrite and white iron pyrite have identical chemical composition (Fe – 46.6 %, S - 53.4 %), color (in reflected light-metallica, goldish), petrophysical properties (specific electric resistance – 10^{-5}–10 Ohm.m·, density- 4.88–5.2 gr/cm^3) [10]. These minerals differ only in their crystal structure: pyrite-cubic (i.e. cubic syngony), while white iron pyrite has tabular and / or spear-like crystals, as well as concretions with radial structure (rhomboid syngony) [2]. Melnicovite (collodial bisulphide iron) is a finely-dispersed matter with specific gravity of 4.1 gr/cm^3 and composed of fine pyrite and white iron pyrite grains (which was identified by X-ray diffraction analysis) [2].

The sulphides in oil-saturated reservoirs are finely-dispersed, having pseudomorphic and idiomorphic crystals, spherulites, irregular formations, veins and seams. The cement- basalt, porous and film-like. Sulphide formations as fines on grains and oils, as well as peripherally to the hydrocarbon-filled pores are of more interest.

Taking into account the fact that the petrophysical properties of both pyrite and white iron pyrite are identical and only their crystal structure differs makes it possible to solve the stated target without considering these two factors. Thus, for avoidance of any doubt, concerning the name of the above-mentioned minerals, one and the same name-pyrite is used according to Betekhin classification, in which pyrite group is included in the sulphide class [2].

Pyrite crystallization is probably due to iron oxide and sulphate reduction in formation water. Increasing SO_4^{-2} ion content (37−113 gr/liter) to (3.7−2.0gr/liter or less) is observed in formation water of Upper Jurassic sediments with anomaly geological description (31 and 49, respectively).

Oil and pyrite association is confined to the lower section of the layers where connate water content is increasing. Pyrite and connate water inclose oil occurrences and develop well-closed path transmitting electric current. Specific electrical resistance decrease can be registered on electrical logging diagrams.

2. Results and discussion

Survey results of geophysical parameters and well samples from SE Western Siberian petroleum province revealed the fact that identifying oil-saturated reservoirs based only on resistivity logging images is relatively unreliable. In this case, additional geophysical methods should be applied to determine other petrophysical rock properties, which, in its turn, would establish the electrically-conductive minerals within these rocks.

According to potential induced polarization values, electrically-conductive minerals are reliably identified. Measurements are conducted in Borkun samples [3] after extraction, i.e. hydrocarbon extraction. Electrical resistivity of semi-conducting minerals sharply increases at growing temperature within formation conditions [8]. In this respect, proposed Borkun nomograph determining reservoir saturation behavior, including electrically-conductive minerals and based on potential induced polarization measurement data [3], should be correlated to well logging. However, it should be stated that this method is not applied.

Another informative method in well logging identifying rock density properties is gamma-ray density logging. It is well-known [1] that in existing host rocks with density of (2.4-2.7)·10^{-3}kg/m^3,sulphides and other ores with a density of (4.5-5.1)·10^{-3}kg/m^3 reflect decreased radiation intensity on logs.

Volume density, electrically-conductive mineral content and effective porosity relationship identified in 300 samples provides data to plot an evaluation nomograph of the reservoir saturation parameters (Fig. 2).

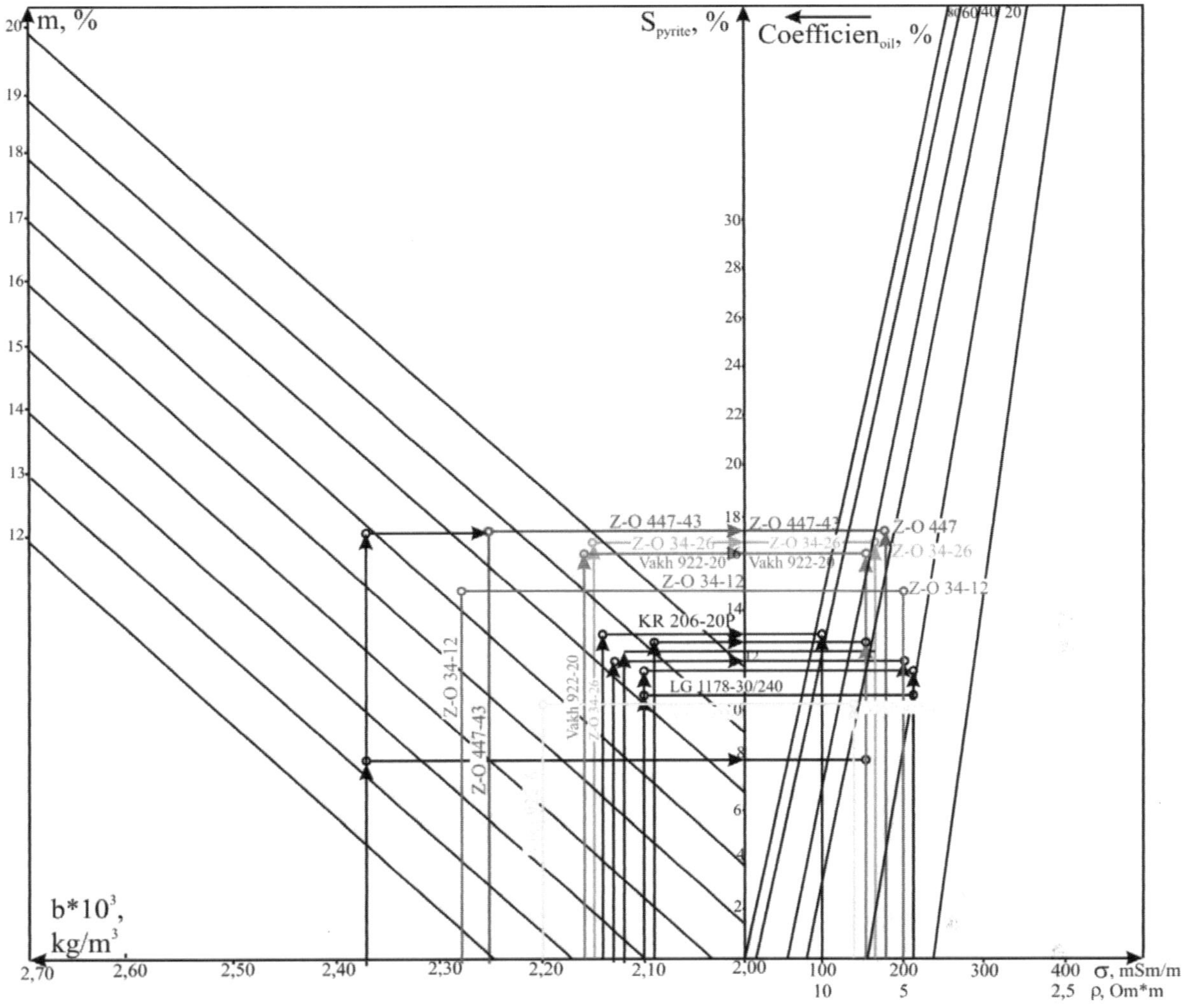

Figure 2. Nomograph of evaluating oil-saturated Upper Jurassic low-resistivity reservoirs

The first section of Borkun nomograph [3] plotted for Upper Jurassic reservoirs of Surgut anticline fold involves a relationship diagram of rock electric conductivity (σ), rock saturation coefficient (K_{sat}) and pyrite content in rock (S_{pyrite} %). Electric conductivity in wholly water- saturated sample without electrically-conductive minerals is 266 mS/m (R_{por} = 3.75 Ohm·m) at formation temperature of 90°C. Electric conductivity of pyrite at above-mentioned temperature is 1.5 Ohm·m.

The left nomograph section plotted by A.V. Ezhova [5] is a diagram indicating the content of electrically-conductive minerals based on rock volume density data (δ) including rock porosity (m) [5].

Application procedure of nomograph:

– rock volume density is determined based on gamma-gamma ray logging data on density;

– rock porosity coefficient is determined based on radioactive logging data;

– from point, corresponding to rock volume density (δ), a perpendicular is erected extending until it intersects the projection of the horizontal line, corresponding to rock porosity value (m). The projection of this point on Y-axis determines the pyrite content in rock (S_{pyrite} %);

– based on inductive electric- magnetic logging data, rock electric conductivity(σ) is determined by erecting a perpendicular extending until it intersects the projection of the horizontal line through Y-axis (S_{pyrite} %). Located point on the slanting line indicates oil saturation behavior in investigated reservoir beds.

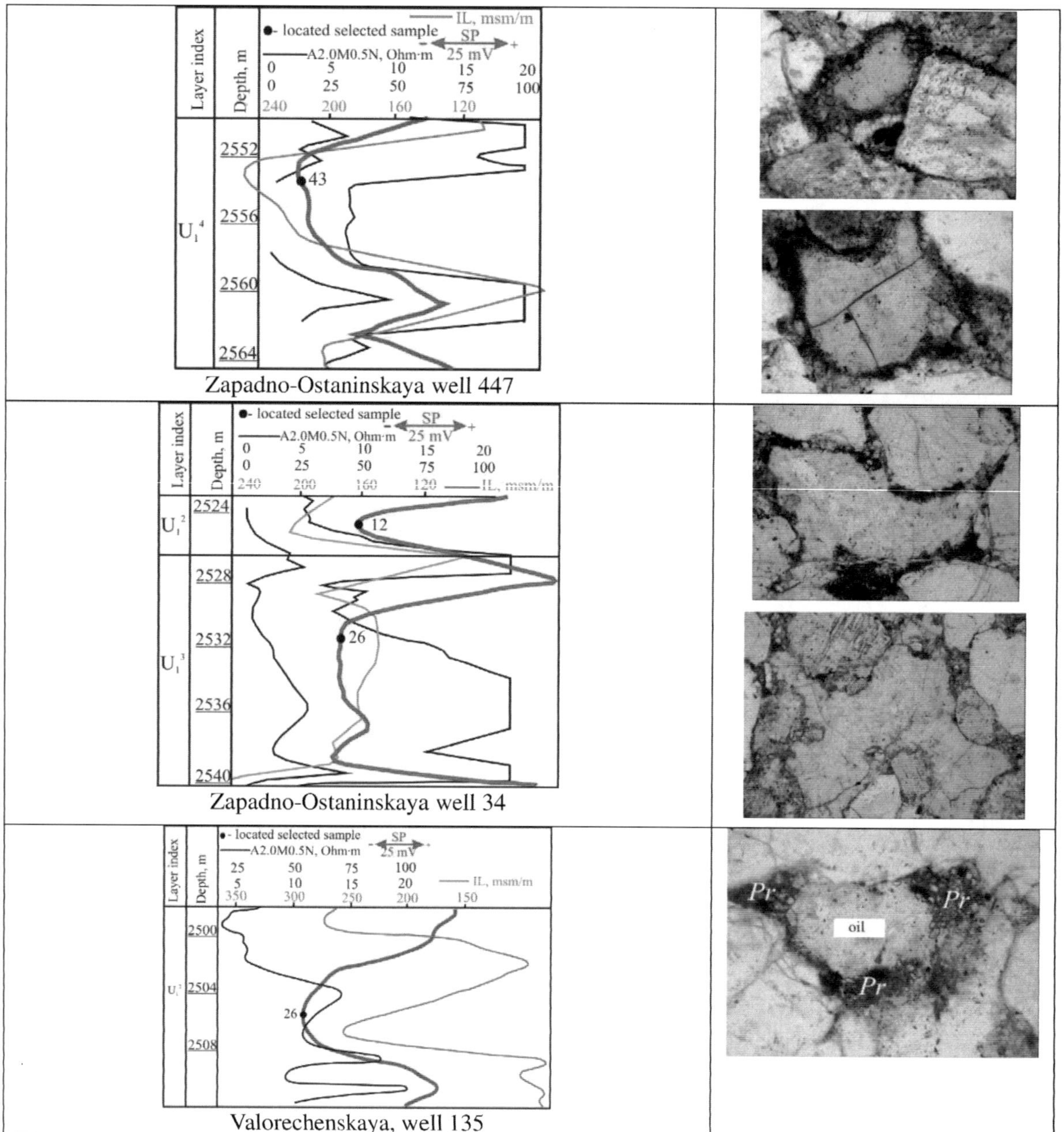

Figure 3. Geophysical characteristic anomalies and photos of thin sections, oil-saturated reservoirs

3. Conclusion

Nomograph example is based on the data of several investigated rock samples. According to geophysical data these selected samples were from water-saturated reservoirs. Occurrence of oil substance within pores was observed in thin sample sections (Fig. 3). Nomograph could be applied in GIS data interpretation to identify oil saturation degree in low-resistivity reservoirs [5].

References

[1] Artsibashev B, Ivanukovich G 1975 *Moscow Atomizdat.* Plotnostnoj gamma-gamma-karotazh na rudnyh mestorozhdenijah. pp. 72.

[2] Betekhtin A 1961 *Kurs mineralogii Moscow Gosgeoltechizdat*. pp. 540.

[3] Ezhova A, Libina V, Borkun F, Salnikova N 1991. Tomsk Polytechnic University. Vlijanie sul'fidnyh

IOP Conf. Series: Earth and Environmental Science **43** (2016) 012054 doi:10.1088/1755-1315/43/1/012054

obrazovanij na petrofizicheskie svojstva kollektorov gorizonta. pp. 162.
[4] Dakhnov V 1985 *Moscow Nedra*. Geofizicheskie metody opredelenija kollektorskih svojstv i neftegazonasyshhenija gornyh porod. pp. 310.
[5] Ezhova A 2006 Izvestiya Tomsk Polytechnic University. *Metodika ocenki neftenasyshhennosti nizkoomnyh kollektorov v jurskih otlozhenijah jugo-vostoka Zapadno-Sibirskoj plity. Vol.* **309 (6)** pp. 23-26.
[6] 1988 *Spravochnik Edited by Dobpinin Moscow Nedra. Interpretacija rezul'tatov geofizicheskih issledovanij neftjanyh i gazovyh skvazhin. pp.* 476
[7] Komarov V 1980 *Leningrad Nedra*. Jelektrorazvedka metodom vyzvannoj poljarizacii. pp. 391.
[8] Latishova M, Vendelshtein B, Tuzov V 1990 M*oscow Nedra*. Obrabotka i interpretacija materialov geofizicheskih issledovanij skvazhin. pp. 312.
[9] Vakhromeev G, Erofeev L, Kanaikin V, Nomokonova G 1997 *Tomsk Polytechnic University*. Petrofizika: Uchebnoe posobie dlja vuzov. pp. 462

"Digital petrophysics" in studies of porosity properties of low-permeable reservoirs

A A Ponomarev[1], M D Zavatsky[1], O V Veduta[1], R S Shulga[2], K O Vyatkina[2] and I A Koryashov[3]

1 Tyumen Industrial University, Tyumen 625000, Russia
2 Siberian Research Institute of Oil Industry, Tyumen 625013, Russia
3 Tomsk Polytechnic University, Tomsk 634050, Russia

E-mail: ponomarev94@mail.ru

Abstract. The article focuses on study in the properties of low-permeable reservoirs. The object of the research is the Bazhenov formation rocks taken from a well drilled in an oilfield of the West Siberian basin. A series of experiments were carried out to determine porosity of rocks by different methods: the helium saturation, nuclear magnetic resonance (NMR), microcomputed tomography (mCT). A comparative analysis was conducted to assess the efficiency of different methods. The obtained results have shown that common methods of petrophysical surveys are insufficient for carbonate-argillaceous Low-permeable rocks, containing a great amount of hand extractable organic matter. The additional data received by the mCT method gave an opportunity to increase sufficiently the reliability of the research results. We have concluded that the most efficient method for the study of low-permeable rocks is a combined use of the NMR and the mCT.

1. Introduction

Nowadays, special attention is given to hydrocarbon accumulations belonging to the category of hard recoverable reserves located in unconventional and structurally complex reservoirs. This is particularly important for Western Siberia where the majority of large oil deposits discovered in the 1960s-80s are nearly exhausted. One of the challenges for the development of oilfields is that low-permeable reservoirs with high oil viscosity cannot be explored by conventional petrophysical methods. Therefore, it is necessary to develop innovative exploration techniques for the study of reservoir properties.

In Western Siberia, hydrocarbon accumulations belonging to the category of hard recoverable reserves mainly occur in Upper Jurassic and Lower Cretaceous reservoirs: the Tyumen, Bazhenov, and Achimovsky formations. The deposits of these formations may be of different geological structures, but they all have the same peculiarity. These reserves are considered to be hard recoverable because of the following characteristics: low reservoir porosity and permeability, lithologic variation, inhomogeneity of spatial structure, and higher formation pressure gradient of productive horizons [1]. In such conditions, the conventional methods of determining reservoir properties using core data may be ineffective. While preparing samples (extracting hydrocarbons) it is hard or impossible to recover the organic matter completely, the residues of bitumen in micro- and nanopores distorts the survey results. Therefore, it seems advisable to introduce new techniques in the petrophysical studies of unconventional reservoirs. The mCT seems to be the most promising method allowing 3D visualizing the pore space structure of a rock sample.

2. Theoretic characteristic of porosity determination methods

The main task of petrophysical studies is to determine porosity and permeability of a reservoir. Any conventional method of open porosity determination by pore space saturation with liquid or gas requires oil pre-extraction from core-samples with subsequent water drying off. After that, pore space is saturated with liquid or gas, and open porosity is calculated. In case of unconventional reservoirs (low permeability, high oil viscosity), accurate calculations are impossible as liquid organic matter cannot be completely extracted without rock destruction, which leads to the significant errors in porosity determination.

An alternative approach to the reservoir properties determination is the nuclear magnetic resonance (NMR) method, i.e. resonance absorption of electro-magnetic waves by nuclei, occurring as a result of change in their spins vectors direction. The NMR takes place in the samples placed in the strong constant magnetic field and simultaneously exposed to the perpendicularly directed alternating magnetic radio-frequency field [4].

As a rule, the NMR method is more efficient for analysis of low-permeable reservoirs than conventional techniques to determine porosity by extraction and saturation. However, it is difficult to determine porosity of argillaceous and highly pyritized rocks by this method. There are also some challenges related to correct interpretation of the combined signal in time allocation of transverse relaxation (T_2).

The disadvantages of the NMR method are described in a number of scientific works devoted to the study of reservoir properties. For example, Zubkov Yu M and Potapov A G wrote, "… the use of the term "porosity" by the NMR is not quite correct because, in fact, we do not determine the porosity of samples by a relaxometer, but the total volume content of hydrocarbons which form a part of the fluids saturating a certain sample. Drying samples with the standard technique at the temperature of 105 ℃ does not allow extracting hydrogen containing fluids completely, so when afterwards we saturate the samples with water or kerosene, we receive overestimated porosity values by the NMR in comparison with the weight method. The errors are especially large when we study argillaceous samples and samples containing liquid hydrocarbons as, for example, bituminous sediments of the Bazhenov formation. On the contrary, the samples containing mixed lattice constituents with swelling components possess specific properties: when molecules of hydrogen containing fluids run into the interlayer space of these clay minerals, they are so tightly held by them that the time of relaxation of hydrogen atoms becomes too short to be fixed by the NMR device. Therefore, the porosity values of these samples received by the weight method are higher than the porosity values received by the NMR method" [3].

Shumskajte M J and Glinskih V N argue that the amount and type of clay minerals influence the results of the NMR measurements; they lead to the variation of amplitude, shift of spectra, and determine decisive influence of clay bound water on the total bound fluid at remaining saturation. [2]

The most reliable data on the porosity properties of low-permeable rocks can be obtained by the microcomputed tomography (mCT) method. The principle of this method is to develop a 3D model of the internal structure of a sample using a set of X-ray photographs. The main advantages of this method are nondestructive testing, assessment of the sample quality for further investigations, complex data on physical-mechanical and structural-textural properties of rocks, and study of samples with irregular geometric forms.

The mCT method also has some disadvantages. They include limited camera resolution, difficulties in distinction of minerals similar in X-ray density as well as challenges with fluid and kerogen segmentation within the rock. Hence, it is necessary to select carefully scanning and reconstruction modes according to the rock lithotype. Due to the challenge of formalizing the requirements for measurement parameters, there are no federal standards for determination of mineral composition, porosity, and permeability by the mCT method. However, there is a number of research experiments, including correlations between the mCT data and information gathered by conventional techniques, the optical microscopy, in particular. The comparative analysis shows a high correlation degree of the obtained results [1, 4-8].

3. Experiment

The object of the research is the Bazhenov formation rocks taken from a well drilled in an oilfield in the eastern part of the West Siberian basin.

Experimental instruments include: microcomputed tomograph Skyscan 1172 (the Academic Geochemical Laboratory, Tyumen Industrial University); NMR-spectrometer Khromatek-20M made by JSC Special Design Bureau "Khromatek"; porosimeter-permeameter AP-608 made by the company "Coretestsystems"; microscope OLYMPUS BX53; X-ray diffractometer DRON-7 with mounted solid state energy-dispersive detector (Siberian Research Institute of Petroleum Industry).

The experiment consists of an assessment of the porosity determination effectiveness in studies of low permeable bituminous reservoirs. A complex analysis of two Bazhenov formation samples was conducted using X-ray phase analysis to determine mineral composition; helium saturation method, NMR, and mCT to determine porosity.

The samples of cylindrical form with the diameter of 4 mm were studied by the mCT. The scanning parameters: voltage across the tube – 100 kV; turning angle – 0.3°; the number of photographs – 5; camera resolution – 1.6 μm/pxl. The software – CTan, CTvox, Nrecon. The results of the complex analysis of the samples are presented in figure 1 and table 1.

Table 1. The porosity of the samples.

№	The nuclear magnetic resonance method	Helium method	micro-CT method
1	8.39	3.46	18.20
2	13.70	11.79	17.70

The helium saturation method shows very low porosity values that can be explained by insufficient extraction of organic matter from capillary and subcapillary porous channels. The lowest porosity values are observed in sample №1 that is composed of biomorphic clay rock (silicon content 50%, clay content 35%), enriched in organic matter.

The size analysis of equivalent-diameters of porous channels was carried out by the mCT method. The results are shown in figure 2. The pore space of sample №1 consists primarily of channels with the diameter less than 8 μm, but the maximum diameter is about 15 μm. Sample №2 is characterized by interbedding of phosphatic calcareous oil-saturated rock (with high content of P2O5 – more than 17%) and argillite enriched in organic matter. It has channels with the diameter of up to 13 μm, and the maximum diameter is about 32 μm. It allows concluding that the size of porous channels influences the extraction of organic matter from the rock. It is advisable to carry out additional experiments in order to identify sizes of porous channels with organic matter which is hard for extraction without rock destruction. These data would be very useful for updating effectiveness criteria for organic matter extraction from low-permeable rock samples.

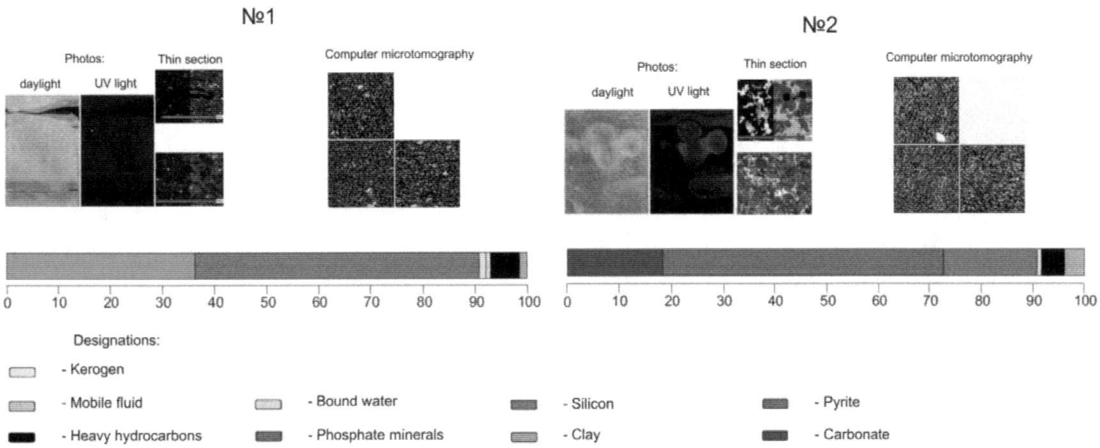

Figure 1. The results of complex analysis of two Bazhenov formation samples.

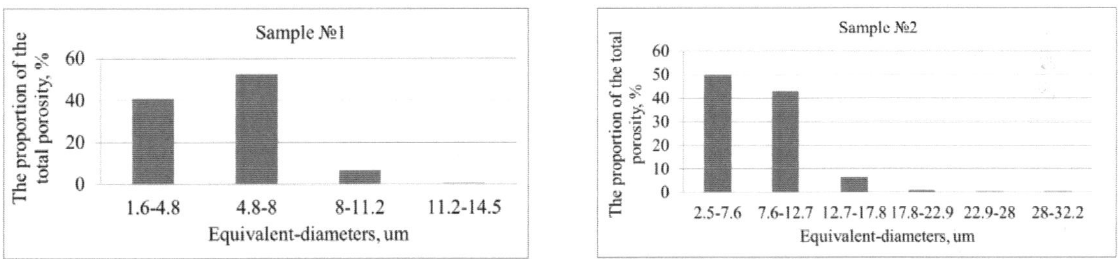

Figure 2. The distribution of equivalent-diameter by the mCT method.

4. Conclusion

The NMR method shows the results similar to those of the mCT while studying sample №2 which does not contain clay matter. While studying sample №1, where clay matter content is about 35%, the NMR method shows erroneously low results.

The mCT method shows the highest porosity values. However, it is important to note that this method determines organic matter including kerogen as a pore space. To receive accurate data it is necessary to subtract the kerogen content from the total porosity found by the mCT method. As a result, we have received the following porosity values: sample №1 – 16.2%; sample №2 – 12.7%.

Based on the comparative analysis of methods for determining porosity of reservoirs with low permeability and high content of heavy hydrocarbons the following conclusions are made:

- The NMR method is effective to determine porosity of rocks with low clay matter content.
- In this case, the method of determining porosity by helium saturation is ineffective. It is explained by the impossibility to extract petroleum from capillary and subcapillary porous channels without rock destruction. To identify the sizes of porous channels it is necessary to conduct additional research as well as to take into account the heavy oil holdup in the rock.
- The mCT method has shown the highest porosity values of rocks. It should be noted that kerogen possesses low X-ray density so the content data in the rock are required. It is necessary to subtract the kerogen content from the obtained total porosity values.
- The complex use of the NMR and mCT methods provides the most reliable data on the permeability and porosity properties of low-permeable reservoirs.

References

[1] Vahrusheva I, Gilmanov Ya and Patrakov D 2015 *Proc. 5th Conf.Supercomputer Technologies in Oil and Gas Indastry. Mathematical Methods, Software and Hardware* (Moscow: Moscow University Press). pp. 112-116.
[2] Shumskajte M and Glinskih V 2015 *Karotazhnik.* Vol. **7** **pp.** 56-62.
[3] Zubkov Ju and Potapov A 2014 *Karotazhnik.* Vol. **8** pp. 3-32.
[4] Ivanov M, Kalmykov G, Belokhin V, Korost D and Hamidulin R 2008 *Petrophysical Methods of Core Material Research* (Moscow: Moscow University Press)
[5] Ponomarev A and Zavatsky M 2015 *J. Higher educational institutions news. Oil and gas.* Vol. **3(15)** pp. 29-33.
[6] Parfenov V, Nikiforov A and Ponomarev A 2014 *Proc. 9th Int. Conf. Geology and Petroleum* Potential of the West Siberian basin. Vol. **2** pp. 276-278 .
[7] Korost D 2012 *Moscow University Press.* Heterogeneity of Terrigene Reservoir Structure and the Type of Structures of the Void Space.
[8] Stuart R. S 2009 *Boca Raton: CRC Press.* Microcomputed Tomography: Methodology and Applications.

IOP Conf. Series: Earth and Environmental Science **43** (2016) 012056 doi:10.1088/1755-1315/43/1/012056

Detection of attenuation zones in a time section based on running window filtration

A Islyamova, M Nemirovich-Danchenko, D Terre

National Research Tomsk Polytechnic University, Tomsk, Lenin Avenue, 30, Russia

Abstract. When developing oil reservoirs composed of carbonate rocks and those characterized by complex structure, the production well flow rate is largely determined by reservoir fracturing/ porosity rather than structure. These reservoir properties can often be reflected in a time section as significant attenuation of a seismic signal. The running time window spectral analysis has been proposed in the previous research to detect fractured zones. The calculations of seismic field diffraction due to a single pore and pore ensemble effects were made. The present research indicates that reservoir fracturing or porosity can cause qualitatively similar behavior of reflected signal amplitude spectra. Based on this finding, a rejection filter was constructed and applied to a real time section of the field in Tomsk Oblast, Prony and Fourier spectra being tested.

1. Problem setting

In [1, 2] we have proposed and further developed the method of running time window spectral analysis to determine fractured zones. The technique for amplitude spectrum analysis of seismic fields was based on the results of numerical finite-difference modeling of forward problems in mechanics of deformable solids in fractured bodies [3] and Gregory's laboratory test published in [4].

The unnormalized spectra obtained in the laboratory test and numerical computation are given in figure 1 a and b. Their comparison indicates that a section of high frequencies drops drastically within the spectrum of a wave reflected off the fractured zone; consequently, it distinctly differs from the wave spectrum in a continuous medium.

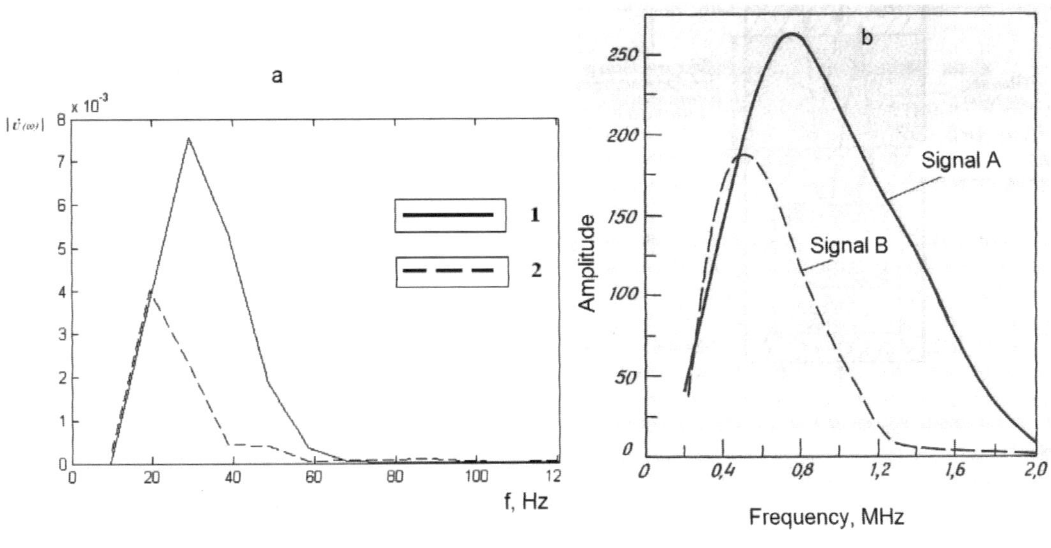

Figure 1. *a – unnormalized spectra of an incident (1) and a reflected (2) wave in a fractured layer [1,2], b – figure from [4]; spectra of an incident (A) and reflected (B) signals in the sandstone sample*

In [5] we solved the problem of plane-wave signal propagation through a medium which contains a single pore or a layer with a random ensemble of pores.

The numerical simulation processing resulted in calculation of the spectra of wave propagation before penetrating a porous layer, and while traveling through a porous medium. Amplitude spectrum moduli are aligned in figure 2.

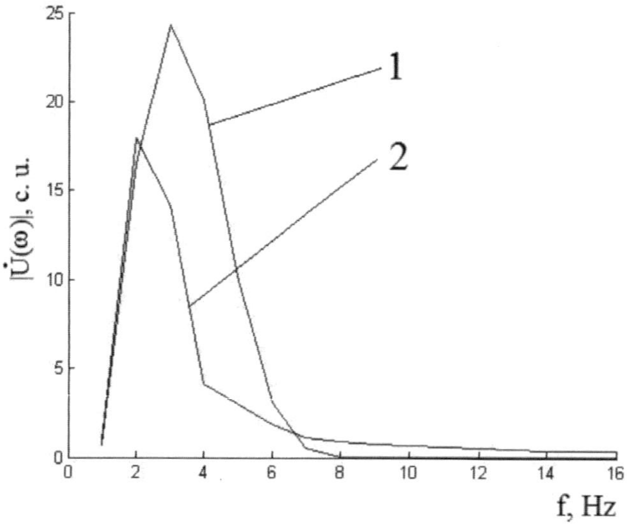

Figure 2. *Incident (1) and refracted (2) wave spectra in a porous layer*

It should be noted that in the course of our research we have observed that spectrum behavior models of seismic wave propagating fractured (figure 1, a) and porous (figure 2) media are similar.

The differences in spectral characteristic of an incident wave and a wave which penetrated fractured or porous zones are obvious. Consequently, the next stage of the research is to detect attenuation and spectrum change zones using real seismic data sets. Therefore, running window Prony and Fourier transforms [6] have been used in further real time section processing.

2. Processing technique

A real section analysis requires numerical filtration technique. Moreover, restrictions connected with sampling theorem and imposed due to final sizes of processing units do not allow extracting harmonic components of any pre-determined frequency. To bypass this restriction, we should apply Prony transform [7] which is not a spectral transformation in its pure form, but is a method to estimate discrete data based on linear combination of exponential functions (like least squares technique). Furthermore, it is possible to calculate spectral energy density (PSD) for any Prony frequency.

Prony method estimates *n*-element of *y(1), ...,y(N)* sampling based on exponential *p*-order model:

$$y(n) = \sum_{k=1}^{p} h_k z_k^{n-1}, \tag{1}$$

where factors h_k и z_k are generally complex and determined by the following formulas:

$$h_k = A_k exp(j\theta_k), \ z_k = \exp[(\alpha_k + j2\pi f_k)\, T].$$

where A_k – amplitude, α_k- attenuation coefficient (dimension - c^{-1}) of *k*-complex exponent, f_k(Hz), $\theta_k(rad)$ – frequency and *k*-sinusoid initial phase.

Figure 3 shows the first four terms of a series for values N=64 and p=15 (1). The first term (p1, the thickened line) is the attenuation exponent; the three remaining terms are damped sinusoids. For the majority of ordinary signals (including acoustic ones), the series (1) converges much quicker than Fourier series.

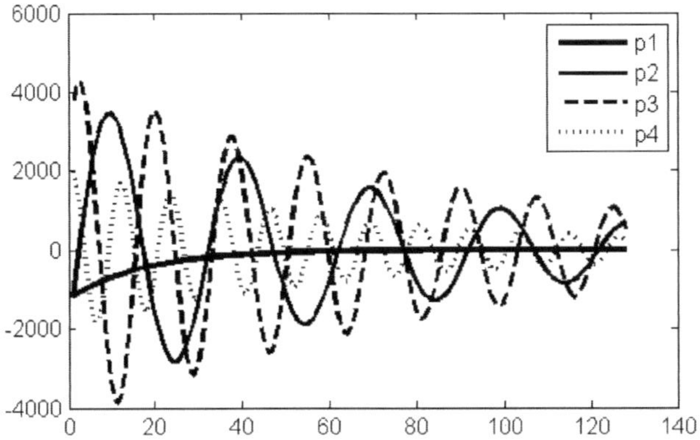

Figure 3. *The first terms of Prony series*

Having calculated h_k and z_k value arrays, it is possible to determine spectral energy density for any f_k frequency from $-(2\Delta t)^{-1} \leq f_k \leq (2\Delta t)^{-1}$ frequency range, where $(2\Delta t)^{-1}$ is an analogue of Nyquist frequency[8, 9].

For the chosen window [10] size N in case of Fourier transform Nyquist frequency is equal to f_N=N/2 (a nondimensional value which defines the amount of harmonics, i.e. events); whereas for the given reference value Δt, the cutoff frequency is fc=1/(2Δt) (a dimensional value, Hz). For example, if the window size is N=64 and Δt=0.002 s, then f_N=32, f_{th}=250Hz, while one harmonic of the window corresponds to 250/32=7.8125 Hz. For the chosen window and fixed reference value this number 7.8125 Hz remains the same.

In case of Prony transform the method of frequency selection from $-(2\Delta t)^{-1} \leq f_k \leq (2\Delta t)^{-1}$ frequency range predetermines variability of frequencies with change in signal parameters. Furthermore, each number («harmonic») in a Prony series will correspond to a certain bandwidth, but not to a particular Hz-frequency. The behavior of these bands is of particular methodological interest for geophysical data processing and interpretation.

3. Results and discussion

Figure 4 below shows the results of Prony notch filter application to «C» area «№95» time section windowing.

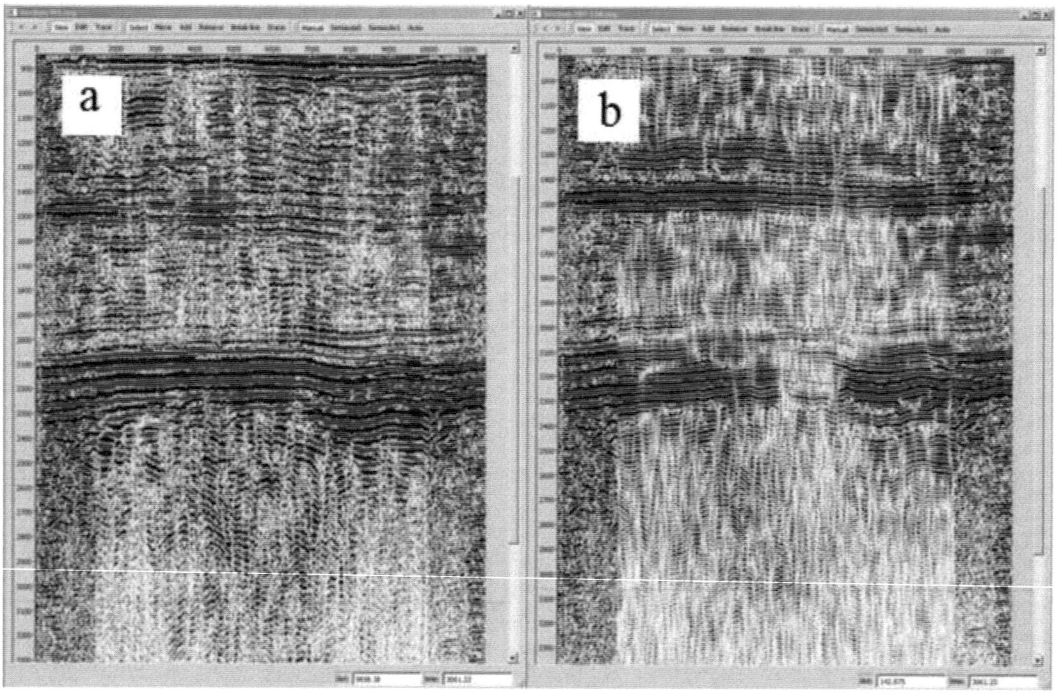

Figure 4. *Initial «C» area seismic section «№95» before (a) and after (b) Fourier transform being applied*

The proposed section processing sequence being applied, higher signal attenuation zones (probable increased fracturing/porosity and/or fluid saturation) are distinguished within the section due to lesser wave amplitude in comparison with the surrounding traces. This effect can be observed in the central section of the profile 95 in time sample of 2100-2300 ms, where the zone with violated phase sequence and low wave amplitudes can be detected among distinct phases of horizontally bedded Jurassic deposits of the sedimentary rocks.

It should be noted that the revealed zone corresponds to the actual pay interval.

The effect of porous/ fractured layer on a seismic signal has been studied in the investigation. In case of a reflected wave Prony and Fourier spectra are analysed. It has been indicated that amplitudes of certain frequencies decrease substantially in case of both porous and fractured layers. The similar spectrum change was observed previously in laboratory modeling of seismic wave propagation through the sandstone sample. In general it can be stated that zones of increase in attenuation can be displayed in a certain frequency band of the spectra. This can identify a procedure for filtration of real time sections to detect potential zones of increased signal attenuation. The presented technique can be used at the stage of processing seismic time sections in the laboratory, interpreting acoustic logging and mapping productive areas.

References

[1] Shatskaya A A, Nemirovich-Danchenko M M, Terre D A 2014 *IOP Conf. Series: Earth and Environmental Science.* Modeling of inclined fracture network and calculation of fracture effect on seismic signal spectrum. Vol. **21** pp. 012016.

[2] Nemirovich-Danchenko M M 2013 *Phys. Mesomech.* Detection of multiple fracturing of solid medium based on evaluation of spectral density of reflection energy. Vol. **16 (1)** pp. 105-110.

[3] Stefanov Y P 2000 *Theoretical and Applied Fracture Mechanics.* Wave dynamics of cracks and multiple contact surface interaction. Vol. **34** pp. 101-108.

[4] Payton C (ed.) 1977. Seismic stratigraphy: applications to hydrocarbon exploration. AAPG Memoir. Vol. **26** pp. 516 p.

[5] Shatskaya A A, Nemirovich-Danchenko M M, Terre D A 2015 *IOP Conf. Series: Earth and Environmental Science*. Modeling of seismic field in porous medium: Simulation study of single pore and pore ensemble effects. Vol. **27** pp. 012027.

[6] Marple S. 1987 Digital Spectral Analysis with Applications, Prentice-Hall Signal Processing Series (Alan. V. Oppenheim, Editor), Prentice-Hall, Englewood Cliffts, N. J.
[7] Marple S. 1982 *Acoustics, Speech, and Signal Processing, IEEE Int. Conf. on ICASSP.* Spectral line analysis via a fast Prony algorithm. Vol. **82** pp. 1375-1378.
[8] Nyquist H 1928 *Transactions of the American Institute of Electrical Engineers.* Certain Topics in Telegraph Transmission Theory. Vol. **47**(2) pp. 617-644 doi: 10.1109/T-AIEE.1928.5055024
[9] Oppenheim A V, Schafer R W 1989 *Prentice-Hall, Englewood Cliffts, N. J.* Discrete-time signal processing.
[10] McClellan J H *et al* 1998 *Prentice-Hall, Upper Saddle River, N. J.* Computer-Based Exercises for Signal Processing Using MATLAB5.

Integral analysis of geological and field data for selection of oilfield development strategy

A A Milke[1] and K V Tsivelev[2]

National Research Tomsk Polytechnic University, 30 Lenin Avenue, Tomsk, 634050, Russia
E-mail: [1]aleks_milke@mail.ru, [2]tsivelevkirill@mail.ru

Abstract. The reservoir development plan is a complex process and should be analysed from different points of view. The process was analysed in terms of geology, petrophysics, modelling, production technology and economics. Therefore, different methods should be used for the project.

1. Introduction

The oilfield A is located in Kaymysovky oil region next to Dvurechenskoe and Krapivenskoe oilfields. There are 9 exploration wells drilled in field A, five of which have core data. The base properties, such as porosity, permeability and water saturation are determined on the basis of core data. As the main material the following logs data are used: gamma ray log (GR), spontaneous potential log (SP), resistivity logs, neutron logs, potential logs, and so on. There were 270 samples taken from core of different production intervals. All data were analyzed separately for two production intervals: $U_1^{1\text{-}2}$ and $U_1^{3\text{-}4}$. The properties distribution was constructed on the basis of core data.

2. Analysis of geological and field data

First of all, the intervals, which could be potential sources of hydrocarbons, were determined on the basis of SP and GR logs for all wells [1]. The shaliness should be calculated for the effective porosity analysis. Shale normally contains radioactive bearing minerals and gamma ray log could be used for shale identification. Shaliness was analyzed using different models, the best result was achieved by the Larionov model. On the basis of this method the effective porosity curve was calculated for each well. The comparison of core porosity and log porosity for $U_1^{1\text{-}2}$ and $U_1^{3\text{-}4}$ was made separately. Then on the basis of the «base well» concept, the effective porosity curves were built for wells without core data. The «base well» was determined using the following criteria: wells lithology similarity and the lateral distance between the wells (well A2 was chosen).

The permeability was measured with the use of nitrogen gas, so the core permeability data were corrected on the Klinkinberg effect, provided that there was a slippage effect of gas molecules along the grain surface. Then the dependency between porosity and permeability was calculated based on core data. The effective porosity curves were used for this reason. It was decided to use unique dependency for each development object. Despite the fact that the exponential type of correlation was obtained, the determination coefficient was high, due to log and core porosity similarity. The average log derived parameters for $U_1^{1\text{-}2}$: φ=15%, k=7.7 mD, S_w=0.48; for $U_1^{3\text{-}4}$: φ=17.2%, k=176 mD,

S_w=0.43. The average core derived parameters for U_1^{1-2}: φ=13.2%, k=8.8 mD, S_w=0.56; for U_1^{3-4}: φ=18%, k=162 mD, S_w=0.37.

One of the main objectives of property estimation is correct calculation of Stock Tank Oil Initially in Place (STOIIP). However, additional step should be made in order to eliminate the values which are of minor importance for reservoir field development. The criteria for this elimination include several steps: water saturation, porosity, permeability and shaliness cut-off criteria [2]. The cut-off criteria are estimated separately for each productive formation. The cut-off criteria are defined for U_1^{1-2}: S_w=0.76, φ=0.116, k=0.88 mD, V_{sh}=0.24 and for U_1^{3-4}: S_w=0.73, φ=0.1202, k=1.06 mD, V_{sh}=0.27.

The interpreted parameters such as porosity, lithology logs and well picks of formation boundaries were used as an input data. The structures of geological model were constructed by offsetting Bazhen bottom structure.

The next step was to build up of 3D structural grid. Using modeling software, separate grids were constructed for U_1^{1-2} and U_1^{3-4}. Geological model was build using cell size of 100 m by 100 m for U_1^{3-4} and U_1^{1-2}. The size was chosen to optimize calculating time and obtain accurate model. The number of layers were selected so that the model fully describes vertical heterogeneity typical for regional depositional environment. The STOIIP (Stock Tank Oil Initially In Place) was U_1^{3-4}=7.92 mln m^3 and U_1^{3-4}=34.81 mln m^3.

The lithological characteristics were distributed by means of indicator modeling. In terms of lateral trends variogram from similar rocks was used and vertical lithology was distributed by vertical proportional curves for each layer separately [3]. Variogram parameters for lateral distribution were: azimuth – assumed direction of sediment deposition, long section rank – 4000 m, cross-section rank – 2000 m. Vertical variogram parameters: 1 m for U_1^{1-2} and up to 3 m for U_1^{3-4}. In terms of input data the pointwise interpretation of porosity log was used and then it was scaled into cells. Then the porosity parameter was distributed by kriging interpolation method. The same azimuth of variogram was used for lithology distribution.

The estimation of hydrocarbons volume is based on statistic data correlation results of petrophysics and core analysis [4]. The STOIIP estimation is conducted by three primary methods: deterministic, stochastic or probabilistic, and geo modeling [5]. Using the deterministic method, the STOIIP for U_1^{1-2} is 7.49*10^6 m^3 and for the U_1^{3-4} is 33.9*10^6 m^3. Using the stochastic method, the STOIIP for U_1^{1-2} is 7.6*10^6 m^3 and for the U_1^{3-4} is 34.2*10^6 m^3. Using the geological model, the STOIIP estimation is 7.92*10^6 m^3 and 34.81*10^6 m^3, respectively.

The simulation model was based on geological model [6]. Upscaling process was implemented to reduce the number of cells and optimize calculating timing. The lateral dimension of a cell remained unchanged; however, vertical cell thickness was scaled up from 0.8 m to 2 m. The reservoir properties were scaled up to a coarser cell. The STOIIP of geological and simulation model were 42.73 mln m^3 and 41.39 mln m^3, respectively, for both layers. The dynamic processes were defined by single relative permeability result provided with the core data [7].

The simulation model was produced by Tempest «Roxar» software. Static parameters, such as geological model porosity, permeability, and saturation were used as initial parameters and also PVT properties (Pressure, Volume, Temperature) were used being approximated by specific correlations [8].

Economic analysis of the project was based on evaluation of several potential scenarios of field development. The main variation parameters were drilling pattern and distances between wells, rate of fluid extraction and water injection, changing pattern orientation, hydraulic fracturing, horizontal wells, separate and unified development of both production intervals [9]. All these scenarios were evaluated by the economic model and the most profitable scenario was 5-point pattern with 500*500 in low permeability-thickness product (kh) zones and 1000*1000 in high kh zones.

The choice of formation pressure maintenance was defined by type of formation, the size of the formation and its oil-bearing zone, the presence of gas cap, formation oil viscosity, type of reservoir rock and its permeability, the level of formation heterogeneity, the presence of tectonic failure, and others. The presence of two objects of development with similar properties resulted in evaluating two distinct variants of development: joint and separate.

Two zones of different kh values were defined during reservoir properties evaluation. The kh varies from 1000 to 5000 mD*m at north-west block, whereas at the south-east block the kh values are less than 1000 mD*m.

The largest oil recovery index of 51.1% was demonstrated by 5-spot with the 500 meter spacing between production and injection wells. However, comparing economic interpretations, it was shown that, according to kh maps, the most efficient approach was to develop two production zones separately: 500*500 m between production and injection wells within zones of low kh and less concentrated pattern of 1000*1000 m in zones of higher kh values. The case was considered to be the most economically viable with recovery factor 50.6%, which was less than previously mentioned pattern (with recovery factor 51.1%) by 0.5%.

Also, the variant with natural depletion mechanism was simulated. Initially, the case had shown recovery factor 2%, whereas after simulation modeling it revealed recovery factor 9%, which indicated that the aquifer was not included in calculations. As one of the potential pattern of development, horizontal well pattern was simulated. The main challenge in this case was to justify the bottomhole pressure on production wells [10].

It is assumed that the construction is started in 2015 and it is to be continued to first quarter of 2018 when the production commences. The estimate economic life of the field is 8 years with a payback occurring between year 3 and 4. The total production of oil recovered during the project life is 21.2 mln tons of oil and 712.1 mln m^3 of gas (used for power generation). The economic oil recovery index (0.45) achieves in 8 years and technical oil recovery index (0.50) achieves in 25 years.

Depreciation of assets was performed using Declining Balance method with 25% rate. There were additional funds accounted for miscellaneous (5% from total Capital Expenditures (CAPEX), excluding drilling cost) and for environmental reclamation (5% from total CAPEX). Also, to account for uncertainties, the contingency fund of 25% from total CAPEX was established. There was some exploration cost included in Well development section of capital expenditure.

The revenue will be generated from sales of oil. Contained gas volumes are not in sufficient marketable quantities. Taxation represents 78%, a significant portion of total expenses on the project. Tax model consists of various federal and regional, labor taxes and royalties.

Sensitivity analysis was carried out on the following parameters by changing one parameter at a time between ±30% at 10% intervals while maintaining the rest of the following parameters constant. The Net Present Value (NPV) of field A is the most sensitive to the taxes and exchange rate and less sensitive to Operating Expenditures (OPEX). The Internal Rate of Return (IRR) period of project was the most sensitive to taxes and CAPEX and, secondly, to oil price and less sensitive to OPEX.

The project will produce marketable Urals brand oil which will be sold to local transfer oil pipeline located 30 km from field A. The oil will be treated and analyzed on site before releasing for sale.

3. Conclusions

As a result of the study the development model was constructed and the final variant was chosen. The best variant has no technological limits and the NPV is much higher than in other variants.

References

[1] Merkulov V P 2008 Production well logging. – Tomsk Polytechnic University, 139

[2] Lalanne B, Massonnat G 2004 Impacts of Petrophysical Cut-Offs in Reservoir Models//SPE Annual Technical Conference and Exhibition. – Houston, Texas

[3] Guirillot D 2002 Integration of geological models and reservoir simulators//17th World Petroleum Congress. – Rio de Janeiro, Brazil

[4] Ross J G 1997 The Philosophy of Reserve Estimation//SPE Hydrocarbon Economics and Evaluation Symposium. – Dallas, Texas

[5] Denney D 1997 Integrated Deterministic/Probabilistic Reserve Estimation//SPE Annual Technical Conference and Exhibition. – San Antonio, Texas
[6] Gregg J M, Rey T L and Wang J J 2009 Constructing A Well Engineered Simulation Model//PSIG (Pipeline Simulation Interest Group) Annual Meeting. – Galveston, Texas
[7] Lomeland F, Ebeltoft E and Hammervold Thomas W 2005 A New Versatile Relative Permeability Correlation//Reviewed Proceedings of the 2005 International Symposium of the SCA (Securities and Commodities Authority). – Abu Dhabi, United Arab Emirates
[8] Hemmati M and Kharrat R 2007 A Correlation Approach for Prediction of Crude Oil PVT Properties//SPE Middle East Oil and Gas Show and Conference. – Manama, Bahrain
[9] Litvak M L, Gane B R and Angert P F 2007 Field Development Optimization Technology//SPE Reservoir Simulation Symposium. – Houston, Texas
[10] Yildiz T 2005 Multilateral Horizontal Well Productivity//SPE EAGE (European Association of Geoscientists and Engineers) Annual Conference. – Madrid, Spain

Changes in composition and pore space of sand rocks in the oil water contact zone (section YU_1^{3-4}, Klyuchevskaya area, Tomsk region)

N Nedolivko[1], T Perevertailo[2], and T Pavlovec[3]

[1, 2, 3]Department of Geology and Mineral Exploration, Institute of Natural Resources, National Research Tomsk Polytechnic University Tomsk, Russia
Email: [1]nedolivkonm@yandex.ru, [2]ptg@tpu.ru, [3]tatyanapavlovec@mail.ru

Abstract. The article provides an analysis of specific features in changes of rocks in the oil water contact zone. The object of study is the formation YU_1^{3-4} (J_3o_1) of Klyuchevskaya oil deposit (West Siberian oil-gas province, Tomsk region). The research data allow the authors to determine vertical zoning of the surface structure and identify the following zones: oil saturation (weak alteration), bitumen-content dissolution, non-bitumen-content dissolution, cementation, including rocks not affected by hydrocarbon deposit. The rocks under investigation are characterized by different changes in composition, pore space, as well as reservoir filtration and volumetric parameters. Detection of irregularity in distribution of void-pore space in oil-water contact zones is of great practical importance. It helps to avoid the errors in differential pressure drawdown and explain the origin of low-resistivity collectors.

1. Introduction

The goal of research is to study the main character of changes in void-pore spaces and composition of terrigenous reservoir rocks in the oil water contact zone (OWC zone) on the basis of section YU_1^{3-4}, Upper-Jurassic producing horizon YU_1, Klyuchevskaya area, 207 Well (Tomsk region, West Siberian oil-gas province).

2. Background

The processes connected with rock alteration during their conversion into hydrocarbons in the oil-water contact (OWC) zone was the discussion topic for many researchers, among which were many Russian [4, 5, 9] and foreign [1–3, 7, 10] authors and the author of the paper [6]. Due to mass exchange, the trap at the water – oil contact zone is not completely filled with oil and hydrocarbons are located at the top, while water is displaced to the bottom. Oil-water contact is the intensive alteration zone of oil, porous water and country rock where the reservoir during long – term OWC formation acquires a zonal structure.

3. Research Methods

The research focuses on sandstone rocks in colored thin sections, when studied in transmitted and polarized light. For this purpose, the microscope Polam 213 was used. The researchers applied X-ray phase analysis to determine mineral composition of the cement. Laboratory data on porosity and permeability were used. The researchers compared the sandstone rocks with the identical median diameters to reduce the influence of sedimentogenic factors.

4. Object

Sandstone rocks in the oil water contact zone, section $YU_1^{3\text{-}4}$ (J_3o_1), are located in the low part of the Upper Jurassic producing horizon YU_1, Upper Vasyugan Subsuite (Klyuchevskaya oil deposit, West Siberian oil-gas province, Tomsk region). The formation under study is located above marine clayey deposits of Lower Vasyugan Subsuite (J_3c_3) and related to regressive longshore bars [8]. The formation is in depths ranged from 2693.95 to 2708.3 m based on the analysis data from the Well No. 207 (Figure 1). Sandstones have feldspar-quartz composition with insignificant amount of clastics (cilica rocks, average and acid effusive rocks, pegmatites) and minerals (mica, chlorite). The cement is composed of clay minerals (kaolinite, chlorite, illite), carbonates (calcium, siderite), quartz, and pyrite.
The degree of sandstone conversion is related to average catagenesis.

5. Key findings

The research data allow the authors to determine vertical zoning of the surface structure and identify the following zones: I – oil saturation (weak alteration), II – bitumen-content dissolution, III – non-bitumen-content dissolution, IV – cementation, V – including rocks not affected by hydrocarbon deposit (Fig. 1).

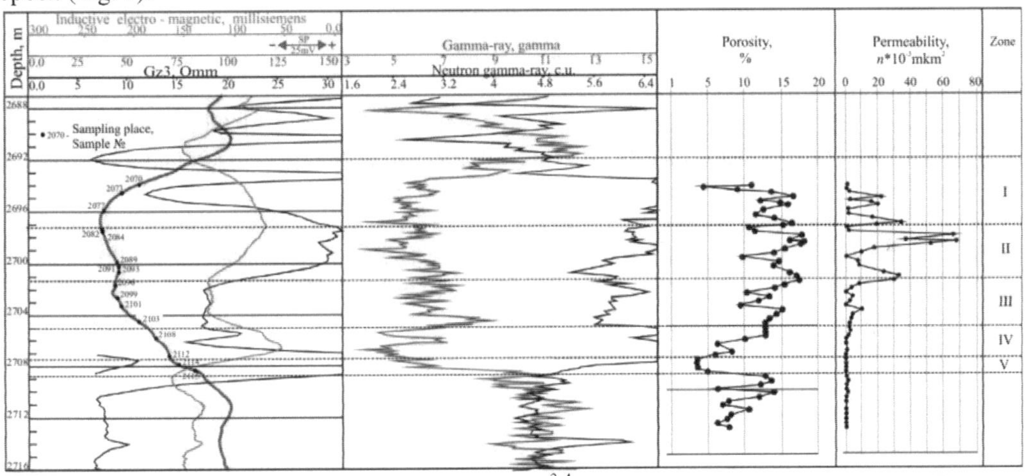

Figure 1. Vertical zoning of section $YU_1^{3\text{-}4}$. Klyuchevskaya area, Well No. 207

In ***oil saturation (weak alteration) zone*** (2692.0–2697.2 m) the sandstones contain significantly high component integrity, with weak corroded quartz grains, sometimes regenerated. Potassium feldspar and plagioclase either remain unchanged or undergo pellitization and sericitization along the cleavage fractures and grain edges. Cement has polymineral chlorite-hydromica-kaolinite composition, sometimes with compound impurities, such as hydromica-montmorillonite, including siderite, pyrite, rare dispersed relict calcite and ferriferous dolomite. Recrystallization of cement components is non-uniform, having selective, spotty features, where pelitomorphic mineral modifications are adjacent to micro-fine crystallized minerals. Kaolinite is pelite (initial cement) non-uniform fine- microcrystallized (in mica) – up to 0.05 mm and uniform microcrystallized (secondary cement) – 0.001mm in voids modifications. Capacity space in oil-containing sandstones include: primary residual sedimentogenic intergranular voids; secondary dissolution voids: micro-cavities on the fragment surfaces; intragranular voids; intercrystalline voids. Intergranular voids formed from several grain contours are predominate in the sandstones of this zone. Void walls are even or irregular. In regenerated sections with weak dissolved fragments the intergranular void walls show distinct relative even surfaces and vice versa, in sections with the dissolved fragments, the void walls show irregular micro-cavernous surface. Intragranular voids in oil-saturated sandstones are insignificant; they are confined to effusive feldspar fragments. Intragranular porosity includes fracture voids of catagenesis compaction and, besides, the following: in plagioclase – slit-like voids along cleavage and twin seams; in kalifeldspath – regular, often isometric voids along perthite growths; in clastics (silic rocks, effusive) – irregular cavity voids with sinous boundaries, sometimes forming thin cellular patterns. Micro-cavernous dissolution voids form separate areas on the fragment surfaces. Such voids in feldspar are found in the altered grains or on the end-parts of prismatic fragments, where lateral prism face surface shows no traces of dissolution. Combination of surface dissolution in several sections and regeneration in adjacent grain sections or contacting enriched quartz grains are found in quartz grains, as well as, acid effusive and flinty rock fragments. Intercrystalline micro-voids are formed in recrystallization sections during primary cement kaolinitization or muscovite. They are located between kaolinite crystals. Due to confined kaolinitization in saturation zone and weak kaolinite recrystallization, these void types have limited and irregular distribution. Thus, the specific features of void-pore space in oil-saturated zone are the following: intergranular voids predominace, weak dissolution resulting in even void wall surface.

In ***bitumen-content dissolution zone*** (2697.2–2701.2 m) formed in the environment containing oil and water, the dissolution of grains as well as the feldspar pelitization and sericitization is increased, kaolinite cement is crystallized, the bitumen and pyrite are formed. Void space in sandstones from this zone, as in oil-saturated sandstones include combination of intergranular, intragranular, micro-cavernous voids on the fragment surfaces and intercrystalline voids, however, the relation between different void types and their peculiarities significantly change. Furthermore, secondary dissolution voids as surface and intergranular voids are widely developed. The following factors were determined: intergranular void size and connectivity degree increase due to fragment and primary cement dissolution; most grains in sandstones from this zone show intensive developed surface micro-cavity, which in its turn, significantly complicates not only void wall surface, but also the intergranular void configuration itself; not only does the number of intragranular voids increase, but also their sizes. Often surface voids transform into intragranular voids, and these on their turn, merge together, while remaining relic cellular grains (in this case, feldspar) are practically dissolved; significantly increases the role of intercrystalline voids due to mass kaolinitization [5, 6] of primary kaolinite-chlorite-hydromica cement and formation of monomineral coarse-flaky (0.06–0.1 mm) aggregates from tabular kaolinite grains with complete tricline lattice. Intercrystalline voids (size 0.01–0.02 mm) have relative uniform distribution. Intergranular voids, micro-voids on inside and surface grain and intercrystalline voids form uniform associated void-pore space which is partially or completely filled with oxidized or pyrite bitumen. Bitumen forms films on the rock fragments, fills small voids and micro-voids in kaolinite cement, covers large void walls and completely seals small intergranular voids.

Newly formed void space structures show significant increase of filter-capacity rock properties, such as: porosity increase up to 15.4–17.6 %, permeability –up to $33.08–52.68 \cdot 10^{-3}$ mkm^2.

In non-bitumen-content dissolution zone (2701.2–2705.0 m) the sandstone composition transformation results in further substitution of pelite and sericite by feldspar, kaolinite –mica minerals; disappearance of primary polymineral clay cement and formation of monomineral kaolinite cement in the void spaces. Structure transformations (as in bitumen-content zone) are connected with the intensive fragment and cement dissolution, intergranular void renewal, micro-porosity development on the inside and surface fragments, and occurrence of intercrystalline voids in kaolinite cement. Porosity and permeability in this zone in some samples can be compared to that in the bitumen-content zone.

Cementation zone (2705.0–2707.4 m) includes sandstone with the secondary quartz regenerated fragments and calcite corroded cement, substituted by rock fragments and primary cement. Quartz [1, 6, 7, 10], secondary in amount (total mineral content is insignificant and has low percentage), has incomplete regenerated boundaries with thickness from 0.006 to 0.02 mm. Calcite completely fills in practically all the remaining voids and rocks lose their reservoir properties: rock porosity and permeability decrease.

If there is ***no*** **hydrocarbon** ***deposit influence*** (2707.4–2708.0 m), rocks contain polymineral, weakly recrystallized kaolinite-chlorite-hydromica cement; include thin-dispersed pyrite, pelite-morph and microgranular siderite and porous calcite. Void space is non-uniformly distributed; intergranular voids are of insignificant sizes (0.02–0.1 mm), weakly associated, and often separated by non-permeable compacted cement sections or non-cement combination of several grains.

Specific features of composition and structure sandstone transformation in oil saturation and oil-water contact zones (Tab. 1) can be explained as following.

Table 1. *Geochemical, mineralogical and structural transformation of sandstones under conditions of incomplete oil infilling of trap*

Zone	Specific features of transformation	Changes of fragments and cement	Porosity types
Oil saturation of weak dissolution	Dissolution of reservoir by oil oxidation products at contact with interstitial water; limited mineral-formation, disrupted oil displacement by interstitial water	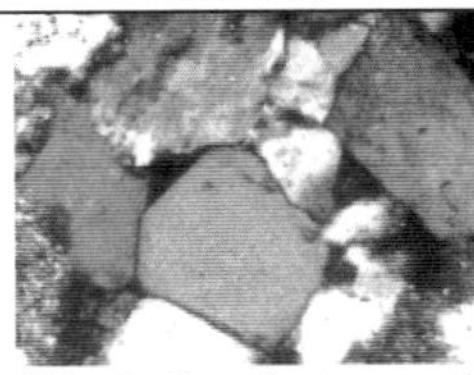 Weak dissolution and substitution of fragments, polymineral cement	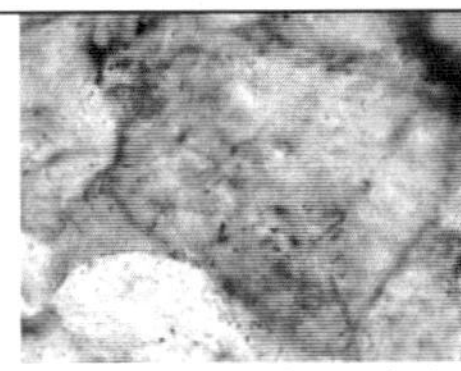 Intergranular voids of simple morphology, rare – intragranular voids
Bitumen-content dissolution	Oil oxidation, bitumen formation, sulphate restoration, pyrite and kaolinite synthesis, dissolution of fragments and cement produced through oil oxidation.	 Dissolved and substituted fragments, kaolinite cement, of bitumen filled	 Intragranular, intercrystalline and micro-cavernous voids of complex morphology
Non-bitumen-content dissolution	Dissolution through acid aggressive solutions from bitumen-content zone, kaolinite synthesis	 Dissolved interfaces and substitutions of fragments, kaolinite cement	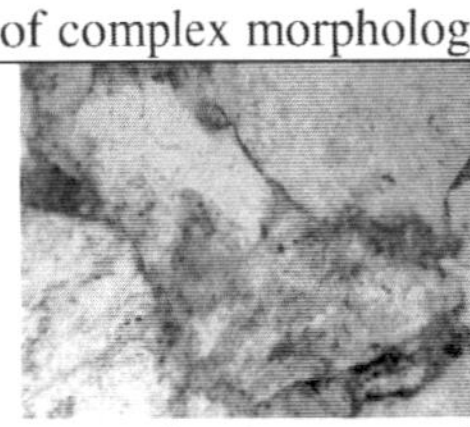 Intragranular, intercrystalline and micro-cavernous voids of complex morphology
Cementation	During acidity decrease quartz and calcite synthesis from solution produced in dissolution zones.	 Regeneration of quartz grains	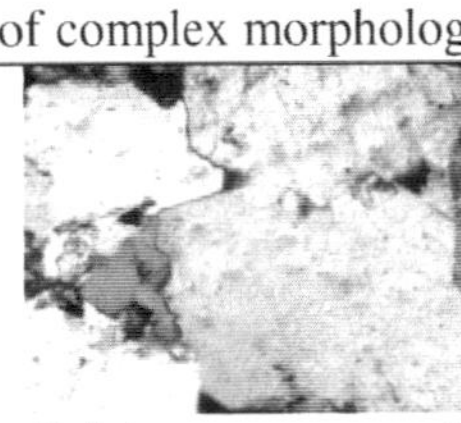 Calcite cement, small isolated residual voids

6. Results and Discussion

In oil saturation zone the oil migrating to the reservoir is in imbalance to void solutions and reservoir mineral phase as a whole. When oil, as any organic substance, interacts with interstitial water it undergoes oxidation due to bacteriological and chemical processes. These oil oxidation products in void solutions stimulate their acidity increase. Previously dissolved elements (silicium, aluminum and others) have precipitated on the newly-formed acid (void-solution) boundary resulting in secondary quartz and kaolinite formations. Unstable minerals, namely, carbonates, aluminum silicates, silicates, etc, at pH decrease are dissolved, forming secondary surface micro-cavernous and intragranular voids. Under conditions of hampering water exchange aluminum silicate and silicate dissolution (feldspar, mica and others) is accompanied by partial displacement of one mineral by another (sericitization and kaolinitization of feldspar, kaolinitization of mica, etc.). As a result of rapid oil displacement from void solutions, mineral and structure transformation processes have a limited nature, while reservoir dissolution is indistinct. Carbonates can be completely dissolved, while their leaching products are carried out beyond the dissolution zone [4]. Transformation processes of oil saturation zones in reservoirs have a limited nature and are, to a certain extent, determined by the amount of interstitial water in sandstones and hydrocarbon infilling velocity in the trap. The composition-structure transformations in the oil saturation part of the reservoir cease during gradual hydrocarbon infilling of the trap and water displacement from void space.

Active substance exchange occurs ***on the oil-water contact zone*** where oil, subjected to bacteriological oxidation, generates organic acids (CnH_2n +1COOH or CnH_2nO_2 type), spirits, aldehydes, ketones and other organic combinations; void solution acidity increases, and, at the same time, their aggressive properties sharply increase; while surrounding (country) rock is subjected to restoration and intensive dissolution. Biochemical and chemical process indicators, occurring in the OWC zone, are mineralogical and structure transformations, reflecting specific processes and their sequence, as well as, fixed space zonality of OWC structure. Rock mineralogical changes at OWC zone include secondary mineral formations, not typical for rocks and structural changes – in the dissolution of rock constituents and secondary porosity formation.

Bitumen-content zone is the zone where oil-water contact immediately occurs, and where oil oxidation takes place, resulting in the formation of solid bitumen, pyrite and other oxidation products (aggressive solvents, such as fatty acid, which in its turn, forms acid environment). Pyrite formation through oxidized bitumen is as following [4, 9]:

$$CaSO_4 + HCO_3 \rightarrow H_2S + CaCO_3 + CO_2;$$
$$Fe + H_2S \rightarrow FeS_2$$

As a result of sulphate reduction the sulphuric and hydrogen boundaries are formed where pyrite precipitation and organic substance pyritization occur in harsh restoration environment. Pyrite formation ceases further from oxidizing oil beyond sulphate reduction. Calcite dissolution in acid environment is as follows:

$$CaCO_3 + CO_2 \leftrightarrows Ca^{2+} + 2HCO_3^-$$

through simple leaching, up to complete mineral dissolution and component removal beyond acid solution reaction. Thus, calcite is not present in the dissolution zone of OWC. During dissolution of aluminum silicate and quartz, component removal is partial as dissolution includes only insignificant mobility of cations, then further formation of new mineral phases: either the replacement of one mineral for another (aluminum silicate) or mineral synthesis from solutions (quartz, kaolinite). During dissolution of feldspar and quartz, cations Na^+ (feldspar kaolinitization) and K^+, Na^+, Fe^{2+}, Fe^{3+}, Mg^{2+}, Ca^{2+} (plagioclase albitization) and anions HCO_3^- pass into void solutions, while Ti^{4+}, Si^{4+} appear during kaolinitization of hydromica, chlorite, montmorillonite and mica. As a result of rock dissolution, the void connectivity degree, rock porosity and permeability are increased. Kaolinite synthesis occurs in acid environments at pH=5–7 from those solutions circulating in the voids.

Geochemical processes ***in non-bitumen-content zone*** are also connected with rock dissolution under the influence of aggressive solutions which proceed from bitumen-content zone. These processes occur during acidity decrease to the horizon bottom, as acid inflow in this direction decreases, and solutions are enriched by alkalines and alkaline-earth elements. Aluminum loses its mobility, that results in mass synthesis of kaolinite.

The ***cementation zone*** is formed in the weak acid environment of the lower dissolution zone, where geochemical solution discharge, enriched by chemical elements, occurs. Rearrangement of dissolution products occurs when environment alkalinity increases to the reservoir bottom, resulting in cementation zone that has zonality nature. First, quartz precipitation, then carbonate precipitation (Fe-Ca-Mg carbonates) occurs. Due to these processes, void-space decreases, and the reservoir is sealed.

7. Conclusions

Detection of vertical zonality of sandstone rocks in oil-water contact zones is of great practical importance that allows explaining electrical resistance drop on the diagrams of induction logging of the low resistivity reservoirs. The electrical resistance decrease can be resulted from the presence of pyrite, current-conducting mineral, in sandstone rocks of Bitumen-content zone, from decrease of capillary-adhesive water around pelletized feldspar grains, in micro-voids in both bitumen-content and non-bitumen-content zones of dissolution. High specific surface of micro-cavernous voids on the surfaces of fragments, intergranular and intercrystalline voids provides expanding the film of capillary-adhesive water which forms the network of interconnected channels. The uniform water space formed by the retained capillary water conducts the current well that is reflected by the induction logging curves.

References

[1] Barclay S A and Worden R H 2000 *Spec. Publ. Int. Ass. Sediment.* Petrophysical and petrographical analisis of quartz cement volumes across oil-water contacts in the Magnus Field, northern North Sea. Vol. **29** pp. 147–161.

[2] Cerepi A, Durand Cl, Brosse Et. 2002 *Journal of Petroleum Science and Engineering.* Pore microgeometry analysis in low-resistivity sandstone reservoirs. Vol. **35(3)** pp. 205-232.

[3] Kalsen L A, Skeil J E, Owe K B, Bjørlikke K, Ostad R, Berge K, Cecchi M, Vik E and Sheafer R G 2004 *Understanding Petroleum Reservoirs: Towards an Integrated Reservoir Engineering and Geochemical Approach.*Petroleum migration, faults and overpressure. Part II. Case history. The Haltenbanker Petroleum Province, offshore Norway Published by Geological Society. Special Publication. Vol. **237** pp. 305–364.

[4] Lebedev B A 1992 *Moscow. Nedra.* Geochemistry of epigenetic changes. pp. 126.

[5] Melnik I A 2012 *Geophysics.* The technique to identify oil and gas objects in epigenetically regenerated reservoirs of Western Siberia. Vol. **1** pp. 31–36.

[6] Nedolivko N 2003 *International Geological Congress, Resumes 33 – Norway, Oslo, International Union of Geological Sciences.* Problem of low-ohm reservoirs in view of rock transformation in ancient water-oil contact zones. pp. 2090577.

[7] Oxtoby N H 1993 . *Petroleum Geology of Northwest Europe: Proceedings of the 4th Conference (ed. PARKER J R). Geological Society of London.* The link between petroleum emplacement and sandstone cementation. pp. 1395–1402.

[8] Perevertailo T, Nedolivko N and Dolgaya T 2015 *IOP Conference Series: Earth and Environmental Science. Scientific and Technical Challenges in the Well Drilling Progress.* Vasyugan horizon structure features within junction zone of Ust-Tym depression and Parabel megaswell (Tomsk Oblast). Vol. 24 pp. 012023.

[9] Sakhibgareev R S 1989 *Leningrad. Nedra.* Secondary changes of collectors in the process of formation and destruction of oil deposits. pp. 260

[10] Worden R H, Oxtoby N H and Smalley P C 1988 *Petroleum Geoscience.* Can oil emplacement prevent quartz cementation in sandstones. Vol. **4** pp. 129–138.

Dynamics of water separation in destruction of water-in-oil emulsions

E A Kuzmenko[1], N V Usheva[1], O E Moyzes[a1], K A Polyakova[1]

[1]National Research Tomsk Polytechnic University, 30 Lenin Avenue, Tomsk, 634050, Russia

E-mail: [a]moe@tpu.ru

Abstract. The dynamics of drop formation and settling processes in breaking water-in-oil emulsions of the West Siberian oil fields was experimentally studied. The investigation results of drop formation in the water-in-oil emulsions were presented. The residual water content in oil was determined after the settling process at the varied initial water content, temperature, and hydrodynamic conditions of emulsion formation.

1. Introduction

Oil treatment in the field involves separation of oil from water, mechanical impurities, and dissolved gases. The processes of dewatering and desalting are complicated. They include drop formation and settling stages and depend on the conditions of emulsion formation and the effectiveness of their breaking [1, 2].

Currently, mathematical modelling is effectively used to study and design petroleum refining processes applying simulated dynamic models and modeling systems [3–9].

The lack of effective methods for estimation of residual water content in oil makes modelling and designing of crude oil treatment units more difficult. It is virtually impossible to consider the influence of water-in-oil emulsions breaking processes in the modelling of dewatering process for systems with real-time fluids without experimental data. Studying the dewatering process dynamics ensures estimating the time of steady-state regime attainment in the system under certain technological conditions.

In the works [10-12] it is shown that studying the water-in-oil emulsions breaking processes is possible by using the real-time oil samples as the objects of research instead of model fluids. Therefore, the aim of this research is to study the water-in-oil emulsions formation and breaking processes of West Siberian oil fields.

2. Object and methods of research

The objects of research were the oil samples of West Siberian fields, the physical-chemical properties of which were determined with the standard methods in the laboratory "Natural Fuel and Utilities Lab" of Tomsk Polytechnic University (table 1).

Table 1. Physical-chemical properties of oil samples

Parameters	Oil sample 1	Oil sample 2	Oil sample 3	Oil sample 4
Density at 20 °C, kg/m^3	814.6	826.5	830.0	862.2
Kinematic viscosity				
at 20 °C, m^2/s	2.75	3.70	4.72	11.44
at 50 °C, m^2/s	1.29	2.30	2.36	4.86
Paraffins content, mass%	0.03	3.5	1.27	0.98
Asphaltene and tar contents	0.01	2.65	0.76	1.03
	0.19	7.57	16.0	14.33

The experimental technique is specified in the work [13]. Emulsions were obtained by mixing oil and tap water with dissolved sodium chloride (40 g/l), 10, 20, 30 % by volume. Each sample of the water-in-oil emulsion in the amount of 100 ml was put into a 200 ml conical flask and manually mixed for 10 min. After that the emulsion was mixed by the automatic paddle mixer ER10 for 10 min.

Two mixing regimes were applied: 1000 rpm (v_1) and 2000 rpm (v_2), which ensured different water-in-oil emulsion dispersion degree.

The obtained water-in-oil emulsions were poured in the 100 ml measuring flasks and put into the thermostat for maintaining temperature regime (20 and 50 °C). Further, the amount of settled water was traced after certain time periods. The measurement error was 1 vol.%.

The process of drop formation was studied with the microscope "LOMO" R2U42. It is a standard measuring microscope with vertical lighting and 100 times amplification.

3. Results and discussion

The study of initial water content influence on the settling process, intensity of emulsion mixing, and temperature was carried out for the selected West Siberian oil samples (figures 1-3, tables 1, 2).

The research results of water separation from oil (sample 2) with various water content at two mixing regimes and temperature 20 and 50 °C are presented in figures 1, 2.

As it can be observed, at the first mixing regime, the time of emulsion breaking decreases from 40 to 18 min. with an increase in temperature to 50 °C (figure 1a and 1b). The speed of water separation decreases with an increase in mixing intensity, the time of steady-state regime attainment for settling process rises and reaches nearly 50 min. at 20 °C (figure 2a and 2b).

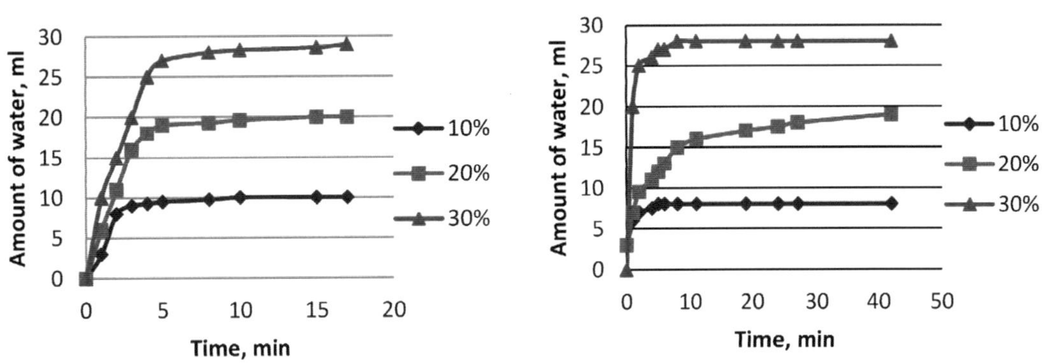

a) b)

Figure 1. Dependence of settled water amount on time at various water content in oil (v_1=1000 rpm, a) T=50 °C, b) T=20 °C)

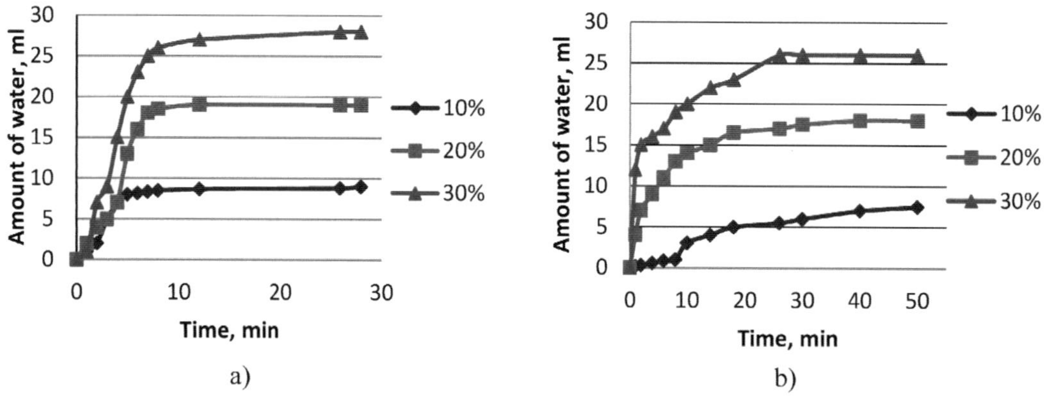

Figure 2. Dependence of settled water amount on time at various water content in oil (v_2=2000 rpm, a) T=50 °C, b) T=20 °C)

The selected oil samples (table 1) vary in their physical-chemical properties. The density varies from 815 to 862 kg/m^3, whereas kinematic viscosity increases continuously and reaches its maximum value of 11.44 m^2/s at 20 °C.

In figure 3 the dynamics of water separation is presented for the various oil samples under the same conditions of emulsion formation.

It is shown that for the oil samples 1 and 2, which have relatively low values of density and viscosity, the amount of separated water is virtually equal, and time of steady-state regime attainment is 15 and 25 min. respectively (table 2). It may be noted that high speed of water separation is observed for these oil samples during the first 10 min. With an increase in density and viscosity of oil (samples 3 and 4), the speed of water-in-oil emulsion breaking declines significantly, the amount of unseparated water increases, and the system attains steady-state regime within the period of more than 50 min.

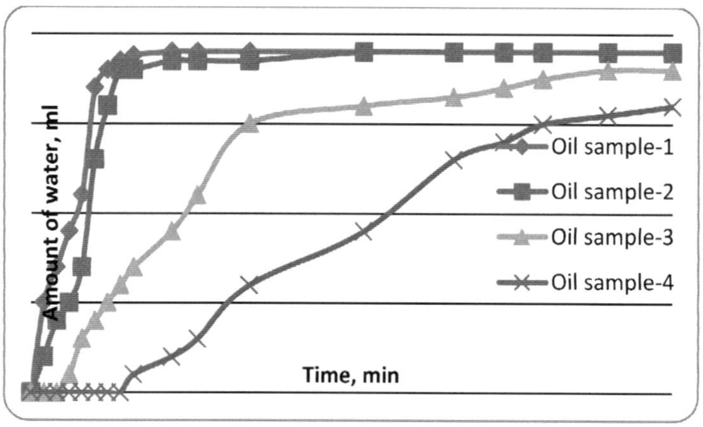

Figure 3. Dependence of settled water amount on time for various oil samples (v_2=2000 rpm, T=50 °C)

Table 2. Results of dewatering process for various oil samples (water content is 20 vol.%, v_2=2000 rpm, T=50 °C)

Parameters	Oil sample 1	Oil sample 2	Oil sample 3	Oil sample 4
The amount of settled water, vol.%	19	19	18	16
The time of settling process, min	15	25	45	

> 50

The settling process generally depends on the settling speed of water drops of certain diameter, which is formed depending on the process operating conditions. Therefore, one of the research stages is the observation of drop formation process and drop sizing.

The most probable particle size in the dispersed phase depends on physical-chemical properties of oil, oil-field water, multiphase flow regime, temperature, and other factors. Dispersing content of water-in-oil emulsion is constantly changing in the processes of its formation and breaking. The features of water drop formation for the various oil samples are studied with visual methods under similar conditions of emulsion formation.

The pictures of drops for various fresh water-in-oil emulsions are presented in figure 4. The diameters of forming water drops under different hydrodynamic conditions are estimated. For example, maximum drop diameters are within the range of 120–350 μm for the oil sample 2, and for heavy oil (sample 4) drops have much lower sizes (≈80 μm), which corresponds to the dynamics of water separation from oils with various physical-chemical properties (figure 3) and conforms to the theoretical laws of drops formation and settling [1, 2, 13].

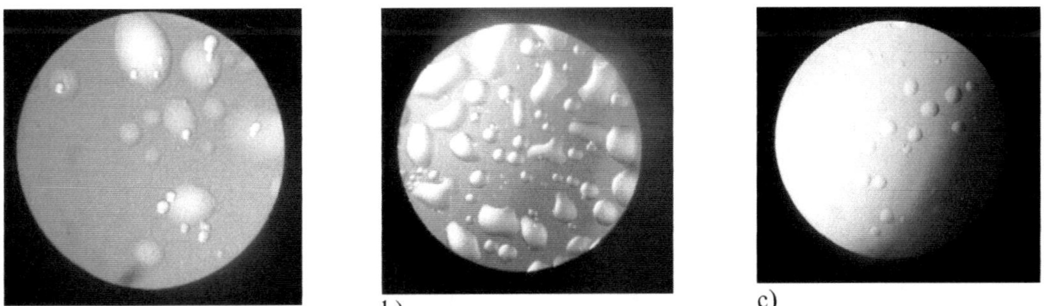

a) b) c)

Figure 4. Pictures of water drops in emulsions (water content is 10 %, mixing regime corresponds to v_2; a) oil sample 1, b) oil sample 2, c) oil sample 4

4. Conclusion

It is shown that physical-chemical properties of oils and content of dispersed phase significantly influence the residual water content and time of water separation process.

As a result of experimental research it is found out that the more stable finely-dispersed systems are formed with an increase in water-in-oil emulsion mixing intensity, dewatering of these systems is possible at higher temperature.

Thus, the obtained results can be used for analysis and selection of process conditions for dewatering of oil with various physical-chemical properties and for modelling crude oil treatment technology [14].

References

[1] Tronov V P 2000 *Oilfield treatment* (Kazan: FEN) p 416

[2] Lutoshkin G S 2005 *Collecting and treatment of oil, gas and water: Textbook for higher education institutions. 3 ed.* (Moscow: Al'yans) p 319

[3] Kim S F, Usheva N V, Samborskaya M A, Moyzes O E and Kuzmenko E A 2013 *Fundamental Research* vol **8** pp 626-629

[4] Kim S F, Usheva N V, Moyzes O E, Kuzmenko E A, Samborskaya M A and Novoseltseva E A 2014 *Procedia Chem.* vol **10** pp 448-453

[5] Usheva N V, Moyzes O E, Kuzmenko E A, Kim S F, Khlebnikova E S, Gizatullina S N and Filippova T V 2015 *IOP Conference Series: Earth and Environmental Science* vol **27** pp 1-5

[6] Frantsina E V, Ivashkina E N, Ivanchina E D and Romanovskii R V 2014 *Chem. Eng. J.*

(Amsterdam, Neth.). vol **238** pp 129-139

[7] Dolganova I O, Dolganov I M, Ivashkina E N, Ivanchina E D and Romanovskiy R V 2012 *Pol. J. Chem. Technol.* vol **14 (4)** pp 22-29

[8] Krivtsova N I, Gaga S G, Desiatnichenco A A, Popok E V and Zaitceva EV 2014 *Procedia Chem.* vol **10** pp 441-447

[9] Belinskaya N S, Ivanchina E D, Ivashkina E N, Frantsina E V and Silko G Y 2014 *IOP Conference Series: Earth and Environmental Science* vol **21 (1)** pp 1-7

[10] Tronov V P 2002 *Gas separation and decrease in oil loss* (Kazan: FEN) p 407

[11] Sadriev A R, Mirgalev I R, Grechuhina A A and Morozov G A 2009 O*il and gas technologies* **1** pp 28-31

[12] Sadriev A R, Grechuhina A A and Hamidullin R F 2008 *Oil and gas technologies* **3** 28-31

[13] Ermakov S A 2007 *Oil and Gas Bussines* **5** pp 102-118

[14] Tronov V P 2000 *Systems of oil-and-gas gathering and hydrodynamics of the main technological processes* (Kazan: FEN) p 416

Influence of sulphide Cu (I) promoting additives concentration on acid and catalytic properties of high-silica zeolites in straight-run gasoline conversion

I S Khomyakov , V I Erofeev and Fan Kuok Khan

Department of Geology and Oil Field Development, National Research Tomsk Polytechnic University, Russia
Khomyakov_i.s@mail.ru

Abstract. In present article the influence of Cu_2S promoting additives concentration on acid and catalytic properties of high silica MFI-type zeolites is investigated in the process of conversion of straight-run gasoline fractions of gas condensate into high octane components of motor fuels. It was shown that zeolite modified with 1 % of Cu_2S nanoscaled powder possesses the highest acid centers concentration and highest catalytic activity.

1. Introduction

Research and development of high-yield advanced stock processing is an challenge of oil and gas chemistry. Within catalytic thermal processes antiknock can sufficiently improve the properties of straight-run gasoline fractions of oil and gas condensates.

One of the modern trends of rational nature management is a conversion of light hydrocarbons into high octane motor fuels component by means of zeolite-containing catalysts [1-3]. MFI-type zeolite is widely used in zeolite-based catalytic processes. It was stated that zeolite can be used to catalyze lots of acid-base hydrocarbons conversion reactions [4-10]. Due to the unique structure that facilitates its molecular-sieve selectivity zeolite sorbs only size-defined substances. Also, zeolite possesses unique acid properties and is related to supersacids. All its features facilitate the application of zeolite in petrochemical industry. Moreover, zeolite is more active catalyst than convenient catalytic systems. In present article we compared acid properties and catalytic activity of MFI-type zeolite and zeolite modified with Cu_2S.

2. Experimental procedure

High-silica zeolites (HSZ) were produced by hydrothermal synthesis from alkaline aluminosilica gels at 175–185 °C in the course of 2–4 days [11]. After the crystallization was complete, the resulting zeolites were washed with water, dried at 110°C, and calcined at 600°C for 6 h. The HSZ were converted to the hydrogen active form by treatment with a 25% NH_4Cl solution at 90 °C for 2 h, with subsequent drying at 110 °C and calcination at 600 °C for 6 h (the content of Na_2O in decationized zeolites was less than 0.01%). The HSZ catalysts obtained were identified by IR spectroscopy (Nicolet 5700 IR Fourier spectrometer) and X-ray phase analysis (DRON-3 X-ray installation, Mo anode, Ni filter). According to X-ray phase analysis data, all the HSZ samples obtained belong to zeolites of the MFI type.

A certain amount of Cu_2S nanoscaled powder was added to the origin zeolite. Mechanochemical activation was carried out with vibratory ball mill KM-1 for 12 hours at 25 °C. Nanoscaled particles of

Cu_2S were obtained within very high-temperature synthesis. The described method was applied for the synthesis of HSZ modified by 0.5, 1 and 3 % (mass) of Cu_2S.

The acid properties of HSZ were studied in a thermal-desorption installation on the basis of ammonia adsorption in a flow of carrier-gas helium in the temperature range 50–650°C at a linear heating rate of 10 deg min^{-1}. To eliminate diffusion hindrances and the effect of the physical form of ammonia adsorption on, HSZ ammonia was adsorbed at a high flow rate of the carrier-gas (110 cm^3 min^{-1}) at 100 °C in the course of 1 h. After that, the reactor with a sample under study was cooled to 50 °C and ammonia was desorbed, with a katharometer serving as detector. Helium of A brand (99.995 vol %) and ammonia of pure grade were used in the experiments. The concentration of acid centers (μmol per gram of catalyst) in the samples under study was determined from the amount of ammonia contained in desorption peaks (forms), with the determination accuracy of adsorbed ammonia by gas chromatography being ±2.5%.

Transformations of the straight-run gasoline fraction with a boiling onset point of 170 °C from the gas condensate were studied on zeolite-containing catalysts on a flow-through catalytic installation with a fixed catalyst bed (reactor volume 10 cm^3) in the temperature range 350–425 °C at a volumetric flow rate of the raw material of 2 h^{-1}, atmospheric pressure, and experiment duration of 1 h at each fixed process temperature.

The gas condensate from the Myl'dzhinskoe deposit has the following fraction composition: boiling onset point 31 °C, 82 vol % of the condensate boils away at 200 °C, end boiling point 297 °C, loss and residue 10 vol %. As regards the group hydrocarbon composition, the straight-run gasoline fraction with a boiling onset point of 170 °C from the gas condensate has the following composition (wt %): n-alkanes 22.7, isoalkanes 33.2, naphthenes 39.9, and arenes 4.2. The octane number of the straight-run gasoline fraction with boiling onset point of 170 °C from the gas condensate of the Myl'dzhinskoe deposit is 60 points by the research method (RM) scale.

Gaseous hydrocarbons were analyzed on a stainless steel column (length 3 m, inner diameter 3 mm) packed with 5% NaOH on Al_2O_3, and liquid hydrocarbons, on a quartz glass capillary column (100 m × 0.25 mm × 0.25 μm) with a supported fixed ZB-1 phase. Gaseous and liquid products formed in conversion of straight-run gasoline fractions of the gas condensate were quantitatively analyzed by gas chromatography on a hardware-software complex based on a Khromatek-Kristall 5000 v. 1 gas chromatograph, with Khromatek-Analitik processing software. The research octane numbers were calculated by using the Khromatek-Analitik processing software from results of a gas-chromatographic analysis of the hydrocarbon composition of the starting raw material and liquid products formed in conversion of straight-run gasoline on the zeolite-containing catalysts under study. The error of the gas-chromatographic determination of gaseous and liquid hydrocarbons is ±2.5%.

3. Experiments and results

3.1. Influence of Cu_2S promoting additives on acid properties of zeolite-containing catalysts

Acid properties of catalysts were estimated by means of temperature-programmed ammonia desorption. Temperature desorption profiles of all catalysts have two peaks. Low-temperature peak (120–250 ºC) of initial zeolite with T_{max}= 185 ºC is mainly related with ammonia desorption from weak acid centers (form I) that are coordinative unsaturated aluminium ions. High temperature peak (300–550 ºC) with T_{max} = 400 ºC is mainly related with ammonia desorption from strong acid centers (form II) that are hydrogen ions of hydroxylic bridge groups. From the data presented in table 1 it can be seen that initial high-silica zeolite possesses 383 μmol/g of weak acid centers and 128 μmol/g of strong acid centers. Within Cu_2S modification the increase of total acid centers concentration takes place due to the increase in weak acid centers concentration while the concentration of strong acid centers remains approximately the same. It should be noted that samples modified by 0.5 % and 1 % of sulphide Cu(I) nanoscale powder possess the highest acid centers concentration (weak acid centers concentration is respectively 415 μmol/g and 410 μmol/g). Also, Cu_2S additions lead to the rise of ammonia desorption temperature. For example, for weak acid centers it increases at 5–12 ºC and at 10–20 ºC for strong centers.

Table 1. Acid properties of zeolite-based catalysts

Catalyst	T_{max}, °C		a.c.c, µmol/g		
	I	II	I	II	common
HSZ	185	400	382	128	510
0.5 % Cu_2S / 99.5 % HSZ	197	425	415	130	545
1 % Cu_2S / 99 % HSZ	196	420	410	130	540
3 % Cu_2S / 97 % HSZ	190	410	390	130	520

3.2. Influence of Cu_2S promoting addition on catalytic properties of zeolite-containing catalysts

The investigation of temperature influence shown that with its rise from 350 to 425 °C and with space feed rate 2 h^{-1} high octane gasoline yield for all investigated samples decreases since the degree of initial stock hydrocarbons conversion increases. The results of the corresponding investigations are presented in table 2.

Firstly the yield of gas products, primarily due to C_3-C_4 paraffines, increases from 9.7 % (350 °C) to 33.7 % (425 °C). The yield of C_6-C_9 liquid aromatic hydrocarbons increases from 9.3 % to 24.8 %. Toluol and xylol take the main part of aromatics. With the increasing temperature the yield of benzene and C_{5+} oleffines increases from 0.4 % and 1.6 % (350 °C) to 1.5 and 2.9 % (425 °C), respectively. But the yield of naphthenic, n- and isoparaffinic C_{5+} hydrocarbons decreases with the increasing temperature (table 2). Gas products are mainly presented by propane and butane.

The dependence of obtained liquid products octane number on temperature is presented in figure 1 (a).

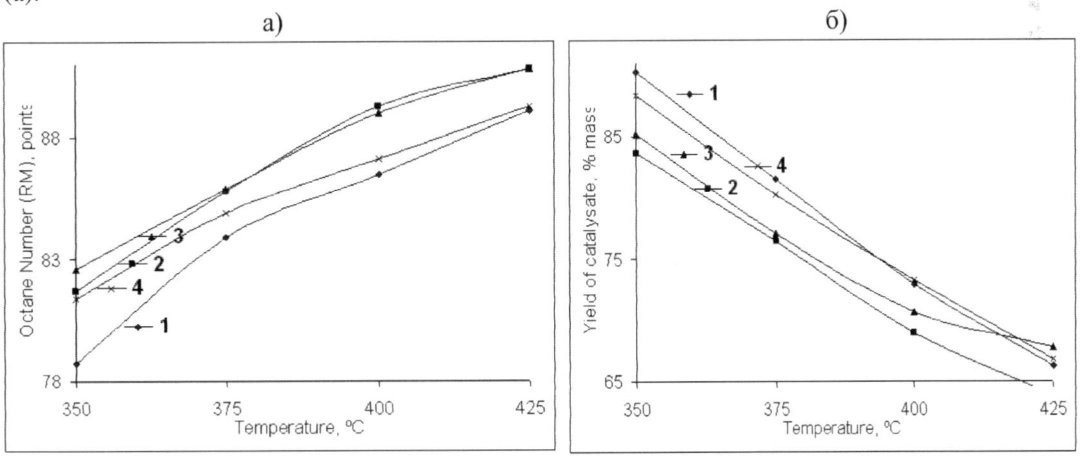

Figure 1. The influence of straight run gasoline conversion temperature with zeolite-containing catalyst on the obtained catalysate octane number (a) and catalysate yield (b). 1 — HSZ, 2 — 0.5 % Cu_2S / 99.5 % HSZ , 3 — 1 % Cu_2S / 99 % HSZ, 4 — 3 % Cu_2S / 97 % HSZ

Table 2. The composition of straight run catalytic gas condensate gasoline conversion products: 1 — HSZ, 2 — 0.5 % Cu_2S / 99.5 % HSZ, 3 — 1 % Cu_2S / 99 % HSZ, 4 — 3 % Cu_2S / 97 % HSZ

	Catalyst											
		1			2			3			4	
Temperature, °C	375	400	425	375	400	425	375	400	425	375	400	425
Gas phase, % mass	18.5	27.1	33.7	22.9	29.3	32.2	19.7	26.7	33.2	23.5	31.0	36.2
Liquid phase, % mass	81.5	72.9	66.3	77.1	70.7	67.8	80.3	73.3	66.8	76.5	69.0	63.8
Gas phase composition, % mass												
Methane	0.2	0.4	0.6	0.2	0.4	0.7	0.2	0.4	0.6	0.2	0.4	0.7
Ethane	1.1	1.5	2.3	1.1	1.6	2.5	1.1	1.6	2.3	1.1	1.6	2.5
Ethylene	1.5	1.7	2.4	1.0	1.4	2.1	1.4	1.7	2.1	1.1	1.5	1.9
Propane	50.6	50.5	50.1	52.7	52.2	52.4	52.0	51.9	52.0	51.8	52.2	52.8
Propylene	3.8	4.2	5.7	2.7	3.6	4.8	3.6	4.3	4.9	2.5	3.5	4.1
Isobutane	22.1	21.5	19.5	22.2	21.2	19.1	21.2	20.4	19.2	22.5	21.2	19.5
N-butane	17.7	17.1	15.3	18.0	17.0	15.1	17.6	16.7	15.6	18.5	16.9	15.4
Isobutene	2.9	3.1	4.0	2.1	2.7	3.2	2.8	3.1	3.3	2.2	2.7	3.0
Liquid phase composition, % mass												
Aromatic	16.5	20.9	24.8	19.4	24.7	28.9	18.1	21.5	26.6	19.5	24.7	28.9
Benzene	0.7	1.1	1.5	0.8	1.4	1.7	0.8	1.2	1.6	0.9	1.4	1.7
Isoparaffines	38.3	38.2	37.0	37.9	37.1	36.0	38.0	37.6	35.4	38.3	37.1	36.0
Naphthenes	27.4	24.9	23.1	28.1	24.9	21.9	26.7	24.6	23.4	25.7	23.3	21.6
Paraffines	15.3	13.3	12.2	11.6	11.0	10.7	14.9	13.6	11.9	14.3	12.7	11.3
Olefines	2.6	2.7	2.9	2.2	2.3	2.5	2.3	2.6	2.8	2.2	2.2	2.2
Octane number (research method)	83.9	86.5	89.1	85.8	89.3	90.3	85.9	89.0	90.8	84.9	87.1	89.3

With this figure we can conclude that the promoting addition of copper sulphide increases the octane number of obtained catalysate in comparison with the catalysate obtained with the origin HSZ. The highest octane number was obtained with samples 2 (0.5 % Cu_2S / 99.5 % HSZ) and 3 (1 % Cu_2S / 99 % HSZ). It varies from 82 (350 °C) to 91 (425 °C) points of RM.

The influence of temperature of straight-run gasoline conversion with zeolite-containing catalyst on catalysate yield is shown in figure 1 (b). Also, we can see that the highest gas product yield takes place with the origin zeolite. Promoting additions increase the activity of raw hydrocarbons conversion. This facilitates the increasing of liquid catalysate yield if compare to the origin zeolite. It should be pointed out that at all process temperature the sample with 1 % of Cu_2S and 99 % of HSZ performs better yield than the sample with 0.5 % of Cu_2S and 99.5 % of HSZ. So the yield of catalysate increases at 2–4 %.

4. Conclusion

Acid properties of initial high-silica zeolite and zeolite-containing catalysts modified with nanoscale Cu_2S powder were investigated. Introduction of 0.5 % mass and 1 % mass of Cu_2S promoting addition facilitated the increase of weak centers concentration at 9 %.

Catalytic properties of zeolite-containing catalysts in straight-run gasoline fractions conversion into high octane components of motor fuels were investigated. It is stated that the yield of aromatic hydrocarbons increases with increasing temperature for all investigated catalysts. Also, increasing temperature leads to the decrease of main product yield due to the increase in degree of stock conversion. High octane catalysate mostly consists of isoparaffins, arenes and naphthens. Gas products mostly consist of propane and butane.

From all the proposed catalysts the most active samples contain 0.5 % Cu2S / 99.5 % HSZ and 1 % Cu2S / 99 % HSZ. These catalysts facilitate the obtaining of products with maximum octane numbers: 82 points at 350 °C and 91 points at 425 °C (research method). However, in the temperature range 350–425 °C catalysate yield with the sample, modified with 1 % of nanoscaled powder of Cu_2S, is higher than the one obtained with the sample, modified with 0.5 % of nanoscaled powder, at 2-4 %, respectively. Thus, it was shown that the sample containing 1 % Cu_2S / 99 % HSZ performs the best activity and desired reaction product ratio.

References

[1] Wang J, Kang M, Zhang Z and Wang X 2002 *Journal of Natural Gas Chemistry.* Propane Aromatization over Mo/HZSM-5 Catalysts. Vol. **11** pp. 43–50.
[2] Raddi de Araujo L R, Schmal M 2002 *Appl. Catal. A: General.* Pt-Cr/ZSM-5 catalysts for propane and cyclohexane conversions. Vol. **235** pp. 139–47.
[3] Erofeev V I, Medvedev A S, Khomyakov I S and Erofeeva E V 2013 *Russian Journal of Applied Chemistry.* Conversion of gas-condensate straight-run gasolines to high-octane gasolines over zeolite catalysts modified with metal nanopowders. Vol. **86** pp. 979–85.
[4] Cheng Y, Liao R H, Li J S, Sun X Y, Wang L J 2008 *Journal of Materials Processing Technology.* Synthesis research of nanosized ZSM-5zeolites in the absence of organic template. Vol. **206** pp. 445–52.
[5] Ha Vu T T, Tiep Le V, Meriaudeau P, Naccache C 2002 *J. Mol. Catal. A: Chemical.* Aromatization of methane over zeolite supported molybdenum: active sites and reaction mechanism. Vol. **181** pp. 283–90.
[6] Wei Li, Yu S Y, Meitzner G D and Iglesia E 2001 *J. Phys. Chem.* Structure and Properties of Cobalt-Exchanged H-ZSM-5 Catalysts for Dehydrogenation and Dehydrocyclization of Alkanes B. Vol. **105** pp. 1176–1184.
[7] Na K and Somorjai G A 2015 *Catal. Lett.* Hierarchically Nanoporous Zeolites and Their Heterogeneous Catalysis: Current Status and Future Perspectives. Vol. **145** pp. 193–213.
[8] Galadima A and Muraza O 2015 *Journal of Natural Gas Science and Engineering.* From synthesis gas production to methanol synthesis and potential upgrade to gasoline range hydrocarbons: A review Raddi de Araujo L R and Schmal M 2002 Pt-Cr/ZSM-5 catalysts for propane and cyclohexane. Vol. **25** pp. 303.
[9] Dapsens P Y, Mondelli C and Pérez-Ramírez J 2013 *ChemSusChem.* Highly selective Lewis acid sites in desilicated MFI zeolites for dihydroxyacetone isomerization to lactic acid. Vol. **6** pp. 831–839.
[10] Giannetto G, Monque R and Galiasso R 1994 *Catal. Rev. – Sci. Eng.* Transformation of LPG into aromatic hydrocarbons and hydrogen over zeolite catalysts. Vol. **36** pp. 274–304.
[11] RF Patent *2236289.*

Changing the composition of the group hydrocarbons of diesel fractions in the process of hydrotreating

N Krivtcova [1,a], K Baklashkina[1,b], Sh Sabiev[1,c], E Krivtsov [2,d] and A Syskina[1e]

[1] 634050, Russia, Tomsk, Lenin Avenue 30, National Research Tomsk Polytechnic University
[2] 634021, Russia, Tomsk, Akademichesky avenue 4, Institute of petroleum chemistry SB RAS
E-mail: [a] krivtcova@mail.ru, [b] ksenija.baklashkina@gmail.com, [c] sabiyev.shyngys@gmail.com, [d]john@ipc.tsc.ru, [e]annesyskina@yahoo.com

Abstract. Change in group composition of sulfur compounds and structural group composition of the diesel fractions in the course of hydrotreating is presented in the paper. The removal degree of sulfur compounds is shown to comprise 95.8 % rel. The homologs of benzothiophenes are removed for 93.9 % rel., ones of the dibenzothiophenes are for 90.7 % rel. A considerable change in group composition of diesel fraction is established in the course of hydrotreating. Hydrogenation degree of aromatic hydrocarbons is 24.4 % wt., the amount of saturated hydrocarbons has increased by 20.4 % wt.

1. Introduction

The main industrial process for sulfur removal is catalytic hydrodesulphurization in which sulfur compounds (SC) break down forming hydrogen sulfide, and the hydrocarbonic part of raw material molecules is restored and remains as a part of the target liquid products [1]. However, tougher ecological requirements to gasolines and diesel fuels [2-5] dictate the need of increase in capital and operational expenditures for production of fuels with low sulfur content. Generally, an increase in production costs is caused by an increase in hydrogen partial pressure in the system and decline in reactor capacity [6-8]. [1] It is believed that it is possible to obtain the motor fuel that meets the strictest standards via improving catalysts and reactors, optimizing technological processes. So, temperature increase of the process causes increase in hydrogenolysis depth of SC, and leads to deeper hydrogenation of aromatic and olefin hydrocarbons. However, the undesirable reactions of hydrocracking leading to formation of gasoline and merkaptan increasing the total sulfur content in the products are also possible.

The paper is aimed at studying influences of temperature and time on the content of sulfur compounds and structural group composition of the diesel fraction in the course of hydrotreating.

2. Experiment

2.1. Diesel hydrotreating

Hydrotreating of diesel fuel with the total sulfur content of 1.430 % wt. was carried out on the laboratory catalytic reactor, the flow-through system with the evaporator, the reactor on the special metal gauze of which the catalyst is placed, the condenser, and the hydrogenate receiver.

Hydrotreating was performed with the aluminum-cobalt-molybdenum catalyst GKD-202, the granule size is 1-2 mm, the catalyst volume in the reactor is 10 ml.

The raw materials and hydrogen came to the reactor from top downward: the diesel fraction was supplied with the high pressure dosing pump; hydrogen supply was regulated with the automatic batcher. The reactor temperature in the air heating oven was set and maintained by the temperature regulator providing the heat accuracy of ±0.5 °C. The thermocouple placed in the pocket in the center of the reactor is used for temperature measurement in the catalyst layer. System pressure was measured by means of the high pressure reducer and controlled according to the indications of the standard pressure gauge.

The products of reaction came to the water cooler-condenser. The liquid products of reaction were gathered for the analysis by discharge in the receiver. The gaseous products via the throttle gate came to the six-port valve by means of which gas is dumped into the atmosphere.

Diesel fuel was hydrotreated under the following experimental conditions: the volume rate for liquid raw materials was 2 h^{-1}, pressure was 3,5 MPas, hydrogen/raw materials ratio was 300/1, temperature was 340, 360 and 380 °C.

2.2. Methods of analysis for initial diesel fuel and hydrotreating products

The method of gas-liquid chromatography (GLC) was used to identify the sulfur compounds. GLC was conducted with use of the chromatograph "Kristall 2000M" which includes 25 m × 0.22 mm quartz capillary column with SE-54 stationary phase, gas carrier is helium. The sulfur- selective flame photometric detector (FPD) was used for analysis of sulfur compounds. There was a linear increase in column heating oven temperature from 50 to 290°C, the heating rate was 4 degrees per minute. The qualitative composition of sulfur compounds was determined by comparing the retention times of the individual compounds (Di-n-hexyl sulfide, benzothiophene, dibenzothiophene) and the analyzed components. Also the literature data was used. The quantitative composition of sulfur compound groups and individual isomers was determined by analyzing the space of chromatographic peak areas, obtained by calculating the space of peak areas (% rel.) in relation to the total sulfur content in the sample (% wt.).

To determine the total sulfur concentration in the initial diesel fuel and the hydrogenate the spectral photometer SPECTROSKAN-S was used.

The group hydrocarbonic composition of the initial distillate was determined by means of the liquid adsorption chromatography on the activated aluminum oxide of the II activity degree according to the method of Brockman (the certified method № 1216-2006 of Institute of petroleum chemistry SB RAS). The chromatographic column, which is 75 × 1.8 cm in size, was filled with the sorbent in the "wet" method for removal of sorbent wetting heat and reduction of channel forming in the sample-to-sorbent ratio of 1:80 by weight. The quantity of the sample was mixed with hexane and evenly distributed on all the surface of the sorbent. Then the fractions of saturated, mono- and bi-aromatic hydrocarbons were separated by n-hexane, the tri-aromatic hydrocarbons were eluted by the mixture of hexane and benzene (3:1 by volume).

The structural group analysis of the diesel fraction was carried out by the n-d-M method which is widely used for research of medium and heavy fractions of oil so far. Hydrocarbons mixture of the studied fraction is presented in the form of one "average molecule" the properties of which are defined in the ratio of arene and cycloalkane fragments and alkane chains.

For calculation of structural group composition of the fractions by the n-d-M method, it is necessary to experimentally define the following physical constants:

1. The index of refraction within the accuracy of ± 0.0001 was defined with the IRF-22 refractometer at 20 °C.
2. Density within the accuracy of ± 0.0002 g/ml was determined by the picnometer method at 20 °C according to GOST 17310-2002.
3. Molecular weight within the accuracy of ± 3% rel. was found according to the tables, for this reason viscosity was defined at 20 °C, 50 °C, 100 °C in accordance with GOST 33-2000.

3. Results and discussion

In the course of hydrotreating of the diesel fraction with the total sulfur content of 1.430 % wt., the maximum degree of SC removal was 94.5, 95.0, 95.8% at the process temperature of 340, 360 and 380 °C, respectively. Change in the total sulfur content in the course of hydrotreating of diesel fuel depending on time of the process at a temperature of 380 °C is presented in table 1.

Table 1. Change in the total sulfur content in the course of hydrotreating at a temperature of 380 °C

Process duration, h	$S_{tot.}$, % wt.	Removal degree S_{tot}, %
0	1.430	–
2	0.094	93.92
4	0.079	94.47
6	0.070	95.10
8	0.060	95.80

Using the formalized reaction of sulfur compounds (SC) transformation in the environment of hydrogen (H_2) into hydrocarbon (HB) and hydrogen sulfide (H_2S):

$$SC + H_2 = HB + H_2S \quad (1)$$

it is possible to determine transformation rate and to compose the kinetic model considering concentration change in sulfur compounds in the course of hydrotreating:

$$W = k \cdot C_{SC} \cdot C_{H2},$$

$$\frac{dC_{sc}}{dt} = -kC_{sc}C_{H_2},$$

where W is chemical reaction rate (1); k – reaction rate constant; C – substance concentration.

After calculation of the inverse kinetic equation, the rate constant of sulfur compounds (SC) transformation on the GDK-202 catalyst was 3.02 s^{-1}.

It is established that the initial fraction contains a wide set of SC – sulfides, homologs of benzo-and dibenzothiophenes (table 2). There are 41% of BT homologs, about 14 % rel. of DBT homologs in the initial fraction. The hydrotreating products contain a limited set of compounds in the composition; they generally are homologs of benzo-and dibenzothiophenes. In the hydrotreating products C the percent of dibenzothiophene sulfur increases up to 48, 47, and 27% of the total sulfur at 340, 360, and 380 °, respectively. According to the data of the chromatographic analysis, the strongest compounds are 4-methyl-DBT and 4.6-dimethyl-DBT the content of which falls no more, than 20% rel.

Table 2. Sulfur content in the sulfur compound groups in the initial fraction and hydrotreating product (8 h)

Fraction	Composition, % wt.					
	C_1BT	C_2BT	C_3BT	DBT	C_1DBT	C_2 BT
Initial	0.0340	0.2020	0.3510	0.0360	0.0880	0.0800
After hydrotreating (340 °C)	0.0017	0.0128	0.0222	0.0236	0.0036	0.0095
After hydrotreating (360 °C)	0.0019	0.0125	0.0219	0.0220	0.0030	0.0071
After hydrotreating (380 °C)	0.0020	0.0152	0.0228	0.0028	0.0065	0.0073

The kinetic researches [9, 10] have shown that these molecules are rather slowly hydrogenated to the dihydrothiophene derivatives. These unstable intermediate products will easily be transformed by two stage hydrogenation into tetrahydro- and hexahydrothiophene compounds or desulphurized with obtaining biphenyl derivatives.

Change in the group hydrocarbonic composition and structural group composition of the initial diesel fraction and hydrotreating products is presented in tables 3 and 4. The general tendency is decrease in the total number of aromatic compounds and, as a result, increase in a percent of saturated compounds.

Table 3. Change in the group composition of the initial fraction and hydrotreating product

Fraction	Composition, % wt.			
	SHB	mono-arenes	bi-arenes	tri-arenes
Initial	53.8	28.2	13.8	2.2
After hydrotreating	64.8	21.1	10.4	1.9

Table 4. Structural group composition of the initial fraction and hydrotreating products

Process duration, h	C_A	C_C	K_A	K_T	C_N	C_P	K_N
0	19.68	60.08	0.50	1.46	45.40	40.90	0.96
2	18.00	59.10	0.50	1.54	41.00	40.90	1.14
4	23.00	54.00	0.50	1.50	31.00	46.00	1.00
6	24.90	52.70	0.60	1.45	26.30	47.30	0.85
8	24.91	30.00	0.60	2.00	14.91	70.00	1.40

Ca - the carbon content in the aromatic compounds; Cc - the carbon content in the cyclic compound (aromatic + naphthenic); K_A – the average number of aromatic rings in a molecule; Kt – the total number of rings (aromatic + naphthenic); Cn - the carbon percent in the naphthenic compounds, %; Cp - the carbon percent in the paraffin compounds, %; Kn - the average number of naphthenic rings in a molecule.

Hydrotreating leads to decrease in quantity of bi- and tri-aromatic hydrocarbons more than twice, the content of monoaromatic hydrocarbons decreases from 28.2 to 21.1 % wt. in the products. There is a significant increase in a percent of saturated hydrocarbons from 53.8 to 64.8 % wt. These changes favorably affect commodity properties of the obtained product. It is known that tendency to hydrogenation of aromatic compounds changes in the following row: tri-aromatic > bi-aromatic > monoaromatic. It is established that the use of the GKD-202 catalyst in the course of hydrotreating leads to approximately equal removal (in (% rel. from the initial quantity) of all groups of aromatic compounds.

According to the structural group analysis of the diesel fraction before and after the process of hydrotreating, there is an increase in paraffin compounds and decrease in the number of aromatic compounds. The average number of aromatic rings also reduces.

4. Conclusion

It is established that the use of the GKD-202 catalyst in the course of hydrotreating of the diesel fraction allows removing more than 95.8 % rel. from the initial quantity of sulfur in the raw materials. It is shown that the removal degree of sulfuraromatic compounds falls with increase in quantity and size of alkyl substitutes. Dibenzothiophenes derivatives are removed much worse than the homologs of benzothiophenes.

The analysis of the initial fraction SC and hydrotreating products has shown that the compounds 4-methyl-DBT and 4.6-dimethyl-DBT are stable and are practically not exposed to hydrogenolysis with the GKD-202 catalyst.

It is established that there is a change in group and structural group composition of the fraction in the course of hydrotreating. The degree of aromatic hydrocarbons removal is 24.4% while the quantity of saturated hydrocarbons increases 20.4%.

5. References

[1] Chertkov Ya, Spirkin V 1971 *Moscow chemistry.* Sulfur and oxygen compounds of oil distillates. pp. 307.
[2] Babich I and Moulijn J 2003 *Science and technology of novel processes for deep desulfurization of oil refinery streams.* a review Vol. **82 (6)** pp. 607-631.
[3] Mitusova T, Polina E and Kalinina M 2002 *Moscow Equipment.* Modern diesel fuels and its additives. pp. 63.
[4] Kaminsky E and Havkin V 2001 *Moscow Equipment.* Deep oil refining: technological and ecological aspects. pp. 384
[5] Danilov A 2003 *Moscow Equipment.* Introduction to chemmotology. pp. 464.
[6] Elshin A, Grishanov G and Mikishev V 2003 Oil processing and petrochemistry. Vol. **8** pp. 26.
[7] Kaminsky E, Melik-Akhnazarov T and Havkin V 1998 *Science and technology of hydrocarbons.* Vol. **1** pp. 68.
[8] Kaminsky E, Havkin V, Kurganov V 2000 *World of oil products.* Vol. **2** pp. 9.
[9] Meille V, Schulz E, Lemire M and Vrinat K 1997 *J.Catal.* Hydrodesulfurization of Alkyldibenzothiophenes over a NiMo/Al_2O_3 Catalyst: Kinetics and Mechanism. Vol. **170** pp. 29-36.
[10] Krivtsova N, Krivtsov E, Ivanchina E and Golovko A 2013 *Basic Researches.* Kinetics of hydrodesulfurization of diesel fraction. Vol. **8-3** pp. 640-644.

Intensification and forecasting of low-pour-point diesel fuel production via modelling reactor and stabilizer column at industrial unit

N S Belinskaya, E V Frantsina, E D Ivanchina, N V Popova, I V Zyryanova and E V Averyanova

Department of Fuel Engineering and Chemical Cybernetics, National Research Tomsk Polytechnic University, 30 Lenin Avenue, Tomsk, 634050, Russia
E-mail: belinskaya@tpu.ru, evf@tpu.ru

Abstract. In this work forecast calculation of stabilizer column in the technology of low-pour-point diesel fuel production was modelled. The results of forecast calculation were proved by full-scale experiment at diesel fuel catalytic dewaxing unit. The forecast calculation and full-scale experiment made it possible to determine the ways of mass transfer intensification, as well as to increase the degree of hydrogen sulphide removal in the column, and thereby to decrease corrosiveness of the product stream. It was found out that maintenance of the reflux rate in the range of 80–90 m^3/h and injection of additional vapourizing streams, such as stable naphtha from distillation unit (in the volume of 10–22 m^3/h) and hydrogen-containing gas (in the volume of 100–300 m^3/h), ensure complete elimination of corrosive hydrogen sulphide from the product stream. Reduction of stream corrosive activity due to suggested solutions extends service life of equipment and pipelines at industrial catalytic dewaxing unit.

1. Introduction

Petroleum refining is a unique sector of industry in respect to corrosion. Virtually all known types of corrosion occur there due to the vast variety of operating conditions [1–5]. Every second failure at petroleum refineries is caused by corrosion of equipment [6].

The objective of this work is to decrease corrosive activity of stable naphtha, which is one of the products of diesel fuel catalytic dewaxing.

2. The subject of research and raw materials characteristics

Catalytic dewaxing unit is intended to produce summer, winter, and arctic diesel fuels (figure 1). Main production stages involve hydrotreating and dewaxing of the feedstock (mixture of atmospheric gas oil, straight-run diesel fractions, and visbreaking gasoline), stabilization of the product, and distillation with diesel fractions and stable naphtha production [7,8].

After unit start-up, high content of hydrogen sulphide in product stable naphtha was observed. It was discovered that hydrogen sulphide is formed in the bottom of the stabilizer column, as well as in the intermediate equipment at high temperature according to the reaction of mercaptane decomposition. Mercaptanes, in turn, are formed in the dewaxing reactor from olefins, which are the products of cracking reaction on dewaxing catalyst. The occurrence of mercaptane formation reaction in the dewaxing reactor was proved using thermodynamic analysis (the change in Gibbs free energy (ΔG) of this reaction is equal to –69,1 ÷ –76,2 kJ/mol).

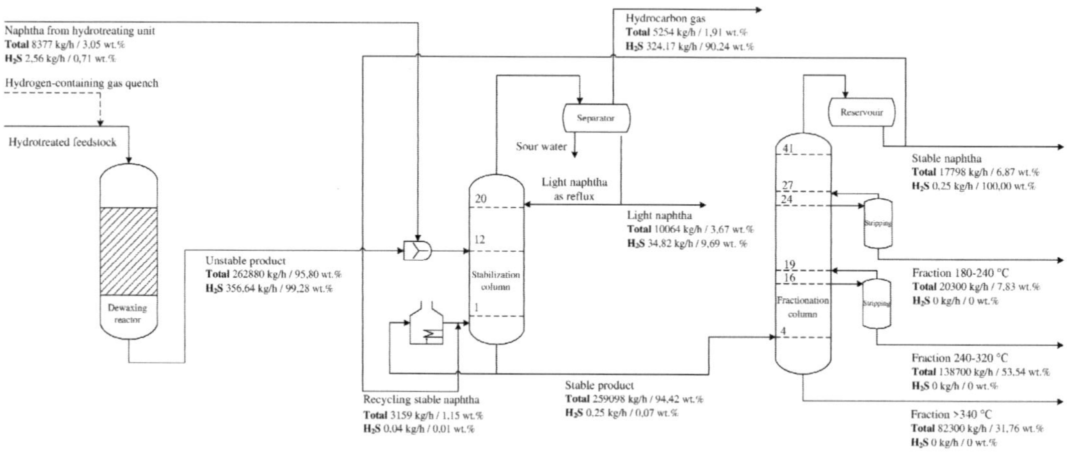

Figure 1. Industrial diesel fuels catalytic dewaxing unit

The bulk of hydrogen sulphide (96–99 wt.%.) enters into the stabilizer column with unstable hydrogenate, which is the product of the reactor section. The rest of hydrogen sulphide (1–4 wt.%) enters into the stabilizer column with naphtha-distillate of hydrotreating unit.

It is known, that sulphur-containing compounds are distributed among the fractions according to chemical structure of hydrocarbons. Aliphatic compounds (low-molecular-weight mercaptanes, sulphides, disulphides) are presented in low boiling fractions. Thus, hydrogen sulphide concentrates in gas, light and stable naphtha. However, in compliance with the project of the unit, stable naphtha should be free from hydrogen sulphide.

The presence of hydrogen sulphide in product stable naphtha is harmful, since it causes corrosion of equipment located after distillation. The latter decreases equipment service life and enhances the probability of failure inception in the unit operation due to damage of equipment as a consequence of corrosion [9].

Hydrogen sulphide must be removed completely as a component of gas and light naphtha at the stabilization stage. However, the actual mode of stabilizer column operation does not provide total removal of hydrogen sulphide.

To achieve the objective of the work, the research on mass transfer intensification and hydrogen sulphide removal in the stabilizer column, as well as forecasting of low-pour-point diesel fuel production were conducted via modelling of the reactor and stabilizer column.

Study of any complex multi-stage petroleum refining process requires taking into account all production stages because of the fact that raw materials composition and technological parameters of previous stages influence significantly on the results of subsequent stages. The most effective method of research, forecasting, and optimization of petroleum refining processes is the method of mathematical modelling [10-12], applied in this work.

3. Results and discussion

To evaluate the influence of flow rates of reflux and recycled stable naphtha, entering the stabilizer column, the operation of industrial diesel fuel dewaxing unit was monitored.

Initially recycled stable naphtha was sent into the column through the furnace before November 2015. This limited its flow rate to 5 m^3/h because of the pump power. As a result of column re-piping (change of recycled stable naphtha direction for straight into the column by-passing the furnace) recycled stable naphtha flow rate was increased to 22 m^3/h.

Thus, stable naphtha flow rate into the bottom of the column was increased from 2.2–4.2 to 17.0–22.0 m^3/h after re-piping, which ensured decrease in hydrogen sulphide content in product stable naphtha from 0.0009–0.0015 to 0.0004-0.0008 wt.% (figure 2).

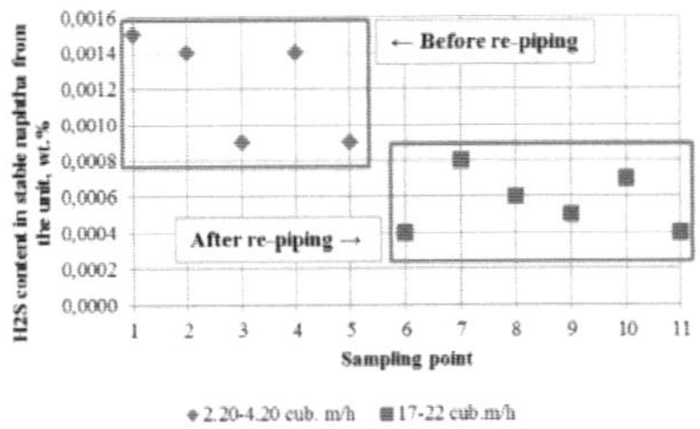

Figure 2. Impact of recycled stable naphtha flow rate increasing due to re-piping on hydrogen sulphide content in product stable naphtha

The monitoring of industrial dewaxing unit provided determination of dependency of hydrogen sulphide content in product stable naphtha on reflux and recycled stable naphtha flow rates. Thus, in case of reflux flow rate equal 76 m^3/h (table 1), hydrogen sulphide content in product stable naphtha varies from 0.0009 to 0.0014 wt.% under change in recycled stable naphtha flow rate from to 3.5 to 3.8 m^3/h. The similar dependence is observed when dewaxing unit operated after re-piping. Thus, in case of reflux flow rate equal to 78 m^3/h (table 2), hydrogen sulphide content in product stable naphtha varies from 0.0006 to 0.0008 wt.% under change in recycled stable naphtha from 20 to 22 m^3/h.

Table 1. Hydrogen sulphide content in product stable naphtha before re-piping

Date	Recycled stable naphtha flow rate, m^3/h	Reflux flow rate, m^3/h	H_2S in stable naphtha, wt.%
13.01.2015	2.20	79	0.0015
03.02.2015	4.20	72	0.0014
05.02.2015	4.20	71	0.0009
11.02.2015	3.80	76	0.0014
17.02.2015	3.50	76	0.0009

Table 2. Hydrogen sulphide content in product stable naphta after re-piping

Date	Recycled stable naphtha flow rate, m^3/h	Reflux flow rate, m^3/h	H_2S in stable naphtha, wt.%
23.11.2015	18	82	0.0004
25.11.2015	20	78	0.0008
26.11.2015	22	78	0.0006
27.11.2015	22	81	0.0005
30.11.2015	20	90	0.0007
07.12.2015	17	88	0.0004

Moreover, analysis of industrial data showed that hydrogen sulphide content in product stable naphtha depends on reflux flow rate. Thus, under constant recycled stable naphtha flow rate equal to 4.2 m^3/h, and reflux flow rate equal to 71 and 72 m^3/h respectively, hydrogen sulphide content varies from 0.0009 to 0.0014 wt.% (table 1). The similar correlation is observed when dewaxing unit operated after re-piping. Thus, under constant recycled stable naphtha flow rate, equal to 22 m^3/h, and reflux flow rate equal to 78 and 81 m^3/h respectively, hydrogen sulphide content varies from 0.0005 to 0.0006 wt.% (table 2).

The developed mathematical model of diesel fuel dewaxing process, brought into coincidence with the model of stabilizer column (figure 3), ensured quantitative evaluation of reflux and recycled stable naphtha flow rates influence on hydrogen sulphide content in product stable naphtha.

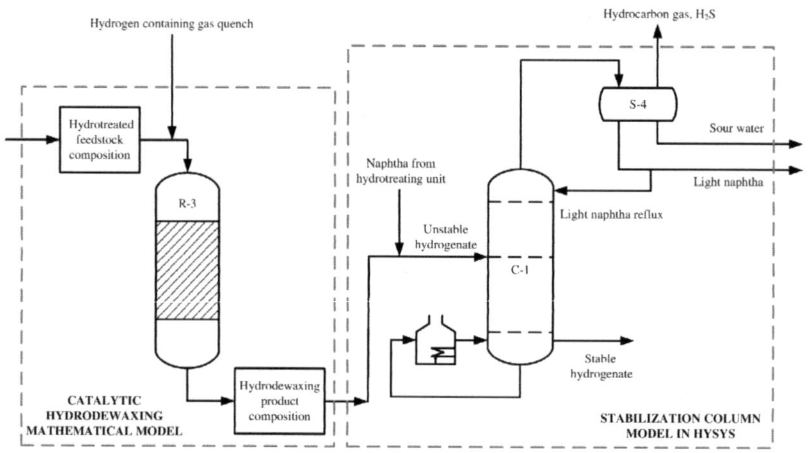

Figure 3. Schematic representation of reactor and stabilizer column joint operation for dewaxing process modelling

Calculation results are presented in figures 4-6.

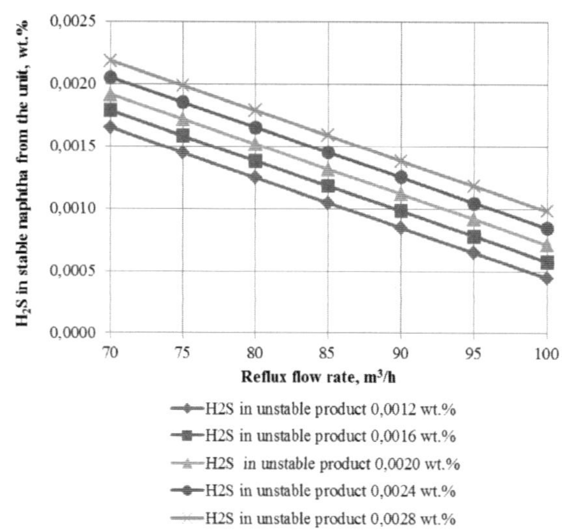

Figure 4. Impact of reflux flow rate on hydrogen sulphide content in product stable naphtha

As figure 4 shows, under increase in reflux flow rate into the column from 70 to 100 m^3/h the content of hydrogen sulphide in product stable naphtha (stable naphtha from the unit) significantly declines from 0.0022 to 0.0005 wt.% depending on hydrogen sulphide content in the column feedstock (unstable product).

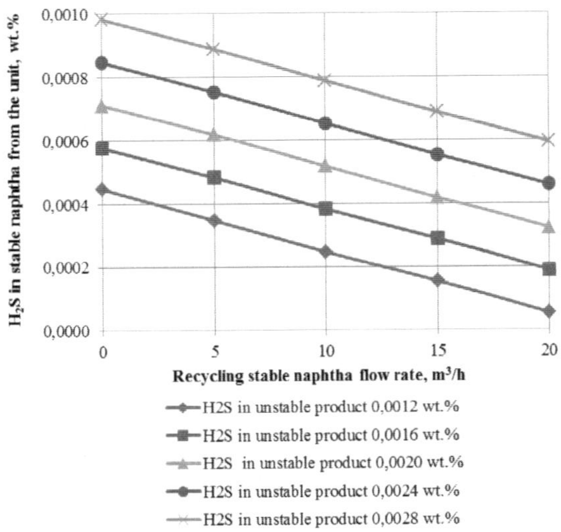

Figure 5. Impact of recycled stable naphtha flow rate on hydrogen sulphide content in product stable naphtha

As figure 5 shows, when recycled stable naphtha flow rate into the column increases from 0 to 20 m^3/h, the content of hydrogen sulphide in product stable naphtha falls gradually from 0.0010 to 0.0001 wt.% depending on hydrogen sulphide content in the column feedstock.

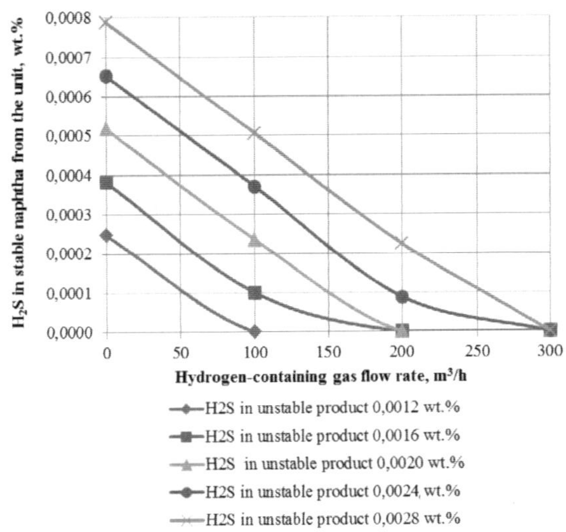

Figure 6. Impact of hydrogen containing gas flow rate on hydrogen sulphide content in product stable naphtha

As it can be seen from figure 6, increase in hydrogen-containing gas flow rate into the column from 100 to 300 m^3/h causes decrease in hydrogen sulphide content in product stable naphtha from 0.0008 wt.% to complete removal depending on hydrogen sulphide content in the column feedstock.

4. Conclusion

As a result of the performed research on mass transfer intensification and hydrogen sulphide removal in stabilizer column, the dependencies of reflux and recycled naphtha flow rates on hydrogen sulphide content in product stable naphtha and its corrosiveness were established. Thus, with increase in reflux flow rate and flow rate of recycled stable naphtha into the bottom of the column, hydrogen sulphide content in product stable naphtha decreases. However, these features do not provide complete removal of hydrogen sulphide. This becomes possible by injection of hydrogen-containing gas into the column as additional vapourizing agent, which ensures increase in partial pressure of hydrogen sulphide. The latter provides complete removal of hydrogen sulphide as a part of overhead product.

The revealed correlations made it possible to determine optimal intervals of column operating conditions. Maintenance of reflux rate in the range of 80–90 m^3/h and injection of additional vapourizing agents into the column, namely recycled stable naphtha from distillation stage (in the volume of 10-22 m^3/h) and hydrogen-containing gas (in the volume of 100–300 m^3/h), ensure attainment of hydrogen sulphide absence in product stable naphtha. Reduction of stream corrosive activity due to suggested solutions extends service life of equipment and pipelines at industrial catalytic dewaxing unit.

References

[1] Hinds G, Wickström L, Mingard K and Turnbull A 2013 *Corros. Sci.* Impact of surface condition on sulphide stress corrosion cracking of 316L stainless steel. Vol. **71** pp. 43–52.

[2] Roffey P and Davies E H 2014 *Eng. Failure Anal.* The generation of corrosion under insulation and stress corrosion cracking due to sulphide stress cracking in an austenitic stainless steel hydrocarbon gas pipeline. Vol. **44** pp. 148–157.

[3] Zheng S, Chen L, Chen C. 2013 *Eng. Failure Anal.* Failure analysis of an A333Gr6 pipeline after exposure to a hydrogen sulfide environment. Vol. **35** pp. 516–523.

[4] Kim S J, Jung H G, Park G T, Kim K Y. 2014 *Appl. Surf. Sci.* Effect of Cu and Ni on sulfide film formation and corrosion behavior ofpressure vessel steel in acid sour environment. Vol. **313 pp.** 396–404.

[5] Fellah M F. 2016 *Fuel Process. Technol.* Adsorption of hydrogen sulfide as initial step of H_2S removal: A DFT study on metal exchanged ZSM-12 clusters. Vol. **144** pp. 191–196.

[6] Medvedeva M L 2005 *Petroleum and Gas.* Corrosion and protection of petroleum and gas processing equipment: Textbook for higher education universities of petroleum and gas profile. pp. 312.

[7] Belinskaya N S, Ivanchina E D, Ivashkina E N, Frantsina E V and Silko G Y 2014 *Procedia Chem.* Optimal technological parameters of diesel fuel hydroisomerization unit work investigation by means of mathematical modelling method. Vol. **10** pp. 258–266.

[8] Belinskaya N S, Ivanchina E D, Ivashkina E N and Silko G.Y. 2014 *Procedia Chem.* Effect of feed composition changing at naphtha catalytic reforming unit due to involvement of gasoline fraction obtained by diesel fuels hydrodewaxing into the processing. Vol. **10** pp. 267–270.

[9] Belinskaya N S, Ivanchina E D, Ivashkina E N, Chuzlov V A and Faleev S A 2015 *Procedia Eng.* Mathematical modeling of the process of catalytic hydrodewaxing of atmosheric gasoil considering the interconnection of the technological scheme devices. Vol. **113** pp. 68–72.

[10] Frantsina E V, Ivashkina E N, Ivanchina E D, Romanovskii R V, 2014 *Chem. Eng. J.* Developing of the mathematical model for controlling the operation of alkane dehydrogenation catalyst in production of linear alkyl benzene. Vol. **238** pp. 129–139.

[11] Dolganov I M, Pisarev M O, Ivashkina E N, Dolganova I O, 2015 *Petroleum and Coal.* Modeling of liquid separators work in gas and gas condensate preparation unit in low-temperature separation technology. Vol. **57** pp. 328–335.

[12] Dmitriev V M, Gandga T V, Dolganov I M, Pisarev M O, Dolganova I O, Ivashkina E N, 2015 *Petroleum and Coal.* Structure of network simulator for training and retraining of operators of controlled technological objects of oil and gas industry. Vol. **57** pp. 691–695.

IOP Conf. Series: Earth and Environmental Science **43** (2016) 012063 doi:10.1088/1755-1315/43/1/012063

Influence of dispersing additive on asphaltenes aggregation in model system

A M Gorshkov[1,4], L V Shishmina[2,5], A Z Tukhvatullina[3], Yu R Ismailov[2] and G A Ges[2]

[1] Petroleum Learning Centre, National Research Tomsk Polytechnic University, Russia

[2] Geology and Oil Field Exploitation Department, National Research Tomsk Polytechnic University, Russia

[3] Arbuzov Institute of Organic and Physical Chemistry, Kazan Research Center of Russian Academy of Sciences, Russia

[4]gorshkovam89@mail.ru, [5]shishmina@tpu.ru

Abstract. The work is devoted to investigation of the dispersing additive influence on asphaltenes aggregation in the asphaltenes–toluene–heptane model system by photon correlation spectroscopy method. The experimental relationship between the onset point of asphaltenes and their concentration in toluene has been obtained. The influence of model system composition on asphaltenes aggregation has been researched. The estimation of aggregative and sedimentation stability of asphaltenes in model system and system with addition of dispersing additive has been given.

1. Introduction

Over the last decade, the share of unconventional Russian reserves has increased significantly, particularly, in high viscosity oil and natural bitumen. Today the commercial development of heavy oil is actively conducted in the Republic of Tatarstan. One of the main problems is the high content of asphaltenes, which can reach 20% in such oils [1, 2].

Asphaltenes are usually classified as the most polar fraction of crude oil that are soluble in aromatic solvents but insoluble in normal alkanes. Asphaltenes are identified as molecules having polyaromatic and polycyclic rings with short aliphatic chains and heteroatoms, such as nitrogen, oxygen and sulfur. Also asphaltenes contain some metals (vanadium, nickel, iron, etc.). The asphaltene molecule has diameter between 10 and 20 Å [1]. High polarity, aromaticity, and heteroatom content in asphaltenes causes a strong tendency to self-association and the formation of supramolecular aggregates even in very dilute organic solvents. In this regard the prevention of asphaltenes sedimentation stability loss is an important scientific and practical task.

One of the most common methods of asphaltenes stabilization in crude oils and hydrocarbon solutions is the use of amphiphilic surfactants, which inhibit or delay the phase separation process of asphaltenes in hydrocarbon solutions. Many scientists are trying to identify mechanisms of the additive structure influence on asphaltenes stability in organic model systems and crude oils. This research is focused on the influence of chemical additives due to their particular structure, composition, and mechanism of influence [1, 3-5]. A great interest is displayed towards dispersing additives among amphiphilic substances. Effect of dispersing additives has been investigated for diesel fuel and oils [6, 7]. However, effect of dispersing additives in terms of hydrocarbon asphaltene solution has not been studied yet.

The aim of this work is to address the influence of dispersing additives on asphaltene aggregation in model system by the photon correlation spectroscopy method.

2. Experimental procedure

2.1. Materials

Asphaltenes have been extracted from heavy oil from one of the oil fields in the Republic of Tatarstan. Asphaltenes aggregation has been investigated in the asphaltene–toluene–heptane model system which is the most common system used to investigate the process of asphaltenes aggregation [8–11]. Chromatographic grade toluene and *n*-heptane of the same grade have been used as a solvent and as precipitant respectively. All experiments have been carried out on freshly prepared solution of asphaltenes in toluene. Initiation of asphaltenes aggregation is performed by adding a certain amount of *n*-heptane.

Alkenylsusinimide C-5A is used as dispersing additive, which is 40% concentrate of alkenylsussinimide group in mineral oil and unreacted polybutylene. The additive is a viscous liquid, which has been dissolved in toluene with mass ratio 1:1.

In the beginning of every experiment, the cuvette with the studied solution is dispersed for a 15 second period by ultrasound wave to mix all components. Cuvettes have a cylinder form with the diameter of 15 mm.

2.2. Method and equipment

The photon correlation spectroscopy (PCS) method allows determining the diffusion coefficient of colloidal particles by measuring the autocorrelation function of light scattering intensity fluctuations. If the particles take a spherical shape, their radius can be calculated by Stokes-Einstein equation [8–11]. Besides, PCS allows measuring the particle size in dispersion system in real time during its growth.

PhotoCor Complex system is based on the PCS method and is used to measure the size of asphaltenes associates and investigate aggregation process [8, 11]. A semiconductor laser beam with a wavelength $\lambda = 654$ nm is used as a radiation source. The main measuring tools of PhotoCor Complex device are two photodetector sensors, operating in photon counting mode. Light scattering intensity is observed in order to calculate the onset point of asphaltenes using a photodetector device Photocor-PC3 which is more sensitive. As a result, it becomes possible to track small changes in dispersion system. Photocor-PC3 is set at an angle of 90°. The light scattering intensity is measured for a short period of time (60 seconds). To research asphaltenes aggregation and particle size determination, second sensor (Photocor-PC1) is used. It is less sensitive but more effective for study of the process. Photocor-PC1 is set at an angle 40°. To avoid influence of the «thermal lens» effect (occurrence of convection flows as a result of the liquid heating by laser beam) on validity of particle size determination, the neutral glass filter has been selected for each experiment. The filter is installed on optical path of laser radiation. All experiments are carried out at the temperature of 27 °C and at atmospheric pressure.

3. Experiments and results

3.1. Onset point of asphaltenes

In this work, onset point of asphaltenes is determined from the dependence of the light scattering intensity on the amount of *n*-alkane in model system [11].

The PCS method has limitations due to the fact that highly concentrated asphaltenes solutions in toluene act as strongly absorbing media. The dependence of the light scattering intensity on the concentration of investigated asphaltenes is presented in figure 1.

The increase in asphaltenes concentration in toluene to 0.15 g/L leads to the growth of light scattering intensity to a maximum value of 218,000 cps. Further addition of asphaltenes to the solution

IOP Conf. Series: Earth and Environmental Science **43** (2016) 012063 doi:10.1088/1755-1315/43/1/012063

results in an intensity drop, indicating the onset of light absorption by the disperse system (figure 1). When concentration of asphaltenes in toluene reaches value of 1 g/L, the scattering intensity drop comes to a minimum value of 15,500 cps. Four solutions with different concentrations of asphaltenes in toluene have been chosen as objects of research: 0.05 g/L (concentration in range of transparent solutions), 0.15 g/L (concentration, corresponding to the maximum value of scattering intensity), 0.4 g/L and 0.625 g/L (concentrations in range of opaque solution) (figure 1). The threshold concentration of *n*-heptane (the onset point of asphaltenes) is determined by PCS method for each concentration of asphaltenes.

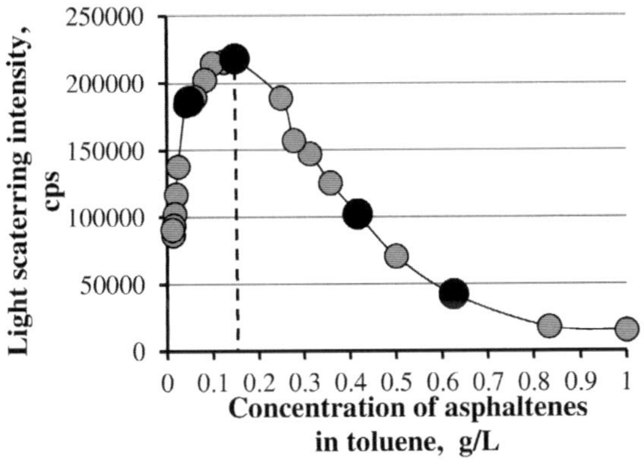

Figure 1. Dependence of the light scattering intensity on the asphaltenes concentration in toluene

The dependence of the light scattering intensity (*Ils*) on the amount of *n*-heptane (*C*) for different concentrations of asphaltenes in toluene is presented in figure 2.

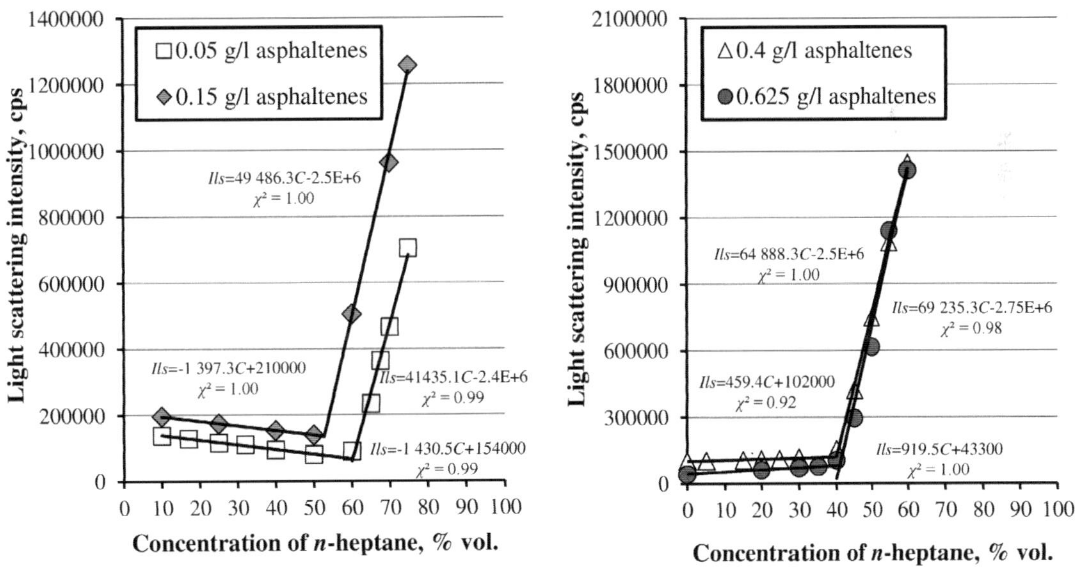

Figure 2. Identification of the asphaltenes onset point in model systems

Based on figure 2, it can be concluded that the dependence of the light scattering intensity on the amount of added *n*-heptane is approximated by linear functions (reliability of approximation $\chi2\geq0.92$) for all concentrations of asphaltenes in toluene. The cross point of these functions is the "onset" parameter of asphaltenes in investigated concentrations. It is worth mentioning that scattering intensity decreases with increasing amount of *n*-heptane to the threshold value for the transparent solution (concentrations 0.05 g/L and 0.15 g/L), while light scattering intensity rises in directly proportional manner with amount of precipitator to the threshold value for the absorbing solution (concentrations 0.4 g/L and 0.625 g/L). This light intensity decreases due to the increasing fraction of *n*-heptane, scattering intensity of which is less than that of pure toluene in case of transparent solutions. In the second case, intensity rises due to the increase in solution transparency. Sharp growth of light scattering intensity is obtained when amount of *n*-heptane is above the threshold value. It indicates the transition of model solution from the molecular-dispersed to the colloidal state [11]. The growth rate of scattering intensity rises with increasing concentration of asphaltenes in toluene (figure 2). Findings once more confirm potential of using the PCS method to determine the onset point of asphaltenes.

The dependence of the onset point of asphaltenes on their concentration in toluene is presented in figure 3.

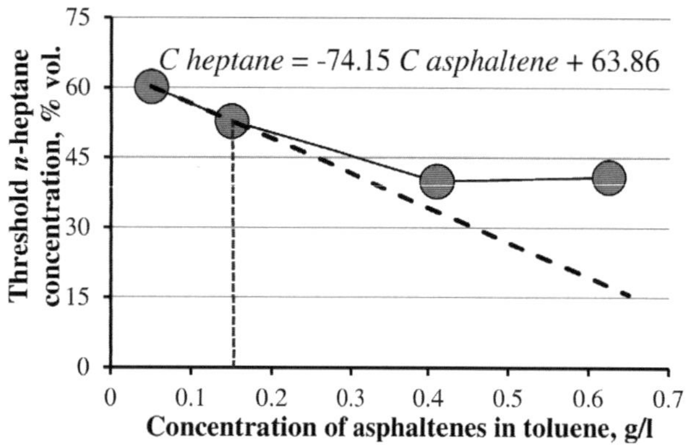

Figure 3. Dependence of the threshold *n*-heptane concentration on concentration of asphaltenes in toluene

Based on this figure, it can be concluded that the onset point of asphaltenes decreases linearly with increasing concentration of asphaltenes to the value of 0.4 g/L, after which it stabilizes at 40% vol. of *n*-heptane. However, asphaltenes aggregation is identified for the concentration of asphaltenes in toluene 0.4 g/L with lower addition *n*-heptane than the threshold value. A possible explanation for this phenomenon is that concentrated solutions of asphaltenes in toluene (0.4 g/L and 0.625 g/L) are strongly absorbing media, for which optical methods, including the PCS method, have their limitations. Thus, the PCS method allows precise defining of asphaltenes onset point only in the range of transparent solutions (for investigated asphaltenes up to concentration of 0.15 g/L). During the experiment, it has been assumed that in order to determine the onset of asphaltenes aggregation in absorbing media it is necessary to extrapolate the curve obtained for transparent solutions (figure 3, dotted line). Figure 3 shows that the more concentrated solution of asphaltene is the more deviation of experimental data from the approximating curve is observed. Thus, for asphaltenes concentrations in toluene 0.4 g/L and 0.625 g/L, the aggregation starts with *n*-heptane addition of 34% and 18% vol. respectively.

3.2. Process of the asphaltenes aggregation

Process of aggregation, i.e. time variation of particles number and degree of their aggregation, is described by the Smoluchowski equation. In the articles [8, 10, 11] it is shown that process of asphaltenes aggregation passes in accordance with the diffusion-limited aggregation mechanism in most cases.

An investigation of the asphaltenes aggregation process is conducted in toluene solution of asphaltenes with concentration of 0.4 g/L, which is an opaque medium (figure 1). The onset point of the asphaltenes concentration in the solution is assumed to be equal to 34% vol. of *n*-heptane.

The investigation results of the asphaltenes aggregation process during the addition of various *n*-heptane volumes are presented in figure 4. Solid curves are the result of the experimental data treatment to the power law dependence in accordance with the diffusion-limited aggregation mechanism by least-squares method.

Figure 4 displays that asphaltenes in toluene stay in the colloidal-dispersed state without the addition of *n*-heptane, with the average radius of particle *R* being approximately 225 nm. The asphaltenes aggregation process in all model systems is described by the power law dependence (reliability of approximation $\chi 2 \geq 0.89$), which suggests DLA mechanism. The only exception is the sample containing 35% vol. of *n*-heptane. In this case, the asphaltenes particles grow to a size of about 310 nm and remain stable throughout the experiment. This behavior is possible because the *n*-heptane concentration value is close to the threshold value (34% vol.) and the amount of *n*-heptane is not enough for a fast asphaltenes aggregation.

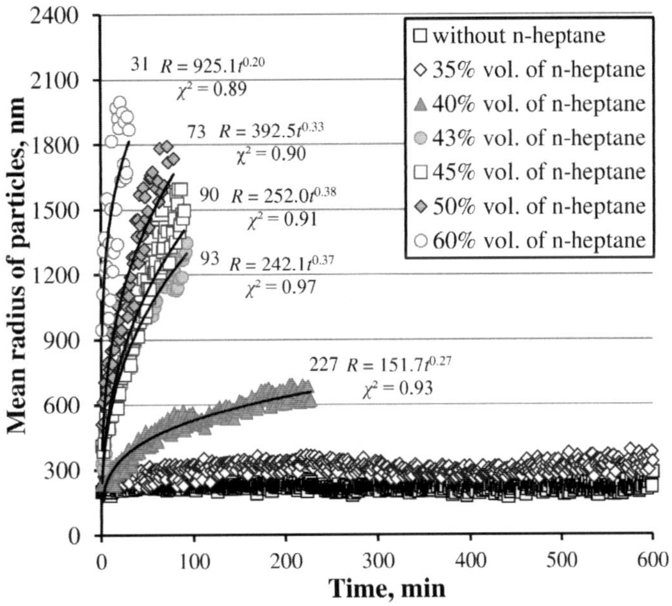

Figure 4. The asphaltenes aggregation process during the addition of various *n*-heptane volumes

In this paper an aggregation time *tagr* has been defined. It is the time during which the particles grow to critical sizes in the solution, and after reaching these sizes, aggregates commence to precipitate and measurements become no longer valid (figure 4). This parameter can be used to evaluate the sedimentation stability of asphaltenes.

Figure 4 displays that increasing amount of added *n*-heptane facilitates the aggregation rate because initial size of the particles fixed by the device is increased and the asphaltenes aggregation time is decreased. Ultimately, these facts lead to earlier precipitation of asphaltenes particles.

3.3. Influence of dispersing additive on the asphaltenes aggregation process

An investigation of the dispersing additive influence on the asphaltenes aggregation process is carried out on the asphaltenes–toluene–heptane model solution. The concentration of asphaltenes in toluene is 0.4 g/L. The amount of added *n*-heptane is 43% vol. for all additive concentrations.

The results of the asphaltenes aggregation process research in solution without additive and with addition of various dispersing additive volumes are presented in figure 5.

The dispersing additive efficiency is evaluated by two parameters: aggregative and sedimentation stability of the asphaltenes solutions. If the additive stabilizes the asphaltenes aggregates and sedimentation is not observed, the aggregative stability of asphaltenes is evaluated by the value of mean particle's radius in the solution. The sedimentation stability is determined by the asphaltenes aggregation time (*tagr*). The results of aggregative and sedimentation stability evaluation of the investigated asphaltenes is presented in table 1.

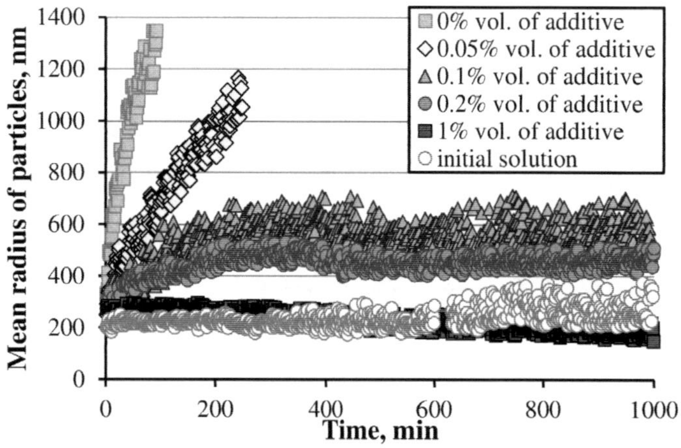

Figure 5. The asphaltenes aggregation process during the addition of various volumes of additive C-5A

Table 1. Influence of the additive C-5A on asphaltenes stability in model systems

Concentration of additive, % vol.	Time of aggregation *tagr*, min	Mean radius of particles *R*, nm
0	93	growth from 300 nm to 1300 nm
0.05	249	growth from 250 nm to 1130 nm
0.1	>1000	590±110
0.2	>1000	455±50
1	>1000	225±65
Solution of asphaltenes in toluene without *n*-heptane	>1000	225±35

With figure 5 and table 1 it can be concluded that addition of the additive in small amounts (up to concentration of 0.05% vol.) leads to the increase in the asphaltenes aggregation time, which ultimately increases the sedimentation stability of asphaltenes. With the increase in additive concentration by more than 0.05% vol. there is the decrease in mean radius of particles in the model solution. When 1% vol. of the C-5A is added to the solution, the additive prevents asphaltenes aggregation completely, as the size of particles in the system coincides with the size of particles in the initial solution of asphaltenes in toluene. These facts indicate the dispersing effect of the additive C-5A on the asphaltenes aggregates.

4. Conclusion

In this paper the onset point of asphaltenes in toluene solutions with different concentrations has been determined by the photon correlation spectroscopy method. It is shown that the PCS method allows reliably determining the onset point for asphaltenes only in transparent solutions. For opaque solutions to determine the asphaltenes aggregation onset, it is necessary to use the linear relationship obtained in this research. This assumption conforms to the article [12] where an inverse relationship between the concentration of asphaltenes in toluene and the amount of precipitant, which is necessary to reach the onset point of aggregation, are experimentally shown.

The influence of model system composition on the asphaltene aggregation process has been researched. It has been identified that investigated asphaltenes in toluene with a concentration of 0.4 g/L stay in the colloid-dispersed state and have a mean particle's radius of 225 nm. The asphaltenes aggregation obeys DLA mechanism.

In this research the estimation of asphaltenes aggregative and sedimentation stability in model system with addition different amounts of additive has been given. The dispersing effect of the additive C-5A on asphaltenes aggregates has been identified.

5. References

[1] Kraiwattanawong K, Fogler H S, Gharfeh S G, Singh P, Thomason W H and Chavadej S 2009 Effect of asphaltene dispersants on aggregate size distribution and growth *Energy and Fuels* **23** 1575–82

[2] Sakhabutdinov R Z, Kosmacheva T F, Sudykin S N, Ismagilov I Kh and Gubajdulin F R 2008 Methods of treatment of superviscous oils of OJSC «Tatneft» oil fields *Oil industry* **7** 86–89

[3] Hashmi S and Abbas F 2010 Effect of dispersant on asphaltene suspension dynamics: aggregation and sedimentation *The Journal of physical chemistry* **114** 15780–88

[4] Lamia G and Abbas F 2004 Effect of resins and DBSA on asphaltene precipitation from petroleum fluids *AIChE Journal* **50** 470–479

[5] Almusallam A S, Shaaban M, Nettem K and Fahim M A 2013 Delayed aggregation of asphaltenes in the presence of alcohols by dynamic light scattering *Journal of dispersion science and technology* **34** 809–817

[6] Hlavaty O L 1989 *Physic chemistry dispersing oil additives* (Kiev: Naukova Dumka) p 184

[7] Rudnick L R 2009 *Lubricant additives: chemistry and applications* ed. L R Rudnick (Chemical industries; 124)

[8] Mal'tseva E V, Gorshkov A M, Chekantseva L V, Shishmina L V and Yudina N V 2013 Influence of aggregation of asphaltenes on the rheological properties of oil *Russian Journal of Applied Chemistry.* **86** 1370–75

[9] Yudin I K, Nikolaenko G L, Gorodetskii E E, Markhashov E L, Frot D, Briolant Y, Agayan V A and Anisimov M A 1998 Universal behavior of asphaltene aggregation in hydrocarbon solution *Petroleum Science and Technology* **3** 401–02.

[10] Yudin I K and Anisimov M. A. 2007 Dynamic light scattering monitoring of asphaltene aggregation in crude oils and hydrocarbon solutions *Asphaltenes, heavy oils, petroleomics* ed. O C Mullins, E Y Sheu, A Hammami and A G Marshall (Berlin: Springer) pp 439–468

[11] Gorshkov A M, Shishmina L V and Roslyak A T 2014 New method for the onset point determination of the petroleum asphaltene aggregation *IOP Conf. Ser.: Earth Environ. Sci.* **21** 012027

[12] Ianguzova A P and Safieva R Z 2011 Study of behavior of asphaltenes in model systems and crude oils in early stages of aggregation *Proc. of VI Int. scientific-technical Conf.* ed. O F Glagoleva and E A Chernysheva (Moscow: Technique) pp 17–19

Features of non-oxidative conversion of methane into aromatic hydrocarbons over Mo-containing zeolite catalysts

A A Stepanov, L L Korobitsyna, A V Vosmerikov

Institute of Petroleum Chemistry, Siberian Branch, Russian Academy of Sciences, Tomsk, 634055, Russia
Email : stepanov153@yandex.ru, kll@ipc.tsc.ru , pika@ipc.tsc.ru

Abstract. The results of study of methane conversion under non-oxidative conditions over molybdenum containing zeolite catalysts prepared by solid-phase synthesis using nanosized molybdenum powder are presented. The kinetic mechanisms of the process behavior under different conditions of methane dehydroaromatization are determined. It is shown that non-oxidative conversion of methane can occur both in the external diffusion and kinetic regions, depending on the methane flow rate. It is found out, that the optimum temperature of the methane conversion is 750 °C. It is shown that increased methane conversion is observed at the feed space velocity of methane decreasing from 1500 to 500 h^{-1}.

1. Introduction

Rational processing of hydrocarbon gases, whose main component is methane, is important in the petrochemical industry for production of valuable chemical products. Currently, most of these gases are used as fuel to produce heat and electricity or burned in flares at oil production sites. A most promising process for methane conversion is oxygen-free methane dehydroaromatization into aromatic hydrocarbons over zeolite catalysts modified with various transition metals. The possibility of methane conversion into aromatic hydrocarbons was shown in [1] in the 80-90-s. The conversion was performed in a pulsed mode at 740 °C over Pt- Zn-, and Ga/ZSM-5 catalysts. In 1993, Wang [2] proposed an effective molybdenum-containing zeolite catalyst based on HZSM-5. Anoxic conditions of this process provide a higher selectivity towards formation of aromatic hydrocarbons, while in the presence of oxygen the main reaction products are CO and CO_2.

From the viewpoint of thermodynamics, the non-oxidizing methane conversion ($6CH_4 \leftrightarrow 9H_2 + C_6H_6$) is not reasonable, because of the high power of C-H bond in a methane molecule (413 kJ/mol). However, despite the fact that $\Delta G = 0$ only at 1075 °C in this reaction [3], significant amounts of benzene can already form at 700-800 °C [1].

Various metals – Mo, W, Fe, Cr, Mn, Pt, etc. and a large number of ZSM-5, ZSM-8, ZSM-11, HMCM-22, HMCM-41, TNU-9, and other zeolites have been investigated as the catalyst carriers for methane conversion [4]. Currently, the Mo/ZSM-5 catalyst is the most studied. A significant disadvantage is its rapid deactivation in the course of non-oxidative conversion of methane. In order to increase the stable performance of the catalyst, use is made of modification of the support surface, addition of promoters and small amounts of oxidants, and changes in the number and strength of the zeolite acid sites by varying the silica modulus [5]. Mo/ZSM-5 catalysts are generally prepared by impregnating of zeolite with an aqueous solution of $(NH_4)_6Mo_7O_{24} \cdot 4H_2O$ [6]. Also, Mo-containing catalysts are prepared by solid-phase synthesis and by mixing the zeolite with a molybdenum oxide or molybdenum salts. It is shown that the catalysts prepared by solid phase synthesis could be either less or more active in the course of conversion of methane into aromatic hydrocarbons compared to the catalysts manufactured by impregnation. Differences in the activity of Mo/ZSM-5 catalysts prepared by different methods are due to different localization of molybdenum on the zeolite surface [7]. In [8], the authors reported the use of nanosized molybdenum powder to prepare Mo/ZSM-5 catalyst characterized by improved activity and stability compared to those prepared by impregnation or mechanical mixing of the zeolite with a molybdenum oxide or salt.

It is known that Mo/ZSM-5 catalysts are bifunctional [4,8], in the cases where methane is activated on the active sites of molybdenum to form ethylene – the initial reaction product, while its further oligomerization and aromatization occur on the acid sites of the zeolite to form benzene and naphthalene.

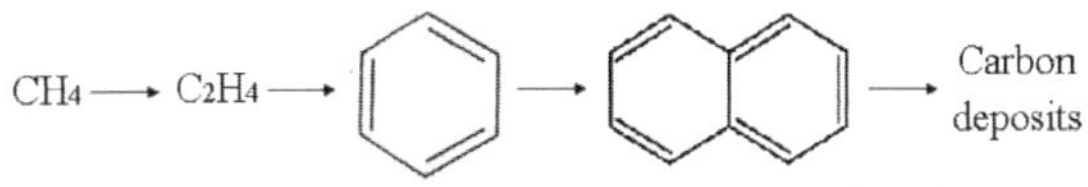

It has been found out that Mo2C is an active site of methane activation [4,9], which is formed during the induction period. However, the excessive formation of coke on the catalyst active sites results in its lower activity and stable operation in the course of non-oxidative conversion of methane into aromatic hydrocarbons.

The aim of this work is to study the effect of the conditions of methane conversion in the presence of Mo/ZSM-5 catalyst, prepared by solid-phase synthesis using nanosized Mo powder, on composition and yield of the resulting products.

2. Experimental

A zeolite with $SiO_2/Al_2O_3 = 40$ was prepared via hydrothermal synthesis using hexamethylenediamine as a template. The degree of crystallinity of the resulting zeolite was 100%. A molybdenum-containing catalyst was prepared via solid-phase synthesis by mechanical mixing of HZSM-5 zeolite with a nanosized powder of molybdenum (Mo NSP) in a vibratory ball mill for 2 hours. The resulting mixture was calcinated at 540 °C for 4 hours in air. The Mo NSP was prepared by the method of electric wire explosion in argon atmosphere (average particle size 70-100 nm) [10]. The content of molybdenum in the catalyst was 4.0 wt%.

The process of methane conversion was conducted in a laboratory flow-type quartz reactor at atmospheric pressure, the reaction temperature 600-750 °C and the feed space velocity of methane from 500 to 1500 h^{-1}. A 0.5-2.0 cm^3 volume of the catalyst (particle size 0.5-1.0 mm) was placed on the quartz reactor lattice to heat in a helium stream to the reaction temperature and maintained at this temperature for 10 min, the supply of helium to reactor was then stopped and methane (purity 99.9 %) was let in. The reaction products (ethane, ethylene, benzene, toluene, and naphthalene) were analyzed by gas chromatography. In order to estimate the catalytic activity of the samples, the degree of propane conversion was estimated. The yield and selectivity towards the formation of the gaseous and liquid products were also calculated.

3. Results and Discussion

To determine the optimal conditions for non-oxidative methane conversion, the effect of temperature and feed space velocity of methane was investigated.

Figure 1 presents the data on the yield of aromatic hydrocarbons versus the methane flow rate at different contact times. These data show that the kinetic curves of the yield in methane conversion products on the flow rate for different contact times are similar. When the flow of methane increases, the yield in aromatic hydrocarbons also increases to a certain value, after which the arene yield remains practically unchanged. This indicates that in the cases where little methane is consumed, the process runs in the external diffusion region, while as the methane flow rate increases the conversion occurs in the kinetic region. We observed that in the latter case different methane flow rates correspond to different contact times.

IOP Conf. Series: Earth and Environmental Science 43 (2016) 012064 doi:10.1088/1755-1315/43/1/012064

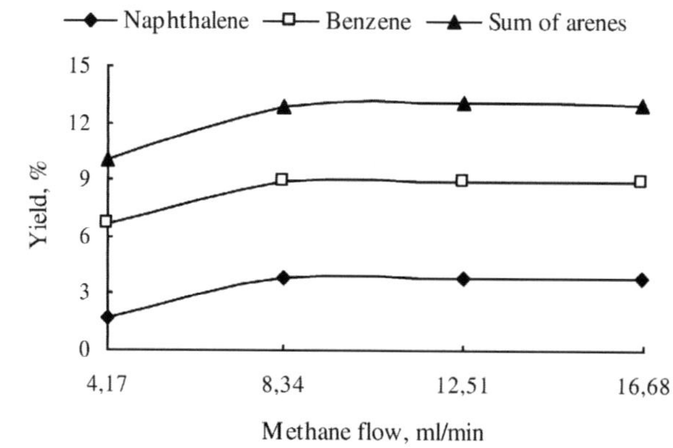

(a)

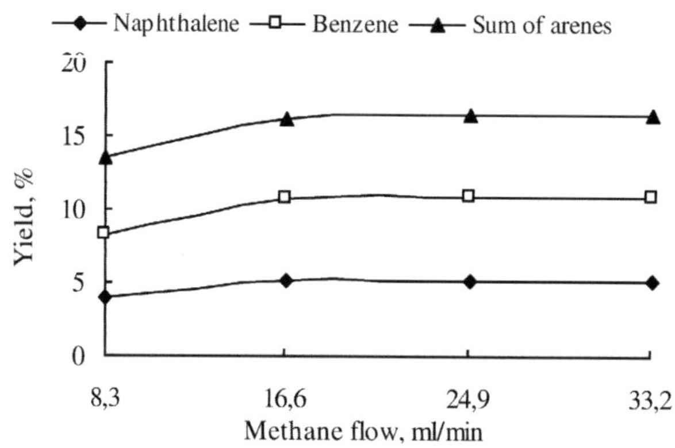

(b)

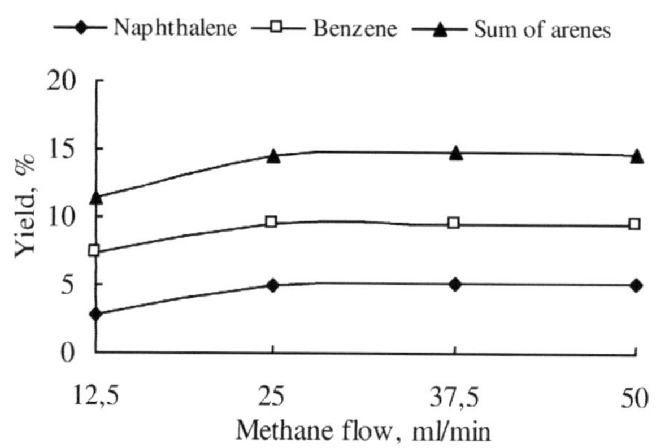

(c)

Figure 1. Yield in aromatic products of methane conversion vs the methane flow rate for contact times: 7.2 s (a), 3.6 s (b), and 2.4 s (c).

Thus, when the contact time is reduced from 7.2 to 2.4 s, the transition into the kinetic region is observed at higher methane flow values. Therefore, in further experiments we selected the methane flow rates to rule out the external diffusion inhibition and to make the process run in the kinetic region.

The data on the effect of reaction temperature on the yield and selectivity towards formation of aromatic hydrocarbons are shown in Table 1.

Table 1. Effect of temperature on the process of methane conversion over Mo/ZSM-5 catalyst (contact time 3.6 s).

Temperature, °C	Yield in products, %			Selectivity towards arenes, %	Conversion of methane, %
	Benzene	Toluene	Naphtalene		
600	0.07	0.01	0.26	78.7	0.3
650	1.40	0.04	1.06	80.3	2.4
700	5.79	0.09	2.31	79.7	7.8
750	10.48	0.21	4.90	79.8	15.2

They show that an increase in the reaction temperature from 600 to 750 °C results in an increase in the methane conversion from 0.3 to 15.2%, which also increases the product yield. For instance, the benzene yield increased from 0.07 to 10.48% due to an increase in the reaction temperature to 750 °C, while selectivity towards formation of aromatic hydrocarbons is practically unchanged and remains at the level of 79-80%. This seems to be due to the thermodynamic characteristics of the process, wherein the increase in the reaction temperature favors the highest rate of selective conversion of methane into aromatic hydrocarbons. The increase in the reaction temperature to 800-850 °C results in a transient increase in methane conversion at the beginning of the process but then the catalyst activity decreases sharply due to an increased surface coking. .

The data on the changes in the activity of Mo-containing zeolite catalyst are shown in Figure 2 versus the time of the process for the methane feed space velocity 500-1500 h^{-1}.

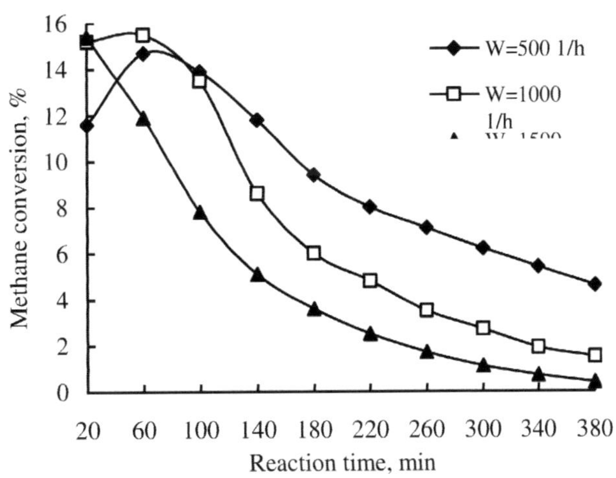

Figure 2. Dependence of methane conversion on the reaction time for methane feed space velocity 500-1500 h^{-1}.

It is evident from Figure 2 that a decrease in the methane feed space velocity results in an increased methane conversion. However, the conversion of methane at the space velocity 500 h^{-1} is lower

(11.6%) within the first 20 minutes of the reaction at 1000 h^{-1} (15.2%) and 1500 h^{-1} (15.4%), which is due to a longer induction period at low methane feed space velocities. An increase in a space velocity to 1500 h^{-1} results in a sharper decrease in the methane conversion and shorter period of stable performance of the catalyst as a result of caused by increased catalyst loading and formation of large amounts of carbonaceous deposits with a high degree of condensation.

Figure 3 shows the yields of basic products of non-oxidative conversion of methane versus the reaction time. An analysis of the gaseous products of non-oxidative methane conversion over Mo/ZSM-5 catalyst demonstrated that the major products are ethylene and ethane. For all feed space velocities of methane under study the yield in ethane is less than 0.16%. It is known that ethylene is the primary product of non-oxidative methane conversion, which is oligomerized on Bronsted acid sites of the zeolite to benzene and naphthalene. Therefore, the lowest C_2H_4 concentration is observed within the first minutes of the reaction. During the reaction of methane dehydroaromatization, the active and acid zeolite sites carbonize, giving lower benzene and naphthalene yields, hence an increase in concentration of ethylene.

The liquid products of methane conversion contain benzene and naphthalene. The fraction of benzene is higher than that of naphthalene. For all feed space velocities of methane under study the highest amount of benzene is formed within the first 20-60 minutes of the reaction, after which the concentration is gradually reduced due to catalyst carbonization. The maximum yield in benzene in the first 20 minutes of the reaction was 7% at feed space velocity of methane under study 1500 h^{-1}. This is due to the fact that at high space velocities, the induction period is shorter. The highest amount of naphthalene is produced at the beginning of dehydroaromatization reaction at the feed space velocity of methane 1000 h^{-1}. As the reaction runs, the yield in naphthalene decreases for all feed space velocities of methane.

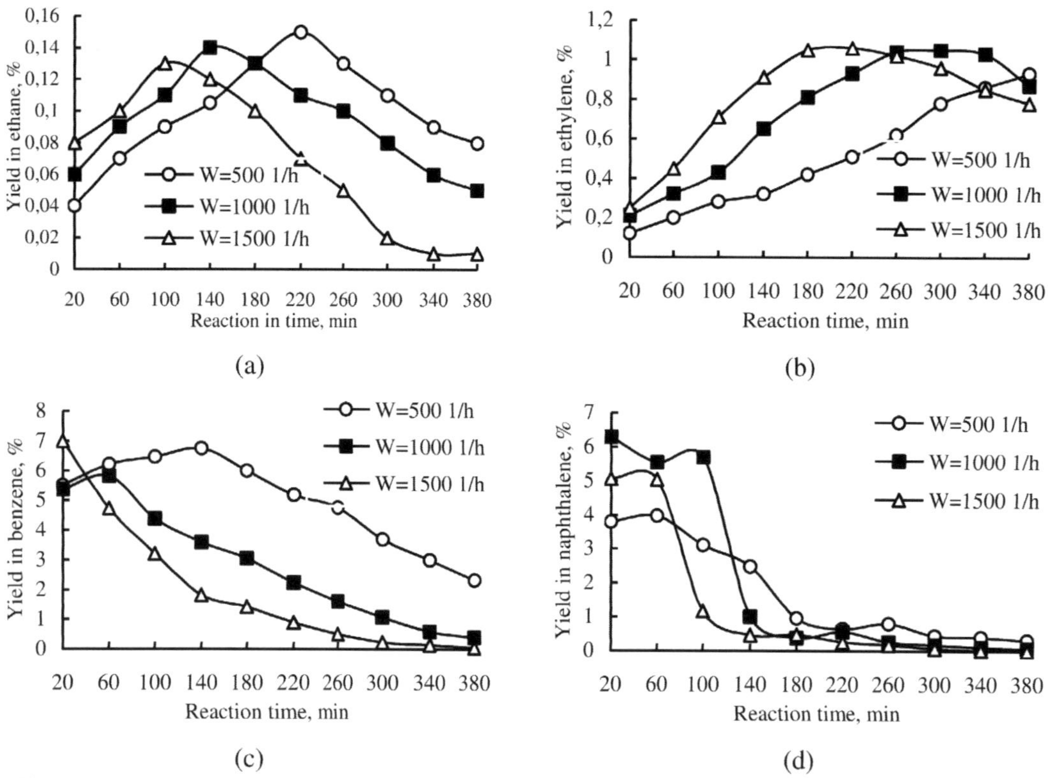

Figure 3. The yield in ethane (a), ethylene (b), benzene (c) and naphthalene (d) over Mo/ZSM-5 catalyst in the course of non-oxidative methane conversion.

Based on the experimental data obtained and available information on conversion of hydrocarbons over ZSM-5 zeolites [9,11], in Table 2 we summarize the chemical reactions representing the basic patterns of non-oxidative methane conversion.

Table 2. Set of chemical reactions occurring in the course of non-oxidative methane conversion.

Reaction	Product	Reaction	Product
$2CH_4 \rightarrow C_2H_2+3H_2$	acetylene	$2C_6H_6 \rightarrow C_{10}H_8+C_2H_4$	naphthalene
$2CH_4 \rightarrow C_2H_4+2H_2$	ethylene	$C_{10}H_8+C_2H_6 \rightarrow C_{12}H_{12}+H_2$	methyl naphthalene
$2CH_4 \rightarrow C_2H_6+H_2$	ethane	$C_6H_6+CH_4 \rightarrow C_7H_8+H_2$	toluene
$6CH_4 \rightarrow C_6H_6+9H_2$	benzene	$C_7H_8+CH_4 \rightarrow C_{11}H_{10}+H_2$	methyl naphthalene
$3C_2H_6 \rightarrow C_6H_6+H_2$	benzene	$C_{11}H_{10}+CH_4 \rightarrow C_{12}H_{12}+H_2$	dimethyl naphthalene
$C_2H_6 \rightarrow C_2H_2+2H_2$	acetylene	$CH_4 \rightarrow C+2H_2$	coke
$C_2H_6 \rightarrow C_2H_4+H_2$	ethylene		

4. Conclusion

The process of non-oxidative conversion of methane over Mo/ZSM-5 catalysts prepared via solid phase synthesis using nanosized molybdenum powder has been investigated. The kinetic mechanisms of methane conversion have been studied and optimal conditions have been identified. It has been shown that the optimum temperature of non-oxidative methane conversion, at which the conversion efficiency is the highest and so is the selectivity towards benzene formation (15.2 and 10.48%, respectively), is 750 °C at the methane feed space velocity 1000 h^{-1}.

Referencies

[1] Bragin O V, Vasina T V, Preobrazhenskii A V and Minachev K M 1989 *Bulletin of the academy of sciences of the USSR. Division of chemical science. Moscow.* Aromatization of methane on pentasil-containing catalysts. Vol. **38** pp. 680.

[2] Wang L,Tao L, Xie M and Xu G 1993 *Catal. Lett.* Dehydrogenation and aromatization of methane under non-oxidizing conditions. Vol. **21 pp.** 35.

[3] Scurrell M S 1987 *Appl. Catal.* Prospects for the direct conversion of light alkanes to petrochemicalfeedstocks and liquid fuels – a review. Vol. **32** pp. 1.

[4] Ma S, Guo X, Zhao L, Scott S and Bao X 2013 *J. Energy Chem.* Recent progress in methane dehydroaromatization: from laboratory curiosities to promising technology. Vol. **22** pp. 1.

[5] Abdelsayed V, Shekhawat D and Smith M W 2015 *Fuel.* Effect of Fe and Zn promoters on Mo/HZSM-5 catalyst for methane dehydroaromatization. Vol. **139** pp. 401.

[6] Wang D, Lunsford J H and Rosynek M P 1997 *J. Catal.* Characterization of a Mo/ZSM-5 catalyst for the conversion of methane to benzene. Vol. **169** pp. 347.

[7] Weskhuysen B M, Wang D, Rosynek M P and Lunsford J H 1998 *J. Catal.* Conversion of methane to benzene over transition metal ion ZSM-5 zeolites: I. Catalytic Characterization. Vol. **120** pp. 338.

[8] Spivey J J and Hutchings G 2014 *Chem. Soc. Rev* Catalytic aromatization of methane. Vol. **43** pp. 792.

[9] Karakaya C, Zhu H and Kee R J 2015 *Chem. Eng. Sci.* Kinetic modeling of methane dehydroaromatization chemistry on Mo/zeolite catalysts in packed-bed reactors. Vol. **123** pp. 474.

[10] Gusev A I 1988 *Nano-crystalline materials: methods of obtaining and properties* (Yekaterinburg UrO RAN Publ). pp. 200.

[11] Li L, Borry R W and Iglesia E 2002 *Chem. Eng. Sci.* Design and optimization of catalysts and membrane reactors for the non-oxidative conversion of methane. Vol. **57** pp. 4595.

Composite fuel based on residue from tyre and secondary polymer pyrolysis

V Popov[1,a], A Papin[1,b], A Ignatova[1,c] and A Makarovskikh[2]

[1]Kuzbass State Technical University, 28 Vesennaya St., Kemerovo, 650000, Russia
[2]Tomsk Polytechnic University, 30 Lenin Ave., Tomsk, 634050, Russia
E-mail: [1,b]papinandrey@rambler.ru

Abstract. The article presents the analysis of obtaining high-quality molded solid fuel from waste that is a carbonaceous residue obtained by pyrolysis of automobile tyres and secondary polymers. Preliminary waste preparing, blending and briquetting have been carried out; fuel samples have been obtained; their strength characteristics have been studied; technical analysis has been carried out.

1. Introduction

Nowadays one of the factors contributing to the pollution of the environment is the increase in the amount of automobile tyres waste and this is a global issue all over the world. These wastes are very specific, are not susceptible to rotting and decomposing. They cover land area, polluting settlements, water basins and air. At the same time, automobile tyres waste is a valuable source of secondary raw materials and can be used to produce an alternative source of fuel.

Secondary polymers are considered to be used in the same way. Every year the production of plastics increases by 5-6%, and by 2020 the production output can reach 300 million tons. Plastics consumption in developed countries is about 100 kg per capita per year. Accordingly, the plastic waste volume increases.

The purpose of the research is to develop a technology to make composite fuel from waste of automobile tyre recycling that is a carbonaceous residue of automobile tyres and secondary polymers pyrolysis.

The research objectives are to study feed composition and to assess its adequacy for practical application; to choose a method of concentration for the carbonaceous residue; to select and study a binder; to obtain briquetted fuel and assess its quality.

The novelty of the research is the development of new alternative methods used for preparing a low-quality carbonaceous residue resulted from automobile tyres pyrolysis, which allows obtaining high-quality low-ash briquetted fuel. The development of the technology allows recycling waste into new products and has practical applicability.

There are a large number of various technologies to recycle used tyres [1-9]. The most promising method is low-temperature pyrolysis [7]. Tyres decompose at a temperature of about 450°C in the reactor. Decomposition results in intermediate products such as black oil, gas, carbonaceous residue, and metal cord. The advantage of pyrolysis is that it is environmentally-friendly. However, solid residue from the pyrolysis, that is low-quality carbon, is characterized by high ash and sulfur content, and cannot be practically applicable accumulating on the industrial site of the factory.

2. Materials and Methods

A solid carbonaceous residue obtained by pyrolysis of automobile tyres, taken from LLC "Ekoshina", Novokuznetsk, was used as an object of the research. Technical analysis of the solid carbonaceous residue obtained by pyrolysis of automobile tyres was carried out. The output of volatile substances was determined according to GOST 6382-2001 [10], while ash content according to GOST 11022-95 [11].

The obtained data are presented in Table 1. It has been identified that the carbonaceous residue has a high value of ash content and volatile substances and cannot be used without recycling.

Table 1. Results of technical analysis of solid carbonaceous residue obtained by automobile tyres pyrolysis

Test object	Analyte	Content, % wt.
Low-grade carbon	Moisture content: W^a	0.68-2.1
	Ash: A^d	10-18
	Volatile substances output: V^{daf}	8.8-12
	Sulfur content: S^d_t,	4-8

Therefore, a concentrate based on the carbonaceous residue was obtained by the method of concentration. Calorific value of the concentrate was determined in accordance with GOST 147-95 [12], sulfur content was determined according to GOST 2059-95 [13], as well as moisture content was determined according to GOST 11014-10981 [14].

The method of concentration was used as it is based on oil agglomeration. Other existing concentration methods turned to be ineffective due to the low process selectivity because of the high ash content and fine materials. The essence of oil agglomeration method consists in different wettability of solid carbonaceous particles with liquid hydrocarbons in water. As a result of turbulence in the pulp there is a selective formation of the aggregates which are thickened structurally transforming into solid spherical granules.

The solid residue obtained by automobile tyres pyrolysis was crushed to the particles of 0.1 mm size and concentrated using a pilot unit by oil agglomeration. As a result concentrates were obtained. Discharge engine oil was used as a reagent for the process of concentration.

3. Results and Discussion

At the first stage of the experiment a solid carbonaceous residue obtained by automobile tyres pyrolysis was mixed with water for 1-2 min. using a paddle stirrer connected to a motor. Then the discharge engine oil in amounts of 8.0-9.0 % wt. was added to the solid carbonaceous residue and stirred for 5-8 min. As a result, a concentrate with characteristics presented in Table 2 was obtained.

Table 2. Characteristics of the concentrate based on the solid carbonaceous residue obtained by automobile tyres pyrolysis

A^d, % wt. (ash)	4.0-5.5
W^a, % wt. (moisture content)	8.5-10.5
V^{daf}, % wt. (volatile substances output)	8.9-12.0
Q_s^r, kcal/kg (combustion heat)	6600-6700
S^d_t, % wt. (sulfur content)	0.5-1.0

Secondary polymers are suggested to be used as a binder during briquetting. In our opinion, the most suitable binder for a composite fuel from solid residue obtained by automobile tyres pyrolysis is secondary polymers wastes that are polyolefins (high-density polyethylene, low-density polyethylene, polypropylene) and polyethylene terephthalate.

Polyolefins are high-molecular compounds which are widely used in different production sectors and are easily processed. [15-17].

Polyethylene terephthalate (PET) is a complex thermoplastic polyester of terephthalic acid and ethylene glycol. Its main advantages are strength and stiffness [15-17]. Today, it is used for the production of various packaging. It is one of the most common polymers in the world, while PET recycling is underdeveloped and requires innovation.

The obtained concentrate was mixed with a secondary binder polymer (polyethylene), 8-9% by weight of the initial concentrate. After that it was put into a die mold, then placed in a muffle furnace and heated for 30 min. at a temperature of 170°C to complete melting of the binder polymer. Next, the die mold was placed under the press and the load was brought up to 100 kgf/cm² to obtain a solid briquette (Figure 1).

Figure 1. Briquetted fuel sample

Prepared samples of briquetted fuel were further tested for mechanical strength. The mechanical strength in the processes of abrasion in the drum, compression and dropping was determined in compliance with GOST 18132-72 [18] and 21289-75 [19]. Test data are shown in Table 3.

Table 3. Test results of mechanical strength of the samples

P, % (mechanical strength)	99
Compression strength, kg/cm^2	50.0
Dropping strength, % content, pieces size> 25 mm	85-99
Abrasion, % content, pieces size> 25 mm	80-90

Next, technical analysis of the samples was carried out. The results are shown in Table 4.

Table 4. Results of technical analysis of the obtained briquette

Test object	Analyte	Content,% wt.
Briquetted fuel on the basis of a solid carbonaceous residue obtained by pyrolysis of automobile tyres and secondary polymers	W^a, % wt. (moisture content)	2.5-3
	A^d (ash)	4.5-7
	V^{daf} (volatile content)	8.0-12.0
	Q_s^r, kcal/kg (combustion heat)	6800-7500
	S^d_t, % wt. (sulfur content)	0.4 -1.0

The dependence of briquette strength on the binder content was studied (Figure 2).

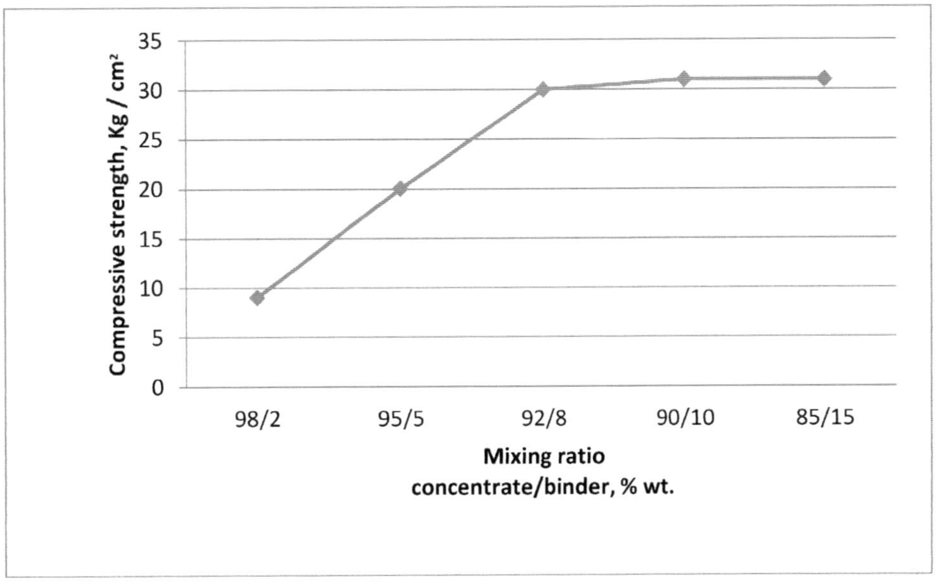

Figure 2. Dependence of briquette strength on the binder content

Briquette strength was optimal when adding a binder from 8.0 to 10.0% wt. The dependence of briquettes abrasion on the binder content is shown in Figure 3.

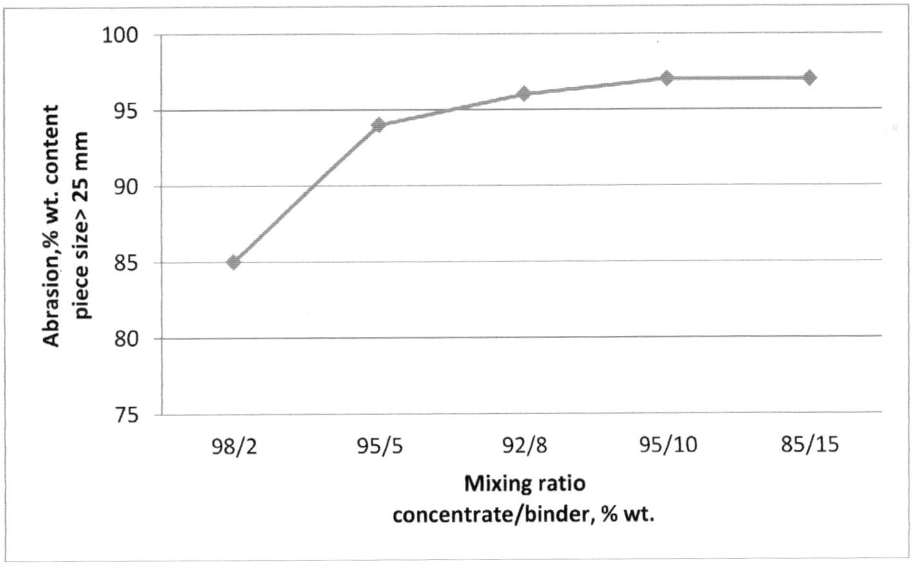

Figure 3. Dependence of briquettes abrasion on the content of binder

Briquettes abrasion strength was optimal when adding a binder from 8.0 to 10.0% wt.

4. Conclusion

The studies have shown the possibility to concentrate a low-grade solid carbonaceous residue obtained by automobile tyres pyrolysis and to get high-quality molded fuel.

Compared to its analogues, the obtained composite fuel has a number of advantages, and therefore, it is considered to be promising and competitive.

Disposal of tyres waste is suggested to be more environmentally-friendly and to expand energy resource base through the use of alternative source of fuel.

Scientific and technical research products are as follows:

1. composite fuel;
2. technology of obtaining composite fuel from solid residue by pyrolysis of automobile tyres with a secondary polymer binder;
3. high-quality carbon.

The fuel can be applied in power economy sector, for domestic boiler houses, and used by private consumers.

Thus, the research aimed at expanding application areas for carbonaceous residue obtained by automobile tyres pyrolysis is relevant. It can solve the problem of automobile tyres and secondary polymers waste disposal, which currently is one of the priorities in terms of economy and ecology, as well as an upcoming trend of business development.

References

[1] Sapronov V A 1982 *Proceedings "Recycling of used tyres" (Moscow: TSNIITENEFTEHIM)*. Economic and environmental importance of the problem of used tyres processing.

[2] Tarasova T F 2006 *Bulletin of the Orenburg State University*. Ecological value and solution of the problem of worn out tyres processing. Vol. **2-2** pp. 130-135.

[3] Wolfson S 2011 *Bulletin of Kazan Technological University*. Methods of disposal of tyres and rubber products. Vol. **1** pp. 74-79.

[4] Valueva A V 2012 *Proceedings SWORLD*. Prospects of automobile tyres recycling in Kuzbass. Vol. **7-1** pp. 19-20.

[5] Boravsky B V 2007 Worn out tyres: methods of solid waste recycling. Vol. **4** pp. 4-5.

[6] Demyanova V S 2011 *Scientific Bulletin of Voronezh State Architecture and Construction University. Construction and architecture*. Prospects of automobile tyres recycling. Vol. **4** pp. 74-79.

[7] Maltsev V M 1995 Actual problems of used tyres recycling *Chemical Industry of Ukraine* **3** 57-58

[8] Volynkina E P 2004 *Ecology and Industry of Russia*. Use of tyres waste. Vol. **1** pp. 40-44.

[9] Strakhov V M 2005 *Coke and Chemistry*. Coke from rubber waste. Vol. **2** pp. 35-43.

[10] GOST 6382-2001 2001 Solid mineral fuel. Methods of determining volatile substances output (Moscow: Publishing House of Standards)

[11] GOST 11022-95 1995 Solid mineral fuel. Methods of determining ash (Moscow: Publishing House of Standards)

[12] GOST 147-95 1995 Determination of the highest combustion heat and calculation of the lowest combustion heat (Moscow: Publishing House of Standards)

[13] GOST 2059-95 1995 Solid mineral fuel. Method of determining total sulfur burning at a high temperature (Moscow: Publishing House of Standards)

[14] GOST 11014-1981 1981 Coal and peat, anthracite and oil shales. Rapid method of determining moisture (Moscow: Publishing House of Standards)

[15] Bazunova M V 2012 *Chemistry*. Multifunctional composites based on ultradisperse carbon materials with polyethylene as a binder. Pp. 57-59.

[16] 1964 *Proceedings Moscow*. Polyethylene and other polyolefins. pp. 214.

[17] Shifrina V S 1961 Polyethylene (Goskhimizdat) p 176

[18] GOST 18132-72 1972 Peat briquettes and half- briquettes Method for determining mechanical strength (Moscow: Publishing House of Standards)

[19] GOST 21289-75 1975 Coal Briquettes Methods of determining mechanical strength (Moscow: Publishing House of Standards)

Production of activated carbon from rice husk Vietnam

V V Korobochkin[1], N V Tu[2], N M Hieu[3]

[1, 3]Department of Chemical Engineering, Tomsk Polytechnic University, Tomsk
[2]School of Materials Science and Engineering, Hanoi University of Science and Technology, 1 Dai Co Viet, Hanoi, Viet Nam
E-mail: [1]vkorobochkin@tpu.ru, [2] tu.nguyenvan@hust.edu.vn,[3]kqhak@yandex.ru

Abstract. This work is dedicated to the production of activated carbon from rice husk from Delta of the Red River in Viet Nam. At the first stage, carbonization of a rice husk was carried out to obtain material containing 43.1% carbon and 25 % silica with a specific surface area of 51.5 m^2/g. After separating of silica (the second stage), the specific surface area of the product increased to 204 m^2/g and the silica content decreased to 1.23% by weight as well. The most important stage in the formation of the porous structure of the material is the activation. The products with the high specific surface area in the range of 800-1345 m^2/g were obtained by activation of carbonized product with water vapour or carbon dioxide at temperatures of 700 °C and 850 °C, with varying the flow rate of the activating agent and activation time. The best results were achieved by activation of carbon material with water vapour at the flow rate of 0.08 dm^3/min per 500 g of material and the temperature of 850 °C.

1. Introduction

Recently the use of activated carbon as an adsorbent has increased rapidly. Resources base of activated carbon is quite varied, from crop waste to lignite and coal. Activated carbons obtained from these resources have high ash content and low mechanical strength. In addition, the inventories of these raw materials are reduced over time. Therefore, finding the alternative sources of raw materials for the production of activated carbon is an actual problem. However, using rice husk to produce activated carbon is significantly cheaper than other precursors, such as synthetic polymers and brown coals.

In 2014, the rice production in Viet Nam was 45 million Tons, and this continued to increase as a result of application of more advanced technologies in the process of cultivation of rice [1]. During the processing of rice, a huge amount of rice husk were formed, these waste products must be utilized. Combining Observance of environmental norms with the possibility of obtaining valuable products from rice husk is an actual task. According to [2], rice husk consists of a large amount of carbon in the composition of the natural polymer - cellulose and lignin. Besides carbon, these polymers contain sufficient amounts of hydrogen and oxygen which contribute to receiving activated carbon with the developed porous structure.

An efficient processing of rice husk into products (silica, activated carbon) is based on thermal decompression including washing, drying, pyrolysis of raw material, condensing the vapour mixture and obtaining a solid residue, followed by activation.

Production of activated carbon from rice husk has been the subject of many studies [3–5], which shows that it is possible to obtain activated carbon from rice husk with a high specific surface area.

However, these studies failed to demonstrate the efficient way to utilize a significant amount of amorphous silica dioxide, which is contained in the ashes.

The technology of processing includes the following three basic steps; First, carbonization of rice husks with an optimum temperature to remove moisture and volatile organic substances, and for the prior formation of porous structure; secondly, leaching of carbonized rice husk using sodium hydroxide is followed by precipitation of silica and filtration and third, activation of carbon material. The study in the field of thermal treatment of rice husk using pyrolysis and combustion present an interesting process. In these processes, different amount of the evolved gasses and vapors is generated, which affects the yield of carbon residue.

The purpose of this work: is to investigate the effect of heat treatment methods on the content of carbon and silica in the carbonized products in order to predict the composition of precursor for next activation stage and to carried out the process of activation of carbon residue.

2. Experimental

The rice husk from the Red River Delta in Viet Nam was used as the object of study. The process of carbonization of a rice husk was carried out in the Flow Reactor with a volume capacity of 500 cm^3 at 600 °C. Aprioristic information about the process was obtained by Differential Thermal Analysis, which is conducted on the device STD Q600 V8.1, in a stream of air or argon. The elemental composition of initial raw material, products of carbonization, products after separation of silica, products of activation and electron micrographs were obtained with the device Scanning Electron Microscope (SEM)-JED-2300 Analysis Station (JEOL). Characteristics of the porous structure (specific surface area, pore volume, pore size) were determined by nitrogen adsorption isotherms (BET method) at 77 K on the device NOVA Station, 2.11. Before measurements, the samples were maintained in the environment of nitrogen within 17 hours at a temperature of 150 °C.

3. Results and discussion

The elemental composition of rice husk, products of carbonization at 600 °C and products after separation of SiO_2 were shown in Table 1.

Table 1. The elemental composition of rice husk, products of carbonization and products after separation of SiO_2

Material	Content,% wt.								
	C	H	O	N	K	Si	Ca	Mg	Al
Rice husk	39,20	5,40	46,00	0,38	1,27	7,00	0,28	0,15	0,06
Products of the carbonization	43,10	0,01	29,33	0,03	1,57	25,00	0,39	0,29	0,08
Products after separation of silica	70,74	0,01	25,30	0,03	0,58	1,23	1,68	0,35	0,12

During carbonization stage, the decomposition and synthesis reaction of the organic part with recombination of removing gasses and vapors occur, which as a result lead to accumulation of flat formations of hexagons – precursors of graphene. Carbonization temperature is a key parameter, which affects the properties of the resulting carbons.

Differential Thermal Analysis (DTA) was conducted to determine the pre-carbonization conditions rice husk in an air flow and an inert gas (Argon). As shown in Figure 1, at a temperature below 200 °C in air flow, decomposition of a hemicellulose and removal of adsorbed moisture were observed. In the range of 280–540 °C there is a decomposition of cellulose, lignin, and other organic compounds with releasing of water and carbon dioxide. Consequencely, at up to 1000 °C a significant amount of carbon (total mass of weight loss. 89.3%) was burned.

For obtaining information about the process of decomposition of rice husk by pyrolysis, DTA in an inert gas flow (Argon) has been carried out. Results of DTA are shown in Figure 2.

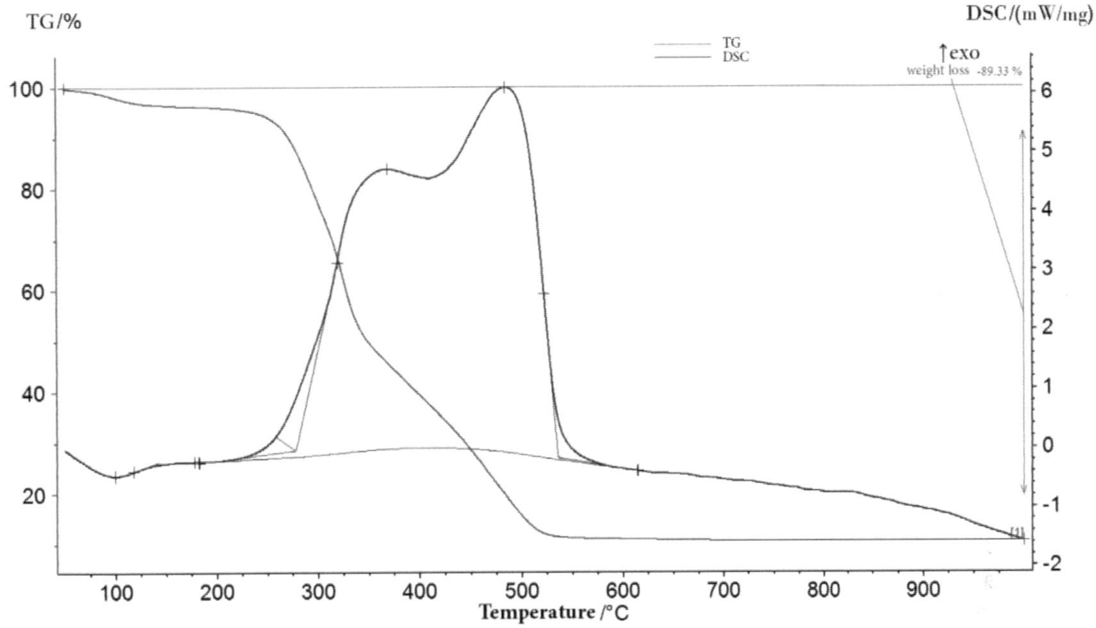

Figure 1. The results of DTA of rice husk in an Air flow

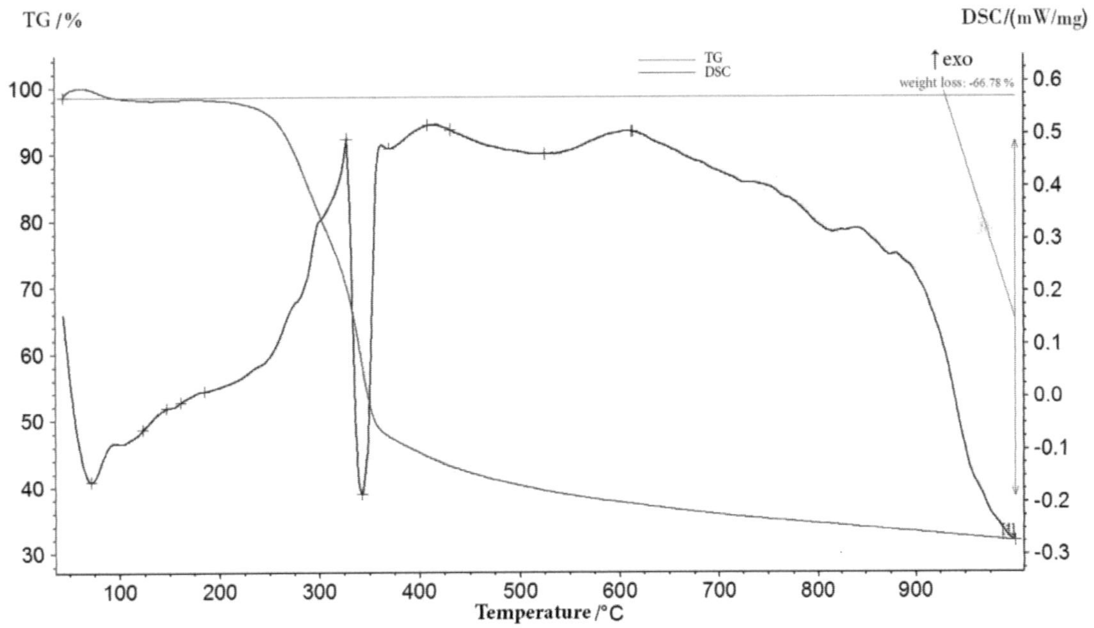

Figure 2. The results of DTA of rice husk in a flow of Argon

As shown in Figure 2, during pyrolysis, an intensive decomposition of cellulose and lignin with preferential releasing of H_2O, CO and CO_2 were observed in the temperature range of 320-380 °C. When temperature of heating upon to 1000 °C, the weight loses is 66.8%. After the temperature of 650 °C, the processes of interaction of H_2O and CO_2 with formed active carbon has begun occurred by following reactions:

$$CO_2 + C = 2CO, \quad (1)$$
$$H_2O + C = CO + H_2. \quad (2)$$

This interaction is evidenced by the DSC curve and a slight weight loss (4.8%) at temperatures above 650 °C. The reaction between carbon and carbon dioxide is a reversible reaction. Entropy factor provides the reaction moves to the right while enthalpy factor provides the reaction moves to the left. At temperatures below 400 °C, the equilibrium is almost completely shifted to the left and at a temperature above 1000 °C to the right (towards the formation of CO) [6].

Results of study [7] have shown that, at a temperature of carbonization above 850 °C the part of a pore space of the obtained carbon material is lost due to condensation of resinous substances in pores. Therefore, carbonization is recommended to be carried out at temperatures of 500-550 °C.

Carbonized material is exposed to leaching by sodium hydroxide to create favorable conditions for the formation of pores in the next stages of activation, as well as for the separation of silica dioxide [8].

Textural characteristics of products of carbonization of a rice husk and products after separation of SiO_2 are provided in Table 2.

Table 2. Pore Structure Characterization of products of processing rice husk

Material	S_{BET}, m^2/g	V_Σ, cm^3/g	V_μ, cm^3/g	Pore Diameter, nm	$S_{BET,\mu}$, m^2/g
After carbonization at 600 °C	51,5	0,0062	0,007	4,8	10,9
After separation of SiO_2	204,4	0,2248	0,057	4,4	97,8

Table 2 shows that the carbonized product has a small specific surface area and a low pore volume. However, after removing the SiO_2 significantly increases the specific surface area (up to 204 m^2/g) and an adsorption capacity. At that, in the leaching process, the carbonaceous matrix is formed with a high content of micropores, which accounts for 50% of a specific surface area of material.

Research Results by BET well illustrate micrographs of the obtained materials (Figure 3)

a) b) c)

Figure 3. Micrographs of materials obtained after carbonization (a) separation of SiO_2 (b) and activation (c)

Micrographs revealed that the structure of a product after separation SiO_2 has changed. The particles have a form of aggregates composed of laminated structures, their size decreases. This result is due to the process of leaching of silica.

Material after separating of SiO_2 contains more than 70% carbon and has the specific surface area more than 200 m^2/g. For improvement of the adsorptive properties, the product was exposed to activation.

There are two main groups of methods of activation of carbon materials: namely, physical activation and chemical activation [9]. In this work, the physical activation using carbon dioxide and water vapor was carried out. The process of activation is described by reactions (1) and (2). During experiments, the effect of temperature, gas (water vapor) flow rate, and the activation time on the value of the specific surface area has been investigated.

In order to reduce loss of the carbon material, activation was carried out in the Flow Reactor without access of air with external electric heating. The mass of loaded material was 500g. The results of experiments are given in Table 3.

Table 3. Results of Activation of carbon obtained from rice husk

T, °C	Flow rate of H_2O, dm^3/min	S_{BET}, m^2/g	T, °C	Flow rate of CO_2, dm^3/min	S_{BET}, m^2/g
Activation time 1h					
700	0.08	1113	700	0.1	891
850	0.08	1345	850	0.1	1054
700	0.16	1089	700	0.5	924
850	0.16	1239	850	0.5	1097
Activation Time 5h					
700	0.08	1092	700	0.1	780
850	0.08	1198	850	0.1	1023
700	0.16	1198	700	0.5	816
850	0.16	1102	850	0.5	1064

The table shows that for same activation time and close value of gas (water vapor) flow rate, the maximum effect of activation is observed at a temperature of 850 °C. The specific surface area of material prepared with water vapor is higher than the one prepared with carbon dioxide. Increasing the activation time up to 5 hours, as well as increasing the values of gas (water vapor) flow rate is not effective. The best results are achieved by activation of carbon material using water vapor at the flow rate of 0.08 dm^3/min and the temperature of 850 °C.

Conclusion

Production of activated carbon from rice husk from Delta of the Red River in Viet Nam is carried out in three steps: carbonization of a rice husk, leaching of silica from carbonized material and depositing it in the form of SiO_2; activation a carbonaceous material using water vapor or carbon dioxide.

During carbonization, the material with a carbon content of 43.3% by weight and silica content of 25.9% by weight is generated. The material has a small (51.5 m^2/g) specific surface area. After separation of silica, the carbon material contains 70.74 % by weight carbon and only 1.23 % silica. The specific surface area of material significantly increases up to 204.4 m^2/g due to the formation of micropores with a diameter less than 2 nm. Micropores accounts for a half (97.8 m^2/g) of the total specific surface area.

Activation by water vapor and carbon dioxide at temperatures of 700 and 850 °C allows obtaining a material with a specific surface area in the range of 800–1345 m^2/g. The best results are achieved by activation of carbon material with water vapor at the flow rate of 0.08 dm^3 /min per 500 g of material and at the temperature of 850 °C.

Reference

[1] General Statistics Office Of Viet Nam. [Electronic resources]. - Available at: https://gso.gov.vn/default.aspx?tabid=717

[2] 2008 *Bioresource Technology.* Production of activated carbon from bagasse and rice husk by a single-stage chemical activation method at low retention time, Dimitrios Kalderis. Vol. **99 (15)** pp. 6809–6816.

[3] Huaxing Xu, Biao Gao, Hao Cao 2014 *Journal of Nanomaterials.* Nanoporous Activated Carbon Derived from Rice Husk for High Performance Supercapacitor. Article ID 714010, doi:10.1155/2014/714010. pp 7.
[4] Liou T , Wu S, Hazard J 2009 *Mater.* Vol. 171 pp. 693–703.
[5] Dang Nhan Thong, Nguyen Manh Hieu, Korobochkin V V 2014 *Scientific and technical and industrial magazine Nanoengineering.* Analysis of the combustion process in the production activated carbon.
[6] Hieu N M, Korobochkin V V, Tu N V 2015 *Procedia Chemistry*. pp. 308 – 312.
[7] Kinle H, Bader A 1984 *Chemistry.* Active coals and their industrial application. pp. 216.
[8] Nekrasov B. V. 1973 *Chemistry.* Basics of general chemistry. pp. 495.
[9] Gridneva T B, Soroka P I, Smirnova E S, Belaya A A, Ryabik P V 2012 *Questions of chemistry and chemical technology*. Investigation of the effect of preparation methods for rice husk chemical composition of the raw material in the production of silicon dioxide. Vol. **3** pp. 50–53.

Single-stage technogy for granulated foam glass production based on the composition of tripoli and technogical microsilica

O Kazmina, A Volkova, V Vereschagin, I Rymanova

National Research Tomsk Polytechnic University, Tomsk, Russian Federation
E-mail: kazmina@tpu.ru

Abstract. The possibility of foam glass production by means of one-stage technology based on the natural tripoli and technogenic silica is determined. 45 % sodium hydroxide solution is used for the synthesis of foam glass. The addition of microsilica as an extra component in an amount of 10 to 50 wt. % in batch component increases the strength of the obtained material to 4 MPa and reduces foaming temperature from 870 to 830 °C. The increased mechanical strength of the material is due to the residual quartz dissolution and cristobalite crystallization. The samples obtained from batch with 30 wt. % microsilica have maximum strength.

1. Introduction

Foam glass is considered to be one of the most environmentally friendly, durable and efficient thermal insulators. Its effectiveness is due to a unique combination of properties, such as low water absorption, frost resistance, resistance to aggressive media, relatively high strength. Currently, the field of foam glass application has been rapidly developing. Granulated foam glass is used for fill insulation, and as an effective filler of various composites, for example, in the production of warm plasters as light concrete and cement slurries.

Glass cullet or finely crashed glass melted in a glass-melting tank is a raw material for foam glass production. In view of the limited resources of waste glass the raw material base development for foam glass production is relevant. Granulate melting is associated with high energy and capital intensity of glassmaking. European researches are mainly focused on the use of various types of glass cullet, for example, glass of electron - ray tubes and computer screens, etc. [1-7]. Some researchers consider not only glass cullet but also various types of natural and technogenic materials as raw material [8-11].

The objective of this study is to obtain foam glass material by means of one-stage technology based on natural silica component in the form of tripoli at a foaming temperature not exceeding 850 ° C.

2. Materials and Experimental method

Tripoli and microsilica were chosen as objects to study siliceous materials of natural and technogenic origin. Chemical and granulometric composition of silica raw materials are shown in Table 1. The chemical composition of tripili differed from the standard silica sand used in glass production, as it had a reduced content of glass- forming oxide and an increased content of SiO_2 Al_2O_3, Fe_2O_3. Silica fume, which is a byproduct of metallurgical production in melting of ferrosilicon and its alloys, corresponds to the sand grade used for the production of foam glass, cans and semiwhite glass bottles.

Along with chemical composition the high dispersion with average particle size of less than 100 microns was the main requirement to the raw material. Initial tripoli contains coarse particles in the form of clay aggregated in an amount of 3 - 7 wt. %. Therefore, tripoli was previously dried and then milled in a ball mill. The results of the grain-size composition of the prepared tripoli were presented in the following table. Microsilica was mainly presented in a form of particles of less than 0.5 microns without additional preparing. According to the results of X-ray analysis it was determined that microsilica refers to X-ray amorphous material. Crystal silicon (d = 0,313; 0,192 nm), carborundum (d = 0.251; 0.217; 0.154 nm), and graphite (d = 0,335) practically without a crystal phase were presented in the results. Small amount of amorphous halo (ring) reflection maxima were shown on the X-ray photograph of tripoli, which corresponded to different modifications of silica, ie, material was amorphous crystalline. A few of diffraction maxima specific for albite were also presented.

Table 1. Chemical and и grain-size composition of silica components

	Chemical composition					
Material name	Content of oxides, wt. %					
	SiO_2	Al_2O_3	Fe_2O_3	CaO	MgO	Δm
tripoli	76,16	7,52	4,10	1,05	0,75	8,73
microsilica	91,70	0,50	0,40	1,20	-	2,00
	Grain-size composition					
Size of particles,	Percentage of fractions, %					
μm	> 0,5	0.5 – 1	1 – 5	5 – 10	10 - 50	50 – 100
tripoli	17,0	30,8	26,4	6,6	18,4	0,8
microsilica	73,0	8,0	1,0	1,0	1,0	16,0

In the study, granular sodium hydroxide, a white crystalline powder with the composition of NaOH - 99.5 wt. %; Na_2CO_3 - 0.5 wt. %; NaCl, Na_2SO_4, Fe_2O_3 – traces, was used as an alkali component. The study of phase composition of the raw and obtained materials was carried out using a X-ray diffractometer DRON-3M in copper radiation. Quantitative X-ray analysis was performed using the Match program. Macro- and microstructure analysis of porous samples was performed on a scanning electron microscope (JCM – 6000).

3. Experimental results

Preparation of foam glass was carried out according to the scheme shown in Figure 1. First, the source raw material was dried, then sieved, and, if necessary, milled. The prepared materials were mixed in a bowl of the mixer-granulator, then granulated with binding substance in an amount of 30 wt. 2 %. The alkaline component adding in a form of solution did not affect the batch homogeneity obtained in the stage of component mixing. At the same time it resulted in intimate contact of the mixture components through the alkaline solution, which led to more complete silication reactions in each pellet. The resulting raw pellets were dried at a temperature not exceeding 200 °C to final moisture content of less than 1.5 %. Foaming was carried out in an electric-tube furnace at different temperatures.

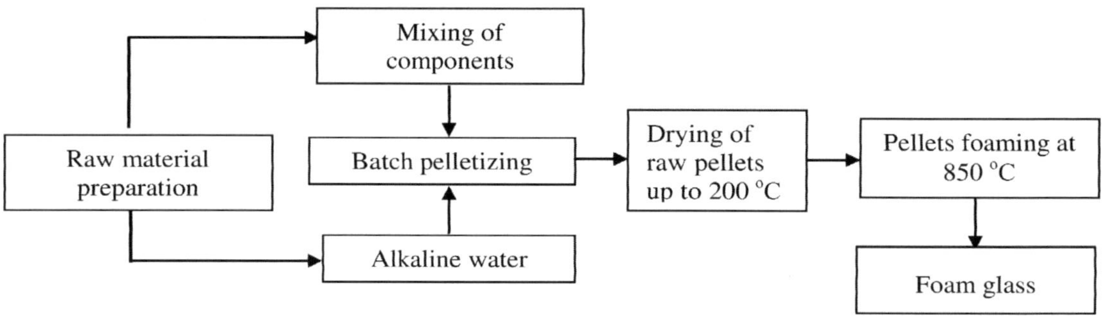

Figure 1. One-stage production of foam glass

Use of only microsilica as a silica component showed that microsilica was completely dissolved in alkali due to structural features and high dispersion. In addition, the pellets foaming occurred at a temperature of 200 °C. However, the foam density exceeded 800 kg/m^3, which didn't comply with the requirements of thermal insulation materials. Therefore, microsilica can be used only as an additional component. Its amount in the batch was varied from 10 to 50 % by weight in 10% increments. The batch compositions were presented in Table 2.

Table 2. Chemical and compositional analysis of batch

№ compound	Compositional analysis of batch, wt. %			Oxide composition of glass phase, wt. %					
	tripoli	microsilica	45 % NaOH	SiO_2	Na_2O	CaO	MgO	Al_2O_3	Fe_2O_3

1.1	100	0	23	73,67	13,36	1,02	0,73	7,26	3,96
1.2	90	10	23	74,86	13,31	1,03	0,65	6,56	3,59
1.3	80	20	23	76,05	13,25	1,03	0,57	5,88	3,22
1.4	70	30	23	77,24	13,19	1,04	0,5	5,17	2,86
1.5	60	40	23	78,41	13,14	1,05	0,43	4,48	2,49
1.6	50	50	23	79,57	13,09	1,06	0,36	3,79	2,13

Self-heating of the moisturized batch up to 80 °C was observed in the stage of preparation and granulation of the batch with formation of sodium hydrosilicate and silica gel in a colloidal state. Interaction of the alkaline solution with the amorphous silica component of the batch occurred due to the following reaction:

$$3SiO_2 + 2NaOH + 3H_2O \xrightarrow{80^\circ C} Na_2O \cdot 2SiO_2 \cdot 3H_2O + SiO_2 \cdot H_2O$$

Silication reactions occurred due to dehydration of crystalline sodium silicate upon heating the granulated material up to 200 ° C at the drying stage:

$$Na_2O \cdot 2SiO_2 \cdot 3H_2O \xrightarrow{200^\circ C} Na_2O \cdot 2SiO_2 + 3H_2O$$

According to X-ray analysis it was determined that reflection peaks corresponding to different sodium silicate were presented on the X-ray photographs of the dried granules (up to 200 °C) of all compositions (Fig. 2). Diffraction peak intensity increased with an increase of microsilica in the batch, which indicated more intensive silication development.

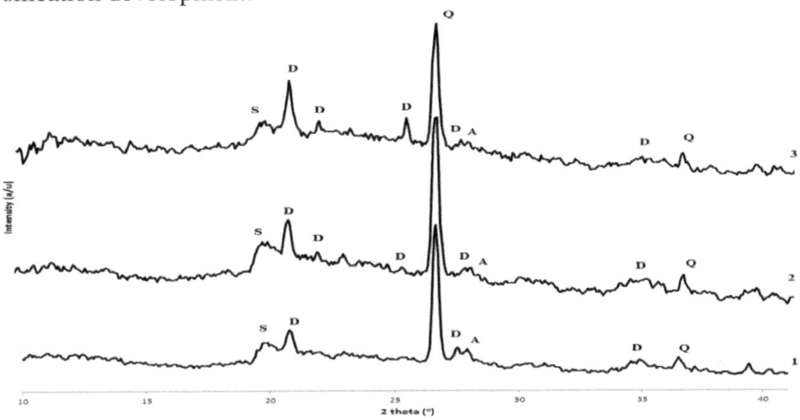

Figure 1. The XRD patterns of the dried pellets due to the content of microsilica in the batch of wt. %: 1 - 10; 2 - 20; 3 - 50 (Q - quartz, D - sodium disilicate; A - albite; S - $Na_2Si_3O_7$)

When the granulated material was heated up to 750 °C and above, a glass foaming process was observed at the foaming stage, and the obtained porous material had a vitrified surface. The process proceeded in stages. The first phase was obtaining of eutectic melt by ternary eutectic melting formed among $Na_2O{\bullet}2SiO_2$ and SiO_2 and albite, at a temperature of 740 °C and the binary eutectic among $Na_2O{\bullet}2SiO_2$ and SiO_2 at a temperature of 793°C. In the second stage, the residual silica was dissolved in the primary melt, then, the granules were cooled to form sodium- alumino-silicate glass at foaming temperatures (800-900 °C).

Quantitative X-ray analysis of the synthesized foam glass showed that the amount of the glass phase in the obtained product depended on the content in the starting batch of microsilica. The glass phase content increased from 74 % to 78 % with microsilica increasing from 10 to 20 %, respectively. The amount of microsilica in the starting material of 30 % resulted in a decrease up to 64 % of glass phase, and when the content of microsilica was 40% and 50 %, the amount of the glass phase increased up to 69 %. Different content of the glass phase in the final product was due to the glass crystallization processes occurring at the stage of pellets foaming. The X-ray photographs of all samples presented reflection peaks corresponding to quartz, and new reflection peaks corresponding to cristobalite appeared (Fig. 3). With an increase of silica in the batch the intensity peaks of cristobalite increased.

On the electronic micrographs of the samples obtained with the addition of microsilica, it can be seen a new crystal phase in foam glass. The picture showed needle-like crystals, which were absent in the samples without microsilica (Fig. 3). According to the X-ray diffraction analysis and the needle shaped crystals one can determine the cristobalite crystallization.

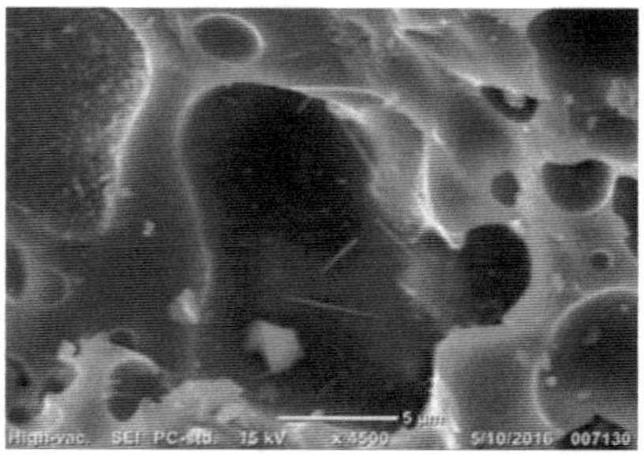

Figure 3. SEM micrograph of the foam glass obtained with addition of 50% microsilica

The granular material strength values specified by the standard techniques showed the following: with an increase of the content of microsilica in the batch up to 30 % the mechanical strength of the granules increased to 4 MPa, the strength reduced to 3.3 MPa with more content of microsilica, but the strength of the granulated foam glass without microsilica addition of 1.5 MPa became higher. The increased strength of the foam glass was due to crystallization processes that occurred at the foaming stage. At a density of 500 kg/m^3 the strength of foam glass was two times larger than the strength of the foam glass obtained by one-stage technology based on tripoli, and was 1.3 times larger than the strength of the foam glass obtained from the glass cullet.

4. Conclusions

The use of highly fine natural amorphocrystalline tripoli and amorphous technogenic microsilica provides obtaining of foam glass material by one-stage technology with 45 % sodium hydroxide solution as an alkaline component at foaming temperatures of 830-850 °C. Addition of microsilica as a minor constituent in the batch in the amount of 10 to 50 wt. % increases the strength of the obtained material to 4 MPa and reduces the foaming temperature from 870 to 830 °C. The increased mechanical strength of the foam glass is due to dissolving of residual quartz and cristobalite crystallization. The samples obtained from the batch containing 30 wt. % of microsilica have the maximum strength of 2 MPa.

References

[1] Chul-Tae Lee 2013 *Journal of Industrial and Engineering Chemistry.* Production of alumino-borosilicate foamed glass body from waste LCD glass. Vol. **19** pp. 1916–1925.

[2] Fernandes H, Andreola F, Barbieri L, Lancellotti I, Pascual M J, Ferreira JMF 2013 *Ceramics international*. The use of egg shells to produce Cathode Ray Tube glass foams. vol. **39** pp. 9071–9078.

[3] Mueller A, Schnell A, Ruebner K 2015 *Constr. Build. Mater.* The manufacture of lightweight aggregates from recycled masonry rubble. Vol. **98** pp. 376–387.

[4] Guo H W, Gong Y X, Gao SY 2010 *Materials letters.* Preparation of high strength foam glass-ceramics from waste cathode ray tube. Vol. **64**. pp. 997 – 999.

[5] Petersen R R, König J, Smedskjaer M M, Yue Y 2014 *J. Non-Cryst. Solids.* Effect of Na_2CO_3 as foaming agent on dynamics and structure of foam glass melts. Vol. **400 (15)** pp. 1–5.

[6] König J, Petersen R R, Yue Y 2015 *Ceramics international.* Fabrication of highly insulating foam glass made from CRT panel glass. Vol. **41** pp. 9793–9800

[7] König J, Petersen R R, Yue Y 2016 *J. Non-Cryst. Solids.* Influence of the glass particle size on the foaming process and physical characteristics of foam glasses. Vol. **447** pp. 190–197

[8] Yatsenko E A, Smolii V A, Kosarev A S, Vilbitskaya N A 2011 *Glass and Ceramics.* Application of mathematical modeling for studying the strength properties of foamed slag glass. Vol. **68 (3-4)** pp. 91-94

[9] Volland S, Brötz J 2015 *Constr. Build. Mater.* Lightweight aggregates produced from sand sludge and zeolitic rocks. Vol. **85** pp. 22–29.

[10] Wanga X, Jin Y, Wang Z, Nie Y, Huang Q, Wanga Q 2009 *Waste Manag* .Development of lightweight aggregate from dry sewage sludge and coal ash.Vol. **29** pp. 1330–1335.

[11] Chen H J, Yang M D, Tang C W, Wangaa S Y 2012 *Constr. Build. Mater.* Producing synthetic lightweight aggregates from reservoir sediments. Vol. **28** pp. 387–394.

The technology improvement and development of the new design-engineering principles of pilot bore directional drilling

A Shadrina, L Saruev, S Vasenin

National Research Tomsk Polytechnic University, Tomsk, Russia
E-mail: avshadrina@rambler.ru, saruevla@tpu.ru, vasenin8@gmail.com

Abstract. This paper addresses the effectiveness of impact energy use in pilot bore directional drilling at pipe driving. We establish and develop new design-engineering principles for this method. These principles are based on a drill string construction with a new nipple thread connection and a generator construction of strain waves transferred through the drill string. The experiment was conducted on a test bench. Strain measurement is used to estimate compression, tensile, shear and bending stresses in the drill string during the propagation of elastic waves. Finally, the main directions of pilot bore directional drilling improvement during pipe driving are determinated. The new engineering design, as components of the pilot bore directional drilling technology are presented.

1. Introduction

Large-scale application of trenchless technology in underground construction has motivated the proposed research. Generally, rotary well drilling rigs are used for these purposes [1-5].

The disadvantages of these drill rigs are as follows: firstly, the inefficiency of drill rigs in medium-hard and harder rocks, and secondly when drilling tools have to deal with similar hard materials, in particular concrete slabs or other construction debris.

In response to this problem, powerful percussive mechanisms are added in drill rigs used for pilot bores and for hardpan penetration [6].

The technologies using impact energy are promising. They allow influencing objects (rocks) with huge efforts. However, their efficiency can be maximised only by understanding the drill string dynamic process and by studying the regularities of strain waves propagating through the drill strings.

An elastic wave transfers the percussive energy and contributes to the penetration of the drilling tool into the rock.

The aim of research is to establish and develop a new design-engineering principle for the directional drilling of pilot bores during pipe driving, which is based on the application of a new drill pipe construction designed with a new nipple thread connection and the construction of a strain wave generator transfered through the drill string.

2. Methodology

The experimental method allowed us to estimate the effectiveness of wave propagation along the drill string. Moreover, our experimental findings proved the effectiveness of the new nipple connections hidden inside the tube in load distribution during the operation [7].

The experiment was conducted on the test-bench developed by scientists of Tomsk Polytechnic University. It consisted of concrete blocks with through holes. These holes simulated an artificial borehole 42 mm in diameter. The drill string with new enclosed-type nipple connections was located in the artificial borehole.

Impact impulses in the drill string are generated by a pendulum impact machine with a fixed cylindrical hammer.

Furthermore, the drill string was loaded at a constant torque of 265 N·m by means of a lever with weight. The hammer pre-impact velocity was adjusted by changing its drop height. The percussive energy ranged from 31.25 to 207.35 J.

To measure the axial deformation in the new thread connection during static and dynamic loads, lateral grooves were formed in the surface of the nipple. Wire strain gauges with a nominal length of 10 mm and a resistance of 100 Ohm were bonded in these grooves (Fig. 2).

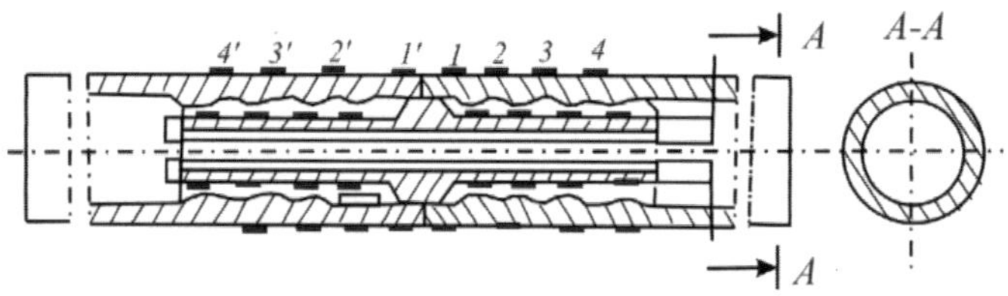

Figure 2. Schematic view of sticking resistance strain gauges for recording compression, tensile and bending stresses

To measure the bending and shear stresses, the same wire strain gauges were used, and stuck in the middle of the nipple (Fig. 3).

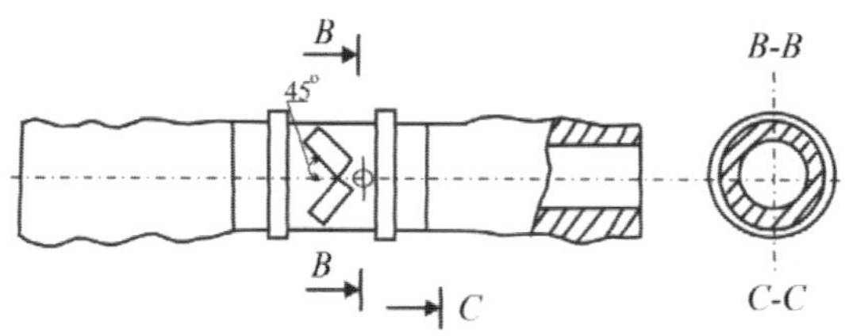

Figure 3. Schematic view of sticking resistance strain gauges for recording shear stresses

Finally, the power pulses were fixed by oscillograph S1 – 117 in five positions of the drill string. Furthermore, oscillograms of compression, tensile, bending and share stresses in elements of the thread connection were obtained and discussed.

The up-to-date hydraulically-driven drilling machines currently replace those with percussion mechanisms supplied by the pneumatic power.

Hydraulic hammers used abroad possess the sufficient capacity, however require a powerful oil-pumping station allowing the hydraulic fluid to flow in high-pressure and drain hoses. The significant energy losses caused by the hammer reciprocal motion, and friction losses caused by the fluid flow in high-pressure and drain hoses equal to approximately 10% of inertial losses. These energy losses are eliminated in the original hammerless hydraulic mechanism due to the formation of power pulses in a closed-loop system having a non-linear elastic element (a high-pressure hose) practically having no free fluid [8].

Power pulses are generated in a closed loop with a non-linear elastic element practically having no free fluid.

The proposed hydraulic mechanism includes two vibrating systems. The first is sinusoidal pressure fluctuations of the trapped fluid induced by the vibration generator. The fluid is pushed by the periodical movements of the plunger into the closed loop, thereby creating the pressure pulses. The second is a conventional vibrational contour, i.e. a mass with an elastic element.

The pressure pulse generator comprises the pulsator hydraulically connected to the driver, high-pressure hose, and hydraulic cylinder with a piston (Fig. 4).

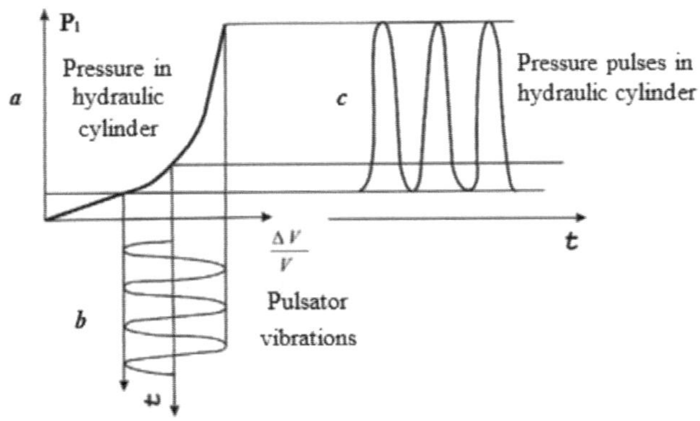

Figure 4. Pressure pulses generated in hydraulic cylinder: *a* – non-linear behavior of fluid depending on the volume change $\Delta P=f(\Delta V/V)$, where ΔV is trapped fluid pushed by the plunger into the closed loop; V is trapped fluid in the high-pressure hose and hydraulic cylinder; *b* – the diagram of relative volume change in the high-pressure hose; *c* – pressure pulses generated in the hollow hydraulic cylinder; P_l – initial pressure load induced by a feed thrust of the hydraulic mechanism with a drill string

3. Results and discussion

Widely used drill pipe joints, such as tool joints, sleeve joints refer to inefficient engineering designs for percussive drilling. Their cyclic tensile stresses have a low energy transfer coefficient.

Compared with other drill pipe joints, the suggested construction allows the transfer of the strain wave energy with minimum dissipation (Fig. 5).

In the beginning, when the drill string was assembled, the nipple connection became deformed. This was due to the effect of normal force on the pipe joint. The tightening force acting on the enclosed-type nipple connection resulted in drill pipe compression and nipple tension.

An elastic impact produced by the interaction between the hammer of the percussive mechanism located outside the borehole and the drill string, resulted in strain wave propagation (Fig. 5, line *1*). In this case the tightening force in the drill pipe is increased to power pulse value (about 15-18 tons) (Fig. 5, line *2*). Furthermore, tension force in the nipple is decreased (Fig. 5, line *3*).

When the nipple is completely unloaded, a gap between adjacent threads is observed both in the drill pipes and the nipple. This results in additional nipple longitudinal displacement in relation to the drill pipe due to the torque. It also leads to additional and equivalent nipple tension (Δl_n) and drill pipe compression (Δl_p).

Nipple and drill pipes have different rigidity. Rigidity refers to the ability to resist deformation in response to an applied force. Rigidity extent is the stiffness. In our research the pipe stiffness was $C_p = 1.6 \cdot 10^9$ N/m and the nipple stiffness was $C_n = 0.44 \cdot 10^9$ N/m.

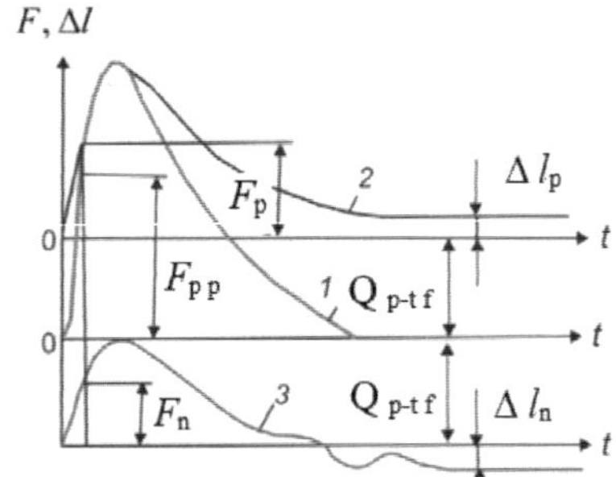

Figure 5. The influence of power pulse on the new nipple joint elements of drill pipes: 1 – power pulse propagation in enclosed-type nipple joint; 2 – the compressive force behavior in the drill pipes; 3 – the tension force behavior in the nipple; $F_{p\,p}$ – power pulse value in the pipe segment before the thread; F_p – power pulse value at pipe end; F_n – the reduced tension force value in the middle of the nipple; $F_{p\,p} = F_p + F_n$ – forces value at some instant; $Q_{p\text{-}t\,f}$ – pre-tightening force

Note that in this case the thread segment of the drill pipe has repeated-load cycle, but the nipple has alternate asymmetric load.

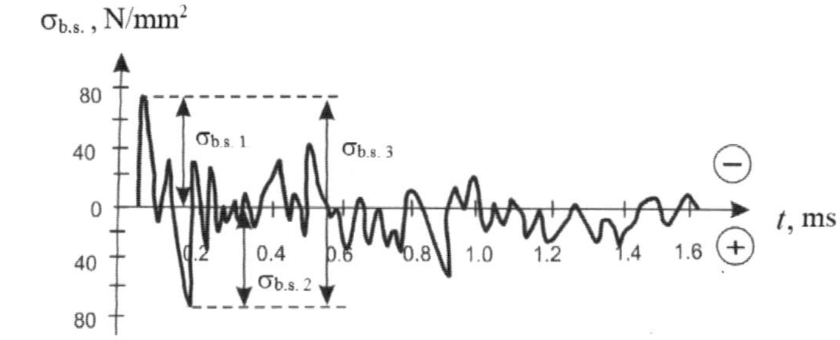

Figure 7. The bending waves propogation in the nipple
$\sigma_{из.1}$, $\sigma_{из.2}$ – maximum of bending stress; $\sigma_{из.3}$ – variation range of bending stresses

The modification of shear stresses is observed in the nipple as a result of a series of percussions generated by the hammer at a constant pre-impact velocity of 5.1 m/s.

IOP Conf. Series: Earth and Environmental Science **43** (2016) 012068 doi:10.1088/1755-1315/43/1/012068

The combined effect from the coupling torque and strain wave generated by the longitudinal impact, the drill pipes tightening occurred while in operation. It results in 2 or 3-fold reduction in the shear stress. Being located inside the drill pipe, the nipple was unloaded from the most of transverse and longitudinal loads, while 70% of coupling torque was transmitted through the drill pipe joints. Experiments show that pressure pulses are formed due to the plunger strokes and inertial mass movements at the compression of trapped fluid. In turn, the formation of power pulses is oriented only towards the borehole. It does not necessitate protecting the drilling machine from power pulses reflected from the pipeline connections and the borehole.

The largest values of the strain wave amplitude and energy in the drill string are observed at the smallest length of the high-pressure hose and its largest diameter. The time dependences of the strain wave amplitude are presented in figure 8. This is due to the fact that a longer high-pressure hose connecting the hydraulic pulsator with the cylinder, provides the lower $\Delta V/V$ ratio at the same parameters of hydropulsator. This results in lowering the pressure pulse amplitude in hydraulic cylinder and, as consequence, the power pulse in the drill string. Moreover, the decrease in the diameter of the high-pressure hose increases its rigidity. This also results in a decrease in the amplitude and energy of the power pulse.

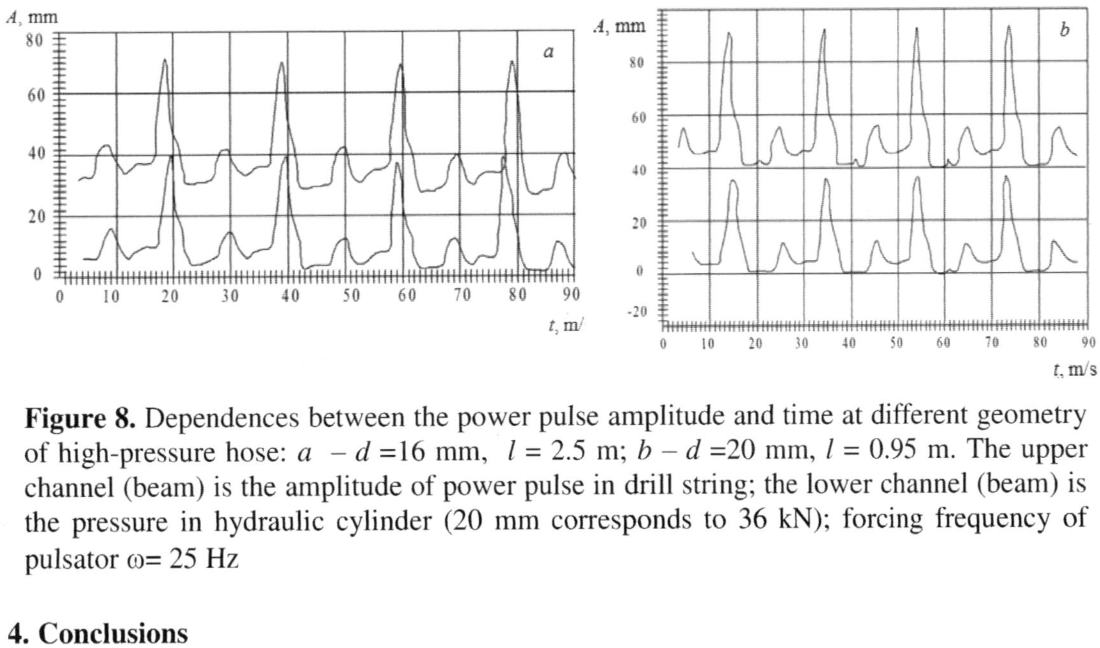

Figure 8. Dependences between the power pulse amplitude and time at different geometry of high-pressure hose: a – d =16 mm, l = 2.5 m; b – d =20 mm, l = 0.95 m. The upper channel (beam) is the amplitude of power pulse in drill string; the lower channel (beam) is the pressure in hydraulic cylinder (20 mm corresponds to 36 kN); forcing frequency of pulsator ω= 25 Hz

4. Conclusions

This study has proposed new engineering solutions as part of the pilot bore directional drilling technology. Using the new drill string construction and the hydropercussion device without a hammer, leads to the fact that power pulses with optimal parameters can be obtained and transferred to a downhole with minimum losses.

The enclosed-type nipple connection of drilling pipes is most efficient and durable as it takes into account wave processes in the drill string. The research has revealed the regularities of compression and tension stress changes in the elements of drill pipe joints.

The novel design solution proposed for the method of power pulse generation in a drill string is a promising research development. Hammerless hydromechanical systems having oil-pumping stations with pressure and drain pipelines simplify the flow chart of the drilling machine, considerably increase the efficiency of hammerless hydraulic mechanisms in comparison with all other up-to-date hydraulic percussion mechanisms.

5. References

[1] Stangl G A, Levings R B 2012 *Proceedings of 30th International No-Dig Conference and Exhibition No-Dig Sao Paulo.* Horizontal Directional Drilling (HDD) systems for pilot bore drilling in mixed soil conditions and rock. pp. 345-366.

[2] Hair J D 1989 *Pipeline and Gas.* Vol. **216 (1)** pp. 29-35.

[3] Allouche E N, Ariaratnam S T and Lueke J S 2000 *Journal of Construction Engineering and Management.* Vol. **126 (1)** pp. 68-76.

[4] Ariaratnam S T, Allouche E N 2000 *Practice Periodical on Structural Design and Construction.* Vol. **5 (4)** pp. 142-149.

[5] Burkov P V, Burkova S P and Kravchenko A N 2015 *IOP Conference Series: Materials Science and Engineering. Proceedings of VI International Scientific Practical Conference on Innovative Technologies and Economics in Engineering.* Finite Element Model of Trenchless Pipe Laying. Vol. **91** pp. 1-7.

[6] Danilov B B, Smolyanitsky B N, Sher E N 2014 *Journal of Mining Science.* Vol. **50 (3)** pp. 484-490.

[7] Shadrina A V and Saruev L A 2014 *International Journal of Mining Science and Technology.* Vol. **24 (2)** pp. 245–249.

[8] Kazancev A A, Shadrina A V, Saruev L A, Saruev A L 2009 RF Patent 79924. Tomsk: RF.

IOP Conf. Series: Earth and Environmental Science **43** (2016) 012069 doi:10.1088/1755-1315/43/1/012069

Developing the drilling tool for trenchless pipeline construction

S Povarnitsyn[1], A Martynuk[2], O Brusnik[3]

[1] OOO RN-KrasnojarskNIPIneft
[2, 3] National Research Tomsk Polytechnic University, Tomsk, Russia
E-mail: [1]povarnitsynsergey@gmail.com, [2]kr_nas_sh@tpu.ru , [3]brusnikov@tpu.ru

Abstract. The most promising methods of underground pipeline construction are discussed. To increase the efficiency of the drilling tool for underground pipeline construction the following research methods have been used in the article: optimization method of Evolutionary strategy, finite element modeling. The influence of geometric parameters of the drilling tool on force characteristics of enlarging the drill hole has been investigated.

1. Introduction

The trenchless methods of installing an underground pipeline, for example, horizontal directional drilling (HDD), microtunneling, pipe jacking and pipe ramming are currently widespread in many fields of construction.

The most promising method of underground pipeline construction is horizontal directional drilling, but this method has a number of limitations, the main of which is the high cost due to the use of drilling fluid. The increase in this method efficiency can be achieved by several ways: using a multistage enlarging method to enlarge a hole; using drilling tools for drilling a pilot hole and enlarging a hole without using drilling fluid.

The objective of the article is to increase the efficiency of trenchless pipeline construction based on improving the drilling methods and the drilling tool.

2. Research methods

A characteristic feature of the drilling tool and its analogues is that in the course of drilling the soil is not removed from the drill hole to the surface, but is pressed into the surrounding soil.

Since the drilling tool consists of a helical auger segment and displacement elements, in the course of drilling the constituent elements of the tool perform operations on destruction, displacement and soil compaction.

The numerical modeling method has been widely used for defining the parameters of interaction of drilling tools and the soil [1].

The Finite Element modeling of a solid body and elasto-plastic soil interaction is presented in the works by V. V. Aleshin, V. E. Seleznev, G. G. Boldyrev, G. Lacey, E. Susila, S. I. Woo, D. Sheng, J. Walker.

Bolton M. D. studies the possibility of using the software ANSYS, and other programs for solving problems considering the elasto-plastic behavior of the soil [2].

The experiments made by G. G. Boldyrev let us deduce an inference that a better description of the qualitative and quantitative characteristics of the real behavior of the soil can be obtained using the soil model of Drucker-Prager for the static and dynamic models [3].

The works by E. Susila are devoted to substantiation of the finite element model simulating interaction between a conical tip of piles and the soil penetrated. The author suggests using the Drukera-Prager's model to describe physical properties of the soil [4, 5, 6].

In the present work the optimization method of Evolutionary strategy (ES) is used [7]. The mathematical formulation of the optimization problem of the drilling tool can be written as follows:

$$\min F(x), x = (x_1, \ldots, x_N) \in X, \quad (1)$$

under the constraint:

$$X = \{x : x_{\min,i} \le x_i \le x_{\max,i}\},$$

x – the vector of parameters defining the geometry of the tool; F – the objective functional; X – geometric constraints; $x_{\min}$, $x_{\max}$ – minimum and maximum values of geometric constraints.

The HDD method involves drilling a pilot hole on a designed path and further enlarging this drill hole. For the optimization problem the stage of enlarging the drill hole diameter from 110 to 150mm is analyzed.

According to the optimization problem the initial geometry of the drilling tool as well as its mode of operation are defined. It is required to determine the optimal geometric parameters of the drilling tool with relation to the chosen criterion of quality that meets the accepted constraints. The schematic construction of the drilling tool is shown in figure1.

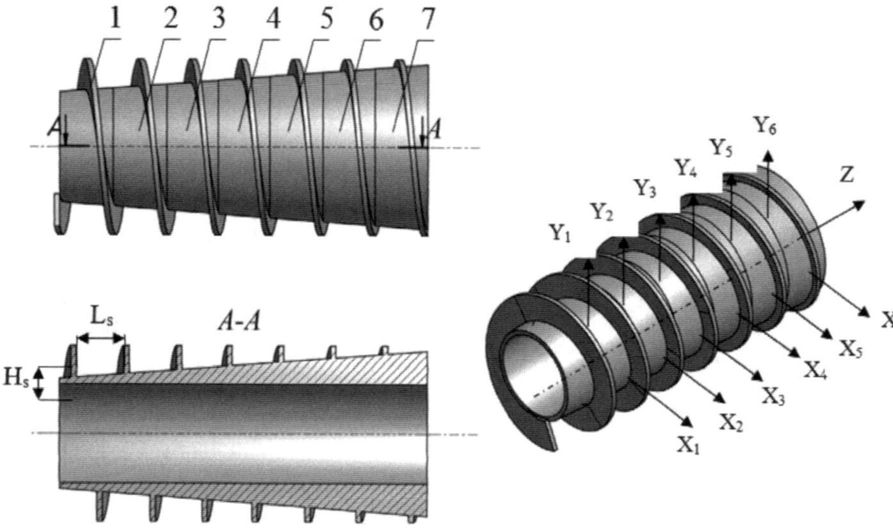

Figure 1. Parameters of the drilling tool: 1 – helical auger segment; 2, 3, 4, 5, 6, 7 – displacement elements

Having studied the interaction of the drilling tool and the soil we can propose a number of geometric parameters to be optimized: helical auger segment (Hs – height of spiral, Ls – length of spiral); the position of displacement elements are relative to the axes X,Y.

In the course of drilling a horizontal hole the drilling tool experiences various loads, which depend on geological conditions and the trajectory of the hole path. In order to accurately determine the force characteristics depending on geometry of the drilling tool, the following simplifications have been specified: the soil material is isotropic, the position of the drilling tool is vertical.

The geometry of the mathematical model is shown in figure 2. The soil is represented by the volume of 500×500×650 mm (x×y×z).

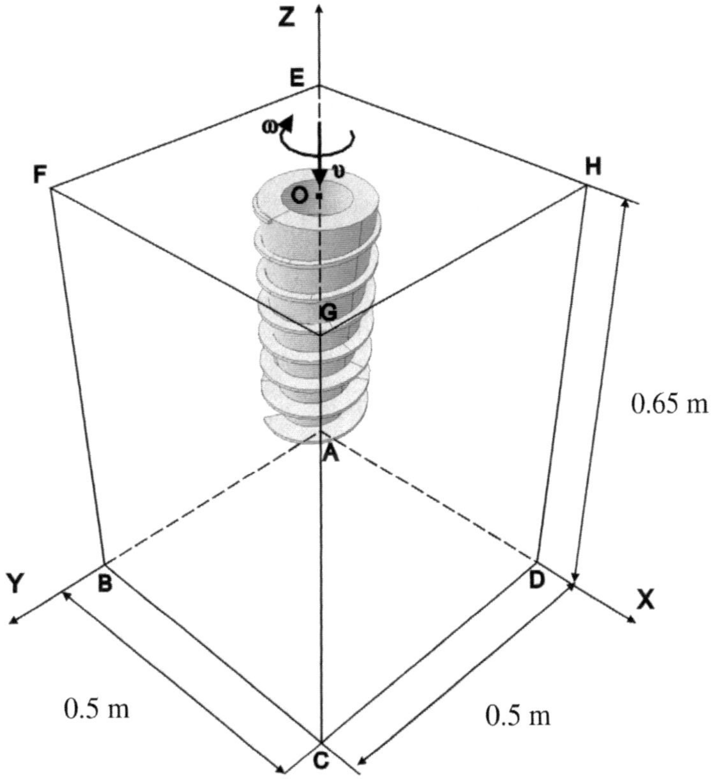

Figure 2. Design scheme of the model

The initial geometry of the drilling tool consists of 1895 finite elements with an average size of a grid x,y,z – 9 mm. The reference point of the finite element grid of the drilling tool (node O) is set on the Z-axis, with constant linear velocity υ – 0.1 m/s and angular velocity ω – 120 rpm.

The output parameter accepted by the corresponding task is the reaction force projected onto the three coordinate axes and torques, Ri and Mi symbols, respectively.

3. Results and discussion

The sample of the drilling tool with the output parameters Rz = 1.54 kN, Mz = 308 N×m and F = 1.14 has the least value of the objective functional from the all sample variants of the drilling tool. A 3D model of the drilling tool with a geometry corresponding to the optimal variant is given in figure 3.

IOP Conf. Series: Earth and Environmental Science **43** (2016) 012069 doi:10.1088/1755-1315/43/1/012069

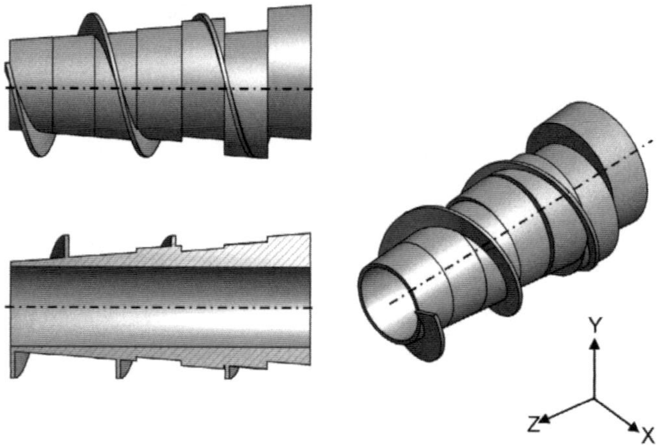

Figure 3. Geometry of the optimal sample

4. Conclusions

1. Based on the study of interaction of the drilling tool and the soil, a number of geometric parameters of the drilling tool are suggested to be optimized.
2. The geometry of the drilling tool providing minimum force characteristics for enlarging the drill hole in the soil by the HDD method has been defined.
3. The influence of geometric parameters of the drilling tool on force characteristics of enlarging the hole has been investigated.

5. References

[1] Zienkiewicz O C and Taylor R L 2005 *The Finite Element Method for Solid and Structural Mechanics* (Oxford: Elsevier) pp. 631.
[2] Bolton M D 1986 *Geotechnique.* Vol. **36 (l)** pp. 65–78.
[3] Boldyrev G G and Nikitin E V 1987 *Foundation, foundations and soil mechanics.* Vol. **1** pp. 26–28.
[4] Jalali J and Vogt N Numerical 2008 *ANSYS Conference & 26th CADFEM Users' Meeting. Darmstadt.* Modelling of Pile Foundation of Noise Protection Walls in Slopes using. pp.1–5
[5] Susila E and Hryciw R D 2003 *International Journal for Numerical and Analytical methods in geomechanics.* Vol. **27** pp. 585–602.
[6] Jalali J Numerical 2008 *GeoDACH-Treffen Grainau Deutschland.* Modelling of Single Piles and Pile Groups under Horizontal Loads In: Tagungsband 10. pp. 51
[7] Moiseev N N, Ivanilov Yu N and Stolyarova E M 1978 *Moscow Nauka. Methods of optimization.* pp. 352

Impact of hydrocarbon drilling mud on mud motor elastomers at different temperatures

A V Epikhin, V V Melnikov, L N Nechaeva and O S Ulyanova

[1]National Research Tomsk Polytechnic University, 30 Lenina Ave., Tomsk, 634050, Russia
E-mail: epikhinav@mail.ru

Abstract. The paper describes the experimental research of hydrocarbon drilling mud impact on engineering parameters of mud motor elastomer samples. It is believed to be urgent due to an increase in using mud motors in oil and gas well construction now, and the issue of intense exploitation is currently topical. The test results of elastomer IRP-1226 dependent on the temperature are shown in the paper. It is proved that the hydrocarbon drilling muds have a significant impact on wearing of mud motors elastomers under the condition of a temperature increase.

Introduction

Mud motors have been widely used in the process of oil and gas wells drilling since the middle of the 20th century. This was connected with the development of directional and horizontal drilling, renewal of production by sidetracking. Besides, mud motors are used for workover and well service [1].

Long-term operation of mud motors under various geological conditions has revealed low service life of elastomer (rubber component of stator). This is confirmed by a rate of mud motors accident encountered by contractors. On average mud motors breakdown cause 12 equipment downtime per a year, which increases drill time by that to 5-10 %. In Western Siberia, it increases the total cost of up to 3 million rubles per one well. According to the experience of various drilling intervals, the key factor of mud motors service life is a type of drilling mud. For instance, when hydrocarbon drilling mud is used, mud motors wearing increases by 250 – 350 % [2-3].

Mud Motors Manuals emphasize the impact of high temperature in bottom-hole zone on elastomer durability and it consequently decreases service life of a motor. Drilling mud comprising hydrocarbon-based reagents causes swelling of a rubber component (elastomer) of the stator and leads to premature motor's breakdown. The differential pressure increasing the recommended working value of mud motor's gear, decreases service life of the stator [4].

According to the information mentioned-above, it was decided to carry out experimental research of the impact of diesel fuel as one of the types of hydrocarbon drilling mud on durability of IRP-1226 rubber samples used as a material for rubber components of mud motors stator; and to identify the dependence of the size on temperature.

1. Materials and Methods

The impact of different mud types on rubber samples IRP-1226, which is oil-resistant rubber with improved wear-resistance and a temperature range from - 20 to $+100^0$ C, has been studied. Its nominal strength is equal to 9.8 MPa, failure elongation is 125% and Shore A hardness number is 65 – 95.

The samples were made in the form of cylinders with a diameter up to 43 mm and a thickness up to 11.5 mm. They were kept in plastic containers throughout the experiment with total immersion in the mud under atmospheric pressure. To verify the experimental results, 3-5 samples were analyzed for each type of the drilling mud. The first batch of samples was kept in the diesel fuel under temperature of 25^0C (room temperature), while the other three subsequent batches of samples were fully immersed in the muds in a cabinet dryer under temperature 50^{0}C, 75^{0}C and 90^{0}C, respectively. The time of the experiment was 20 days for each batch. This is due to the fact that the mud motor operation time is equal to 200 – 600 hours, with 240-300 hours on average [5-6].

It is necessary to note that the previous experiments made by researchers clearly showed elastomer instability in the drilling mud causing swelling or a decrease in the sizes [1,7 - 9].Therefore, the evaluation was made based on geometric dimensions of the rubber sample.

2. Results and discussion

According to the experimental data, the hypotheses about elastomers instability in regard to drilling muds causing swelling or the decrease in the sizes [10] were proved. Therefore, the experimental results were divided into two categories according to elastomers instability: 1 – a decrease in sizes (fig. 1 – 2); 2 – sample swelling (fig. 3 – 4).

At temperature 25^0 – 50^{0}C, a decrease in the size of the rubber sample was observed. Furthermore, the diameter of the sample reduced by 1.41 mm per 480 hours at temperature 25^{0}C, while at temperature 50^{0}C this parameter was 1.59 mm. At temperature 50^{0}C within the period of 0-48 hours, a decrease in the sample sizes is more intense than at room temperature. At the same time, throughout 100 – 480 hours at room temperature a decrease in a sample size is more intense than at temperature 50^{0}C. This is due to the high volatility of diesel fuel with the increase in temperature.

During the experiment splitting of samples in diesel fuel was observed in all batches after 24 hours (fig.2). Mud gets insignificantly dark at temperature 50^{0}C, which confirms indirectly the possibility of drilling mud impact on elastomer wearing even without dynamic loading.

Intense change of the sample parameters in the form of swelling was observed at temperatures 75^0 and 90^{0}C. The following results were registered: at temperature 75^0 C diameter of the sample increased by 1.42 mm, while at temperature 90^{0}C swelling of the sample is 1.3 mm per 480 hours.

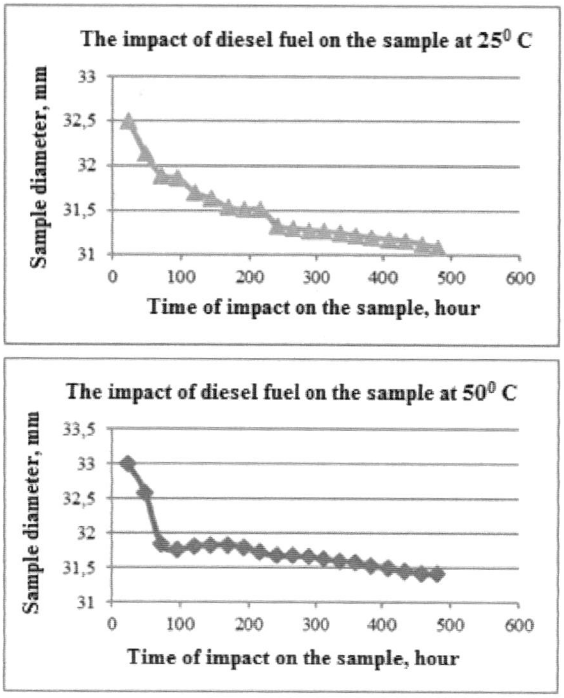

Figure 1. The impact of diesel fuel on the samples at 25^0 and 50 ^{0}C

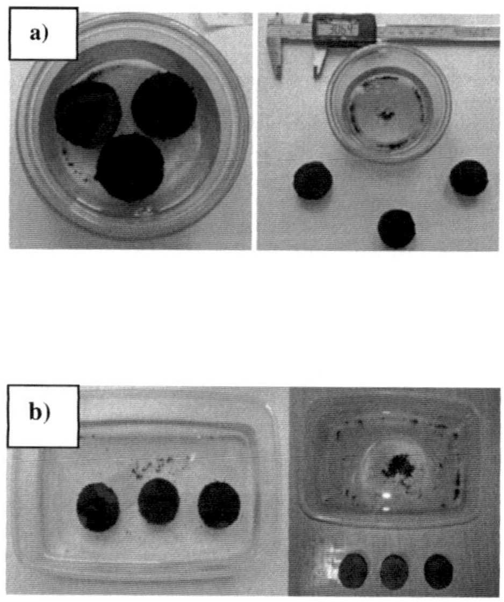

Figure 2. The samples and diesel fuel throughout 24 hours (left) and 480 hours (right), a) at 25 ^{0}C; b)at 50 ^{0}C

It should be noted that the degree of mud impact to the sample is determined not only by the temperature but also the duration of the experiment. At temperature 90^{0}C throughout 0-72 hours swelling of the sample is more intense than at temperature 75^{0}C. At the same time, throughout 144 – 148 hours more intense swelling is registered at temperature 75^{0}C.

When studying the temperature range of 75^0 – 90^{0}C, an increase in sample size was obvious even within the first days of the experiment. This is opposite to the experimental data registered in the temperature range of 25^0 – 50^{0}C. Splitting of samples was less intense but it mainly occurred at temperature 75^0 C (fig.4). The color of diesel fuel rapidly darkened (within 24 hours), and at temperature 90^{0}C it became close to black.

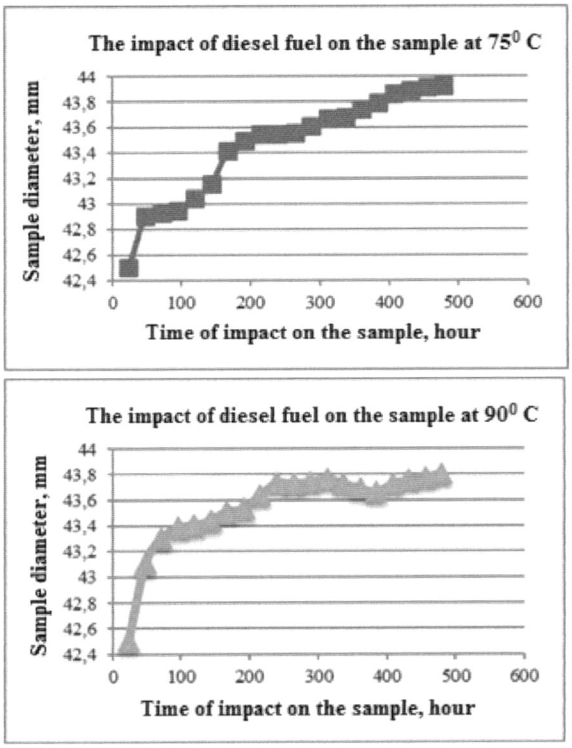

Figure 3. The impact of diesel fuel on the samples at 75^0 and 90^{0}C

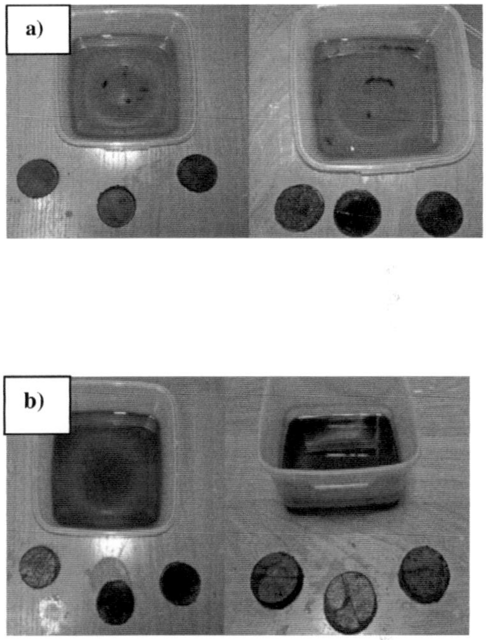

Figure 4. Figure 2. The samples and diesel fuel throughout 24 hours (left) and 480 hours (right), a) at 75 ^{0}C; b)at 90 ^{0}C

Conclusions

Evaluating the results of the research, it can be noted that the impact of diesel fuel as a type of drilling mud on elastomers' parameters is confirmed. In all studied temperature ranges intense deformation and splitting of the samples were observed. In further research it is necessary to pay particular attention to the temperature range of 75^0 – 90^{0}C, since according to the experimental data; it is within this range, where more intense elastomers deformation is registered.

Acknowledgement

We gratefully acknowledge the financial support given by The Russian Foundation for Basic Research (RFBR), project *№16-38-00701 m_a).*

References

[1] Korotaev Yu A 2003 *Ph.D. thesis in Engineering Science. Perm.* Study and design the multiple-thread set screw gerotor mechanisms downhole motors production. pp. 386.

[2] Anyanwu O N, Klotz C, Labrecque D, Ulrich C 2012 *Society of Petroleum Engineers - Coiled Tubing and Well Intervention Conference and Exhibition.* Optimized downhole mud motor delivers outstanding performance improvement in Alaska coiled tubing drilling. pp. 88-100.

[3] Herbig C, Hohl A, Oueslati H, Reckmann H, Heuermann-Kühn L, Baumann J, Heinisch D 2014 *SPE/IADC Drilling Conference, Proceedings.* Data-driven drilling optimization: Using historical data to improve future performance. Vol. **1** pp. 83-94.
[4] Design and operation of mud motors (in Russian) [Electronic resource] URL: http://www.gazpb.ru/ekspluatatsiya-turbinnoj-tekhniki/105-ustrojstvo-i-rabota-vintovyx-zabojnyx-dvigatelej.html.
[5] Ignatov D N 2012 *Youth and Science 7th Russian scientific and technical conference of students, under graduates and young scientists, proceedings (Krasnoyarsk) Siberian Federal University.* Reliability indexes of mud motors. URL: http://conf.sfu-kras.ru/sites/mn2012/section18.html.
[6] Yadav A, Ali A 2014 *Society of Petroleum Engineers - SPE Bergen One Day Seminar.* Drilling technology taking drilling operations to new frontiers. pp. 210-217.
[7] Varela R, Guzman F, Cruz D, Atencio N, Iturrizaga F, May R, Solano R, Perez R 2015 *Oil Gas European Magazine.* First application of turbodrill and hybrid bit to optimize drilling times in cretaceous formations with high chert content in Mexico South Region Vol. **41 (1)** pp. 32-33.
[8] Baldenko D, Korotaev Yu 2012 *Journal "Burenie i neft"* The current status and Russian PDM perspectives of development [Electronic resource] URL: http://burneft.ru/archive/issues/2012-03/1.
[9] Baldenko D, Baldenko F, Gnoevyh A 2007 *Downhole drilling motor. Moscow.* Screw hydraulic engine. Vol. **2** pp. 470
[10] Epikhin A V, Ushakov A V, Barztaikin V V, Melnikov V V, Ulyanova O S *IOP Conference Series: Earth and Environmental Science* 2015 Experimental research of drilling mud influence on mud motor mechanical rubber components Vol. **27 (1)** pp. 012051.

Synthesis of Carboxymethyl Starch for increasing drilling mud quality in drilling oil and gas wells

K M Minaev[1], D O Martynova[2], A S Zakharov[1], R R Sagitov[1], A A Ber[1] and O S Ulyanova[1]
[1] Tomsk Polytechnic University, 30 Lenina prospect, 634050, Tomsk, Russia
[2] Tomsk State University, 36 Lenina prospect, 634050, Tomsk, Russia
E-mail: minaevkm@bk.ru, martynova1847@gmail.com

Abstract. This paper describes the impact of carboxymethyl starch preparation conditions on physicochemical properties of polysaccharide reagent, widely used as fluid loss reducing agent in drilling mud. Variation of the main parameters of carboxymethylation is researched in the experiment. The following conditions such as temperature and reaction time, amount of water, as well as ratio of NaOH to monochloracetic acid define the characteristics of carboxymethyl starch. The degree of substitution is defined for polysaccharides, as well as the characteristics of samples have been studied by infrared spectroscopy. Rheological characteristics and fluid loss indicator have been investigated to study the impact of the reagents on drilling mud quality.

1. Introduction

Being natural polysaccharide, starch is characterized by unique characteristics, such as high biocompatibility and annual renewability of raw material resources (potato plant, corn, rye, wheat, cassava, pea, etc.) that distinguish it from cellulose derived from wood, which period of maturing is 18 – 20 years for fast growing wood. In addition, another obvious advantage of starch is the ability to control the properties of polysaccharide by chemical, physical, bacteriological, biotechnological or mixed affect.

Due to its unique characteristics as natural polymer, starch is used as a base for production of a wide range of products. Chemically modified starch is produced by large companies, such as Aкzo Nobel and Avebe (Netherlands), Raisio Chemical, Grain Processing Corp. (the USA), INQUIL® (Brazil), Henkel (Germany) Aloja Starkelsen (Latvia), Rolnas (Poland), etc. All of them have produced a wide range of starch. One of the main directions of application of their chemically modified starch production is the petroleum industry [1].

Starch was the first polymer reagent used for the drilling muds. In 1939 it was added to salt muds in the western part of Texas to regulate the filtration properties. However, with the introduction of polymer reagents based on cellulose ethers, especially carboxymethylcellulose, widespread use of starch started to reduce. This was primarily due to the need to use bactericides and low thermostability of starch reagents [2].

Currently, the natural starch is not used in drilling muds. Instead of it, the following types of modified starch are applied to regulate filtration and rheological characteristics of drilling muds as carboxymethyl, hydroxyethyl, starch hydroxypropyl, and phosphates of starch as well as oxidized starch. Carboxymethyl starch (CMS) is the most commonly used starch among all modified starch reagents. Providing the required filtration properties of drilling mud in free-salined mud and saline-based mud, carboxymethyl starch can be used as a cheaper alternative of carboxymethyl cellulose. It is characterized by good water solubility, resistance to high content of salt, high durability to thermal and bacteriological affect.

Despite the fact that the carboxymethyl starch is available on an industrial scale, it is urgent to research the technology of its production with the required characteristics and its application as a component of drilling mud. For this purpose various syntheses of carboxymethyl starch are developed. The most important methods to obtain carboxymethyl starch are dry method, extrusion, aqueous method and organic solvent slurry method.

Dry method of obtaining carboxymethyl starch is the easiest one, but it requires a period of time from several hours to several days and, the given carboxymethyl starch is characterized by low degree of substitution that restricts its further application.

As for extrusion method, it is characterized, on the one hand, by some technological advantages, in particular, high productivity, energy efficiency, but, on the other hand, it contains many additives, which require additional purification. The use of water as a reaction medium may be accompanied by gelling carboxymethyl starch, which leads to low degree of substitution and difficulties in the process of its extraction from a reaction vessel, as well as increase in the time of drying [3]. carboxymethyl starch with a high degree of substitution and a low content of additives can be obtained by the suspension procedure with a mixture of water and organic solvent as a reaction medium. It is suggested to add such amount of water that is sufficient for swelling starch grains at increased temperature, but not sufficient for gelling and agglomeration of polysaccharide. Swelling starch grains promotes rapid access of substances to a reaction zone, but their destruction impedes this process [4]. Therefore, the retention of starch grains is a key factor that determines industrial importance of the procedure that simplified considerably synthesis of carboxymethyl starch [5]. In addition to this procedure, excess swelling of starch grains can also be controlled by addition of cross-linking agents prevented full destruction of starch grains due to formation of intra- and intermolecular covalent bonds between polysaccharide chains. This treatment also increases the stability of grains to pH medium and high shear pull impact [6]. The proportion of chemical bonds is generally low in relation to the total weight of starch and the number of anhydroglucose units in grains in total, therefore to increase the efficiency of the procedure, the substances containing some active groups, for example, dialdehydes are used as cross-linking agents [7, 8]. Glyoxal refers to such substances, but being widely used as a cross-linking agent for starch and cellulose in pulp and paper industry, it is not enough studied. It might be suggested to get efficient conditions of carboxymethylation for modification of starch by cross-linking agents due to alkaline solution with pH = 7.5 –12 [9].

The main parameters defining the efficiency of carboxymethylation by suspension procedure are the following: a type of solvents, amount of alkali, fluid, temperature, and reaction time [10]. Therefore, the purpose of the paper is to define the optimal conditions of carboxymethyl starch obtaining.

2. Materials and methods

Carboxymethyl starch was prepared by the following stages. Firstly, 13.8 gm of starch and 5.2 ml of sodium hydrate saturated solution were added to monochloracetic acid solution, and then, if it was necessary, cross-linking agents in a form of 40% solution were injected. The mixture was heated to the reaction temperature and stirred for 1 hour. The resulting product was filtered, purged several times by ethanol, and dried at temperature of 50^0C.

In the paper the ratio of alkali to monochloracetic acid in carboxymethylation agent, the type of starch, the volume and type of solvent, water mass fraction in solvent content, amount of glyoxal as a cross-linking agent, as well as temperature and reaction time are varied (Table 1).

IOP Conf. Series: Earth and Environmental Science **43** (2016) 012071 doi:10.1088/1755-1315/43/1/012071

Table 1. Content of reaction mixture and reaction condition of carboxymethylation

Sample	Type of starch	Ratio m(NaOH)/m (*monochloracetic* acid)	Solvent	V(solvent), ml	Temperature, ^0C	Reaction time, min	Glyoxal percent by weight, %.	Degree of substitution
CMS-IV	potato starch	0.88	water	25	40	10	-	60
CMS-cl-VI	potato starch	0.88	isopropanol	40	30	60	5	44
CMS-VI	potato starch	0.88	isopropanol	40	30	60	-	36
CMS-VII	potato starch	0.88	butanol-2	40	30	60	-	30
CMS-8	corn starch	0.88	isopropanol	40	30	60	-	30
CMS-10	potato starch	1.75	isopropanol	40	30	60	-	23
CMS-14	potato starch	3.50	isopropanol	40	30	60	-	11
CMS-11	potato starch	0.88	isopropanol	40	45	60	-	30
CMS-cl-11	potato starch	0.88	isopropanol	40	45	60	1	31
CMS-12	potato starch	0.88	isopropanol	40	55	60	-	26
CMS-cl-12	potato starch	0.88	isopropanol	40	55	60	1	33
CMS-13	potato starch	0.88	isopropanol	40	65	60	-	33

The chemical composition of samples was studied by FTIR spectroscopy using ATR method (attenuation total reflection) by Nicolet 6700 (Thermo Scientific) in a range of 400 – 4000 sm^{-1}. The degree of substitution of carboxymethyl starch was defined by the formula

$$\gamma = 162 \cdot \omega \cdot 100/(2300-80 \cdot \omega), \quad (1)$$

taking into account sodium content in the samples, obtained by atomic absorption,

where γ – the degree of substitution, showing number of OH groups, substituted by carboxymethyl in a chain of 100 anhydroglucose units;

ω – sodium mass fraction, % wt.

To evaluate viscosity fluctuation in the studied samples of carboxymethyl starch, the rotor viscometer OFITE 800 was used. The apparent viscosity (AV), mPa·s was calculated by the formula "test value of viscometer at 600 rpm"/2.

$$\frac{\text{test value of viscometer at 600 rpm}}{2} \quad (2)$$

Wall-plastering properties of drilling mud were determined by the filter press OFITE. To investigate the influence of carboxymethyl starch on drilling mud, a model system, consisting of fresh drilling mud (5% solution of modified bentonitic drilling mud powder) and saline-based mud (5% solution of modified bentonitic drilling mud powder+400 gr/dm^3 NaCl) was used.

3. Results and discussion

3.1. Reaction temperature

Evaluating the influence of carboxymethylation reaction temperature of potato starch on its characteristics was carried out in the temperature range from 30^0 to 65^{0}C. It is shown that by using a cross-linking agent, the ratio of drilling mud viscosity to the temperature of carboxymethylation is characterized by 45^{0}C maximum temperature, while in the case of a cross-linking agent absence, this parameter increases at higher temperature up to 55^{0}C (fig.1).

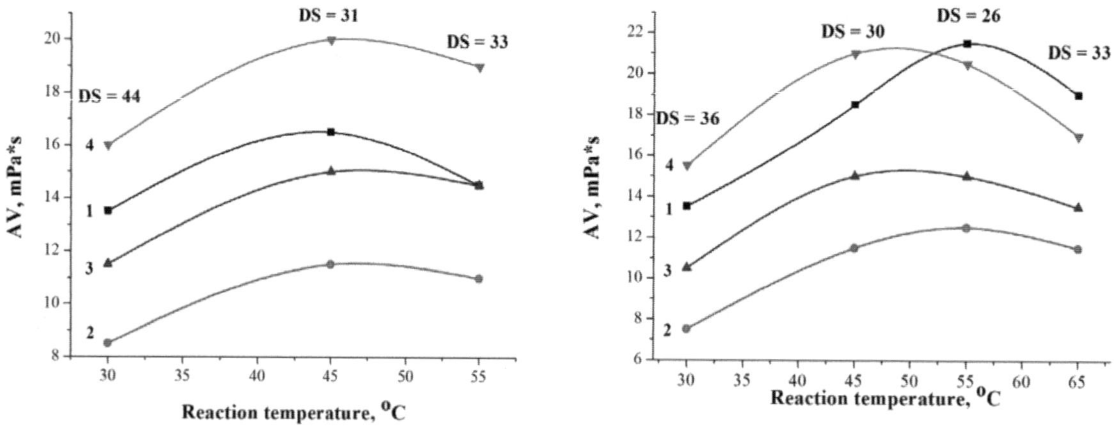

a b

Figure 1. The dependence of drilling mud viscosity on temperature of carboxymethylation a) with a cross-linking agent; b) without a cross-linking agent.
1 – fresh drilling mud; 2 – salt mud (5% - carboxymethyl starch); 3 – salt mud (1% - carboxymethyl starch);4 – salt mud (1,5% - carboxymethyl starch).

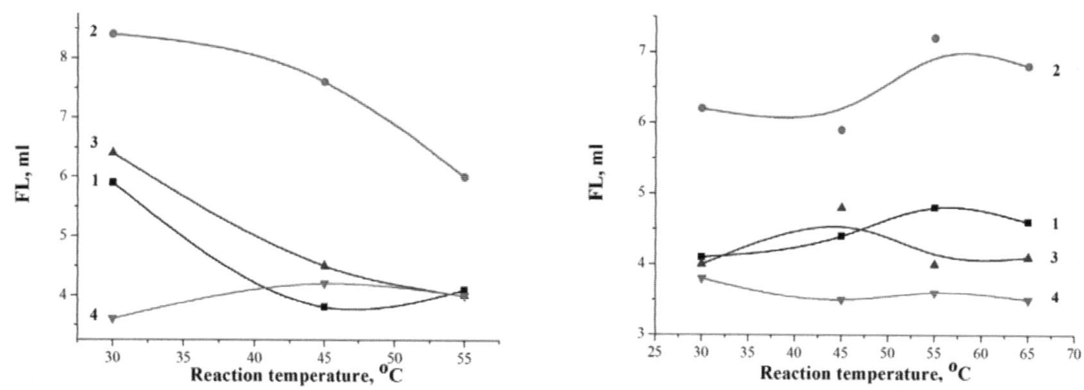

a b

Figure 2. The dependence of reaction temperature of carboxymethylation on the fluid loss indicator a) with a cross-linking agent; b) without a cross-linking agent.
1 – fresh drilling mud; 2 – salt mud (5% - carboxymethyl starch); 3 – salt mud (1% - carboxymethyl starch);4 – salt mud (1,5% - carboxymethyl starch).

An increase in the reaction temperature of starch carboxymethylation, accompanied by the introduction of the cross-linking agent results in lowering the fluid loss indicator (fig. 2a). Despite this fact the degree of substitution is maximum at low temperature of synthesis. It is known that lowering filtration in a medium of polymer reagents is due to absorption into positively charged clay particles with solvation shell formation, and increasing viscosity of a dispersion system due to partial or complete free fluid fixation. Therefore, it can be concluded that viscosity of a dispersion system influences significantly drilling mud filtration in low carboxymethyl starch muds. However, inverse relation is observed in high carboxymethyl starch mud (fig. 2a), where the lowest fluid loss indicator corresponds to maximum degree of substitution and is a result of stronger adsorptive influence of carboxymethyl starch with clay particles due to more numbers of OH-groups in a polymer chain.

When a cross-linking agent is absent, synthesis temperature has little effect on fluid loss properties of fresh drilling mud and salt mud with synthesized samples of carboxymethyl starch (fig. 2b). The fluid loss indicator was more subjected to a change in reaction temperature for the samples with a cross-linking agent. At the same time, the samples of carboxymethyl starch without a cross-linking agent provide less fluid loss properties under the equal conditions.

*3.2. Ratio m(NaOH)/m(*monochloracetic acid*)*

Changing the fluid loss properties of fresh drilling mud is not registered due to changing the ratio m(NaOH)/m(monochloracetic acid), because all obtained samples demonstrate sufficient degree of substitution to solve carboxymethyl starch and act without aggressive mineral influence (fig. 3a).

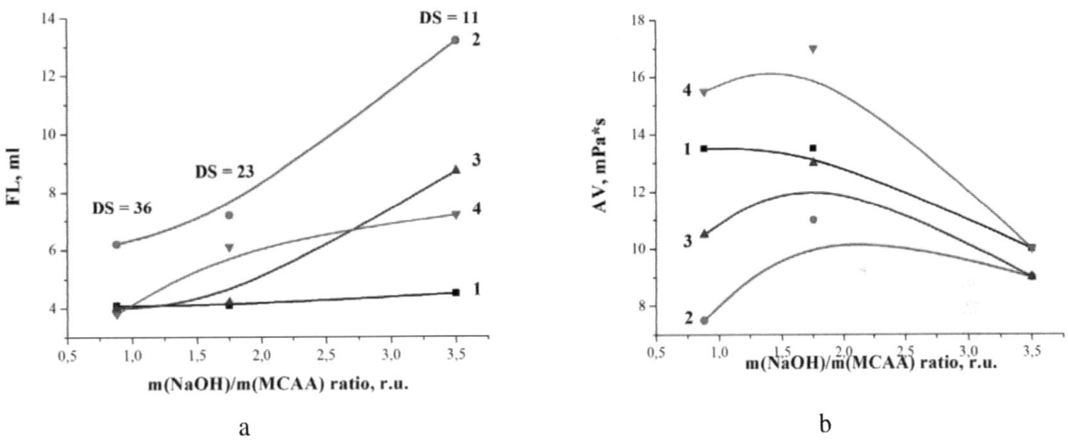

Figure 3. The dependence of ratio m(NaOH)/m(monochloracetic acid) on the fluid loss indicator and apparent viscosity a) with a cross-linking agent; b) without a cross-linking agent.
1 – fresh drilling mud; 2 – salt mud (5% - carboxymethyl starch); 3 – salt mud (1% - carboxymethyl starch);4 – salt mud (1,5% - carboxymethyl starch).

When ratio m(NaOH)/m(monochloracetic acid) has become more, it is observed a significant decrease of degree of substitution caused an increase in the velocity of hydrolysis side reaction of *monochloracetic acid* and, respectively, a decrease in alkylating agent used in carboxymethylation of starch As a result, a fluid loss indicator increases sharply, but apparent viscosity decreases in salt muds with high m(NaOH)/m(monochloracetic acid*)* (fig. 3b).

3.3. Solvent

The samples obtained in water are characterized by the highest viscosity in comparison with carboxymethyl starch obtained in suspension media (fig. 4a). Such behavior can be explained by the homogeneous conditions of carboxymethylation in water due to destruction of starch kernels, that provides higher viscosity. A fluid loss indicator of the sample does not differ from samples obtained in isopropanol and butanol-2 (fig. 4b).

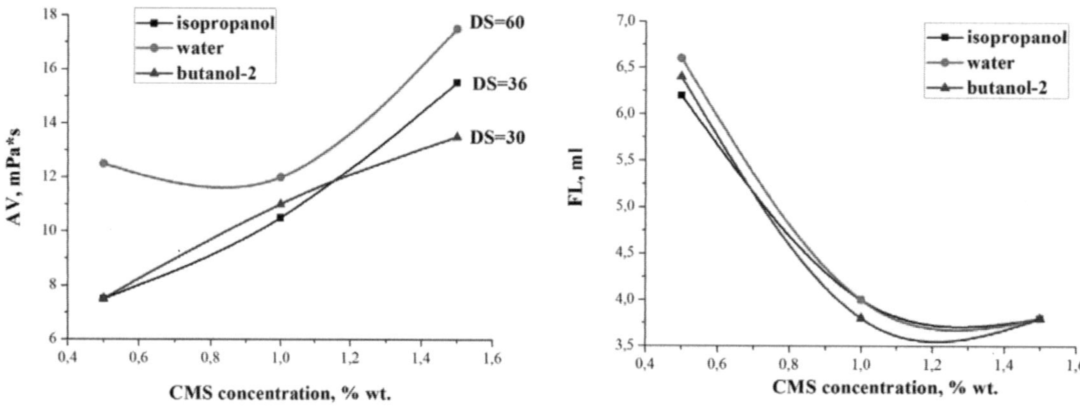

a b

Figure 4. The impact of a solvent on viscosity (a) and a fluid loss indicator (b) of saline-based mud with different concentration of polysaccharide.

3.4. Types of starch

It is proved that viscosity and a fluid loss indicator of salt mud with carboxymethyl starch on a basis of corn is higher than potato starch (fig. 5a). It can be explained by the fact that corn starch is more resistant to chemical and physical influences and has higher molecular weight, so, that provides a higher polymerization number and a lower degree of substitution.

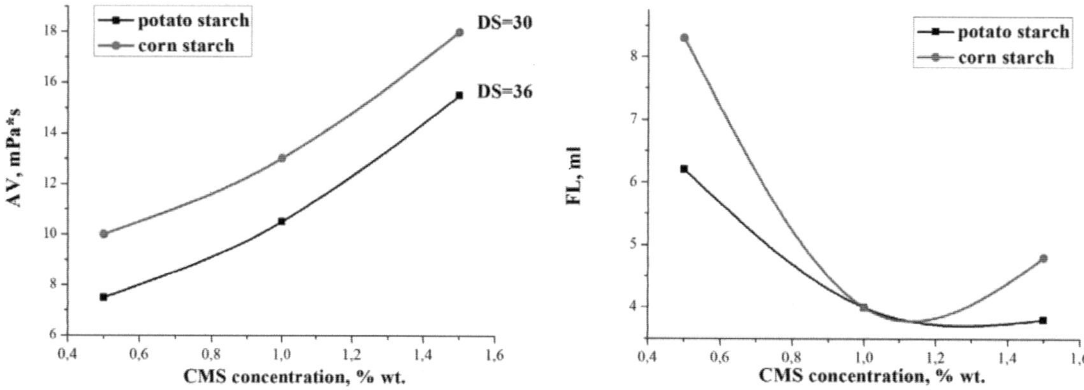

a b

Figure 5. The impact of a type of starch on viscosity (a) and a fluid loss indicator (b) of saline-based mud with different concentration of polysaccharide.

In spite of this, a fluid loss indicator of the potato starch sample is lower as compared with corn starch, because of the determined impact of higher degree of substitution of the carboxymethyl starch.

3.5. Research of carboxymethyl starch samples by FTIR spectroscopy

A broad absorption band is observed on IR -spectra of the samples in 3400 cm^{-1} corresponding to stretching vibrations of OH -groups, as well as intra- and intermolecular hydrogen bonds [11]. An absorption band corresponds to vibrations of isolated OH-groups. Decreasing the intensity of these bands indicates reducing the fraction of OH-groups in a result of substitution by carboxymethyl in case of substitution of the native starch by carboxymethyl starch. The bands at 2930 cm^{-1} and 2878 cm^{-1} refer to stretching vibrations of C-H bonds. During starch carboxymethylation, primary and secondary OH-groups of starch are subjected to esterification, and as a result glycoside bonds are partially destroyed, it leads to decrease in molecular weight of starch. It is proved by a new peak at 1700-1750 cm^{-1} showing an ester group formation. the bands about 1600 cm^{-1} corresponding to carboxymethyl ether groups are registered

At spectra of the substituted samples. In 1200-1500 cm-1 intra deformation vibrations of -CH2, -CH, -COH groups are observed and it is sensitive with respect to a change in conformation of -CH2OH group. In particular, an absorption band at 1423 cm-1 shows scissoring vibrations of –CH2 group. An absorption band about 1082 cm-1 corresponds to valence vibrations -OCH-O-CH2 groups. Low-intensity bands in zone 500-700 cm-1 correspond to out-of-plane vibrations of OH-group of glucopyranose ring.

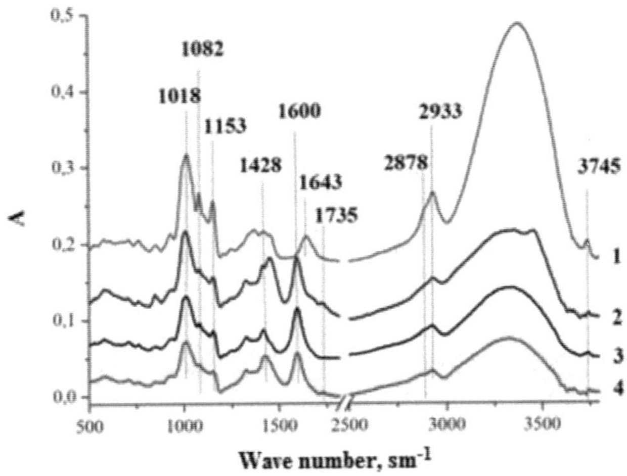

Figure 6. IR -spectra of the samples of the native potato starch (1), carboxymethyl starch in butanol-2 (2), carboxymethyl starch in isopropanol (3), carboxymethyl starch in water (4).

4. Conclusion

To sum up, during the experiment described in the paper:

1. the samples of carboxymethyl starch have been obtained by suspension procedure in organic solvents and water for drilling mud;
2. the influence of synthesis condition of carboxymethyl starch on fluid loss and rheological properties of fresh drilling mud and salt mud has been studied;
3. it is shown that the highest viscosity of drilling mud is observed for carboxymethyl starch obtained in water;
4. the change in infra-red spectra of the carboxymethyl starch samples in carboxymethylation has been evaluated.

Acknowledgment

This work was supported by the Federal Target Program «Research and development on priority directions of scientific-technological complex of Russia for 2014-2020» (RFMEFI57815X0119)

References

[1] Kryazhev V N, Romanov V V, Shirokov V A 2010 *Khimija Rastitelnogo Syrja* Vol. **1** pp. 5.
[2] Grey J R, Darly G S 1985 Composition and properties of drilling agents (flushing fluids) *Nedra M.* pp. 509.
[3] Bhandari P N, Hanna M A 2011 *Industrial & Engineering Chemistry Research* Vol. **50** pp. 12784.
[4] Villwock V K, BeMiller J N 2005 *Starch/Stärke* Vol. **57** pp. 281.
[5] Spychaj T, Wilpiszewska K, Zdanowicz M 2013 *Starch/Stärke* Vol. **65** pp. 22.
[6] Kaur L, Singh J, Singh N 2006 *Journal of the Science of Food and Agriculture* Vol. **86** pp. 1945.
[7] Uslu M-K, Polat S Vol. **87** pp. 1994.
[8] Minaev K, Martynova D, Knyazev A, Zaharov A, Shenderova I 2014 *IOP Conference Series: Earth and Environmental Science* 21, 012035.
[9] Kittipongpatana N, Janta S, Kittipongpatana O 2011 *Pakistan Journal of Pharmaceutical Sciences* Vol. **24** pp. 415.
[10] Bhattacharyya D, Singhal R S, Kulkarni P R 1995 *Carbohydrate Polymers* Vol. **21247** pp. 253.
[11] Joshi G, Naithani S, Varshney V K, Bisht S S, Rana V, Gupta P K 2015 *Waste Management* Vol. **38** pp. 33.

Vertical section construction of wells at Kuyumbinsky oil field via percussive-rotary drilling with DTH hammer.

K V Buzanov, L N Nechaeva and O S Ulyanova

National Research Tomsk Polytechnic University, 30 Lenina Ave., Tomsk, 634050, Russia
E-mail: byzanovaltay@mail.ru

Abstract. The article presents the analysis of sequential use of mobile percussive-rotary drilling sets with DTH hammer and bottom-hole cleaning by foam mud in construction of vertical sections along with at Kuyumbinsky oil field. On the basis of the analysis, an engineering solution is proposed to prevent disastrous mud loss that is the key factor of efficiency in implementation of resource-saving technologies.

1. Introduction

The analysis of the current situation in Russian oil and gas sector demonstrates a considerable shift in its development towards the regions of Eastern Siberia and Yakutia. During 2006-2014, JSC Rosneft discovered six fields at the license blocks of Irkutsk Region: 2 oil-gas condensate fields (Severo-Danilovskoye and Savostyanov fields); 1 gas-oil field (Lisovsky field); 2 oil fields (Sanarskoye and Mazur fields); 1 gas field (Yuzhno-Danilovskoye field).

On January 1, 2015 the total reserves of the above-mentioned fields amount to 477 million tons of oil and condensate and 192 billion m^3 of gas [1]. At present, Irkutsk and Yakutia gas production centers, with PJSC Gazprom as the main investor, are being built. To ensure the efficiency of the implemented projects, particularly at times of difficult macroeconomic situation, there is an urgent need to use resource-saving technologies [2].

One of the perspective engineering solutions for oil and gas drilling with mud loss is a sequential use of mobile percussive-rotary drilling sets with fixed drilling rigs and drilling mud circulation.

2. Material and Methods

The analysis of modern experience, including foreign (the USA, China, Canada), has shown high technological efficiency of applying air, nitrogen, fog, foam and gas-liquid muds in similar geological conditions [3].

In 1999-2015, 963 wells were drilled in China using gaseous fluids, including 676 wells in Szechuan, in partnership with experts from Chuanqing Drilling Engineering Company (CCDC) [4].

Bottom-hole cleaning with gaseous muds was applied at various depths. In particular, drilling of 508 mm, 426mm, and 324mm sections with percussive-rotary drilling was considered the most successful in Shangluo field with the casing depths of 60m, 350m and 600m respectively. Depending on the size of water inflow and reservoir properties it is advisable to drill 245mm and 168mm sections to the depths of 1,450m and 2,800m using foam, gas-liquid mud or aerated mud.

In 2014, LLC Intes initiated pilot works on the construction of vertical 426mm and 324mm sections along with down-the-hole drilling of water wells and bottom-hole cleaning with foam mud at Kuyumbinsky oil field (Krasnoyarsk Region).

The chosen design and engineering solution was caused by specific characteristics of the geological structure, namely incompatibility of drilling conditions within the field marked by disastrous mud loss in the interval ranging from 6 to 540 meters [5].

Investigating drilling conditions incompatibility alongside with lost circulation and factors contributing to their origin, as well as nature of existence and statistics of elimination, allowed experts of LLC Intes to make the following conclusions:

1. Disastrous mud loss was observed in the gross interval of the Evenkiiskaya Suite (6-540 meters) characterized by fractured dolomites (75%), laminated marls and marlstones (15% and 10% respectively) with high alteration and heterogeneity of rocks through the section.
2. Well deepening accompanied by lost circulation begins at the required capacity of mud pumps (36 hp) thus ensuring high-quality bore-hole cleaning and stability of PDM motors DRU-240.

3. Application of various wall packing materials and fractions, profile packers, reduction of the drilling mud specific weight, as well as introduction of visco-elastic mud did not bring any positive results within the drilling interval under study. Only repetitive cement plugs are productive (on average 3-5 per interval).
4. Intensive mud loss while deepening a well with surface rotor drive lead to drilling rate limits of up to 2-3 m/h, including due to presence of hard streaks. In case of DDM drilling it is necessary to provide bottom-hole continuous supply of industrial water, taking into account 50 m^3/hour absorption [5].

Development of the individual drilling program for vertical intervals of wells of Kuyumbinsky oil field with the use of DTH hammer became one of the key measures in design and implementation of the aforesaid technology: the analysis of drilling patterns for studied sections, justification of the chosen equipment and development of recommended drilling parameters.

Geological structure of the Evenkiiskaya Suite determines the presence of well walls taluses and collapses in poorly cemented shattered terrigenous deposits along with mud loss. For design purposes it was decided to consider 35 meters as the instability boundary of terrigenous deposits. Besides, the statistical analysis allowed the specialists to reveal common lost circulation patterns and sharp drilling rate decrease within the range of 60-80m in depth. Water inflow of up to 12 m^3/h observed within the studied interval was also taken into account.

Relevance and novelty of engineering solutions that ensure borehole stability, maintenance of the required drilling rate and implementation of the most suitable well construction program are caused by the construction of 426mm vertical sections 35m in depth and 324mm conductors to the depth of 550m by applying the mobile drilling complexes with DTH hammers.

Practical feasibility of the project challenged engineers and researchers to use tools, which were not applied earlier in Russian oil and gas industry, to solve the problem of mud loss, drilling rate, and optimization of well deepening in geological conditions of oil fields within Eastern Siberia.

The drilling rig "COOPER 550 LTO" was chosen as a plant capacity stock unit aimed to build emergency intervals for vertical sections and conductors. The loading capacity of the rig amounted to 106.5 tons that ensured lowering the 324mm casing to the depth of 550 meters [6].

In order to pass through quaternary deposits it was planned to drill the first ten meters using a 520mm drill bucket KFB-K. In fact, drill bucket footage made 1.2 meters at a time because it required further lifting to clean the drill cuttings. In doing so, the following drilling mode was chosen: axial loading - 1 ton, rotation frequency – 60 rpm, torque – 0.5 kN·m. Table 1 shows the bottom-hole assembly in drill bucket drilling.

Table 1. Bottom-hole assembly within 0-10 m.

Item	Description	Length, m
1	Drill bucket KFB-K 520 No.2606	2.34
2	Drill collar UBT-203 No.05	1.5
3	Drilling tool substitute 3-117x3-121L	0.4
4	Kelly VBT-112	11.4

When reaching 10m bottom-hole and depth of 35m the design implied drilling with a crushing-chipping tool from surface rotor drive with further air-blast cleaning of a well. Drilling without an air hammer in this interval was justified by substantial risks of well walls taluses and collapses. To ensure air-blast cleaning the drilling rig was equipped with three SULLAIR Combo 1150XHH/1350XHDL compressors with 38.2 m^3/min rated capacity and 2.4 MPa developed pressure. The designed bottom-hole assembly is indicated in Table 2.

Table 2. Bottom-hole assembly for surface casing drilling

Item	Description	Length, m
1	Drill bit EBXT1GS III 508 mm s\n 11767592	0.7
2	Drilling tool substitute M 3-177/M 3-152 No.22707	0.56
3	Stabilizer III 444 mm No.512-10	2.28
4	Drilling tool substitute H 3-152/M 3-mm No.104	11.4
5	Drill collar UBT-203 mm	4.5
6	Drilling tool substitute 3-117x3-121L	0.4
7	Kelly VBT-112 mm	11.4

Drilling mode parameters: axial loading - 2 tons, rotation frequency – 40 rpm, torque – 0.5 kN·m.

The air hammer and drill bit selection procedure offered by the Halco Company laid the basis for developing a program section containing technical and engineering solutions for drilling within the interval of 324mm (from 35 to 550 meters) [7, 8]. Figure 1 shows the drill bit selection pattern.

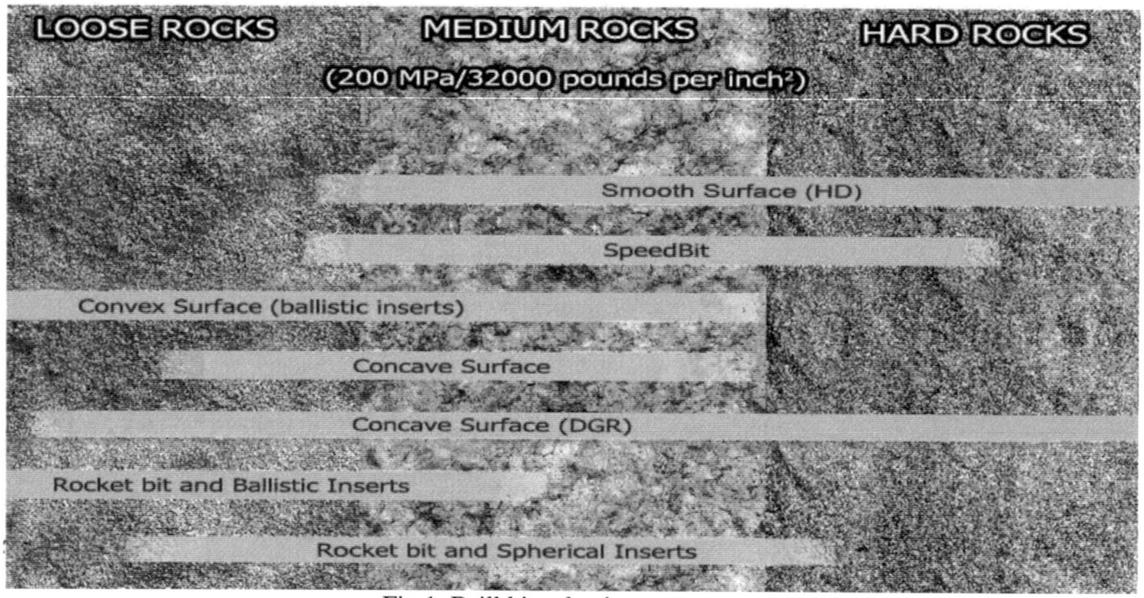

Fig.1. Drill bit selection pattern

A concave drill bit without double gauge teeth (Concave) was selected for standard operating conditions with the maximum control of a wellbore vertical position for rocks of the Evenkiiskaya Suite, mainly for medium rocks. Spherical substrates ensuring the best operational characteristics and length of service at all designed modes recommended by the Halco were selected for drilling extremely hard and abrasive rocks. The diameter of the drill bit was 381mm.

Two fundamental requirements were satisfied while following Halco air hammer selection methodology: the required speed of rock lifting and the maximum similarity of the outer diameter with the diameter of the well. Tables 3 and 4 show the highest lifting speed achieved when applying the air hammer Mach 122/132/142 [2].

Table 3. Air flow rate for Halco hammers (m^3/min.)

Hammer type	0.7 MPa	1.05 MPa	1.2 MPa	1.4 MPa	1.7 MPa	2.4 MPa
Dominator 100	1.6	2.6	3.0	3.5	4.5	-
Mach 20	2	3.8	4.8	5.8	7.6	-
Mach 303	2.8	4.4	5.0	5.9	7.3	10.0
Dart 350	2.5	4.7	5.6	7.0	9.3	13.8
Dominator 350	2.3	4.4	5.2	5.5	8.5	12.5
NT4	4.3	5.5	8.8	11.0	13.2	15.5

Mach 44	4.8	7.0	8.0	9.5	12.0	17.2
Dominator 400/ S-Dominator 400	5.2	7.5	8.5	9.9	12.2	17
NT5/NT5 HD	5.6	9.4	13.2	17.0	20.8	24.6
NT5E	5.6	11.2	15.8	20.4	25.0	29.5
Mach 50	5.7	7.2	8.2	11.0	14.9	23.4
Dominator 500/ S- Dominator 500/ S- Dominator 550	5.5	8.5	9.9	12	15.6	22.65
NT6/NT6HD	5.8	10.8	14,.8	18.8	22.8	26.8
NT6E	8.8	13.5	18.2	22.9	27.5	32.2
Dominator 600/ S- Dominator 600/ S- Dominator 650	-	9.5	12	14.1	19.5	30
Mach 60	5.1	7.9	9.1	12.5	16.4	25.5
Dominator 750	9.9	15.7	18	21.5	27.3	38.5
Dominator 800	5.5	10.8	13.9	15.1	19.3	27.9
Dominator 880	10.9	16.5	18.9	22.4	28.1	39.5
Dominator 880 DeepWell	14.6	23.5	27.1	32.5	41.5	59.4
Dominator 850	11	16.8	19	22.5	28.2	39.5
Dominator 1000	15.6	24	32	40.3	48.5	56.5
Mach 120	-	-	24.35	28.3	35.4	48.1
Mach 122/132/142 S-Dcminaccr 1200	19.8	29.45	33.4	39.05	48.1	70.8

Table 4 illustrates the speed rates (m/min) per each cubic meter of air passing through the hammer for every match of the drilling pipe and the drill bit diameters.

Table 4. Speed rates for 1 m^3 passing through the hammer.

Bit gage (inches/mm)	Drilling pipe diameter							
	2 1/8" 54 mm	2 3/4" 70 mm	3" 76 mm	3 1/2" 89 mm	4" 102 mm	4 1/2" 114 mm	5" 127 mm	5 1/2" 140 mm
2 3/4"/70 mm	642							
3 1/3"/85 mm	295	548	880					
3 1/2"/90 mm	254	420	607					
3 15/16"/100 mm	180	249	304					
4 1/8"/105 mm		208	243	410				
4 1/4"/108 mm		188	218	340				
4 1/3"/110		177	201	305	751			

mm								
4 1/2"/115 mm		153	171	240	451			
4 3/4"/120 mm		134	148	196	319	907		
5"/127 mm		113	123	155	222	407		
5 1/8"/130 mm		106	114	142	196	326		
5 1/2"/140 mm		87	92	109	138	193	367	
6"/152 mm		70	73	83	99	125	179	
6 1/8"/156 mm		66	69	78	92	114	158	
6 1/4"/159 mm		63	66	74	86	105	140	
6 1/2"/165 mm		57	59	66	76	89	115	
6 3/4"/171 mm		52	54	59	67	78	96	
7"/178 mm				54	60	69	82	105
7 1/2"/191 mm				45	49	55	63	90
7 7/8"/200 mm				40	43	47	53	62
8'/203 mm				38	41	45	51	58
8 1/2"/216 mm				33	35	38	42	47
8 7/8"/225 mm				30	32	34	37	41
9 5/8"/245 mm				24	26	27	29	32
10"/254 mm				22	24	25	26	28
11"/279 mm				18	19	20	21	22
11 7/8"/300 mm				15	16	17	17	18
12 1/4"/311 mm				14	15	15	16	16
13"/330 mm						13	14	14
14 3/4"/375 mm						10	10	11
15"/381 mm						10	10	10
17 1/2"/445 mm						7	7	7
18 1/2"/470 mm						6	6	6
20"/508 mm						5	5	5
22"/559 mm						4	4	4

The required air flow rate for Mach 122 amounts to 70.8 m^3/min under 2.4 MPa air-gauge pressure. Similar diameters (drill bit – 381mm and drilling pipe – 114mm) require 10 m/min per each cubic meter of air passing through Mach 122. Thus, when using a 381mm drill bit, hammer Mach 122 and a 114 drilling pipe the lifting speed for an average particle taken from the bottom-hole makes 708 m/min or 11.8 m/s [2].

It should be noted that the required speed of rock lifting from the well bottom is one of the crucial and complex parameters to be defined during air-blast percussion-rotary drilling. B.B. Kudryashov gave a sound assessment of this parameter, its dependence on the rock type, form and size of cuttings [6].

The outer diameter of Mach 122 makes 273mm, which is the closest to the diameter of wells, according to Halco dimensions range. Bottom-hole assembly for conductor drilling at Kuyumbinsky oil field is presented in Table 5.

Table 5. Bottom-hole assembly for conductor drilling

Item	Description	Length, m
1	Drill bit 381 mm CONCAVE SD 12 BIT No.345732	0.38
2	Hammer MACH 122 No.1100	6.6
3	УБТ-229 mm	9.94
4	Drilling tool substitute H 3-171/M 3-152	0.46
5	Stabilizer OD-374 mm	1.5
6	Drilling tool substitute H 3-152/M 3-133	0.36
7	Heavy-weight drill pipe SBT-114 mm	as needed
8	Check valve	0.5
9	Kelly VBT-112	11.4

Drilling mode: axial loading – 3 tons, rotation frequency – 20 rpm, air-gauge pressure – 1.7 MPa and air flow rate – 76.4 m^3/min.

The implementation of the developed technical and engineering solutions demonstrated the following results. At 118m bottom-hole and 100 m/h drilling rate in well No. 401 there was a sticky hole situation, which resulted in the losing of drill string mobility.

It became possible to eliminate the emergency only by water injection. To ensure artificial footage constraints and prevent the risk of emergency, further drilling in advance was carried out by MDR rotor drive without a hammer and the drill bit was replaced by a DSD tool. Experts of LLC Intes decided to clean the bottom-hole with industrial water. This entailed lost circulation in well No.402 when passing through 10-81m interval and forced RIH operations of dual selective packer set with bypass and safety valves with subsequent DRILL HLUG hydrogel pumping. A similar hazardous situation occurred when drilling all four subsequent wells.

3. Results and discussion

To avoid the above problems we propose the following methodology: during the analysis and scientific justification of the new design and engineering approach to the solution of the complex problem of failure-free drilling of the specified intervals it is necessary to perform a complete set of calculations based on fundamental laws of rock fracture physics and aerodynamics with regard to well cleaning with gaseous fluids.

In particular, a real success of introducing the technology of surface casing drilling at Kuyumbinsky oil field will be achieved by applying the methods aimed at determining the desired pumping speed of drilled solids and calculating the air supply necessary for high-quality bore-hole cleaning developed by B.B. Kudryashov. Some aspects of practical application of such methods are described in other research works [10 - 12].

The expected recalculation of the flushing mode parameters will result in additional equipment of a mobile drilling rig with 3.5 MPa developed pressure compressors and booster stations to compensate back pressure from water inflow, as well as change of the drilling mode parameters. Modern mobile drilling rigs manufactured by Schramm, Atlas Copco, KERUI are equipped with top power spinner which rotates a column more effectively and prevents substrates from hitting the earlier chipped bottom-hole sections, and as a result, avoids drilling rate drop.

4. Conclusion

Nevertheless, LLC Intes initiative can be considered a "technological breakthrough" in solving problems of well constructions in Eastern Siberia and the Republic of Sakha (Yakutia) due to its production and engineering advancement. Viability, feasibility and proved applicability of this technology can contribute to its further thorough study.

Acknowledgement

The authors express their sincere gratitude to Andrey V. Petrov, the Director General of LLC Intes, for providing data for the analysis under this research.

References

[1] Official site of CJSC Rosneft [Electronic resource] – URL: http://www.rosneft.ru/Upstream/Exploration/easternsiberia/evenkia/ free, reference date 13.03. 2016
[2] Brilin V I, Ulyanova O S 2015 *IOP Conference Series: Earth and Environmental Science.* Well extension in fractured and broken rocks by wire-line tool (SSK-59). Vol. **24** pp. 012006.
[3] DTH drilling from A to Z 2007 (Western Australia: TEREX Inc. – Malaga) 74.
[4] Zhao Z, Gao D, Zheng D 2010 *SPE Drilling engineering.* Mechanism of well deviation in air drilling and it's control 130201
[5] Individual program of vertical section drilling in well No.401 of Kuyumbinsky oil field 2014 *LLC Intes* (Nizhnevartovsk) 17
[6] Operating Manual for COOPER 550 LTO 1995 (USA: COOPER Corporation) 25
[7] Zhu H, Lin Y, Meng Y, Zhao S, Liu D, Luo F 2010 *SPE Drilling engineering.* Influence of relevant parameters on hole cleaning and pipe string erosion in air drilling 126515.
[8] Trade catalogue – equipment for air hammer drilling 2007 *Atlas Copco Secoroc AB* (Sweden)
[9] Malloy K P, Medley G H, Stone R C 2007 *IADC/SPE Drilling engineering.* Air Drilling in the presence of hydrocarbons: a time for pause 108357
[10] Buzanov K V, Borisov K I, Lavrov A A 2015 *Inzhener-Neftyannik: Scientific and technical journal (Moscow).* Justification and calculation of air drilling parameters for surface casing drilling at Dulisminsky oil-gas condensate field. Vol. **3** pp. 30 – 38.
[11] Buzanov K V, Boyarko. Yu L, Ulyanova O S 2015 *IOP Conference Series: Earth and Environmental Science.* Development of engineering solutions for air drilling at Dulisminskoye oilfild, Irkutsk oblast. Vol. **27** pp. 012049.
[12] Maranuk C, Rodrigues A, Trapasso J, Watson J 2014 *SPE Drilling engineering.* Unique system for underbalanced drilling using air in the Marcellus Shale. Vol. **17** 1024-MS.

IOP Conf. Series: Earth and Environmental Science **43** (2016) 012073 doi:10.1088/1755-1315/43/1/012073

Methods to ensure optimal off-bottom and drill bit distance under pellet impact drilling

A V Kovalyov, Ye D Isaev, A R Vagapov, V V Urnish and O S Ulyanova

National Research Tomsk Polytechnic University, 30 Lenina Ave., Tomsk, 634050, Russia

E-mail: arteka011287@mail.ru

Abstract. The paper describes pellet impact drilling which could be used to increase the drilling speed and the rate of penetration when drilling hard rock for various purposes. Pellet impact drilling implies rock destruction by metal pellets with high kinetic energy in the immediate vicinity of the earth formation encountered. The pellets are circulated in the bottom hole by a high velocity fluid jet, which is the principle component of the ejector pellet impact drill bit. The paper presents the survey of methods ensuring an optimal off-bottom and a drill bit distance. The analysis of methods shows that the issue is topical and requires further research.

1. Introduction

Recently the trend to drill hard rocks has intensified; therefore, new drilling methods are urgent to increase the drilling speed, as well as the rate of penetration. One of the most promising drilling methods is pellet impact drilling currently investigated by the Department of Well Drilling, Tomsk Polytechnic University.

Pellet impact drilling implies rock destruction by metal pellets in the immediate vicinity of the bottom hole encountered. The pellets are circulated in the bottom hole by a high velocity fluid jet, which is the principle component of the ejector pellet impact drill bit. According to the previous experiments and studies [1, 2, 3], the method ensures higher drilling speed, the rate of penetration, and speed per run.

As in the case with conventional drilling methods of drilling, the efficiency of pellet impact drilling depends on many important parameters; one of them is an off-bottom and a drill bit distance. Ensuring an optimal off-bottom and drill bit distance is a complicated issue, which was first studied in the middle of the previous century. Since then different ways to solve this task have been suggested.

2. Review and analysis of the methods ensuring an optimal off-bottom and a drill bit distance

The first method suggested by A.B. Uvakov [1, 2] is based on calculating the average rate of penetration. If it is known, the ejector pellet impact drill bit is run to bottom hole at estimated distance per equal period of time. However, the lack of well logging data may significantly restrict the use of the method. Also, it is difficult to apply it for drilling deep wells.

The second method to ensure an optimal off-bottom and a drill bit distance is the jostle of a drill bit, when it is run to a bottom hole and then lifted up at the distance required and lifted then at specific time [3]. It should be noted that a drill bit is subjected to deformation under the influence of axial load when it contacts with bottom hole. Besides, hydraulic standoff is possible due to shutoff of drill bit's inner valves.

In 1952 L. W. Ledgerwood suggested a new version of the above-mentioned construction [4]. It is based on application of a mechanical feeler 7 with an enlargement 8 (fig. 1). When a critical distance to the bottom hole controlled by the length of a mechanical feeler 7 is exceed, an enlargement 8 partially blocks a primary nozzle 2 enhancing the pressure in a mud system that is a signal for the ejector pellet impact drill bit to be run. However, there are some disadvantages of this version, firstly, it is unreliable due to the high risk to break the feeler, and secondly, it fails to destroy the rock immediately under the feeler of the drill bit. In addition, it is required to use the pellets of smaller diameter, which decreases efficiency of pellet impact drilling.

There is a method to ensure an optimal off-bottom and a drill bit distance via a pellet meter [5]. It is based on the principle of pellets circulation (fig.2), when circulating in a mixing chamber 2 each pellet 8 changes either inductance or capacity of a sensor 5 that is registered by a meter 6. In this case, an optimal off-bottom and a drill bit distance is set by a maximum pellet supply. This version involves some disadvantages too. Firstly, it is necessary to design an additional source to supply and power the pellets that complicates a flow chart of pellet impact drilling. Secondly, making the construction more complicated in general, the version decreased durability and wear-resistance of a drill bit due to wall thickness reduction reducing of walls thickness.

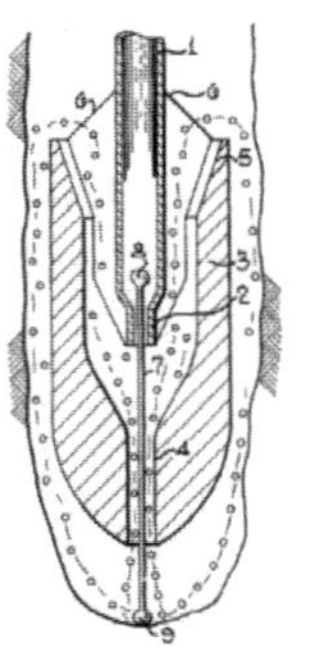

Figure 1.
A device with a mechanical feeler:
1 – a tubular support member ;
2 – a primary nozzle;
3 – a sleeve;
4 – a secondary nozzle;
5 – a hopper;
6 – a web plate element; 7 – a mechanical feeler;
8 – an enlargement;
9 – a spherical enlargement.

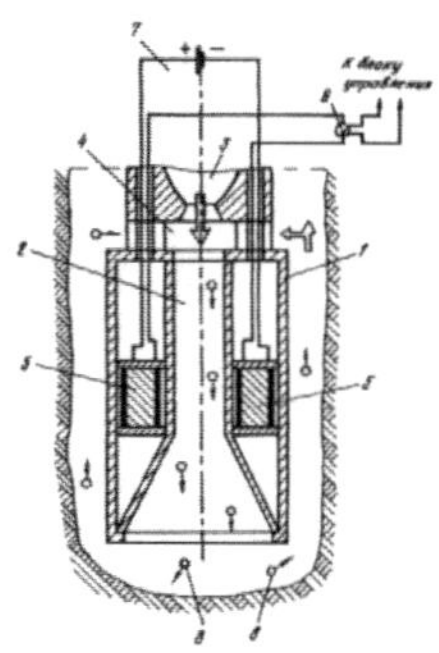

Figure 2.
A device with a pellet meter:
1 – a bit;
2 – a mixing chamber;
3 – a nozzle;
4 – process windows;
5 – a sensor;
6 –a meter;
7 – power source;
8 –pellets.

Other versions have been designed to ensure an optimal off-bottom and a drill bit distance via supports. These designs imply to maintain a fluid jet rigidly in some distance from drill bits destroying a peripheral part (fig. 3) or a central part (fig. 4) of bottom hole via rotary drilling.

However, the first version has its limitations. Firstly, it is difficult to synchronize drill bit operations performed via both rotary drilling and pellet impact drilling [6]. If a central part of the bottom hole is destroyed faster or slower than a peripheral part, it is difficult to control off-bottom and a drill bit distance. Secondly, it is necessary for a drill bit to rotate, which is in contrast to pellet impact drilling, where there is no need for rotation. In addition to above-mentioned limitations, a serious disadvantage of the second version is low wear-resistance because of pellet impact upon the supports.

The third design comprises a tubular support member (fig. 6). The disadvantages are as follows: the ejector pellet impact drill bit should be rotated; it is difficult to synchronize drill bit operations; and pellets of small diameter are to be used.

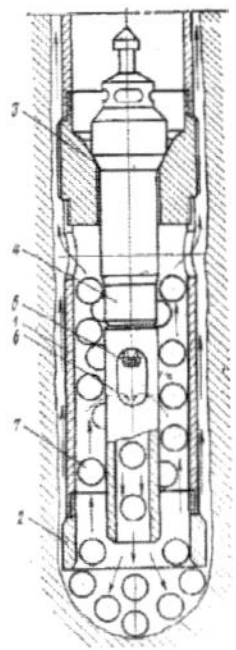

Figure 3.
A pellet impact drill (A B Uvakov and V V Shtrasser design):
1 – a body; 2 – rock-breaking foot; 3 – a cradle; 4 – a fluid jet; 5 – a nozzle; 6 – a mixing chamber; 7 – pellets; 8 – end.

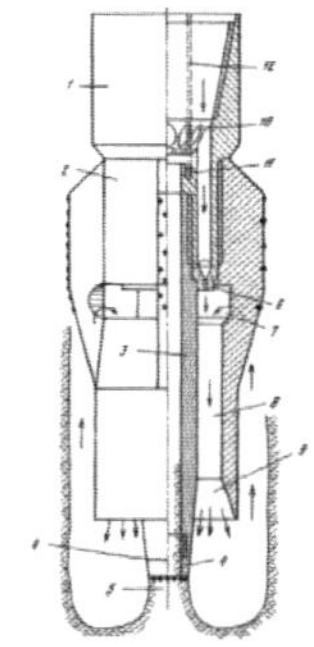

Figure 4.
A pellet impact drill (M M Maylibaev design):
1 – a body; 2 – a ribbed checker; 3 – an inner core tube; 4 – a bore bit; 5 – core; 6 – a nozzle; 7 – ; suction ports 8 – suction port ; 9 – a cone; 10 – canals; 11 – a cradle; 12 – a tube with a core-cutter.

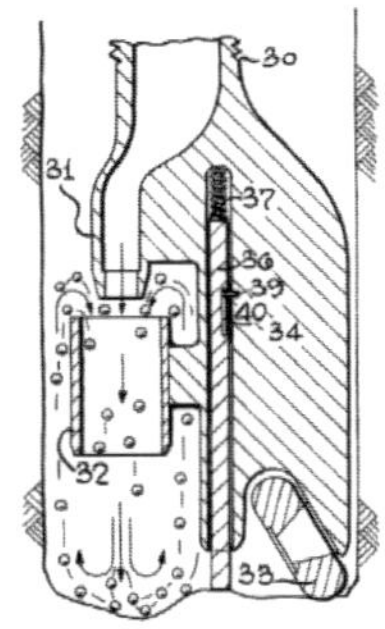

Fiqure 5.
A pellet impact drill (P Williams design):
30 – a tread; 31 – a primary jet nozzle; 32 – a secondary jet nozzle; 33 – a wheel ; 34 – a frame; 36 – a guard deflector; 37 – a spring; 39 – a pin; 40 – a slot.

Other method to ensure required off-bottom distance is based on changing the hydrodynamic conditions in the bottom hole. Figure 6 illustrates a modification of the device. In a position I the pressure at A is less than at B, at point II pressure A equals pressure B, and in a position III the fluid pressure at A is greater than at B. This principle of pressure variation was implemented in McNatt's embodiment (fig.7). [9]. The limitations are complicated design of the ej.p.im.d.b. and costs for adjustment.

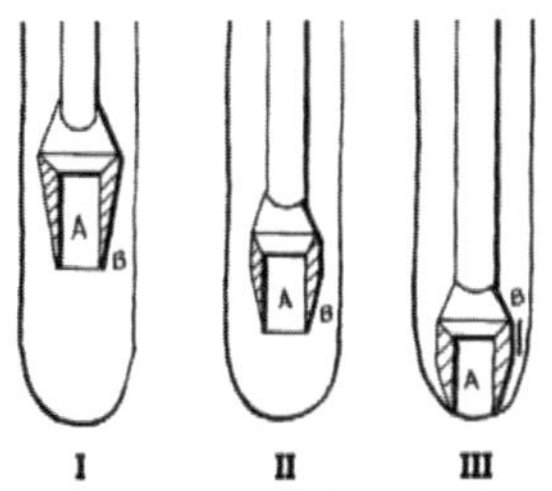

Figure 6.
The principle of the hydrodynamic conditions in the bottom hole

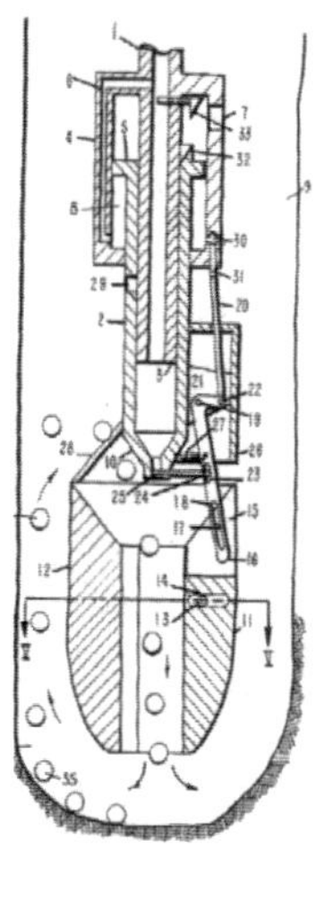

Figure 7. A pellet impact drill (McNatt's design):
1 – a drill string; 2 – a primary nozzle barrel; 3 – a tubular support member; 4 – a cylinder; 5 – a piston; 6 – a channel; 7 – ports; 8 – an annular chamber; 9 – bore hole annulus; 10 – a primary nozzle; 11, 12 – pieces of the secondary nozzle assembly; 13 – pins; 14 – slots; 15 – a slot; 16 – a bell crank; 17 – a pin; 18 – a slot; 19 – a pin; 20 – a connecting link; 21 – web; 22 – a nozzle barrel; 23 – a slot; 24 – a pin; 25 – a member; 26 – a body; 27 – a spring; 28 – web members; 29 – a port; 30 – a bleed port; 31 – a slide valve; 32 – a cam; 33 – a cam; 34 – a slide valve; 35 – a port; 36, 37 – ports; 39 – a controlling port; 40 – a spring; 41 – a plate; 50 – a controlling port; 51 – catalytic fines.

3. Conclusion

The review has revealed a number of limitations in the methods for an optimal off-bottom position. Therefore, the issue is still topical and requires further research.

References

[1] Uvakov A B 1965 *Ph.D. thesis in Engineering Science*. Research and design of pellet impact drilling the directional exploration wells. pp.120.

[2] Uvakov A B 1969 *Pellet impact drilling.* Moscow. pp. 207.
[3] Kovalev A V 2015 *Ph.D. thesis in Engineering Science.* Theoretical and experimental study of pellet impact drilling engineering processes. pp. 143.
[4] Ledgerwood L W 1955 Hydraulic standoff control for pellet impact drilling *Patent US 2724574* [Electronic resource] URL: http://www.freepatentsonline.com/2724574.html
[5] Derbenev L S, Bogolyubov A A, Kapustin A A, Karkashadze G G, YAnchenko G A 1981 Erosion jet drilling and its drilling tool (in Russian) *Patent A. C. № 870705, E 21 C 37/16, E 21 C 21/00, E 21 B 7/18, № 2798122/22-03*
[6] Uvakov A B, Shtrasser V V 1974 Pellet impact drill (in Russian) Patent *A. c. № 417599, E 21 B 7/18, № 1451266/22-3*
[7] Maylibaev M M 1983 A pellet impact drill (in Russian) Patent *A. c. № 1002498, E 21 B 7/18, № 3278854/22-03*
[8] Williams P S 1959 Pellet impact drilling apparatus *Patent US 2868509* [Electronic resource] URL https://www.lens.org/images/patent/US/2868509/A/US_2868509_A.pdf
[9] McNatt E M 1955 Controlling off-bottom position of pellet impact drill *Patent US 2728557* [Electronic resource] URL http://www.google.com/patents/US2728557

Designing gravel pack for uranium ISL wells

A A Ber[1], K M Minaev[1], L M Ber[1], Ye D Isaev[1] and O S Ulyanova[1]

[1]Tomsk Polytechnic University, 30 Lenina prospect, 634050, Tomsk, Russia

e-mail: beraa@tpu.ru

Abstract. The paper describes the improvement of gravel packing technique applied for the production wells. The authors have suggested new design of gravel pack for gravel packing of productive formations. The issue is currently topical because gravel packing at drillhole ISL is less time- and money-consuming. The subject of the research is gravel pack design and content. The purpose defined by the authors is to design the gravel pack and to suggest the composition of gravel cement agent. As a result of the research, the authors have described different designs of the gravel pack, its optimal shape, as well as a choice and justification of cement agents, a hold cover of the gravel pack, and suggested the methods of experimental research.

1. Introduction

Drillhole ISL permits poor uranium ores to be used more efficiently. In this case, it is no need for actual mining, additional ore breaking and reduction, radioactive ore pile stocking and waste storage. Drillhole ISL efficiency depends directly on a pack in a production interval of the bottomhole. The key component of the production well construction for uranium ore drillhole ISL is a pack, irrespectively of well type (injection well or production well), because it influences well integrity and, consequently, the efficiency of drillhole ISL.

The requirements to packs for production wells are as follows:

- high aggressive reagent resistance;
- high mechanical resistance;
- long-term performance within well production;
- open ratio enough for required amount of fluid to pass at low inlet velocity and hydraulic resistance;
- sanding up a pack (acceptable for pumping tests only).

Gravel packs meeting the requirements mentioned-above have considerable advantages over other packs, such as consistency of operation, higher parameters of intake capacity and flow rate. They are particular efficient when there is fine-grained sand in the productive formation. Using gravel packs at drillhole ISL increases flow rate by 30-40 % compared with that when using slotted liners or gauze filters. As a result, there is a twofold increase in flowing time, as well as in round trip equipment capacity [1].

There are two main designs of gravel packs – open hole gravel packs and cased hole gravel packs. The first design - open hole gravel pack, which can be of different types, is inefficient and expensive, tends to colmatage easily, and is difficult to be placed in a well. Besides, after open hole gravel packing, additional packing by a gravitational method is necessary, which increases the costs.

For the second design – case hole gravel pack – the density of gravel pack is recommended to be 1.9 ton/m^3. The methods to place cased hole gravel pack are as follows:

- gravitational method: gravel is intentionally placed into the annular space under the action of gravity (well depth is no more than 100 m);
- hindered settling of gravel by pumping the fluid into the botomhole (well depth is no more than 250 - 300 m).

However, it is difficult to ensure efficient gravel packing by cased hole gravel packs. In addition, it is necessary to increase the diameter of a well and to expand a gravel packer to secure required gravel pack thickness, causing time and financial expenses. To reduce costs and time at gravel packing at drillhole ISL, new technique of gravel packing of producive formations is a critical issue.

The purpose of the paper is to design a gravel pack for uranium ISL wells. The authors have designed a new gravel packing technique. They have suggested using the gravel pack in a form of a solid composition of gravel and cement agents. The gravel pack is placed with a production string into a well, where cement agents are destroyed under the action of acids and permit the gravel pack to evenly fill annular space.

The advantages of this gravel pack design are as follows:

- efficiency of the gravel pack placement in the productive formation since it is installed simultaneously with a pump column and gravel pack equipment;
- low price and simplicity of the construction;
- reducing labor costs of well construction.

The subject of the research is designing the gravel pack and formulation of cement agents. The purposes are as follows:

- to design the gravel pack with considerable durability and low hydraulic resistance over the round trips;
- to prepare the composition of cement agents.

2. Methods and materials

The methods of sample preparation were as follows. To obtain the gravel pack, the gravel-cement agents' composition is prepared and put into metal and plastic frames. The metal rod was installed in a frame, imitating gravel pack equipment through a hole. Then gravel composition was tamped and the samples were put into the drying cabinet ShS-80-01, SNOL under specified temperature over the specified time interval.

Physical and mechanical properties of the gravel pack are defined by the following parameters: time and temperature in the drying cabinet, formulation and concentration of a cement agent, which are varied within the experiment. The sample was placed into a glass cylinder, imitating a well and filled with solvents of various concentrations. The experimental data were registered by a digital camcorder. The test parameters affecting the time of the gravel pack solution were as follows: cement agents' concentration n_{ca}, %; dryout time t_d, h; temperature, T; cement agents' concentration of a hold cover n_{hc}, %; and thickness of a hold cover, h_{hc}, mm.

3. Results and discussion

The gravel pack is a substance consisting of gravel, cement agent and (or) a hold cover, installed on gravel pack equipment, which dissolves within the specified time filling the annular space.

The requirements to the gravel pack are as follows:

- it should dissolve within 5 – 10 hours after its placement in a well;
- it should not be destroyed when running a casing;
- it should ensure minimum hydraulic resistance;
- it should be easily assembled with a production string;
- it should secure homogeneous gravel packing.

Taking into account hydraulic resistance, the following designs of gravel pack have been suggested (fig. 1.): solid cylinder, longitudinally ribbed cylinder, capsule, cylinder with longitudinal holes for fluid to flow, cylinder of changing diameter.

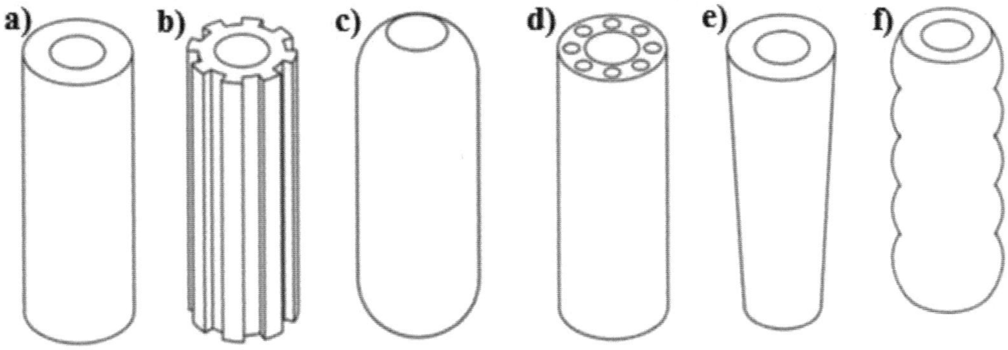

Figure 1 – The designs of gravel pack a) solid cylinder, b) longitudinally ribbed cylinder, c) capsule, d) cylinder with longitudinal holes for fluid to flow, f) and e) cylinders of changing diameter.

Ideal engineering solution should satisfy the following parameters: minimum outer diameter and length of the gravel pack to provide required amount of gravel to a gravel packer.

When cement agents have been identified and analyzed, the possible compositions of the gravel pack have been investigated within the experiment [2], such as:

- the gravel pack with a soluble hold cover;
- the gravel pack, consisted of gravel and the cement agent;
- the gravel pack with cement agents and a soluble hold cover.

4. Conclusion

To sum up, the following results have been obtained within the experiment:

- three designs of the gravel pack have been suggested: with cement agents, with a hold cover, as well as both in combination;
- it has been proved that to reduce the hydraulic resistance and ensure the effective volume of the gravel pack, it is necessary to ensure that the ratio of gravel pack diameter to a well diameter is about 0.8 – 0.9;
- the possible cement agents and a hold cover have been analyzed;
- the methods of experimental research have been suggested;
- the methods of increasing time for the gravel pack to dissolve have been proposed.

Acknowledgments

This work was supported by Federal Target Program «Research and development on priority directions of scientific-technological complex of Russia for 2014-2020» (RFMEFI57815X0119)

References

[1] Brilin V I 2010 Drilling of the uranium ore production wells by via drillhole ISL (Tomsk) p. 218 (in Russian).
[2] Kvesko N G, Chubik P S 2007 Methods and research tools (Tomsk) p. 124 (in Russian)

IOP Conf. Series: Earth and Environmental Science **43** (2016) 012075 doi:10.1088/1755-1315/43/1/012075

Development of small diameter pilot hole directional drilling for trenchless utility installation

A L Saruev, L A Saruev, S S Vasenin

National Research Tomsk Polytechnic University
30, Lenin Ave., 634050, Tomsk, Russia
E-mail: [1]saruev@tpu.ru, [2]saruevla@tpu.ru, [3]vasenin8@gmail.com

Abstract. The paper overviews trenchless utility installation techniques and prospects of further development of horizontal directional drilling technology to drill small diameter pilot holes. The improved design is suggested for the thread connection of drill pipes and hydraulic system to generate power pulses.

1. Introduction

Infrastructure and housing development led to the increase in construction of underground utility and transport systems. Trenchless installation of utility systems is a predominant technology in developed countries. The trenchless technology used for the underground utility systems is based on the formation of subsurface pilot holes drilled in soil with the possibility of further extension up to the required diameter.

So far, worldwide experience has been accumulated in laying pipelines and installing utility systems using both controlled and uncontrolled methods.

Uncontrolled method is based on horizontal directional penetration of pipelines or sheathing pipes due to static or impact loads. The end of a sheathing pipe can be either open or closed. Pipe pushing is one of the first and simple methods of trenchless technology. Originally, the pipe pushing method was mechanical. Later, the invention of hydraulic pipe pushing machines and vibratory hammers allowed improving it greatly. The former implies a high-pressure water-jet to washout soil before using a rock cutting tool; the latter uses spinning counter-weights to create vibration to the pile, which allows it to "cut" soil into the ground.

It should be noted that the pipe pushing can be performed using a pneumatic percussive machine, comprising a hollow casing in which a piston is in reciprocating motion; and a valve assembly through which compressed air is supplied to the piston face. Pipe pushing is limited to 500 mm maximum diameter and 100 mm length of the pipe.

Pipe ramming differs from the pipe pushing by directing the soil into the pipe interior instead of compacting it outside the pipe.

Controlled method of trenchless utility installation is performed by drilling machines and relates to horizontal directional drilling that includes several stages such as pilot hole drilling, hole reaming, and pipeline laying.

The main properties of drilling machines are the drag force, torque, and performance of reciprocating piston pump.

The drilling machines are divided into rotary, percussive, vibratory, and vibratory-rotary drilling machines.

The modern percussive drilling machines provide significant increase in their drilling capacity and drilling rate in severe conditions. The main parameters of the efficient installation of utilities in complicated urban conditions are the unit capacity and its dimensions. The unit capacity is important for the drivage in hard rock, while the unit dimensions influence its mobility and operating capacity in restrained urban conditions.

2. Thread connections of drill pipes

During the percussive drilling of small diameter pilot holes, the thread connections of drill pipes are subjected to the axial load, torque, and transient load. The information on loading effects on thread connections at the design stage, allows increasing their service life.

The practical experience shows that the break of thread connections during the percussive drilling occurs due to their imperfect design revealed under the transient loads.

The insufficient operating capacity of drill pipes and their connections leads to the increase in drill-steel production and consumption and results in great loss of time spent to replace the damaged pipes. Therefore, the pipe connection should be designed to increase the operating capacity of the drill string and the labour capacity while drilling holes.

During the power pulse propagation over the drill string, the strain waveform is changed and the strain wave energy is dissipated. The literature review shows that a safe operation of thread connection is associated, first of all, with the decrease in the strain wave energy when the power pulse travels through it. The strain wave energy is dissipated in thread connections rather than directed to the longitudinal movement of the drill string and rock cutting tool. It is induced by the multiple alternating strains, thread friction, and heating of thread connections [1, 2, 4].

Nowadays, internal upset drill-pipes and pipe-in-pipe system connection are used for directional drilling of a small diameter pilot hole (fig. 1).

In addition to the elastic forces occurring during the power pulse propagation through this type of thread connection, dissipative forces also arise, which requires additional energy.

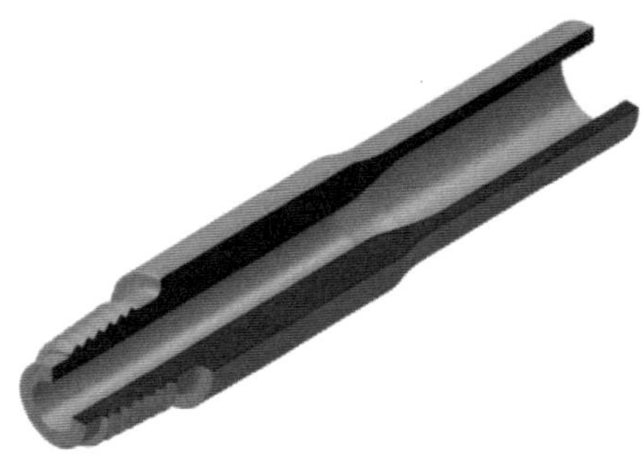

Figure 1. Pipe-in-pipe system connection.

While drilling, the thread connections are subjected to different loads that considerably complicate the analytical determination of dissipative forces induced by the dynamic loads. The quasistatic loading of thread connection is used to investigate the behavior of elastic and dissipative forces observed in thread connections.

The suggested design [3, 6, 7] of nipple thread connection shown in figure 2, is more efficient and allows improving the drill string reliability and cutting transport as well as ensuring the automation of making-up and breaking-out drill pipes.

The nipple connection comprises two drill pipes having a female straight thread and a nipple having a continuous male thread. The nipple has a collar with two annular grooves on its both sides. Collar grooves are cut on the internal side of the drill pipe ends. Non-threaded segment of one of the nipple ends is rigidly fixed in the drill pipe and has longitudinal cuts that divide the non-threaded segment into elastic plates. Elastic plates have collars with a bevel front face which are fixed in the annular groove after the achievement of strain tolerance in the tapered groove. Wrench flats are cut on another non-threaded segment of the nipple end.

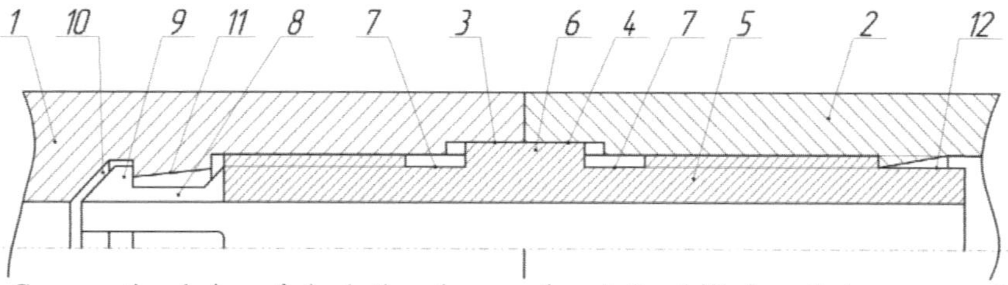

Figure 2. Cross-sectional view of nipple thread connection: 1, 2 – drill pipes; 3, 4 – grooves; 5 – nipple thread; 6 – collar; 7, 10 – annular groove; 8 – elastic plates; 9 – collar with bevel front face; 11 – tapered groove; 12 – wrench flat

The suggested design of the nipple connection [10] has the following advantages:

- Nipple locates inside two butt-coupling pipes providing the minimum energy loss at strain wave transmission;
- In strain wave propagating through the connection, pipe sections are compressed, while the nipple is unloaded from normal tensile and deformation stresses produced by the pre-tightening force;
- The collar and annular grooves cut in the nipple improve the connection durability at bending strain;
- The automation of making-up and breaking-out drill pipes is possible due to the nipple fixation in one of the drill pipes;
- The nipple connection hidden inside the drill pipe provides the consistency of the outer diameter of the drill string, which improves the cutting transport and eliminates the possible drill string jamming in the well.

The suggested nipple thread connection can be used in percussion drilling of medium-hard and harder rocks (f=6…14). The transmission rate of strain wave energy along the drill string increases due to the increase in contact stiffness between thread connections and decrease in the nipple stiffness.

3. Hydraulic system of pulse shaping

The literature review on power pulse generation by horizontal drilling machines reveals the availability of well-developed theoretical foundation of pneumatic equipment. It has facilitated the development and production of effective percussive drilling machines. It should be noted that pneumatic percussion mechanisms has lower efficiency in comparison with hydraulic or electric machines. Electric percussion mechanisms have a number of limitations that impair the invention of more powerful mechanisms. New technical decision is patented by Shadrina *et al.* [8]. It relates to the percussion mechanism that generates power pulses in a drill string. Hammerless hydromechanical system with no oil-pumping stations simplifies the flow chart of the drilling machine and considerably increases the efficiency of the mechanisms in comparison with all other hydraulic percussion mechanisms. However, the hydraulic power cylinder of the suggested percussion mechanism has an imperfect design, which has negative impact on the capacity and efficiency of the mechanism.

The improved design of percussion mechanism is suggested for the percussive drilling machine comprising the hydraulic pulsator connected with the drive; the fluid-pulse shape correcting device; hydraulic power cylinder with piston contacting with the drill string shank. The hydraulic pulsator is represented by cylinder and plunger, while the fluid-pulse shape correcting device is represented by a hollow elastic element with non-linear characteristic. It is filled with a fluid and characterized by a bellows installed in the hydraulic power cylinder and connected to the hydraulic pulsator by means of the hollow elastic element. Thereby, a closed cavity is formed filled with fluid. On one side, the bellows is rigidly fixed to the spring loaded piston, and on the other, with the spring-loaded inertial mass. The cross-sectional view of percussion mechanism is presented in figure 3.

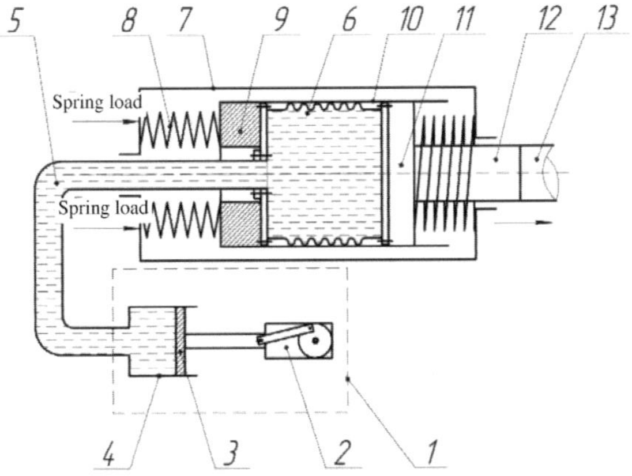

Figure 3. Cross-sectional view of percussion mechanism: *1* – hydraulic pulsator; *2* – drive; *3* – plunger; *4* – hydraulic power cylinder; *5* – elastic element; *6* – bellows; *7* – protective case; *8* – spring; *9* – inertial mass; *10* – end of hydraulic power cylinder; *11* – spring-loaded piston; *12* – shank; *13* – drill string

The hydraulic pulsator is connected with the drive and plunger located inside the hydraulic power cylinder. Its cavities are connected to the bellows by an elastic element and filled with fluid. The protective case is equipped with a spring supported by inertial mass. The inertial mass is fixed to the end of the hydraulic power cylinder connected to the bellows cavity. The bellows cavity is rigidly connected with the spring-loaded piston supported by a drill string shank.

In switching the drive of hydraulic pulsator, plunger starts a reciprocating motion in hydraulic power cylinder. The reciprocating motion produces the fluid pulses the shape of which is characterized by a gradual pulse rise. In turn, these fluid pulses induce forced oscillations of hydraulic fluid in the elastic element. The shaped fluid pulse travels then in the bellows and acts on the spring-loaded piston. Then it passes through the shank and achieves the drill bit having propagated over the drill string [9].

The pressure pulses, generated in the closed volume, propagate in hydraulic power cylinder through the piston in the capacity of the longitudinal elastic strain wave and transfer to the drill string shank. Further, the strain wave propagates at the velocity of sound in the material of the drill string and then travels to the rock cutting tool. Owing to the power pulses, the cutting tool intensifies the rock disintegration, thereby, increasing the drilling penetration rate.

It is worth noting that the suggested percussion mechanism also includes the automated control for the power pulses. In drilling soft rock, the amplitude of power pulses is not high, since it exhibits no resistance to the cutting tool. On the contrary, in hard rocks the amplitude of the power pulse generated by the plunger pair is high, which increases the power capacity of the percussion mechanism.

4. Conclusions

The equipment of drilling facilities with percussion mechanisms is the prospective way of engineering and technological development of small diameter pilot hole directional drilling used for trenchless utility installation. In its turn, the percussion mechanism equipment requires the improved design of the drill string, in particular, the pipe connections that provide the pulse propagation from the percussion mechanism to the rock cutting tool.

The early investigations of the dynamic processes occurring in a drill string during the percussive drilling of small diameter pilot holes have shown that nipple pipe connections have a range of advantages over the pipe-in-pipe or coupling connections. The suggested design of the nipple connection ensures the automation of making-up and breaking-out drill pipes due to the nipple retention in one pipe and its constrained movement inside the thread connection. It led to the decrease in the dynamic loads and stresses in the nipple during the pulse transfer from the thread connection to the cutting tool.

The suggested hydraulic system of pulse generating is a hammerless percussion mechanism that has higher efficiency than those with conventional hydraulic hammers. In the latter ones, the reciprocating motion of a hammer in hydraulic cylinder consumes most of the energy. The experiments showed that power pulses can be generated with a gradual rising edge and longer duration, which will ensure the efficient use of energy. This hydraulic system includes the automated control of the pulse amplitude depending on the rock resistance. With the increase of the rock hardness, the pulse amplitude will be automatically increased.

The implementation of the suggested design solutions patented by the Russian Federation will considerably increase the productivity and reliability of the equipment used for the horizontal directional drilling of small diameter pilot holes.

References

[1] Antonov I S 1999 *Thread connections under impact and cyclic loading.* DSc Abstract. Kurgan
[2] Vlasyuk V I 1999 *Conf. Problems of Scientific-Technical Progress in Hole Drilling.* Scientific problems of small diameter hole directional drilling technology
[3] Dvornikov L T, Puchinyan G S 1972 *Mining Journal.* Thread parameters of rope profile and load distribution by thread turns. Vol. **1**
[4] Rosikhin G B, Akhmetov M M 2004 *Mining Journal Moscow.* Production and operation of drill strings made of Swedish assortment steel. Vol. **5** pp. 89-90.
[5] Saruev A L, Saruev L A 2011 *Tomsk: TPU Publ.* Dynamic processes and stresses in thread connections of drill strings at percussion drilling.
[6] Saruev A L 2004 *Gosgortekhizdat. Sci. Conf. 'Problems of Scientific-Technical Progress in Hole Drilling.* Structural analysis of thread connections of drill strings at percussion drilling of small holes.
[7] Fisher G 1962 *Fracture mechanics of rocks.*
[8] Shadrina AV, Saruev L A, Saruev A L 2009 *Tomsk TPU Publ.* Dynamic processes in drill string during small hole directional drilling.
[9] Vasenin S S, Saruev L A, Kuznetsov I V 2016 Patent Rus. Fed. №159850 Percussion mechanism for percussive drilling machine (in Russian)
[10] Vasenin S S Saruev L A 2015 Patent Rus. Fed. №157331 Nipple connection of drill strings (in Russian)

About determining the wettability of coal surface

V A Arkhipov[1], S A Basalaev[2], N N Zolotorev[3], and A S Usanina[4]
[1,2,3] Department of gas dynamics and burning of combustion of Scientific Research Institute of Applied Mathematics and Mechanics of Tomsk State University, Tomsk State University, Russia
[4] Laboratory of scientific basis of technology of coal benefication, Federal Research Center for Coal and Coal Chemistry of Siberian Branch of the Russian Academy of Sciences, Russia

leva@niipmm.tsu.ru[1] tarm@niipmm.tsu.ru[2] nikzolotorev@mail.ru[3]
usaninaanna@mail.ru[4]

Abstract. A new method of measuring the wettability of coal powders based on determining the mass fraction of wetted powder particles at their interaction with water droplets during the process of gravitational sedimentation in a transparent cell has been proposed. Wettability parameter is calculated from the results of measuring the dependence of spectral transmittance of laser emission in the cell on time of powder sedimentation.

1. Introduction.

The effectiveness of modern methods of coal preparation and processing is achieved by using new energy and environmentally beneficial technical and technological solutions based on background knowledge about the phenomena of chemical and physical and chemical processes taking place on the interfaces (liquid - solid - gas). In particular, in many processes of coal-mining industry the data on the wettability process and contact angle value of coal surface are required. As an example, the process of separating the minerals from draw rock during the coal flotation process at coal benefication [1]. Coal wettability (hydrophobicity or hydrophility) is one of the main parameters that determines the process quality. In the technology of wet dust extraction in coal mines including the water pulverization in the cloud of suspended coal particles the data on powder particles wettability is required for effective particles «enveloping» by liquid and removing them from the air flow.

Since coal is an organic sedimentary rock with a complex structure the question of determining its wettability is complex and controversy. The basic method of determining the wettability of the coal surface is the method of "sitting" drop [2] that consists in determining the contact angle of a drop placed on the surface of the sample. In literature there are three basic ways to prepare the working surface of the coal sample. The first method is based on the briquetting of coal powder particles in tablets [3], the second one is to use a monolithic coal sample [4], and a third method consists in grinding the coal monolith [5].

This paper describes a method for determining the wettability of the fine coal powder without any mechanical action on the surface of coal particles that allows to perform the measurement in a dusty cloud directly [6].

2. Method of measurement.

The proposed method for determining the wettability of the fine powders is based on calculating the fraction of wetted powder particles. A uniform air-born dust is created in a cell with transparent plate walls. The maximum diameter of particles is no more than 5 μm. Using a laser and radiation detector the spectral transmittance of dust suspension is measured (fig. 1). Then, a system of monodispersed droplets with diameter 0.8÷2.5 mm from droppers that is placed uniformly on cross-section of the cell is given in the cell. After that, the spectral transmittance is measured again. The initial concentration of powder particles is taken from the condition условия $T_0 \le 0.2$ (where T_0 is the spectral transmittance before supplying the droplets) and time interval of droplets supplying is determined from the condition $T_k > 2T_0$ (where T_k is the spectral transmittance after supplying the droplets).

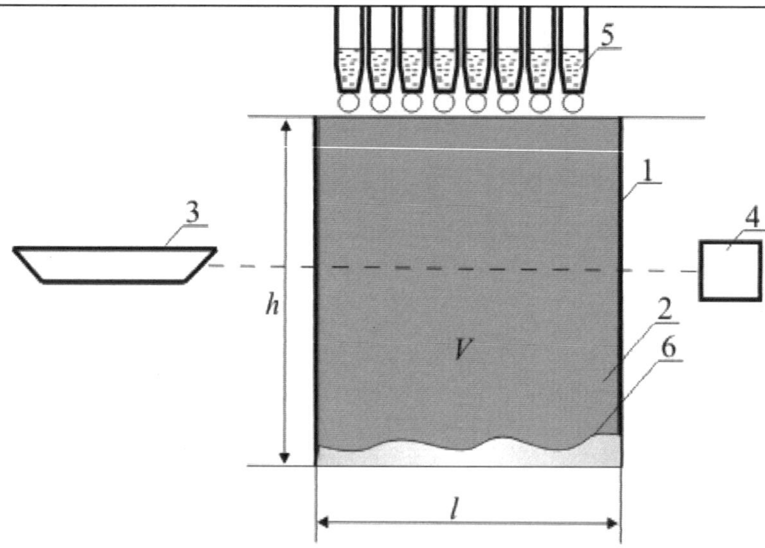

Figure 1. Scheme of experimental setup for determining the wettability of fine powder: 1 – cell; 2 – air-born dust; 3 – laser; 4 – radiation detector; 5 – system of droplets supplying; 6 – liquid

The wettability characteristic is assumed the wettability parameter [2] – mass fraction of dust particles settled on a droplet (wetted particles)

$$\beta = \frac{m_c}{m}, \tag{1}$$

where m_c is the mass of wetted dust particles (settled on a droplet); m is the mass of powder particles that run against the droplet on its free settling in the cell.

In the cell with a height h, with a droplet of a diameter D, m is calculated

$$m = \frac{\pi D^2}{4} hc\eta,$$

where c is the mass concentration of powder particles in the cell; $\eta \le 1$ is the capture coefficient.

Mass of wetted powder particles taking into account (1) is equal to

$$m_c = \beta m = \beta \frac{\pi D^2}{4} hc\eta. \tag{2}$$

The analysis of powder particle settling on a moving droplet is to take into account curving of dust flow lines. The capture coefficient is a ratio of particles that run against the droplet to particles that could run against the droplet if their flow lines never skirted the droplet. As a consequence, not all particles placed in cross-section $S=\pi D^2/4$ (midlength section of the droplet) run against the droplet. The share of collided particles is determined (for potential flow-around) from the Langmuir-Blodgett formula [7]

$$\eta=\left(\frac{\mathrm{Stk}}{\mathrm{Stk}+0.125}\right)^2, \tag{3}$$

where Stk is Stokes number.

For monodispersed powder particles of a diameter D_p, Stokes number is determined using the equation

$$\mathrm{Stk}=\frac{\rho_p D_p^2 u}{18\mu D},$$

where ρ_p is the density of particle; u is the droplet velocity; μ is the coefficient of dynamic viscosity of liquid.

Since a air-born dust is a system of polydispersed particles it is necessary to use average Stokes number [7]:

$$\mathrm{Stk}=\int_0^\infty \frac{\rho_p D_p^2 \varphi(D_p) u}{18\mu D} dD_p=\frac{\rho_p u}{18\mu D}\int_0^\infty D_p^2\varphi(D_p)dD_p=\frac{\rho_p u}{18\mu D}D_{20},$$

where $\varphi(D_p)$ is the differential function of counting size distribution of powder particles; $D_{20}=\int_0^\infty D_p^2\varphi(D_p)dD_p$ is the mean-square diameter of powder particles.

n drop bottles are arranged in upper part of the cell to give the flow of monodispersed droplets (fig. 1). The drop bottles are placed in cross-section. All drop bottles form a droplet of equal diameter D at the frequency f (number of generated droplets per second). Therefore, on a time interval t, N droplets pass the cell:

$$N=nft.$$

Since the mass of powder particles settled on a drop is determined using the formula (2), for N droplets the sum mass of wetted particles is equal to

$$M_c=\beta\frac{\pi D^2}{4}hcN\eta=\beta\frac{\pi D^2}{4}hcnft\eta. \tag{4}$$

It is necessary to take into account the time change of mass concentration of powder particles in the cell to determine the wettability parameter β since some share of particles is settled on droplets. For this reason the equation (4) is written as

$$dM_c(t)=\beta\frac{\pi D^2}{4}hc(t)nf\eta dt, \tag{5}$$

where $dM_c(t)$ is the mass of wetted particles in the time dt.

The particles wetted and droplets are settled on the bottom of the cell so the reduction in sum mass of suspended particles $M(t)$ is:

$$dM(t)=-dM_c(t). \tag{6}$$

Taking into account (6) the equation (5) has the form

$$dM(t)=-\beta\frac{\pi D^2}{4}hc(t)nf\eta dt.$$

Dividing each member of the above equation by the cell volume V yields:

$$dc(t) = -\beta Bc(t)dt, \quad (7)$$

where

$$B = \frac{\pi D^2}{4V} hnf\eta = const. \quad (8)$$

Equation (7) can be written in the form

$$\frac{dc(t)}{c(t)} = -\beta Bdt. \quad (9)$$

Integrating (9) in the range of $t = 0$ to t allows:

$$\ln\left[\frac{c(t)}{c_0}\right] = -\beta Bt.$$

It follows that

$$c(t) = c_0 \exp(-\beta Bt), \quad (10)$$

where $c(t)$ is the mass concentration of powder particles at a time $t > 0$; c_0 is the initial mass concentration of powder particles.

From the equation (10) the wettability parameter β can be defined as:

$$\beta = \frac{\ln[c_0 / c(t)]}{Bt}. \quad (11)$$

Placing (11) in the equation for β from equation (8), we obtain:

$$\beta = \frac{4V \ln[c_0 / c_k]}{\pi D^2 hnft_k \eta}, \quad (12)$$

where c_k is the concentration at time t_k corresponding to water droplets outage.

It is necessary to determine the mass concentration of powder particles in the cell in initial time c_0 (beginning of the droplets supplying) and after settling of the droplets c_k to determine the wettability parameter β using the formula (12).

For this reason a uniform air-born dust is created in the cell with plate parallel transparent walls (for example, glass). Using the source of probe radiation (laser) and radiation detector the spectral transmittance in the cell is measured

$$T = \frac{I}{I_0},$$

where I is the intensity of radiation passed through suspended particles; I_0 is the intensity of radiation of incoming beam.

In accordance with Bouguer's law [8]

$$T = \exp(-\tau),$$

where $\tau = Kcl$ is the spectral optical density of powder particles layer; K is the spectral index of depletion that characterizes the reduction of light intensity by unit volume of medium including independently scattering particles; l is the cell width (the thickness of the powder layer).

For polydisperse particles with distribution function $\varphi(D_p)$ the index of depletion is [8]

$$K = \frac{3c}{2\rho_p} \frac{\int_0^\infty Q(\alpha, \bar{m}) D_p^2 \varphi(D_p) dD_p}{\int_0^\infty D_p^3 \varphi(D_p) dD_p},$$

where $Q(\alpha,\bar{m})$ is the efficiency factor of depletion; $\alpha = \pi D_p / \lambda$ is the dimensionless diffraction parameter (Mi parameter); λ is the wave-length of probing radiation; $\bar{m}$ is the complex refractive index of the material of the particles.

Assuming that the distribution function $\varphi(D_p)$ does not change during the settling of the powder particles on droplet, we can write

$$\frac{c_0}{c_k} = \frac{\tau_0}{\tau_k} = \frac{\ln\left(\frac{1}{T_0}\right)}{\ln\left(\frac{1}{T_k}\right)}. \tag{13}$$

where τ_0 , τ_k is the spectral optical density of powder layer before and after droplets settling, correspondingly.

Placing (13) in (12) we obtain the working formula for calculating β

$$\beta = \frac{4V \ln\left[\left(\ln\frac{1}{T_0}\right)\left(\ln\frac{1}{T_k}\right)^{-1}\right]}{\pi D^2 hnft_k\eta}. \tag{14}$$

3. Example of realization of the method.

The possibility of realization of the method has been determined by means of performing direct calculations of change of concentration and optical density of medium for coal dust at their settling in an air at giving the water droplets for parameters presented in Table 1. The distribution function of the coal particles was obtained using the setup Mastersizer 2000 (MALVERN, UK). Rms particle diameter is $(D_p)_{20} = 1.9$ μm. The obtained distribution function was approximated by a gamma distribution

$$\varphi(D_p) = 15.9 D_p^{5.5} \exp(-3.6 D_p), \tag{15}$$

where $[\varphi(D_p)]$=μm^{-1}; $[D_p]$=μm.

Table 1. Values of parameters for calculation of β

$\rho_p = 1200$kg/m^3	$\mu = 1.808\cdot 10^{-5}$ kg/m·s	n=50	f =2s^{-1}
$\rho = 1.205$kg/m^3	h=0.2 m	V =0.002m^3	η =0.53
$\rho_l = 1000$kg/m^3	D=1.5 mm	$(D_p)_{20} = 1.9$ μm	t_k=100 s

The results of calculation are presented in fig. 2, 3. The dependence of dimensionless concentration of suspended coal particles $\bar{c}(t) = c(t)/c_0$ (where c is the current value of the concentration of particles in the cloud) on time of droplets supplying for the wettability parameter value $\beta = 0.8$ is given in fig. 2. Dependence of optical density on time of droplets supplying in the cell at value β=0.8 is presented in fig. 3. Check the adequacy of the method can be carried out using fig. 2 and fig. 3.

Select the time interval of droplet feed t_k=100s at which the change of the initial concentration of suspended powder is significant and the condition T_k>2T_0 for spectral transmittance occurs (fig. 2, 3). At time values t_k=100s the concentration of suspended particles and ratio of optical densities are equal to $\bar{c}_k$ =0.47, τ_0 / τ_k =2.1, correspondingly.

Placing found values in the formula (14) for calculating the parameter β, we obtain $\beta = 0.8$.

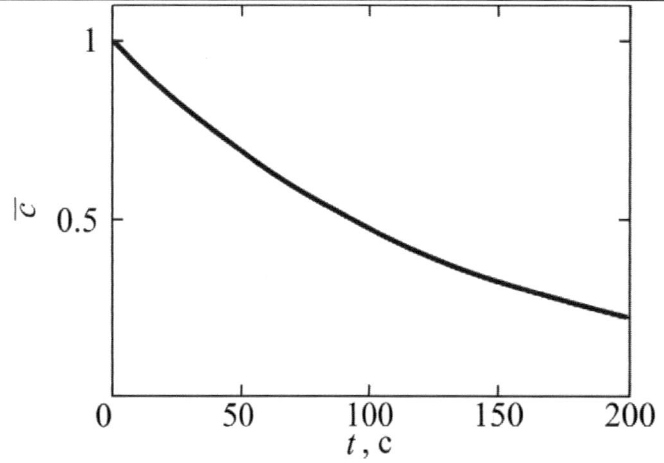

Figure 2. Dependence of concentration of coal dust particles with distribution (15) on time of droplets supplying for the wettability parameter value β =0.8 calculated using formula (10)

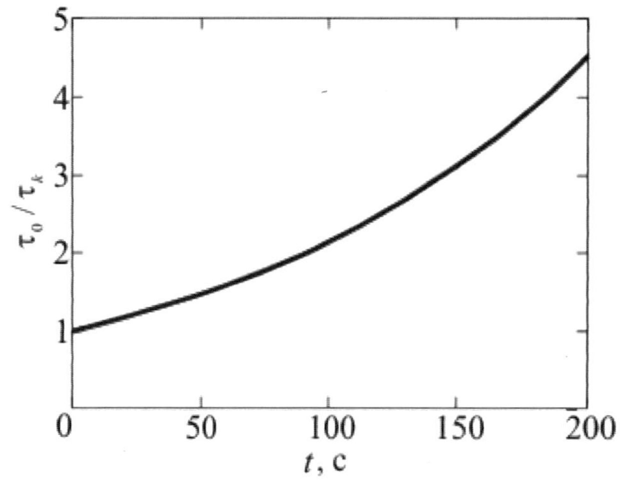

Figure 3. Dependence of optical particle densities τ_0 and τ_k ratio having the distribution (15) on time of droplets supplying for the wettability parameter value β =0.8 calculated using formula (13)

As we can see from the above example, specified and calculated values of wettability parameter coincide with each other ($\beta = 0.8$). Similar results are obtained for any value of the wettability parameter in the range $\beta = 0 \div 1.0$.

4. Experimental working out the method

Working out the method for measuring the wettability of coal powder is carried out using the experimental setup (fig. 4) comprising of the cell, system for creating of suspended powder, system for droplets supplying in the cell, and system for measuring the spectral transmittance.

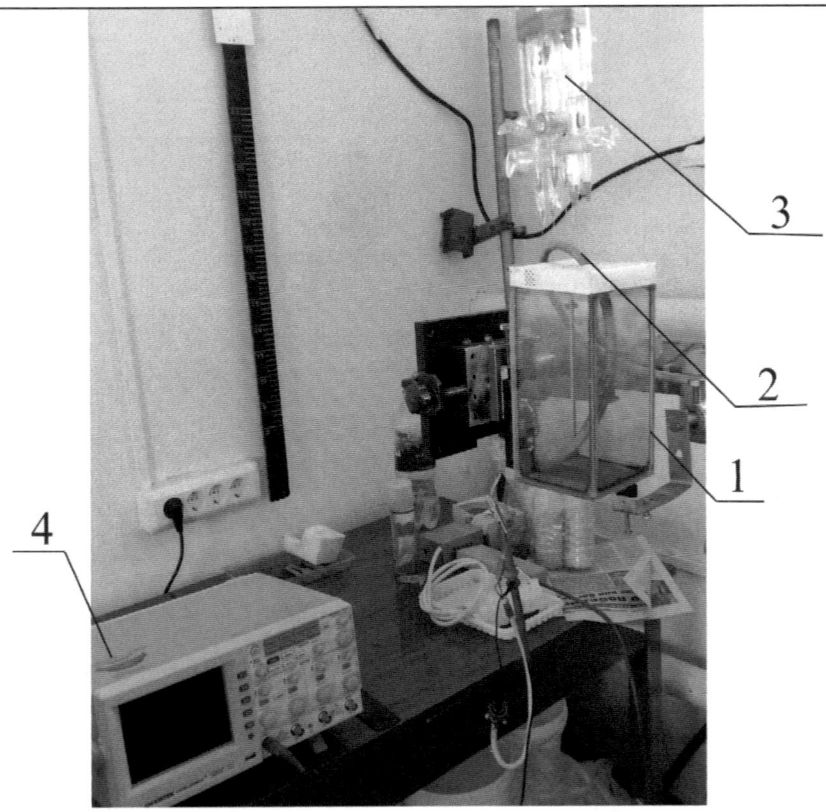

Figure 4. Photograph of experimental setup: 1 – cell; 2 – tube; 3 – drop bottles; 4 – oscilloscope

System for monodispersed droplets supplying – nine of uniformly placed drop bottles – is arranged in upper part of the cell by size of 10×10×20 sm and with walls from optical glass. The variation of the diameter of the capillary and the height of the liquid column in the drop bottle provides to obtain a stream of droplets of distilled water of the spherical shape and predetermined size D = 1÷2.5 mm with a predetermined frequency of separation f = 0.5 ÷ 5 Hz.

System for creating the uniform suspended coal particles includes the source of compressed air and tube placed along lateral and bottom side of the cell. The specified mass of coal powder is placed in the ell to obtain the cloud of suspended particles. Air is fed through the holes in the tube, located at the bottom of the cell. During the air supplying, the cell is closed with lid and suspended coal particles cloud is created.

System for measuring the spectral transmittance includes the source of probing radiation - helium-neon laser LG-78 with power of 1 mW and a wavelength of λ = 0.6328 μm and radiation detector - germanium transistor FTT-4 with an area of the sensor element of 3 mm^2. The radiation detector is located in light-tight housing with pinhole aperture on its butt end to exclude the effect of background illumination. Measurement of spectral transmittance is performed at condition of passing the probing radiation on the distance 12 sm from upper end of the cell. Register signal from a radiation detector is carried by a digital oscilloscope GDS-2064 agreed upon with personal computer.

The visualization of the particle cloud is conducted using the video camera located vertical to probing beam of the laser to control of uniformity of distribution of coal particles.

Working out the method is conducted for coal samples. The characteristics of the samples (maximum particle diameter D_{max}, moisture content W, ash content Z, volatile content V^{daf}) are given in Table 2.

Table 2. Characteristics of coal powder

Sample	D_{max}, mm	W, % wt	Z, % wt	V^{daf}, % wt
1	0.05	0.4	21.86	23.72
2	0.05	0.4	7.0	24.13
3	0.05	0.5	17.05	33.63

The results of working out the method allowed finding the procedural error of experimental setup for realization of the method for determining the powder wettability. In particular, during the atomization of coal powder by compressed air some of the particles located on the cell walls. It results in the error in measuring the spectral transmittance. Relevant modification of experimental setup is necessary to create a uniformly distributed particle cloud isolated from the inner cell walls.

5. Conclusion

A method for determining the wettability of the fine coal powder has been proposed. The proposed method allows improving the accuracy of determining the characteristics of the wettability of the fine powder and carry out the measurement directly in a dusty cloud. This method can be used to study a broad class of organic and inorganic powder materials. As a result of working out the experimental setup, the difficulties associated with the creation of a uniform cloud of suspended particles that is isolated from the inner cell walls have been found.

6. References

[1] Abramov A A 2008 *Moscow State university.* Flotation methods of benefication.
[2] Zimon A D 1974 *Moscow. Chemistry.* Liquid adhesion and wettability.
[3] Li Man Xu Hai-Hong Shu Xin-qian 2008 *Journal of coal science and engineering*. Study on coal dust wettability measurement using cold briquetting technique. Vol. **14** № 4 pp. 571-574
[4] Arkhipov V A, Korotkikh A G, Patrakov Yu F, Usnaina A S 2015 *Journal Tuva Staty university. Tekhnicheskie and fiziko-matematicheskie nauki.* Experimental study of the effect of method of sample preparation on the coal surface wettability. Vol **3**. pp. 95-104
[5] Drelich J, Laskowski J S, Pawlik M 2000 *Coal Preparation.* Improved sample preparation and surface analysis methodology for contact angle measurements on coal (heterogeneous) surfaces. Vol. **21** № 3 pp. 246-275
[6] Arkhipov V A, Zmanovskiy S V, Paleev D Yu, Patrakov Yu F, Usanina A S 2014 *The patent of the RF* 2522805. Method for determining the fine powder wettability / Publ. 20.07.2014, № 20
[7] Shilyaev M I, Shilyaev A M 2003 *Publishing house Tomsk State University of Control Systems and Radioelectronics.* Air dynamics and heat-and-mass transfer of gas dispersed flows.
[8] Arkhipov V A, Bondarchuk S S 2010 *Tomsk State University.* Optical methods of diagnostics of heterogeneous plasma of combustion products.

Genetic Classification of ore-forming processes

I Kucherenko, E Sinkina , R Abramova

[1] Institute of Natural Resources, Tomsk Polytechnic University, Tomsk, Russia
E-mail: kivr@tpu.ru vea1986@mail.ru

Abstract The absence of a genetic classification of ore-forming processes is the result of a rather slow-paced accumulation of knowledge in ore-forming processes, which, in its turn, could be explained by objective conditions such as the fact that during the last few decades the genetic classification included only the description of mineral deposits. The existing genetic mineral deposit classifications has developed and embraced the genetic classification of ore-forming processes in accordance with system theory, structured taxons revealing the nature of the processes and the basis- information source of these processes. The base involved the developed model system of ore mineralization- ore formations (geological formations with syngenetic mineralization) within poly-component and mono-component subformations; for convergent mineralization- geological types of deposits. The latter, as well as non-convergent subformations has accumulated all data about initiating and conditioning ore formation within wide-scaled geological processes. The geological mineralization types within ore formations and subformations are included in the genetic classification of ore-forming processes. This, in its turn, makes it possible to forecast the functions and provide a regular transition into the geological-genetic classification of ore-forming processes in accordance to above-mentioned matrix-structure, and, simultaneously, further the development of the existing geological-genetic theory of ore formation.

1. Introduction

The classification of natural processes (phenomena) based on fundamental information and the structure and principles of which have been developed within the framework of contemporary theory system reflects the maturity of the existing theory. In the absence of a theory, a classification has been developed to describe either genetically natural objects and ore mineral deposits if it concerns ore geology. However, on this case, such precise and accurate classifications would completely depend on the information (data) of ore formation conditions.

In the first half of the 20th century such geologists as V Lindgren [8] and V Obruchev [4] considered that mineral deposits should be genetically-based classified and according to the geological processes. Nevertheless, they understood that it was impossible to put forward such an "ideal" genetic classification of ore-forming processes, due to the fact that there is no justifying theory to prove this.

This, all in all, is the result of a rather slow-paced accumulation of knowledge in ore-forming processes, which, in its turn, could be explained by objective conditions such as the fact that during the last few decades the genetic classification included only the description of mineral deposits.

The authors proposed a transformation of existing genetic mineral deposit classification into a more sophisticated genetic classification involving the ore-forming processes. This classification could be based on an updated ore-formation theory and include the application of system theory approach in describing the natural objects and processes. This proposed classification will be described further.

2. Genetic classification of mineral deposit structure and composition

V Lindgren, the author of the earliest mineral deposit classification, wrote "....the only reasonable classification of deposits into classes could be the following: 1) deposits formed from pre-existing minerals by mechanical concentration or 2) deposits formed as a result of some reaction in solutions", while ".......the present classification describing the complex formation of ore deposits is the only possible one" [8]. It was this fact that became the basis of Lindgren classification:

Deposit I includes clastic rocks, clays and placer. **Deposit II** embraces the existing Russian classification i.e. exogenous (weathered **rocks**, sedimentary) and endogenous.

Simultaneously, another genetic classification of mineral deposits was developed by V Obruchev (Mining Department, Tomsk Technological Institute- 1901-1912) [4].

Genetic classification of mineral deposits (Obruchev [4])

GROUP A. ABYSSAL DEPOSITS (ENDOGENOUS)

Category 1. Magmatic
- ***Class 1.*** Evamagmatic
  - *Type a.. Segregated*
  - *Type b. Liquated*
- ***Class 2.*** **Injection**
- ***Class 3.*** Pegmatitic

Category 2. Emanation
- ***Class 1.*** Contact
- ***Class 2.*** Pneumatolytic
- ***Class 3.*** Exudates and sublimates

Category 3. Hydrothermal
- ***Class 1.*** Hypothermal
- ***Class 2.*** Mesothermal
- ***Class 3.*** Epithermal

(Classes 1–3: *Type a. Packed void*; *Type b. Metasomatic*)

GROUP B. SURFACE DEPOSITS *(EXOGENOUS)*
- ***Class 1.*** Sedimentary
- ***Class 2.*** Infiltration
- ***Class 3.*** Residual
- ***Class 4.*** Clastics or placers

GROUP C. ALTERED DEPOSITS (METAMORPHIC)
- ***Class 1.*** Pyrometamorphic
- ***Class 2.*** Dynamometamorphic
- ***Class 3.*** Hydatometamorphic
- ***Class 4.*** Complex genesis (proto-magmatic, proto-contact, metamorphised proto-sedimentary, for example, ferruginous quartzites)

It should be noted that V Obruchev considered metamorphic (metamorphogenic) deposits as deposits of different origin and minerals, existing before metamorphism and transformed differently during the metamorphism itself, but preserving the initial minerals. Although the genetic classifications of deposits by Lindgren and Obruchev include authentic data information revealing the ore-forming processes, they are different in form and content.

According to Obruchev classification, clastic and placer deposits conforming with the concentration behavior of commercial minerals in deposits are comparable to Lindgren classification "the mechanically- formed concentrations." And, according, to Lindgren classification the chemically-concentrated deposits are described as taxons in Obruchev classification. It should be stated that in the first and second cases both classifications are rather distinct and comparable. Obruchev borrowed from Lindgren classification the concept of differentiated hydrothermal deposits as hypo-, mezo- and epithermal. The latter two terms are applicable even today.

The basic difference between these two classification is the description sequence of deposit formation processes, i.e by Lindgren- from exogenous to endogenous and visa versa from endogenous to exogenous by Obruchev. This difference is expressed as prioritization in evaluating the relevant factors determining ore-forming processes and mechanism of mineral concentration within Lindgren classification, while energy source needed for deposit formation within Obruchev classification.

All the ensuing numerous domestic genetic mineral deposit classifications only extended Obruchev classification, but this included insignificant and or significant alterations or additions. This could be due to either the accumulation of advanced data or even be motivated by the desire of some authors to make one's own contribution in upgrading the base classification.

The first significant alteration in above-mentioned Obruchev classification was notable after the publication of the following classification [1]: in Group B "altered deposits" have been renamed into "metamorphogenous" in Group C which included metamorphic deposits formed during metamorphism and being the result of metamorphism itself. Thus, two associated deposits in the upper level third taxon were combined: metamorphosed (metamorphic, metamorphogenic, altered, consequently existing before metamorphism, according to Obruchev) and metamorphic, not existing before metamorphism, but formed by metamorphism itself. This addition is extant only in all classifications developed during the last few decades [6]. The existing base classification involves another insignificant alteration: hydrothermal taxon is divided into hydrothemal deposit categories- carbonate, skarn, albite and greisen.

3. Results and discussion

This brief descriptive history of the development and design of the genetic mineral deposit classification reflects the evidence of its periodicity in the research of generation-to-generation geologists. It is well-known that the results characterizing the ore-forming processes prove their reliability and relevant scenario which reflect how nature influenced this or that deposit.

The 100-year accumulated direct knowledge could be sufficient enough to build the backbone of the theory – genetic ore-forming deposit theory – which would describe the processes occurring within the developing deposits [5–7]. One should also mention the considerable achievements in developing the ore-forming theory from the fundamental geological (metallogenic) aspect so as to reconstruct and describe (decipher) rather wide-scaled geological processes in the earth crust and/or mantle trriggering and conditioning ore-formation.

The geological basis of this theory, concerning exogenous and endogenous processes which are aggravated by multiple-diversified geological (magmatic, metamorphic) processes within one and the same genetic deposit type and in one specific deposit and which, in its turn, stipulates ore formation. The major problem arises in reconstructing the geological factors involved in the formation of hydrothermal deposits, due to limited accessibility in investigating the sources of energy, fluids and ore matter being located in significant hypothetical depths.

In this case, these achievements in developing the genetically ore-formation theory determine the practicability and completeness of transforming the genetic mineral deposit classification into a more sophisticated classification of ore-forming processes, the structure of which would compile to the requirements of system theory [2]. Obruchev classification is the basis in the development of this updated classification. According to the system theory the classification should include subdominant taxons which differentiate the processes and direct foundation, i.e. information source which, in its turn, is the basis in detecting these processes.

The classification is designed in accordance to: 1) transferability principle, providing the transition from the immediate base to the processes and reversely; 2) entity principle identifying the processes as internal dissected units; 3) use-principle of one feature by autonomy of each peer-to-peer taxon; 4) hierarchical principle of leveled process structure as system constituents, defining distinct taxonomy level hierarchically interconnected so as one taxon is the basis for the next one; 5) principle of taxon discreteness.

Genetic classification of ore-forming processes

ENDOGENOUS GROUP
Category Magmatic
Class Liquation
Class Crystallization
Subclass Early crystallization
Subclass Late crystallization
Category Fluid-magmatic
Class Plutonogenous
Ore formation Rare-metal-*ornamental* ***granite*** *pegmatite*
Ore subformation Rare-metal
Ore subformation Ornamental stones
Class Ultra-metamorphogenous
Category Hydrothermal
Class Magmatogenous
Subclass Plutonogenous
Ore formation *Ferrum-phosphous- rare metal* ***carbonates***
Ore subformation Ferrum-phorsphous
Ore subformation *Rare-metal- rare earth*
Ore subformation *Polymetallic*

Ore formation *Gold-uranium-polymetallic* ***beresite***
Ore subformation *Gold*
Geological type
Ore subformation *Uranium*
Geological type
Ore subformation *Antimonite*
Geological type
Ore subformation *Polymetallic*
Subclass **Volcanogenous**
Class Metamorphogenous
Subclass Greenschist
Subclass Epidote-amphibolite (amphibolite)
Subclass Granulite
EXOGENOUS GROUP
Category Hydragenous
Class Residual
Subclass Siallites
Subclass Laterites
Class Infiltrated
Category Sedimentary
Class Mechanical substance sedimentation
Subclass Continental

Subclass Marine
Class Chemical substance sedimentation
Subclass Continental
Subclass Marine
Ore formation *Ferrum-manganese-carbonate-sandy-**argillaceous***
Ore subformation *Iron ore*
Ore subformation *Manganese*
Class Biochemical substance sedimentation
Subclass Continental
Subclass Marine

POLYGENOUS GROUP
Category Hydrothermal-sedimentary (kuroko type)
Category Metamorphised
Category Unconformity type

According to this classification the metamorphogenic group is replaced by polygenic group when subjected to exogenous and endogenous processes. Carbonate, skarn, albite and greenschist deposits were formed as a result of hydrothermal processes, consequently, these constituents represent all hydrothermal associations, including fenite, beresite, argillite, alkaline metasomatites, propylite, and others. Gold, uranium, antimonite and other minerals in metamorphogenic deposits are displaced from earlier metamorphogenic group to endogenous hydrothermal group.

Accumulating type- models (repeating in numerous features) for each genetic feature type are preferable to stand-alone deposits with their specific features. These models are ore formations providing that the system of ore formations is not based on fragmentary metal mineral ore composition only (as earlier stated), but based on both geological and ore-forming processes of the substance within the geological formation itself [3]. In this case, ore formation is the geological formation with syngenetic mineralization.

Endogenous (magmatic, metasomatic) and sedimentary formations are classified on the composition-genetic basis excluding the geodynamic formation regimes, while metasomatic formations – without correlation with magmatism. It is advisable to detect geodynamic formation regimes, which, in its turn, proves their correlation with wide-scaled geological processes not at the stage of segregation, but within the segregated geological formations. In this case, the geological formation processes coupled with their composition-genetic basis are intended to characterize the earth crust formation regularities This approach minimized and / or excluded all possible discussion concerning the autonomous formation typification of rock associations and being applied in the formation classification of mineral deposits could eliminate all insuperable difficulties in implementing the metal-ore-mineral approach [3].

To define the nature laws that form this or that type of mineral deposit it is proposed to differentiate the poly-component ore formations into mono-component subformations.

Following observations, in some cases the geological conditions (geodynamic regimes, conditioned by the priming ore formation of more wide-scaled geological processes) in ore formation convergence embraces geological types, the number of which equals the number of convergence conditions. The geological mineralization types within ore formations and subformations are included in the genetic classification of ore-forming processes. This, in its turn, makes it possible to forecast the functions and provide a regular transition into the geological-genetic classification of ore-forming processes in accordance to above-mentioned matrix-structure [3], and, simultaneously, further the development of the existing geological-genetic theory of ore formation.

4. Conclusion

In the first quarter of the XX century geologists from different countries remodeled practically all well-known mineral formation processes. The obtained information was adequate enough to classify the deposits terming their formation processes. Most described terms, except for outdated ones, are still used in the modern classification. However, each term embraces incomparably more deeper and extensive knowledge which was accumulated in the past decades.

The first mineral deposit classifications developed by such well-known geologists as V. Lindgren, V. Obruchev, Matthew Emmons, Paul Niggli and Hanz Schneiderhöln were conceptually different. The above-described genetic mineral deposit classification by Obruchev more or less conforms with the classification principles of scientific units and phenomena being based on the existing system theory. In this case it is appropriate as a classification base.

But contemporary classifications could be exceptions to the rules. For example, endogenous metamorphic deposits are not included in the endogenous group, but are included in the metamorphogenic group. Some complexes of hydrothermal deposits are excluded from the hydrothermal category, but still possess equal taxon status with the latter (distributed in one-rank level taxon).

To convert the deposit classification into the classification of processes it is necessary to introduce corresponding alterations in the base classification structure itself, i.e. ranking within three taxons of the base where "geological types" taxon could accumulate new data of geological processes conditioning ore formations and revealing the principles of distribution and formation of mineralization.

Thus, the genetic mineral deposit classification transforms naturally into the genetic ore-forming process classification, while the latter- into geological-genetic classification reflecting the ore-forming theory completely.

References

[1] Betekhtin A, Ershov S, Zverev V. Ivanov G et al 1938 *Leningrad.* Kratkij kurs mestorozhdenij poleznyh iskopaemyh. pp. 473

[2] Blaubergi. Udin E 1973 *Moscow Nauka.* Stanovlenie i sushhnost' sistemnogo podhoda. pp. 270

[3] Kucherenko I 2004 *TPU Izvestya.* Teorija i praktika formacionnogo metoda v rudnoj geologii. Vol. **307 (6)** pp.25-30

[4] Obruchev V 1934 *Ore deposits* 2nd edition Novosibirsk. pp. 596

[5] Smirnov V 1989 *Moscow Nedra.* Geology and Mineral Resources .pp. 326

[6] Starrostin V, Ignatov P 2004 *Geology and Mineral Resources,* Moscow. pp. 512

[7] Strakhov N 1963 *Osnovy teorii litogeneza.* Moscow. Vol. **1, 2, 3.**

[8] Lindgren Waldemar 1928 *Mineral Deposits.* New York and London. Vol. **1** pp. 196

Operation effectiveness of wells by enhancing the electric-centrifugal pump

P N Zyatikov, I N Kozyrev and V S Deeva

Department of geology and oil field development, National Research Tomsk Polytechnic University, 30 Lenin avenue, Tomsk, Russia
zpavel@tpu.ru[1], veradee@mail.ru[3]

Abstract. We present the method to improve the operation effectiveness of wells by enhancing the electric-centrifugal pump. Some of the best ways to extend the electric-centrifugal pumps operating lifetime is using today's techniques as well as additional protective equipment as a part of the electric-centrifugal pump. In paper it is shown that high corrosiveness of formation fluid (a multi-component medium composed of oil, produced water, free and dissolved gases) is a major cause of failures of downhole equipment. Coil tubing is the most efficient technology to deal with this problem. The experience of coil tubing operations has proved that high-quality bottom hole cleaning saving the cost of operation due to a decreased failure rate of pumps associated with ejection of mechanical impurity.

1.Introduction

Nowadays, the use of electric-centrifugal pumps (ECPs) is the most advanced and efficient method of artificial oil lifts. High corrosiveness of formation fluid (a multi-component medium composed of oil, produced water, free and dissolved gases) is a major cause of the failures of downhole equipment operated using this method [1]. Lifetime of the electric-centrifugal pumps can be increased by using additional protective equipment as a part of the ECP. Many devices and techniques to develop the oilfield including artificial lift do not always satisfy requirements of the oil companies. Thus, it is important to propose a new technique and approach to improve the production efficiency, optimise plant and equipment uptime, and reduce product costs. The ECP can be used in producing well stock where oil, liquids and gases are present. To lower energy costs per ton of oil produced more profitable to use the ECP rather than the sucker rod pump. Therefore, the relevance of the theme is not in doubt.

2. The basic principle of the calculation of the ECP parameters

To select a borehole pump it is convenient to use the method of artificial lift. The basic principle of this method is based on the following information: the well productivity index (based on the hydrodynamic research findings), the inclinometer survey, the gas/oil ratio, the formation pressure, the saturation pressure, the well water cut, and the concentrations of entrained solids [2]. JSC «Tomskneft» uses the programm «WellFlo» to select ECP in one producing well. Calculation algorithm for ECP include: receiving information about productive formation and well geometry, choosing the ECP setting depth and required lift, calculating the gas separation at the pump output, adjustment of the pump service properties, and selecting the electric motor, cable, transformer, and control station.

The ECP setting depth is determined in accordance with requirements for the minimum bottom-hole pressure. The gas/oil ratio recommended value into a highly dispersed phase at the pump inlet is 35%. The optimum value of the gas/oil ratio at the ECP inlet is 30–40%. Taking into account this fact, it is necessary to find the range of the borehole having this value in the pressure line and further to find the pump-setting depth with allowance for bend angle of the borehole.

However, as EPC service treatment shows, with so much free gas onsite the pump, the time interval between overhauls increases by 10–15% [3].

The checking calculation of the pump-setting depth is done separately for 2–3 choices for equipment design as well as for another calculation.

It should be noted that checking calculation of pressure change along borehole in the depth higher than pump output and pressure change along oil-well tubing (OWT), one must have regard to gas separation at the ECP inlet. The separation index for the boreholes with pump, is found from [4,5]

$$\delta = \frac{1}{(1+0.6\cdot(Q_{f.st} / W_0 \cdot F_h)}, \qquad (1)$$

where W_0 is the relative velocity bubble gas floating-up, m/s; F_h is the cross-section area of the annular space between the casing pipe and electrical submersible motor (ESP), m^2.

To coordinate the pump characteristic with wells characteristic and thus, to determine the specific energy, transferring to gas-liquid mixture and fluid withdrawal rate maintenance in order to choose pump-setting depth, one could plot a head-flow characteristic graph for the well $Q=f(H_{well})$

$$H_{well} = H_{well.dn.} + P_{whp} / (p \cdot g) + h_{fr} - H_g, \qquad (2)$$

where $H_{well.dn.}$ is the depth of the well dynamic level in which the fluid withdrawal holds, m; $P_{whp}/(p\cdot g)$ is the well-head pressure under average density of the gas-liquid mixture at the range «pump-well-head» into borehole, meter of liquid column; h_{fr} is the friction head loss, m H_2O.

If we don't have the data about the dynamic level, we could estimate it approximately:

$$H_{dnp} = L_l - \frac{P_{ar} - Q_f / K}{\rho_{vo} \cdot g} \qquad (3)$$

where ρ_{vo} is the density of the oil-water mixture at the pump output.

The friction head loss caused by the hydraulic friction into the OWT can be found as for Newtonian fluid:

$$h_{fr} = \lambda \cdot \frac{H_{well} \cdot \omega^2}{2 \cdot g \cdot D}, \qquad (4)$$

where λ is the hydraulic resistance coefficient; ω is the linear flow rate, m/s.

$$\omega = \frac{Q_н \cdot vh + Q_b \cdot v_v}{86400 \cdot F_{pct}}, \qquad (5)$$

We can approximately calculate the pressure relevant to the gas lift effect into buried pipes:

$$H_g = 4 \cdot D \cdot G_{o.act} \cdot (1-(P_w / P_{pump})^{1/3} \cdot \left[1 - B(\overline{P})\right], \qquad (6)$$

where D is the outer diameter of the OWT in inches; P_{pump} is the saturation pressure after the gas separation; $B(\overline{P})$ is the average well water cut in the eductor under average pressure condition $P = 0.5 \cdot (P_{in} / P_w)$.

In order to plot a graph of the head-flow characteristic, the several values of the production rate (as a rule 5 – 6 points, starting from 0) are taken. And then one can plot point by point $Q = f(H)$ the formation heat line H_{well} of the well.

Further, the graph of the head-flow characteristics should be superimposed on the graph of the pump characteristics to find the point of intersection. This point indicates the well production rate that is equal to pump capacity. Selection of the OWT diameter should be determined as a function of production rate:

Table 1. Selection of the OWT diameter (for ECP)

Production rate, m^3 per day	less than 150	150–300	more than 300
The inside diameter of the oil-well tubing, mm	50.3	62	76

Taking into account the typical head-flow characteristics of the ECP, production rate, and required lift, we can choose several pumps, which meet drainage requirements for operation under conditions:

$$0.6 \le Q_f / Q_{w.b} \le 1.2,$$

where $Q_{w.b}$ is the pump capacity measured for water flow in optimum performance.

The point of intersection of these graphs indicates the pump capacity measured for water flow.

In practice, the discharge liquid characteristics differ from the water characteristics caused by occurrence of the oil-water mixing. Therefore, if the pressure at the pump output is smaller than the saturation pressure, the free gas occurs in pump. Due to this fact, we need to adjust the pump characteristics to the discharge liquid and free gas to improve accuracy.

The dependence of the pump head, the pump capacity and coefficient of efficiency on the viscosity of the discharge liquid are considered by using the specific coefficients. When the viscosity of the pumping step is raised, the following indexes are also increasing: the flow resistance, the energy loss in liquid in which wheel disk rotates, as well as the friction of the wheel rotor pin. It leads to reduction of the pump head, pump capacity, and efficiency coefficient by while the power consumption.

When the gas/oil ratio at the ECP inlet is 5–7 % and less, the impact of the gas can be negligible. Further re-calculation of the pump characteristic measured by water to the pump characteristic measured by viscous liquid is made by the P.D. Lyapkov-Maksimov graphical diagram. It can be used for viscous liquid with viscosity less than 0.03–0.05 cm^2/s under formation conditions. If the viscosity of the liquid exceeds this value, it is essential to make adjustments of the pump operating characteristics [6].

Additionally, the kinematic apparent viscosity (m^2/s) is using as a viscosity rating of the well production. It is calculated by the formula

$$\nu_{ap} = \mu_{ap} / \rho_{in}, \tag{7}$$

where μ_{ap} is the apparent viscosity under the appropriate temperature and the fluid shear rate, ρ_{in} is the average density of the production fluid in device channel, kg/m^2.

$$\rho_{in} = \rho_o \cdot \beta + \rho_w \cdot \beta_w, \tag{8}$$

where β and β_w are the volume ratio of the oil and volume ratio of the water correspondingly.

It could be mentioned, that the dependence of the pump head, the pump capacity and coefficient of efficiency on the viscosity of the discharge liquid could be estimated by using the coefficients

$$K_{w,Q} = H / H_w = Q / Q_w, \quad (9)$$

$$K_\eta = \eta / \eta_в, \quad (10)$$

where H_w, Q_w and $\eta_в$ are the pump head, the pump capacity and coefficient of efficiency measured by water under standard conditions; H, Q, η are the same parameters measured by viscous liquid.

The coefficients K_η, Q and K_η depend on the Reynolds Number in ECP channels

$$\mathrm{Re}_w = \frac{4.3 + 0.816 \cdot n_s^{0.274}}{n_s^{0.575}} \cdot \frac{Q}{\gamma_{ex}} \cdot \sqrt[3]{\frac{n_1}{Q_{w.opt.}}}, \quad (11)$$

where n_s is the pumping step power-speed coefficient, n_l is the pump speed, 1/s.

$$n_s = 193 \cdot n_1 \cdot Q^{0.5}{}_{w.opt} \cdot \frac{(g \cdot H_{w.opt})^{-0.75}}{z_n}, \quad (12)$$

where $Q_{w.opt}$ and $H_{w.opt}$ are the pump capacity and pump head measured by water that will give optimal performance, z_n is the number of the pumping step.

We can find the coefficient K_w, Q, K_η using the special graphs from Lyapkov [6] and the value Re_w whereat we can re-calculate the characteristic of the ECP measured by water to characteristic of the ECP measured by viscous liquid. Moreover, we can use the approximating formulas to find these coefficients apart from the graphical method.

For laminar flow:

$$K_{w,Q} = \mathrm{Re}_w / (\mathrm{Re}_w - 50 + 200 * (Q_w / Q_{w.opt});$$
$$K_\eta = 0.485 * \lg \mathrm{Re}_w - 0.63 - 0.26 * (Q_w / Q_{w.opt}).$$

For turbulence flow:

$$K_{w,Q} = 1 - (3.585 - 0.821 \cdot \lg \mathrm{Re}_w) \cdot (0.027 + 0.485 \cdot (Q_w / Q_{w.opt})),$$
$$K_\eta = 0.274 \cdot \lg \mathrm{Re}_w - 0.06 - 0.14 \cdot (Q_w / Q_{w.opt}),$$

where Q_w is the pump capacity measured by water flow under required condition, m^3/s.

The power consumption can be found from formula:

$$N = 10^{-3} \cdot g \cdot Q_w \cdot H_w \cdot \rho_{wo} \cdot K^2{}_{H,Q} / (\eta_w \cdot K_\eta). \quad (13)$$

When the gas/oil ratio at the ECP inlet is under 7 % it could be negligible. When the gas/oil ratio increases, the head-flow characteristic shifts left while the ECP coefficient of efficiency falls. In practice, to avoid the deleterious effect of the gas it is helpful to install the equipment with special gas separator designed by P.D. Lyapkov [5].

3. Use of the coil tubing technology to stimulate wells after hydraulic fracturing.

Coil tubing is the most efficient technology for post-hydraulic fracturing proppant flowback. The experience of coil tubing operations has proved that high-quality bottom hole cleaning and well stimulation with nitrogen save the cost of ECP operation due to decreased failure rate of pumps

IOP Conf. Series: Earth and Environmental Science **43** (2016) 012078 doi:10.1088/1755-1315/43/1/012078

associated with ejection of mechanical impurities. Using coil tubing, firstly, one can lower a flexible pipe into a well very rapidly (up to 50 m/min). Secondly, nitrogen can be injected through it to form a very light fluid column. During the cleaning, the clean-up flow goes to the well and perfectly cleans a wellbore.

Moreover, the coil tubing provides the possibility to achieve an earlier payback and save on costs of further well operation. Quick payback is associated with a number of factors, such as a shorter period of well stimulation; reduction of oil losses due to saving time of well stimulation; longer service life of the well after commissioning; reduction of negative impact on a formation during treatment; and increase in revenue from oil sales [7].

4. The use of a flexible ball joint as a part of ECPs.

As known [8], when ECPs are operated in areas with a high bend angle induced by lateral forces, cases and shafts are often subjected to off-design stresses. This causes one-sided wear of parts and reduction time between overhauls. When running ECP of a certain lateral dimension, the maximum allowable bend angle is defined by allowable elastic strain of the materials of which ECP is made. According to regulations of Russian and foreign manufacturers, the maximum allowable bend angle of the well is equal to 2 per 10 meters. According to the same regulations, ECP should be installed in the well where it is not subjected to deflection or at least fits into a well segment.

The use of the flexible ball joint (FBJ) as a part of ECP makes it possible to achieve the well potential; to prevent torque retention loss because of asymmetrically tightened bolts of ESP intersectional flange connections by removing bending loads acting on ECP when it passes through intervals with a high bend angle [8].

When operating in area with a bend angle above the allowable level, i.e. in a stress-strain state, the ECP fitted with a FBJ is well fit in a deviated borehole and, therefore, can be operated with a higher stability. The maximum bend angle of FBJ is 5° or 10°, depending on a version [9].

5. Conclusion

The reasonable selection of pumping equipment and optimization of its operation allow to improve the operational efficiency of wells, to evaluate the equipment, and to calculate rational operating parameters.

The maximum effect of energy efficiency improvement can be achieved by enhancing the performance of electrical submersible pumps. For this purpose, it is necessary to use a balanced approach to select each element of ECP (cable, pump, submersible motor, etc.) and also to optimize the pump operation after its commissioning. In general, the introduction of the proposed methods into production will significantly increase time between overhauls of wells fitted with ECPs for artificial lift and, therefore, increase daily output of mineral resources.

6. References

[1] Ageev S R, Grigiryan E E, Makienko G P 2007 *Press-Master.* Russian equipment fitting with wing pump for petroleum production and their treatment. pp. 645.

[2] Ivanovsky V N, Daricshev V I, Sabirov A A 2002 *Oil and gas.* Oil-well pumping unit for petroleum production. pp. 824.

[3] Standing orders to submersible equipment fotted with ECPJSC «Tomskneft» EOC

[4] Vasilevsky M V, Zyatikov P N, Burykin A Y, Deeva V S 2015 *IOP Conference Series: Earth and Environmental Science.* Stability of the liquid particles separation in the apparatus of oil and gas systems. Vol. **27** pp. 1–6.

[5] Tronov V P 2002 *Fan.* The gas separation and loss oil reduction. pp. 408.

[6] Lyapkov P D 1955 *UNRIOil.* Russian Thesis in candidacy for a degree "additional properties analysis of the electric submersible pump design and exploitation for the petroleum production and also calculation methods of their operating device".

[7] Zemlyanoy A A, Dolgushin D A, Shatalov D A 2013 *Drilling and oil.* pp. 44–46

[8] Polyakov V S, Barbash I D, Ryahovsky O A 1979 *Engineeringindustrye.* Guide of the ring joint. pp. 344.
[9] Gepshtein F S, Dyachyk I A, Shayakberov V F 2002 *Flexile ball joint.* The patent of the RF 2230233.

Measures to limit subsidence of underground oil pipeline in insular permafrost

N Chukhareva[1,a], A Rudachenko[1,b], A Chekhlov[1,c]
[1]Tomsk Polytechnic University, Tomsk, Russia

[a]Natasha@tpu.ru, [b]ktxng@tpu.ru, [c]alex_chehlov@mail.ru

Abstract. In this paper optimal solutions to limit the subsidence of underground oil pipeline in insular permafrost are proposed.

1. Introduction

The advance of Russian oil and gas companies on the Asia-Pacific Region markets requires development of new fields and construction of gathering and trunk pipelines in Eastern Siberia and the Far East. Most of these areas are characterized by vast zones of insular permafrost that has an adverse effect on the stress-strain state of underground structures. Therefore, this fact should be taken into account in pipeline design and construction.

The insular permafrost is characterized by discontinuous occurrence of permanently frozen layer of small thickness and temperature close to 0 °C. Therefore, such a negative process as thermokarst is a rare case. In addition, as oil pipelines operate under positive temperature, which is conditioned by the need to reduce oil viscosity by heating, frost heaving is also not distinctive.

Soil settlement that occurs during formation of thawing halos represents a real threat to pipeline operation in insular permafrost conditions. To ensure safe operation of underground oil pipeline under these conditions, it is necessary to undertake a complex of actions to limit the subsidence of oil pipeline.

The goal of this research is to propose optimal solutions to limit the subsidence of underground oil pipeline laid in the frozen soil of different properties: thawing rate and depth of thawing halos [1].

2. Materials and methods

There are several technological concepts to limit the subsidence of pipeline in the permafrost environment: use of various types of thermal insulation, removal of icy soil from the trench, application of expansion joints (doglegs) and soil thermo-stabilizers. Nowadays, to ensure safe operation of underground oil trunk pipeline in insular permafrost, Russian pipeline engineering companies use polyurethane foam heat insulation and supplementary polystyrene heat insulation laid in the trench, as well as replace frozen soil by dry soil in the trench [2]. The choice of the definite engineering solution is based on soil settlement modeling and should meet pipeline strength requirements.

The choice of the measures that limit the severity of pipeline subsidence rests on the calculation of ultimate stress limit occurred within the subsidence area. The design ultimate stress limit of pipe steel is determined according to [3]. The magnitude of stress caused by internal pressure and temperature difference is deducted from this value. The ultimate stress defines the maximum pipeline subsidence.

The most intensive subsidence-induced stress occurs between frozen soil and thawed soil, as well as in permafrost areas characterized by different properties. Being the most dangerous, these zones are considered in calculating maximum allowable subsidence according to the method proposed in [4]. The pipeline is regarded as a restrained beam, and pipeline deflection is determined. Stress-induced deflection that corresponds to ultimate one is the maximum allowable subsidence of pipeline which is calculated by the following formula (1).

$$S_{\lim} = \frac{6 \cdot \sigma_{\lim}^2 \cdot I}{r^2 \cdot q \cdot E}, \quad (1)$$

where, S_{lim} – the maximum allowable subsidence;
σ_{lim} – stress limit;
I – centroidal moment of inertia;
r – pipe radius;
q – pipeline load;
E – modulus of elasticity.

According to [1], the soil settlement occurs due to thawing of ground ice and soil compaction caused by pipeline and backfill weight. The assumption that soil settlement occurs due to ice thawing only has been accepted. This assumption allows calculating the maximum allowable depth of soil thawing at given thawing rate by the following formula (2).

$$H_{\lim} = \frac{S_{\lim}}{A}, \quad (2)$$

where, H_{lim} – the limit depth of thawing halo;
A – thawing rate.

Calculating the depth of permafrost thawing under the pipeline is done by the method proposed in [1], which is based on the solution of the soil heat equation. The depth of thawing halos is determined for bare pipes, preinsulated pipes and pipes with supplementary insulation. The obtained values are compared with the maximum allowable depth of thawing halo. If the calculated thawing halo exceeds the maximum allowable depth, it is necessary to replace frozen soil by dry soil in the trench. In this case, the depth of soil to be replaced is determined by the difference between design and the maximum allowable depth of thawing halo.

The next step is to calculate the cost of measures aimed at minimizing subsidence of a pipe section (11 meters). Based on the calculation results, the optimal engineering solution (an action or set of measures) is selected. The chosen solution should guarantee allowable subsidence of underground pipeline at the lowest cost.

The choice of measures that limit the subsidence of underground oil pipeline was made in accordance with described algorithm. The pipeline specifications were as follows: outer diameter – 720 mm, wall thickness – 8 mm, internal pressure – 6.0 MPa, oil temperature – 30 °C. The modes of pipeline construction in permafrost characterized by different characteristics were examined. Based on the data [5-7], the depth of thawing halos is assumed to be from 3 to 11m, thawing rates – from 0.1 to 0.4, which corresponds to soils of second and third subsidence categories.

3. Results and discussion

According to the data [8], the optimal measures to limit the subsidence of underground oil pipeline in permafrost were selected (table 1).

Table 1. Measures to limit underground oil pipeline subsidence.

Depth of thawing halo (m)	Thawing rate			
	0.1	0.2	0.3	0.4
3	Bare pipe	Factory-insulated pipe	Factory-insulated pipe + 0.6m soil replacement	Factory-insulated pipe + 0.8m soil replacement
5	Factory-insulated pipe	Factory-insulated pipe + 0.9m soil replacement	Factory-insulated pipe + 1.5m soil replacement	Supplementary insulation + 1.0m soil replacement
7	Factory-insulated pipe	Supplementary insulation + 0.6m soil replacement	Supplementary insulation + 1.1m soil replacement	Supplementary insulation + 1.4m soil replacement
9	Factiry-insulated pipe + 1.0 m soil replacement	Supplementary insulation + 1.9m soil replacement	Supplementary insulation + 2.5m soil replacement	Supplementary insulation + 2.8m soil replacement
11	Supplementary insulation + 0.6 m soil replacement	Supplementary insulation + 2.2m soil replacement	Supplementary insolation + 2.8m soil replacement	-

IOP Conf. Series: Earth and Environmental Science **43** (2016) 012079 doi:10.1088/1755-1315/43/1/012079

4. Conclusion

Based on the obtained results, it is possible to draw a conclusion that the selected measures could guarantee the allowable subsidence of underground pipeline in all examined soils, except if thawing rate is greater than 0.3 at thawing depth being more than 9 m. Under these conditions, soil replacement is difficult to carry out due to the technical limitations of excavation equipment. In this case, construction of above-ground pipeline can be recommended.

References

[1] SNiP 2.02.04-88 2012 *Bases and foundations on permafrost. (Moscow: MinRegion)* p 118
[2] Lisin Y V, Soschenko A E, Surikov V I, Pavlov V V, Zotov M Y 2014 *Science and technology of pipeline transport oil and oil products* **1** 24-28
[3] SNiP 2.05.06-85* 2005 *Trunk pipelines (Moscow: FSLUE LAC)* p 60
[4] Mustafin F M 2008 *Building construction of oil and gas facilities (St. Petersburg: Nedra)* p 780
[5] Sokolov S M 2008 *Oil industry* **10** 126-127
[6] Wang Y, Jin H, Li G 2016 *Cold Regions Science and Technology* **126** 10-21
[7] Wang T, Zhou G, Wang J, Zhao X 2016 *Applied Thermal Engineering* **99** 591-598
[8] Local costing standards [Electronic resource]: *Official web-site Krasnoyarsk Region* URL: http://minstroy.krskstate.ru/tsn (reference date 24.04.16)

Stress-strain analysis of pipelines laid in permafrost

P Burkov[1,2], Van Yan'nan'[1], S Burkova[1]

[1] National Research Tomsk Polytechnic University, 30, Lenin Ave., Tomsk, Russia
[2] Tomsk State University of Architecture and Building, 2, Solyanaya Sq., Tomsk, Russia
E-mail: [1]burkovpv@mail.ru

Abstract. Increasing reliability of pipelines becomes a real challenge at all stages: design, construction and operation of pipeline systems. It is very important to determine the behaviour of the constructed pipeline under the operational and environmental loads using the design model in accordance with that one adopted in the rules and regulations. This article presents the simulation of pipeline in permafrost. The evaluation of the stress-strain state is given herein and the areas of the stress concentration are detected with the account for different loads occurred during the pipeline operation. Information obtained from the assessment of the stress-strain state of the pipeline allows determining sections in pre-emergency state (even before damages) and take all the necessary measures for eliminating them, thus increasing the pipeline system reliability. It is shown that the most critical pipeline cross-section is observed at the point of transition from one environment to another. The maximum strains decrease the level of the pipeline reliability. The finite element model is presented to determine the pipeline sections in pre-emergency state.

1. Introduction

The total length of the main pipelines in Russia is currently about 50 thousand kilometers. Despite the fact that the problem of the pipeline safety is constantly the focus of attention, over 40 thousand pipeline malfunctions and accidents occur annually in Russia, the losses being over 3% of the total volume of oil and gas production. A considerable part of accidents is conditioned by beyond-design loads received by pipelines. Thus, the assessment of technical conditions of pipelines must include monitoring of their stress-strain state during the operation. One of the main factors that define the operating reliability of underground pipelines is the soil-pipe interaction system.

The pipeline is subjected to soil effects, such as load, temperature, moisture, chemical and biological attacks, corrosion, and others. In turn, the pipeline affects the soil either through dead loads (pipe weight, pipe pressure, etc.) and live loads produced by temperature changes and the internal pressure of the pumped product [1-7].

The additional loads which occur during the pipeline operation, result in a rapid lifespan exhaustion. Beyond-design loads induced by landslips or seasonal soil movement are the most prevalent and dangerous and impossible to be accounted in designing. Beyond-design loads can be detected at the stage of construction and during the operation by measuring the stress-strain state of the pipeline.

As it is known, the main oil fields in Russia are located in West Siberia and the Far North. Therefore, the main pipelines inevitably cross the permafrost areas. Presently, the investigation and forecasting of the stress-strain states of the underground pipelines is a relevant problem nowadays.

Laying of the pipelines in permafrost conditions is usually carried out when the soil is frozen, since a seasonal thawing makes it impossible to move the construction machinery along the pipeline route. Herewith, the original state of the underground pipeline is determined by the position of the pipes laid in a frozen soil, and their stress-strain state should be analyzed with the account of the mechanical-and-physical properties of frozen soil [8].

2. Results and discussion

With the beginning of exploitation, the pipeline heats the frozen soil under the pipe which melts at the oil temperature of 0°C and higher. Due to the change of the mechanical-and-physical properties of frozen soil along the pipe, the area of thawing out is different at different pipe sections. This leads to bending, sagging and large spatial deflections (often with buckling) of the pipeline sections and, sometimes, to its destruction. The strength analysis allowing for the real operating conditions, is one of the main procedures that assure the high reliability of the pipeline. The strength analysis is mostly focused on the stress-strain state of the pipeline modified by the external loads [9-11]. It is conducted using the ANSYS finite element program. The pipeline section under research is laid in terrains with the temperature range from –46 to –49 °C of most cold five-day period. Thus, the northern-type pipes are used having the following characteristics: K56 strength class; 820 mm diameter; 14 mm wall thickness; 9.2 MPa internal pressure; 12 m pipeline section length; 09G2FB1[1] steel type.

The elastic properties of the pipe material are described by the Young's modulus and Poisson number. The analysis of the stress-strain state includes the circular stresses induced by the internal pressure and longitudinal or axial stresses due to all external loads. A solid model is designed using the ANSYS finite element (FE) program, is subjected to the boundary conditions and loads in accordance with the original data.

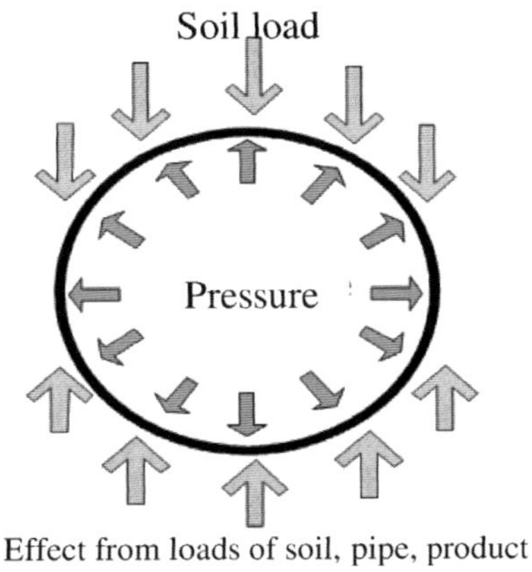

Figure 1. Pipeline walls exposed to loads

The three-dimensional models are obtained for the distribution of the stress-strain state observed in the most critical areas of the pipeline, that present all loads affecting the given structure. The FE models are shown in Figures 1–6.

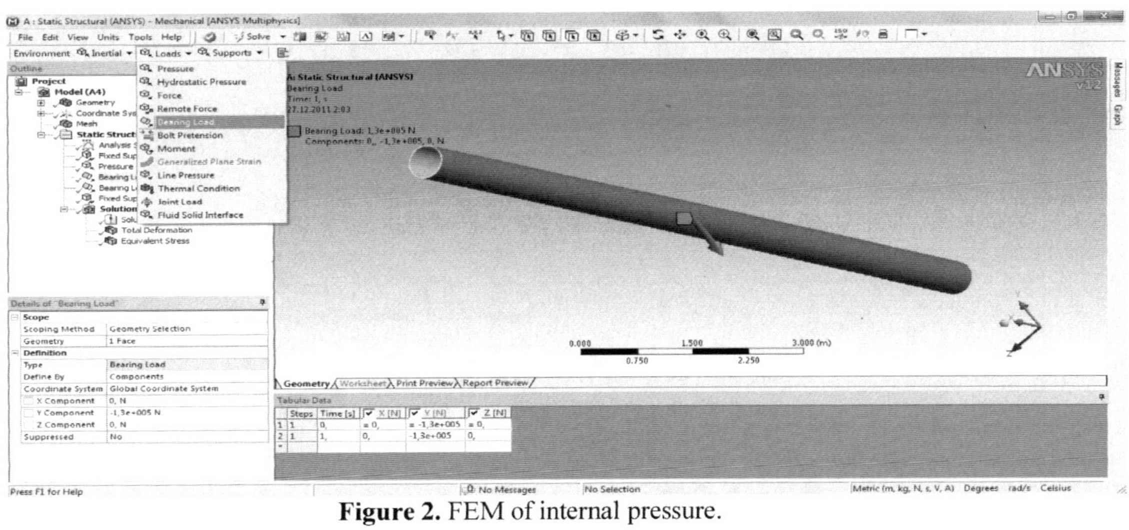

Figure 2. FEM of internal pressure.

[1] Steel composition: 0.09% carbon; 2% manganese; 1% silicon, 1.5% niobium, 0.1% vanadium.

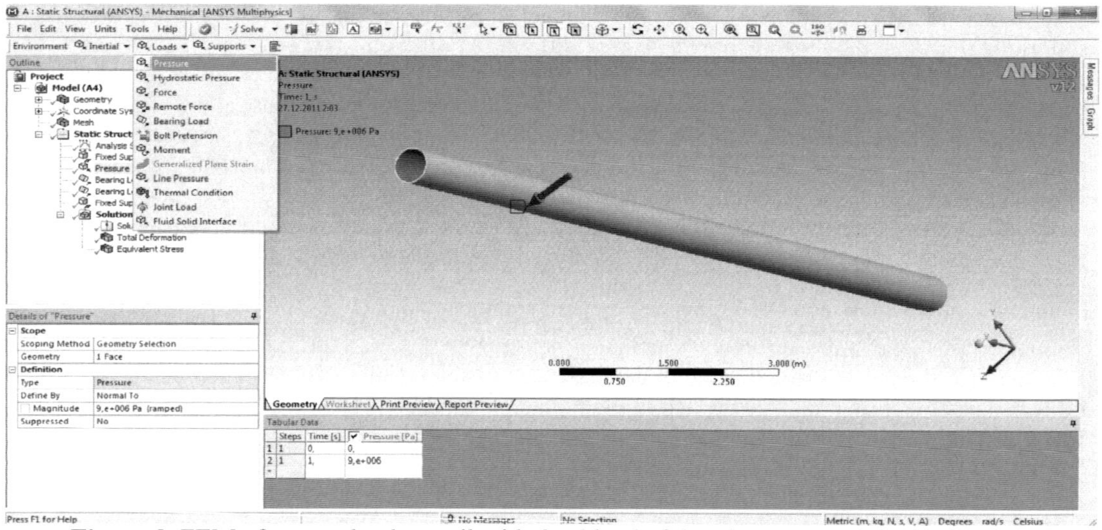

Figure 3. FEM of current load on soil with dead load of the pipe and product weight.

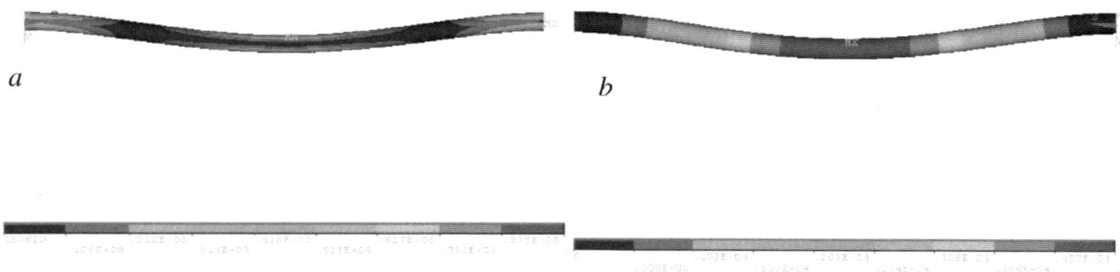

Figure 4. FEM of pipeline section in permafrost conditions:
a – maximum von Mises stress; *b* – deformation.

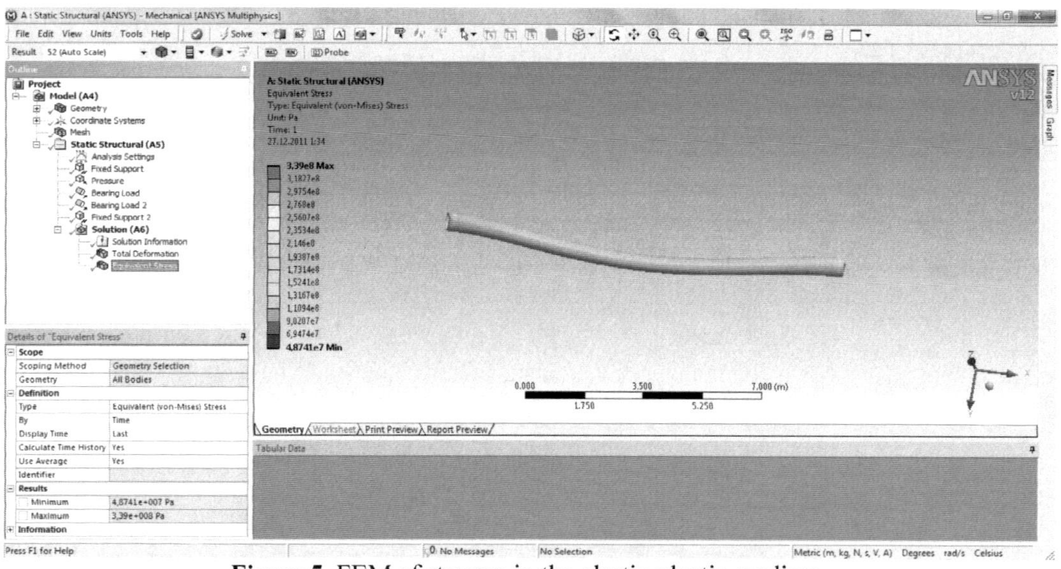

Figure 5. FEM of stresses in the elastic-plastic medium.

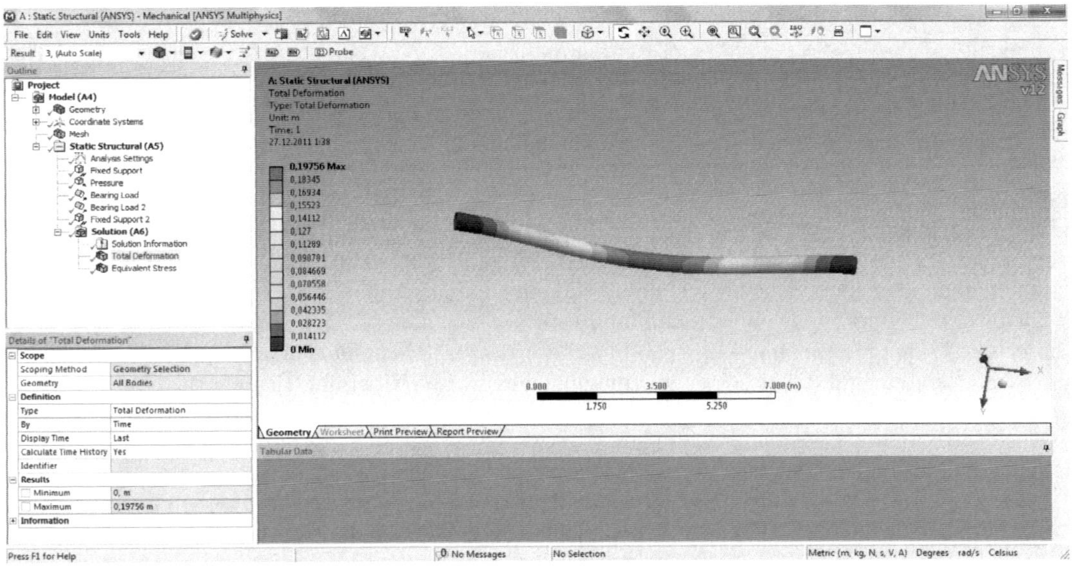

Figure 6. FEM of deformations in the elastic-plastic medium.

The FE models show no bending in places of the pipeline fixation. The obtained information allows constructing the diagram of the normal bending shown in figure 7.

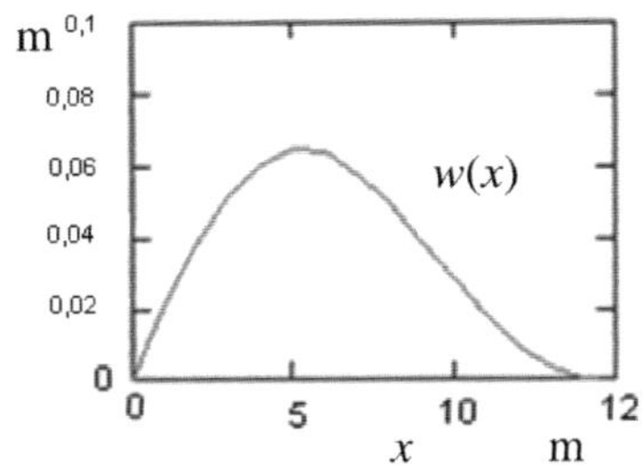

Figure 7. Normal bending of pipeline section in permafrost conditions: x – length; $W(x)$ – bending.

The maximum value of the pipeline bending corresponds to its 5 m removal from the conjunction edge. The bending is approx. 61 mm. To estimate the strength of the pipeline, the diagram is constructed for the axial and circular stresses depending on the pipe section length, figure 8.

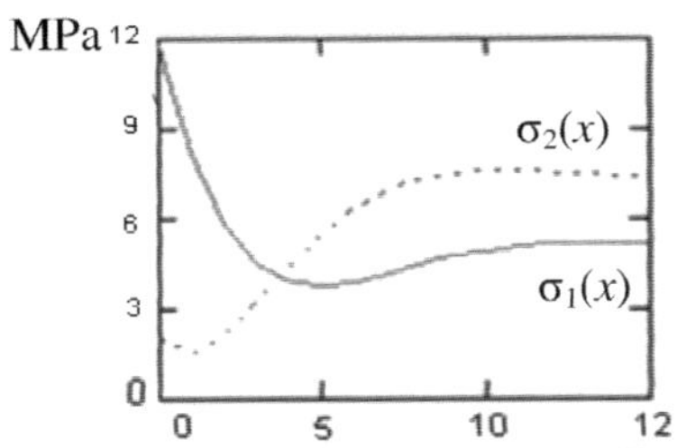

Figure 8. Distribution of axial and circular stresses in pipeline section in permafrost conditions: σ^1 – axial stress; σ^2 – circular stress.

The larger value of the axial stress corresponds to places of the pipeline fixation, i.e. the axial stress value nearby the internal wall of the pipeline exceeds the yield strength of the pipe material. The load-bearing capacity of the pipeline in places of its fixation is exhausted when the pressure in the pipeline exceeds its limit value at which the whole transversal section will be subjected to the plastic deformation.

The FEM analysis allows detecting the pre-emergency conditions of the main pipeline linear sections including those without defects and, taking measures for their elimination, increase the system reliability.

3. Conclusions

The strength analysis of the pipeline showed that buckling or sagging of its sections resulted in unallowable stresses. Larger pipeline sagging produce plastic deformations. Thus, the estimation of the stress-strain state and pipeline sections mostly exposed to loads, including those static and dynamic, is the important factor in providing a safe operation of the pipeline.

In order to increase the pipeline serviceability and estimate its technical conditions, it is advisable to timely conduct the technical diagnostics and strength analysis using the finite element method.

4. References

[1] Khasanov R R 2010 *Oil and Gas Business*. Расчет напряженно-деформированного состояния тройников штампосварных (ТШС) Vol. **2** URL: http://ogbus.ru/authors/KhasanovR/KhasanovR_1.pdf

[2] Kurochkin V V, Malyushin N A, Stepanov O A 2001 *Nedra-Biznestsentr*. Working life of pipelines. pp. 231.

[3] SNiP 2.05.06-85* (in Russian)

[4] Shammazov A M, Mugallimov F M, Nefedova N F 2000 *Nedra-Business-Center*. Main Oil-Pipelines Underwater Crossings. pp. 237.

[5] Basov K A 2002 *Computer Press*. ANSYS in Examples and Problems. pp. 224.

[6] Chigarev A V, Kravchuk A S, Smalyuk A F 2004 *Mashinostroyeniye*. *Ansys* for Engineers: Reference Book. pp. 512.

[7] Methodological Instructions on Examination and identification of Steel Tubes for Oil and Gas Pipelines. Industrial Guideline 39-1.11-014-2000

[8] Burkov P V, Kalmykova K G, Burkova S P, Do T T 2014 *IOP Conf. Ser.: Earth and Environ. Sci.* Research of stress-deformed state of main gas-pipeline section in loose soil settlement. Vol. **21** pp 012039

[9] Antropova N A, Krets V G, Luk'Yanov V G, Baranova AV 2015 *IOP Conf. Ser. Earth and Environ. Sci.* Reliability assessment of tunneling flow charts. Vol **24** pp. 012019.

[10] Chigarev M A, Kravchuk A S, Smalyuk A F 2004 *Mashinostroenie-1*. ANSYS for engineers. pp. 512.

[11] Butov V G, Nikulchikov V K, Nikulchikov A V, , Yashchuk A A 2016 *Proc. 16th International Multidisciplinary Scientific GeoConference SGEM*. Study of the stress-strain behavior and strength characteristics of sealer devices for oil-trunk pipelines.

IOP Conf. Series: Earth and Environmental Science **43** (2016) 012081 doi:10.1088/1755-1315/43/1/012081

Underground pipeline laying using the pipe-in-pipe system

N Antropova, V Krets, M Pavlov

National Research Tomsk Polytechnic University, Tomsk
E-mail: antropova@tpu.ru; krets@tpu.ru

Abstract. The problems of resource saving and environmental safety during the installation and operation of the underwater crossings are always relevant. The paper describes the existing methods of trenchless pipeline technology, the structure of multi-channel pipelines, the types of supporting and guiding systems. The rational design is suggested for the pipe-in-pipe system. The finite element model is presented for the most dangerous sections of the inner pipes, the optimum distance is detected between the roller supports.

1. Introduction

Methods of trenchless pipelay technology have a number of benefits in comparison to trenching, such as:

- low ecological damage;
- lack of necessity to develop and backfill bottom trenches, reinforce the coasline;
- reliability improvement due to the pipelay in a non-fractured rock to deeper depths;
- reduction of operating costs due to the lack of necessity for underwater inspections and the works on preventing periodical pipeline flooding;
- wild landscape conservancy, and many others.

The main drilling methods are directional drilling, method of curved pipes, Geokhod tunneler, screw-driven tunneling, microtunneling, tunneling, and pipe-in-pipe system.

2. Methodology

Trenchless pipeline technologies are used, in particular, for the main oil and gas pipeline laid under the artificial and natural barriers including the underwater crossings (see table 1).

Table 1. Main trenchless pipeline technologies

Trenchless pipeline	Field of application, benefits	Limitations
Directional drilling (325-1720 mm pipe diameter)	crossings under water reservoirs, roads, buildings, and other constructions; natural barriers, watersides; used when necessary to avoid surface damages.	Impossibility of pipeline repair and reconstruction works in case of emergencies.
Method of curved pipes (530 -1420 mm pipe diameter)		Restricted crossing length; impossibility of pipeline repair and reconstruction works in case of emergencies

Geokhod tunneler. Mining diameter up to 3 m; weight – 19 t		In development (Tomsk Polytechnic University)
Screw-driven tunneling		In development (Tomsk Polytechnic University)
Microtunneling (Herrenknecht tunneling system). Mining diameter up to 4 m; laying length – 1500 m		It is restricted by geological conditions and requires large space for assembly area
Shield tunneling		High construction cost. Minimum pipe diameter – 2 m.
Pipe-in-pipe method	used in multi-channel pipelines; simplicity of pipeline repair and reconstruction; drag force reduction in tunneling	Insufficient structural reliability of supporting and guiding systems

Except for the pipe-in-pipe methods, all other methods presented in table 1 are provided with the inverted syphon comprising coaxially welded pipes and representing a single-channel pipeline.

The inverted syphon is a multi-channel pipeline comprising the outer pipe and, at least, one inner pipe installed in the internal space of the outer pipe using the supporting and guiding systems.

There are several types of multi-channel pipelines with different design features:

- comprising at least one inner pipe installed in the outer pipe using plastic spring frames (fig. 1 a), [4];
- comprising at least one inner pipe installed in the outer pipe using supporting centering devices in the form of roller supports (fig. 1 b);
- comprising the outer pipe and at least two inner parallel pipes fixed to each other by the elastic compensating mechanism (fig. 1 c), [6].

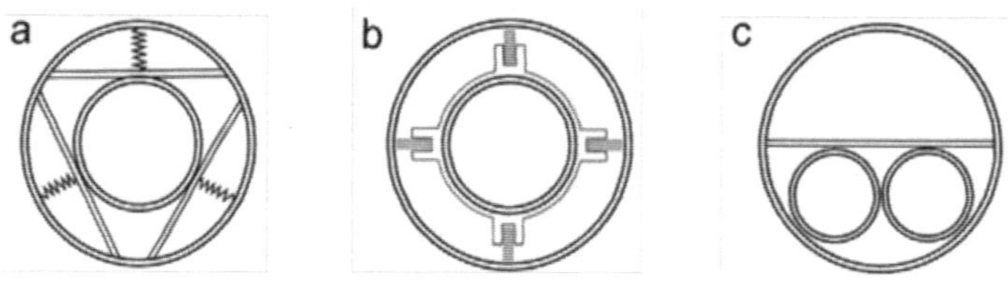

Figure 1. Types of multi-channel pipelines

The pipe lining can be also provided with the variety of wood and polymer lining materials. There is a range of the supporting and guiding systems that facilitate the pulling and centering of the inner pipeline and form a two- and multi-channel inverted syphon.

Several types of supporting and guiding systems are currently used for the pipeline arrangement. They are as follows:

- dielectric rings or spacers (fig. 2 a) are made of two or three semi-rings connected to one ring by metalware (bolts, nuts and washers) and fixed to the pipe. Spacer is made of the types 276-73 (83, 84) polyethylene in compliance with GOST 16338 [3].

- roller guide ring (fig. 2, b) are made of two or three semi-rings connected to one ring by metalware (bolts, nuts and washers) and fixed to the pipe. The dielectric rollers made of polyamide blocks are fixed to the outer pipe and facilitate the pipeline pulling. These polyamide blocks serve as a dielectric to prevent the contact between the inner and outer pipes [3].

- guide rings of PMTD type (fig. 2, d) are made of two semi-rings connected to one ring by bolt connection. The roller sliding supports manufactured in compliance with GOST 10589-87 of the dielectric material (polyamide-610), are installed and mounted to the outer surface of the PMTD ring. To reduce the drag force when pulling the pipeline, the sliding supports are mounted to the upper semi-ring, while the roller supports are mounted to the lower semi-ring [3].
- guide rings with a rocking roller (fig. 2, c) are made of three metal semi-rings with bolt connections. The lower support element has two rollers and rotates to a certain vertical angle [5].

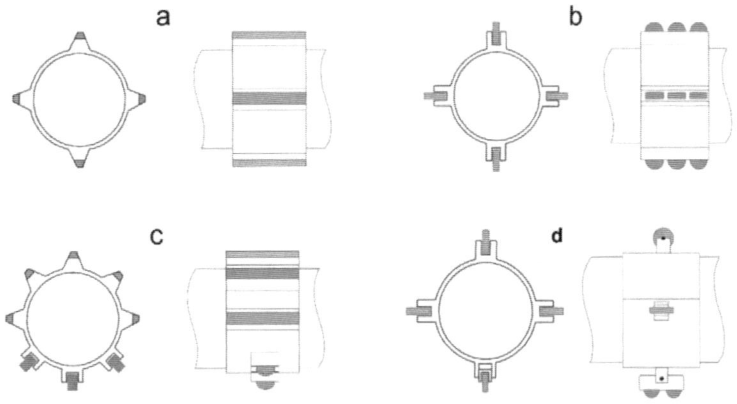

Figure 2. Types of supporting and guiding systems

The roller guide rings and the type PMTD rings are the most advanced guiding systems that allow reducing the drag force and the improvement of the structural reliability of the multi-channel pipeline.

The multi-channel pipeline with the roller guide rings was selected for the finite element modeling (FEM). The pipe section between the roller supports was analyzed and the pipeline axial displacements and stresses were identified using the ANSYS finite element program. FEM of the pipeline displacement is presented in figure 3.

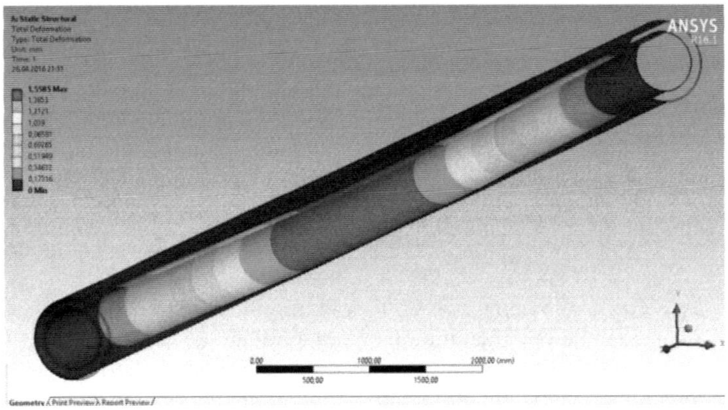

Figure 3. FEM of axial displacement of pipeline (distance between supports – 15 m)

3. Results and discussions

The dependence between the estimated von Mises stress and the pipeline displacement relative to the distance between roller supports is presented in table 2. The pipeline is subjected to plastic deformation tests using the finite element modeling. The task conditions are as follows:

- 530 mm pipe diameter;
- 8 mm wall thickness.

Table 2. Estimated von Mises stress and pipeline displacement relative to the distance between roller supports

N	Distance between roller supports, m	Dead load, N	Axial displacement, mm	von Mises stress, MPa
1	10	10202.4	0.32	5.7
2	**15**	**15303.6**	**1.5**	**13.3**
3	20	20404.8	4.8	20
4	25	25506	11.7	34.5

According to table 2, the most dangerous pipeline section is in the centre and the interface of the roller support and the outer pipe, i.e. the maximum stresses are observed, which reduces the pipeline reliability. The maximum axial displacements of the pipeline are observed in the central part between the roller supports. Depending upon the maximum axial displacements, the optimum distance between the roller supports is 15 m.

4. Conclusions

1. The structural analysis of the multi-channel pipeline and the types of the supporting and guiding systems showed that the most efficient technology is the pipe-in-pipe system.
2. The multi-channel pipeline was designed using the supporting and guiding systems in the form of the roller guide rings. FEM analysis of the inner pipe section between the roller supports showed that the optimum distance between them is 15 m since the pipeline bending relative to its axis tends to zero.

5. References

[1] Rybakov A P 2005 *Moscow, PressByuro Publ* . Principles of trenchless technologies (theory and practice). pp. 304.

[2] STO NOSTROI Underground utilities systems laid using directional drilling 2015 Moscow, BST Publ. pp. 145.

[3] Guide rings. URL: http://rosngsk.ru/oporno-napravlyayushchie-koltsa. Reference date: 08.04.2015.

[4] Erenkov O Yu, Bogachev A P and Yagubov E Z 2013 RF Patent 125228.

[5] Miklush S Ya, Minaev V N and Vylevko V V 2013 RF Patent 124762.

[6] Tsyplakov O G, Tskhadaya N D, Neredov V N, Yagubov Z Kh and Yagubov E Z 1999 RF Patent 2140605.

[7] Antropova N A, Krets V G, Luk'yanov V G and Baranova A V 2015 *IOP Conference Series: Earth and Environmental Science*. Reliability assessment of tunneling flow charts. URL: http://dx.doi.org/10.1088/1755-1315/24/1/012019. Reference date: 26.05.2016. Vol. **24.**

[8] A. Shadrina, T. Kabanova, V. Krets and L. Saruev 2014.*IOP Conference Series: Earth and Environmental Science. XVIII International Scientific Symposium in Honour of Academician M. A. Usov: Problems of Geology and Subsurface Development.* A study of specific fracture energy at percussion drilling. URL: http://iopscience.iop.org/1755-1315/21/1/012036. Reference date: 26.05.2016. Vol. **21**.

[9] Shadrina A V and Saruev L A 2015 *IOP Conference Series: Earth and Environmental Science In: Problems of Geology and Subsurface Development.* The behavior of enclosed-type connection of drill pipes during percussive drilling. URL: http://dx.doi.org/10.1088/1755-1315/27/1/012057 Vol. *27.*
[10] Shadrina A V and Saruev L A. 2014 *Conf. Mechanical Engineering, Automation and Control Systems.* Thread connection simulation and experimental analysis of strain wave energy loss. URL: http://ieeexplore.ieee.org/xpls/abs_all.jsp?arnumber=6986921.

IOP Conf. Series: Earth and Environmental Science **43** (2016) 012082 doi:10.1088/1755-1315/43/1/012082

The analysis of repeated failures of pipelines in Kal'chinskoe oil field

E N Shavlov[1,2], O V Brusnik[1,3], V G Lukjanov[1,4]
[1]National Research Tomsk Polytechnic University, 30, Lenin Ave., 634050, Tomsk, Russia
[2]shavlov@tpu.ru, [3]brusnikov@tpu.ru, [4]lukjanov@tpu.ru

Abstract. The paper presents the chemical analysis of oilfield water and hydraulic analysis of the liquid flow in Kal'chinskoe oil field pipeline that allow detecting the causes of the internal corrosion processes. The inhibitor protection is suggested to reduce the corrosion rate in the pipelines of Kal'chinskoe oil field. Based on the analysis of the pipeline failures, it is suggested to replace steel pipes by fiberglass pipes.

1. Introduction

The system of oil gathering and transportation depends on the natural climatic conditions (continental climate, bogginess); cluster-well development; method, amount, and rate of the oil production; water-cut oil; physico-chemical properties of oil, gas, water and parameters of the reservoir production [1].

Kal'chinskoe oil field was found in 1990 and is today the southmost field in the Tyumen region. The field site lies within the Middle-Ob' Plain and is represented by a dissected lacustrine-alluvial plain. The hydrographic network of this oil field includes the Irtysh basin and its feeders. Owing to the proper drainage of surface and groundwater, a thick forest is found nearby the river valleys. The bogginess is 81% of the area, including 4% of open bogs. The peat bog layer achieves 6-8 cm. The field site is represented by loams, clayey and sandy soils of different size. In accordance with the SNiP 2.01.01-82, the field site refers to the 1st climatic area, D subarea. And in accordance with the SNiP P-7-81 'Seismic construction', it refers to a seismic area with the earthquake intensity of 6. The main volume of oil extracted from Kal'chinskoe field, falls to Achimovskaya band.

Kal'chinskoe oil field is characterized by 86% water-cut oil. Oilfield waters of Kal'chinskoe oil field are represented by complex solutions comprising inorganic salts, gases, soluble organic substances. Among the substances dissolved in oilfield water organic salts predominate, such as chlorides, sulfates, and alkali and alkali-earth metal carbonates. Gas dissolved in oilfield water mostly contains methane. The total composition of salts dissolved in oilfield water is called mineralization. Oilfield waters of Kal'chinskoe oil field relate to the sodium bicarbonate type of water. The water mineralization ranges between 12.21 and 14.21 g/l upon the average [1].

2. Hydraulic analysis of the liquid flow in Kal'chinskoe oil field pipeline

Table 1 presents the original data on the physical properties of fluids pumped through the pipeline and its hydrodynamic parameters. Using these data, the fluid flow rate and the critical rate of transition from stratified to non-stratified oil–water flows were obtained for the pipeline section in Kal'chinskoe oil field.

The condition of corrosion resistance is satisfied in case the flow rate is higher than the critical rate. If this condition is not satisfied, intense localized corrosion frequently occurs on the metallic surface of the pipeline [2].

The calculations are carried out in compliance with the methodology given in *The Design and Operation Manual for Corrosion-Resistant Protection of Pipelines in West Siberian Oil Fields* (Regulatory Document 390147323-339-89-R) with a view to determine the modes of fluid control and the maximum rate of corrosion.

Table 1. Physical properties of controlled fluid and hydrodynamic parameters of Ø114 pipeline

Original data	
Parameters	**Values**
Average pressure, [Pa]	1060000
Volumetric flow rate, [m^3/day]	212
Inner diameter of pipeline, [m]	0.098
Oil delivery temperature, [K]	316
Gas factor, [m^3/m^3]	45
Water content, [%]	72
Oil/water interfacial tension, [N/m]	0.03
Oil dynamic viscosity, [Pa*s]	0.008
Saturation pressure, [Pa]	11300000
Oil density, [kg/m^3]	884
Density of bottom water, [kg/m^3]	1005
Ion composition of bottom water	
Parameters	**Values**
Cl, [g/l]	3.3
HCO3, [mg/l]	2554
Ca, [mg/l]	3.3
pH	7
Design values	
Parameters	**Values**
Oil density at delivery temperature, [kg/m^3]	869.4732
Volume flow rate, [m^3/s]	0.002453704
Volume flow rate in operation conditions, [m^3/s]	0.002450865
Volume gas discharge in operation conditions, [m^3/s]	0.002739484
Volume flow rate of oil-gas-water mixture in operation conditions, [m^3/s]	0.005190349
Consumption gas content	0.527803418
Density of oil-in-water emulsion, [kg/m3]	908.2
Oil dynamic viscosity in operation, [Pa*s]	0.006968444

The hydraulic analysis of the liquid flow in the pipeline of Kal'chinskoe oil field includes the calculation of the estimated flow rate V_f and absolute critical rate V_c of oil-gas-water mixture. The estimated flow rate V_f is obtained from

$$V_f = \frac{4Q}{\pi D^2} = \frac{4 \cdot 0.005190349}{3.14 \cdot 0.098^2} = 0.69\, m/s\,,$$

where Q is the volume flow rate of oil-gas-water mixture in operation conditions; D is the inner diameter of pipeline.

The absolute value of the critical rate V_c can be found from

$$V_c = 6.69 \frac{D_0^{0.268} \sigma^{0.171} |g(\rho_w - \rho_e)|^{0.366}}{\rho_w^{0.536} \nu^{0.073} \left|\left(-10.96\beta^2 + 9.94\beta + 1\right)^{0.659}\right|} = \frac{0.098^{0.268} 0.03^{0.171} |9.8(1005 - 908.2)|^{0.366}}{1005^{0.536} 0.000001^{0.073} \left|\left(-10.96 \cdot 0.52^2 + 9.94 \cdot 0.52 + 1\right)^{0.659}\right|} =$$

$$= 6.69 \frac{0.5365 \cdot 0.55 \cdot 12.29}{40.659 \cdot 0.364 \cdot 2.18} = 0.76\, m/s,$$

where σ is the oil/water interfacial tension; g is the gravity factor; ρ_w is the density of bottom water; ρ_e is the density of oil-in-water emulsion; ν is the kinematic coefficient of viscosity; β is the consumption gas content.

The hydraulic analysis shows the stratified structure of the flow since the flow rate of the oil-gas-water mixture is 0.69 m/s, which is lower than 0.76 m/s of the critical rate and leads to the corrosion conditions.

The chemical analysis of oilfield water and the hydraulic analysis of the liquid flow in Kal'chinskoe oil field pipeline show that the localized corrosion occurs due to the oilfield water attack and corrosion processes induced by the flow conditions.

3. Corrosion inhibition of pipelines

The internal surface of pipelines is exposed to a corrosive environment. Thus, the steel pipes, welded joints and pipeline facilities must be protected from the effects of internal corrosion. At a rate of corrosive environment of 0.2 mm per year and higher, nonmetallic pipe materials should be used for the pipes with the protective internal coating. In steel pipes without the protective coating corrosion inhibitors are used to slow down galvanic corrosion processes. Corrosion inhibitors are used in pipelines in accordance with the developed technology [1].

In Kal'chinskoe oil field, the pipelines are protected by chemicals subject to such physico-chemical properties as

- name, percentage, hazard effect;
- appearance, color, smell;
- solubility;
- specific weight/density at 15-20 °C (g/cm^3);
- vapor pressure at 20°C (millibar);
- pH of dilute solution (concentration, mole);
- viscosity (cS);
- flash point (°C);
- dissolving temperature (°C);

- boiling temperature (°C);
- chilling temperature (°C).

In Kal'chinskoe oil field, Azole 5010 corrosion inhibitor has found a use. It refers to nitrogen-based cationic surfactants and is used to protect steel pipes in oil-and-gas industry. Azole 5010 is produced in two types, i.e. oil-soluble A type and water-soluble B type. Additionally, the available type A and the type B inhibitor concentrates contain 60–70 and 80 % active agents, respectively. Azole 5010 inhibitor concentrates can be used to produce both commodity forms and active-matrix conservation materials and thin-film coatings to protect pipelines from atmosphere corrosion. While storing and transporting, Azole 5010 retains its aggregative stability within the temperature range from –45° to 50°C. With the constant monthly dosage of 25–50 g/t, Azole 5010 has no negative effect on commercial oil, prevents steel from corrosion cracking, inhibit the growth of sulfur reducing bacteria. The type B Azole 5010 is more efficient in oil-gathering systems and pressure pipelining. It possesses low flow rates at stratified flow patterns and also in reservoir pressure maintenance systems. The optimum inhibition dosages of Azole 5010 are 100–150 g/t (shocking) and 15–25 g/t (working). The protective effect of Azole 5010 film in oil-water emulsions comes to 85–95% [3]. Figure 3 presents the time/rate dependence of the pipeline corrosion.

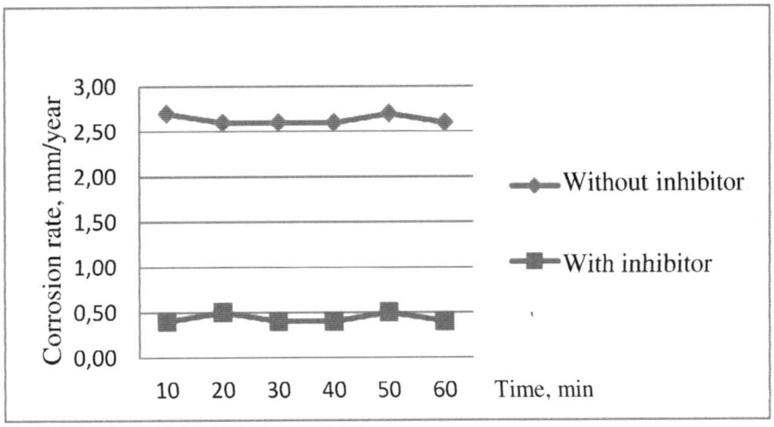

Figure 1. Time dependence of corrosion rate

4. The analysis of the pipeline failures

The operational data are used to analyze the pipeline failures in Kal'chinskoe oil field. This analysis shows that during the time period from 2011 to 2015, three failures were registered as shown in Figure 2. Despite the corrosion protection provided by inhibition, a greater number of failures are registered in the pipelines that have been in operation for not more than five years. This is probably, due to mechanical impurities presenting in the fluid that are involved in hydroerosion of the protective film formed by inhibitors.

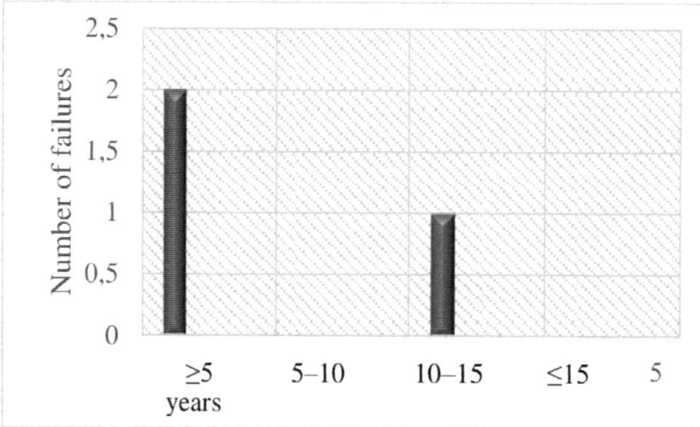

Figure 2. The dynamic failure analysis in Kal'chinskoe oil field for the years of 2011–2015

5. Conclusion

To avoid the localized corrosion processes occurring on the internal surface of pipelines, it is advisable to apply composite or fiberglass pipes. The implementation of fiberglass pipes in Kal'chinskoe oil field will prevent the pipelines from the repeated failures occurred due to the internal corrosion. The use of fiberglass pipes will provide a longer operation of the pipeline extending its service life up to 25 years that, in turn, will result in the reduction of their maintenance costs and elimination of emergencies.

References

[1] Pipeline technology regulations for the formation pressure maintenance system of Kal'chinskoe oil field 2012 (Tyumen) (in Russian)
[2] Regulatory Document 390147323-339-89-R The Design and Operation Manual for Corrosion-Resistant Protection of Pipelines in West Siberian Oil Fields (in Russian)
[3] Kotlasskii chemical plant [official website] URL: http://kchz.ru/item/azol5010

Strength characteristics of sealer devices at design stage

V G Butov[1], K A Golubeva[1], V K Nikulchikov[1] and A A Yashchuk[2]

[1] National Research Tomsk Polytechnic University, 30, Lenin Ave, Tomsk, Russia
[2] National Research Tomsk State University, 36, Lenin Ave, Tomsk, Russia
E-mail: [1] nikulchikov@tpu.ru, [2] rainbow@niipmm.tsu.ru

Abstract. The paper deals with strength characteristics of several sealer devices used to seal the inner cavity of an oil pipeline during the replacement of a pipeline section. Construction calculations revealed the stresses related to the pressurized rubber-cord sealing element rupture resulted from an emergency situation. It was concluded that it is necessary to test the operating parameters before applying the sealers. Estimation of the safety factor of existing sealer devices designs was conducted and recommendations for its increase were proposed.

1. Introduction

Nowadays the environmental policy of petroleum industry enterprises aims at improving environmental management, environmental protection and environmental safety with regard to the requirements of ISO 14001:2004 standard.

The priority tasks of the enterprises include the reduction of risk of pipeline accidents having environmental consequences and timely repair of pipelines, during which the sealer devices needs to be used for sealing the inner cavity of pipeline.

The design of sealer devices requires structural strength assessment taking into account the stresses arising from the rupture of rubber-cord sealing member in a possible emergency situation.

Structural scheme of the sealer device for covering the inner cavity of DN 400/500/800/1000/1200 pipelines is shown in figure 1 [1].

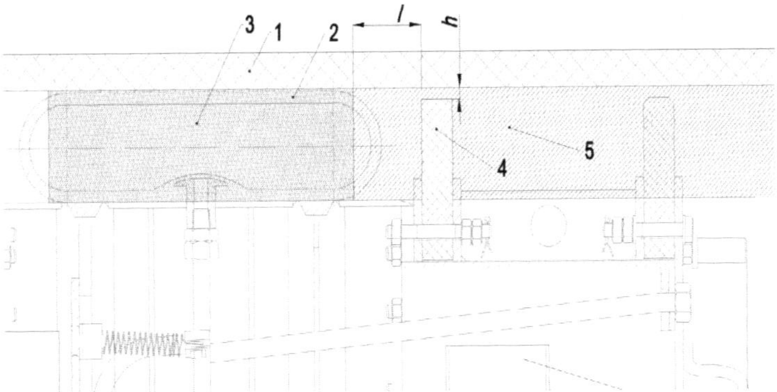

Figure 1. Structural scheme of oil pipeline sealer device and FVM model domains: *1* – pipeline; *2* – sealing member; *4* – aligning bearing; *3* – domain with initial operating pressure; *5* – domain with atmospheric pressure.

IOP Conf. Series: Earth and Environmental Science **43** (2016) 012083 doi:10.1088/1755-1315/43/1/012083

2. Research approaches

The finite element method was used to calculate the stress-strain state of the sealer device body under the pressure pulse caused by sealing member rupture [2, 3]. The ANSYS Mechanical software was applied in numerical calculations.

The finite element model was built with second order integration elements. Due to the symmetry only 1/6 part of the device body needs to be analyzed. The model takes into account only those structural elements and welded joints which have a considerable influence on the stress-strain behavior of the body. T-bar and butt welds connecting body structural elements with the flange are the most stressed elements in case of the pressure pulse from sealing element rupture. Therefore, the stress-strain state of these joints was given particular attention when analyzing the calculation results.

The following boundary conditions were applied to the model:

– the edges of holes for bolts securing body to the core was rigidly fixed;

– pressure on aligning bearing side facing the core.

The following physical properties of steel were used [4]: Young modulus of 210 GPa, Poisson's ratio of 0.3, density of 7900 kg/m^3 and yield stress of 245 MPa.

According to [5], the following allowable stresses were used when analyzing the results:

– nominal allowable stress of the parent material 140 MPa;

– nominal allowable stress of butt weld 126 MPa.

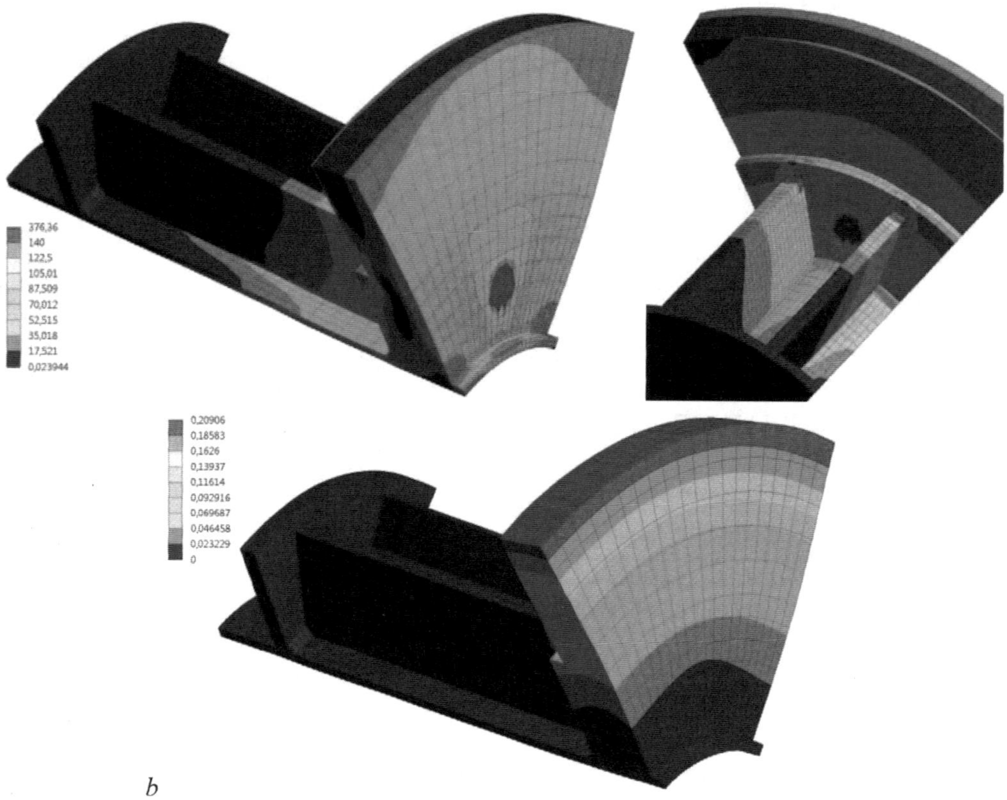

a

b

Figure 2. Stress-strain behavior of DN 400 sealer device under static load: *a* – von Mises stress, MPa; *b* – total displacements (scaled up), mm.

2.1. Elastic approach

First, we assess the strength of the body according to standard engineering procedures. In this approach, the model of the elastic body material and the maximum possible excess pressure of the shock wave accounting for its dynamic nature were used.

The maximum overpressure caused by the shock wave can reach 1.95 MPa, under the operating pressure of 2.05 MPa in the sealing member.

As this overpressure is applied to the support almost instantaneously (the natural period of the body is approximately (0.5-3) $\cdot 10^{-3}$ s while upslope time of the shock wave is ~$5 \cdot 10^{-5}$ s), it is necessary to take into account the impact effect of the load. Theoretical dynamic amplification factor of the instantaneous load is 2 [6], in practical calculations it is recommended to use the value of not less than 1.5 [5]. Thus, the overpressure was set to 3.9 MPa.

The results of elastic approach are presented in figures 2 and 3. In the equivalent stress contour plots the red color indicates regions with stresses exceeding the values allowed by [5]. It should be noted that the evaluation based on engineering approach suggests the possible breakage of the device.

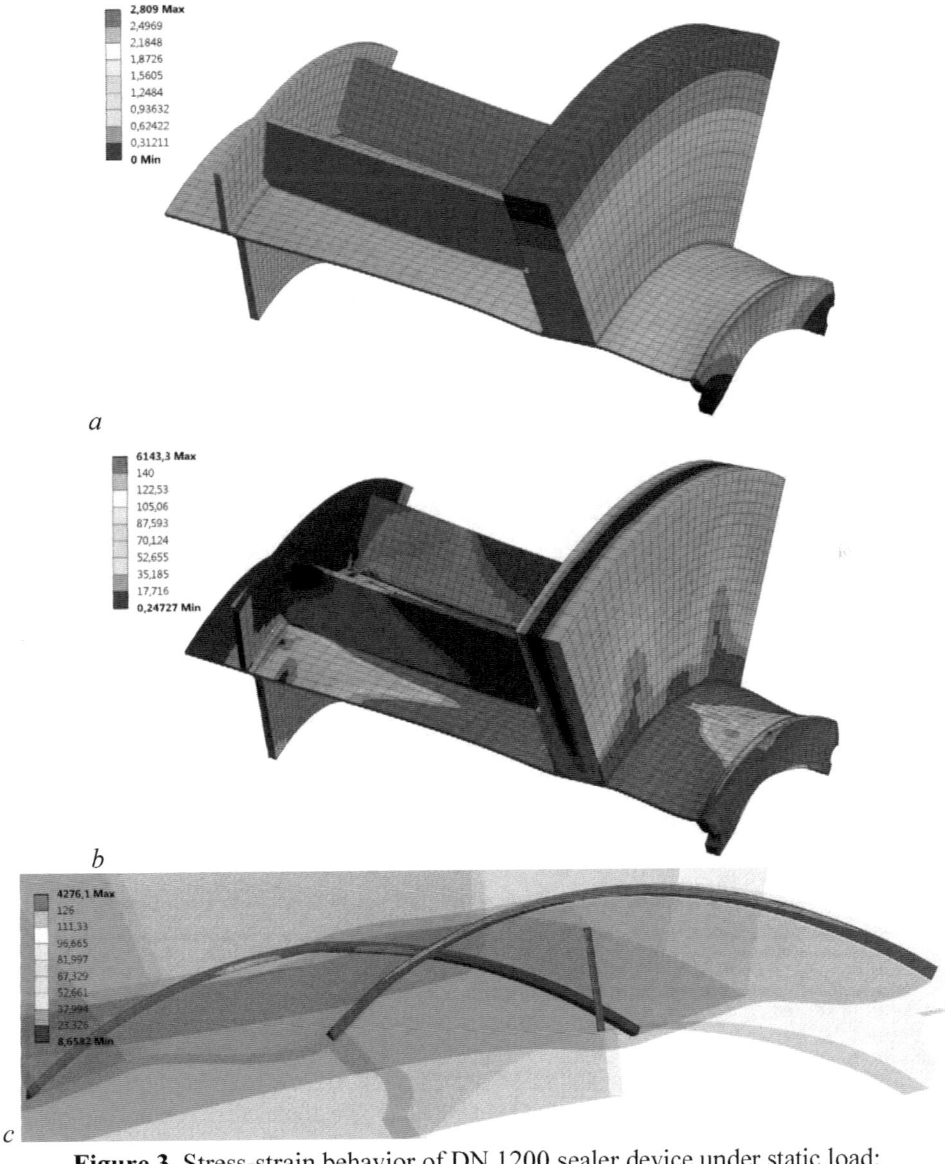

Figure 3. Stress-strain behavior of DN 1200 sealer device under static load:
a – total displacements (scaled up), mm; *b* – von Mises stress, MPa;
c – von Mises stress in most loaded welded joints, MPa.

2.2. Elastoplastic approach
More rigorous and close to real-life calculations were based on elastoplastic material model [7]. Multilinear isotropic hardening material model used the following additional steel properties: strength limit of 370 MPa and reduction of area at fracture of 55% (figure 4).

The results of elastoplastic calculations for DN 400 sealer device are presented in figure 5. The figure shows that there are no stresses close to the strength limit, thus the DN 400 sealer device is capable of withstanding the pressure pulse caused by the sealing member rupture.

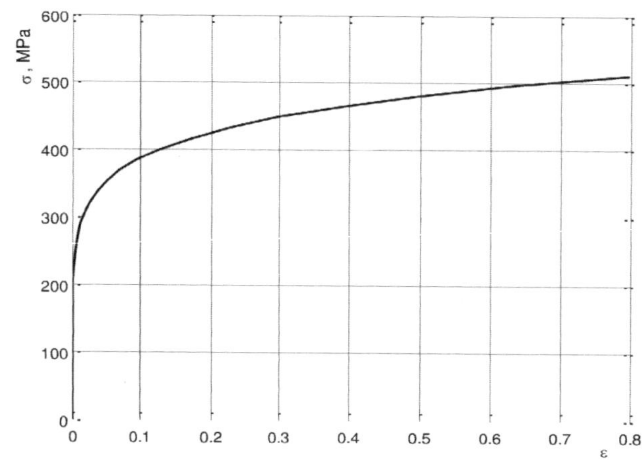

Figure 4. The stress-strain curve for steel.

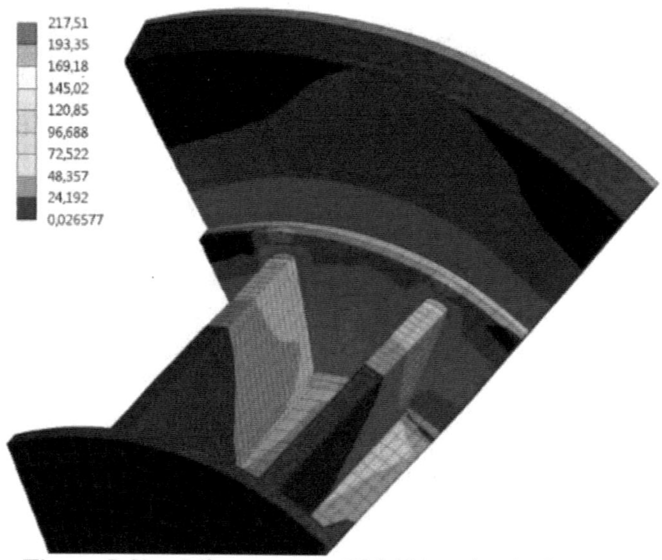

Figure 5. Von Mises stress in DN 400 sealer device, MPa.

The numerical solution for DN 1200 sealer device with the overpressure of 39 atm is non-converging, apparently indicating breakage. The results of elastoplastic calculations for DN 1200 sealer device with 15 atm overpressure (maximum value for which the numerical solution could be obtained) are presented in figure 6. As in the figures herein above, the red color indicates regions with stresses and strains exceeding the allowed values. The results show that plastic strains in most of the aforementioned welds exceed the allowable limit.

IOP Conf. Series: Earth and Environmental Science **43** (2016) 012083 doi:10.1088/1755-1315/43/1/012083

3. Results and discussion

The results of the calculations should be taken into account when designing oil pipeline sealer devices with rubber-cord sealing element.

1. The results based on elastic approach using the upper estimate of overpressure give the equivalent stresses in the body of DN 400 sealer device exceeding the maximum allowable values according to [5]. For the DN 1200 sealer device the equivalent stresses in all the welds connecting the body with the flange exceed the maximum allowable values.

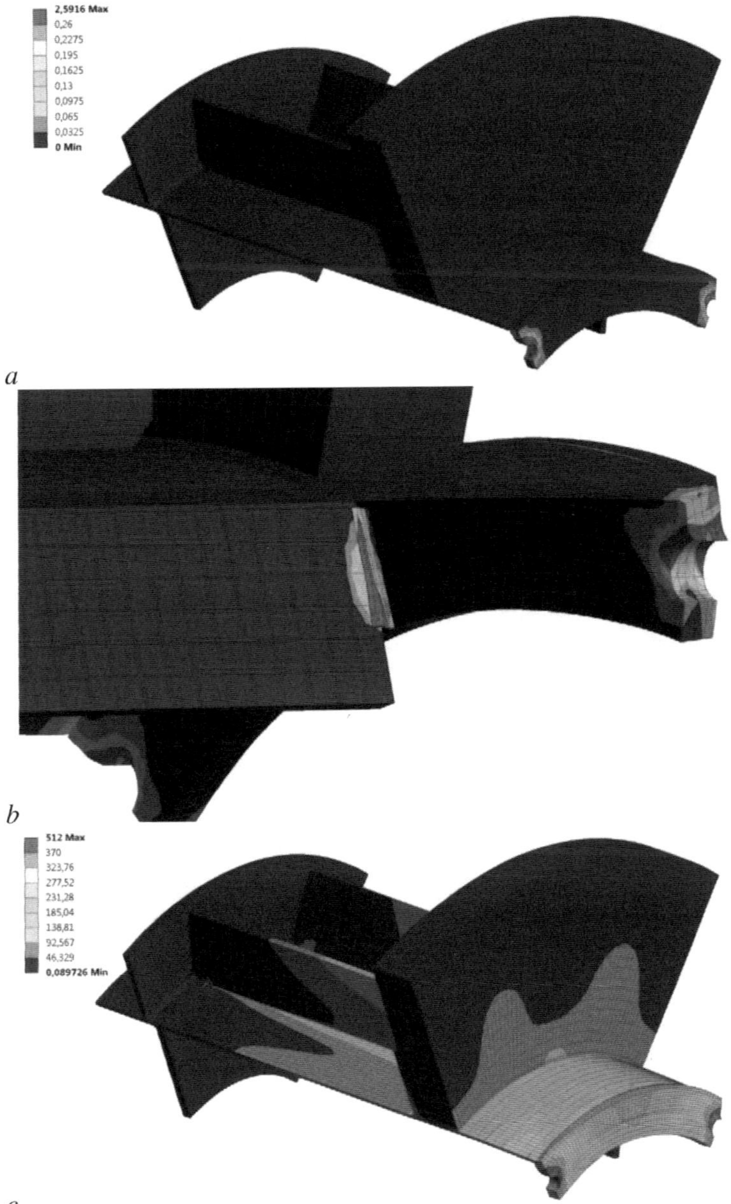

Figure 6. Elastoplastic stress-strain behaviour of DN 1200 sealer device: a – equivalent plastic strain; b – equivalent plastic strain in most loaded regions; c – von Mises stress, MPa.

2. Numerical calculation based on elastoplastic approach reveals that stresses occurring in the DN 400 sealer device exceed yield stress, but are less than the strength limit. Thus, in an emergency situation there will be an irreversible deformation of the device without its destruction. Therefore, the DN 400 sealer device has sufficient strength to ensure safety in a case of emergency (the case of a pressurized sealing member rupture).

3. Numerical calculation based on elastoplastic approach reveals that the stresses occurring in the DN 1200 sealer device exceed the strength limit, thus the device will break down in an emergency situation.

4. Conclusions

The results for the DN 1200 sealer device suggest that the safety factor is less than 1. In order to increase the safety factor to a safe level it is necessary to ensure the strength of the connections of the sealer elements:

1. Increase the strength of the T-bar welded joints connecting the ribs with the flanges.
2. Increase the strength of the butt welded joints connecting the body with the flanges.
3. If necessary, increase the strength or number of bolts connecting the flanges to the core.
4. Consider other ways of connections used in the sealer design. For example, it is possible to use additional connection of the flanges by stud-bolts and nuts instead of T-bar welded joints. Another option is to extend the bolts connecting the body with the core to the flange and thus replace the stud-bolts proposed above.

5. References

[1] Brezgin A E, Denisov F M and Siciak M M 2004 Device for reusable sealing of the pipeline's open end, RF Patent no. 2240466

[2] Zienkiewicz O C 1977 *The Finite Element Method* (London: McGraw-hill)

[3] Burkov P V, Burkova S P and Kravchenko A N 2015 *IOP Conference Series: Materials Science and Engineering.* Finite Element Model of Trenchless Pipe Laying. Vol. **91** pp. 012052.

[4] Anuryev V I 2001 *Moscow. Mashinostroenie.* A Handbook for Designers and Mechanical Engineers. Vol. **1**

[5] Laschinsky A A and Tolchinsky A R 1970 *Leningrad. Mashinostroenie.* Fundamentals of Design and Calculation of Chemical Equipment.

[6] Darkov A V and Shapiro G S 1975 *Strength of Materials* (Moscow: Higher School)

[7] Gokhfel'd D A, Getsov L B and Kononov K M 1996 *Reference book. Ekaterinburg: Ural Branch of Russian Academy of Sciences.* The Mechanical Properties of Steels and Alloys in Unsteady Loading.

Financial Aspects of Technological Concept for Energy Efficiency Enhancement during Stripper Wells Development in Tomsk Region

I Sharf[1], M Sundetov[2], I Shenderova[3], L Grinkevich[4]
[1,2,3] Institute of Natural Resources, National Research Tomsk Polytechnic University, Tomsk, Russia
[4]Tomsk State University, 36 Lenin Av., Tomsk, 634050, Russia

E-mail: irina_sharf@mail.ru

Abstract. The issue of operating costs cutting in terms of falling oil prices on the world market actualizes the challenge to find technological solutions to reduce electricity consumption during well operation. This is especially important for stripped-wells of small deposits in Tomsk region. The correlation analysis between the cost of oil production, electricity, heat and fuel consumption during the extraction of one ton of oil allowed the authors to focus on the financial aspect of such technological solutions like periodic well operation in the Shinginskoye field as well as to recommend the application of this method at the other fields in Tomsk region.

1. Introduction

Cost of oil production is one of the basic parameters determining the threshold of oil companies' financial stability. This issue has become even more relevant in the conditions of world energy prices fall on the world market [1, 2]. As a consequence, the issue of one oil barrel cost (Table 1) has become one of the most important stimuli for the technological innovations development [3]. As it could be seen from the Table 1, there is a significant difference in the amount of operational expenditures. In particular, operational expenditures in the Great Britain are twice higher than in Norway.

Table 1. Overall cost, capital and operational expenditures to produce one barrel of oil

	Overall cost	Capital expenditure	Operational expenditure
United Kingdom	52.5	21.8	30.7
Brazil	48.8	17.3	31.5
Canada	41.0	18.7	22.4
United States	36.2	21.5	14.8
Norway	36.1	24.0	12.1
Angola	35.4	18.8	16.6
Colombia	35.3	15.5	19.8
Nigeria	31.6	16.2	15.3
China	29.9	15.6	14.3
Mexico	29.1	18.3	10.7
Kazachstan	27.8	16.3	11.5
Libya	23.8	16.6	7.2
Venezuela	23.5	9.6	13.9
Algeria	20.4	13.2	7.2
Russia	17.2	8.9	8.4
Iran	12.6	6.9	5.7
UAE	12.3	6.6	5.7
Iraq	10.7	5.6	5.1

Saudi Arabia	9.9	4.5	5.4
Kuwait	8.5	3.7	4.8

The reducing operating expenditures were the major factor for shale oil production rise in the USA. Indirect evidence of this is the fact that the number of patents related to the introduction of innovations in the production industry has more than doubled from 2005 to 2010. Besides, the reduction of operating costs contributed to the fact that electricity prices in the United States for industrial companies, including oil producers, are almost one and a half times lower than for the residential sector. Compare – in 2014 the cost of electricity for the citizens was 12.5 cents per 1 kWh, while for industry 7.01 cents per 1 kWh [4].

It should be noted that the oil cost in Russia is rising due to a number of factors, such as: inflation, complexity of production, infrastructure and geological conditions (Table 2).

Table2. Oil production cost in Russia, rub/t [5]

	2012	2013	2014	2015
I quarter	7476.9	7420.5	8655.5	9812.2
II quarter	7194.9	7312.5	8666.4	10123.8
III quarter	7600.7	7853.4	8845.2	9665.1
IV quarter	7695.1	8307.9	8246.3	8784.6

The figures below illustrate the tendency of heat, electric and fuel energy rising in reliance on one ton of produced oil and oil production cost (Figure 1, 2, 3).

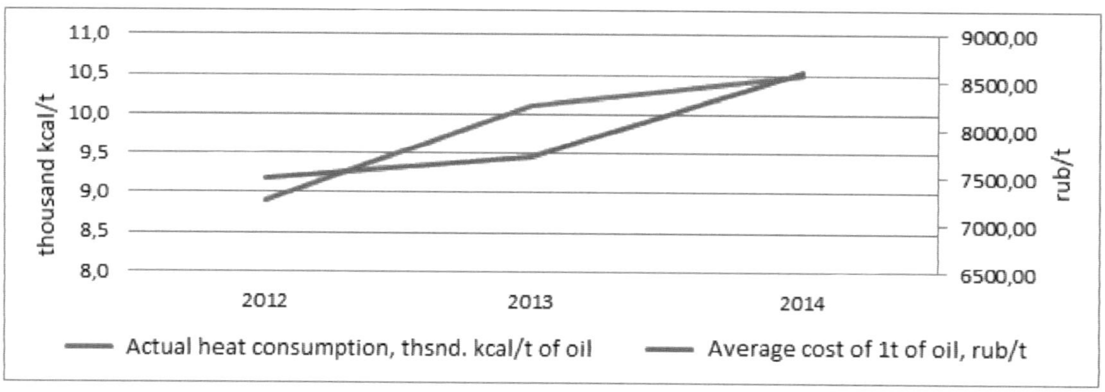

Figure 1 Correlation of actual heat consumption per 1 ton of oil and average oil production cost (rub/t).

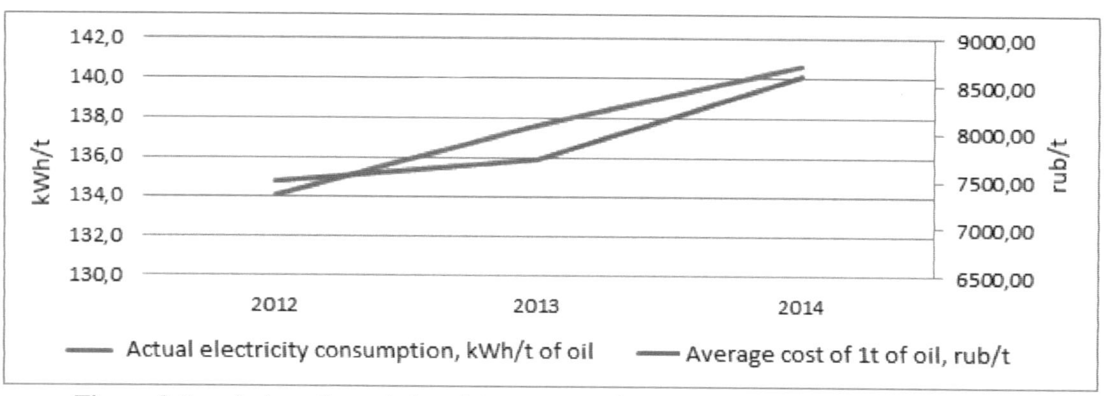

Figure 2 Correlation of actual electricity consumption per 1 ton of oil and average oil production cost (rub/t).

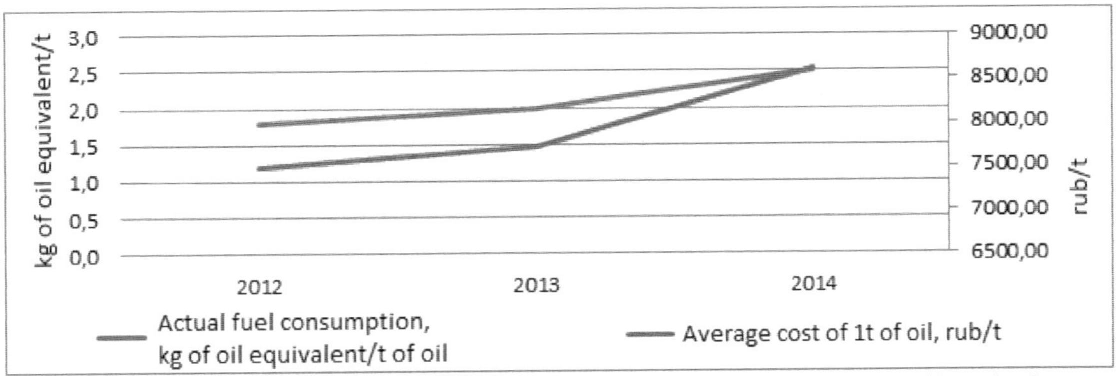

Figure 3 Correlation of actual fuel consumption per 1 ton of oil and average oil production cost (rub/t).

As it can be seen from these figures, there is a difference between the growth rates of oil production cost and energy consumption costs, which defines the objectives of the companies in the development of programs to improve the energy efficiency of the production processes.

In addition, the data in Table 3 shows mixed dynamics in electricity consumption for different production methods [5].

Table 3. Actual electricity consumption per one unit of specific product

	Electricity consumption per one ton of produced oil, kWh		
	2012	2013	2014
Oil production including gas condensate - total	134.1	137.6	140.6
- steam treatment	69.3	67.7	54.2
- gas lift	154.6	153.2	3.1
- other	139.3	142.9	148.0

This issue is extremely considerable for small low production rates fields. In Russia, small deposits are mainly developed by small producing companies, which, due to the scale of its activities, credit and tax state policy, have significant financial difficulties for the realization of innovative investment programs. For example, Tomsk region, the mineral and raw material base of which is characterized by a large number of small and medium-sized fields. According to Tomsknedra, 131 deposits of hydrocarbons were opened in the region, among which 102 are oil, 21- oil and gas, 8 - gas condensate. From 131 fields, 112 are referred to small, while 6 are referred to large. 34 production companies are operating on the territory of the region, but only 15 are involved in the production of hydrocarbons [6]. Therefore, there is an attempt to implement such technological innovations that will improve the profitability at minimal financial input [7, 8].

2. Periodic well operation

One of the most effective ways to solve the above mentioned issue is the introduction of periodic well operation (PWO), which allows significantly to reduce costs, in particular via reduction of energy consumption during oil extraction [9]. Such way of stripped wells exploitation is based on the cycles of oil extraction and oil accumulation in the deposit. Accumulating cycle varies from 30 min to 2 hours.

The aim of the study is a comparative analysis of the influence of short-term well operation modes on energy consumption and the definition of energy efficiency of such method under the conditions of Tomsk region, in particular through the example of Shinginskoye field, which is developed by OOO "Gazpromneft-Vostok". After the exploration in 1971 and reaching of the flow rates of 5-15 m^3/day, it was considered as cost-ineffective. Later in 2007, however, after the fracturing and reaching of inflows of about 150 m^3/day, this deposit was put into operation.

The features of such operation mode are as follows:

1. on-operation well workover;

2. shifting of well to PWO mode during servicing and workover;
3. putting of new wells into production with the potential of less than 30 m³/day, initially in PWO mode [10].

3. Comparative analysis of PWO economic efficiency in the context of energy consumption

The following well operation methods were viewed to carry out the comparative analysis:
1. Nonstop oil production from the whole well stock.
2. Application of PWO technology for high-rate wells, i.e. at $q \geq 80$ m³/day.

The calculation of economic efficiency in monetary term was based on the technical characteristics of electric centrifugal pumps of various capacities applied during oil production. The results are given in Table 4.

Table 4 Results

q_{rated}, m³/day	35	50	80	100	124	For pumps of different power
Pump rate, m³/day	20-60	35-70	60-110	70-140	95-160	
η, %	39	50	55	60	61	
Power, kW	20.35	22.63	33.02	38.29	46.88	
1st Method						
Number of wells, n	25	40	30	25	12	
n, unit fraction	0.37	0.48	0.23	0.21	0.10	
q_{actual}, m³/day	35.00	50.00	18.72	21.27	12.87	
ΣQ, m³/day	875.00	2000.00	561.49	531.72	154.44	4122.65
w, kWh/day	488.40	543.12	792.48	918.96	1125.12	
W, kWh/day	12210.00	21724.80	23774.40	22974.00	13501.44	94184.64
W/Q, kWh/m³	13.95	10.86	42.34	43.21	87.42	
2nd Method						
Number of wells,n	25	40	26	23	18	
n, unit fraction	0.37	0.48	0.52	0.57	0.58	
q_{actual}, m³/day	35.00	50.00	8.95	12.21	15.39	
Q, m³/day	875.00	2000.00	232.76	280.77	277.01	3665.54
w, kWh/day	488.40	543.12	169.72	196.81	240.96	
W, kWh/day	12210.00	21724.80	4412.79	4526.64	4337.34	47211.57
W/Q, kWh/m³	13.95	10.86	18.96	16.12	15.66	
ΔW, kWh			209.32	330.64	1104.41	
ΣΔW, kWh			5442.44	7604.76	19879.46	32926.67
Δ W/Q, %/m³			0.55	0.63	0.82	0.65
ΔPq, rub/m³			93.53	108.34	287.06	150.61
ΣΔPq, rub/t			74.82	86.67	229.64	120.48
ΔPt, rub/day			837.30	1322.57	4417.66	
ΣΔPt, rub/day			21769.78	30419.05	79517.86	131706.68

Where:

q_{rated}- Pump rate

η – pump efficiency;

$q_{actual}=q\cdot\eta$ – actual production rate considering electric centrifugal pump efficiency;

$Q_{35}=\sum_{i=1}^{n_{35}} q_{35}$ – total production rate for all pumps of one type;

W – electric energy consumed by one pump;

$W_{35}=\sum_{i=1}^{n_{35}} w_{35}$ – electric energy consumed by all pumps of one type;

$W/Q=\frac{\sum_{i=1}^{n_{35}} w_{35}}{\sum_{i=1}^{n_{35}} q_{35}}$ – electric energy consumption per 1m³ of produced crude oil;

$\Delta W=(W/Q_{I}-W/Q_{II})\cdot q_{actual}$ – electric energy saving by one electric centrifugal pump;

$\Delta W/Q=\frac{W/Q_{I}-W/Q_{II}}{W/Q_{I}}$ – percent increase of pump power rate;

$\Delta Pq=(W/Q_{I}-W/Q_{II})\cdot k$ – saving on electricity cost per one unit of produced product (k– cost of 1 kWh),($\Sigma\Delta Pq$ – for all wells of one type);

$\Delta Pt=(W/Q_{I}-W/Q_{II})\cdot q_{actual}\cdot t$ - saving on electricity cost for a definite period of time ($\Sigma\Delta Pt$ – for all wells of one type)

The obtained results show that shifting to periodic well operation will allow reducing electric energy consumption by all electric centrifugal pumps in average on 65%, whilst the cash saving on 1m³ and 1 ton of produced oil in average will be 150.61 RUB and 120.48 RUB respectively.

4. Conclusion

1. The successfully introduced and applied PWO mode shifted Shinginskoye field from low to high cost-effective fields.
2. The application of this technological mode is recommended for application during oil production in the other fields in Tomsk region.

References

[1] Commodity Special Feature from WORLD ECONOMIC OUTLOOK International Monetary Fund. April 2015. [Electronic resource] *International monetary fund* [website] URL: https://www.imf.org/external/np/res/commod/pdf/WEOSpecialAPR15.pdf (reference date 25.04. 2016).

[2] Jackson J K U.S. Trade Deficit and the Impact of Changing Oil Prices. Congressional Research Service [Electronic resource] *Federation of American Scientists* [web-site] URL: https://www.fas.org/sgp/crs/misc/RS22204.pdf (reference date 25.04. 2016).

[3] Patterson R Oil Production Is Going To Drop And Oil Prices Are Likely To Increase [Electronic resource] *Peak oil barrel* [website] URL: http://peakoilbarrel.com/oil-production-is-going-to-drop/ (reference date 25.04. 2016).

[4] Many factors influence electricity prices [Electronic resource] *U.S. Energy Information Administration* [website] URL: https://www.eia.gov/energyexplained/index.cfm?page=electricity_factors_affecting_prices (reference date 25.04. 2016).

[5] Cost of oil production in the Russian Federation [Electronic resource] *Federal State Statistics Service* [website] URL: http://www.gks.ru/wps/wcm/connect/rosstat_main/rosstat/ru/statistics/economydevelopment/# (reference date 15.02.2016).

[6] Sharf I, Filjushin V, Shenderova I, Kochetkova O 2015 *IOP Conference Series: Earth and Environmental Science.* Production operation of small petroleum enterprises in Tomsk region. Vol. **27.**
[7] Ishchuk T, Ulyanova O, Savchits V 2015 *IOP Conference Series: Earth and Environmental Science.* Approaches of Russian oil companies to optimal capital structure Vol. **27.**
[8] Konovalov V, Pogharnitskaya O, Rostovshchikova A, Matveenko I 2015 *IOP Conference Series: Earth and Environmental Science* Potential of renewable and alternative energy sources Vol. **27.**
[9] Lapshin A 2014 *Inzhenernaya praktika.* Stripped well exploitation by UETSN in periodic operation mode Vol. **11** (in Russian).
[10] Aptykayev G, Suleymanov A 2011 *Inzhenernaya praktika.* Intensification of oil production and enhancement of well operation equipped with UETSN by method of short-time periodic operation: production-technical. Vol. **4** (in Russian).

Gold as a Tool for Hedging Financial Risks

T L Ischuk[1], D V Zhilkin[2], T Yu Aikina[3]

[1, 2] National Research Tomsk State University, 36 Lenin Ave., Tomsk, 634050, Russia
[1, 2, 3] National Research Tomsk Polytechnic University, 30 Lenin Ave., Tomsk, 634050, Russia

E-mail: [1]tana.itl@mail.ru, [2]postdvz@yandex.ru, [3]aikina@tpu.ru

Abstract. The article discusses gold as a protective asset, which claims to be a high-efficiency tool for hedging financial risks. In the introductory part the general characteristics of hedging as a method of full or partial risk elimination is given, and the main known types of risk hedging typical for a financial asset portfolio holder are considered. Further, dynamics of the world prices for gold is analyzed in a historical retrospective, whereby the conclusion is drawn on a tendency of this asset to grow during the periods of financial instability, and also if new financial assets appear. In the final part of the article the assessment of gold as a tool for hedging financial risks is given.

Historically, gold plays a special role as an instrument for hedging financial risks. Many investors hold the view of gold as a universal protective asset and often connect rise in its value with temporary instability of the world markets.

It is known that hedging is the investment aimed at making profit from a situation, prevention or compensation of another possible risky or out-of-control situation [15]. Thus, purchase of gold can be both a direct investment under rise in its value, and a preventive mechanism (hedging) in case of demand growth and decrease in offer [1]. Purchase of gold is also hedging against consequences of inflation [2], since gold holds its value, when dollar loses in cost [10].

Transactions are hedged against probable price change by taking an opposite position in the parallel (forward or exchange) market. For example, real goods with a certain delivery time are purchased /sold alongside with purchasing /selling forward contracts with the same time of performance on the stock exchange. Thereby, change in price for real goods is compensated by a reverse transaction, which enables to avoid heavy losses involved with unfavourable price movement [14].

While hedging protects against probable losses, on the other hand, it reduces probability of receiving unplanned profit at favorable movements in the asset value. However in most cases the role of additional profit is not as important as risk of loss in target profit, when financial flows and obligations are planned. For this reason hedging is widely applied as an exposure-driven approach [13]. As a matter of actual practice, various hedging options have been formed (Table1).

Table 1. Types of hedging.

№	Types of hedging	Characteristics
1.	Classical (pure hedging).	When a transaction with real delivery in the commodity market is made, parallel opposite positions in forward markets open (in options market or futures market).

2.	Proactive	A market contract replaces a real delivery contract during the period, which divides contracting in the forward market and trading with real delivery in the commodity market.
3.	Full or partial	Full hedging definitely excludes both probability of losses and probability of extra profit from the change in insurable asset price. Both risk of insuring losses and probability of earning an additional profit from a transaction uninsured part remain with partial hedging.
4.	Cross hedge	As an insurance tool the underlying asset is applied, which is close to the really traded one but not coinciding with it. For instance, an export-import transaction involving physical gold can be hedged by a gold futures contract, while an actual stock purchase can be hedged with the help of a stock index futures.
5.	Selective	This type is the most risky and sometimes considered equivalent to exchange speculation: based on the analysts' predictions about presumable price movement, only a part of transaction amount (part of currencies or goods) is insured. Besides, contracting and trading time in the commodity and forward markets may not coincide.

The most widespread hedging tool is a futures contract; futures options or forward contracts are used less often [6]. Selecting this or that instrument for conducting a hedging operation is defined by the type of hedging chosen, taking into account specific terms of transaction. Thus, a gold futures contract can help an investor to avoid losses under a volatile political, financial and economic situation [8].
To get a fair view of the gold entity, as well as its adequate assessment as a protective asset, it is necessary to refer to the history of this precious metal.
Since 1867 Paris currency system has operated in the world. Within this system gold served as international means of payments [12]. Therefore, gold substantively operated as world money. As observed from the graph below, gold value practically did not change within that period, the balance was disturbed only in 1920 and gold reached 20.68 dollars per troy ounce. The world community required a new currency system.

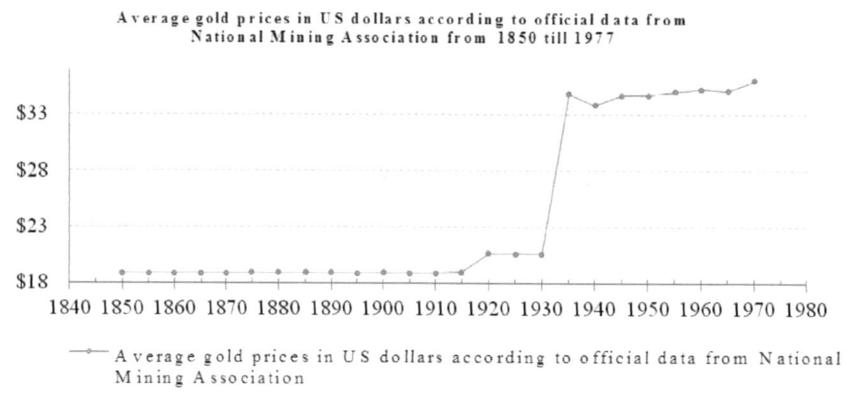

Fig. 1. Average gold prices between 1840 and 1980 [18].

Since 1922 Genoese currency system has begun to function. Gold ceased to be international money and became the international measure of value. Domestic-currency rates were measured by means of gold parity in international payments [12]. Currencies could be freely converted into gold before the Great Depression (1929-1933) and, as a consequence, World War II.
In 1944 the agreement was reached on transition to Bretton currency system which came into operation in 1946. That new currency system involved using US dollar rather than gold as the international measure of value. US dollar, in its turn, could be converted into gold at the fixed rate 35 dollars per troy ounce. The USA undertook the commitment to provide gold parity [12].
However, the given obligation was broken in 1971, when USA repudiated the dollar-gold exchange at a fixed rate.

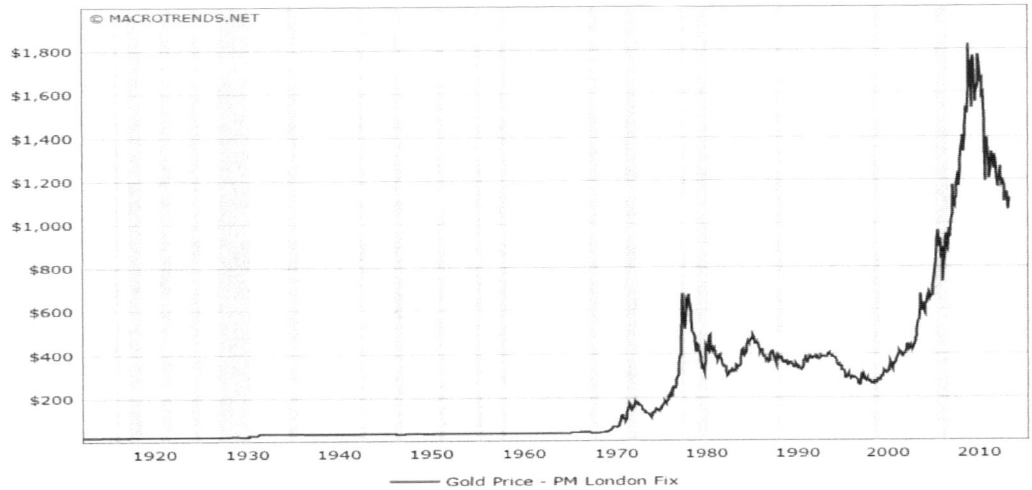

Fig. 2 Long-term graph of gold price in current prices at London fixing.

It is clearly seen from the given graph that the beginning of gold value intensive growth was observed around 1971, when gold guarantee of dollar was suspended. The growth spurted after 1976, when US dollar forfeited gold guarantee completely as a result of transition to Jamaican currency system assuming floating exchange rates. Gold volatility immediately reached unprecedented scale for the first time after 1971.

Introducing Euro – Common European Currency – was marked by a new round of sharp increase in gold value after 1999 [4].

Beyond all doubt, demand for gold as a protective asset increases due to investors' negative expectations concerning the situation in the developed foreign exchange markets and the capital markets [7], at least, for the reason that famous and reputable financiers whose judgments are taken into account keep telling about unique protective financial properties of gold. Moreover, gold possesses the status of "the asset of final instance" i.e. is the asset investors rely on, when the developed world capital markets are not capable to provide desirable profitability [9]. To confirm this statement we will consider the time series of price values of a large-cap stock index (US Broad Market Index S&P 500 is taken as an example), gold itself for the same period, Brent crude oil and dollar index that reflects dollar value towards the basket of the most liquid and high-demand world currencies.

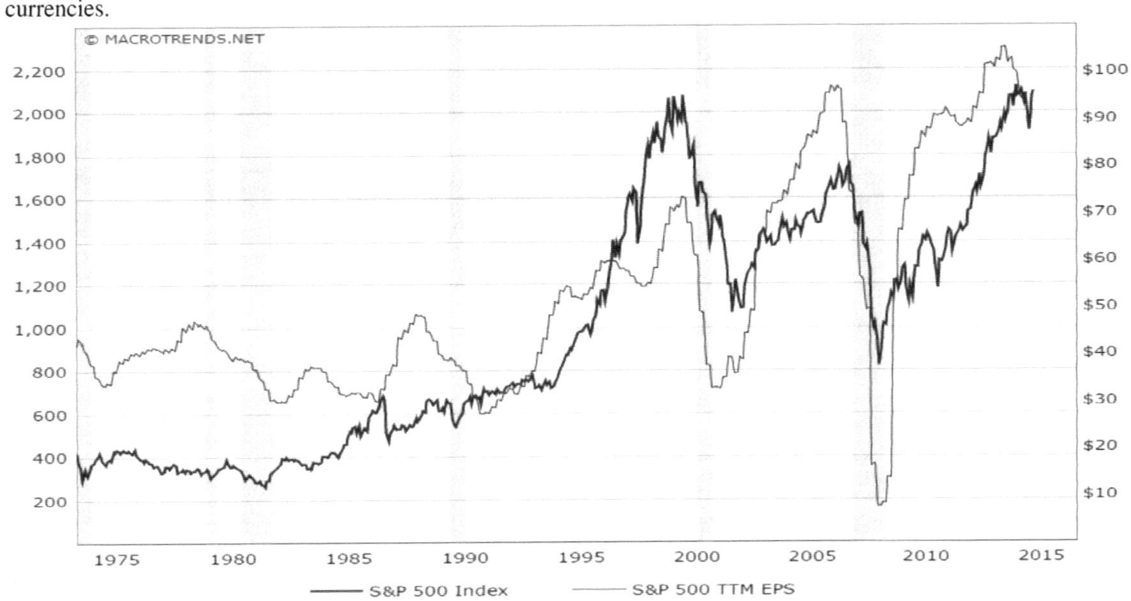

Fig. 3. Time series of price values of US indexes whose basket includes 500 US selected incorporated companies.

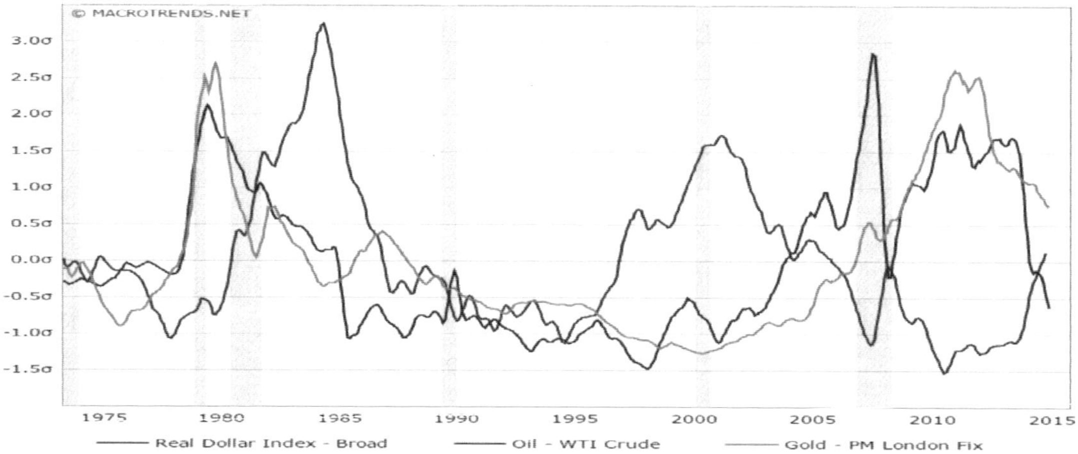

Fig. 4. Time series of price values of gold stock index, Brent crude oil, and dollar index.

As may be inferred from the graphs (Fig. 3, 4), gold correlates with both index of the developed capital market and oil and dollar. It is easy to note that the crash of US Broad Market Index S&P 500 in 1999 coincided with the moment when Common European Currency was introduced [3].

At present, non-deliverable futures and option contacts for high karat gold in bars are elaborated in the Russian Trading System [11]. The underlying asset of a futures contract is high karat gold in bars, the value of which is fixed during the morning fixing in London spot market of actual metal – "The London Gold Market Fixing Limited". The amount of one futures underlying asset is equal to one troy ounce. Calculations are made in money based on values of London fixings. Futures price in the course of stock exchange trading is fixed in US dollars per troy ounce of the underlying asset accurate within 0.1 US dollar. The underlying asset of an option contact for high karat gold in bars is the abovementioned futures contract for high karat gold in bars.

Besides, it is important to emphasize that gold as a financial asset and derivative financial instruments on its basis can be applied in practical terms for hedging not only against currency or system risk, but also against price risk tangentially related to gold of goods. In particular, E.E. Spiridonov in the article "Futures contracts: calculating the efficiency of procurement activities hedging of enterprises" considers futures for high karat gold in bars as a market risk hedging tool at construction organizations which is connected with movements in prices for raw and construction materials. The article also discusses hedging against risk of changing prices for such specific construction materials as pipes and hollow profiles of foundry-iron with the help of futures contacts, the underlying asset of which is high karat gold in bars [16]. Thus, gold that has a negative correlation with US dollar value towards the basket of the major world currencies and a positive correlation with other commodity prices is quite a multifunctional instrument for hedging a non-uniform risk spectrum.

According to the report from US Commodity futures trading commission as of 02.02.2016 (Table 2) 40 376 long and 99 861 short positions on futures and option contacts for gold have been opened by companies hedging their risks in COMEX section of New York Mercantile Exchange.

Table 2. Total futures and option positions of hedgers in the gold market according to the report from US Commodity futures trading commission) as of 02.02.2016 [17].

Trading participants	positions	
	long	short
All participants in the market, executed contracts	**314 142**	**314 142**
Hedgers, executed contracts	40 376	99 861
Hedgers, market share	13%	32%

Thus, hedgers' positions occupy a substantial share in the major North American gold market. The observed imbalance of hedgers' long and short positions gives evidence of prevailing undesirability of decrease in gold prices for the given market participants group.
Studying the gold value movement, some foreign researchers point out the following peculiarity: there is a stable relation between the dynamics of the world gold market and that of PR China stock market, which is expressed in regular capital movements between them. This is also confirmed the by type of interdependence between these markets, which is expressed not only in prices, but also in volatility [5]. The aforesaid again fortifies the role of gold in the world financial market as a protective asset, which investors tend to resort to because of negative expectations concerning future profitability from both developed and emerging capital markets.

Conclusion
Gold is an excellent protective asset. It is pertinent to consider rise in gold value as market reaction to the emergence of new financial assets (or major change in the status of old ones), which causes a temporary condition of uncertainty.

References

[1] Queralto A 2013 A Model of Slow Recoveries from Financial Crises *Board of Governors of the Federal Reserve System International Finance Discussion Papers* 1097
[2] Farlow A 2013 Crash and Beyond: Causes and Consequences of the Global Financial Crisis *Oxford: Oxford University Press*
[3] Breuss F 2011 Global Financial Crisis as a Phenomenon of Stock Market Overshooting *Empirica* **Vol.** 38 1 pp. 131-152
[4] Pisani-Ferry J 2011 The Euro Crisis and its Aftermath *Jean Pisani-Ferry New York : Oxford University Press*
[5] Arouri M 2015 World Gold Prices and Stock Returns in China: Insights for Hedging and Diversification Strategies *Mohamed El Hedi Arouri, Amine Lahiani, Duc Khuong Nguyen Elsevier* **Vol.** 44 pp. 273-282
[6] Morris N 2014 Capital Failure: Rebuilding Trust in Financial Services *Nicholas Morris, David Vines, Oxford, United Kingdom : Oxford University Press*
[7] Du W and Schreger J 2013 Local Currency Sovereign Risk *Board of Governors of the Federal Reserve System International Finance Discussion Papers* 1094.
[8] Boboshko N M 2014 Financial Credit System: Manual for Graduate Students in Specialty "Accounting, Auditing and Analysis" and "Finance and Credit", "Law Enforcement" *N M Boboshko, S M Proyava M.: UNITI-DANA* 239 p
[9] Jagerson J 2013 All about Investing in Gold *John Jagerson, Wade Hansen M.: Mann, Ivanov and Ferber* 248 p
[10] Dirgin A 2010 Risk Hedging with the Help of Derivatives Market *Stock Market* **Vol.** 4 pp. 28-29
[11] Lakhno Yu V 2013 To issue of Russian Derivatives Market's Development *Finance and credit* **Vol.** 15 (542) pp. 49-54
[12] Lewis N 2011 Gold: The Once and Future Money *Nathan Lewis M.: Grafika.Ru* 296 p
[13] Mishkin F S 2011 The Economics of Money, Banking, and Financial Markets 8th Edition Translated from English *M.: LCG "I.D. Williams"* 880 p
[14] Ilyin V V et al 2011 Financial Management *M.: Omega-L* 560 p
[15] Finance, Money Circulation and Credit 2012 *M.: Yurait, Yurait Publishing* p. 720
[16] Yankina I A 2014 Futures Contracts: Calculating the Efficiency of Procurement Activities Hedging of Enterprises *I A Yankina, E E Spiridonov Finance and Credit* **Vol.** 44 (620) pp. 10-17
[17] US Commodity Futures Trading Commission. [Electronic resource] URL: www.cftc.gov/dea/options/other_lof.htm (accessed: 09.02.2016)
[18] Information resource NUMISMASTER.COM *Krause Publications* [Electronic resource] URL: www.numismaster.com/ta/inside_numis.jsp?page=history-gold-prices-historical-charts (accessed: 09.02.2016)
[19] Vazim A A, Romanyuk V B, Akhmadeev K N, Matveenko I A 2015 Associated Petroleum Gas Utilization in Tomsk Oblast: Energy Efficiency and Tax Advantages (Article number 012078) *IOP Conference Series: Earth and Environmental Science* **Vol.** 27 (DOI: http://dx.doi.org/10.1088/1755-1315/27/1/012078)

Financial planning as a policy tool in the petroleum industry (the case study: ojsc "SURGUTNEFTEGAS")

Vera Romanyuk[1], Anna Karyakina[2], Elena Vershkova[3], Larisa Grinkevish[4] and Galina Pozdeeva[5]

[1,2,3,5]Institute of Natural Resources, National Research Tomsk Polytechnic University, 30 Lenin Avenue, Tomsk, 634050, Russia

[4]Institute of Natural Resources, National Research Tomsk State University, 36 Lenin Avenue, Tomsk, 634050, Russia

E-mail: [1]romavb@tpu.ru, [2]karyakinaan@gmail.com, [3]vem@tpu.ru, [4]nasty_saturn@mail.ru, [5]galyunya-tom@yandex.ru

Abstract. The article deals with the financial planning of oil and gas company activities including capital structure optimization. One of the main tasks of up-to-date financial management is to optimize the capital structure of an organization and minimize the weighted average cost of capital. The applied method in capital structure optimization affects the research quality results, as well as management decisions. The study was conducted on the basis of OJSC "Surgutneftegas" financial statements.

Keywords: financial planning, capital structure, weighted average cost of capital, optimization.

1. Introduction

Under modern conditions the financial planning, financial policy process analysis and futures analysis, is an integral part of the financial policy of any company.

Within the framework of any enterprise activities financial planning is reduced to the establishment of rigid proportions in monetary resources expenditure, control over the implementation of adopted budgets, as well as strategic and tactical plan integration.

In today's severe conditions oil and gas business is challenged with the need of reducing expenditure and efficient use of debt and equity capital. In recent years, due to economic decline conditions it has become more difficult for oil and gas enterprises to raise borrowed funds. Reduction of oil value, rouble currency weakening and the US dollar strengthening triggered a significant increase in interest rate on credit. Oil and gas sector companies are faced with the rigid need to optimize its capital structure. Therefore, financial planning, one of the most effective tools of financial policy, is very significant for oil and gas companies.

2. Materials and Methods

The target tool of the company's financial planning is primarily providing necessary financial funding, of production, investment and financing activities, as well as identifying the rational resources use. Financial planning has a range of methods.

The most widely accepted method of financial planning is budgeting. Budgeting is a process of its development and execution on the basis of separate division budgets. This method application is due to several advantages. In contrast, budgeting involves the resources allocation according to responsibility accounting, as well as a special budget execution monitoring system and interconnection of financial plans with operational and investment budgets [1]. This information system integrates planning results, accounting, value control and analysis of the company activities.

Budgeting of a company is based on final financial budgets: income and expenditure budget; budget's balance sheet.

These budgets embrace findings that are executive decision-making based on the financial resources use of the enterprise. Therefore, financial management is carried out in the format of these basic accounts, which form a specific frame for evaluating the financial results of the enterprise activities.

3. Results and Discussion

According to accounting statements of the third quarter, the company possesses cash flow in the volume of 33 billion rubles. The volume of investments is 1.74 trillion rub. [2].

OJCS Surgutneftegaz policy tool in financial planning involves the balance sheet and income and expenditure budget to develop the budget itself.

When developing these budgets, the company is guided by identical balance sheet item findings for the prior period, as well as analyses of the external indicators: interest rate level on deposits for free cash placement (short-term and long-term investments), the market's value of fixed assets planned for purchasing, exchange rate fluctuation, petroleum products value in the world market.

Based on the proposed budget, all indicators increase from 7 to 64% . This indicates the existing resource base of the company resulting in the ability to carry out activities without raising debt funds. The company has a high purchasing power and regularly pays dividends to its shareholders.

The budget balance sheet shows that the conservative financial policy via financial planning methods has allowed the company not to raise long-term debt funds in order to finance its activities, even during the crisis period of 2014-2015.

Table 1 Consolidated income and expenditure budget of OJCS Surgutneftegas

Budget item	30.09.2015 г.			
	Planned	Actual	Changes	
			mil. rub.	%
Sales receipts	678 117	754 181	76 064	11
Cost value	474 395	508 687	34 292	7
Profit contribution	203 721	245 498	41 772	20
Business expenses	48 956	59 965	11 009	22
Before-tax profit	154 764	185 527	30 763	19
Financial costs	42 592	72 915	30 323	71
Income tax	68 022	99 746	31 724	46
Other interest income	1 122 960	1 845 494	722	64
Net income	357 407	498 437	141	39

The budget also shows the fixed assets growth up to 4% due to the company's own funds. Rate of cash on hand and current accounts is increased up to 6%.

Intangible assets increased up to 6% due to the development and further implementation of new technologies in oil and gas exploration and production in different climatic zones.

Due to the fact that recent development technologies in oil and gas recovery has been invested by available stocks, and not by raised external resources, the company's capital stock has declined to 9 815 mil. rubles.

Long-term financial investments of OJSC Surgutneftegas have increased by 35% compared to 2014. It should be noted that OJSC Surgutneftegas has completely fulfilled all financial obligations to credit institutions by 30.09.2015 and, at present, the company has no debt to the banks.

The company's credit debts significantly decreased which indicates high solvency of this 'oil major'.

Retained earnings of the company in 2015 demonstrated a 2% increase in comparison to the previous year. Taking into consideration the above-mentioned facts, the application of the budgeting method in the conservative financial policy framework allowed OJSC Surgutneftegas financial stability and leading position in the oil and gas industry market for many years.

Let's consider the calculating methodology for the key indicator of capital optimization - average cost of capital, defined by the formula:

$$CBC_K(WACC) = \sum_{i=1}^{n} k_i \cdot d_i \quad (1)$$

where, WACC - weighted average cost of capital,%;

d_i - share of i-th source in the total amount of funds;

k¿ - the cost of ¿-th financial source, % [1].

In business practice three models of optimal capital structure determination are used:

- calculation of minimum weighted average cost of capital;
- calculation of maximum financial return (by maximum return on equity);
- calculation of maximum market value of equity: application of the CAPM method (Capital Asset Pricing Model), as well as the P/E multiplier.

According to the calculation method of minimum weighted average cost of capital, determining the optimal capital structure, it should be based on its total value of the individual sources. The optimal ratio is the minimum overall achieved cost of capital. Thus, the value of the weighted average cost of capital determines the optimal capital structure (Table 2).

The price of stockholder equity of OJSC Surgutneftegas includes income tax and shareholders dividends. The return of OJSC Surgutneftegas shareholders was 18.5% in 2015 [3].

Table 2 Optimal capital structure determination of OJSC Surgutneftegas by calculating the minimum weighted average cost of capital

Indicators	Options of capital structure				
	1 (basic)	2	3	4	5
Equity per share	100	90	80	70	60
Equity to debt ratio (leverage)	0	10	20	30	40
Price of debt capital, %	0	18	18.2	18.5	19
Price of equity, %	38.5	38.8	39	39.5	40.5
WACC (Weighted Average Capital Cost) , %	38.5	36.7	34.8	33.2	31.9

According to this model, the optimal capital structure is a ratio of debt to equity of 60 to 40; since such a ratio weighted average cost of capital is minimal. The model is easy to calculate, but it has a significant drawback - it directly depends on credit interest rates, which may change in current conditions.

In order to form an optimal funding structure, multiple financial calculations are carried out via the second model, varying the leverage ratio proportion (Table 3).

Table 3 Optimal capital structure determination of OJSC Surgutneftegas in accordance with the maximum return on equity criteria

Indicators	Options of capital structure				
	1 (basic)	2	3	4	5
1.Total capital bln. rub., including:	3 392.8	3 392.8	3 392.8	3 392.8	3 392.8
2. Stockholder equity: - percentage; - bln. rub.	100% 3 392.8	90% 3 053.5	80% 2 714.2	70% 2 374.9	60% 2 035.7
3. Debt capital: - percentage; - bln. rub.	0	10% 339.3	20% 678.6	30% 1 017.8	40% 1 357.1
4. Equity to debt ratio (leverage)	0	0.11	0.25	0.43	0.67
5. Loan charge, %	0	18	18.25	18.5	18.75
6. Profit contribution, bln. rub.	245.5	245.5	245.5	245.5	245.5
7. Gross profit margin, %	7.24	7.24	7.24	7.24	7.24
8. Short-term loan charge, bln. rub.	0	6.1	12.4	18.8	25.4
9. Gross profit after reduction of short-term loan charge	245.5	239.4	233.1	226.7	220.1
10. Income tax	20	20	20	20	20
11. Net profit	225.5	219.4	213.1	206.7	200.1
12. Return on equity%	6.65	7.19	7.85	8.7	9.83

According to this model the maximum return on equity is the equity base of 60% and debt capital of 40%. The disadvantage of this model is its one-sidedness. Applying only one criteria – return on equity, objectively, it is difficult to determine the optimal capital structure of the oil and gas company.

The valuation model of capitalized assets (CAPM method) is based on the coefficient β for determining the return of the company's equity stock:

$$Rs = RFR + \beta * (AMR - RFR), \qquad (2)$$

where, *Rs* - return of the company's ordinary share,%
RFR - risk-free return securities%
AMR - average return of the securities market % (average market return)
β - covariance coefficient between market securities return on and company securities return

The model of capital structure optimization via P/E multiplier is based on the cash flow generated at a certain ratio of debt to equity. It is based on the company's figures of multiple calculations. Further, the stock price is determined in accordance with required dividend on market shares calculated by the P/E ratio [4].

Assuming that the risk-free portfolio return (RFR) of OJSC SNG is 6%, the average return of the market securities (AMR) is 20%, the market price per share of OJSC Surgutneftegas is 44.05 rubles (Table 4).

Table 4 Optimal capital structure determination of OJSC Surgutneftegas applying CAPM method + P/E multiplier

Leverage ratio %	Cost of debt %	Earning per share, rub	P/E multiplier		CAPM			WACC
			Price-earnings multiple per share	Estimated price, rub	Determined coefficient β	Required rate of return on shares, % $Rs = RFR + \beta * (AMR - RFR$	Estimated price, rub	
1	2	3	4	5	6	7	8	9
0	0	29.3	1.5	43.95	1.4	25.6	114.45	38.5
10	18	29.5	1.49	43.95	1.4	25.6	115.23	36.7
20	18.25	29.8	1.47	43.8	1.45	26.3	113.3	34.8
30	19	31	1.4	43.4	1.5	27	114.81	33.2
40	20	31.2	1.41	44	1.5	27	115.6	31.9

The table shows that the maximum value of the company's stock due to the presented models is debt to equity ratio, i.e. 60 to 40%. The advantage of these models is fairly accurate determination of profitability of the company's price/earnings ratio. However, this calculation is time-consuming and requires significant reliable statistical data.

Conclusion

Thus, based on the given research it is much more profitable for the company to raise debt funds up to 40% of the structure than to finance its activities based only on its own resources. To determine the objective optimal capital structure, it is equitable to use all presented models. This makes it possible for the company to maximize the market value of its shares and profitability in the current economic conditions.

IOP Conf. Series: Earth and Environmental Science **43** (2016) 012086 doi:10.1088/1755-1315/43/1/012086

References

[1] Zlotnikova L G, Kolyadov L V, Tarasenko P F 2005 *Publishing house «Neft' i gaz»*. Finansovyj menedzhment v neftegazovyh otraslyah. pp. 456 s.

[2] 2014 and 2015 Annual reports of OJSC Surgutneftegas for. URL: http://www.surgutneftegas.ru/ru/investors/reports/, free

[3]. Sharf I V, Kuznetsova L P, Chukhareva N V 2014 *IOP Conference Series: Earth and Environmental Science.* Financial and tax risks at implementation of "Chayanda-Lensk" section of "Sila Sibiri" gas transportation system construction project Vol. **21** pp. 1-6

[4]. Tsibulnikova M R, Pogharnitskaya O V, Strelnikova A B 2015 *IOP Conference Series: Earth and Environmental Science.* Designing economic and legal mechanism of land management in oil and gas companies. Vol. **24** pp. 1-5

Energy efficiency, low-carbon energy production, and economic growth in CIS countries

A Vazim[1], O Kochetkova[2], I Azimzhamov[3], E Shvagrukova[4], N Dmitrieva[5]
[1,2,3,4] Institute of Natural Resources, National Research Tomsk Polytechnic University, 30 Lenin Av., Tomsk, 634050, Russia

[2] Department of Geology and Geophysics, Novosibirsk State University
Novosibirsk, Russia
2 Pirogova str., Novosibirsk, 630090, Russia

E-mail: [1] vazim@tpu.ru, [2] kochetkova@tpu.ru, [3] Ilgos-tpu-11@mail.ru, [4] shvagrukova@tpu.com,

Abstract. The paper studies the peculiarities of energy efficiency increase in national economy and decrease of carbon dioxide emission for CIS countries. The conditions that allow achieving parameters of sustainable development are determined according to indexes of GDP energy intensity and carbon intensity. Focusing on the indexes of GDP energy intensity and carbon intensity dynamics as well as on carbon intensity of energy production, a real movement towards implementation of program conditions presented by international organizations is analyzed, namely, economic conversion to the model of sustainable development. The examples demonstrate both the presence of significant differences between 12 countries and the lack of fatality in these differences. At determining dependencies linear models are preferred to non-linear ones, with the explanation of reasons in each particular case. Attention to success of these countries may help to understand the advantages of conversion to the model of sustainable development and also it helps to decrease demands in terms of costs for this conversion.

1. Introduction

Russia has supported the concept of sustainable development, and this fact has determined the direction of economy competitiveness increase via decline in GDP energy intensity. WTO accession has defined timescales and the amount of this decline – by 40% during 2007-2020. However, despite significant achievements in recent years, this index is still staying almost 2.5 times higher than a worldwide average level. It may turn out that the given amount of decline in GDP energy intensity will not be achieved at the abovementioned timescales. So, the Minister of Energy of the Russian Federation A.V. Novak, during the session in April, 9, 2014, noted that the rate of decline in GDP energy intensity in 2011-2013, as compared to 2007, was sustainably lower than intended indicators. Under the conditions, when the opinion about energy reduction of world economy dominates among the experts, this lag seems to be the inefficient economy of Russia [1]. Maintenance of energy-efficiency measures will lead to decline in GDP energy intensity by 45% to 2040, in comparison with 2010 (by 2% per year) in the Base Case, and twofold (by 2.3% per year) in the Case of «Alter Asia» («Drugaya Aziya»). Forecasted rates of decline in GDP energy intensity in Russia will become 2.8-2.1 times lower than those achieved in 2007-2008, when they amounted 5.8% on average, per year. The Base Case assumes the absence of any significant technological revolutions and breakthroughs. «Alter Asia» is the Case, where the world deals with tough resource deficiency. In this case, very favourable opportunities are created for the countries that produce energy resources, including Russia [2]. However, price decline for hydrocarbons was planned neither in the Base Case, nor in the alternative one. Therefore, the measures aimed at energy efficiency increase and environment load decrease have been delayed unintentionally.

In British Petroleum Case the results of energy efficiency policy and decline in carbon intensity are compared during 1994-2014 and 2014-2035 (Fig.1)[3].

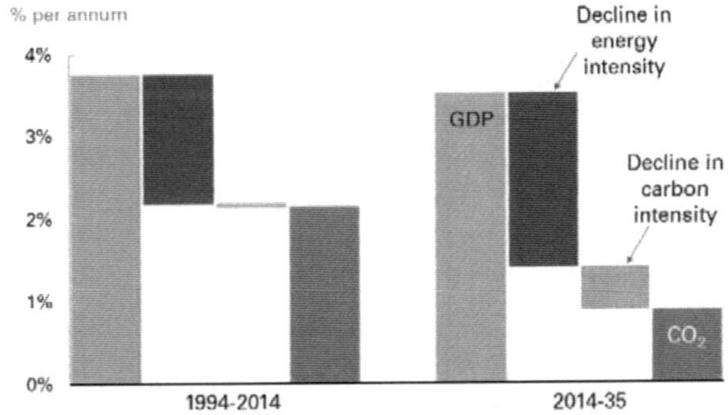

Figure 1. Separation of carbon intensity growth from GDP growth

Figure 1. shows the increase of intensive factors in primary energy production and a sharp increase of CO2 emission decline. The conditions of CO2 emission decline include transition to low-carbon fuel for automobiles and heating as well as the changes in the structure of power system in Russia.

Further let us consider trend data facilitating sustainable development and ecosystem stability: energy efficiency, carbon dioxide emissions in relation to GDP and the energy used, and also transition to low-carbon energy [4].

2. Decline in energy intensity

Energy intensity of a country is often used as an index of country energy efficiency. This is the ratio of total primary energy supply (TPES) to gross domestic product (GDP) of the country. According to the data of Organization for Economic Cooperation and Development (OECD), in 2000-2013 GDP energy intensity in the world economy, in general, reduces from 0.22 to 0.16 of total primary energy supply per GDP unit, which is calculated as an amount of toe (tones of oil equivalent) per GDP unit (1000 US dollars $, 2005)[5]. At the same time, GDP energy intensity in Russia reduces from 0.49 to 0.33, in Canada – from 0.25 to 0.19. In Norway it does not change - 0.13. In contrast, in China the decline occurs from 0.29 to 0.22, in the USA – from 0.20 to 0.15, in European Union (28 countries) – from 0.14 to 0.11, in the countries of Organization for Economic Cooperation and Development – from 0.16 to 0.13, in Japan – from 0.14 to 0.11. See Table 1 [6].

An absolute leader in GDP energy intensity decline is China – 0.31. The second important country, which has achieved success, is India – 0.43. Then, follow the former socialist countries such as Estonia (0.39), Poland (0.42), and Slovakia (0.44). However, these countries are not comparable with Russia either according to population and territory, or social and economic policy (the Eastern-European members of EU take a relaxed look at the sharp decrease in manufacturing and emigration of young staff from the countries).

Among CIS countries, Armenia is the leader in decline of energy intensity (decline by 80%). In the 1990-'s this country faced the blockade of energy supplies, and it had to provide energy-efficiency measures, regardless of economic costs. The second place in terms of energy reduction is taken by neighbouring Azerbaijan – 0.31.

Table 1. Production of toe per GDP unit according to PPP in 1990, 2000, and 2013 in CIS countries

Countries	Years			
	1990	2000	2013	2013/1990
Canada	0.28	0.25	0.19	0.68
Czech Republic	0.29	0.22	0.16	0.55
Estonia	0.61	0.30	0.24	0.39
Finland	0.25	0.22	0.19	0.76
France	0.16	0.14	0.12	0.75
ФРГ	0.17	0.13	0.11	0.65
Iceland	0.34	0.36	0.48	1.41
Japan	0.13	0.14	0.11	0.85
Norway	0.15	0.13	0.13	0.87
Poland	0.33	0.20	0.14	0.42
Slovakia	0.34	0.25	0.15	0.44
Sweden	0.22	0.17	0.14	0.64
	0.16	0.12	0.09	0.56
the USA	0.24	0.20	0.15	0.63
EU–28	0.17	0.14	0.11	0.65
OECD	0.19	0.16	0.13	0.68
Brazil	0.13	0.11	0.11	0.85
China	0.70	0.29	0.22	0.31
India	0.30	0.19	0.13	0.43
Russia	0.47	0.49	0.33	0.70
Armenia	0.74	0.29	0.15	0.20
Azerbaijan	0.42	0.35	0.10	0.24
Belarus	0.62	0.38	0.19	0.31
Georgia	0.36	0.22	0.14	0.39
Kazakhstan	0.40	0.28	0.24	0.60
Kyrgyzstan	0.55	0.26	0.25	0.46
Moldavia	0.47	0.38	0.21	0.45
Tajikistan	0.31	0.33	0.14	0.45
Turkmenistan	0,64	0.69	0.42	0.66
Uzbekistan	0.83	0.93	0.32	0.39
Ukraine	0.52	0.63	0.34	0.65
World, in bulk	0.22	0.19	0.16	0.73

Although this country is a manufacturer and exporter of energy, however, Azerbaijan has become a leader in attracting foreign investment capital, and it has managed to reduce its GDP energy intensity. The third place is taken by Belarus – 30.6. This country is limited in its resources, and it tends to develop machine building and the other branches, which give a significant amount of added value [7].

Turkmenistan demonstrates the worst results among CIS countries – 0.66. This country, like Azerbaijan, is a manufacturer and exporter of energy supplies. However, the lack of investment leads to restrictions in energy intensity decline. Ukraine, similar to Belarus in terms of energy supply resources, nevertheless, appears to be at the end of the list of energy intensity decline, at the same level as Turkmenistan – 0.65. It is caused by the lack of investments in new technologies. Kazakhstan takes the third place with the index of 0.60.

Therefore, having chosen a trend for decline in primary energy intensity, it is necessary to consider possible reserves for energy efficiency increase, from the view of branch.

3. Carbon dioxide emission

Reduction in greenhouse gas emissions may serve as an additional factor to energy efficiency, because the main gas among these gases is carbon dioxide (CO2). CO2 emissions depend on energy intensity (the amount of energy per GDP unit) and carbon content in the energy balance (carbon per energy unit)[8]. Table 2 presents the volumes of CO2 released in CIS countries: Azerbaijan, Belarus, Kazakhstan, Russia, Turkmenistan, Ukraine, and Uzbekistan [9].

Table 2. CO2 emissions in 1985-2014 in CIS countries, million tons (CO2)

	1985	1990	1995	2000	2005	2010	2014	2014/1990
Azerbaijan	53.6	59.0	36.7	30.5	34.7	25.6	33.5	56.8%
Belarus	105.4	113.1	61.9	61.0	64.4	67.3	76.7	67.8%
Kazakhstan	232.3	250.9	167.8	124.1	149.4	168.5	188.6	75.2%
Russia	2284.5	2356.2	1714.7	1557.9	1594.5	1646.1	1657.2	70.3%
Turkmenistan	33.9	35.7	24.5	38.1	49.8	65.4	78.1	218.8%
Ukraine	686.5	755.4	382.8	342.2	336.5	300.5	243.3	32.2%
Uzbekistan	116.8	122.9	112.8	122.4	111.0	101.6	120.5	98.0%

The indexes presented in Table 2 show that the leader of reduction is Ukraine (32.2%), the second place is taken by Azerbaijan (56.8%), the third – Belarus (67.8%). In this case, the growth of emissions is observed in Turkmenistan – 218.8%. Russia has a middle position with the amount of 70.3%.

Being tabulated, the results obtained in CO2 emission reduction do not take in account many factors, from the influence of GDP dynamics to the whole complex of measures on carbon intensity decline in the energy production of the country.

The following table characterizes dependence of CO2 emissions on production output determined by purchasing power parity in dollars, 2005 (shortly, CO2/GDP PPP). It is measured in CO2 kg per US dollar, at 2005 values.

The analysis of Table 3 allows making the following conclusion. Firstly, it is observed that index CO2/GDP PPP varies significantly for these countries. So, in 2013 the leader of the countries with low carbon intensity is Tajikistan (0.19), the second is Azerbaijan (0.21), and the third – Georgia (0.24). Let us consider another pole. The country with the highest level of GDP carbon intensity is Turkmenistan (1.04), and Ukraine (0.77) takes the second place. Then follows the group of countries with a small variation in indexes: Kazakhstan (0.72), Uzbekistan (0.71), and Russia (0.70) [10]. It indicates serious differences in the economic structure of the chosen countries. Russia has great reserves for reduction of CO2 emissions.

Russia's position behind the leading group seems to be unsurpassable, but the previous successes in this index decline in recent years present prospects for future [11].

Table 3. Dynamics of GDP carbon intensity according to PPP, 1999-2013, in CIS countries, CO2 kg per US dollar, at 2005 values

	1999	2004	2009	2010	2011	2012	2013	Trend
Armenia	0.44	0.28	0.25	0.24	0.26	0.28	0.26	y = -0.0138x + 0.4193
Azerbaijan	0.90	0.58	0.20	0.18	0.21	0.22	0.21	y = -0.0548x + 0.8879
Belarus	0.88	0.64	0.45	0.45	0.41	0.41	0.41	y = -0.0339x + 0.8509
Georgia	0.35	0.19	0.24	0.21	0.24	0.25	0.24	y = -0.0045x + 0.2765
Kazakhstan	0.99	0.76	0.76	0.78	0.77	0.73	0.72	y = -0.0084x + 0.8496
Kyrgyzstan	0.54	0.48	0.48	0.45	0.50	0.67	0.56	y = 0.0057x + 0.4528
Moldavia	0.96	0.76	0.63	0.63	0.59	0.58	0.47	y = -0.0294x + 0.9428
Russia	1.26	0.93	0.75	0.78	0.76	0.71	0.70	y = -0.0385x + 1.2056
Tajikistan	0.42	0.26	0.17	0.16	0.16	0.17	0.19	y = -0.0141x + 0.3515
Turkmenistan	1.75	1.91	1.22	1.27	1.20	1.12	1.04	y = -0.0569x + 1.9954
Uzbekistan	2.17	1.69	1.01	0.91	0.92	0.86	0.71	y = -0.1120x + 2.3473
Ukraine	1.52	1.03	0.81	0.83	0.83	0.81	0.77	y = -0.0523x + 1.4420

4. Conclusion

By 2013, Russia has stayed the country with a high level of GDP energy intensity, provided that the decline of this index is the lowest among those in CIS countries, but it is comparable with the world level of decline. It is necessary to range the following countries in descending order according to the index of energy intensity decline: Armenia, Azerbaijan, Belarus, Georgia and Uzbekistan, Moldavia and Tajikistan, Kyrgyzstan, Ukraine, Turkmenistan, and Russia. As to the level of CO2 emissions per US dollar, 2005 (GDP PPP), Russia takes the middle position. The leader in CO2 emission reduction in 1990-2014 is Ukraine – 32.2%, and the «anti-leader» is Turkmenistan – 218.8%. According to the rate of gas emission reduction, Russia has the fourth place. Uzbekistan, Turkmenistan, and Azerbaijan are placed gradually ahead of Russia.

References

[1] Tsibulnikova M R, Kupriyanova O S and Strelnikova A B 2015 Economic assessment of environmental impact in the course of oil field development and production IOP Conf. Ser.: Earth Environ. Sci. 27 012076 DOI:10.1088/1755-1315/27/1/012076 (20th April, 2016).
[2] Forecast on energy production development in Russia and abroad to 2040 **P. 123** [web-site] URL: http://www.eriras.ru/files/forecast_2040.pdf (20th April, 2016).
[3] BP Energy Outlook 2016 edition p. 48 [web-site] URL: http://www.bp.com/content/dam/bp/pdf/energy-economics/energy-outlook-2016/bp-energy-outlook-2016.pdf (20th April, 2016).
[4] *Technological overview of world energy production to 2050* [web-site] URL: http://old.rgo.ru/wp-content/uploads/2011/02/Kartina-mira.pdf (20th April, 2016).
[5] *OECD.org.statistics. Total primary energy supply per unit of GDP* [web-site] URL: http://dx.doi.org/10.1787/888933028045 (12th April, 2016).
[6] *International Energy Agency>Statistics> Statistics Search>Report* [web-site] URL: http://www.iea.org/statistics/statisticssearch/report/?country=RUSSIA&product=indicators&yea

r=2013 (20th April, 2016).
[7] Matugina E G, Pogharnitskaya O V, Petrova K O and Strelnikova A B 2015 Intensive use of energy resources regarding sustainable cities IOP *Conf. Ser.: Earth Environ. Sci.* **27** 012071 doi:10.1088/1755-1315/27/1/012071 (20th April, 2016).
[8] The International Energy Agency Energy and Climate Change. World Energy Outlook Special Report. 67 p. [web-site] URL: http://www.iea.org/publications/freepublications/publication/WEO2015SpecialReportonEnergyandClimateChange.pdf
[9] BP Statistical Review of World Energy 2015 [web-site] URL: http://www.bp.com/content/dam/bp/excel/energy-economics/statistical-review-2015/bp-statistical-review-of-world-energy-2015-workbook.xlsx
[10] Tsibulnikova M R, Ospanov A T, Salata D V and Strelnikova A B 2015 The comparative analysis of payments for negative environmental impact in Russia and Kazakhstan *IOP Conf. Ser.: Earth Environ. Sci.* **27** 012077 DOI:10.1088/1755-1315/27/1/012077 (20th April, 2016).
[11] Tsibulnikova M R, Pogharnitskaya O V and Strelnikova A B 2015 Designing economic and legal mechanism of land management in oil and gas companies *IOP Conf. Ser.: Earth Environ. Sci.* **24** 012032 doi:10.1088/1755-1315/24/1/012032 (20th April, 2016).
[12] Konovalov V, Pogharnitskaya O, Rostovshchikova A and Matveenko I Potential of renewable and alternative energy sources *IOP Conf. Ser.: Earth Environ. Sci.* **27** 012068 DOI:10.1088/1755-1315/27/1/012068 (20th April, 2016).

Tariffs Formation on oil transportation

Glyzina T.S.[1,3], Kolbysheva Yu.V.[1], Grivtsova I.S.[1], Dmitrieva N.V.[2]

[1]Tomsk Polytechnic University, 30 Lenin Ave., Tomsk, 634050, Russia
[2]Novosibirsk State University, 2 Pirogova str., Novosibirsk, 630090, Russia

e-mail: [3]inosine@mail.ru

Abstract. Oil transportation via trunk pipelines is an important part of the oil industry's activity. The main instrument of tariff regulation is the method of tariffs formation. Three methods of tariffs formation such as the method of economically justified costs (the Cost plus method), the method of economically justified return on investment capital (the RAB method), and the method of tariffs indexation were considered.

Key words: oil, oil transportation, tariffs, methods for calculating tariffs on oil

1. Introduction

Oil transportation via trunk pipelines is one of the most significant oil industry components. Oil is mainly produced in Western Siberia, which is located far away from places of its consumption. Tariff regulation of natural monopolies' activity is one of the state tools, which can influence the price changes in this sector and national economy as well. Moreover, the state focuses on the use of traditional energy sources.

The main aim of tariff regulation is the balance between the essential inflow of investments in the sector of natural monopolies, promoting the maintenance of reproductive processes, and structural changes in other sectors of economy, including the orientation of manufacturers on the use of energy-saving technologies. In recent years the method of economically justified return on investment capital is gaining popularity. This method is considered to be an alternative to the method of economically justified costs, which is currently used.

2. Materials and Methods

According to the state regulation of tariffs and their cap levels on services related to the transportation of oil and petroleum products via trunk pipelines and provided by natural monopoly entities, the following methods of tariffs formation on oil transportation are permitted:

a) method of economically justified costs (Cost plus method);
b) method of economically justified return on investment capital (RAB method);
c) method of tariffs indexation.

This paper presents the advantages and disadvantages of these methods.

3. Results and Discussions

The indexation method is applied in the case of changes in the conditions of the natural monopoly management and deviation from targets within the projected inflation rate of revenue. To set the tariff according to this method, the functioning tariff should be multiplied by the index of change, which is determined by the regulatory authority.

The index of tariff changes is determined by the regulatory authority on appeal of a natural monopoly.

The advantages of this method are [1]:

• simplicity in use. This method allows changing the tariff according to the inflation rate, skipping over the stage of expenditures investigation of a certain company carrying out the oil transportation.

• form of incentives. Within the framework of the indexation method more emphasis is given to increase the productivity, rather than regulation based on the service cost.

• protection of supplier from the excessive growth of tariffs exceeding the inflation.

The simplicity of this method lies in the focus on the entire industry prices. The differences between the cost changes, which are reflected in the index, and changes in the cost of certain oil pipeline companies which can't escape from using this method due to the fact that the index sets only the cap level of tariff rather than its value. However, the indexation method is implemented in the Russian legislation in such a way that its use is impractical.

The basic indicator for the method of economically justified costs (the cost plus method) is the target tariff revenue. Due to it, justified costs, related to the services rendering and taxes payment, are covered, as well as the required profit margin is ensured.

Economically justified costs, incurred in the previous period, but not included in tariff, refer to expenditures that should be offset or excluded:

- expenditures incurred due to the objective and unplanned increase in prices on services and products, consumed by the organization;

- expenditures incurred due to servicing the borrowed funds required to cover these expences. These expenditures are taken into account to determine income required for setting the tariff in the next accounting period.

Funds used for certain articles of expenditures, which have not been used during the preceding period, should be included in tariff in the next period.

According to the formula for calculating the planned tariff revenue, the following components can be distinguished [2]:

1. operating

• expenditures related to the cost;

• income tax;

• balance of accrued and paid interests.

2. investment

• expenditures on reconstruction and technical re-equipment (except the sum of depreciation charges);

• expenditures on filling the pipeline with oil;

• expenditures on financing investment projects (part included in tariff);

3. financial

• expenditures on repayment of debt;

• dividends.

This method does not prompt the economic entity to reduce the costs due to the fact that the company's costs are fully included in the tariff. Moreover, this method does not set the relationship between the tariff on services and their rendering quality.

The method of economically justified return on investment capital (the RAB method) is relatively new. It was developed in the 1990s in the UK and used to regulate electricity tariffs. Later this method spread in Western Europe, Canada, the USA, and Australia. Since 2002 the European Union established a mandatory application of this method to regulate tariffs for industry-monopolies in Eastern Europe. In the Russian Federation this method is used in power industry. In practice this method does not apply for setting the tariff on oil transportation. However, it is legally permitted.

This method provides the possibility of annual tariff indexation including macro-economic indicators. Also the regulatory authority can encourage or punish the company by reducing or increasing the tariff. In this case the percentage of investment program execution, reliability of rendering services, and their quality are assessed.

The investment costs are not included in the required gross revenue (RGR). They are considered to be the basis of the investment capital. The return on investment capital is included in tariff. Thus, the investment costs will return to the investor through accrued income within a period of 40 years. The tariff formation by the RAB method is presented in figure 1.

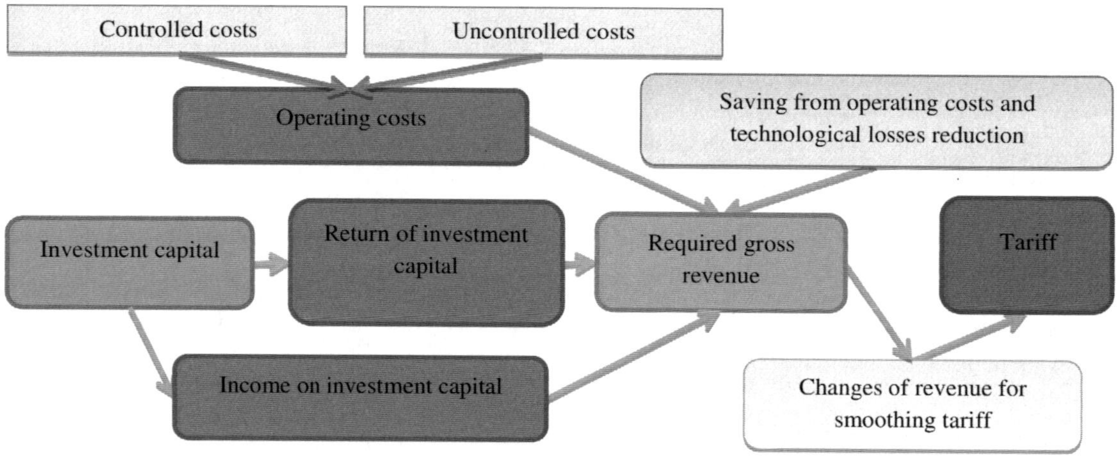

Figure 1. Tariff formation by the RAB method

Features of certain indicators of the RAB method are represented in table 1.

Table 1. The RAB method

Basic indicator	Required Gross Revenue (RGR)
Components of RGR	return of investment capital (depreciation of new and initial capital within 40 years) income on investment capital operating costs
Considered factors	number of assets efficiency index of operating costs period of investment capital return investment capital dimension rate of return of investment capital
Regulatory period	long-term
Objective deviations adjustment	adjustments exist
Incentives	costs cut and creation of conditions for attracting foreign investments
Incentives to reduce current costs	sum of saving is at the disposal of a company during tariff validity
Sources of investment capital	owned and borrowed capital of a company, which is paid by consumers for 40 years
Value determination	according to the normative method, the actual cost of capital may have a different value
Incentives to improve rendering services quality	investment program and RGR are connected with the levels of reliability
Procedure of funds inclusion on investment	investment costs are included in the investment capital base and accounted in tariff

At present the method of economically justified costs (the Cost plus method) is used to set the tariff on trunk pipelines oil transportation. In this regard a comparison of this method with the method of economically justified return on investment capital (RAB method) was conducted.

Table 2 presents the comparison of methods of Costs plus and RAB with the characteristics of the planned tariff growth, long-term management and incentives to reduce costs.

Table 2. Comparison of methods of Cost plus and RAB

	Costs plus	RAB
Smooth tariff growth	-	+
Long-term regulation	-	+
Incentives to reduce costs	-	+
Incentives to improve the services quality	-	+
Inclusion of the credit value in tariff	+	-
Accurate assessment of the actual capital cost	+	-
Developed techniques	+	-

According to table 2 the RAB method has several advantages over the Cost plus method. It provides a smooth tariffs growth and more accurate prediction of tariff rate, as well as its dynamics.

Moreover, the RAB method includes incentives to reduce costs of companies together with the increase in services quality. In recent years this method is gaining popularity at the state level. It is used to set tariffs in electricity sector. According to researches of the Department of Economics, the RAB method is an alternative to the Cost plus method which is currently used [3-5].

In this case the state is free from approving the programs of energy-saving and resource-efficiency of an oil company as the RAB method involves incentives to reduce costs.

Implementation of this method is beneficial both for the of oil pipeline company, that receives a fixed rate of return and investment projects financing, as well as consumers of this company. The latter can predict costs on transportation, as the tariff is set for a period from three to five years.

Thus, the RAB method can be considered as one of the alternative methods to regulate tariffs on oil transportation through pipelines.

It should be taken into account that according to researches, it is advisable to introduce the RAB method after 2030, when a number of capital projects in constructing oil pipelines are completed [6-8].

The use of the RAB method for tariff regulation in the pipeline transport will require the development of appropriate techniques. Currently these techniques are not developed and approved by the Federal Tariff Service of Russia [9-10]. Preliminary regulation of return rate and investment capital rate, their coordination, as well as experimental testing of the method with further indicators adjustment will need.

4. Summary and Conclusions

The method of tariff indexation has several advantages, including its simplicity of use. Nevertheless, the Russian indexing method consists in the fact that the application of the method leads to a large discrepancy between the tariff rate and actual price. Now in Russia the establishment of tariff on oil transportation via trunk pipelines is carried out by the Cost plus method. However, the method is considered obsolete, as it does not create incentives to reduce the costs in the economic entity.

The RAB method is appropriate to consider as an alternative method of tariff regulation on the oil transportation by trunk pipelines. International experience of tariff regulation has shown that the RAB method encourages companies to reduce costs in the long term. As a result the gradual decrease of tariffs simultaneously with the increase of investment inflow occurs. Consequently, this method is considered as the most appropriate for industries such as electricity, water, communication systems, as well as oil transportation by trunk pipelines.

References

1. Serikov P Yu 2010 Network strategy *Pipeline oil transportation* **2** 20-25 (in Russian)
2. Serikov P Yu 2012 Vicissitudes of Method. How the tariffs on oil transportation in the long term will change *Pipeline oil transportation* **6** 6-23 (in Russian)
3. Scarf I V, Glyzina T S and Ochirov S E 2013 Tariff regulation as a factor in investment possibilities of monopolies of pipeline oil transportation *Fundamental researches* **11(8)** 1689-1692 (in Russian)
4. Sharf I V, Malanina V A and Kamynina L A 2014 Features of the marketing strategy of oil and gas companies in exploration drilling *IOP Conf. Ser.: Earth and Environ. Sci.* **21** 012047

5. Sizov A V, Tretjyakov K N, Boyarko G Y and Shenderova I V 2015 Liability of the supervisor under petroleum drilling contract *IOP Conf. Ser.: Earth and Environ. Sci.* **24** 012029

6. Sharf I, Borzenkova D and Grinkevich L 2015 Tax incentives as the tool for stimulating hard to recover oil reserves development *IOP Conf. Ser.: Earth and Environ. Sci.* **27** 012079

7. Vazim A A, Romanyuk V B, Akhmadeev K N and Matveenko I A 2015 Associated petroleum gas utilization in Tomsk Oblast: energy efficiency and tax advantages *IOP Conf. Ser.: Earth and Environ. Sci.* **27** 012078

8. Ishchuk T L, Ulyanova O S and Savchits V A 2015 Approaches of Russian oil companies to optimal capital structure *IOP Conf. Ser.: Earth and Environ. Sci.* **27** 012066

9. Konovalov V, Pogharnitskaya O, Rostovshchikova A and Matveenko I 2015 Potential of renewable and alternative energy sources *IOP Conf. Ser.: Earth and Environ. Sci.* **27** 012068

10. Tsibulnikova M R, Ospanov A T, Salata D V and Strelnikova A B 2015 The comparative analysis of payments for negative environmental impact in Russia and Kazakhstan *IOP Conf. Ser.: Earth and Envir. Sci.* **27** 012077

Petroleum taxation: a comparison between Russia and Kazakhstan

M R Tsibulnikova[1,2] D V Salata[1] V V Drebot E A Vorozheykina

1 Tomsk Polytechnic University, 30 Lenina street, Tomsk, 634050, Russia;
e-mail: tsibulnikova2011@yandex.ru
2 Tomsk State University,30 Lenina street, Tomsk, 634050, Russia;

Abstract
The paper compares mineral resource recovery taxes for oil to be paid in Kazakhstan and the RF. It provides a case study on an average Kazakh oil and gas company and presents tax calculations as an example. To compare the taxation systems in Kazakhstan and the RF, the situation is modelled as if the field was located in the RF and the relevant calculations are carried out in compliance with national laws and regulations.
Key words: tax, mineral resource recovery, oil field

Introduction

Being one of the leading sectors of many national economies, oil recovery is obviously taxed. There is a global challenge to develop an adequate system of taxation to maximize the resource rent [1,2]. However, oil companies should get the profit enough to enhance production and conduct geological survey [3,4]. Taxation can either encourage or repel investments from the economic sector, which, in its turn, either contributes to or hinders the development of the industry [5,6]. The development of an adequate taxation system is currently a topical issue and it often becomes a cause for agreement reviews or a subject of disputes and controversies under international jurisdiction [7]. The research on taxation system and plausible effects is rather important since it allows identifying and implementing the best practices [8].

Materials and Methods

The present paper describes a case study of an oil and gas company of Kazakhstan in terms of the resource rent paid in Russia and Kazakhstan in compliance with relevant national laws and regulations. The comparison was made through calculations based on the data from JSC Mangistaumunaigas report over 2014. The petroleum tax calculations were carried out regarding a Kazakhstan oil field. The petroleum tax to be paid in compliance with Russian laws and regulations was calculated as well.

Results and Discussion

The mineral resource recovery tax (MRRT) is effective in the RF from January 1, 2002, with enactment of Chapter 26 of the Russian Tax Code. Taxpayers of the mineral resource recovery tax are individual entrepreneurs and organizations involved in mineral resource recovery and having license for the use of subsurface resources. The tax is comprised of a percentage rate, which regularly increases, multiplied by a coefficient depending on global oil price and dollar exchange rate.

In 2014 MRRT on hydrocarbon raw material accounted for 98% of the total tax and all regular payments for the use of natural resources. The MRRT on oil input into the national budget was 2463.6 billion rubles, which made up 87% of MRRT on hydrocarbon raw material and 17% of all national revenues in the RF, respectively [9].

Compared to the taxes effective before the MRRT enactment (payments for mineral reserve replacement and the right to use subsurface resources), the MRRT has a number of advantages. The major advantage is simplicity of tax administration for both taxpayers and the regulatory authority. The simplicity is achieved through transparent systems of tax assessment and determination, as well as the absence of tax concessions. Moreover, the tax allows national revenues from both the resource rent and the differential oil rent.

However, there are some limitations of MRRT to be pointed out regarding its effect on the economic sector and taxpayers' finance. The major disadvantage is dependence of the tax rate (and the amount of payments) on global oil prices. The dynamics of average oil prices (Urals crude oil, USD/bbl) and the range of price index (Ip) values over the period 2002–2015 are shown in table 1. The direct dependence of the price index Ip, which is used to determine the amount of MRRT in the RF, on global oil prices is obvious.

Table 1. Dynamics of MRRT on oil

Year	Price USD/bbl	Dollar exchange rate	Ip	Tax rate with regard to Ip (USD)	Oil price, USD/ton	Percentage of tax in price per 1 ton, %
2015	51	60.98	8.26	103.7	367.2	28.2
2014	97.6	38.4	11.83	151.8	702.72	21.6
2013	107.96	31.83	11.34	167.4	777.3	21.5
2012	110.58	31.09	11.36	162.9	796.2	20.5
2011	109.55	29.39	10.63	151.6	788.8	19.2
2010	78.05	30.37	7.34	101.2	561.96	18
2009	60.74	31.75	5.47	72.4	437.3	16.6
2008	94.04	24.85	7.95	133	677.1	19.6
2007	69.43	25.58	5.89	96.6	499.9	19.3
2006	60.89	27.19	5.4	83.3	438.4	19
2005	50.2	28.28	4.47	66.2	361.4	18.3
2004	34.51	28.81	3.03	36.5	248.5	14.7
2003	27.13	30.69	2.36	26.1	195.3	13.4
2002	23.72	31.34	1.96	21.3	170.8	12.5

Over the time period from 2002 to 2008, the percentage of tax in price increased from 12.5 to 19.6, and then in 2010 decreased to 18%. From 2011 till 2015 MRRT rose again. The particular increase in tax load is characteristic for 2015. As table 1 shows, the tax burden peaks in the periods of economic crises. Within these periods the tax withdraws oil price differential that is due to high global prices for energy sources. When MRRT and custom duties are both effective, there is double taxation system: in both cases, the tax basis depends on the amount of oil and export prices. Another MRRT limitation is flat taxation scale, which fails to be applied with due regard to geological, economic and geographic, infrastructural issues, as well as to the particularities in the field development and exploration, reserve structure and quality, etc. Therefore, additional benefits are actual for the taxpayers whose sites are located in better fields: the recovery is profitable due to the simple methods applied for resource extraction. Therefore, national oil production increases due to intense exploitation of highly profitable wells neglecting both new field exploration and development, as well as new well drilling [10].

In Kazakhstan, the issue of adequate natural resource taxation, which maximizes national revenues and ensures the profit of subsurface resource users, has always been topical. The taxation system is developed in compliance with the Constitution of the Republic of Kazakhstan, which implies that the state holds a monopoly over the use of mineral reserves. This corresponds to the international principles of income taxation from natural resources. Over a short time period, the basic principles of petroleum taxation have been determined and the legislative framework has been developed. It is noteworthy that in Kazakhstan the taxation systems of leading oil-exporting countries are intensively studied and best global practices are applied to ensure petroleum sector development and to reach the aim set by the President of the Republic of Kazakhstan, namely, to be in top 50 most competitive countries [11].

The Law on Subsoil and Subsoil Use creates a legislative framework for granting, exercising, transferring, and terminating the rights of a subsurface resource user. In compliance with the Tax Code, taxes and other mandatory payments are regulated by the Tax Code only. These taxation principles have nothing to do with the agreements between the government and particular companies [12].

In the Republic of Kazakhstan, subsoil user taxation includes corporate income tax, oil export rent tax, commercial discovery bonuses, and mineral extraction tax. The only exception is production sharing agreements, which became effective before January 1, 2009, and special contracts approved by the President of the Republic of Kazakhstan.

The mineral resource recovery tax (MRRT) is analogous to the royalty, which is based on the recovery amount, and is applied for crude oil, gas condensate and natural gas. The rate ranges between 0.5% and 18% depending on the recovery amount. There are different rates and bases which are determined by the following factors: what is produced and whether production is exported or not. In 2013, the share of MRRT in the Republic of Kazakhstan budget revenues accounted for 12%.

The taxation system in Kazakhstan is inefficient due to uncollected tax debt. There should be a system of government monitoring over implementation of enhance recovery methods and rational utilization of feedstock, as well as environmental safety [13]. The above-mentioned methods will be efficient in case there are amendments to tax legislation which will allow using all available funds to ensure social and economic development of petroleum producing regions [14].

Let us compare the taxes for crude oil recovered under the same conditions in Kazakhstan and Russia. We assume that Kazakh company recovered 15,000 tons of crude oil within the first quarter of 2014, with 80000 tons transported to the refinery for 35000 tenge (equal to 7446.8 rubles) per ton, 5000 used towards production and the rest sold. Based on the accounting data, the production cost is 30000 tenge (equal to 6250 rubles) per ton. The global oil price over the first quarter of 2014 was 18400 tenge (3833.3 rubles) per barrel. The barrels to tons conversion coefficient is 7.2314. As the recovery amount planned for 2014 was 450000 tons, which is less than 50000 tons, the tax rate is 7%. As for the oil transported to the refinery and used towards production, the tax rate is applied with decreased coefficient 0.5, i.e. 7%*0.5=3.5% (in compliance with article 332 (2) of the Tax Code of the Republic of Kazakhstan). The MRRT for crude oil over the first quarter of 2014 is 708662808 tenge (147638085 at the exchange rate in 2014):

$$\text{MRRT} = ((80\,000\ \times\ 35\,000\ +\ 5\,000\ \times 30\,000)\ \times\ 3.5\%)\ +\ (65000\ \times 7.2314\ \times 18\,400\ \times\ 7\%) = 708\,622808 \text{ tenge} = 146\,638\,085 \text{ rubles}$$

Let us calculate the tax amount as if this oil (150,000 tons) was recovered in Russia (in compliance with article 342 of the Tax Code of the RF). On average the coefficient of

IOP Conf. Series: Earth and Environmental Science **43** (2016) 012089 doi:10.1088/1755-1315/43/1/012089

world oil price in the first quarter of 2014 was 12.3841 and the tax rate was 493 rubles per ton. The MRRT calculation is as follows:

$$\text{MRRT} = 150000 \times 12.3841 \times 493 \times 1 \times 1 \times 1 = 915\ 804\ 195 \text{ rubles}$$

(we assume that the coefficient of field depletion (Cfd), the index of reserve stock level (Irs) and the degree of recovery complexity (Drc) are equal to 1).

The calculation results indicate that under the same conditions the tax amount in Russia is six times as much as that in Kazakhstan. This explains higher budget revenues in the Russian Federation compared to those in the Republic of Kazakhstan.

Conclusion

There is a significant difference between the mineral resource recovery taxes in the RF and Kazakhstan. In Kazakhstan, a progressive taxation scale is applied, while in the RF a flat taxation scale is used. The progressive taxation in Kazakhstan encourages exploration and development of minor oil and gas fields by small companies. As for big companies, the tax payments are bigger since they recover greater amounts of oil.

In the RF, the same approach is applied for both big and small companies. The differences in the rate are connected with the type of the field, in other words, the decreasing coefficient is applied where the reserves are difficult to extract or if the field tends to be depleted. As a result, with oil prices falling, the number of small independent companies involved in resource recovery in the RF is decreasing, which strengthens the monopoly within the petroleum sector. In this regard, the MRRT in Kazakhstan is more flexible and appropriate regarding the taxpayers. Therefore, the progressive minerals taxation system of Kazakhstan might be applied in the RF.

References

[1] Weijermars R 2015 *Natural Resources Research.* Natural Resource Wealth Optimization: A Review of Fiscal Regimes and Equitable Agreements for Petroleum and Mineral Extraction Projects. Vol**. 24 (4)** pp. 385-441.

[2] Smith J L 2013 *Resources Policy.* Issues in extractive resource taxation: A review of research methods and models. Vol. **38** Issue 3 pp. 320-331.

[3] Boadway R, Keen M 2010 *Problems and Practice.* Theoretical perspectives on resource tax design The Taxation of Petroleum and Minerals: Principles. pp. 13-74.

[4] Konovalov V, Pogharnitskaya O, Rostovshchikova A I, Matveenko I A 2015 *IOP Conference Series: Earth and Environmental Science.* Potential of renewable and alternative energy sources. Vol. **27** pp. 5.

[5] Demirmen F 2010 *SPE Hydrocarbon Economics and Evaluation Symposium.* Win-win upstream fiscal systems: What they are and how to achieve them. pp. 189-196.

[6] Romanyuk V B, Baitova Y S, Safronova E V, Pozdeeva G P 2015 *IOP Conference Series: Earth and Environmental Science.* Regulatory framework of pricing and estimate standards in the petroleum sector (Article number 012073). Vol. **27** pp. 1-5.

[7] Costa Lima G A, Gaspar Ravagnani A T F S, Schiozer D J *Conference Proceedings 2.* Proposed Brazilian fiscal system for pre-salt production projects: A comparative study of gain and loss of government and companies SPE Latin American and Caribbean Petroleum Engineering. pp. 1478-1486.

[8] Bugorskaya M V 2009 *Mineral resource recovery tax.* Methodological aspect Tax Policy and Practice. Vol. **3-1**. pp. 25–27.

[9] The Federal Treasury (official website) [Electronic resource] URL: http://www.roskazna.ru reference date 04.05.2016

[10] Panskov V 2008 *Petroleum sector taxes.* Changes are essential, but not sufficient Tax Policy and Practice. Vol. **11** pp. 12–14.

[11] Vygon G V 2004 *OilGasLaw 2.* Challenges in oil rent assessment and oil recovery taxation mechanisms.

[12] Taxation of Subsoil users in Kazakhstan [Electronic resource] URL: http://www.investkz.com/journals/80/901.html

[13] Akhmetov B 2014 *Vestnik of Revenue Service of the Republic of Kazakhstan.* Analysis of budget revenues in the Republic of Kazakhstan over 2013. Vol. **3** pp.7–8.

[14] Statsurina Yu A 2015 *Topical Economic and Management.* Issues for Engineering and Oil and Gas Enterprises under Innovation Economy (in Russian) Vol. **1** pp. 86–94..

Systematic approach in petroleum personnel competence assessment

Vera Romanyuk[1], Evgeniya Nekhoda[2], Andrey Dmitriev[3], Dmitriy Khudyakov[4], Galina Pozdeeva[5]

[1,3,4,5]Institute of Natural Resources, National Research Tomsk Polytechnic University, 30 Lenin Avenue, Tomsk, 634050, Russia
[2]Institute of Natural Resources, National Research Tomsk State University, 36 Lenin Avenue, Tomsk, 634050, Russia
E-mail: [1]romavb@tpu.ru, [2]dom-hors@mail.ru, [3]dmitrievau@hw.tpu.ru, [4]KhudyakovDV@hw.tpu.ru, [5]galyunya-tom@yandex.ru

Abstract. The article is devoted to professional competence improvement of personnel in the petroleum industry. The technique for competence assessment optimization in oil and gas well drilling is developed. The specification for the oil and gas industry competence profiles has been provided.

Introduction

Research in the systematic approach has been conducted for several decades in Russia. The systematic theoretical foundation description and analysis of the most significant fundamental management schools include human relations, modern organizational theories, industrial sociology, as well as classical management theory which have been represented in the works of D. M. Gvishiani, S. Popkov, V. N. Sadowski [1]. Systematic approach involves planning, engineering design, evaluation, etc. Considering the professional competence assessment within the organization management system, the assessment of both a separate system element and a subsystem itself with its relationships, elements and principles are defined.

1. Materials and Methods

A. A. Damadaran, leading scientist in assessment field (as a philosophical category) identifies three important rules for any assessment [2]. Developed by high-qualified and experienced experts this system is influenced to a certain degree by bias of the resulting assessment itself. That can be explained by high availability level of different information in our modern world. It is possible to reduce bias influence on the assessment in two ways:

- before assessment, radical public opinion should be excluded which could possibly lead to its undervaluation or revaluation;
- before assessment, personal convictions should be minimized concerning how overvalued or undervalued the object is.

Systematic approach characterizes organization activities and identifies patterns and relationships with a view to more effective use. Fundamentally, this approach is a problem-solving method, as well as a goal setting method. It is more objective than subjective.

Today, the systematic approach is used widely in economic research. For instance, the academician K. A. Bagrinovsky School developed a systematic approach methodology in strategic business development planning. Systematic studies are absolutely specific where general systems theory is associated with the work objectives and the nature of the object being studied. To optimize the assessment system, the test item numbers for each competence was correlated to the competence matrix. This was based on:

- the number of *i-th* questions to the complexity level of the *j-th* competence issues;
- the percentage of the *j-th* competence complexity level.

Within the assessment system, the maximum number of tests was given to the competencies having the largest testing percentage. For instance, practically all the organization office and service specialists have been tested for such competencies as "Drilling", "Oil and gas well drilling technology", "Oil and gas well installation and operation". Consequently, the number of these competence tests was the largest in the assessment system. In other words, the assessment system is optimized on the basis of the competence importance.

The competence percentage depended on the *j-x* competences number of *i-th* complexity level, according to the assessment system specifications and codifiers and total number of test items.

The number of questions for each competency was determined by the following formula:

$$m_{ij} = \left(d_{ij} \cdot \sum_{i=1}^{n} m_{ij} \right) / 100 \quad , (1)$$

where, the number of questions the *i-th* level of complexity for the *j-th* competence;
dij - the proportion of the *j-th* competence of the *i-th* level of complexity.
The competence percentage was defined as:

$$d_{ij} = \left(k_{ij} / \sum_{i=1}^{n} k_{ij} \right) \cdot 100, \quad (2)$$

where, *kij* - the *j-th* competence number of *i-th* complexity level.

2. Results and Discussion

Individual specialist specifications include a set of competencies, taking into account the complexity level, the number of questions based on their complexity level, the maximum number of points, and the test time. The higher a specialist position in the enterprise structure, the greater range of competencies is included in his /her specification (Table. 1).

Table 1. Specification of department supervisor "Capital workover management"

Capital workover management		Competence level	Number of questions	Total score	Test time
Department supervisor					
Fundamental Management	1. Fundamentals of Management	3	3	9	
	2. Fundamentals of Oil and Gas-Field Engineering	3	2	6	
	3. Fundamentals of legal studies	2	2	4	
Business organization	4. Documents, policies and procedures	1	3	3	
	5. Institutional environment	2	2	4	

Oil and gas well drilling	6. Technology and technique of oil and gas well drilling	2	2	4	120 min.
	7. Repair operations	2	2	4	
	8. Petrophysics	3	3	9	
	9. Installation and maintenance of drilling equipment	2	2	4	
	10. Pipeline engineering, construction and operation and oil and gas storage	2	6	12	
IT skills	11. *SAP Repairs and maintenance of fixed assets	3	3	9	
	12.Internet/Intranet	2	2	4	
Competence level	1 – beginning	1	3	3	
	2 — user	7	18	37	
	3 – experienced	4	11	32	
Total		12	32	72	

After tests, codifiers and specifications, the trial testing was conducted according to the proposed competencies modules [3].

Due to the final personnel testing results, the career development program was established.

Based on these techniques, in some oil and gas companies personnel effectiveness analysis was launched and established - for primary and after career development training testing. Table 2 shows the assessment system of drilling engineers in the petroleum industry.

The results of this study can be summarized as follows:

- based on the competence model and accounting and economic services structure, the general competencies profile was formed;
- codifiers and specifications were designed;
- the test items bank were examined;
- the strengths and weaknesses in specialist training were analyzed;
- recommendations on the advanced training programs were formulated.

Table 2. Test questions specification on competence profiles petroleum specialists (Oil and gas well drilling)

Model elements	Competence levels				k_{ij}	d_{ij}	m_{ij}
	Beginning	User	Experienced	Expert			
Fundamentals of Management							
Enterprise Economics and Management (in oil and gas industries)	25	25	20	5	75	3.7	37
Fundamentals of Management	20	15	10	5	50	2.5	25

Fundamentals of Oil and Gas-Field Engineering	30	25	20	5	80	4.0	40
Fundamentals of legal studies	15	10	10	5	40	2.0	20
Knowledge of business organization							
Incorporation documents, policies and procedures	25	15	5	5	50	2.5	25
Institutional environment	40	25	15	5	85	4.2	42
Business trends	15	15	10	5	45	2.2	22
Oil and gas well drilling							
Geology	25	15	15	0	55	2.7	27
Field development	50	25	25	10	110	5,5	55
Well drilling	50	40	30	5	125	6.2	62
Oil and gas recovery	60	50	15	6	131	6.5	65
Production operations metrology support	20	15	10	5	50	2.5	25
Field gathering system and preparation of production fluid	20	15	15	5	55	2.7	27
Repair operations	50	25	15	10	100	5.0	50
Oilfield chemistry	25	15	10	5	55	2.7	27
Petrophysics	25	10	10	5	50	2.5	25
Pipeline engineering, construction and operation and oil and gas storage	35	15	10	5	65	3.2	32
Mechanics of solid rock	20	15	10	5	50	2.5	25
Oil and gas well drilling technology	35	30	15	5	85	4.2	42
Drilling equipment installation and maintenance	50	30	15	5	100	5.0	50
Geophysical well logging	25	10	10	5	50	2.5	25
Horizontal drilling	45	25	15	5	90	4.5	45
IT skills							
SAP Logistics	15	10	5	5	35	1.7	17
SAP Repairs and maintenance of fixed assets	15	10	5	5	35	1.7	17
Well drilling production process automation	22	15	5	5	47	2.3	23
Lotus Notes	14	38	7	5	64	3.2	32

Electronic archive system	20	21	5	5	51	2.5	25
MS Office (Word, Excel, Power Point)	25	52	17	5	99	4.9	49
Internet/Intranet	12	53	10	5	80	4.0	40
Overall competencies k_{ij}	346	564	147	5	2007	*	*
Percentage of competence questions d_{ij}, %	32.6	53.11	13.84	0.47	100	100	*
Number of questions m_{ij}, total	313	510	133	5	-	-	1000

This assessment algorithm of professional competence has been successfully tested to motivate personal and career growth of its employees, in such Russian oil and gas companies as: JSC "Gazprom", OJSC "Rosneft", OJSC "Surgutneftegas", OJSC "GazpromNeft", "Lukoil West Siberia", OJSC "NOVATEK", LLC "TomskNIPIneft". The system is targeted for Chief Geologist services (geology; field development, oil production, drilling, workover); Chief Engineer services (oil production, production, refining, power generation, automated control system, mechanics).

After the analysis of the personnel effectiveness of primary and after advanced training testing, the dynamics of growth was revealed. The system is constantly being improved: new directions are being developed due to the customer requests an the latest methods of personnel assessment are being applied, etc. Moreover, thanks to the system of competence assessment a number of new short-term courses were developed and more than 400 corporate employees attend them annually.

The study has solved the following tasks:

- perspective systematic approach, methodology questions and personnel competence assessment methods;
- different theories and personnel professional competence assessment models;
- assessment model (codifiers and specifications, personnel competency matrix);
- professional competence assessment method for oil and gas company specialists;
- professional competence assessment system of oil and gas service companies specialists and
- professional competence assessment system of oil and gas companies specialists via developed test programs [4].

3. Conclusion

The professional competence assessment improvement requires a systematic approach - firstly, from the standpoint of the methodology application, equipment and technology applied; secondly, taking into account the testing methods and further assessment examination. The assessment system examination should be conducted by the organization leading experts and external auditors, experienced in knowledge and skills of assessment examination and implementation.

References

[1] Zlotnikova L G, Kolyadov L V, Tarasenko P F 2005 *Neft' i gaz. M. Publishing house.* Financial management in oil and gas industries. pp. 456.

[2] Annual reports of OJSC "Surgutneftegas" for 2014 and 2015. [Electronic resource] URL: http://www.surgutneftegas.ru/ru/investors/reports/, free, reference date 01.02.2016

[3] Sharf I V, Kuznetsova L P, Chukhareva N V 2014 *IOP Conference Series: Earth and Environmental Science*. Financial and tax risks at implementation of "Chayanda-Lensk" section of "Sila Sibiri" gas transportation system construction project. Vol. **21 (1)** pp. 1-6.

[4] Tsibulnikova M R, Pogharnitskaya O V, Strelnikova A B 2015 *IOP Conference Series: Earth and Environmental Science.* Designing economic and legal mechanism of land management in oil and gas companies. Vol. **24** pp. 012032.

Strategic planning features of subsurface management in Kemerovo Oblast

V Romanyuk, A Grinkevich, K Akhmadeev, G Pozdeeva

Institute of Natural Resources, National Research Tomsk Polytechnic University, 30 Lenin Avenue, Tomsk, 634050, Russia
E-mail: romavb@tpu.ru, nasty_saturn@mail.ru, ahmad1994@mail.ru, galyunya-tom@yandex.ru

Abstract. The article discusses the strategic planning features of regional development based on the production and subsurface management in Kemerovo Oblast. The modern approach - SWOT analysis was applied to assess the regional development strategy. The estimation of regional development plan implementation was given for the foreseeable future.

1. Introduction
Research and practice of the raw material industry management indicate qualitative changes in enterprise development and operating conditions. Globalization and raw material sector boundaries change business processes beyond individual enterprises. Enterprises are becoming strategic regional networks with changing participants from different economy sectors. Effective strategic management becomes critically important for competitive enterprises. Under current circumstances, it is becoming more interdisciplinary, integrating concepts, models and economics and decision-making tools, marketing and finance, sociology and psychology. Strategy, management and business processes are closely interconnected and form each company integrity. The company's business model efficiency depends on these elements, concurrence, as well as correlation to each other.

2. Materials and Methods
Kemerovo region is a major ore mining region in the Russian Federation. The region consists 20 towns, 19 rural districts, 23 urban-type settlements, and 167 village councils. The largest towns are Novokuznetsk (population of 563.5 thousand) and Kemerovo (population of 521.2 thousand), and the towns of Prokopyevsk, Belovo, Leninsk-Kuznetsky, Kiselevsk, Mezhdurechensk with a population of over 100 thousand, respectively. Kemerovo region share in the total industrial output of the Russian Federation is about 2.5-3.0%. In particular, Kemerovo region share accounts for 58 percent of the coal mining and 76 per cent of coking coal in Russia. The region produces 12 percent of rolled ferrous metal products, 12 percent of steel, 7 percent of chemical fibers and yarns. Kemerovo Oblast with an area of 95.5 thousand sq. km is a great mineral and raw potential development area which is becoming one of the leading industrial regions of Russia [1]. It is the major mineral resource producer and consumer at the same time. The region mainly exports coal and natural raw materials rich in aluminum (urtites). Many other types of raw minerals are imported: from white-burning clay, glass-making sand, alloying additions for steel production to natural gas. There are virtually all types of mineral raw material objects: from liquid and gaseous

hydrocarbons to construction materials and mineral underground waters. Undoubtedly, coal and iron are primary minerals and ferrous metallurgy base of the region. Plenty of minerals are intensively being developed such as: coal, iron ore, construction materials, polymetals, gold, nepheline ore, underground drinking and mineral water.

As in any other regions, the regional mineral developers have a number of issues:

- fast growth of core product cost development. For instance, mining cost per a tonne of coal is increasing 20% annually;
- high transport expenses of producers, as primary consumers of coal and metallurgic production/output are at distance of 4-5 thousand km from Kemerovo Oblast;
- limited homegrown raw material base for ironworks;
- inadequacy of professional development system of the market demands.

To increase the regional development efficiency including the base material sector, 'Kemerovo Oblast: Strategy for social and economic development through 2025' document has been developed.

The strategy target is to increase the regional economy competitiveness and the welfare growth of the population. The main trends should be taken into account when developing the goals and their further implementation, i.e.:

- complex extracted raw materials processing and its effective implementation;
- regional consumer product consumption development; Kemerovo region manufacturers' entry into new markets;
- reducing infrastructure limitations in the development of the regional economy coal sector;
- economical recovery provision;
- personnel training system development, imbalance elimination in labor market development;
- innovative centers and product development [2].

These trends are of contradictory significance for the region. On the one hand, they create the possibility of market players capable of implementing large-scale investment projects and resistant to market price fluctuations. On the other hand, they reduce potential impact of the regional and municipal authorities on large companies. In particular, the region already suffers from the fact that major companies sell their products regionally manufactured at transfer prices and conduct transactions at market price through external trading companies. According to the regional administration estimates through the transfer pricing and tolling (foreign raw materials processing and further finished product export), the volume economy income tax payment is 3-4 bln. rub. per year for large companies. The regional development target implementation is its economy competitive recovery and population welfare growth. This in its turn, closely interrelates sub-targets solutions and makes it necessary to implement a number of major strategic projects, including stable growth provision in the economy core sector:

- economy sector raw material base development;
- basic sector product consumption increase within the country and entry to new markets;
- infrastructural limitations ease affecting the regional basic sector development and its economy enterprises;
- personnel training system development, inbalance elimination in labor market and settlement system development.

At the same time, the strategy identifies a number of risks that could change regional growth directions into long-term perspectives. They are:

- economic dependence risk on the regional public sector on world coal and metal commodity markets environment.
- resource base depletion risk
- environmental risks
- lack in technology and others.

Each region has its own features that are the key in the economy itself. In Kemerovo Oblast there are coal and steel industries, the share of these two sectors account for 70% of the total volume of shipped goods, and over 40% of the total volume of gross regional product. Kuzbass possesses the necessary objective prerequisites for further increasing production volumes and maintaining a leading position of the country's coal basin. There is 265 bil. coal within Kuzbass area, which today reflects the most favorable conditions for the development and account 44.5% of similar CIS countries reserves. In Kuzbass, within these years, more than 6.5 bln. tons of coal were mined. On that basis and on conditions of the market economy restructuring, a more reasonable approach in determining coal mining volume has appeared and there are good reasons to believe that its favorable mining reserves will be enough for tens of centuries.

3. Results and Discussion

Kemerovo Oblast is the major developed regions in Siberia, which have a pronounced raw-material specialization. It is the leading economy sector. Its energy contribution to the GRP - 49.8%, which is much higher than the average for the SFD and the Russian Federation. In the national-wide and international division of labor the region is involved in the raw materials and output production of its primary processing.

Table 1. SWOT - analysis of Kemerovo Oblast development

S (strengths)	W (weaknesses)
1.80% of the Russian Federation mines are concentrated in Kemerovo region; 66.4% of Russia's coal reserves, 57% of metal ores large stocks and mining. The main R & D centers in the coal and mining industry. 2. 14% of the Russian rolled products, 17%of steel, 80% of the mains and 100% of tram rails. 3. Region is focused on export. 4. The largest regional system of urban settlements in the Urals (22 cities, including 2 with a population of over 500 thousand., 5 with a population of from 100 to 250 thousand, 5 with a population from 50 to 100 thousand.).	1. Innovation and technological developments backlog compared to the best world practice. 2. Infrastructure limitations of the basic economy sectors. 3. Strong dependence on the export conditions in the world markets. 4. Depletion of available natural resources, significant investment needs for the resource base improvement. 5. Poorly developed urban environment (this discourages investment flows in the high-tech and management sectors).
O (opportunities)	T (threats)
1. Changing technology platform in the coal and mining industry of the Russian Federation, center displacement of technological and support services. New large-scale projects in Siberia and the Far East. 2. The growth of demand for coal (growth share in the country fuel balance) and the high demand for ferrous metals. 3. Increase of oil and gas cost is a prerequisite for coal and chemistry high profitability. 4. Major investment projects in the region.	1. Improvement of competition from the world's largest producers. 2. Decrease of the attractiveness of the primary industry employment, growth of personnel deficiency risks. 3. High risk of environmental condition deterioration due to the growth of coal mining and other mineral production. 4. Increase in transportation and raw materials delivery costs, outrunning the growth of prices of the basic sector products. 5. Complete homegrown engineering replacement by import.

4. Conclusion

The development strategy of Kemerovo region for the period of up to 2025 will be implemented in most areas. The main source of further social and economic development of Kemerovo Oblast should be economic efficiency improvement and its core activities based on its productivity growth. It is necessary for the region to manage more efficiently its unique natural and mineral resources, capital and labor potential. The current regional economy framework is most likely will be a long term perspective. The main source of growth in Kuzbass will remain the primary industry sectors (metallurgy, coal and chemical industries). There is a need to introduce new innovative projects in these sectors. This difficult task implementation is possible only via joint business, regional authorities and society efforts.

References

[1] Territorial office of Federal State Statistics Service of Kemerovo Oblast [Electronic resource] URL http://kemerovostat.gks.ru/wps/wcm/connect/rosstat_ts/kemerovostat/ru/, free, reference date 01.02.2016

[2] Federal State Unitary Enterprise A.P. Karpinsky Russian Geological Research Institute (FGUP VSEGEI) [Electronic resource] URL: http://www.vsegei.ru/ru/, free, reference date 01.02.2016

[3] Laws and regulations of the Russian Federation. [Electronic resource] URL: http://bazazakonov.ru/, free, reference date 01.02.2016

Prospects of supervising service development as the tool of input quality control

A. Sizov[1], K. Tretyakov[2], G. Boyarko[3], I. Shenderova[4], I. Ostranitsyn[5]

[1,2,3,5] Department of Natural Resources Economics, Institute of Natural Resources, National Research Tomsk Polytechnic University

[4] Department of Foreign Languages, Institute of Natural Resources, National Research Tomsk Polytechnic University

E-mail: [1]Sizov_aleksey@rambler.ru, [2]tkn1@tpu.ru, [3]Gub@tpu.ru,
[4] inna-shenderova@yandex.ru,[5] Iostr_93@sibmail.com

Abstract. Supervising provides a foothold in the Russian Federation domestic market of oilfield services. Despite the rapid growth of supervising services market, there is still a definite demand in developing this domain. The authors consider the implementation of supervising in Russian oil and gas industry sector, as well as the possible execution paths of its improvement and development.

1. Introduction

Supervising has proved its efficiency in the oil and gas industry sector. It is obvious that the incoming quality monitoring of oil and gas well construction and repair activities has positively affected the performance of different service organizations. Therefore, one of the main trends in supervising development is transferring positive experience and its further implementation as a basic tool of incoming quality monitoring of contracting in other activity spheres and industry sectors. The authors suggest transferring existing successful supervising experience into the following spheres: core drilling in geological solid mineral exploration, design and construction of water supply wells, engineering surveys, implementing environmental plans. Legal gaps, lack of legal norms and regulations in the Russian Federation Standard Laws reveal the awkwardness and diversity of contract relation conditions, which, in its turn, in most cases results hindering difficulties and uncertainty in the organization of supervising service within the Russian Federation. This leads to the need in redeveloping existing legal regulatory to create a supervising market.

2. Research Methods

The research methods include the following: (1) method of abstraction, which involves the abstraction of the supervising service demand level on market in the Russian Federation oil and gas industry sector. The authors assume that supervising services are in demand on the Russian Federation oil and gas industry service market. (2) method of analogy, i.e. the authors project the main economic supervising mechanisms into various industries.

3. Introducing supervising into other activity spheres and production industries

Oil-gas well construction, design and construction of water supply wells, core drilling in geological solid mineral exploration and engineering surveys involve complex industrial process. The specific feature is that the basic technological operations are carried out under the conditions of final result uncertainty. This stipulates an extremely wide variety of working conditions and probabilistic nature of the factors due to the lack of reliable information on geological-mining and technical production conditions. The first production operation results could be accurate, however, further work performance may be distorted.Under the conditions of uncertainty, these issues could be solved by introducing supervising, as the basic tool of technical and technological quality monitoring of work performance. Thus, conceptual similarity of the presented activity domains associated with the oil and gas well construction and repair embraces the fact that the production process itself could lack final result uncertainty and, this in itself, reveals the fact these processes are volatile indicating that operations accuracy production technology will be of fundamental importance and will be the main consolidating factor in the issues of implementing supervising into the other activity domains.

The set methods and goals in the water well construction, prospecting works in core drilling for solid minerals, as well as functional requirements in engineering survey are similar in content. In this case, there exist all the necessary prerequisites for introducing accumulated supervising experience in the oil and gas industry sector. This could be an effective management and monitoring tool. Supervising services are not regulated by the existing Civil Laws, and the provision of such services is based on the basic principle of contract freedom, which makes supervising more adaptive to any activity in compliance with the needs and desires of this or that customer. The supervisor provides external monitoring, while the motivated supervision involves honest execution of one's obligations through fines and penalties established by the contract relations and charged for performance non-identified shortcomings and defects of the contractor [1,2].

Supervising services are more essential when the supervisor is present throughout the work performance in order to monitor the process quality, and it is less effective if it involves only single one-time supervisory control. In this case, it is more difficult for the supervisor to "focus" the operation results on the object under supervision. For example, a serious challenge could be the reliability assessment of sample recovery values, the main indicator of geological information of subsurface structure during core drilling. This information could be cases of fraud and falsification of the core material. As drilling crew workers are focused only on the implementation of planned targets, i.e. core sample recovery at any price, the core material could be falsified in the following cases [3]:

1) slurry up-filling from the sump to the drilled core– slurry is composed not only of drilled out material from current footage, but also from borehole wall fragments over its entire length (the most frequent case);
2) adding core from the drilled wells into the core of the current footage;
3) falsification of coal or ore core- boring cylinders from matching materials on lathe machine.

Involving a supervisor in water well construction is obvious during the experimental pumping. This operation should also be carried out in compliance with the industrial standards, which state that the total duration of the pumping should be 1-2 days on each reduction after establishing permanent dynamic level for a given production rate. The duration of such an operation could be up to 3 months. Although this operation is characterized by a high automation level, still there are some cases of well test result falsification of the performing organization itself [4].

Without proper monitoring of engineering survey, the specialists relying on previous materials or indirect evidence could not be able to perform the necessary studies fully. For example, they can increase the distance between actual mine workings, which, in its turn, could lead to inevitable missing or mistaken contouring areas and further development of hazard geological processes. Other possible factors could be decrease of true vertical parametric well depth, gapping and/or no soil samples, inadequate identification of the engineering-geological elements and ground water level and etc. Contractors often use such approaches, especially if they have extensive prospecting experience and simultaneously try to adjust the findings on its face. In this case, the customer could receive unrepresentative or falsified survey results, which would certainly affect further performance on site [5].

In all above - mentioned cases of work performance result falsification the main task of the supervisor would be to improve the quality of the representative results. This would embrace an exclusively applied monitoring of any technology in accordance with the technical specifications and internal regulations of the customer during work performance. For example, a solid core bar is extracted from the core barrel during continuous fracture free solid rock drilling, which is composed of mineral grains embracing similar hardness and abrasion properties. In this case, reduction could be governed by such geological and technical factors as abrasion of ends and the diameter of combined core fragments during the rotation in the receiving core barrel, loose soil and sand rocks, ultra-soft and soluble minerals, dispersing fine particles in dense silty rocks, alternating thin layers of hard and soft rocks, sharp transition zones from solid rocks in soft formations. There could also be a feasible core output reduction up to 0% during karst massif drilling in case of cavities and voids or high ice content in soil (at room temperature soil volume "dries out" sometimes up to 60-70%). For each factor causing core destruction and further output reduction, the following drilling technologies were developed: decreasing rotation; shorter runs; dry run (without drilling mud); applying double coring shells (inside a non-rotating tube); "salt" dense drilling muds at drilling, sealant packing of inner pipe and etc. These factors, influencing standard core recovery should be examined by the supervisor, who should first analyze and then properly follow the core drilling process, eliminating any cause-effect technology violations during production and exploration activities, as well as analyzing the recovered core or confirming its absence.

The main condition for the water well construction is its location. The location must comply with sanitary regulations, and should not be located in flood, landslide areas, near highways and past cattle burial grounds. These conditions are fundamental excluding possible aquifer contamination. Faulty detection of such violations could lead to aquifer contamination, even if exploited aquifers have not been drilled. Bad annulus plugging, resulting in poor aquifer overlapping, is a serious violation of the water well construction technology. Failure in annulus isolation technology could result in water flow between developed and non- developed aquifers. This gradually erodes the soil and increases gaps between the casing and the formation [6].

Since contaminated water can enter the developed formation, degrading water quality. The supervisor during work performance, could register above -mentioned facts and give a preliminary conclusion on the technical condition of the object, its location and accurate maintenance water well construction technology. Supervisor could confirm the fact of fixed pipe connections and lack of pipe clearance. There are cases of well casing without fixing and application of infringing materials. Contractors restore previous lifting equipment, by using pipes, filters, defective materials contrary to sanitary regulations. These facts can not be identified as labeling and/or accompanying documents are falsified, which, in its turn, hinders possible identification. Applying anti-falsification incoming control, the supervisor could not only improve the quality of construction, but also significantly reduce the risk of accidents.

Engineering survey results are influenced not only by technological factors but also such factors on-site infrastructure, geographic (landscape, climate, tectonic) and geological.These factors are probabilistic and directly dependent on the object under the study. These factors not only influence the engineering survey quality, but also the cost of such work, namely, production cost increase during the survey. Engineering survey executives quite often try to hide the fact that further implementation is not practical, in order to increase funding by increasing work volume and convincing the customer that there is no other existing option. Supervisor analyzes the primary surey results and suggests relative changes in the planned work program. This results in proving economic justification of for the expenditure and policy for implementation of each program phase.

At present, quality control of solid mineral core drilling, engineering survey and water well construction are mainly performed by the customer. Public authority expertise and technical supervision have definite results, but are still ineffective as the production process remains in the shadows. Supervisory authorities conduct only one-time inspection control, which could only reflect the present on-site situation. In this case, it is the customer who is more interested in the quality of performed work. Currently, supervision and control organizations are established by the customers, such as Internal Oversight Services, but again they are inefficient [7,8]. Internal quality control inspector does not always work autonomously as he/she is also dependent on the authorities who are often focused on the implementation of specific targets. This results in the concealment of the facts, deviations and violations committed during work performance. Such control is ineffective, as the inspector is subjected to outside pressure which, in its turn, excludes possible objectivity in assessing the quality of both the results and production process. This fact defines a certain need for accurate quality evaluation of the work performance independent of external control. These functions could be performed by supervisors in the sphere of oil and gas well construction and repair. Supervisor is motivated and highly-responsible for his/her performance which could be the key quality factor in rendering supervising services. It should be noted that the supervisor's report has an evident value in legal disputes confirming or denying the correctness of actions, as well as the order and the quality of work implementation.

4. Conclusion

It is obvious that introducing supervising into solid minerals core drilling, water well construction and engineering survey domains is effective which is defined by anti-falsification control of work performance results, as well as the supervision and control proper production technology execution. This would not only improve the production control quality, but also proper performance of one's responsibilities. It has been proved that the most promising aspect would be the fact of increasing the work performance quality results in different industrial sectors.

References

[1] Sizov A, Tretyakov K, Boyarko G and Shenderova I 2015 Liability of the supervisor under petroleum drilling contract [Electronic resource] *IOP Conference Series: Earth and Environmental Science* **Vol.** 24 pp. 1-5

[2] Sizov A, Tretjyakov K, Boyarko G, Shenderova I, Shakirov A 2015 Supervising development in petroleum industry of the Russian Federation [Electronic resource] *IOP Conference Series: Earth and Environmental Science* **Vol.** 27 pp. 1-6

[3] Sizov A, Boyarko G 2015 Supervising during core drilling within the geological-prospecting works *Izvestiya Vysshikh Uchebnykh Zavedeniy. Geologiya i Razvedka* **Vol.** 6 pp. 50-53

[4] Sizov A, Boyarko G 2015 Technical supervision during construction of water wells *Prospect and protection of mineral resources* **Vol.** 2 pp. 35-38

[5] Sizov A, Boyarko G 2014 Engineering survey supervising *Engineering survey* **Vol.** 13-14 pp. 24-29

[6] Vazim A, Romanyuk V, Akhmadeev K, Matveenko I 2015 Associated petroleum gas utilization in Tomsk Oblast: energy efficiency and tax advantages [Electronic resources] *IOP Conference Series: Earth and Environmental Science* **Vol.** 27

[7] Sizov A, Boyarko G and Shenderova I 2014 Development of well construction and workover supervising in Russian Federation [Electronic resources] *IOP Conference Series: Earth and Environmental Science* **Vol.** 21 pp. 1-5

[8] Zolotenkov Ya, Boyarko G 2015 Legal and economic problems and solutions in geological exploration *Gornyi Zhurnal* **Vol.** 2 pp. 4-8

Economic tools for realization of methane production project on Kuzbass coal deposits

I Sharf[1], M Sokolova[2], O Kochetkova[3], N Dmitrieva[4]

[1,2,3] Institute of Natural Resources, National Research Tomsk Polytechnic University, Tomsk, Russia
[4]Department of Geology and Geophysics, Novosibirsk State University, Novosibirsk, Russia

E-mail: irina_sharf@mail.ru

Abstract. Environmental issues and, above all, issues related to the release of greenhouse gases into the atmosphere, such as coal bed methane, actualize the challenge of searching a variety of options for its disposal. The difference in the macroeconomic, industrial, geological and infrastructural features determine the need to choose the most cost-effective option for using of methane emitted from the coal deposits. Various economic ways to improve the profitability of production are viewed on the basis of the analysis of methane production project from Kuzbass coal deposits, Kemerovo region, Russia.

1. Introduction

Coal mine or coalbed methane is methane, which is released into the surrounding rocks. It is explosive at concentrations of 5-15%. This fact may lead to economic, environmental and social consequences. Furthermore, the release of methane into the atmosphere contributes to the greenhouse effect [1]. Its impact is many times greater than the impact of carbon dioxide. According to the US Environmental Protection Agency, in the global human impact on the atmosphere, the share of coal-bed methane is 8-10% [2, p.1]. For example, in the USA in 2014 the share of coalbed methane was 9% [3]. The impact of human activity on the quality of the atmosphere is the basic factor for alternative energy industry development [4].

One of the positive characteristics of methane is its energy value, which is used to reduce methane concentration in the coal beds and to eliminate negative impact of methane on the environment.

Production of methane from coal beds, as an independent mineral resource, has been carried out for more than 40 years in the world. The USA and Australia are the leaders in methane production. Currently methane is used:

- in electricity generation for gas engines and turbines;
- in pipelining;
- as a boiler fuel for heating;
- as a heat source for the coal mine air ventilation and coal drying.

Utilization of coalbed methane depends on the quality of gas (concentration of methane and the presence of impurities), and the economic efficiency of variety of ways of its use. Thus, during methane transportation the gas concentration should be at least 95%. Also such methane can be used as a motor fuel. For electricity generation the concentration should be of more than 25% [5]. Therefore, issues of cost-effectiveness of different options for methane use are of particular importance in the world [6, p.21].

2. Project realization aspects of methane production from coal deposits in Kuzbass

The project "Methane production from Kuzbass coal deposits" (Kemerovo region) implemented by "Gazprom Dobycha Kuznetsk" LLC is the first project of such kind in Russia. This project has a double purpose: production of an important raw material for gas provision to Kemerovo region and subsequently the creation of safe conditions of miners' work by advance degassing of coal deposits.

The development and exploitation of CBM wells, due to the specific features, requires a forced pumping of associated formation water. Only after complete drainage of simultaneously developed groups of productive coal deposits maximum production rates of wells are achieved.
As a rule, at well completion, the productive reservoirs (up to 8 production facilities in one well) are subjected to additional stimulation by hydraulic fracturing. In this regard, the development of CBM wells has to be carried out in two stages.
The first, development stage, allows indirectly to determine the production potential of reservoir fluid and gas, to choose the most suitable size of pump equipment and minimize the risks of downhole equipment failure, due to the clogging of operating facilities by hydraulic fracturing products and coal slurry. The experience of well operation allowed setting the most optimal depth of downhole pumping equipment (DPE) running at the first stage of development - 50 m above the perforated interval of upper operating facility. At the second stage the assembly runs to the projected depth below the last perforated interval [7, p. 15].
At reducing the dynamic level of the reservoir fluid the moment of gas desorption should be controlled (increasing of pressure in the annulus) as well as the physical properties of the pumped formation fluid. If possible, the operation of DPE should be excluded.

3. Analysis of equipment operation during project realization.

Currently, 23 downhole units are operating in the wells of Taldynskoe and Naryksko-Ostashkinskoe deposits: sucker-rod screw pumps, electrical submersible pump units of different size and balanced pumping units.
Sucker-rod screw pumps were more optimal under the conditions of CBM deposits development.
The analysis of sucker-rod screw pump exploitation for the period from 2009 to 2015 (Figure 1) has shown that operating time in 2015 on well stock was 640 days, and the maximum operating time was 1150 days.

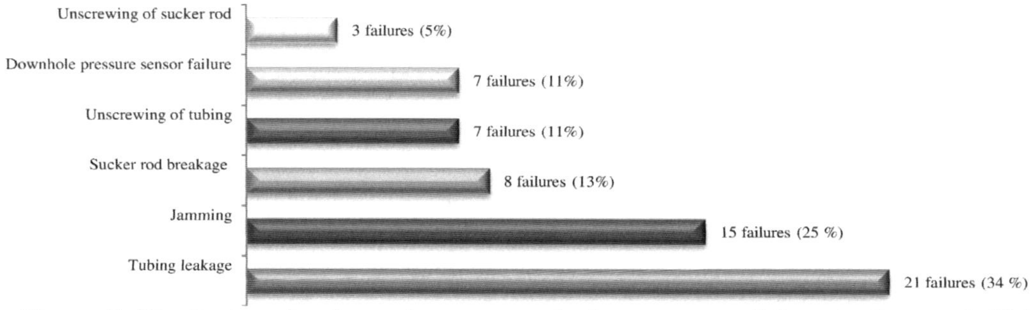

Figure 1. Distribution of sucker-rod screw pump basic components failure (on January 1, 2016).

As it can be seen from the diagram the major causes of components failure are: tubing leakage (due to the design features which were later eliminated by the increase in the number of centralizers on the rod string) and high content of mechanical impurities (proppant, coal slurry, mud) in the pumped liquid.
Electrical submersible pump units also showed good operating results. These pumps are having extended turnaround interval for CBM wells at the established operating mode.
The analysis of electrical submersible pump units exploitation for the period from 2009 to 2015 (Figure 2) has shown that operating time in 2015 on well stock was 1090 days, and the maximum operating time was 1704 days.

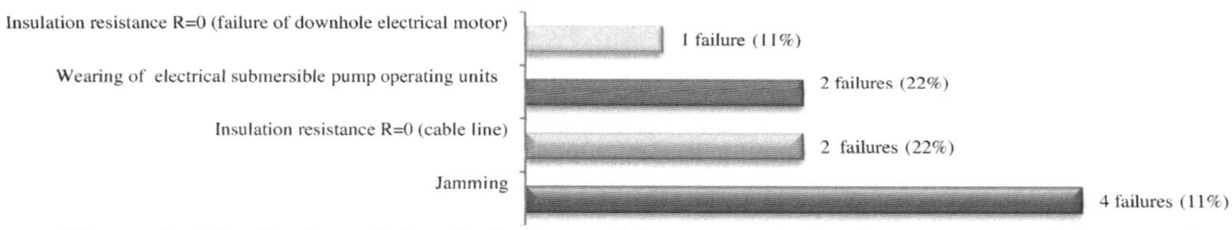

Figure 2. Distribution of electrical submersible pump units basic components failure (on January 1, 2016).

The cause of 85% of failures is the presence of the abovementioned mechanical impurities in the pumped fluid.
In 2013 balanced pumping units were introduced into CBM deposits development.
The analysis of balanced pumping units' exploitation for the period from 2013 to 2015 has shown that operating time in 2015 was 317 days on well stock, and the maximum operating time was 751 days. Only two failures occurred during the whole period of operation due to the pump plunger jamming.

4. Comparative analysis of downhole equipment economic efficiency

To evaluate the effectiveness of the three main types of downhole equipment under discussion, a comparative analysis of purchase cost, service and repair works is provided.
The results are shown in Figure 3.

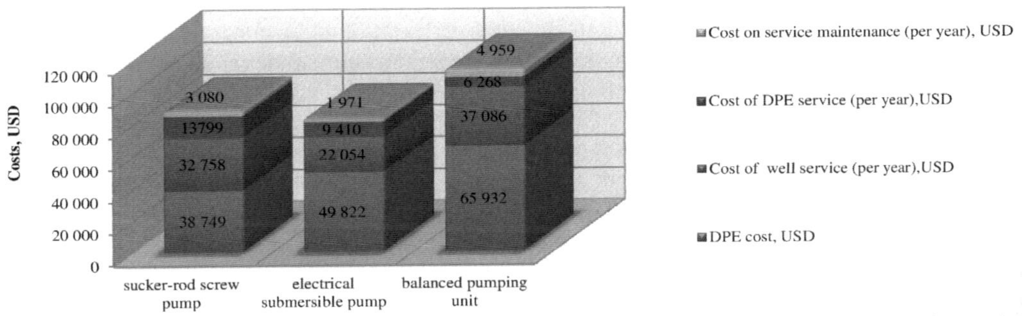

Figure 3. Correlation of downhole pumping equipment cost and costs on service and repair.

Considering the analysis of downhole pumping equipment efficiency during the development and exploitation of CBM wells and the results of cost correlation analysis, the following conclusion could be drawn:

1. The main cause of failure of all types of DPE, which were tested on wells, is the clogging of operating facilities by mechanical impurities (81%) during pumping of formation fluid, due to the fact that the developed CBM reservoirs were subjected to additional stimulation by fracturing. The first phase of development was marked by intense proppant and coal slurry lifting. This further resulted in the need of bottom-hole cleaning at each well servicing;
2. At balanced pumping unit exploitation basic costs (58%) fall on the purchase of equipment and on DPE service and repair (42%) due to the complexity of assembly work, service maintenance and the need for special service crew.
3. During sucker-rod screw pump unit exploitation the equipment purchase costs will be minimal (45%) if compared with the electrical submersible pump unit (60%), balanced pumping unit (58%) and major costs (55%) – the cost on downhole equipment service and repair.
4. During electrical submersible pump unit application, the costs on DPE service and repair are minimal (41%) due to the large turnaround interval if compared with the sucker-rod screw pump unit (56%) and balanced pumping unit (42%);
5. For operation on CBM wells, the most optimal types of downhole equipment are sucker-rod screw pump unit and electrical submersible pump unit.

In spite the fact that costs on electrical submersible pump unit and sucker-rod screw pump unit are practically the same, for more efficient exploitation it is recommended:

- At the first stage to use sucker-rod screw pump unit due to its appliance flexibility and stability of operation at the presence of high quantity of mechanical impurities in the pumped reservoir fluid;
- At the second stage on the steady-mode, under the equal conditions of exploitation (at low water inflow), it is more rational to use electrical submersible pump.

5. Economic ways to optimize costs for methane production from coal deposits

Due to the fact that an important component of success and effectiveness of project functioning is the optimal use of downhole pumping equipment and ground equipment, as well as the rational use of its resources and the increase of its operating period, the issues related to the optimization of the equipment purchase and maintenance costs are of particular relevance.

Currently the majority of oil and gas companies implement a rental scheme (lease) of downhole equipment, which includes the provision of full range of services in equipment selection, operation, maintenance and repair:

- calculation, fitting of equipment for a definite well;
- analysis of exploitation results with the recommendations for further increase of operation rates;
- incoming control and transportation of equipment to the well (from the well);
- supervision of pumping unit installation and reaching the steady-mode of operation;
- service maintenance and repair;
- storage and safekeeping of Customers equipment.

At the same time, companies by transferring functions to the third party contractor receive the following benefits:

- absence of need to insure the property, which in its turn minimizes the tax liabilities and reduces the risks during storage and disposal of non-repairable equipment (optimization of costs for environmental protection);
- the possibility to reduce personnel maintenance costs, labor costs, insurance premiums, communication costs, transportation costs, clothing costs, certification costs, training costs, etc.;
- lack of necessity to withdraw money from circulation on purchasing oil-field equipment, thus there appears an opportunity for other projects financing, for example, new wells construction and putting into operation.

In addition, there is an opportunity for company's management team to focus on the main form of business, thus not to be distracted by non-core activities. As a result, the company reduces its capital costs.

Let's consider the case of business process transformation to outsourcing and the shifting to downhole pumping equipment rental scheme [8, 9, 10].

The major criteria, in the market monitoring process of service companies, which provide rent (lease) of DPE with complex service maintenance, are: geographical remoteness of deposit from the production bases providing rental services and equipment supply.

Calculation of rental costs was realized on the basis of commercial proposals of two companies (from Tomsk and Perm).

To determine the economic efficiency of rental scheme introduction, the results of two options were compared (Figure 4).

The first option - exploitation of wells by sucker-rod screw pump unit, purchased at company's own expense, alongside with service maintenance considering amortization period. The second option - the application of DPE rental scheme (lease) with the service maintenance provided by the contractor (Tomsk company), due to the nearest geographic location and therefore lower service costs.

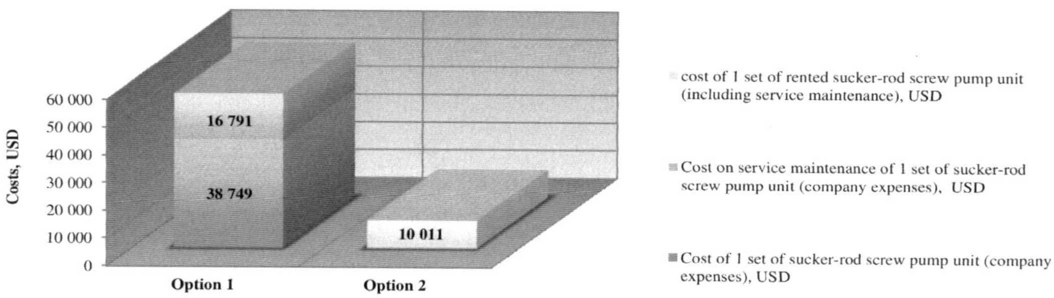

Figure 4. Cost correlation of one set of sucker-rods crew pump unit.

The following formula determines the feasibility of shifting to equipment rental scheme and the E_a design ratio:

$$E_a = \sum_{t=1}^{m} C_s \,/\, \sum_{t=1}^{n} C_r = \frac{2{,}516+0{,}200+0{,}896}{0{,}650} = 5{,}5 \quad (1)$$

where C_s — cost of equipment purchase and service maintenance (company expenses),USD;
C_r — rental cost and service maintenance (contractor's expenses), USD/year.

If the designed ratio is E_a>1, the outsourcing project (shifting to DPE rental scheme) will be economically efficient, and if E_a<1 is economically inefficient thus, the transformation of business-process to outsourcer will not bring such economic benefit to the company as the cost cutting.
As it is seen from the calculation, the ratio is E_a>1, thus outsourcing project (rent of DPE) is economically efficient.
The economic efficiency from shifting to DPE rent will be:

$$E = (C_s/A - C_r) * N = 0{,}554 \quad (2)$$

Where E – economic efficiency, USD;
C_s –cost of 1 set of sucker-rod screw pump unit (company expenses) including service maintenance, USD;
A – amortization period, year;
C_r – cost of 1 set of rented sucker-rod screw pump unit (including service maintenance), USD/year;
N – demand in DPE for the period of one year, set.

The demand for DPE for the year 2016 is 4 sets of sucker-rod screw pump units. Thus shifting to the rental scheme will bring the economic efficiency of 33 882,33 USD.

6. Conclusion

Thus, the conclusion could be drawn that in order to increase the efficiency and cost optimization in general, "Gazprom Dobycha Kuznetsk" LLC needs to select the option of shifting to DPE rental scheme with service maintenance rendered by the Contractor company and to consider the geographical remoteness.

References

[1] Tsibulnikova M, Ospanov A, Salata D, Strelnikova A 2015 The comparative analysis of payments for negative environmental impact in Russia and Kazakhstan *IOP Conference Series: Earth and Environmental Science* **Vol.** 27
[2] Coal Mine Methane Recovery: A Primer. U.S. Environmental Protection Agency September 2009 EPA-430-R-09-013 [Electronic resource] *U.S. Environmental Protection Agency* [website] URL: https://www3.epa.gov/cmop/docs/cmm_primer.pdf (reference date: 01.04.2016)
[3] Overview of Greenhouse Gases. [Electronic resource] *United States Environmental Protection agency* [website] URL: https://www3.epa.gov/climatechange/ghgemissions/gases/ch4.html (reference date: 01.04.2016)
[4] Konovalov V, Pogharnitskaya O, Rostovshchikova A, Matveenko I 2015 Potential of renewable and alternative energy sources *IOP Conference Series: Earth and Environmental Science* **Vol.** 27
[5] Best Practice Guidance for Effective Methane Drainage and Use in Coal Mines *Ece energy series* **Vol.** 31 [Electronic resource] URL: http://www.unece.org/fileadmin/DAM/energy/se/pdfs/cmm/pub/BestPractGuide_MethDrain_es31.pdf (reference date: 01.04.2016)
[6] Su S, Ren T, Balusu R, Beath A, Guo H and Mallett C 2006 Development of Two Case Studies on Mine Methane Capture and Utilisation in China [Electronic resource] *Commonwealth Scientific and Industrial Research Organisation (CSIRO)*, Australia [website] URL: https://www.globalmethane.org/documents/china_case_study_report.pdf (reference date: 01.03.2016)
[7] Regulations on the development and commissioning of the test well operation mode for the extraction of methane from coal deposits, Moscow 2012
[8] Dubinkin Ye 2014 Service instead of purchase: pump leasing in the petroleum industry *Sibirskayaneft'* **Vol.** 113 (in Russian)
[9] Yengalycheva A 2014 Outsourcing: Actual and professional *Spravochnik ekonomista* **Vol.** 1 (in Russian)
[10] Sharf I, Malanina V, Kamynina L 2014 Features of the marketing strategy of oil and gas companies in exploration drilling *IOP Conference Series: Earth and Environmental Science* **Vol.** 21

Outlook for the Development of Oil and Gas Industry in Vietnam

M R Tsibulnikova[1, 2], V A Pham[1], T Yu Aikina[1]

[1, 2] National Research Tomsk Polytechnic University, 30 Lenin Ave., Tomsk, 634050, Russia
[2] National Research Tomsk State University, 36 Lenin Ave., Tomsk, 634050, Russia

E-mail: [1]tsibulnikova2011@yandex.ru

Abstract. Oil and gas industry makes a significant contribution to Gross Domestic Product of Vietnam. In 2015 it appeared to be in an intricate situation under the conditions of fall in oil prices caused by excess of supply over demand in the oil market. On the one hand, low prices for oil enable Vietnam as an importer to purchase more oil in the world market. On the other hand, the state company PetroVietnam loses a part of its customers in the domestic market. This tendency can lead to stagnation of the oil and gas industry in Vietnam under weak oil prices if the government does not take timely action.

1. Introduction

Oil and gas industry is one of the priority fields in the development strategy in Vietnam. Oil production is not only the source of foreign currency for the country, but also the most important motive power for the current economic development. After more than 30 years of construction and development, oil and gas industry in Vietnam has achieved remarkable progress. According to the law on foreign investments, dozens of oil companies are producing oil and gas. The total amount of investments is about 7 billion US dollars. Many new oil and gas fields have been discovered; oil and gas output has been grown rapidly. This has facilitated the country's turnaround in the late 80-s of the 20th century and brought Vietnam into the list of oil-exporting countries which has improved the national energy security [1].

2. Method and Materials

The analysis of oil and gas industry in Vietnam has been conducted on the basis of studying annual statistic report from OPEC, oil companies British Petroleum and PetroVietnam. Inventory and production status of oil and gas in Vietnam has been considered, assessment of Vietnamese industry needs has been conducted, and industry growth dynamics has been estimated.

3. Results and Discussion

Geologic exploration of oil and gas in Vietnam started in 1959. At the behest of the Vietnamese government, the Soviet Government sent experts to assist in oil geology development of Vietnam. From 1959 till 1961 Soviet and Vietnamese experts completed the summary report "Geology and oil-and-gas potential in the Democratic Republic of Vietnam" [2].

The national petroleum corporation PetroVietnam dominates in the oil and gas industry of Vietnam, it operates under the direction of Ministry of Trade and Industry and controls this branch. The oil industry of Vietnam includes three segments: Upstream (Exploration and production), Midstream (Transportation) and Downstream (Processing and distribution).

Oil and gas industry in Vietnam is currently a relatively young branch and cannot satisfy the demand in the country [3]. Insufficient number of qualified specialists obstructs the branch development. The country mainly exports crude oil and imports oil products for local demand. The existing oil refinery plants Dung Quat and Nghi Son meet only about 35% of domestic demand. ORP Dung Quat is going to start reconstruction and planning to increase further output up to 9.5 million tons per year, which will make it possible to satisfy 50% of domestic demand.

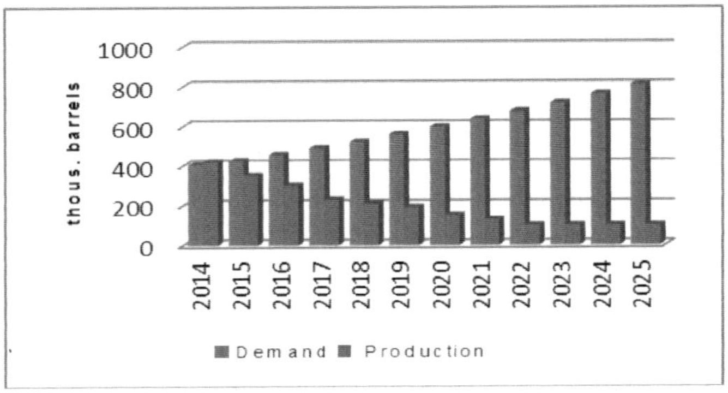

Fig. 1. Oil demand and production forecast in Vietnam.

At the same time, increase in demand for oil products both in short term and in the long term is predicted. This is due to the population growth and development of industry, transport sector in particular. According to OPEC, demand for oil fuel rapidly grows, especially in developing countries, and deliveries will not satisfy the demand [4] according to forecasts by 2025.

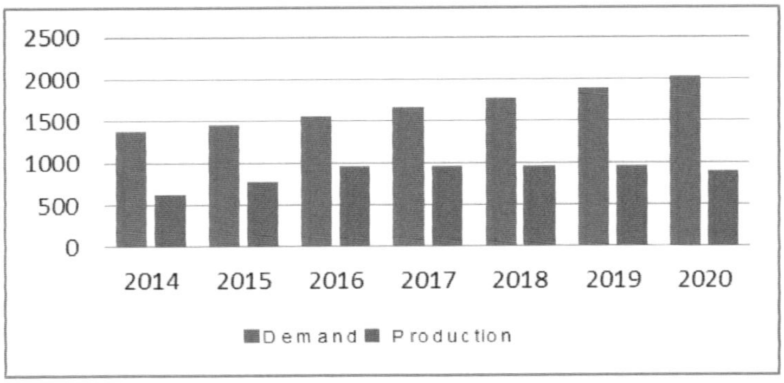

Fig. 2. Gas demand and production forecast in Vietnam.

At present the annual oil and gas production is on the average about 24 mln. tonnes of oil equivalent. Vietnam ranks the 4th place in oil reserves and 7th – in gas reserves in the Asia-Pacific Region (according to the data from British Petroleum, 2010); therewith it takes the 25th and 30th places in the world. The ratio between reserves and production (R / P) is 32.6 in oil and 66 in gas. This indicates a significant potential of the branch future development [5].

Proven reserves of hydrocarbon raw materials in sedimentary basins in Vietnam are considerable (about 4600 million tons of oil equivalent, where gas makes about 50%), they are mainly widespread in the continental shelf. Oil and gas ultimate reserves produced in 11 fields are about 1200 million tons

of oil equivalent [6]. Potential of unexplored oil and gas reserves in the rest of area is rather high. This is a valuable asset and basis for future development of the oil and gas industry.
According to the data from BP, oil reserves of Vietnam in 2013 were 0.3% of total volume in the world, average annual growth was 8.5% between 2000 and 2012.
Oil and gas potential of Vietnam is mainly concentrated in 7 basins: Kyulong, Con Son, Hong He, Malay Tho Chu, Fu Han, The Paracel Islands and Truong Sa. Five of them are being developed, while two are at exploration stage (The Paracel Islands and Truong Sa bassin). Oil and gas basins of Vietnam are generally located in the southern Vietnam, they are difficult for exploration. In particular, two basins The Paracel Islands and Truong Sa are located at a great depth in the South China Sea, thus they require key investments. The basin Kyulong was the first in Vietnam to be exploited, and it is considered to be the largest by oil reserves. However, this basin has been being developed for 23 years and now it shows evidence of decrease in production. Malay Tho Chu also used to have a substantial oil and gas potential, but at present time the reserves in the basin Hong He decreased [7, 8].
Further development of Vietnam keeps on being associated with strategic partners, large oil companies abroad. Thereby system generation of oil and gas ports is one of important trends. Oil ports in Vietnam are located in Hai Phong, Quang Ninh, Nghe An, Da Nang, Binh Dinh, Ho Chi Minh City, Dong Nai, and Vung Tau. Oil ports in Vietnam can receive ships of 60 - 120 thousand tonnes. These ports will allow fuel importers to search for new sources of oil and gas which are cheaper than the available sources in this area, reduce costs for discharge and losses during cargo handling operations.
To consider the branch development, we have conducted a swot-analysis of oil and gas industry in Vietnam:
Strengths: the market share of domestic oil and gas makes 35% due to the development plan and rational expansion of production; the activities in oil industry were synchronized from exploration and exploitation, distribution to services connected with oil and gas production.
Weaknesses: total dependency on global oil prices; low degree of flexibility due to the government control; human resources capability and the applied technology cannot fully satisfy the demands of the branch.
Opportunities: state participation entitles to a lot of privileges; there are significant prospects of oil production during approximately 60 years; there are no alternative energy sources since they require great investments, but their efficiency is not high, using nuclear energy sources is rejected on account of radioactive waste toxicity.
Threats: oil reserves are decreasing due to the fact that production ratio exceeds exploraction ratio; exploraction extension and development of deep-water areas will be costly and risky; the restructuring plan for the state company PetroVietnam affects the condition of the whole branch.

4. Conclusion

According to the presented analysis, the strategy of oil and gas complex development in Vietnam must include the following aspects.
Increasing volumes of searching and exploring oil and gas resources; supporting investigation of new sites, field development; improving domestic infrastructure of oil and gas transportation and processing. The government control must be maintained so as to guarantee growth in domestic supply and meet the demand for the country's development, lowering the dependence on deliveries from abroad.
Increasing investments for scientific research in oil and gas sector, enhancing training of high-qualified specialists in order to find new energy sources (alternative energy forms, green power), emerging development technology, reducing expenditure in the oil sector.
Legislative development, tax regulations and state funding policy should be changed in order to contribute to increasing income of oil companies, creating sufficient financial capacity, encouraging them to develop new energy sources and improve competitiveness in the free market.
National environmental policy is required to conform to international safety requirement and meet demands for energy sources.

Increasing strategic oil reserve to prevent delivery interruptions by advancement of high offshore technology in the South China Sea that will make it possible to increase raw hydrocarbon production output and promote economic growth in the republic.

References

[1] Nguyen C P and Richard J 2011 *Economic Transition and Accounting System Reform in Vietnam. European Accounting Review* **Vol.** 20 4 pp. 693–725

[2] Knutsen H M 2015 Transfer of 'Western' Institutions to a Transition Economy Context: Experiences from the Vietnamese Petroleum Sector *Forum for Development Studies* **Vol.** 42 1, 2 pp. 65-87

[3] Thinking about the Oil and Gas Industry of Vietnam [Electronic resource] URL: http://oilgas.vn/suy-nghi-ve-nganh-dau-khi-viet-nam.t1868.html (accessed: 01.03.2016)

[4] OPEC Annual Statistical Bulletin. [Electronic resource] URL: https://www.google.ru/url?sa=t&rct=j&q=&esrc=s&source=web&cd=5&cad=rja&uact=8&ved=0ahUKEwihwuyQ8p_LAhVDEpoKHUCpAIkQFgg5MAQ&url=https%3A%2F%2Fwww.opec.org%2Fopec_web%2Fstatic_files_project%2Fmedia%2Fdownloads%2Fpublications%2FASB2014.pdf&usg=AFQjCNEtHkXVPvA7JCoZoVtZRP2OGD1CbQ&sig2=dtqGqoJ2g8Dueiwuqpg-vw (accessed: 20.01.2016)

[5] Vietnam Oil and Gas Industry: Great Potential, High Growth [Electronic resource] URL: http://tapchitaichinh.vn/thi-truong-tai-chinh/chung-khoan/nganh-dau-khi-viet-nam-tiem-nang-lon-tang-truong-cao-14143.html (accessed: 22.07.2015)

[6] Oil and Gas Vietnam [Electronic resource] URL: http://www.mofahcm.gov.vn/vi/mofa/tt_baochi/nr041126171753/ns060824151244 (accessed: 21.09.2015)

[7] Oil and Gas Basins in the Continental Shelf of Vietnam [Electronic resource] URL: http://nguoidongbang.blogspot.ru/2014/08/ban-o-tham-do-khai-thac-dau-khi-tren.html (accessed: 30.05.2015)

[8] Offshore Oil and Gas Basins of Vietnam [Electronic resource] URL: http://nguoidongbang.blogspot.ru/2014/08/ban-o-tham-do-khai-thac-dau-khi-tren.html (accessed: 30.05.2015)

Education in the system of today's employee values

N Makasheva, T Ischuk, Yu Makasheva and D S Kalashnikova

Institute of Natural Resources, Tomsk Polytechnic University, Tomsk, Russia

E-mail: yuliam@pptsu.ru

Abstract. Establishing the new society is accompanied by a global intellectualization of the entire economic system influencing the development of information, scientific and innovative branches of the indusry. The key factors of the society develpment are information and knowledge. An important quality of a modern employee is the ability for continuous training. Currently, the interest to education has significantly increased as well as the demand for education and the educational level of employees. Education is one of the most vital values, meeting the natural needs of humans in the knowledge of the external world and themselves. The growth of social demand for education in the evolution of social development has led to the gradual transformation of the elite educational institutions into the institutions for everyone. For most people nowadays, higher education has become not only sustainable, but one of the most significant values. The successful development of knowledge-based economy could be provided by the employees not only having a high degree, but having a desire for the continuous lifelong knowledge development.

1. Introduction

The present stage of social development is characterized by the global changes which are the result of the industrial society transformation into the society of the new type. The most important characteristics of this society are the intellectualization of labor and the introduction of information technology into the industrial process of the society. Besides the natural wealth, the accumulation of knowledge, information and experience are the main sources and the key factors of the society development.

In contrast to the industrial, the "new economy" is an economy of intangible factors of production. Alexander Repiev, the president of "Mekka Advertising", said that it is a world where people are working using their minds, not hands [1]. T.Sakayya identified these processes as "... a new stage of civilization where values are the driving force generated by knowledge" [2].

The leading role of information and scientific knowledge is reflected in the concept of "knowledge-based economy" or intellectual economy. The phrase "the knowledge industry" in the early 60s of the 20th century was used by F. Machlup in his research, later, in the late 60s of the last century Peter Drucker introduced a new scientific term "knowledge-based economy." The main capital of such an economy is human resources, and their values as the main productive force are characterized by the extent of their mastery of the modern knowledge [3]. An essential quality of the today's employee is the ability for continuous training.

1.1. The increase of public demand on education in the conditions of flexible postindustrial economy

Investigation of the world around to adapt to its environment for further comfortable existence is one of the essential human needs. In the process of social evolution development, the need of humans in the knowledge has transformed into the existing type of activity – educational system. Incentives to education

at different stages of human civilization development were different, which was largely determined by the level of society development and economy demands. The economy of post-industrial society has accelerated the knowledge growth. Thus, the activities related to the production, storage, transmission and use of knowledge are becoming crucial. According to the research of the first decade of the 21st century, the share of "educated employees" is about one-third from the total amount of employees involved in the economy development [4].
The growth of social demand for education has led to the gradual transformation of the educational institution. At present, the education is affordable for everyone and the demand for it is growing steadily. The interest to education is significantly increasing throughout the world, as well as to the educational level of employees [5]. According to OECD data, currently in France, Great Britain, Norway, more than 60% of national income growth is provided by the employees having higher education. The world's leading trend is the transition to universal higher education. For example, in Japan and South Korea the vast majority of the population aged 25-34 is having higher education. In 15 most developed countries of the world 26% of the employed population is having higher education [6].

1.2. The increase of the role of education in the system of values of the contemporary employee

The authors of the Report on Human Development in the Russian Federation (2014) noted that the formal feature of Russian population was in being of one of the most educated country in the world. In the share of tertiary education Russia is the leading country, and in the share of population possessing higher education Russia ranks second in comparison with ten most developed countries [4]. Education in Russia was and remains a primary value. Good education is considered to be a guarantee of successful socialization and self-realization, as well as one of the most important requirements for successful employment [7]. According to the survey, 53% of Russians over the age of 18 consider that getting of at least a university degree is crucial for self development, and 9% among the surveyed noted the need in further postgraduate education [8].
The system of values related to the obtaining of higher education is determined by a guarantee of getting a highly paid job, prestigious profession and the possibility of self-realization and career development. Higher education is associated with the growth of such consumer features as: the ability to keep a family (65%), to have a car (50%), to travel abroad (50%), to own a house (15%) [8]. The authors of this study marked that the differentiation of consumer expectations of the university, college and vocational school students are almost identical with the differentiation of consumer abilities of the employees with higher, specialized secondary and primary vocational education.
The successful development of the knowledge-based economy could be provided by the employees who have the ability to constant knowledge acquiring, searching and finding ways of problem solving, as well as having the skill of applying their knowledge in practice. In today's knowledge-based economy, the education obtained in the best universities, becomes insufficient very quickly [9]. The new economy requires from the employees a continuous improvement of the educational level [10]. The motto "lifelong education" is changing into "education throughout the life" [11]. This requirement is reflecting in the most important trends of recent years – an increase of the number of able-bodied population involved in further training (non-formal education in terms of International Studies). This could be clearly seen through the correlation: the higher the extent of able-bodied population involved in further education, the higher is the level of economic development of the country. In 2012, 41% of the OECD countries population aged 25-64, were involved in non-formal education. In Russia, only a small part of the labor force is involved in the additional vocational training (about 17%) [4].

2. Conclusion

For most people, higher education has become a stable value and the most important value orientation. The value orientation is a special social attitude of people towards certain values, which is represented in the mind as an object, which a person needs in order to meet the requirement. The value of education is associated with the needs of each individual and the society in whole. In terms of personality, two values of education could be highlighted – the terminal, which is determined by the internal desire to learn and the instrumental, when education is considered as an opportunity to implement the consumer expectations at a higher level.

In contemporary Russia there is a special attitude to higher education. For the majority of school graduates (and their parents) the most significant is the fact of entering a university. The specifics of the modern rush for higher education in Russia (coefficient of admission to higher education institutions, the total number of first-year students and 17-year-olds, is more than 90%) lies in the receiving of basic higher education, i.e. "education for a diploma" that does not require much effort. Getting of the diploma of higher education is considered as a necessary condition for the success in life.

The large-scale study of students' life orientations (late 20th - early 21st century) has showed that education was and remains a priority instrumental value in the mass consciousness of young people. Thus, there is an increase of importance of education as the main tool to achieve life objectives [12]. The terminal value of education was marked by many researchers, but, according to the same survey, 60% of students have showed a lack of desire to continue their education after graduation.

The main consumers of higher education, on the one hand, form the demand for education, but on the other hand, they are guided by the existing demand and the prestige of a particular specialty. Demand characterizes the amount of people willing to improve their educational status and ready to invest in their education. Identifying of the demand trends for higher education is possible through the study of public opinion [13].

Nowadays, according to the theory of generations, the representatives of last three generations of the past century are dominating on the labor market: the baby boomers (1943-1963), the unknown (or "generation X" – 1963-1983) and the millenials (or "generation Y" - 1983- 2003).

The value system of "generation Y" or "generation me" is freedom, creativity and mobility, visible results and immediate payment, the possibility of training and retraining. Such employees are occupying the labor market nowadays. For them, not only traditional educational programs are relevant, but also remote ones as well as non-formal education, master programs, including the obtaining of another qualification and profile training [14]. The next generation could be conventionally referred to as "generation Z" (2003-2023). Its system of values could be only predicted now. Within 8-10 years, the system of values of this generation will determine their behavior in the market of educational services.

References

[1] Rep'ev A Do we have a place in the "New Economy"? (in Russian) [Electronic resource] - URL: http://lib.rus.ec/b/98589/read (reference date: 02.11.2014)

[2] Sakaiya T 1999 The cost created by knowledge or the future history - The new industrial wave in the West: An Anthology Moscow, *Akademia* pp. 337-371 (in Russian)

[3] Syupova M, Bondarenko N 2014 The role of higher education in the formation of an innovative economy *Scientific notes PNU* **Vol.** 3 158-165 (in Russian)

[4] Grigorieva L, Bobilev S 2014 Report - Human Development in the Russian Federation Moscow, *Analytical Centre under the Government of the Russian Federation* pp. 204 (in Russian)

[5] Marginson S 2016 The worldwide trend to high participation higher education: dynamics of social stratification in inclusive systems *Higher Education* **Vol.** 2 pp. 1-22

[6] Plaksiy S 2014 Higher education: challenges and answers, Moscow *Izdadel'stvo nacional'nogo instituta biznesa (The National Business Institute Publishing)* pp. 603 (in Russian)

[7] Fedurina K 2016 Formation of values in today's youth in Russia. - *Collection of materials of the International scientific-practical conference. - "The value system of modern society"* **Vol.** 44 pp. 139-142

[8] Bondarenko N Higher Education in the system of personal values and market guide (in Russian). [Electronic resource] URL: http://magazines.russ.ru/oz/2013/4/24b.html (reference date: 06.02.2016)

[9] Salmi J 2009 Creating a world-class universities Moscow, *All The World Publisher* pp. 132 (in Russian)

[10] Lorenz E, Lundvall B-Å, Kraemer-Mbula E and Rasmussen P 2016 Work Organization, Forms of Employee Learning and National Systems of Education and Training *European Journal of Education. Special Issue: Adult Learning, Adult Skills and Innovation* **Vol.** 51, 2 pp. 154–175
[11] Galimova M 2009 Education throughout life as a successful tool *Life Long Learning as the perspective of university education transformation* **Vol.** 2 pp. 16-19 (in Russian)
[12] Nemirovskiy V 2006 Structure and dynamics of life orientations of student's youth: 1988 -2004. Postnonclassical approach Krasnoyarsk, *RIOKrasGU* pp. 161 (in Russian)
[13] Filatova L 2014 Trends in demand for higher education: conditions, opportunities, risks *Economics of Education* **Vol.** 5 pp. 32-48 (in Russian)
[14] Kostina E, Orlova N 2016 The labor market and building a professional career of students *Higher education in Russia* **Vol.** 11 pp. 28-36

Current situation of oil refinery in Bulgaria

Elena Vershkova[1]**, Petinka Petkova**[2] **Anastasia Grinkevich**[3]

[1,2] Institute of Natural Resources, National Research Tomsk Polytechnic University, 30 Lenin Av., Tomsk, 634050, Russia
[3]Institute of Natural Resources, National Research Tomsk State University, 36 Lenin Av., Tomsk, 634050, Russia

E-mails: [1,2]vem@tpu.ru, [3]nasty_saturn@mail.ru

Abstract. This article deals with the classification approach for oil refineries in international practices. Criteria of refinery estimation group, including its financial status estimation, have been investigated. The analysis object is "Lukoil Neftochim Bourgas" AD (LNCHB) activity. This company is a leading enterprise in Bulgaria. The analysis of LNCHB operating: energy intensity index; index of operating costs and return on investment index have been performed.

1. Introduction

A comprehensive comparison of separate refineries status and their development dynamics in the last 25 years have being carried out by various international rating agencies. Participants receive final state assessment accounts annually, as well as analytical findings that help refinery management in further development processes. In order to ensure confidentiality of the investigation and comparative analysis results, standard peer groups (representative groups) are introduced. According to the same indicators investigated refineries are divided into four subgroups – quartiles. The first quartile includes 25% of all the participating refineries which showed indexes in this group, whereas the fourth quartile included 25% of the refineries with the worst indexes [1]. In case, when the number of participating refineries is not so great, division is 3 or even 2 quartiles. Standard group data comparison provides an opportunity for all refineries to understand what their index level is.

2. Materials and methods

This information provides enterprise managers with sustainable estimation of refinery problems being a matter of priority to shape plans and improvement measures.

The investigation structure involved a large number of indexes being divided into two main groups:

1) index of effectiveness;

2) index of competitiveness [2].

Index of effectiveness includes energy intensity index; maintenance cost efficiency index; personnel efficiency index; maintenance personnel efficiency index; non-maintenance efficiency personnel index; non-energy cost efficiency index; volume gain index; carbon emissions index.

Index of competitiveness includes OPEX - cash operating expenses; NEOPEX – non-energy cash operating expense; maintenance index; personnel index, working hours; PCI - personnel cost index; refinery process utilizations; capital investment index; net cash margin; ROI - return on investment.

Refinery grouping is carried out in two ways (Table 1).

Table 1. Refinery grouping methods in international practices

According to the graphic		According to the refinery scale and complexity	
BAL	Baltic Region (Denmark,	*	Viscracking and thermal

According to the graphic		According to the refinery scale and complexity	
	Sweden, Finland)		cracking
DNX	the Benelux (Belgium, Netherlands)	GOC 1	Refinery with integrated index of less than 6.9
GER	Germany		
FRA	France	GOC 2	Refinery integrated index ranging from 6.9 to 8
UKI	The UK and Ireland		
IBE	South-Western Europe (Spain)	GOC 3	Refinery integrated index in the range of 8 to 9.5
M	Mediterranean Europe (Italy, Greece)		
SEE	South-Western Europe (Bulgaria, Romania)	GOC 4	Refinery integrated index more than 9.5
CEU	Central Europe (Czech Republic, Poland, Austria)		

An integrated indicator represents refinery process capacity as thousand barrels per day.

Leading Bulgarian refinery "Lukoil Neftochim Bourgas" AD (LNCHB) refers to the GOS-2 group (a refinery group with EDC -equivalent distillation capacity- of 1400-2199 thousand barrels per day) and C&SE region (Central and Southern Europe Region) [3,4].

3. Results and discussion

Let's consider the basic LNCHB analysis indexes defining the refineries status from 2008 to 2014. Forecast cost for some 2016 indexes relevant to refinery development plans was calculated.

Energy Intensity Index is the percentage ratio of actual refinery process unit energy consumption to standard refinery process unit energy consumption. Standard energy consumption is determined by the refinery development, i.e. each process involves standard energy consumption per unit of useful efficiency.

The lower the index value, the more efficient the refinery energy consumption. Figures 1 and 2 show the LNCHB EII dynamic pattern comparable to Central and Southern Europe and GOC-2 refinery group EII. Both diagrams clearly demonstrate LNCHB energy efficiency improvement trend for the period from 2008 to 2014 [5]. In this case, the set of implemented activities and projects in this period, were aimed at the LNCHB energy efficiency improvement.

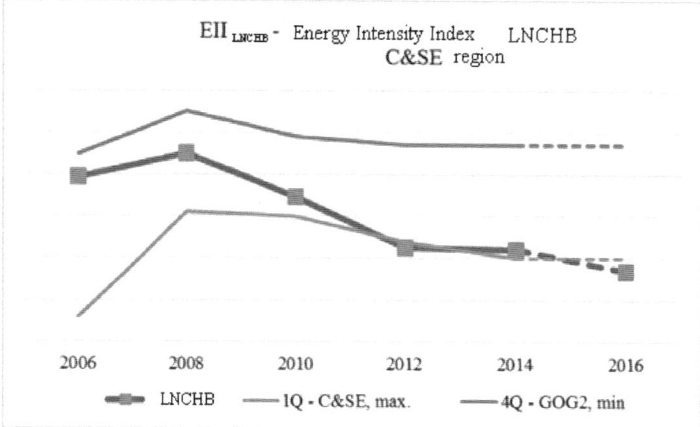

Fig. 1. LNCHB EII (Energy Intensity Index) dynamic pattern comparable to Central and Southern Europe refineries EII, %

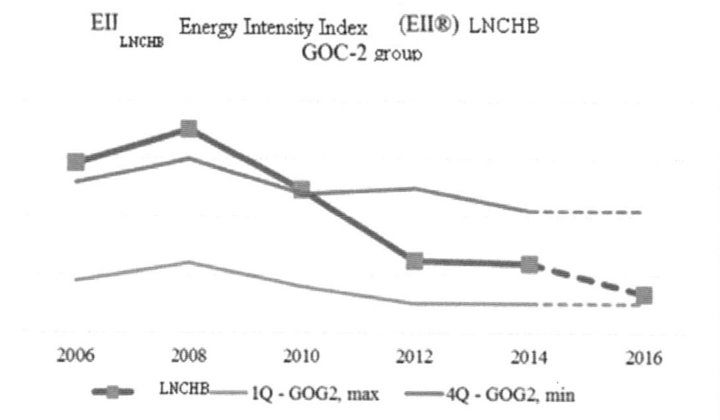

Fig. 2. LNCHB EII dynamic pattern comparable to GOC-2 refinery group EII, %

The diagrams depict that these implemented measures furthered the refinery from the third quartile to the first, as well as the Central and South-eastern Europe refineries from the fourth quartile to the second (for GOC-2 group) [6]. Despite the achieved energy efficiency results, the figures also show that the refinery implementation plans is to continue energy efficiency improvement in 2016. This would be possible due to the planned activities and implementation projects, such as:

- high-radiative coating duplication via LNCHB furnaces (in 2015, the coverage produced in the furnace P-101 AHP-1 (CPF EURO) considerably reduced fuel consumption in the furnace));
- mono ethanolamine gas purification to MDEA solvent (after minor reconstruction) would save significant heat under conditions of hydrocarbon and hydrogen sulfide removal from gas;
- new furnace construction with HRSG (heat recovery steam generator), high efficiency on AD-4 installation (planned construction will be completed in the first quarter of 2017 and would significantly reduce fuel consumption at AD-4);
- other measures (included in the LNCHB energy efficiency program).

OPEX index shows the degree of rationality use of refinery operational costs [7].

Figures 3 and 4 show the LNCHB OPEX dynamic pattern comparable to Central and Southern Europe, as well as GOC-2 refinery group OPEX.

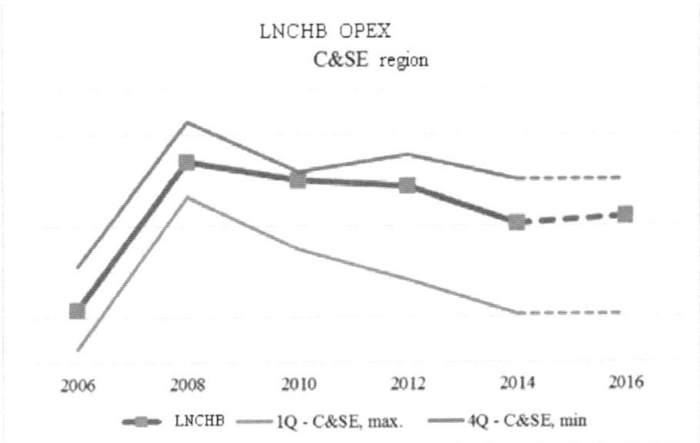

Fig. 3. LNCHB OPEX-index dynamic pattern comparable to refinery OPEX index of the Central and Southern Europe region

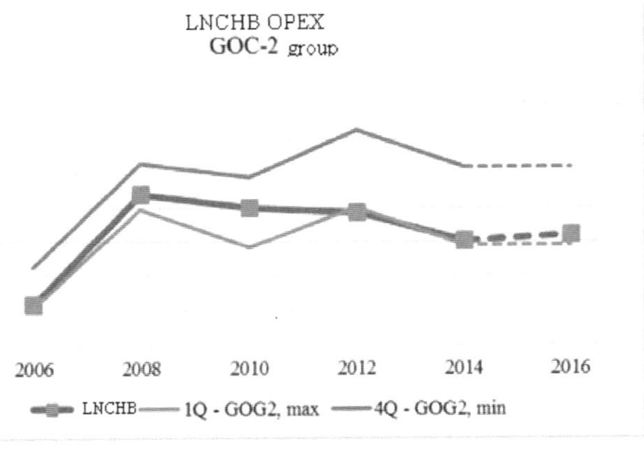

Fig. 4. LNCHB OPEX-index dynamic pattern comparable to refinery group OPEX-index of GOC-2

IOP Conf. Series: Earth and Environmental Science **43** (2016) 012096 doi:10.1088/1755-1315/43/1/012096

The figures clearly show the operating expenses increase in 2008 compared to 2006, for all analyzed participants, i.e. 1/3 energy resource rise in price (specifically for oil) being the major costs (about 2/3 of expenses) in 2008 compared to 2006. Both figures also depict a steady decline in the LNCHB operating costs in the period from 2008 to 2014 and this is most clearly observed in the GOC-2 group, where LNCHB is among the leaders (due to the first quartile OPEX) in 2012 and 2014.

The main reason for such a quartile difference of LNCHB in the GOC-2 group and Central and Southeast Europe Region is the energy sources price (mainly natural gas) [8]. It still continues to be higher in Bulgaria than in Western Europe and the Central and Southern Europe Region. The reduction of OPEX expenses in value (from 2008 to 2014) is due to implemented activities and projects on energy efficiency into the LNCHB, as well as the implementation of a number of measures to minimize hydrocarbons, steam, industrial and drinking water loss etc., related to introduced hydrocarbon vapor disposal technology for final products tank storage; fuel storage and interconnections optimization [9].

Figures 5 and 6 show LNCHB return on investment index (ROI) dynamic pattern comparable to the Central and Southern Europe refineries and GOC-2 refinery group ROI. LNCHB sustains the third quartile ROI level annually exclusively, (as well as for NCM index (net cash margin) comparable to the Central and Southern Europe and GOC-2 group refineries. This is due to the fact that the Return on Investment Index (ROI) is the investment margin ratio to capital employed [10].

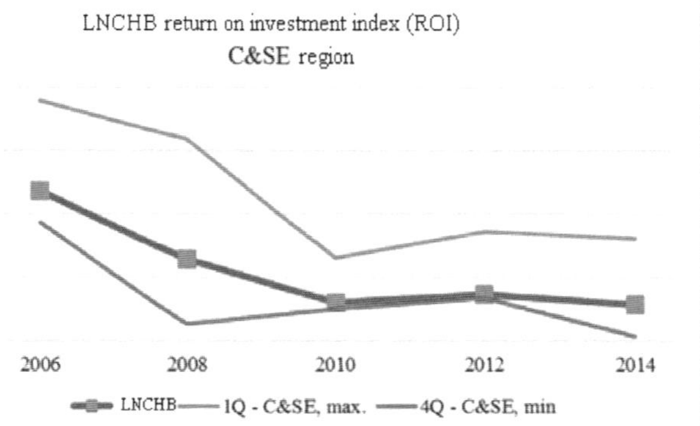

Fig. 5. LNCHB index return on investment (ROI) dynamic pattern comparable to the Central and Southern Europe refineries ROI

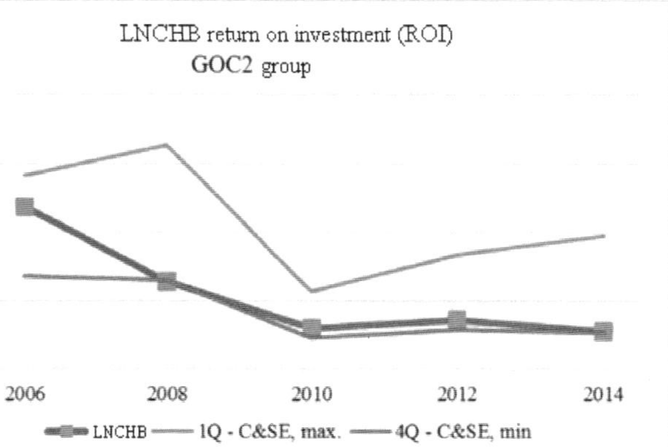

Fig. 6. LNCHB return on investment (ROI) comparable to the GOC-2 refinery group ROI

Based on the above-mentioned methodology personnel efficiency index (PEI) is calculated as productive capability by the formula:

EDC = Σ ((PUC * Pc* PUpt) * Cm)

where, EDCRPUC - equivalent distillation capacity of a refinery process unit;

PUC - process unit capacity, ton per day;

Cf - configuration factor;

Cm - coefficient of multiplicity;

Pc - proportionality coefficient, Pc = 0,01 PUpt

where, PUpt – process unit productive time,% per year.

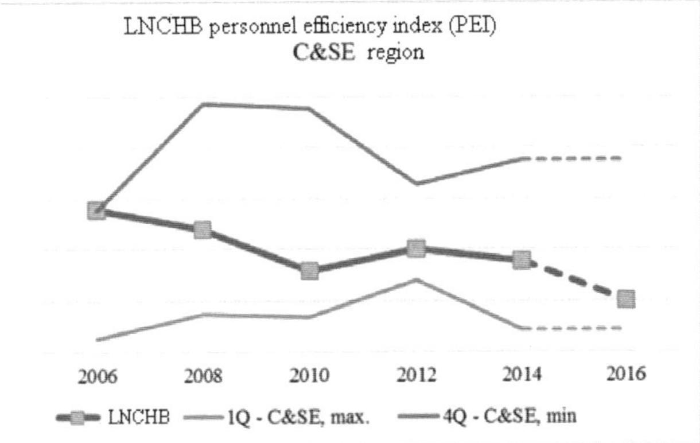

Fig. 7. LNCHB personnel efficiency index (PEI) dynamic pattern comparable to Central and Southern Europe refineries PEI

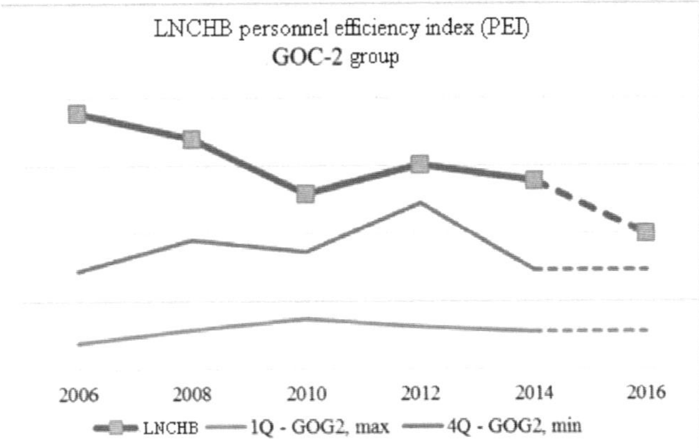

Fig. 8. LNCHB personnel efficiency index comparable to the GOC-2 refinery group PEI

4. Conclusion

It should be noted that it is economically unprofitable to downman personnel as it is the main development resource.

The main difference of the LNCHB quartile in the GOC-2 and Central and Southern Europe group is low-level remuneration of labor in Central and Southern Europe regions.

References

[1] Out of more than 340 oil refineries, a half-dozen that stand out from the crowd [Electronic resource] URL: http://blogs.platts.com/2011/11/01/out_of_more_tha/, free, reference date 01.02.2016
[2] Romanyuk V B, Baitova Y S, Safronova E V and Pozdeeva G P 2015 *Regulatory framework of pricing and estimate standards in the petroleum sector* (Article number 012073) IOP Conference Series: Earth and Environmental Science. **Vol.** 27. pp. 1-5.
[3] Lukoil Neftochim Bourgas AD (LNHB) [Electronic resource] URL: www.neftochim.bg/ru, free, reference date 01.02.2016
[4] Lukoil Refining [Electronic resource] URL: http://www.lukoil.ru/materials/doc/Materials2014/Refining_FB_rus_56-58.pdf, free, reference date 02.02.2016
[5] World Population Prospects [Electronic resource] URL: http://esa.un.org/unpd/wpp/publications/files/key_findings_wpp_2015.pdf, free, reference date 02.02.2016
[6] International Energy Agency [Electronic resource] URL: http://www.iea.org/Sankey/#?c=IEA Total&s=Balance, free, reference date 03.02.2016
[7] Vazim A A, Romanyuk V B, Akhmadeev K N and Matveenko I A 2015 *Associated petroleum gas utilization in Tomsk Oblast: energy efficiency and tax advantages* (Article number 012078) // IOP Conference Series: Earth and Environmental Science. **Vol.** 27. pp. 1-5
[8] Refining Sector Fitness Check Step 1. Solomon Databases [Electronic resource] URL: https://www.concawe.eu/uploads/Modules/Events/2.Symposium2015_Breakout2_REFIT_Solomon_Step1_FINAL.pdf, free, reference date 04.02.2016
[9] Tesla Motors [Electronic resource] URL: https://www.teslamotors.com/supercharger, free, reference date 05.02.2016
[10] Solarcity [Electronic resource] URL: http://www.solarcity.com/, free, reference date 05.02.2016

The policy of import substitution as the basis for economic security and well-being of society

Yu S Makasheva, N P Makasheva, A S Gromova, N V Andreeva and S A Ishtunov

Institute of Natural Resources, National Research Tomsk Polytechnic University, 30 Lenin Av., Tomsk, 634050, Russia

E-mail: e-mail: yuliam@pptsu.ru

Abstract. The study presents the analysis of import substitution opportunities on separate branches of economic activity, preceding the realization of import substitution policy with the aim to support national economic security, which is essential for the contemporary society welfare insurance. Currently, social well-being is considered to be the reflection of economic activity, the instrument of state influence on the society, as well as an indicator of the social security system. Due to the fact that Russia is integrated into the world economy, the foreign-economic policy currently is playing an important role in the development of national security and the state's interest to the spheres of economy considering external and internal threats. Decline in external economic conditions may result in serious consequences for the functioning and development of the country as well as for the trade and investment activities, which will further lead to the decline in export, withdrawal of capital, recession of industrial production, trade and investment sphere, fall of GDP and living standards. Thus, considering the current state of instability in the world economy and the growing political tension in relation to Russian Federation, the measures to increase economic security in the country should be taken. The policy of import substitution is considered to be one of the major solutions nowadays.

1. Introduction

Economic security is traditionally considered to be the most important qualitative characteristics of economic system which defines its ability in the maintenance of high standards of living, steady provision with the resources of national economy development, as well as in the consecutive realization of the national interests [1]. Under the conditions of world economy instability and growing political tension in relation to Russian Federation, the measures to increase the economic security in the country should be taken. The policy of import substitution (IS) is considered to be one of the major solutions nowadays.

2. The need for import substitution policy

There are two main features of IS: competiveness and economic effectiveness. This research is aimed at the revealing of the basic aspects of these features, not only at the detailed theoretical analysis. The development of competiveness, as a means to leave the competitors behind, is a strategic goal and the basis of sustainable socio-economic growth of any country, but the development of "knowledge economy" is the major aspect influencing the growth of the competitiveness. [2] Thus, in order to analyze the IS in terms of economic security it is necessary to investigate not only the level of competiveness in the areas of concern, but also the level of innovations.

Besides, while speaking about the economic efficiency it is necessary to consider the fact that from the point of view of international trade theory, the structure of foreign trade of any country basically depends

on the structure of economy, which in its turn is defined by the features of the country recourses. [3]
The country will have a competitive advantage (mainly in price) in case when it is producing goods from the resources which are in abundance. If the production of the product involves scares and therefore expensive resources, it will be imported. In realization of IS policy it is important to understand that the goods which are producing in accordance with such policy are likely to be more expensive than the imported alternative. The protectionist policy under the conditions of multilateral trade agreements between various organizations such as WTO could lead to the serious consequences. Thus, for the successful realization of the IS policy it is necessary not only to increase the level of competiveness, which will enable to squeeze the competitors from national market, but also to implement the investment policy (basically the state support) at the enterprises producing goods which are analogous to the imported ones. Consequently, while analyzing the IS, apart from the evaluation of the innovation and competitiveness level, it is also important to consider the level of investments into the troubled branches of economic activity. [4]
Thus, *the following analysis will include three main parts: the study of competitiveness level, innovations and investments in accordance with the highly dependent on import branches of economic activity.*
The development of efficient IS policy requires, first of all, objective evaluation of the degree of the country's dependence on import, detection of the most dependent on import branches of economic activity and the degree of readiness to import substitution. This will precede the research of the *competitiveness level, innovations and investment.*
To define the dependence of Russian economy on import, it is necessary to calculate the share of import in different branches of economic activity (table 1). For calculation, the data were taken at current prices [5] according to purchasing-power-parity exchange rate [6].

Table 1. The share of import in domestic consumption on RNCEA, %.

Code		**2005**	**2010**	**2012**	**2013**
	In whole consumption	6.72	9.33	11.77	11.54
01	Agriculture, hunting	11.47	18.76	20.19	20.86
02	Forestry, logging	0.52	0.10	0.18	0.27
05	Fishing, operation of fish hatcheries and fish farms	27.34	62.74	186.60	289.92
11	Extraction of crude petroleum and natural gas	0.11	0.25	0.40	0.24
12	Mining of uranium and thorium ores	0.00	0.01	0.13	0.27
13	Extraction of metallic minerals	13.00	9.75	26.62	29.35
14	Extraction of other mineral resources	4.62	7.76	10.08	9.72
15	Manufacture of food products and beverages	13.04	16.93	18.39	18.38
16	Manufacture of tobacco products	5.30	3.64	2.48	3.41
17	Manufacture of textiles	61.97	65.24	77.78	79.05
18	Manufacture of wearing apparel; dressing and dyeing of fur	67.46	82.59	89.20	90.20
19	Manufacture of leather and leather products, including footwear	31.51	50.23	57.33	61.97
20	Manufacture of wood and wood products, except furniture	10.04	19.53	31.71	40.80
21	Manufacture of pulp, paper and paper products	42.08	44.32	53.50	63.81
22	Publishing, printing	6.83	10.34	12.78	13.30
24	Manufacture of chemicals and chemical products	50.75	69.77	83.37	107.11
25	Manufacture of rubber and plastic products	50.14	51.79	62.70	66.53
26	Manufacture of other non-metallic mineral products	10.62	17.11	19.59	20.92
27	Manufacture of basic metals	14.23	34.00	62.83	88.52
28	Manufacture of fabricated metal products	28.30	37.00	45.98	50.03
29	Manufacture of machinery and equipment	55.30	67.33	76.33	80.78
30	Manufacture of office machinery and computers	84.91	88.30	90.26	92.87

31	Manufacture of electrical machinery and apparatus	45.39	64.44	76.24	80.90
32	Manufacture of radio, television and communication equipment	66.06	73.84	77.44	80.41
33	Manufacture of medical equipment, instruments and appliances for measuring, checking, testing, navigating and other purposes	31.60	39.67	47.58	43.59
34	Manufacture of motor vehicles, trailers and semi-trailers	7.35	7.45	5.51	6.44
35	Manufacture of aircraft and spacecraft and other transport equipment	55.86	59.90	71.13	65.30
36	Manufacture of furniture and goods, which are not included into the other groups	32.42	41.25	57.37	59.76
37	Recycling	5.85	9.20	12.28	11.92
40.1	Production and distribution of electricity	0.37	0.06	0.10	0.17

Before the data analysis, it is necessary to define the criteria explaining the necessity of IS. For this purpose, the current legislation could be applied, where these criteria are registered only for a limited branches of activity. For example, in forestry sector this share will go down to 10 % by 2020; in paper, cardboard and furniture manufacture – to 10.5 % and 10. 7% respectively [7]. In food products the share will go down in average up to 3-15% by 2020 (the share of Russian production must be: in sugar up to 96.7%, in vegetable oil – 84%, meat and meat products – 88.3%, in fish products – 82%, in milk and milk products – 85.3) [8]. Thus, regarding the necessity of IS, the branches of economic activity with the criterion from 15% and higher could be considered. The index share of IS could be more than 1 (100%) if the production requires semi-products or constituting parts which are included in the production of export goods.

3. Innovative parameters and indexes of competiveness

Basing on the analysis, the basic types of troubled economic branches with a different degree of dependence (marked in Table 2) could be defined - the manufacture of metals, chemicals and chemical products, machinery and equipment, textile manufacture, fishing and fish farms, wood manufacture. These branches will further be viewed in more detail in the next analysis.

Three types of branches could be distinguished basing on the indexes of competiveness in Russian national economy:

1) competitive branches: oil and gas industries, forest industry, diamond industry, ferrous and non-ferrous industries, power industry. These branches have powerful raw materials base and they make up approximately 75% of Russia's export.

2) semi-competitive branches: aerospace industry, atomic industry, heavy engineering industry, wood-pulp and paper industry, power machine building, production of metalware, food manufacture, food industry, fish industry.

3) noncompetitive branches: machine-tool industry; manufacture of high-tech_consumer electronics; equipment for metal manufacture, petroleum and chemical manufacture, wood and paper manufacture; agricultural machinery industry, light industry.

The main conclusions on the competitiveness of import dependent branches of economic activity are given in the final table.

The provision of competitive IS requires serious investments in the development of local competitive industries. Deprecation of the main reserves in Russia is about 50% in total (in 2005 it was 45.2%, in 2013 – 48.2%) [9].

The worst indexes, according to the deprecation degree data of basic funds of Russian Federation commercial organizations on the types of economic activity (especially from the point of view of IS), belong to the fish industry and fish farming – in 2010 the deprecation was 60.3%, by the end of 2013 this index lowered, but it is still considerable – 51.3%. In the other branches under the analysis, this index is slightly lower, but on the whole it is close to the average index in Russia. [10]

According to the FSSS data [10], the investment in the analyzed branches was 7.2% (in 2005) from the total value of investments into the main capital all over the Russia. During the next years, the volume gradually increased, but considering the volume of investments in whole, the share of investments in these branches decreased to 5.3% by the end of 2013.
The lag in metallurgical production is more notable (1.5 percentage points in 2013 in comparison with 2005). The share of investment in fish industry and fish farming, in manufacture of textile and clothing, as well as (with little decrease in 2010 and 2011) in manufacture of electrical and optical equipment was not changing over a period from 2005 to 2013.
Only the manufacture of chemicals and chemical products showed a little increase of the investment share by 0.1 percentage point (at the same time, in 2010 and 2011 it also decreased).
In all analyzed branches the main sources of investments are the own company's funds. The highest share of the own funds is in the manufacture of machinery and equipment, in the manufacture of electrical and optical equipment (70.8 and 72.5%% respectively in 2013), in the manufacture of basic metals (66.2 % in 2012 г. and 69.2 % in 2013), in fishing and fish farms (72.7% in 2012 and 51.9% in 2013). [10]
It should be noted that the financial investments of the organizations of the analyzed types of economic activity are mainly the short-term investments [11].
It is a big challenge for the company to cope with IS increase with its own funds due to the fact that the money should be invested not only in the recovery of the deprecated reserves, but also in the modernization. It is essential to develop new innovative technologies [12] for the provision of the competitive production for further replacement of the imported products with the domestic ones, which will be put to the global market.
Russian researches also note Russia's lag in many innovative parameters. For example, insufficient level of national scientific and technical potential and effectiveness of its use. This points at an extremely low level of our economy innovativeness [13].
The share of innovative products is very low in most of the analyzed spheres. [14], [15] Only the manufacture of chemicals and chemical products shows a relative increase of the indexes (from 4.8% in 2010 to 9.2% in 2013). But even these index fall (from 11.9% in 2010 to 9.6% in 2013). According to the FSSS data in 2013 there was an increase in the production of electrical and optical equipment - 10.7%. In other branches it is much lower. Thus, in the manufacture of textiles and clothing it was 2.9% in 2010, but in 2013 it decreased to 2.4%. Compare: the unit weight of new for the innovative products market goods (in total amount of shipped goods of innovation-active enterprises) was 12% in Great Britain, 14% in Hungary, 14.1% in Germany, 15.4% in Portugal [8].
As for the costs on innovative activity, only the metal manufacture has a higher index in comparison with the other branches in the analyzed sector. Its dynamics is negative: 18.9% in 2011, 15.0% in 2012 and 8.2% in 2013 – thus, the decrease is more than 10% in 3 years only! In other branches this index is much lower (from 0.1% in the textiles and clothing manufacture to ≈ 9 % in the manufacture of chemicals and chemical products).

Considering the given indexes, it is very hard to talk about a really effective import substitution as well as to ensure the competiveness product on the branches of economic activity experiencing currently the high import dependence.

Table 2. Analysis of IS possibilities on RNCEA.

	Branches	Import quota in domestic consumption		Competiveness [16],[17],[18]		Innovations		Investments	
		Level	Dynamics	Level	Dynamics	Level	Dynamics	Level	Dynamics
1	Manufacture of metals	high	–	high	–	high	–	low	–
2	Manufacture of chemicals and chemical	high	–	high	–	average	–	low	+

	products								
3	Manufacture of wood and wood products, except furniture	high	–	low	0	very low	–	low	+
4	Manufacture of machinery and equipment	high	–	low	–	averag e	0	low	+
5	Fishing, operation of fish hatcheries and fish farms	high	–	high	–	-	–	low	0
6	Manufacture of textiles and clothing	high	–	low	+	very low	0	low	0
7	Manufacture of electrical and optical equipment	high	–	low	–	high	0	low	0

Basing on the data given in Table 2 it could be noted that the dynamics of import quota is negative for all troubled branches of economic activity as well the level of investments. Taking this into consideration, the branches with the high or average level of competiveness and innovation activity could be referred to the branches of economic activity, which are open for IS and where economic security threat free IS policy could be implemented. These branches are – the metal manufacture (against the negative dynamics), the manufacture of chemicals and chemical products and fish industry. The realization of the IS policy in the other branches is impossible without serious government program and government financing, otherwise this could result in the threat to the country's economic security. The results of the research indicate to the necessity of serious analysis of IS opportunities on separate branches of economic activity, preceding the realization of IS with the aim to support the national economic security.

4. Conclusion

Considering the given indexes, it is very hard to talk about a really effective import substitution as well as to ensure the competiveness of product of the economic activity branches experiencing currently the high import dependence.

To cope with this challenge, it is necessary to attract all possible funds, to involve both budget and extra budgetary funds in the investment process. Regional and municipal authorities could play a significant role by participating in the renewal and development of corresponding branches on their territory. This will allow to reach not only economic targets but also social ones, namely - to rise the employment level, to provide corresponding domestic goods, to rise the income and the standard of living.

References

[1] Aslanova L, Batova B 2014 Geoeconomic position of Russia in the world: problems and prospects. *Modern problems of science and education* **Vol.** 2 pp. 1-9

[2] Kravchenko N, Bobylev G, Valieva O, Fyodorov A 2013 Competitiveness on the Basis of Innovation: The International Position of Russia. *Studies on Russian Economic Development* **Vol.** 5 pp. 90-101

[3] Mukherjee S 2012 Revisiting the Debate over Import-substituting versus Export-led Industrialization. *Trade and Development Review* **Vol.** 5 pp. 64 – 76

[4] Elvin David 2008 An Analysis of Methods for Identifying Local Import Substitution Opportunities to Foster Sustainable Regional Economies. Master's Theses 1896 February 2014. Paper 141
[5] Hand Books of statistics 2014 [Electronic resource]: [Official website] / United Nations Conference on Trade and Development (UNCTAD) – URL: http://unctadstat.unctad.org/ReportFolders/reportFolders.aspx (reference date: 21.06.2014)
[6] World Economic Outlook Database April 2014 [Electronic resource]: International Monetary Fund (IMF) - URL: http://www.imf.org/external/pubs/ft/weo/2014/01/weodata/weorept.aspx?sy=2003&ey=2012&scsm=1&ssd=1&sort=country&ds=.&br=1&c=922&s=PPPEX&grp=0&a=&pr.x=32&pr.y=4 (reference date: 28.09.2014)
[7] About the approval of Russian Federation forestry industry development for the period to 2020 [Electronic resource]: Order of Ministry of Industry and Trade RF № 248, Ministry of Agriculture RF № 482 from 31 October 2008 – URL: http://www.consultant.ru
[8] About the approval of Russian Federation food and recycling industry development for the period to 2020 [Electronic resource]: Decree of the RF Government from 17 April 2012 № 559-r – URL: http://www.consultant.ru
[9] Federal State Statistics Service. [Electronic resource] // URL: http://www.gks.ru (reference date: 17.02.2015)
[10] Investments in Russia 2013 Moscow, *Rosstat* 300 (in Russian)
[11] Investment activity in Russia: conditions, factors, tendencies 2013 (in Russian) [Electronic resource] – URL: http://www.gks.ru/bgd/regl/b13_112/Main.htm (reference date: 17.08.2014)
[12] Supporting Investment in Knowledge Capital, Growth and Innovation 2013 OECD, *OECD* 360
[13] Mindeli L 2012 Sceince and Innovation in Modern Russia. *Energy: economics, technics, ecology* **Vol.** 3 pp. 11-16
[14] Indicators of Innovation in the Russian Federation 2013 Moscow, *National Research University "Higher School of Economics"* p. 472 (in Russian)
[15] Falcman V 2014 Competitiveness of the basic types of Russian goods on the world market *Contemporary Europe* **Vol.** 1 pp. 87-98 (in Russian)
[16] Romanov V 2013 Modern condition and perspectives of forestry complex development *Regional problems of transforming the economy* **Vol.** 2 pp. 37-46 (in Russian)
[17] Tumalanov N, Ivanov V, Tumalanov E 2013 The impact of modernization on the competitiveness of the domestic manufacturers *The Bulletin of KrasGAU* **Vol.** 8 pp. 14-18 (in Russian)

Stimulation of commercial coal seam methane production aimed at improving mining technology

E A Shubina[1], V G Lukyanov[2]

Department of Oil and Gas Complex Engineering Design, Tomsk Polytechnic University, Tomsk

[1]Lena_shubina@mail.ru, [2]lev@tpu.ru

Abstract. The relevance of the current research is due to the urgent need to revise the existing normative bases and procedures involved in intensive development of coal-methane deposits and commercial production of coal seam methane. The article presents the analysis of data on coal production volume and amount of methane emitted into the atmosphere in Kuzbass. There is a need to develop the exploration techniques that would allow implementing pre-mining gas drainage of coal seams and provide the companies with the guidance on coal seam methane drainage in very gassy coal mines. Commercial production of methane should become an integral part of economy and energy balance of the Russian Federation, which, in its turn, would enhance environmental protection due to reducing methane emissions, the largest source of greenhouse effect.

Introduction

The coal seams play a significant role in methane accumulation within the earth's crust and open up new wide horizons in building up hydrocarbon gas resources. Due to the fact that the Russian Federation (RF) is rich in coal deposits, unconventional sources of hydrocarbons have been neglected. And just recently one can notice a fairly obvious interest towards this sector of petroleum industry and, in particular, to the State balance of methane resources.

The predicted methane accumulation within the basic coal basins of the RF accounts for 83.7 trillion cubic meters, which, in its turn, makes up about one-thirds of Russian natural gas resources. Among coal basins of the RF, a special role belongs to Kuzbass which is proved to be the largest and most studied coal-methane basin of the world. Forecast resources of methane in the basin account for more than 13-20 trillion cubic meters [1].

This estimation of methane resources in coal seams corresponds to the depth of 1800–2000 m and provides a great potential for wide-scale production of methane as an independent mineral resource within Kuznetsk coal basin.

In accordance with the reports of the Kemerovo regional administration, 25 coal pits and open-pit coal mines will be closed down by 2025 [2], with 13 coal mines being closed due to resource depletion, while the rest – due to unprofitability. The coal mines are basically located in Kisilyovsk and Prokop'evsk. They were constructed in 1920-1940 and characterized by complex rock and geological conditions. Most coal enterprises closed in the 1990s. Those enterprises that still operate just queue to wait for their turn and do not produce anything. In Kuzbass, the total number of coal mines reduced from 86 to 62 between 1996 and 2013. An increase in coal seam gas content was one of the basic causes for closing coal mines.

Today, coal enterprises that operate in Kemerovo Oblast make up 57% of Russia's total coal production. In accordance with the Instruction on Mine Air Content Monitoring, Gas Content Determination and Mine Classification based on Methane and Carbon Dioxide Concentration [3], coal mines are annually classified based on relative gas content. The relative gas content is a volume of gas released during coal mining over a certain period per amount of coal produced over the same period and it is measured in m^3/t [4].

Thus, the coal mines in Kuznetsk coal basin are classified as follows:

- 9 coal mines are referred to I category, with relative gas content being up to 5 m^3/t of average daily production;
- 7 coal mines are referred to II category, 5-10 m^3/t;
- 11 coal mines are referred to III category 10-15 m^3/t;
- 21 very gassy mines, i.e. more than 15 m^3/t or blows;
- 14 mines are classified as dangerous due to methane outburst.

It is worth noting that relative gas content of very gassy mines is sometimes several times higher than that indicated in the classification and it can reach 100 and even more m^3/t.

To ensure safety of coal mining operations, costly and at the same time inefficient technologies to gasify coal seams have to be applied. This makes coal production costs much higher. The violation of dust-gas regulations leads to outbursts in coal mines which, in this case, should undergo complete reconstruction. The emergency response plans of the coal and gas outbursts are commonly rather expensive and time-consuming involving such issues as development and application of gas drainage techniques to control gas emission after complete underground mine reconstruction. Therefore, those mining companies that do not follow the management plans of outbursts in underground coal mines or do not have any opportunities to implement up-to-date gas drainage techniques have to stop mining activity, shut down coal mines or dissolve mining companies. It is due to the fact that production costs do not ensure either loss-free operation of the mining enterprises or safety of mining activity.

Increase in underground coal mining costs caused by high gas content in coal seams inevitably results in reduction of coal production volume. As an illustration, table 1 shows the data on coal production in Kuzbass between 2004 and 2013 based on materials provided by the State Balance of Minerals [5].

Table 1. Data on coal production in Kuznetsk coal basin between 2004 and 2013

Year	Coal production in Kuzbass basin, million tonnes	Proportion of coal production within Kuzbass basin	
		underground mining	surface mining
2004	141	45	55
2005	149	44	56
2006	156	44	56
2007	162	44	56
2008	162	41	59
2009	158	45	55
2010	161	40	60
2011	166	37	63
2012	175	37	63
2013	178	35	65

The data presented in table 1 reveal the increase in coal production over the past 10 years in Kuznetsk coal basin. Due to construction of new mines, the total number of coal mining companies that have to produce coal by underground mining methods because of unprofitable mine liquidation does not significantly decrease.

On the basis of the data presented in table 1, special attention should be paid to the increase in volume of coal extracted by surface mining methods, i.e. up to 65% of total coal production volume in Kuznetsk coal basin. To make it more obvious the data are given as a diagram (figure 1).

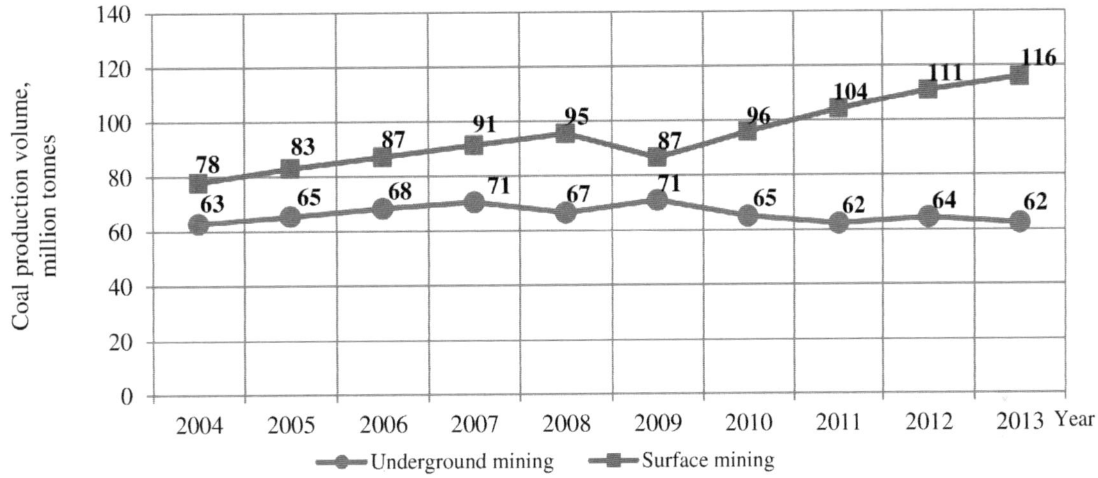

Figure 1. Volume of coal produces by underground and surface mining methods between 2004 and 2013 in Kuznetsk coal basin.

The data illustrated in figure 1 indicate a slight increase in volume of coal extracted by surface mining methods in 2004 in respect to the volume of coal extracted by underground mining methods and steady increase in surface mining over the past 4 years.

For further analysis of coal production volume and methane emission, we collected the data on the amount of methane being emitted into the atmosphere between 2004 and 2014. The data were provided by regional office of the Federal State Statistics Service in Kemerovo Oblast.

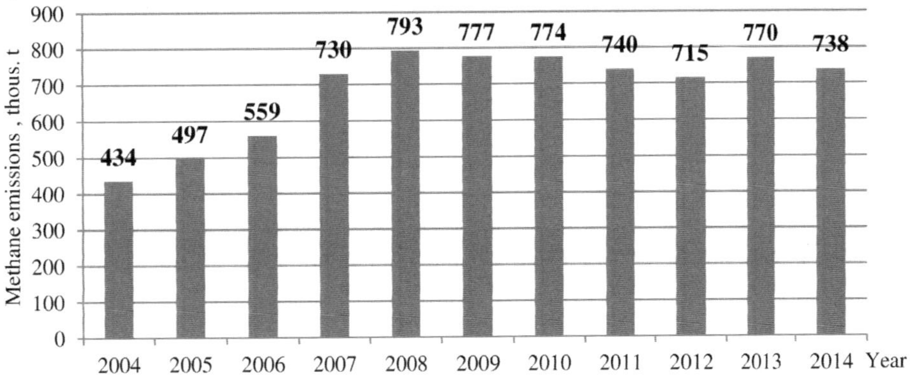

Figure 2 Data on the amount of methane being emitted into the atmosphere between 2004 and 2014 in Kuznetsk coal basin.

Considering the specific features of coal-methane deposits, coal is basically extracted by surface mining methods in gas weathering zones. Therefore, figure 3 presents the comparative analysis of volume of coal extracted by underground mining methods and amount of methane being emitted to the atmosphere.

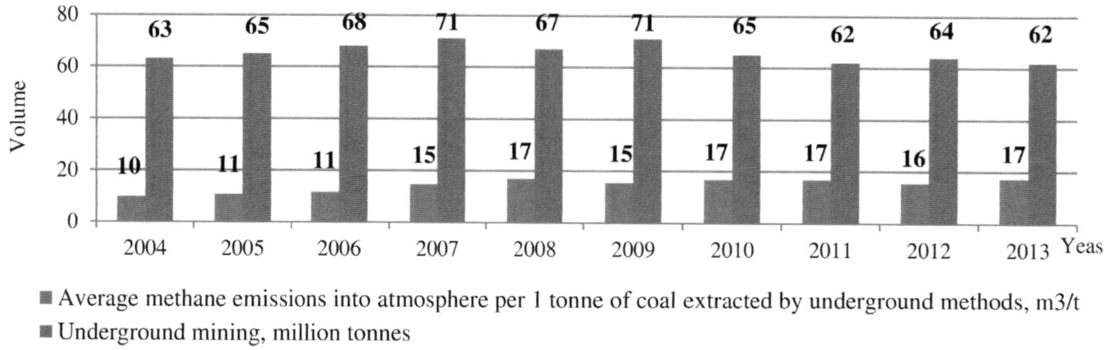

Figure 3 Methane emissions per 1 tonne of coal extracted by underground mining method.

As apparent from the data shown in figure 2, the amount of methane emitted into the atmosphere has increased since 2008. Despite reduction in underground coal production, average number of methane emissions per a tonne of extracted coal is 17 m^3/t. This proves the urgency of the issue concerning increase in coal mine gas.

This situation has been actively discussed at various levels of public authority. In 2014, the government of the RF approved the Coal Industry Development Program up to 2030 [6].

The program is aimed at ensuring high-efficient operation of coal mining industry due to the following steps: complete modernization of Russian coal industry, implementation of up-to-date coal production technologies, development of new coal mining activity centers and complexes intended to convert coal into products with the high share of added value, implementation of the already taken decisions of federal executive authorities on technical, social, and environmental issues related to liquidation of unprofitable underground and open-pits mines and reestablishment of safe living conditions for local civilians.

The program also implies construction of new coal mines. It is worth noting that application of various gas drainage techniques is the only way to provide safe coal mining activity at depth greater than 500 meters, with gas content being 20-30 m^3/t when reported on a dry, ash-free basis. In this context, implementation of pre-mining gas drainage and utilization would ensure the safety of further coal production and reduce methane emissions.

In accordance with the data presented in figure 2, annual emissions of coal mining companies reached 434 thousand tonnes in 2004, while in 2014 – 739 thousand tonnes. Therefore, it is possible to conclude that the amount of methane being emitted into the atmosphere has risen by 70 % over the past 10 years. This proves the urgency of selecting more appropriate coal production method in coal mines characterized by high gas content.

The standard tax rate per tonne of methane within the established standards for permissible emissions is 50 rubles, within the established limits for emissions – 250 rubles [7]. It is noteworthy that the tax on methane emissions into the atmosphere has remained constant for more than 10 years. Due to the growth rate of methane emissions and corresponding increase in coal production costs, the tax reduction is not foreseen. However, the tax rate does not ensure safety of coal mining and guarantee shut-down free operation of high gassy coal mines.

Today, within the program of integrated development of Kuzbass coal deposits, gas drainage is carried out in the following coal mines: "Taldinskaya-Zapandnaya-1,2", "Kotinskaya", "7th November", "Kirova", "Komsomolets", "Polysaevskaya", "Mine named after Lenin", "Chertinskaya-Koksovaya", "Mine № 7". Most of the enumerated coal mines are operated by OJSC SUEK (OJSC Siberian Coal Energy Company)-Kuzbass [8]. In the above-mentioned coal mines, methane is basically extracted and utilized in development and production headings, i.e. directly during coal production. Therefore, the applied technologies have nothing in common with pre-mining gas drainage.

Geographically, coal-rich regions are located a significant distance from natural gas deposits. The designated purpose of wide-scale coal-seam methane production is to supply coal-mining regions of the RF with their own local gas. Being an alternative to natural gas, methane is more available, much cheaper, and more environmentally friendly among most known gases. The coal production costs

IOP Conf. Series: Earth and Environmental Science **43** (2016) 012098 doi:10.1088/1755-1315/43/1/012098
would reduce up to 30 % if coal mining companies are provided with heat and power generated from coal seam methane.

To provide commercial methane production in Kuzbass, in 2001 the RF developed and approved the program "Hydrocarbon Production in Kemerovo Oblast".

Gazprom Dobycha Kuznetsk is the license owner entitled to operate a test site in Taldinskoye and Harykско-Ostashkinskoye areas. The company has already produced almost 40 million m^3 of coal seam methane, in 2014 – 12 million m^3. Today, the production rate is 1.2 million cubic meters per month.

In Russia, coal seam methane reserves were first proved in May, 2011(protocol № 2420–dsp of the State Committee for Mineral Reserves in south-east area of Taldinskoye coal-methane deposit [9].

Based on the results of Taldinskoye and Harykско-Ostashkinskoye site development, it is planned to elaborate legal and regulatory basis to carry out geological study of gas content in coal seams during the exploration stage, to develop methodology to assess geological and recoverable methane reserves and resources and provide the guidance on pre-mining methane drainage and integrated coal-methane deposit development in order to ensure rational use and preservation of mineral resources.

Experience of such countries as USA, Australia, and China shows that wide-scale production of coal seam methane was launched, especially after the Government of these countries began to support the projects by providing the companies engaged in coal seam methane production with significant tax breaks [10].

Implementation of the above recommendations obviously requires significant state support. In Russia, in order to stimulate coal seam methane production, a number of tax breaks have been adopted.

In Kemerovo Oblast, the Congress of People's Deputies adopted tax relief Act N 5-OZ dated 28 January, 2010 "Tax Relief for the Residents of Industrial Clusters and Entities Involved in Investment Activities Engaged in Coal Seam Methane Production in Coal Deposits of Kemerovo Oblast" [11].

Precisely, under article 3 of the above Act, the residents of the industrial clusters and entities involved in investment activities are granted with the following tax concessions:

- the corporate income tax rate to be paid to the regional budget has been reduced from 18 % up to 13.5 %;
- the property tax rate has been reduced from 2.2 % up to 0 %.

In order to ensure coal mine safety, the Government of the RF has granted the companies with tax relief and abolished mineral extraction tax introducing zero import customs duties rates in respect of unique individual types of components parts and parts needed for gas drainage and methane production [12].

Federal Act No. 425-FZ of 28 December, 2010 amended the Russian Tax Code [13]. Under article 325.1, taxpayers are provided with a tax deduction in terms of mineral extraction tax reduction by cost value that is related to provision of coal mining safety, especially in the sites characterized by high gas content and coal seams prone to spontaneous ignition.

Taxpayers may spend the savings from tax deduction as they determine or consider the corresponding costs in calculating the taxation base for the income tax. These amendments have been in force since 1 April, 2011. Besides, article 343.1 amends the procedure of reducing coal severance tax due to the costs related to providing safe working environment.

Pursuant to Federal Act N 278-FZ of 29 December, 2012 on Amendments to Articles 336 and 337, coal seam methane is regarded as independent mineral resource. Moreover, in accordance with Article 336, paragraph 2, coal seam methane is not recognized as an item of taxation. The amendments significantly stipulate coal seam methane production on an industrial scale and provide the basis for coal mine safety due to implementation of various gas drainage techniques [14].

Conclusion

Taking all those considerations into account, therefore, it could be said that there is an urgent need to revise the procedure of underground coal mining in terms of coal seam production. With favorable conditions provided by the RF Government, there is an impetus to the development coal seam methane production on an industrial scale in order to provide coal mine safety. At the same time, the current legislation system does not provide coal-mining companies with exemption from paying for methane emissions, which, in its turn, encourages companies to reflect on their own attitudes towards methane.

Research-and-development organizations and subsurface users are currently implementing the plans for integrated development of coal-methane deposits.

References

[1] http://www.gazprom.ru/about/production/extraction/metan/
[2] Makin M V 2014 *Kemerovo:Standart kachestva.* Kuzbasse budut stroitsya novye vysokoproizvoditelnye predpriyatiya [In the Kuzbass will build new high-performance enterprise] no. **37/38** pp. 8-10.
[3] 2012 *Federal Service for Ecological, Technological and Atomic Supervision of the Russian Federation.* Instruktsiya po kontrolyu sostava rudnichnogo vozdukha, opredeleniyu gazoobilnosti I ustanovleniyu kategoriy shaht po metanu i dioksidu ugleroda [Instructions for monitoring the composition of mine air, determination of the gas content and the establishment of categories of mines for methane and carbon dioxide] (Moscow)
[4] 2011 *Federal Service for Ecological, Technological and Atomic Supervision of the Russian Federation.* Instruktsiya po degazatsii ugolnykh shakht [Instruction on gas drainage of coal mines] (Moscow)
[5] 2004-2013 *Federal Agency on Subsoil Usage «Rosnedra».* Gosudarstvennyy balans zapasov tverdykh goryuchikh iskopaemykh [State balance of stocks of solid fuels] Ugol. (Moscow: Rosnedra)
[6] Approved by the Government of the RF dated 21 June, 2014 № 1099-r Programma razvitiya ugolnoy promyshlennosti Rossii na period do 2030 goda. [The program of development of coal industry in Russia for the period up to 2030] (Moscow)
[7] *Government Act of the RF* dated 12 June, 2003 № 344 *O* normativakh platy za vybrosy v atmosfernyy vozdukh zagryaznyayushshikh veshshestv statsionarnymi i peredvizhnymi istochnikami [On rates of charges for emissions of air pollutants from stationary and mobile sources] (Moscow)
[8] http://m.ria.ru/kem/20131212/983666091.html
[9] *Gosudarstvennaya ekspertiza materialov podschota zapasov metana v ugolnyh plastah Yugo-Vostochnogo uchastka Taldinskogo metanougolnogo mestorozhdeniya [State examination of materials calculation of reserves of methane in coal seams of the Southeast region Taldinskoye CBM field]* Protokol GKZ Rosnedra no. 2420-dsp dated 25 February, 2011 (Moscow)
[10] http://www.gazprominfo.ru/articles/methane-from-coal/
[11] *Act approved in Kemerovo Oblast* dated 28 January, 2010 № 5-OZ O nalogovykh lgotakh rezidentam tehnoparkov i subektam investicionnoy deyatelnosti, osushshestvlyayushshim deyatelnost po dobyche prirodnogo gaza (metana) iz ugolnykh mestorozhdeniy na territorii Kemerovskoy oblasti [On tax benefits to residents of industrial clusters and investment activity entities engaged in activities for the extraction of natural gas (methane) of coal deposits in Kemerovo Oblast] (Kemerovo)
[12] *Act of Council of Federation of the Federal Assembly of the Russian Federation* dated 09 June, 2010 № 218-SF O zakonodatelnom obespechenii bezopasnykh usloviy truda v ugolnoy otrasli [On the legislative provision of safe working conditions in the coal industry] (Moscow)
[13] *Federalnyy zakon* ot 28.12.2010 no. 425-FZ *O* vnesenii izmeneniy v glavy 25 i 26 chasti vtoroy Nalogovogo kodeksa Rossiyskoy Federatsii. [On Amendments to Chapters 25 and 26 of the Tax Code of the Russian Federation] (Moscow)
[14] *Federal Act* dated 29 December, 2012 no.278-FZ *O* vnesenii izmeneniy v stati 336 i 337 chasti vtoroy Nalogovogo kodeksa RF. [On Amendments to Articles 336 and 337 of the Tax Code of the Russian Federation] (Moscow)

Allowable and critical risks of the Arctic development in terms of global climate change

Y Bolsunovskaya[1], D Volodina, A Sentsov[3]
[1,2] National Research Tomsk Polytechnic University, Tomsk, Russia
[3] National Research Tomsk State University, National Research Tomsk Polytechnic Russia Tomsk, Russia

E-mail: [1]ju_al@inbox.ru, [2]volodina.da2014@yandex.ru, [3]sentsov@tpu.ru

Abstract
The Arctic development is accompanied by different high risks which basically arise due to natural and technogenic factors. The changes in the Arctic cryosphere are commonly considered the most serious ones by the international scientific community. In our study we regard the changes in Arctic cryosphere as natural risks. Due to the fact that complex ice conditions, on the one hand, present the serious obstacle to Arctic resources development and, on the other hand, serve as indicator of alarming global climate change, the current research proposes the risk analysis based on the analytical model, with the risks being classified by their level of impact.

1. Introduction

The Arctic issues are becoming increasingly important in the current studies of global climate change. This is due to the fact that the Arctic region which is characterized by complex and rather specific climatic features (ice sheet) secures the balance of global climate system. As for the Arctic region itself, it should be noted that today it presents a vivid example of transforming climatic and environmental problems into economic and political ones. The growth of the Arctic geo-political importance on a worldwide scale due to Arctic shipping potential and the anticipated amount of hydrocarbons and strategic metals opens new avenues for economic activities both of the Arctic states and the states which do not border the Arctic region. The specific geographical and climatic conditions of the Arctic present a challenge to economic activity in the region. In fact, it is these conditions that restrain geophysical exploration of the Arctic continental shelf. Besides, they make higher demand on communications-infrastructure availability, sustainable development of social systems, staff resources trained specifically to work under Arctic conditions, and, most significantly, programs aimed at preserving and protecting the Arctic environment and unique eco-systems. Therefore, development of the Arctic resources is always accompanied by a great number of various risks.

Such a great concentration of risks in the Arctic region is due to different factors which can be conventionally divided into natural and technogenic ones. The natural factors involve geographical location of the region and harsh weather conditions, while technogenic factors imply the impact of hydrocarbon development activities. The combination of the above factors which constantly affect the Arctic ecosystems poses an enormous threat both to the Arctic region itself and the global climate system, in general.

2. Materials and methods

The current research presents the complex analysis of the risks related to the Arctic region development. The effects of regional climate changes, their role in global climate system and relationships among the Arctic countries, including modernization of the already existing models of the Arctic region development were examined. Based on the analysis of natural, social, economic, and political uncertainties, four groups of the Arctic risks – natural, social and ecological, political, transport-related and technological – were distinguished. Each group includes the maximum number of risk-factors divided into catastrophic, critical, and allowed risks for definite areas. This was done to define the likely level of harm on a world-wide scale. The results of such a division are presented in a form of analytical model.

The allowable risks admit effective economic activity in the Arctic region, with risks being properly managed. The group of critical risks is characterized by high danger of potential losses which exceed the expected benefit. The group of catastrophic risks includes the risks which can lead to global irreparable damage (for example, environmental damage, etc.) or substantially affect geopolitical frames. The proposed model could serve as a basis to make strategic decisions on developing international Arctic projects.

2.1. Data sets

Increasing interest in the Arctic region is basically conditioned by its geo-political importance, the Arctic region currently experiences surge in economic activity. Indeed, due to implementing up-to-date hydrocarbons production technologies, it has become possible to extract increasing amounts of oil and gas from hard-to-reach or highly complex reservoirs located in the traditional oil and gas regions. In addition, the shelf areas of the southern seas are also intensively developed. Thus, it can be stated that there is not great urgency in developing hydrocarbon resources of the Arctic region to address growing global demand for energy. Therefore, the crucial issue is what risks the countries are ready to bear in order to develop hard-to-reach Arctic resources. To address this issue, it is essential to reveal the risks related to the Arctic region development, define their possible adverse effects and estimate the anticipated damage. Arctic operations engender a variety of risks. Some risks can have a significant impact on the whole world, while others can be mitigated due to international, state and cooperative initiatives. The current study is based on the idea that any economic activity in the Arctic region is always accompanied by a number of risks. Among the scientists and experts there is no clear agreement on the classification of risks, their adverse impact and anticipated consequences. Besides, the present approaches to classify the risks do not fully describe the nature of the risks related both to the Arctic region itself and the Arctic states. The analysis of the research literature has revealed that the classification of the investment risks related to the Arctic region development is the most debated in the international practice.

We believe that identification of climate risks with investment ones is not quite appropriate due to the fact that such a classification is aimed at evaluating impact of risks only in terms of economic activity in the Arctic region. As a result, despite an increasing number of international initiatives launched to deal with environmental issues, the consequences of global climate changes, social and environmental protection problems are regarded as issues of secondary importance.

2.2. Risk assessment during the Arctic development in the context of global climate change

The basic components of the Arctic cryosphere, such as sea ice and permafrost, are among the most seriously affected by global and regional climate changes. Let us examine each component more closely.

The basic role of the sea ice in the Arctic is to preserve the heat balance between water body and bottom sediments. In summer, sea ice cover reduces the absorption of solar radiation by water surface. In winter, ice cover prevents heat exchange between deep water and surface layers [1]. In the coldest months, the ice extent is up to 16 mln. km^2 in the seas of the Northern Hemisphere. The thickness of the seasonal ice accounts for 2 m, while multiyear ice ranges from 3 to 6 m. The hummocks rise 5-9 m in the open sea and 20 m in the coastal area. In the Arctic seas, the shipping routes through fully consolidated ice with different degrees of compaction reduce by 10-25% in summer, in winter – by 20-40%. The increased ice-free period accelerates the processes of coastal erosion and degradation. In general, the process of ice cover transformation is of cyclic character: in different periods scientists note the stages of ice shrinkage or ice extension [2]. However, in accordance with the various forecasts, the Arctic sea ice will continue its rapid shrinkage [3, 4].

As National Snow and Ice Data Center [4] reported (figure 1), on September 11, the forth lowest Arctic sea ice extent in the satellite record for 2015 was 4.41 million square kilometers. After the date of the sea ice minimum, Arctic ice started increasing. The average Arctic ice sea extent for September 2015 was 4.63 million square kilometers, i.e. the lowest values in the satellite record. This is 1.87 million square kilometers below the 1981 to 2010 average extent; however, it is 1.01 square kilometers above the record low monthly average for September, 2012.

Figure 1. Satellite monitoring data on Arctic sea ice extent in September, 2015. The magenta line shows the average ice extent for the month in the 1981-2010 period [5].

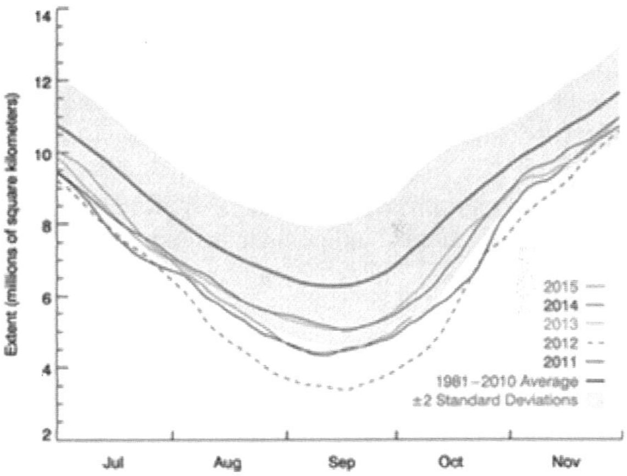

Figure 2. Arctic Sea Ice Extent (Area of ocean with at least 15% sea ice). The graph presents Arctic sea ice extent for 5 October, 2015 compared to daily ice extent data for the 2011 to 2014. The grey area around the grey line (1981-2010) shows two standard data deviation ranges [5].

The findings indicate that despite the seasonal increase (September, 2015) Arctic sea ice extent does not reach the 1981-2010 average for that month. Thus, Arctic sea ice extent changes in the cyclic manner. However, on a global scale, it continues melting. This trend is illustrated in (figure 2) which shows data deviation from the 1981 to 2010 for four previous years.

Thawing or degradation of permafrost and associated release of carbon dioxide (CO_2) and methane (CH_4) present another significant concern. Permafrost degradation is termed as uneven melting of ground ice leading to ground surface subsidence (thermokarst). Unlike other cryosphere components which are subjected to direct climate change impact, permafrost is protected by the surface organic layer. Therefore, it can be affected only if this insulating organic surface layer is completely destroyed. Thus, it can be stated that the most serious damage is basically caused by technogenic activities, while climate changes serve as accelerating agents. However, the intensity and duration of greenhouse gas emission in the Arctic region, as well as the impact of such gases on climate change have not been completely studied yet [6]. The urgency and importance of the above issues contributed to launching a number of international initiatives aimed at monitoring climate change in the Arctic region and forecasting the changes in the global climate system based on modeling principle (AMAP, IPCC, CAFF, etc.).

Methane is considered a more potent greenhouse gas whose impact on the environment is more than several times higher than CO_2. Despite the fact that the amount of methane release into the atmosphere is much lower than that of carbon dioxide, its heat-trapping ability is 25 times higher. Thus, methane contributes substantially to global climate change in the near term (10-15 years). Therefore, quantitative estimates of methane emission in various Arctic regions, on land and sea, constitute a great part of international research [6, 7, 8]. Based on Global Methane Initiative [9] data, methane emissions from oil and natural gas systems for 2010 accounted for 20% of global anthropogenic methane emissions. These emissions are projected to grow approximately by 35% by 2020. Oil and gas emissions are expected to increase the projected amount of emitted methane by 3% annually. Therefore, it can be stated that the effects of oil and gas activities in the Arctic region can be rather crucial and catastrophic. Thus, the specific natural and climatic conditions of the Arctic region pose significant limitations to the resource development perspectives in this region. It is impossible to completely eliminate natural risks which consequences are hard to predict. In addition, it is rather difficult to address these risks.

Oil and gas activity causes substantial environmental damage to the Arctic region, as a whole. The most dangerous and hard-to-eliminate consequences of such an activity are oil and chemical spills that accompany any hydrocarbon field development including drilling operations and oil and gas transportation. The most harmful effect of such accidental spills is a constant degradation of Arctic sea ecosystems. Such adverse effects of anthropogenic activities are examples of ecological risks [10]. The basic danger related to ecological risks is due to the fact that they produce combined impact on the whole region, thus, reducing efficiency of risk mitigation plans. Therefore, high environmental safety of the Arctic region is a key factor of its sustainable development.

Another challenge that impedes sustainable development of the Arctic region is significant social risks. In the Arctic region, social consequences of climate changes and anthropogenic impact are more notable due to the fact that it is place for indigenous peoples to live. Such factors as landscape continuity, maintenance of the traditional biodiversity of the Arctic region are obligatory conditions for indigenous peoples' well-being. However, the rapid warming trend and consequences of industrial activity in this region significantly reduce the possibility of on-time adaptation of local communities to ongoing changes threatening their traditional lifestyle. Besides, there is a growing risk for the health of northern peoples: increase in morbidity and mortality resulting from anomalous high/low temperatures, increased accidents of injury and mortality caused by unpredictable weather conditions (storms, floods, etc.), growing number of infectious, parasitic diseases and cancer, acute and chronic illnesses, reproductive dysfunction [11, 12].

IOP Conf. Series: Earth and Environmental Science **43** (2016) 012099 doi:10.1088/1755-1315/43/1/012099

Efficient and safe development of Arctic resources is also complicated by transport-related and technological problems. Oil and gas activity in the Arctic region inevitably involves the use of stationary drilling and production platforms, ice-class vessels designed for ice navigation, exploration, and rescue operations under harsh Arctic weather conditions. кyport infrastructure that would meet international navigating requirements is also of great importance. Due to projected increase in cargo flows on the Northern Sea Route, it is of great importance to find the ways to reduce the discussed transport-related and technological risks. Here Russia's Arctic strategies are of great value as the Northern Sea Route is the historically formed national single transport communication of the Russian Federation in the Arctic. To guarantee cargo flow security on the Northern Sea Route, it is necessary to use up-to-date engineering facilities and provide highly developed infrastructure in Russian Arctic. However, currently these conditions are not sufficiently fulfilled. In comparison with other polar countries, Russia is characterized by insufficient level of infrastructure development in the northern regions. Therefore, the problems related to the applied technologies and infrastructure should be considered in a wider arena including political background [13] and political risks [14] of the Arctic region. Today, there is no well-developed international Arctic law. The legal status of the Arctic seas is defined by the principles and norms of the international law, precisely by 1958 Geneva Conventions on the Law of the Sea and UN Convention on the Law of the Sea which do not capture many legal aspects of the Arctic issues. There is no unique approach to define the Arctic borders as a geographical region with strict division into political and administrative regions of the Arctic countries and clear mapping of areas of their responsibility. Such an uncertainty stipulates geopolitical competition to claim the right to natural resources found in the Arctic and control of Northern Sea routes. Therefore, the system of international cooperation in the Arctic region is actually twofold. On the one hand, competition among the Arctic nations seeking to assert their influence in the region is getting tougher; on the other hand, no Arctic country appears to have sufficient scientific and technological capacities to implement Arctic resources development projects alone [14].

The international Arctic cooperation has achieved such a scale that it can be carried out on the following bases: bilateral cooperation, multilateral cooperation, and large scale cooperation (including all Arctic countries, as well as a number of non-Arctic nations). It is assumed that such a cooperation could be affective not only for participants, but for the Arctic region, as a whole. In accordance with Oran R. Young [15], improvement of the current Arctic governance systems and development of new ones with due regard to the newly emerged risks are the ultimate priority of the Arctic international cooperation. The focus of the non-Arctic countries has definitely shifted from pure research activity to the intent to exploit Arctic strategic potential, which makes a perspective trend in the Arctic international cooperation. However, there is a potential for international conflicts. The list of non-Arctic countries and the procedure to gain the observer status in the Arctic Council unveil the unseen links between the Arctic and non-Arctic countries. In addition, membership of the Arctic and non-Arctic countries in various international organizations (EU, NATO) can influence their relationships.

3. Results and Discussion

The basic findings of the current research are as follows: relationships of the Arctic countries, as well as the Arctic region itself, are intensively transforming. Due to interacting forces of climate change, globalization, social and economic trends, there is an urgent need to modernize the already existing models of the Arctic region development. The Arctic sustainable development is directly dependent on natural, social, economic, and political uncertainties which contribute to new risks. Based on our estimates and specific geographical and climatic conditions of the Arctic, we assume that uncertainties can emerge within any activity related to the Arctic development. The analysis of ongoing transformations allows us to define four basic groups of Arctic risks which could cause negative effects on a world-wide scale. To predict the likely level of harm, each group of Arctic risks is presented in the form of analytical model, with the risks being divided into catastrophic, critical, and allowable (figure 3). The model demonstrates that catastrophic and critical risks are predominant in each group of Arctic risks. This proves the fact that Arctic resources are impossible to develop by one country.

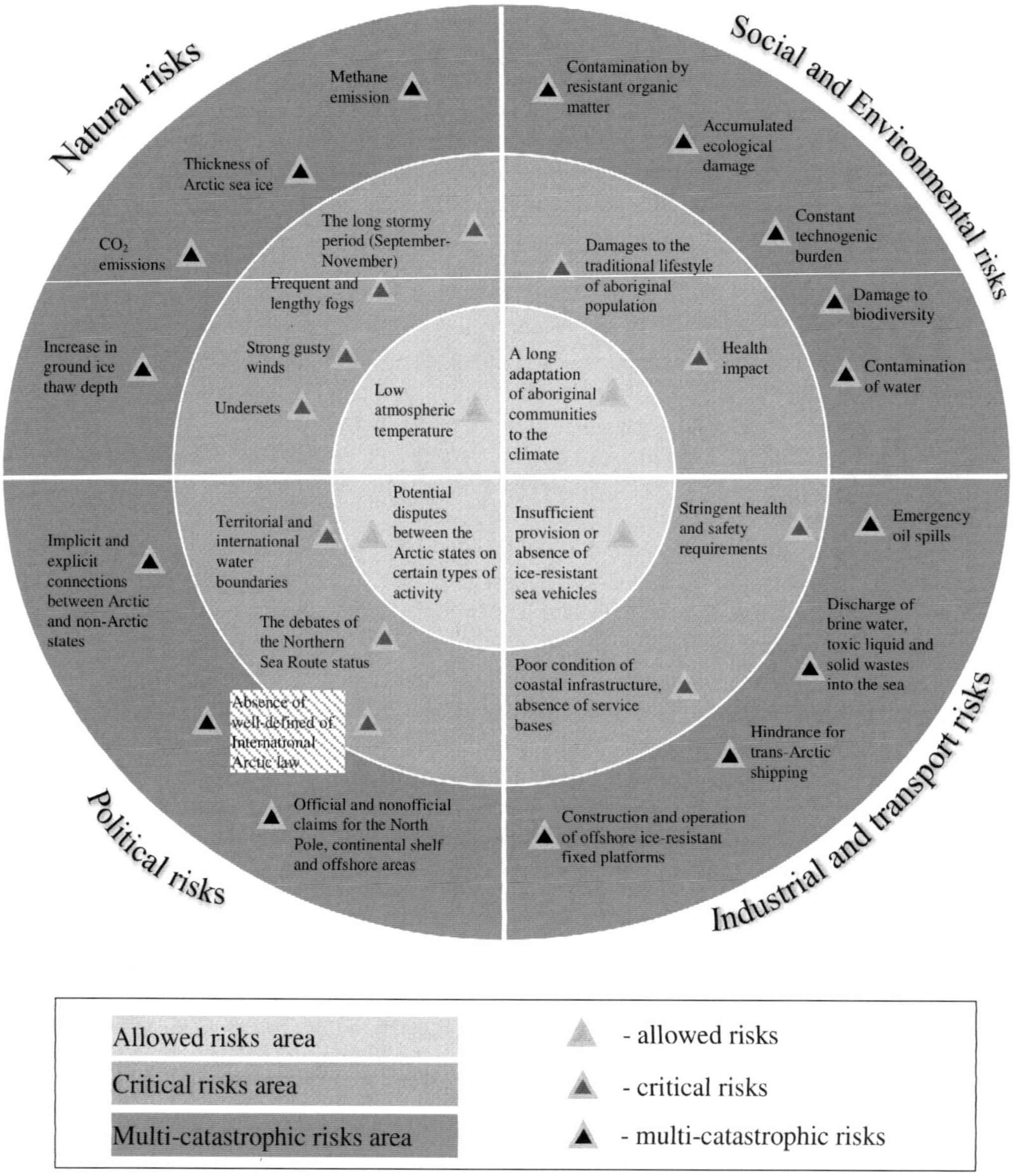

Figure 3. Analytical model of Arctic risks breakdown by the areas of possible harm (Author's edition)

4. Conclusion

Today, it is difficult to predict the dynamics of global climate changes. In addition, it is difficult to define the impact of the Arctic climate changes on the global climate. However, the tremendous expansion of economic activity in the Arctic region involves a growing number of new risks which should be consistently analyzed with regard to natural factors. There is little doubt that the specific geographical and climatic conditions of the Arctic contributed to the shift of non-Arctic countries' interests from research level to the intent to participate in Arctic resource development and control of the Northern Sea routes. This proves the fact that it is of particular importance to establish the partnership among all participants involved in economic activity in the Arctic, to achieve the balance between efficient use and economic feasibility of Arctic resources development. Besides, it can be required to enhance the dialogue with non-Arctic countries in order to foster mutually beneficial cooperation.

References

[1] Olerskiy V A 2014 Razvitie transportnoy sistemy Arkticheskoy zony RF: Processing of IV International Forum «the Arctic: present and future» 15-18
[2] Rosgidromet. Russian Federal Service for Hydrometeorology and Environmental Monitoring. URL: http://www.meteorf.ru
[3] Lindsay R and Schweiger A 2015 Arctic sea ice thickness loss determined using subsurface, aircraft, and satellite observations *The Cryosphere* No. **9** 269-283
[4] 2015 Melt season in review. National Snow and Ice Data Center: Advancing knowledge of Earth's frozen regions. URL: http://nsidc.org/arcticseaicenews/
[5] Tayushchaya krasota. Izmeneniya klimata i ego posledstviya 2009 Heinrich Böll Foundation, Russian Regional Ecological Center
[6] Schuur E A G et al 2015 Climate change and the permafrost carbon feedback *Nature* vol 520 (7546) 171–179
[7] Natali S M et al 2015 Permafrost thaw and soil moisture drive CO2 and CH4 release from upland tundra *Journal of Geophys. Res.* vol **120** (3) 525-537
[8] Fisher R E et al 2011 Arctic methane sources: Isotopic evidence for atmospheric inputs *Geophys. Res. Lett.* vol **38** L21803
[9] Global Methane Emissions and Mitigation Opportunities 2011 Global Methane Initiative. URL: https://www.globalmethane.org/documents/analysis_fs_en.pdf
[10] Bolsunovskaya Y and Bolsunovskaya L 2015 Ecological risk analysis as a key factor in environmental safety system development in the Arctic region of the Russian Federation *IOP Conf. Ser.: Earth and Environ. Sci.* vol **24** 012003
[11] Revich B et al 2008 Climate Change Impact on Public Health in the Russian Arctic. The United Nations Office in Russia Arctic Initiative. URL: http://www.unrussia.ru/sites/default/files/doc/Arctic-eng.pdf
[12] Dudley J P, Hoberg E P, Jenkins E J and Parkinson A J 2015 Climate Change in the North American Arctic: A One Health Perspective *EcoHealth* vol **12** (2) online first articles
[13] Sentsov A, Bolsunovskaya Y and Bolsunovskaya L 2014 Effective Planning of the Future of the Arctic *IOP Conf. Ser.: Earth and Environ. Sci.* vol **21** 012014
[14] Bolsunovskaya Y, Boyarko G and Bolsunovskaya L 2014 Political risks of hydrocarbon deposit development in the Arctic seas of the Russian Federation *IOP Conf. Ser.: Earth and Environ. Sci.* vol **21** 012046
[15] Young Or 2012 Arctic tipping points: governance in turbulent times *Ambio* vol **41**(1) 75-84

The Arctic zone: possibilities and risks of development

A Sentsov[1], Y Bolsunovskaya[2], and E Melnikovich[3]
[1] National Research Tomsk State University, Tomsk, Russia
[2 3] National Research Tomsk Polytechnic University, Tomsk, Russia

E-mail: [1]134-43@mail.ru, [2]ju_al@inbox.ru, [3]melhmelll@gmail.com

Abstract. The authors analyze the Arctic region innovative possibilities from the perspective of political ideology and strategy. The Arctic region with its natural resources and high economic potential attracts many companies and it has become an important area of transnational development. At present, the Arctic region development is of great importance in terms of natural resource management and political system development. However, the most important development issue in the Arctic is a great risk of different countries' competing interests in economic, political, and legal context. These are challenges for international partnership creating in the Arctic zone, Russian future model developing for the Arctic, and recognition of the Arctic as an important resource for the Russians. The Russian economic, military, and political expansion in the Arctic region has the potential to strengthen the national positions. The authors present interesting options for minimizing and eliminating political risks during the Arctic territories development and define an effective future planning model for the Russian Arctic.

Acknowledgment. The present work was realized and subsidized within the framework of the research "Political identity formation in the context of the image of the future modeling" supported by the President of Russian Federation Grant for state support of young candidates of science MK-7447.2016.6.

1. Introduction

Nowadays, the Arctic zone is a very important subject of international geopolitics. In addition, this is an extremely important ecological and environmental subject. Again the Intergovernmental Panel on Climate Change has informed about serious and irreversible effects of global warming. The ice floe melts, the inhabitants run away, the biodiversity suffers, and, often, disappears. So the threat is real [3].

The challenges are numerous: to open a new seaway, to create air stops, to exploit new hydrocarbon and gas deposits, etc. No doubt, the Arctic is a zone of technological innovations and new perspectives and possibilities. For centuries, and even in the twentieth century, the Arctic has remained almost unknown, except for the indigenous peoples living there for millennia. Subsequently, the increasing military build-up in the North, from Alaska to Siberia, has placed the region at the heart of the conflicts of the Cold War and made from it a forbidden zone for almost the second half of the previous century [10].

Also historically, in comparison with European and Asian civilizations, the Arctic, as cooperation territory, in the field of economic and political progress, is almost like a new planet; the region is so young in the global context that has no equivalent in the world.

For the Russian Federation the Arctic is a key zone of potential economic growth and positioning in the international community. The effective modeling of the Russian Arctic future is extremely important for the Russian state.

2. Arctic zone division and cooperation

One of the most important and dangerous issue in terms of international security assurances is the Arctic zone division between different countries (see the Figure 1). Originally, there are five countries of the polar circumference: the United States, Canada, Denmark (via Greenland), Norway, and the Russian Federation. Approximately 25% of the undiscovered reserves of natural gas and oil are supposed to be located in the Arctic [1].

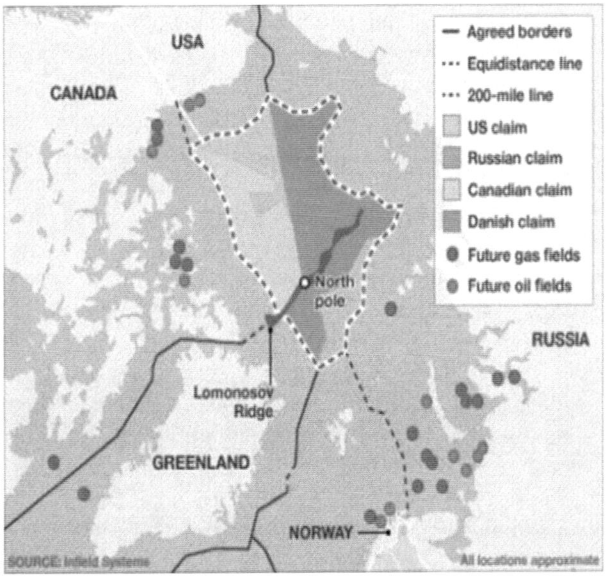

Fig. 1. Five countries of the polar circumference [4]

But in the 1990s these five countries with the participation of another three ones situated in the Arctic zone, namely, Finland, Iceland, and Sweden, created a high-level intergovernmental forum – the Arctic Council.

During the creation of the Arctic Council in the 1990s, the eight member states were still hesitant in their approach and their mandate was very limited, mainly reduced to science and the environment; a rotating presidency and no permanent secretariat.

Nevertheless, the Arctic Council has matured, has turned into an effective tool for negotiating treaties, global agreements and concrete actions. In addition, even more remarkable, since the decision of the Council of Ministers of Kiruna in 2013, more than half of the G20 countries, the main countries of Europe and Asia, will be one way or another at the Arctic table.

The Arctic, yet distant and unknown to the late twentieth century, has become a new field of action where the main economic and political forces advance gradually, ensuring their presence and long-term interests [9].

The intersection of multiple factors can cause the geopolitical conflicts in the Arctic, often under the guise of legal considerations [8]. Land claims are not frozen. They are even essential, especially in the context of the exclusive economic zone extension through the continental shelf enlargement. There are limits of the Committee of the Continental Shelf and of the Committee of London. It is up to them to decide these claims based on rules set by the United Nations Convention on the Law of the Sea. The treatment of territorial disputes seems to take place in serene conditions despite what the media echoes let appear.

In this context, it is very important that the economic interest is a powerful engine of compromise. In this regard, the relationship between Russia and Norway can be a good example. The land border between Russia and Norway was conformed in 1944, at the end of World War II. A treaty was signed between the two countries, to the detriment of Finland, which lost its access to the Barents Sea. Nowadays, Norway and Russia are active and well represented in a multitude of organizations: the Arctic Council, the Euro-Arctic Barents Council, and the Council of Baltic Sea States. This institutional overlapping is also established at the regional level. For example, the county of Finnmark and Murmansk Oblast have cooperation agreements, either economical or institutional. Another example of this stability: economic interdependence.

This interdependence between Oslo and Moscow is also found on the capital and financial plan. The Norwegian pension fund, very powerful, has a varied portfolio of shares in major companies in the world and especially in major Russian companies in the banking and energy sectors.

For example, Tele North Norwegian Telecommunications Company owns 33% of the Public joint-stock company "VimpelCom", one of the three largest Russian telecommunications companies. It should be added that in September 2010 the Treaty of Murmansk was signed, to resolve a long border conflict between Russians and Norwegians. This treaty, signed between Russian President and Norwegian Prime Minister perfectly illustrates this stability [6].

Note, that the Barents region is the most populous of all the Arctic area. There are nearly 3 million between Murmansk Oblast, Arkhangelsk Oblast and the Republic of Karelia. The three most populated cities are Murmansk, 300 000 inhabitants, Arkhangelsk 350 000, and Petrozavodsk, capital of Karelia 270 000. The Russian side represents the largest population. On the Norwegian side, Tromsø is situated, the largest city in northern Norway with 70 000 inhabitants.

Then we may say about indigenous people, Fennoscandia and the Kola Peninsula are home to around 40,000 Sami, Norway is home to half of this population. For Russia, although there are no official figures, the population of the Sami could remain around 2000 inhabitants [6].

The bilateral relationships between some Arctic riparian countries are very good organized. But as for the eight Arctic Council member states one of the basic document that regulates their relationships in the Arctic region is the Agreement on Cooperation on Aeronautical and Maritime Search and Rescue in the Arctic (see the Figure 2), signed on May 12, 2011 in Nuuk, Greenland [5].

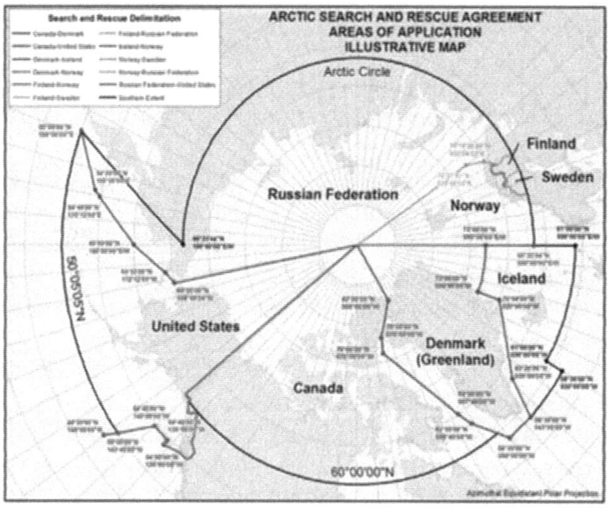

Fig. 2. Map of the Arctic search and rescue areas, the Agreement of 2011 [5]

3. Modeling of the Russian Arctic future

The Arctic zone is a region of high importance for the Russian Federation. Russia has its interests and considerable claims in this region (see the Figure 3). This territory is an extremely significant object of national economic and politics [8]. Russian President Vladimir Putin points out that "our interests are concentrated in the Arctic. And of course, we should pay more attention to issues of the Arctic development and the strengthening of our position" [7].

Fig. 3. Russian claims in the Arctic zone [4]

The modeling of the Russian Arctic future is a vital challenge for the Russian state. No doubt, for the time being "the expansion of Russian economic, military, and political influence in the Arctic region has the potential to strengthen the position of the state in the international community" [10].

The sense of the future is a fundamental part of a human consciousness. This sense is located on the basic axis of coordinates on which a person is oriented in the outside world. It is included in the individual, social, and thus political temporal plan. The time horizon of modern society is determined by the future. Therefore, modeling and programming of the future may be so important in our modern society. The future plays a key role in the value system of a person and a society and now the future plan broadly defines the main directions in the creation of a dynamic society. The modeling of an effective image of the Arctic future is extremely important for our Russian society in the context of national security, sustainable development, and improvement of political, social, and economic wellbeing.

The model of the Russian Arctic future can be effective providing that the future of this Russian region will go hand in hand with the future of the other Arctic territories. The harmonious future and development are possible only in cooperation with the other Arctic states.

4. Interpretation of the results

The modern Arctic is the crossing point of scientific parameters, technical, diplomatic considerations and legal questions. At present, it is accepted that global warming contributes to lower limits of the polar caps, opening up new opportunities in navigation, which would constitute a new ocean permanently ice-free.

It is generally admitted that among the so-called Arctic seaways the most important are the Northern Sea Route along the Russian coast and the Northwest Passage that borrows multiple straits and bypasses in the Canadian archipelago of islands. The only uncertainty concerns the date on which a track is navigable throughout the year. Over the decades, these two Arctic sea-lanes were completely ice-free, but only for very short periods, especially in September. Moreover, it is not sure that the appropriate port infrastructure already exist, including the Canadian side. However, the riparian States have begun to develop these Arctic waterways. Russia uses it more than any other country. In 2010, four ships passed through the Northern Sea Route; the following year thirty-four ships passed along the Russian coast.

Another important issue leads us to emphasize the global dimension of the Arctic fate: the continued protection of its ecosystem [2]. Any human intervention is likely to disturb the balance between plant and animal species. The Arctic is a very vulnerable region. Take the example of migratory species arriving in the Arctic summer, precisely when routes are the most feasible. Ships and animals are concentrated in the same places, and that increases the risks. Now it is known that the disappearance of species in a chain influences others, especially in an environment where the species number is relatively small, that is the case of the Arctic. There is no need to mention the risks of pollution because of the oil and gas development.

5. Conclusion

The above-mentioned examples show that the issues of the Arctic cannot be treated in small groups. It interests the whole international community and not only the riparian states. The most important point in the Russian Arctic future modeling is in this context that the continued active engagement of all the countries is of particular importance. The Arctic zone risks can be transformed into new possibilities not only for the Russian Arctic but also for the whole world. And in this situation the most important and fundamental efforts should be done by the key players of the international geopolitics of our days, namely Russia, the United States, the European Union, and the People's Republic of China.

The cooperation of different countries in the region can influence the economic development as well as the political situation and international relationship in the Arctic zone. The Arctic will play a key part in the modeling of an effective future for the whole world.

References

[1] Alhadeff, I. (2015). United States and Russia in the Arctic Zone. URL: https://iakal.wordpress.com/2015/08/05/united-states-and-russia-in-the-arctic-zone (access date: 02.03.16)

[2] Bolsunovskaya, Y., et al. (2015). Integrated analysis of risks in terms of Russian Arctic zone sustainable development. *IOP Conf. Ser.: Earth Environ. Sci.* **27** 012021, doi: http://dx.doi.org/10.1088/1755-1315/27/1/012021

[3] Climate Change 2014: Impacts, Adaptation, and Vulnerability. (2014). Report, Intergovernmental Panel on Climate Change, 31.03.2014. URL: http://www.ipcc.ch/report/ar5/wg2 (access date: 04.03.16)

[4] Emmerson, C. (2010). *The Future History of the Arctic*. New York: Public Affairs

[5] Farré, A.B., et al. (2014). Commercial Arctic shipping through the Northeast Passage: Routes, resources, governance, technology, and infrastructure. *Polar Geography*. **37** (4), 298 - 324

[6] Isachenkov, V. (2015). Russia to UN: We are claiming 463,000 square miles of the Arctic. *Military & Defense, Business Insider*, 04.08.2015. URL: http://www.businessinsider.com/russia-to-un-we-are-claiming-463000-square-miles-of-the-arctic-2015-8 (access date: 04.03.16)

[7] Nato on high alert as Putin eyes up Arctic for empire. (2015). *Daily Express*. Northern and Shell Media Publications. URL: http://www.express.co.uk/news/world/508019/Nato-high-alert-Putin-eyes-Arctic-empire (access date: 05.02.16)

[8] Sentsov, A., Aleksandrov, O., Bolsunovskaya Yu., Kuimova M. (2015). Modeling of the Future in the Programs of Political Parties. *Proceedings of the International Conference on Research Paradigms Transformation in Social Sciences 2014*. Vol. **166**, doi: 10.1016/j.sbspro.2014.12.568

[9] Sentsov, A., Bolsunovskaya, Yu., and Bolsunovskaya, L. (2014). Effective Planning of the Future of the Arctic. *IOP Conf. Ser.: Earth Environ. Sci.* **21** 012014, doi: http://dx.doi.org/10.1088/1755-1315/21/1/012014

[10] Shcherbinin, A., Danilova, E., Sentsov, A., Bolsunovskaia, L., Bolsunovskaya, Y. (2015). The Russian Arctic: innovative possibilities at the turn of the past and the future. *IOP Conf. Ser.: Earth Environ. Sci.* **27** 012022, doi: http://dx.doi.org/10.1088/1755-1315/27/1/012022